# Fundamentals of AIR POLLUTION ENGINEERING

RICHARD C. FLAGAN

JOHN H. SEINFELD

*California Institute of Technology*

DOVER PUBLICATIONS INC
Mineola New York

*Bibliographical Note*

This Dover edition, first published in 2012, is an unabridged, corrected republication of the work originally published in 1988 by Prentice-Hall, Inc., Englewood Cliffs, New Jersey. For this edition the authors have provided a new Preface and three Appendices.

Instructors who would like the Solutions Manual for the problems in this book can contact the authors at *seinfeld@caltech.edu* or *flagan@caltech.edu.*

*Library of Congress Cataloging-in-Publication Data*

Flagan, Richard C., 1947–
Fundamentals of air pollution engineering / Richard C. Flagan and John H. Seinfeld.
p. cm.
Includes bibliographical references and index.
ISBN-13: 978-0-486-48872-1
ISBN-10: 0-486-48872-1
1. Air—Pollution. 2. Environmental engineering. 3. Environmental protection. I. Seinfeld, John H. II. Title.

TD883.F38 2012
628.5'3—dc23

2012008972

Manufactured in the United States by Courier Corporation
48872101
www.doverpublications.com

# Contents

# Preface to the Dover Edition

Despite major advances in combustion technology and emissions controls since the publication of Fundamentals of Air Pollution Engineering in 1988, the basic principles underlying the study of the generation and control of air pollutants remain unchanged. These include chemical kinetics, thermodynamics, fluid mechanics, and heat and mass transfer. We adopted the approach of providing a fundamental basis for analysis of combustion and gas-cleaning processes. This viewpoint is as timely now as it was in 1988.

This Dover edition of the original volume includes three new appendices that did not appear in the original book. While the book does not cover methods for the measurement of airborne particles (aerosols) per se, a powerful class of aerosol measurement methods involves the manipulation of charged particles flowing in an electric field. The theory of this class of methods, as embodied in the Differential Mobility Analyzer, is an important application of aerosol dynamics and is the subject of Appendix A. Appendix B is an analysis of the deposition of aerosol particles to the wall of a tube in steady laminar or turbulent flow. This problem is both technologically important in particle sampling, control, and measurement, and an outstanding example of the application of the aerosol convection/diffusion equation. The diffusive deposition of particles in laminar tube flow is analogous to the classic problem of heat transfer between the wall and a flowing fluid, the so-called Graetz problem. Appendix B contains the formulation and solution of both the laminar and turbulent flow cases with a numerical example. Appendix C considers the evaporation of diesel-generated particles as an exhaust stream is diluted upon emission to the air. Since the publication of the original text, the importance of the near-source behavior of vehicle particulate emissions has become evident based on ambient sampling immediately downwind of roadways. Traditionally, combustion-generated particles have been considered as non-volatile. These particles, which consist primarily of carbonaceous material, were thought to remain more or less inert in the atmosphere. Laboratory studies have shown that, upon dilution, combustion particles can evaporate, indicating that the components of typical combustion particles are not totally non-volatile. Once evaporated, the organic compounds contained in the particles may undergo atmospheric oxidation to products that are even less volatile than the parent compounds. These low volatility oxidation products can re-condense in the atmosphere to form so-called secondary organic aerosol. This technologically important problem also provides an excellent example of the theory of multi-component evaporation of aerosols. In Appendix C, a simplified example of diesel particle evaporation is formulated and solved.

We are pleased that Dover Publications has reprinted our 1988 text. Finally, we note that a Solutions Manual is available for the problems at the end of each chapter.

Richard C. Flagan
John H. Seinfeld
Pasadena, CA
January 2012

# Preface

Analysis and abatement of air pollution involve a variety of technical disciplines. Formation of the most prevalent pollutants occurs during the combustion process, a tightly coupled system involving fluid flow, mass and energy transport, and chemical kinetics. Its complexity is exemplified by the fact that, in many respects, the simplest hydrocarbon combustion, the methane–oxygen flame, has been quantitatively modeled only within the last several years. Nonetheless, the development of combustion modifications aimed at minimizing the formation of the unwanted by-products of burning fuels requires an understanding of the combustion process. Fuel may be available in solid, liquid, or gaseous form; it may be mixed with the air ahead of time or only within the combustion chamber; the chamber itself may vary from the piston and cylinder arrangement in an automobile engine to a 10-story-high boiler in the largest power plant; the unwanted by-products may remain as gases, or they may, upon cooling, form small particles.

The only effective way to control air pollution is to prevent the release of pollutants at the source. Where pollutants are generated in combustion, modifications to the combustion process itself, for example in the manner in which the fuel and air are mixed, can be quite effective in reducing their formation. Most situations, whether a combustion or an industrial process, however, require some degree of treatment of the exhaust gases before they are released to the atmosphere. Such treatment can involve intimately contacting the effluent gases with liquids or solids capable of selectively removing gaseous pollutants or, in the case of particulate pollutants, directing the effluent flow through a device in which the particles are captured on surfaces.

The study of the generation and control of air pollutants can be termed *air pollution engineering* and is the subject of this book. Our goal here is to present a rigorous and fundamental analysis of the production of air pollutants and their control. The book is intended for use at the senior or first-year graduate level in chemical, civil, environmental, and mechanical engineering curricula. We assume that the student has had basic first courses in thermodynamics, fluid mechanics, and heat transfer. The material treated in the book can serve as the subject of either a full-year or a one-term course, depending on the choice of topics covered.

In the first chapter we introduce the concept of air pollution engineering and summarize those species classified as air pollutants. Chapter 1 also contains four appendices that present certain basic material that will be called upon later in the book. This material includes chemical kinetics, the basic equations of heat and mass transfer, and some elementary ideas from probability and turbulence.

Chapter 2 is a basic treatment of combustion, including its chemistry and the role of mixing processes and flame structure. Building on the foundation laid in Chapter 2, we present in Chapter 3 a comprehensive analysis of the formation of gaseous pollutants in combustion. Continuing in this vein, Chapter 4 contains a thorough treatment of the internal combustion engine, including its principles of operation and the mechanisms of formation of pollutants therein. Control methods based on combustion modification are discussed in both Chapters 3 and 4.

Particulate matter (aerosols) constitutes the second major category of air pollutants when classified on the basis of physical state. Chapter 5 is devoted to an introduction to aerosols and principles of aerosol behavior, including the mechanics of particles in flowing fluids, the migration of particles in external force fields, Brownian motion of small particles, size distributions, coagulation, and formation of new particles from the vapor by homogeneous nucleation. Chapter 6 then treats the formation of particles in combustion processes.

Chapters 7 and 8 present the basic theories of the removal of particulate and gaseous pollutants, respectively, from effluent streams. We cover all the major air pollution control operations, such as gravitational and centrifugal deposition, electrostatic precipitation, filtration, wet scrubbing, gas absorption and adsorption, and chemical reaction methods. Our goal in these two chapters, above all, is to carefully derive the basic equations governing the design of the control methods. Limited attention is given to actual equipment specification, although with the material in Chapters 7 and 8 serving as a basis, one will be able to proceed to design handbooks for such specifications.

Chapters 2 through 8 treat air pollution engineering from a process-by-process point of view. Chapter 9 views the air pollution control problem for an entire region or airshed. To comply with national ambient air quality standards that prescribe, on the basis of health effects, the maximum atmospheric concentration level to be attained in a region, it is necessary for the relevant governmental authority to specify the degree to which the emissions from each of the sources in the region must be controlled. Thus it is generally necessary to choose among many alternatives that may lead to the same total quantity of emission over the region. Chapter 9 establishes a framework by which an optimal air pollution control plan for an airshed may be determined. In short, we seek the least-cost combination of abatement measures that meets the necessary constraint that the total emissions not exceed those required to meet an ambient air quality standard.

Once pollutants are released into the atmosphere, they are acted on by a variety of chemical and physical phenomena. The atmospheric chemistry and physics of air pollution is indeed a rich arena, encompassing the disciplines of chemistry, meteorology, fluid mechanics, and aerosol science. As noted above, the subject matter of the present book ends at the stack (or the tailpipe); those readers desiring a treatment of the atmospheric behavior of air pollutants are referred to J. H. Seinfeld, *Atmospheric Chemistry and Physics of Air Pollution* (Wiley-Interscience, New York, 1986).

We wish to gratefully acknowledge David Huang, Carol Jones, Sonya Kreidenweis, Ranajit Sahu, and Ken Wolfenbarger for their assistance with calculations in the book.

Finally, to Christina Conti, our secretary and copy editor, who, more than anyone else, kept safe the beauty and precision of language as an effective means of communication, we owe an enormous debt of gratitude. She nurtured this book as her own; through those times when the task seemed unending, she was always there to make the road a little smoother.

R. C. Flagan
J. H. Seinfeld

# 1

# Air Pollution Engineering

The phenomenon of air pollution involves a sequence of events: the generation of pollutants at and their release from a source; their transport and transformation in and removal from the atmosphere; and their effects on human beings, materials, and ecosystems. Because it is generally either economically infeasible or technically impossible to design processes for absolutely zero emissions of air pollutants, we seek to control the emissions to a level such that effects are either nonexistent or minimized.

We can divide the study of air pollution into three obviously overlapping but somewhat distinct areas:

1. The generation and control of air pollutants at their source. This first area involves everything that occurs before the pollutant is released ''up the stack'' or ''out the tailpipe.''
2. The transport, dispersion, chemical transformation in, and removal of species from the atmosphere. This second area thus includes all the chemical and physical processes that take place between the point of emission and ultimate removal from the atmosphere.
3. The effects of air pollutants on human beings, animals, materials, vegetation, crops, and forest and aquatic ecosystems, including the measurement of gaseous and particulate species.

An air pollution control strategy for a region is a specification of the allowable levels of pollutant emissions from sources. To formulate such a strategy it is necessary to be able to estimate the atmospheric fate of the emissions, and thus the ambient concentrations, so that these concentrations can be compared with those considered to give rise to adverse effects. The ultimate mix of control actions and devices employed to achieve the allowable levels might then be decided on an economic basis. Therefore, the formulation of an air pollution control strategy for a region involves a critical feedback from area 3 to area 1. Consequently, all three of the areas above are important in air pollution abatement planning.

A comprehensive treatment of each of these three areas is beyond the scope of a single book, however. The present book is devoted to an in-depth analysis of the generation and control of air pollutants at their source, which we refer to as *air pollution engineering*.

## 1.1 AIR POLLUTANTS

Table 1.1 summarizes species classified as air pollutants. By and large our focus in this book is on the major combustion-generated compounds, such as the oxides of nitrogen, sulfur dioxide, carbon monoxide, unburned hydrocarbons, and particulate matter. Table 1.2 provides a list of the most prevalent hydrocarbons identified in ambient air, and Table 1.3 lists potentially toxic atmospheric organic species.

### 1.1.1 Oxides of Nitrogen

Nitric oxide (NO) and nitrogen dioxide ($NO_2$) are the two most important nitrogen oxide air pollutants. They are frequently lumped together under the designation $NO_x$, although analytical techniques can distinguish clearly between them. Of the two, $NO_2$ is the more toxic and irritating compound.

Nitric oxide is a principal by-product of combustion processes, arising from the high-temperature reaction between $N_2$ and $O_2$ in the combustion air and from the oxidation of organically bound nitrogen in certain fuels such as coal and oil. The oxidation of $N_2$ by the $O_2$ in combustion air occurs primarily through the two reactions

$$N_2 + O \longrightarrow NO + N$$

$$N + O_2 \longrightarrow NO + O$$

known as the Zeldovich mechanism. The first reaction above has a relatively high activation energy, due to the need to break the strong $N_2$ bond. Because of the high activation energy, the first reaction is the rate-limiting step for NO production, proceeds at a somewhat slower rate than the combustion of the fuel, and is highly temperature sensitive. Nitric oxide formed via this route is referred to as *thermal*-$NO_x$. The second major mechanism for NO formation in combustion is by the oxidation of organically bound nitrogen in the fuel. For example, number 6 residual fuel oil contains 0.2 to 0.8% by weight bound nitrogen, and coal typically contains 1 to 2%, a portion of which is converted to $NO_x$ during combustion. (The remainder is generally converted to $N_2$.) Nitric oxide formed in this manner is referred to as *fuel*-$NO_x$.

Mobile combustion and fossil-fuel power generation are the two largest anthropogenic sources of $NO_x$. In addition, industrial processes and agricultural operations produce minor quantities. Emissions are generally reported as though the compound being emitted were $NO_2$. This method of presentation serves the purpose of allowing ready comparison of different sources and avoids the difficulty in interpretation associated with different ratios of $NO/NO_2$ being emitted by different sources. Table 1.4 gives $NO/NO_x$ ratios of various types of sources. We see that, although NO is the dominant $NO_x$ compound emitted by most sources, $NO_2$ fractions from sources do vary somewhat with source type. Once emitted, NO can be oxidized quite effectively to $NO_2$ in the atmosphere through atmospheric reactions, although we will not treat these reactions here. Table 1.5 gives estimated U.S. emissions of $NO_x$ in 1976 according to source category. Utility boilers represent about 50% of all stationary source $NO_x$ emissions in the United States. As a result, utility boilers have received the greatest attention in past $NO_x$ regulatory strategies and are expected to be emphasized in future plans to attain and maintain $NO_x$ ambient air quality standards.

### 1.1.2 Sulfur Oxides

Sulfur dioxide ($SO_2$) is formed from the oxidation of sulfur contained in fuel as well as from certain industrial processes that utilize sulfur-containing compounds. Anthropogenic emissions of $SO_2$ result almost exclusively from stationary point sources. Estimated annual emissions of $SO_2$ in the United States in 1978 are given in Table 1.6. A small fraction of sulfur oxides is emitted as primary sulfates, gaseous sulfur trioxide ($SO_3$), and sulfuric acid ($H_2SO_4$). It is estimated that, by volume, over 90% of the total U.S. sulfur oxide emissions are in the form of $SO_2$, with primary sulfates accounting for the other 10%.

Stationary fuel combustion (primarily utility and industrial) and industrial processes (primarily smelting) are the main $SO_2$ sources. Stationary fuel combustion includes all boilers, heaters, and furnaces found in utilities, industry, and commercial/institutional and residential establishments. Coal combustion has traditionally been the largest stationary fuel combustion source, although industrial and residential coal use has declined. Increased coal use by electric utilities, however, has offset this decrease. $SO_2$ emissions from electric utilities account for more than half of the U.S. total. A more detailed breakdown of U.S. sulfur oxide emissions in 1978 is given in Table 1.7.

### 1.1.3 Organic Compounds

Tables 1.2 and 1.3 list a number of airborne organic compounds. Organic air pollutants are sometimes divided according to volatile organic compounds (VOCs) and particulate organic compounds (POCs), although there are some species that will actually be distributed between the gaseous and particulate phases. The emission of unburned or partially burned fuel from combustion processes and escape of organic vapors from industrial operations are the major anthropogenic sources of organic air pollutants.

A major source of airborne organic compounds is the emissions from motor ve-

**TABLE 1.1** AIR POLLUTANTS

| | Physical properties | Concentration levels[a] | Anthropogenic sources | Natural sources |
|---|---|---|---|---|
| $SO_2$ | Colorless gas with irritating, pungent odor; detectable by taste at levels of 0.3 to 1 ppm; highly soluble in water (10.5 g/100 $cm^3$ at 293 K) | Global background concentration levels in the range 0.04 to 6 ppb; hourly averaged maximum concentrations in urban areas have occasionally exceeded 1 ppm | Fuel combustion in stationary sources; industrial process emissions; metal and petroleum refining | Atmospheric oxidation of organic sulfides |
| $H_2S$ | Colorless, flammable gas; highly toxic; characteristic rotten egg odor | Global background about 3 $\mu g\ m^{-3}$; urban levels have been observed as large as 390 $\mu g\ m^{-3}$ | Kraft pulp mills; natural gas and petroleum refining; rayon and nylon manufacture; coke ovens | Biological decay processes; volcanoes and geothermal activities |
| $\dot{N}O$ | Colorless, odorless gas; nonflammable and slightly soluble in water; toxic | Global background level from 10 to 100 ppt; urban levels have been observed as large as 500 ppb | Combustion | Bacterial action; natural combustion processes; lightning |
| $NO_2$ | Reddish-orange-brown gas with sharp, pungent odor; toxic and highly corrosive; absorbs light over much of the visible spectrum | Global background level from 10 to 500 ppt; urban concentrations have reached values exceeding 500 ppb | Combustion | |
| $NH_3$ | Colorless gas with pungent odor; detectable at concentrations exceeding 500 ppm; highly soluble in water | Global background level of 1 ppb; urban concentrations in range of 5 ppb | Combustion | Bacterial decomposition of amino acids in organic waste |
| $CO_2$ | Colorless, odorless, nontoxic gas moderately soluble in water | Global background concentration has increased from 290 ppm in 1900 to about 345 ppm in 1985 | Combustion of fossil fuels | |

| | | | | |
|---|---|---|---|---|
| CO | Colorless, odorless, flammable, toxic gas, slightly soluble in water | Global average concentration of 0.09 ppm; concentrations in northern hemisphere are about twice those in southern hemisphere; urban levels in the vicinity of heavily traveled roadways can exceed 100 ppm | Combustion of fossil fuels | Atmospheric oxidation of methane and other biogenic hydrocarbons |
| $O_3$ | Colorless, toxic gas, slightly soluble in water | Global background concentrations range from 20 to 60 ppb; polluted urban levels range from 100 to 500 ppb | No primary sources; formed as a secondary pollutant from atmospheric reactions involving hydrocarbons and oxides of nitrogen | Natural tropospheric chemistry; transport from stratosphere to troposphere |
| Nonmethane hydrocarbons (see Table 1.2) | | Global background concentrations range from 10 to 20 ppb; polluted urban levels range from 500 to 1200 ppb | Incomplete combustion; industrial sources | Vegetation |

[a] Two concentration units that are commonly used in reporting atmospheric species abundances are $\mu g\ m^{-3}$ and parts per million by volume (ppm). Parts per million by volume is not really a concentration but a dimensionless volume fraction, although it is widely referred to as a "concentration." Parts per million by volume may be expressed as

$$\text{"concentration" of species } i \text{ in ppm} = \frac{c_i}{c} \times 10^6$$

where $c_i$ and $c$ are moles/volume of species $i$ and air, respectively, at $p$ and $T$. Given a pollutant mass concentration $m_i$ expressed in $\mu g\ m^{-3}$

$$c_i = \frac{10^{-6}\, m_i}{M_i}$$

where $M_i$ is the molecular weight of species $i$ and $c = p/RT$. Thus the "concentration" of a species in ppm is related to that in $\mu g\ m^{-3}$ by

$$\text{"concentration" of species } i \text{ in ppm} = \frac{RT}{pM_i} \times \text{concentration in } \mu g\ m^{-3}$$

Parts per billion by volume (ppb) is just $(c_i/c) \times 10^9$.

**TABLE 1.2** HYDROCARBONS IDENTIFIED IN AMBIENT AIR

| Carbon number | Compound | Carbon number | Compound |
|---|---|---|---|
| 1 | Methane | | 2,3-Dimethylbutane |
| | | | *cis*-2-Hexene |
| 2 | Ethane | | *trans*-2-Hexene |
| | Ethylene | | *cis*-3-Hexene |
| | Acetylene | | *trans*-3-Hexene |
| | | | 2-Methyl-1-pentene |
| 3 | Propane | | 4-Methyl-1-pentene |
| | Propylene | | 4-Methyl-2-pentene |
| | Propadiene | | Benzene |
| | Methylacetylene | | Cyclohexane |
| | | | Methylcyclopentane |
| 4 | Butane | | |
| | Isobutane | 7 | 2-Methylhexane |
| | 1-Butene | | 3-Methylhexane |
| | *cis*-2-Butene | | 2,3-Dimethylpentane |
| | *trans*-2-Butene | | 2,4-Dimethylpentane |
| | Isobutene | | Toluene |
| | 1,3-Butadiene | | |
| | | 8 | 2,2,4-Trimethylpentane |
| 5 | Pentane | | Ethylbenzene |
| | Isopentane | | *o*-Xylene |
| | 1-Pentene | | *m*-Xylene |
| | *cis*-2-Pentene | | *p*-Xylene |
| | *trans*-2-Pentene | | |
| | 2-Methyl-1-butene | 9 | *m*-Ethyltoluene |
| | 2-Methyl-1,3-butadiene | | *p*-Ethyltoluene |
| | Cyclopentane | | 1,2,4-Trimethylbenzene |
| | Cyclopentene | | 1,3,5-Trimethylbenzene |
| | Isoprene | | |
| | | 10 | *sec*-Butylbenzene |
| 6 | Hexane | | α-Pinene |
| | 2-Methylpentane | | β-Pinene |
| | 3-Methylpentane | | 3-Carene |
| | 2,2-Dimethylbutane | | Limonene |

hicles. Motor vehicle emissions consist of unburned fuel,* in the form of organic compounds; oxides of nitrogen, in the form primarily of nitric oxide; carbon monoxide; and particulate matter. Since motor vehicle emissions vary with driving mode (idle, accelerate, decelerate, cruise), to obtain a single representative emission figure for a vehicle, it is run through a so-called driving cycle in which different driving modes are attained

*Gasoline is the 313 to 537 K fraction from petroleum distillation and contains approximately 2000 compounds. These include $C_4$ to $C_9$ paraffins, olefins, and aromatics. Typical compositions vary from 4% olefins and 48% aromatics to 22% olefins and 20% aromatics. Unleaded fuel has a higher aromatic content than leaded fuel.

**TABLE 1.3 POTENTIALLY HAZARDOUS AIR POLLUTANTS**

| Chemical name | Chemical formula | Toxicity[a] | Average concentration[b] (ppt) |
|---|---|---|---|
| Halomethanes | | | |
| Methyl chloride | $CH_3Cl$ | BM | 788 |
| Methyl bromide | $CH_3Br$ | BM | 141 |
| Methyl iodide | $CH_3I$ | SC, BM | 2.7 |
| Methylene chloride | $CH_2Cl_2$ | BM | 978 |
| Chloroform | $CHCl_3$ | SC, BM | 346 |
| Carbon tetrachloride | $CCl_4$ | SC, NBM | 221 |
| Haloethanes and halopropanes | | | |
| Ethyl chloride | $C_2H_5Cl$ | — | 100 |
| 1,2-Dichloroethane | $CH_2ClCH_2Cl$ | SC, BM | 558 |
| 1,2-Dibromoethane | $CH_2BrCH_2Br$ | SC | 32 |
| 1,1,1-Trichloroethane | $CH_3CCl_3$ | Weak BM | 512 |
| 1,1,2-Trichloroethane | $CH_2ClCHCl_2$ | SC, NBM | 29 |
| 1,1,2,2-Tetrachloroethane | $CHCl_2CHCl_2$ | SC, BM | 10 |
| 1,2-Dichloropropane | $CH_2ClCHClCH_3$ | BM | 60 |
| Chloroalkenes | | | |
| Vinylidene chloride | $CH_2{=}CCl_2$ | SC, BM | 19 |
| Trichloroethylene | $CHCl{=}CCl_2$ | SC, BM | 143 |
| Tetrachloroethylene | $CCl_2{=}CCl_2$ | SC | 401 |
| Allyl chloride | $ClCH_2CH{=}CH_2$ | SC | <5 |
| Hexachloro-1,3-butadiene | $Cl_2C{=}CCl{-}CCl{=}CCl_2$ | BM | 5 |
| Chloroaromatics | | | |
| Monochlorobenzene | $C_6H_5Cl$ | — | 280 |
| α-Chlorotoluene | $C_6H_5CH_2Cl$ | BM | <5 |
| *o*-Dichlorobenzene | $o$-$C_6H_4Cl_2$ | — | 12 |
| *m*-Dichlorobenzene | $m$-$C_6H_4Cl_2$ | — | 6 |
| 1,2,4-Trichlorobenzene | 1,2,4-$C_6H_3Cl_3$ | — | 5 |
| Aromatic hydrocarbon | | | |
| Benzene | $C_6H_6$ | SC | 3,883 |
| Oxygenated and nitrogenated species | | | |
| Formaldehyde | $HCHO$ | SC, BM | 14,200 |
| Phosgene | $COCl_2$ | — | <20 |
| Peroxyacetyl nitrate (PAN) | $CH_3COOONO_2$ | Phytotoxic | 589 |
| Peroxypropionyl nitrate (PPN) | $CH_3CH_2COOONO_2$ | Phytotoxic | 103 |
| Acrylonitrile | $CH{\equiv}CN$ | SC | — |

[a]BM; positive mutagenic activity based on Ames salmonella mutagenicity test (bacterial mutagens); NBM, not found to be mutagenic in the Ames salmonella test (not bacterial mutagens); SC, suspected carcinogens.

[b]Average from 2 weeks of measurements in Houston, St. Louis, Denver, and Riverside.

*Source:* Singh et al. (1981).

**TABLE 1.4** NO/$NO_x$ RATIOS IN EMISSIONS FROM VARIOUS SOURCE TYPES

| Source type | NO/$NO_x$ |
|---|---|
| Industrial boilers | |
| Natural gas | 0.90–1.0 |
| Coal | 0.95–1.0 |
| No. 6 fuel oil | 0.96–1.0 |
| Motor vehicle | |
| Internal combustion engine | 0.99–1.0 |
| Diesel-powered car | 0.77–1.0[a] |
| Diesel-powered truck and bus | 0.73–0.98 |
| Uncontrolled tail gas from nitric acid plant | ~0.50 |
| Petroleum refinery heater: natural gas | 0.93–1.0 |
| Gas turbine electrical generator: No. 2 fuel oil | 0.55–1.0[b] |

[a] The lower limit is for idle conditions; the higher for 50 mi/hr (80.5 km $h^{-1}$).

[b] The lower limit is for no load; the higher for full load.

*Source:* U.S. Environmental Protection Agency (1982a).

for prescribed periods. The driving cycle is carried out in the laboratory on a device called a dynamometer that offers the same resistance to the engine as actual road driving.

Three different driving cycles have been employed in emissions testing: the Federal Test Procedure (FTP), a cycle reflecting a mix of low and high speeds; the New York City Cycle (NYCC), a low-speed cycle to represent city driving; and the Crowded Urban Expressway (CUE) cycle, representative of high-speed driving. The average cycle speeds of the three cycles are: FTP—19.56 mi/hr (31.5 km $h^{-1}$); NYCC—7.07 mi/hr (11.4 km $h^{-1}$); CUE—34.79 mi/hr (56.0 km $h^{-1}$). Emissions of all pollutants are generally larger for the lower-speed cycles.

### 1.1.4 Particulate Matter

Particulate matter refers to everything emitted in the form of a condensed (liquid or solid) phase. Table 1.7 gives the total estimated U.S. particulate matter emissions in 1978, and Table 1.8 presents a summary of the chemical characteristics of uncontrolled particulate emissions from typical air pollution sources.

In utility and industrial use, coal and, to a lesser extent, oil combustion contribute most of the particulate (and sulfur oxides) emissions. Coal is a slow-burning fuel with a relatively high ash (incombustible inorganic) content. Coal combustion particles consist primarily of carbon, silica ($SiO_2$), alumina ($Al_2O_3$), and iron oxide (FeO and $Fe_2O_3$). In contrast to coal, oil is a fast-burning, low-ash fuel. The low ash content results in formation of less particulate matter, but the sizes of particles formed in oil combustion are generally smaller than those of particles from coal combustion. Oil combustion particulate matter contains cadmium, cobalt, copper, nickel, and vanadium.

**TABLE 1.5** ESTIMATED ANTHROPOGENIC $NO_x$ EMISSIONS IN THE UNITED STATES IN 1976 ($10^6$ metric tons/yr, expressed as $NO_2$)[a]

| Source category | |
|---|---|
| Transportation | **10.1** |
| Highway vehicles | 7.8 |
| Nonhighway vehicles | 2.3 |
| Stationary fuel combustion | **11.8** |
| Electric utilities | 6.6 |
| Industrial | 4.5 |
| Residential, commercial, and institutional | 0.7 |
| Industrial processes | **0.7** |
| Chemicals | 0.3 |
| Petroleum refining | 0.3 |
| Metals | 0 |
| Mineral products | 0.1 |
| Oil and gas production and marketing | 0[b] |
| Industrial organic solvent use | 0 |
| Other processes | 0 |
| Solid waste disposal | **0.1** |
| Miscellaneous | **0.3** |
| Forest wildfires and managed burning | 0.2 |
| Agricultural burning | 0 |
| Coal refuse burning | 0.1 |
| Structural fires | 0 |
| Miscellaneous organic solvent use | 0 |
| | **23.0** |

[a] One metric ton = $10^3$ kg.

[b] A zero entry indicates emissions of less than 50,000 metric tons/yr.

*Source:* U.S. Environmental Protection Agency (1982a).

**TABLE 1.6** ESTIMATED ANTHROPOGENIC $SO_2$ EMISSIONS IN THE UNITED STATES IN 1978 ($10^6$ metric tons/yr)

| Source category | |
|---|---|
| Stationary fuel combustion | 22.1 |
| Industrial processes | 4.1 |
| Transportation | 0.8 |
| | 27.0 |

*Source:* U.S. Environmental Protection Agency (1982b).

**TABLE 1.7 ESTIMATED ANTHROPOGENIC SULFUR OXIDE AND PARTICULATE MATTER EMISSIONS FROM STATIONARY SOURCES IN THE UNITED STATES IN 1978 ($10^3$ metric tons/yr)**

| Source category | Sulfur oxides | Particulate matter |
|---|---|---|
| Fuel combustion | | |
| Utility | | |
| Coal | 15,900 | 2,350 |
| Oil | 1,720 | 140 |
| Gas | 0 | 10 |
| Industrial | | |
| Coal | 1,890 | 700 |
| Oil | 1,150 | 90 |
| Gas | 0 | 40 |
| Other fuels[a] | 150 | 280 |
| Commercial/institutional | | |
| Coal | 40 | 20 |
| Oil | 900 | 60 |
| Gas | 0 | 10 |
| Residential | | |
| Coal | 60 | 20 |
| Oil | 260 | 20 |
| Gas | 0 | 30 |
| Industrial processes | | |
| Metals | | |
| Iron and steel | 110 | 830 |
| Primary smelting | 1,960 | 480 |
| Iron foundries | 0 | 140 |
| Other | 0 | 120 |
| Mineral products | | |
| Cement | 670 | 780 |
| Asphalt | 0 | 150 |
| Lime | 0 | 150 |
| Crushed rock | 0 | 1,340 |
| Other | 30 | 910 |
| Petroleum | | |
| Refining | 900 | 70 |
| Natural gas production | 140 | 0 |
| Chemical | | |
| Sulfuric acid | 220 | 0 |
| Other | 0 | 190 |
| Other | | |
| Grain processing | 0 | 730 |
| Pulp and paper | 80 | 240 |
| Other | 0 | 60 |
| Solid waste disposal | 0 | 500 |
| | 26,180 | 10,460 |

[a]Primarily wood/bark waste.

*Source:* U.S. Environmental Protection Agency (1982b).

Major industrial process sources of particulate matter include the metals, mineral products, petroleum, and chemicals industries. Iron and steel and primary smelting operations are the most significant emission sources in the metals industry. The iron and steel industry involves coke, iron, and steel production, each of which is a source of particulate emissions. The primary metals industry includes the smelting of copper, lead, and zinc, along with aluminum production. Sulfur in unprocessed ores is converted to $SO_2$ during smelting, with a relatively small portion emitted as particulate sulfate and sulfuric acid. Emissions from the mineral products industry result from the production of portland cement, asphalt, crushed rock, lime, glass, gypsum, brick, fiberglass, phosphate rock, and potash. The particles emitted from crushing, screening, conveying, grinding, and loading operations tend to be larger than 15 $\mu$m.

## 1.2 AIR POLLUTION LEGISLATION IN THE UNITED STATES

The 1970 Clean Air Act Amendments* was a major piece of legislation that in many respects first put teeth into air pollution control in the United States. A major goal of the Act was to achieve clean air by 1975. The Act required the Environmental Protection Agency (EPA) to establish National Ambient Air Quality Standards (NAAQS)—both primary standards (to protect public health) and secondary standards (to protect public welfare). The Act also required states to submit State Implementation Plans (SIPs) for attaining and maintaining the national primary standards within three years.

Automobile emissions were arbitrarily set at a 90% reduction from the 1970 (for CO and hydrocarbons) or 1971 (for $NO_x$) model year emissions to be achieved by 1975 (or 1976 for $NO_x$). Since there was no proven way to achieve these goals when the law was enacted, the industry was in effect forced to develop new technology to meet the standards by a certain deadline. This has been called "technology-forcing legislation." Emissions standards were to be written by the EPA for certain new industrial plants. These New Source Performance Standards (NSPS) represented national standards that were to be implemented and enforced by each state.

The Clean Air Act Amendments of 1977 incorporated a number of modifications and additions to the 1970 Act, although it retained the basic philosophy of federal management with state implementation. In this Act, the EPA was required to review and update, as necessary, air quality criteria and regulations as of January 1, 1980 and at five-year intervals thereafter. A new aspect was included for "prevention of significant deterioration" (PSD) of air quality in regions cleaner than the NAAQS. Prior to the 1977 Amendments it was theoretically possible to locate air pollution sources in such regions and pollute clean air up to the limits of the ambient standards. However, the Act defined class 1 (pristine) areas, class 2 (almost all other areas), and class 3 (industrialized) areas. Under the PSD provisions, the ambient concentrations of pollutants will be

*The original Clean Air Act was passed in 1963.

**TABLE 1.8 CHARACTERISTICS OF UNCONTROLLED PARTICULATE EMISSIONS**

| Source category | Particle size (weight % less than stated size) 15 μm | 2.5 μm | 1.0 μm | Chemical composition: Major elements and compounds | Trace elements (less than 1% by weight) |
|---|---|---|---|---|---|
| Fuel combustion | | | | | |
| Utility | | | | | |
| Coal | 15–90 | 5–70 | 1–15 | Al, Ca, Fe, Si, sulfates, organics | As, B, Ba, Be, Cd, Cl, Co, Cr, Cu, F, Hg, K, Mg, Mn, Na, Ni, P, Pb, S, Se, Ti, V, Zn, Zr |
| Oil | 95 | 70–95 | 5–20 | Al, Ca, Fe, Mg, Na, sulfates, organics | As, Ba, Br, Co, Cr, Cu, K, Mn, Mo, Ni, Pb, Se, Sr, Ti, V |
| Industrial | | | | | |
| Oil | — | — | 65–95 | Al, Fe, Mg, Si, sulfates, organics | As, Ba, Ca, Cd, Co, Cr, Cu, Hg, K, Mo, Ni, Pb, Se, Sr, Ti, V, Zn |
| Gas | — | — | 100 | Cl, Na, sulfates, organics | — |
| Commercial/institutional/residential | | | | | |
| Oil | — | — | — | Al, Ca, Mg, Zn, sulfates | As, Ba, Cd, Cr, Cu, Hg, K, Ni, Pb, Sb, C |
| Gas | — | — | 100 | Cl, Na, sulfates, organics | — |
| Industrial processes | | | | | |
| Metals | | | | | |
| Iron and steel | — | 35–99 | 30–95 | Al, C, Ca, Cr, Fe, K, Mg, Mn, Pb, Si, Zn, sulfates, organics | Ag, As, Br, Cd, Cs, Cu, F, I, Mo, Ni, Rb, Se, Sn, Sr, V, Zr |

| | | | | | |
|---|---|---|---|---|---|
| Primary aluminum | 90 | 75 | 35–45 | Al, C, Ca, F, Fe, Na | — |
| Primary copper | — | 20–95 | 70 | Cu, Pb, S, Zn | Ag, Al, As, Cd, Hg, Sb, Se, Si, Te |
| Primary lead | — | 80 | — | Pb, Zn | As, Cd, Se, Te |
| Primary zinc | — | 90–98 | — | Cd, Fe, Pb, S, Zn | Cu, Hg, Mn, Sn |
| Iron foundries | 70–95 | 65–90 | 65 | — | — |
| Mineral products | | | | | |
| Cement | 80 | 30 | 5–30 | Al, C, Ca, Cl, K, Mg, Na, Si, carbonates, sulfates | Ag, Ba, Cd, Cr, Cu, F, Fe, Mn, Mo, Ni, Pb, Rb, Se, Ti, Zn |
| Asphalt | 10–15 | 1–2 | $<\frac{1}{2}$ | Al, C, Ca, Fe, K, Mg, Si, sulfates | Ag, As, Ba, Cr, Ti |
| Lime | — | 25–50 | 5 | Ca, Fe, Mg, Se, Si, carbonates | — |
| Gypsum | — | — | 20 | Al, C, Ca, Mg, Na, sulfates | As, Ba, Br, Cd, Cl, Cr, Cu, Fe, K, Mn, Mo, Ni, Pb, Se, Sr, Y, Zn |
| Crushed rock | — | — | 1–2 | Ca, Si, P | Ba, Cu, Fe, K, Mn, Sr |
| Petroleum | — | — | 50–90 | Asphalt, coke dust, sulfuric acid mist, fly ash, soot | — |
| Chemicals | | | | | |
| Sulfuric acid | — | 40–95 | 10–55 | Sulfuric acid mist | — |
| Others | | | | | |
| Grain processing | 15 | 1 | 0 | Organics | — |
| Pulp and paper | 90–95 | 70–80 | — | Ca, Mg, Na, carbonates, sulfates | — |
| Solid-waste disposal incinerators | 45 | 35 | — | — | — |

*Source:* U.S. Environmental Protection Agency (1982b).

allowed to rise very little in class 1 areas, by specified amounts in class 2 areas, and by larger amounts in class 3 areas.

The 1977 Amendments also addressed the issue of nonattainment areas: those areas of the country that were already in violation of one or more of the NAAQS. The law appeared to prohibit any more emissions whatsoever and thus seemed as if it would prevent any further growth in industry or commerce in these areas. However, subsequent interpretations by EPA led to a policy known as emissions offset that allowed a new source to be constructed in a nonattainment area provided that its emissions were offset by simultaneous reductions in emissions from existing sources.

Emissions standards for automobiles were delayed, and the standard for $NO_x$ was permanently relaxed from the original goals of the 1970 Act. CO and hydrocarbon standards were set at a 90% reduction from the 1970 model year to 3.4 g/mi for CO and 0.41 g/mi for hydrocarbons to be achieved by the 1981 model year. The required $NO_x$ reduction was relaxed to 1 g/mi by the 1982 model year, representing a reduction from about 5.5 g/mi in 1970. Standards were also proposed for heavy-duty vehicles such as trucks and buses.

Two types of air pollution standards emerged from the legislation. The first type is ambient air quality standards, those that deal with concentrations of pollutants in the outdoor atmosphere. The second type is source performance standards, those that apply to emissions of pollutants from specific sources. Ambient air quality standards are always expressed in concentrations such as micrograms per cubic meter or parts per million; whereas source performance standards are written in terms of mass emissions per unit of time or unit of production, such as grams per minute or kilograms of pollutant per ton of product.

Table 1.9 presents the current National Ambient Air Quality Standards. Some states, such as California, have set their own standards, some of which are stricter than those listed in the table. New Source Performance Standards (NSPS) are expressed as mass emission rates for specific pollutants from specific sources. These standards are

**TABLE 1.9** NATIONAL AMBIENT AIR QUALITY STANDARDS (PRIMARY)

| Pollutant | Averaging time | Primary standard |
|---|---|---|
| Sulfur dioxide | Annual average | 80 $\mu g\ m^{-3}$ |
| | 24 h | 365 $\mu g\ m^{-3}$ |
| Nitogen dioxide | Annual average | 100 $\mu g\ m^{-3}$ |
| Carbon monoxide | 8 h | 10 $mg\ m^{-3}$ |
| | 1 h | 40 $mg\ m^{-3}$ |
| Ozone | 1 h | 0.12 ppm (235 $\mu g\ m^{-3}$) |
| Particulate matter ($PM_{10}$)[a] | Annual geometric mean | 50 $\mu g\ m^{-3}$ |
| | 24 h | 150 $\mu g\ m^{-3}$ |

[a] See the text.

*Source:* 40 CFR (Code of Federal Regulations) 50, 1982.

generally derived from field tests at a number of industrial plants. A separate category of standards for emissions from point sources has been created for hazardous air pollutants, such as beryllium, mercury, vinyl chloride, benzene, and asbestos.

The particulate matter entry in Table 1.9 requires some explanation. After a periodic review of the National Ambient Air Quality Standards and a revision of the Health and Welfare Criteria as required in the 1977 Clean Air Act Amendments, the EPA proposed in 1987 the following relative to the particulate matter standard:

1. That total suspended particulate matter (TSP) as an indicator for particulate matter be replaced for both the primary standards, that is, the annual geometric mean and the 24-hour average, by a new indicator that includes only those particles with an aerodynamic diameter smaller than or equal to a nominal 10 $\mu$m ($PM_{10}$)
2. That the level of the 24-hour primary standard be 150 $\mu g\ m^{-3}$ and the deterministic form of the standard be replaced with a statistical form that permits one expected exceedance of the standard level per year
3. That the level of the annual primary standard be 50 $\mu g\ m^{-3}$, expressed as an expected annual arithmetic mean

EPA also proposed in the *Federal Register* to revise its regulations governing State Implementation Plans to account for revisions to the NAAQS for TSP and $PM_{10}$. Under the Act, each state must adopt and submit an SIP that provides for attainment and maintenance of the new or revised standards within nine months after the promulgation of an NAAQS. The revision authorizes the EPA Administrator to extend the deadline for up to 18 months as necessary.

Table 1.10 gives some selected New Source Performance Standards. The uncontrolled emission rates for a variety of processes can be estimated from the data available in the EPA publication generally referred to as AP-42, "Compilation of Air Pollutant Emission Factors" (U.S. Environmental Protection Agency, 1977).

## 1.3 ATMOSPHERIC CONCENTRATION UNITS

We note from Table 1.8 that two concentration units that are commonly used in reporting atmospheric species abundance are $\mu g\ m^{-3}$ and parts per million by volume (ppm). Parts per million by volume is just

$$\frac{c_i}{c} \times 10^6$$

where $c_i$ and $c$ are moles per volume of species $i$ and air, respectively, at pressure $p$ and temperature $T$. Note that in spite of the widespread reference to it as a concentration, parts per million by volume is not really a concentration but a dimensionless volume fraction.

**TABLE 1.10 SOME NEW SOURCE PERFORMANCE STANDARDS (NSPS)**

| | |
|---|---|
| Steam electric power plants | |
| Particulate matter | 13 g/$10^6$ kJ |
| $NO_x$ | |
| Gaseous fuel | 86 g/$10^6$ kJ |
| Liquid fuel | 130 g/$10^6$ kJ |
| Coal | 260 g/$10^6$ kJ |
| $SO_2$ | |
| Gas or liquid fuel | 86 g/$10^6$ kJ |
| Coal | At least 70% removal depending on conditions |
| Solid waste incinerators: particulate matter | 0.18 g/dscm[a] corrected to 12% $CO_2$ (3-hr average) |
| Sewage sludge incinerators: particulate matter | 0.65 g/kg sludge input (dry basis) |
| Iron and steel plants: particulate matter | 50 mg/dscm[a] |
| Primary copper smelters | |
| Particulate matter | 50 mg/dscm[a] |
| $SO_2$ | 0.065% by volume |

[a] Dry standard cubic meter.

*Source:* 40 CFR (Code of Federal Regulations) 60, 1982.

Given a pollutant mass concentration $m_i$ expressed in $\mu$g m$^{-3}$,

$$c_i = \frac{10^{-6}\, m_i}{M_i}$$

where $M_i$ is the molecular weight of species $i$ and $c = p/RT$. Thus

$$\text{concentration of species } i \text{ in ppm} = \frac{RT}{pM_i} \times \text{concentration in } \mu\text{g m}^{-3}$$

If $T$ is in kelvin and $p$ in pascal, then (see Table 1.15 for the value of the gas law constant $R$)

$$\text{concentration of species } i \text{ in ppm} = \frac{8.314T}{pM_i} \times \text{concentration in } \mu\text{g m}^{-3}$$

**Example 1.1** ***Conversion between Parts per Million and Micrograms per Cubic Meter***

Confirm the relation between ppm and $\mu$g m$^{-3}$ for ozone given in Table 1.9 at $T = 298$ K and $p = 1$ atm $(1.0133 \times 10^5$ Pa)

$$\text{concentration in } \mu\text{g m}^{-3} = \frac{pM_i}{8.314T} \times \text{concentration in ppm}$$

$$= \frac{(1.0133 \times 10^5)(48)}{8.314(298)} \times 0.12$$

$$= 235.6\ \mu\text{g m}^{-3}$$

The 24-hour $SO_2$ NAAQS is 365 $\mu$g m$^{-3}$. Convert this to ppm at the same temperature and pressure.

$$\text{concentration in ppm} = \frac{(8.314)(298)}{(1.0133 \times 10^5)(64)} \times 365$$
$$= 0.139 \text{ ppm}$$

## 1.4 THE APPENDICES TO THIS CHAPTER

Analysis of the generation and control of air pollutants at the source, air pollution engineering, requires a basis of thermodynamics, fluid mechanics, heat and mass transfer, and chemical kinetics. This chapter concludes with five appendices, the first four of which provide some basic material on chemical kinetics, heat and mass transfer, probability, and turbulence that will be called upon in later chapters. Appendix E presents the units that will be used throughout the book.

## APPENDIX A CHEMICAL KINETICS

Chemical kinetics is concerned with the mechanisms and rates of chemical reactions. A single chemical reaction among $S$ species, $A_1, A_2, \ldots, A_S$, can be written as

$$\sum_{i=1}^{S} \nu_i A_i = 0 \qquad \text{(A.1)}$$

where the stoichiometric coefficient $\nu_i$ is positive (by convention) for the products and negative for the reactants. For example, the reaction

$$\nu_1 A_1 + \nu_2 A_2 \longrightarrow \nu_3 A_3 + \nu_4 A_4$$

is written in the form of (A.1) as

$$\nu_3 A_3 + \nu_4 A_4 - \nu_1 A_1 - \nu_2 A_2 = 0$$

If there are $R$ chemical reactions, we denote them by

$$\sum_{i=1}^{S} \nu_{ij} A_i = 0 \qquad j = 1, 2, \ldots, R \qquad \text{(A.2)}$$

where $\nu_{ij}$ is the stoichiometric coefficient of species $i$ in reaction $j$.

Let $R_i$ be the rate of generation of species $i$ by chemical reaction (g-moles $i$ m$^{-3}$ s$^{-1}$), and let $r_j$ be the rate of reaction $j$ (g-mol m$^{-3}$ s$^{-1}$). Then in a closed system,

$$\frac{dc_i}{dt} = R_i = \sum_{j=1}^{R} \nu_{ij} r_j \qquad i = 1, 2, \ldots, S \qquad \text{(A.3)}$$

which also implies that

$$\frac{(R_i) \text{ due to reaction } j}{(R_l) \text{ due to reaction } j} = \frac{\nu_{ij}}{\nu_{lj}} \qquad i, l = 1, 2, \ldots, S; \; j = 1, 2, \ldots, R \tag{A.4}$$

We now define the *extents* of the $R$ reactions, $\xi_1, \ldots, \xi_R$, by

$$c_i - c_{io} = \sum_{j=1}^{R} \nu_{ij} \xi_j \qquad i = 1, 2, \ldots, S \tag{A.5}$$

so that the extents satisfy

$$\frac{d\xi_j}{dt} = r_j(\xi_1, \ldots, \xi_R) \qquad j = 1, 2, \ldots, R; \quad \xi_j(0) = 0 \tag{A.6}$$

Consider the gas-phase chemical reaction

$$A + B \xrightarrow{k} C$$

occurring in a closed reactor, the volume of which $V(t)$ can change with time. The intrinsic reaction rate is $kc_A c_B$. Let us examine the rate of change of the concentrations in the system. The molar concentration of species $A$ at any time $t$ is $c_A = n_A/V$, and the rate of change of the number of moles of species $A$ is

$$\frac{d}{dt}(c_A V) = -Vkc_A c_B \tag{A.7}$$

or

$$\frac{dc_A}{dt} = -kc_A c_B - \frac{c_A}{V}\frac{dV}{dt} \tag{A.8}$$

Thus we see that in a system with volume change, the concentration of a reacting species changes due to the volume change as well as to the reaction itself. It is desirable to be able to express the rate of change of the concentration in a way that depends only on the chemical reaction occurring. In a closed system the total mass $m$ is unchanging, so a concentration based on the total mass in the system rather than on the total volume would seem to satisfy our desires. Let $\rho$ be the overall mass density of the system, equal to $m/V$. Let us replace $V$ by $m/\rho$ in (A.7):

$$\frac{d}{dt}\left(\frac{c_A m}{\rho}\right) = -\frac{m}{\rho} kc_A c_B \tag{A.9}$$

Since $m$ is constant, it can be canceled from both sides, giving

$$\rho \frac{d}{dt}\left(\frac{c_A}{\rho}\right) = -kc_A c_B \tag{A.10}$$

Thus, for a system where volume is changing in time, a quantity that reflects only the concentration change due to chemical reaction is

$$\rho \frac{d}{dt}\left(\frac{c_i}{\rho}\right)$$

We will frequently use this form of writing the rate of a combustion reaction to isolate the effects of chemistry from those due to expansions, heating and cooling.

**Example 1.2** ***Extent of Reaction***

Consider the two reactions

$$N_2O_4 \longrightarrow 2NO_2$$

$$NO_2 + NO_2 \longrightarrow 2NO + O_2$$

We let $A_1 = O_2$, $A_2 = NO$, $A_3 = NO_2$, and $A_4 = N_2O_4$. Assume that at $t = 0$ only $N_2O_4$ is present. The stoichiometric coefficients are $\nu_{31} = 2$, $\nu_{41} = -1$, $\nu_{22} = 2$, $\nu_{12} = 1$, and $\nu_{32} = -2$. We introduce the extents of reaction, according to (A.5),

$$c_1 = \xi_2$$

$$c_2 = 2\xi_2$$

$$c_3 = 2\xi_1 - 2\xi_2$$

$$c_4 - c_{40} = -\xi_1$$

For a closed, uniform system at constant volume,

$$\frac{dc_i}{dt} = \sum_{j=1}^{2} \nu_{ij} r_j \qquad i = 1, 2, 3, 4$$

which can be written in terms of the extents of the two reactions,

$$\frac{d\xi_1}{dt} = r_1(\xi_1, \xi_2) \qquad \frac{d\xi_2}{dt} = r_2(\xi_1, \xi_2)$$

Theory provides expressions for the reaction rates $r_j$ as functions of concentrations and temperature, including certain parameters such as the frequency factor and the activation energy. A reaction as written above is an *elementary* reaction if it proceeds at the molecular level as written. Sometimes a reaction does not proceed microscopically as written but consists of a sequence of elementary reactions. For example, the photolysis reaction

$$H_2 + Br_2 \longrightarrow 2HBr$$

consists of the sequence of elementary reactions,

$$Br_2 + h\nu \longrightarrow 2Br$$

$$Br + H_2 \rightleftarrows HBr + H$$

$$H + Br_2 \longrightarrow HBr + Br$$

$$Br + Br \longrightarrow Br_2$$

The sequence of elementary reactions is called the *mechanism* of the reaction. Aside from the fundamental interest of understanding the chemistry on a molecular level, a reaction mechanism allows us to derive an expression for the reaction rate.

The number of molecules participating in an elementary reaction is its *molecularity*. Customarily, there are monomolecular and bimolecular reactions. Truly monomolecular reactions consist only of photolysis, such as $Br_2 + h\nu$ above, radioactive decay, or a spontaneous transition from a higher to a lower electronic state. Frequently, reactions written as monomolecular, such as isomerizations, are in fact bimolecular because the energy necessary to cause the reaction is provided by collision of the molecule with a background species. Such a background species that acts only as a reaction chaperone is usually designated $M$. There are no true termolecular reactions in the sense that three molecules collide simultaneously; one written as $A + B + C \rightarrow$ is most likely the result of two bimolecular steps,

$$A + B \longrightarrow AB \qquad AB + C \longrightarrow$$

**Example 1.3** ***Independence of Reactions***

Given a chemical reaction mechanism, there is the possibility that two reactions are multiples of each other or that one reaction is a linear combination of two others. Such a reaction tells us nothing more in a stoichiometric sense than the reactions on which it is dependent, since any changes in composition it predicts could equally well be accounted for by the other reactions.

For small numbers of reactions we can frequently determine if they are linearly independent by inspection, observing whether any reaction can be reproduced by adding or subtracting other reactions. In general, however, there is a systematic approach to determining the independence of a set of reactions (Aris, 1965).

Consider the set of reactions

$$H_2 + O_2 \xrightarrow{1} 2OH$$

$$OH + H_2 \xrightarrow{2} H_2O + H$$

$$H + O_2 \xrightarrow{3} OH + O$$

$$OH + H \xrightarrow{4} H_2 + O$$

If we define

$$A_1 = H_2 \qquad A_4 = O$$

$$A_2 = O_2 \qquad A_5 = OH$$

$$A_3 = H_2O \qquad A_6 = H$$

the reactions may be written as

$$\begin{aligned}
-A_1 \quad -A_2 \qquad\qquad\qquad +2A_5 \qquad\qquad &= 0 \\
-A_1 \qquad\quad +A_3 \qquad\quad -A_5 \quad +A_6 &= 0 \\
-A_2 \qquad\quad +A_4 \quad +A_5 \quad -A_6 &= 0 \\
A_1 \qquad\qquad\qquad +A_4 \quad -A_5 \quad -A_6 &= 0
\end{aligned}$$

To test for independence, form a matrix of the stoichiometric coefficients with $\nu_{ij}$ in the $j$th row and the $i$th column, that is,

$$\begin{bmatrix} -1 & -1 & 0 & 0 & 2 & 0 \\ -1 & 0 & 1 & 0 & -1 & 1 \\ 0 & -1 & 0 & 1 & 1 & -1 \\ 1 & 0 & 0 & 1 & -1 & -1 \end{bmatrix}$$

Now take the first row with a nonzero element in the first column (in this case it is the first row) and divide that row by the leading element. This will yield a new row of

$$1 \quad \frac{\nu_{2j}}{\nu_{1j}} \quad \frac{\nu_{3j}}{\nu_{1j}} \quad \cdots \quad \frac{\nu_{Sj}}{\nu_{1j}}$$

where $j$ is the number of the row used. This new row may be used to make all the other elements in the first column zero by subtracting $\nu_{1k}$ times the new row from the corresponding element of the $k$th row,

$$0 \left( \nu_{2k} - \nu_{1k} \frac{\nu_{2j}}{\nu_{1j}} \right) \cdots \left( \nu_{Sk} - \nu_{1k} \frac{\nu_{Sj}}{\nu_{1j}} \right)$$

The matrix for the present example becomes

$$\begin{bmatrix} 1 & 1 & 0 & 0 & -2 & 0 \\ 0 & 1 & 1 & 0 & -3 & 1 \\ 0 & -1 & 0 & 1 & 1 & -1 \\ 0 & -1 & 0 & 1 & 1 & -1 \end{bmatrix}$$

The next step is to ignore the first row and first column and repeat this matrix reduction process for the reduced matrix containing $R - 1$ rows. This yields

$$\begin{bmatrix} 1 & 1 & 0 & 0 & -2 & 0 \\ 0 & 1 & 1 & 0 & -3 & 1 \\ 0 & 0 & 1 & 1 & -2 & 0 \\ 0 & 0 & 1 & 1 & -2 & 0 \end{bmatrix}$$

This reduction process is continued until we have 1's as far as possible down the diagonal and 0's in all elements in rows below the last 1 on the diagonal. Continuing, we find

$$\begin{bmatrix} 1 & 1 & 0 & 0 & -2 & 0 \\ 0 & 1 & 1 & 0 & -3 & 1 \\ 0 & 0 & 1 & 1 & -2 & 0 \\ 0 & 0 & 0 & 0 & 0 & 0 \end{bmatrix}$$

At this point we have three rows with 1's on the diagonal and only 0's in the final row. The number of independent reactions is the number of 1's with only zeros to their left. Alternatively it is the number of reactions minus the number of rows that are entirely zero. In this case, then, only three of the four reactions are linearly independent. The procedure does not tell us which reactions are dependent, but inspection of the set reveals that reaction 3 is the sum of reactions 1 and 4. Thus we need to replace any one of reactions 1, 3, or 4 with another reaction and repeat the test. For example, reaction 4 could be replaced with

$$O_2 \xrightarrow{5} 2O$$

which will be found to be independent.

## A.1 Reaction Rates

Gas molecules can react only when they come close enough to one another for direct energy exchange that can lead to bond breaking. For the di- or triatomic molecules that are important in the latter phases of combustion chemistry, the centers of the two molecules must approach within a few angstroms. From elementary kinetic theory, the frequency of collisions per unit volume of gas of molecules of type $i$ of mass $m_i$ with molecules of type $j$ of mass $m_j$ is (Benson, 1960)

$$Z_{ij} = \left(\frac{8k_BT}{\pi m_{ij}}\right)^{1/2} \pi\sigma_{ij}^2 N_i N_j \quad \text{m}^{-3}\ \text{s}^{-1} \tag{A.11}$$

where $N_i$ is the number concentration of species $i$ ($\text{m}^{-3}$), $(8k_BT/\pi m_{ij})^{1/2}$ is the root-mean-square relative speed of the $i$ and $j$ molecules, $k_B$ is the Boltzmann constant ($1.38 \times 10^{-23}$ J molecule$^{-1}$ K$^{-1}$), $m_{ij} = m_i m_j/(m_i + m_j)$ is the reduced mass, and $\pi\sigma_{ij}^2$ is the cross-sectional area in which interaction can occur.

The characteristic time during which molecules in thermal motion in a gas are close enough to interact is brief, on the order of $10^{-12}$ to $10^{-13}$ s. At ambient temperature and pressure the mean time between molecular collisions can be shown from (A.11) to be the order of $10^{-9}$ s. Thus collisions are short in duration compared to the time between collisions.

Whereas the collision of two molecules is a necessary condition for reaction, sufficient energy must be available to break chemical bonds. Theory indicates that the fraction of collisions involving energy greater than a required energy $E$ is given by exp $(-E/k_BT)$. In this form $E$ has units of energy per molecule. More commonly, $E$ is expressed in terms of energy per mole, and we use exp $(-E/RT)$, where $R$ is the universal gas constant (see Table 1.15). The rate of reaction is expressed in a form that accounts for both the frequency of collisions and the fraction that exceed the required energy,

$$r = A(T) \exp\left(-\frac{E}{RT}\right) c_i c_j \tag{A.12}$$

The preexponential factor $A(T)$ may depend on temperature since the translational kinetic energy and internal degrees of freedom of the molecules influence the probability of reaction in any collision event. The rate of reaction is usually written as $r = kc_ic_j$, where the parameter $k$ is called the *rate constant*,

$$k = A(T)\exp\left(-\frac{E}{RT}\right) \tag{A.13}$$

If $A(T)$ is independent of $T$, we have the *Arrhenius form*, $k = A\exp(-E/RT)$.

The parameter $E$ appearing in (A.13) is the *activation energy*. Figure 1.1 illustrates the energetics of an exchange reaction of the type

$$A + B \longrightarrow C + D$$

The difference in the energies of the initial and final states is the *heat of reaction* $\Delta h_r$. The peak in the energy along the reaction coordinate is associated with the formation of an activated complex $AB^\ddagger$, a short-lived intermediate through which the reactants must pass if the encounter is to lead to reaction. By estimating the structure of this transition state the activation energy $E$ may be estimated (Benson, 1960), although the most reliable estimates of $E$ are obtained by correlating rates measured at different temperatures to the Arrhenius form of $k$.

Most elementary reactions can be considered to be reversible,

$$A + B \underset{k_b}{\overset{k_f}{\rightleftharpoons}} C + D$$

The time rate of change of one of the reactants or products due to this one reaction is

$$\rho\frac{d([D]/\rho)}{dt} = k_f[A][B] - k_b[C][D]$$

where the brackets represent an alternative notation for the species concentration (i.e., $[A] = c_A$) and where we have used the moles per unit mass, $[D]/\rho$, in anticipation of combustion kinetics.

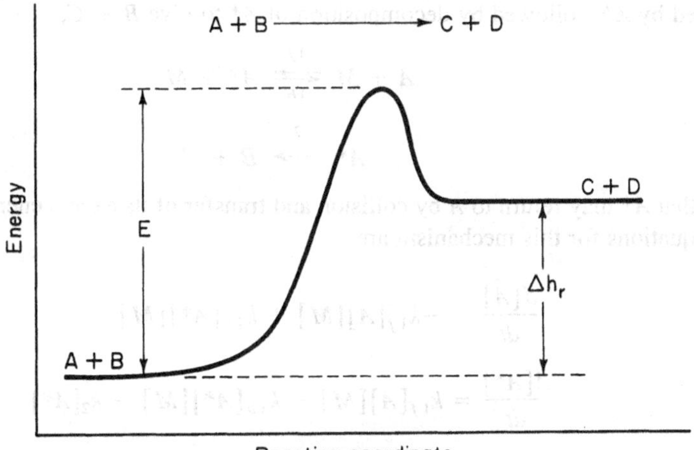


**Figure 1.1** Energetics of an exchange reaction $A + B \rightarrow C + D$.

At chemical equilibrium

$$0 = k_f[A]_e[B]_e - k_b[C]_e[D]_e \tag{A.14}$$

or, rearranging,

$$\frac{k_f}{k_b} = \frac{[C]_e[D]_e}{[A]_e[B]_e} \tag{A.15}$$

The right-hand side is equal to the equilibrium constant expressed in terms of concentrations, $K_c$. Thus we see that the ratio of the forward and reverse rate constants of a reaction is equal to the equilibrium constant, $k_f(T)/k_b(T) = K_c(T)$. This principle of *detailed balancing* is very important in the study of chemical kinetics since it allows one of the two rates to be calculated from the other rate and the equilibrium constant. Often, direct measurements of rate constants are available for only one reaction direction. When measurements are available for both reactions, detailed balancing provides a check on the consistency of the two rates.

## A.2 The Pseudo-Steady-State Approximation

Many chemical reactions, including those occurring in combustion processes, involve very reactive intermediate species such as free radicals, which, due to their very high reactivity, are consumed virtually as rapidly as they are formed and consequently exist at very low concentrations. The pseudo-steady-state approximation (PSSA) is a fundamental way of dealing with such reactive intermediates when deriving the overall rate of a chemical reaction mechanism.

It is perhaps easiest to explain the PSSA by way of a simple example. Consider the unimolecular reaction $A \rightarrow B + C$ whose elementary steps consist of the activation of $A$ by collision with a background molecule $M$ to produce an energetic $A$ molecule denoted by $A^*$, followed by decomposition of $A^*$ to give $B + C$,

$$A + M \underset{1b}{\overset{1f}{\rightleftarrows}} A^* + M$$

$$A^* \xrightarrow{2} B + C$$

Note that $A^*$ may return to $A$ by collision and transfer of its excess energy to an $M$. The rate equations for this mechanism are

$$\frac{d[A]}{dt} = -k_{1f}[A][M] + k_{1b}[A^*][M] \tag{A.16}$$

$$\frac{d[A^*]}{dt} = k_{1f}[A][M] - k_{1b}[A^*][M] - k_2[A^*] \tag{A.17}$$

The reactive intermediate in this mechanism is $A^*$. The PSSA states that the rate of generation of $A^*$ is equal to its rate of disappearance; physically, what this means is that

$A^*$ is so reactive, as soon as an $A^*$ molecule is formed, that it reacts by one of its two paths. Thus the PSSA gives

$$k_{1f}[A][M] - k_{1b}[A^*][M] - k_2[A^*] = 0 \tag{A.18}$$

From this we find the concentration of $A^*$ in terms of the concentrations of the stable molecules $A$ and $M$,

$$[A^*] = \frac{k_{1f}[A][M]}{k_{1b}[M] + k_2} \tag{A.19}$$

This expression can then be used in (A.16) to give

$$\frac{d[A]}{dt} = -\frac{k_{1f}k_2[M][A]}{k_{1b}[M] + k_2} \tag{A.20}$$

We see that the single overall reaction $A \rightarrow B + C$ with a rate given by (A.20) is not elementary because of the dependence on $M$. If the background species $M$ is in such excess that its concentration is effectively constant, the overall rate can be expressed as $d[A]/dt = -k[A]$, where $k = k_{1f}k_2[M]/(k_{1b}[M] + k_2)$ is a constant. If $k_{1b}[M] \gg k_2$, then $d[A]/dt = -k[A]$, with $k = k_{1f}k_2/k_{1b}$. On the other hand, if $k_{1b}[M] \ll k_2$, then $d[A]/dt = -k_{1f}[M][A]$.

One comment is in order. The PSSA is based on the presumption that the rates of formation and disappearance of a reactive intermediate are equal. A consequence of this statement is that $d[A^*]/dt = 0$ from (A.17). If this is interpreted to mean that $[A^*]$ does not change with time, this interpretation is incorrect. $[A^*]$ is at steady state with respect to $[A]$ and $[M]$. We can, in fact, compute $d[A^*]/dt$. It is

$$\frac{d[A^*]}{dt} = \frac{d}{dt}\left[\frac{k_{1f}[A][M]}{k_{1b}[M] + k_2}\right] \tag{A.21}$$

which, if $[M]$ is constant, is

$$\frac{d[A^*]}{dt} = -\frac{k_{1f}^2 k_2[M]^2[A]}{(k_{1b}[M] + k_2)^2} \tag{A.22}$$

The key point is that (A.19) is valid only after a short initial time interval needed for the rates of formation and disappearance of $A^*$ to equilibrate. After that time $[A^*]$ adjusts slowly on the time scale associated with changes in $[A]$ so as to maintain that balance. That slow adjustment is given by (A.22).

**Example 1.4** ***Analysis of Bimolecular Reactions***

When two molecules collide and form a single molecule,

$$A + B \underset{k_{-a}}{\overset{k_{+a}}{\rightleftharpoons}} AB^{\ddagger}$$

the initial collision produces an activated complex that has sufficient energy to overcome an energy barrier and decompose. The lifetime of the activated complex is short, on the order of the vibrational period of the complex (e.g., $10^{-12}$ to $10^{-13}$ s). Unless another molecule collides with the activated complex within this period and removes some of this excess energy, that is,

$$AB^{\ddagger} + M \underset{k_{-s}}{\overset{k_{+s}}{\rightleftharpoons}} AB + M$$

the activated complex will decay back to $A$ and $B$. At ambient temperature and pressure the frequency of collisions of background molecules (e.g., air) with the complex is of the order of $10^9$ s$^{-1}$. Thus only one $AB^{\ddagger}$ complex out of $10^3$ to $10^4$ formed can produce a stable molecule. The actual number may be lower and may depend on the type of third body $M$ involved.

The rate of formation of the stable product, $AB$, is

$$\frac{d[AB]}{dt} = k_{+s}[AB^{\ddagger}][M] - k_{-s}[AB][M]$$

The PSSA can be applied to $[AB^{\ddagger}]$, giving

$$[AB^{\ddagger}] = \frac{k_{+a}[A][B] + k_{-s}[AB][M]}{k_{-a} + k_{+s}[M]}$$

Substituting into the rate equation and grouping terms, we find that

$$\frac{d[AB]}{dt} = \frac{k_{+s}k_{+a}[A][B][M]}{k_{-a} + k_{+s}[M]} - \frac{k_{-a}k_{-s}[AB][M]}{k_{-a} + k_{+s}[M]}$$

At low pressure, $[M] = p/RT$ is small, so $k_{-a} \gg k_{+s}[M]$ and

$$\frac{d[AB]}{dt} = \frac{k_{+s}k_{+a}}{k_{-a}}[A][B][M] - k_{-s}[AB][M]$$

In the high-pressure limit, $k_{-a} \ll k_{+s}[M]$ and

$$\frac{d[AB]}{dt} = k_{+a}[A][B] - \frac{k_{-a}k_{-s}}{k_{+s}}[AB]$$

We see that in the low-pressure limit the forward reaction appears from the rate expression to be a termolecular reaction, whereas the reverse reaction is bimolecular. On the other hand, in the high-pressure limit, because of the high concentration of $M$, the collisional stabilization of the activated complex is very efficient and the forward reaction appears to be bimolecular, whereas the reverse reaction appears to be unimolecular.

## A.3 Hydrocarbon Pyrolysis Kinetics

As a prelude to our analysis of combustion kinetics it will be useful to consider the thermal decomposition or pyrolysis of hydrocarbons. It is generally accepted that the pyrolysis of hydrocarbons occurs by a free-radical mechanism. Free radicals are entities that contain one unpaired electron. They are often molecular fragments formed by the rupture of normal covalent bonds in which each fragment retains possession of its contributing electron. Examples of free radicals are the methyl radical, $CH_3\cdot$, the ethyl radical $CH_3CH_2\cdot$, and the chlorine atom, Cl.

Let us consider the mechanism for the pyrolysis of ethane. The process is initiated by the thermal breakdown of the ethane molecule into two methyl radicals:

$$C_2H_6 + M \xrightarrow{1} 2CH_3\cdot + M$$

The alternative $C_2H_6 \rightarrow C_2H_5\cdot + H\cdot$ has a much higher activation energy than reaction 1 and thus can be neglected. This initiation reaction is followed by the chain propagation steps:

$$CH_3\cdot + C_2H_6 \xrightarrow{2} CH_4 + C_2H_5\cdot$$

$$C_2H_5\cdot + M \xrightarrow{3} C_2H_4 + H\cdot + M$$

$$H\cdot + C_2H_6 \xrightarrow{4} H_2 + C_2H_5\cdot$$

These reactions are called chain propagation reactions since free radicals are continuously propagated by the reactions. Although some radicals are destroyed, others are generated in equal numbers. The termination reactions involve the combination of two free radicals to form a stable molecule or molecules:

$$2H\cdot \xrightarrow{5} H_2$$

$$H\cdot + C_2H_5\cdot \xrightarrow{6} C_2H_6$$

$$H\cdot + C_2H_5\cdot \xrightarrow{7} C_2H_4 + H_2$$

$$H\cdot + CH_3\cdot \xrightarrow{8} CH_4$$

$$CH_3\cdot + C_2H_5\cdot \xrightarrow{9} C_3H_8$$

$$2C_2H_5\cdot \xrightarrow{10} C_4H_{10}$$

Note that if the termination reactions did not occur, it theoretically would be necessary for only one molecule of $C_2H_6$ to decompose by reaction 1 in order for complete conversion of $C_2H_6$ to $C_2H_4$ to occur. All the rest of the $C_2H_6$ would react via reaction 2 and the chain sequence of reactions 3 and 4. If, on the other hand, hydrogen atoms are terminated by any of reactions 5 to 7 as soon as they are generated by reaction 3 and before they can react by reaction 4, each molecule of $C_2H_6$ that decomposes by reaction 1 can generate (via $CH_3\cdot$) at most two molecules of $C_2H_4$. Under these conditions the chain sequence of reactions 3 and 4 is completely suppressed. Actually, an intermediate situation will exist in which propagation and termination reactions compete for radicals. The average number of times that the chain sequence is repeated before a chain-propagating radical is terminated is called the chain length of the reaction.

For hydrocarbons larger than ethane the initial bond rupture may occur at any

C—C bond; for example, *n*-pentane may decompose by

$$C_5H_{12} \longrightarrow CH_3\cdot + C_4H_9\cdot$$

$$\longrightarrow C_2H_5\cdot + C_3H_7\cdot$$

It is generally assumed that the activation energy of the initiation step is approximately equal to the bond energy of the bonds being broken. The reverse radical recombination reactions then have zero activation energy.

The rate at which hydrogen abstraction reactions of types 2 and 4 proceed depends on the location of the hydrogen atom on the carbon chain backbone. A primary carbon atom forms a normal covalent bond with one other carbon; a secondary C atom with two other C atoms; a tertiary with three other C atoms. Consider 2-dimethyl,4-methyl pentane:

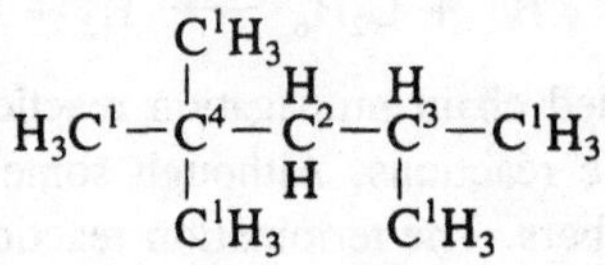

where $C^1$ are primary carbons, $C^2$ is a secondary carbon, $C^3$ is a tertiary carbon, and $C^4$ is a quarternary carbon. The rate of H-atom abstraction from such a molecule generally proceeds in the order $C^3 > C^2 > C^1$, with an approximate difference in activation energy of about 2 kcal (8374 J; see Table 1.15 for conversion factor) between each carbon, that is, $E_3 - E_2 \simeq E_2 - E_1 \simeq 2000$ cal/g-mol (8374 J/g-mol).

Let us now analyze the mechanism for ethane pyrolysis. In doing so, we want to determine which of the termination reactions are most important. To decide which is the dominant termination reaction, we have to consider the relative concentration of the free radicals, the relative rate constants, and the need for a third body. Since recombination reactions can be assumed to have zero activation energy, the frequency factor determines the rate constant. Theory predicts that the frequency factor is lower the larger the molecules. Thus the frequency factor for $C_2H_5\cdot + C_2H_5\cdot$ is lower than that for $C_2H_5\cdot + CH_3\cdot$, and so on. On the other hand, recombination reactions such as $H\cdot + H\cdot$, $H\cdot + CH_3\cdot$, and $H\cdot + C_2H_5\cdot$ require a third body, so in spite of their higher frequency factor, their rate is considerably lower than those of the larger radicals. Finally, and most important for this case, the concentration of ethyl radicals is much higher than those of $H\cdot$ and $CH_3\cdot$.

We can estimate the relative concentrations of free radicals using their PSSA equations, neglecting the contributions from initiation and termination reactions. Neglecting the initiation and termination reactions is valid if the chain length of the reaction is sufficiently long. For example, the PSSA applied to $C_2H_5\cdot$ under these conditions is

$$-k_3[M][C_2H_5\cdot] + k_4[H\cdot][C_2H_6] \simeq 0$$

from which we obtain

$$\frac{[H\cdot]}{[C_2H_5\cdot]} = \frac{k_3[M]}{k_4[C_2H_6]}$$

The ratio is $\ll 1$ since $k_3[M] \ll k_4[C_2H_6]$. Thus we can consider reaction 10 to be the sole termination reaction.

To derive a rate equation for the overall process, we apply the PSSA to $[H\cdot]$, $[CH_3\cdot]$, and $[C_2H_5\cdot]$:

$$0 = k_3[C_2H_5\cdot][M] - k_4[H\cdot][C_2H_6]$$

$$0 = 2k_1[C_2H_6][M] - k_2[CH_3\cdot][C_2H_6]$$

$$0 = k_2[CH_3\cdot][C_2H_6] - k_3[C_2H_5\cdot][M] + k_4[H\cdot][C_2H_6] - 2k_{10}[C_2H_5\cdot]^2$$

We obtain

$$[C_2H_5\cdot] = \left(\frac{k_1[M]}{k_{10}}\right)^{1/2} [C_2H_6]^{1/2}$$

$$[CH_3\cdot] = \frac{2k_1[M]}{k_2}$$

$$[H\cdot] = \frac{k_3[M]}{k_4}\left(\frac{k_1[M]}{k_{10}}\right)^{1/2} [C_2H_6]^{-1/2}$$

The overall rate of disappearance of ethane is

$$\frac{d[C_2H_6]}{dt} = -3k_1[M][C_2H_6] - k_3[M]\left(\frac{k_1[M]}{k_{10}}\right)^{1/2} [C_2H_6]^{1/2}$$

If we neglect the contribution from initiation (assuming long chains), we have

$$\frac{d[C_2H_6]}{dt} = -k_3[M]\left(\frac{k_1[M]}{k_{10}}\right)^{1/2} [C_2H_6]^{1/2}$$

and the overall reaction has order $\frac{1}{2}$.

If, instead of reaction 10, we had specified the principal termination reaction to be reaction 9, the overall reaction can be shown to have order 1.

## APPENDIX B MASS AND HEAT TRANSFER

Virtually every process that we will study in this book involves heat and/or mass transfer. The rates of combustion reactions depend on the rates of mixing of fuel and air, on the rate of removal of energy in the combustion equipment, and in the case of the burning of solid particles or liquid drops, on heat and mass transfer between the particle and the surrounding gas. Several of the most important processes for removal of gaseous pollutants from effluent streams involve contacting of the waste gas stream with a liquid or solid sorbent. In such a process the vapor species must diffuse to the gas–liquid or gas–solid interface, cross the interface, and, in the case of a liquid, diffuse into the bulk of the liquid.

### B.1 Basic Equations of Convective Diffusion

Whereas we assume that most readers will have some familiarity with heat transfer, this may not be the case with mass transfer. Thus in this section we develop the basic equations of mass transfer that will be used subsequently in the book and present some essential material on heat transfer.

Consider a binary mixture of species $A$ and $B$. The mass flux of $A$ (kg m$^{-2}$ s$^{-1}$) at any point in a fluid is $\boldsymbol{n}_A = \rho_A \boldsymbol{u}_A$, where $\rho_A$ is the density of species $A$ (kg m$^{-3}$) and $\boldsymbol{u}_A$ is the average velocity of species $A$ molecules at that point (m s$^{-1}$). If species $A$ is being generated by chemical reaction at a rate $\bar{r}_A$ (kg m$^{-3}$ s$^{-1}$), a balance on species $A$ over an incremental volume of the fluid produces the differential equation of continuity of component $A$ in a binary mixture of $A$ and $B$,

$$\frac{\partial \rho_A}{\partial t} + \nabla \cdot \boldsymbol{n}_A = \bar{r}_A \tag{B.1}$$

The same equation in molar units is

$$\frac{\partial c_A}{\partial t} + \nabla \cdot \boldsymbol{N}_A = r_A \tag{B.2}$$

where $\boldsymbol{N}_A$ is the molar flux of $A$ (g-mol m$^{-2}$ s$^{-1}$), equal to $c_A \boldsymbol{u}_A$, $c_A$ is the molar concentration of $A$ (g-mol m$^{-3}$), and $r_A$ is the molar rate of generation of $A$ (negative if $A$ is being consumed).

To obtain equations strictly in terms of $\rho_A$ or $c_A$ we need to relate the fluxes $\boldsymbol{n}_A$ and $\boldsymbol{N}_A$ to these quantities. The fundamental relation describing mass transfer by molecular diffusion in a binary mixture is *Fick's law*, expressed in mass and molar units as

$$\boldsymbol{n}_A = \frac{\rho_A}{\rho}(\boldsymbol{n}_A + \boldsymbol{n}_B) - \rho D_{AB} \nabla \frac{\rho_A}{\rho} \tag{B.3}$$

$$\boldsymbol{N}_A = x_A(\boldsymbol{N}_A + \boldsymbol{N}_B) - c D_{AB} \nabla x_A \tag{B.4}$$

Notice that the flux of species $A$ in a binary mixture at a point is the sum of two contributions. Let us examine the first terms on the right-hand sides of (B.3) and (B.4). The sum $\boldsymbol{n}_A + \boldsymbol{n}_B = \rho \boldsymbol{u}$, the product of the total density of the mixture and the mass average velocity (i.e., the total mass flux at that point). The quantity $\rho_A/\rho$ is the mass fraction of species $A$, so the first term on the right-hand side of (B.3) is the fraction of the total mass flux of the fluid that is $A$. Similarly, $\boldsymbol{N}_A + \boldsymbol{N}_B = c\boldsymbol{u}^*$, where $c$ is the total molar concentration of the mixture, and $\boldsymbol{u}^*$ is the molar average velocity, $\boldsymbol{u}^* = x_A \boldsymbol{u}_A + (1 - x_A)\boldsymbol{u}_B$. Thus the first terms on the right-hand sides of (B.3) and (B.4) are the contributions to the fluxes from convection. In sum, then, the flux of species $A$ (or $B$) may arise from bulk flow of the mixture and/or molecular diffusion.

Substituting (B.3) and (B.4) in (B.1) and (B.2), respectively, we obtain the binary convective diffusion equations,

$$\frac{\partial \rho_A}{\partial t} + \nabla \cdot (\rho_A \boldsymbol{u}) = \nabla \cdot \rho D_{AB} \nabla \frac{\rho_A}{\rho} + \bar{r}_A \tag{B.5}$$

$$\frac{\partial c_A}{\partial t} + \nabla \cdot (c_A \boldsymbol{u}^*) = \nabla \cdot cD_{AB}\nabla x_A + r_A \tag{B.6}$$

Either of these equations describes the concentration profiles in a binary diffusing system. They are valid for systems with variable total density ($\rho$ or $c$) and variable diffusivity $D_{AB}$.

If $\rho$ and $D_{AB}$ can be assumed to be constant, we can employ the overall continuity equation, $\nabla \cdot \boldsymbol{u} = 0$, and then divide (B.5) by $M_A$ to obtain

$$\frac{\partial c_A}{\partial t} + \boldsymbol{u} \cdot \nabla c_A = D_{AB}\nabla^2 c_A + r_A \tag{B.7}$$

## B.2 Steady-State Mass Transfer to or from a Sphere in an Infinite Fluid

In both combustion and aerosol applications we will be interested in mass transfer to or from a sphere in an infinite fluid. We assume that the gas phase is a binary mixture of a diffusing species $A$ in a background gas $B$ (e.g., air). We wish to determine the steady-state concentration profile of $A$ around the sphere and the mass flux. Although we can in principle solve the steady-state form of (B.7), difficulties arise in specifying the velocity $\boldsymbol{u}$. It is easier to begin with a balance on species $A$ over a thin shell between radii $r$ and $r + \Delta r$ in the gas. If $A$ is not reacting in the gas, such a balance gives

$$(4\pi r^2 N_{A_r})_r - (4\pi r^2 N_{A_r})_{r+\Delta r} = 0 \tag{B.8}$$

which, upon dividing by $\Delta r$ and letting $\Delta r \to 0$, becomes

$$\frac{d}{dr}(4\pi r^2 N_{A_r}) = 0 \tag{B.9}$$

or

$$4\pi r^2 N_{A_r} = \text{constant} \tag{B.10}$$

If species $B$ is not transferring between the gas and the sphere, then $N_{B_r} = 0$ everywhere. Thus (B.4) becomes

$$N_{A_r} = -\frac{cD_{AB}}{1 - x_A}\frac{dx_A}{dr} \tag{B.11}$$

Combining (B.10) and (B.11) and calling the constant $-C_1$ gives

$$4\pi r^2 \frac{cD_{AB}}{1 - x_A}\frac{dx_A}{dr} = C_1 \tag{B.12}$$

Integrating (B.12) gives

$$-4\pi c D_{AB} \ln(1 - x_A) = -\frac{C_1}{r} + C_2 \tag{B.13}$$

One boundary condition is that $x_A = x_{A\infty}$ as $r \to \infty$, from which we find

$$C_2 = -4\pi c D_{AB} \ln (1 - x_{A\infty}) \tag{B.14}$$

The other boundary condition is that the mole fraction of $A$ just above the surface of the sphere is $x_{A0}$ (i.e., $x_A = x_{A0}$ at $r = R$). Applying this condition gives

$$C_1 = 4\pi R c D_{AB} \ln \frac{1 - x_{A0}}{1 - x_{A\infty}} \tag{B.15}$$

and thus, from (B.13),

$$\ln \frac{1 - x_A}{1 - x_{A\infty}} = \frac{R}{r} \ln \frac{1 - x_{A0}}{1 - x_{A\infty}} \tag{B.16}$$

or

$$\frac{1 - x_A}{1 - x_{A\infty}} = \left( \frac{1 - x_{A0}}{1 - x_{A\infty}} \right)^{R/r} \tag{B.17}$$

Let us compute the flux of species $A$ at the surface of the sphere. Since $-4\pi R^2 N_{A_R} = C_1$, we have

$$N_{A_R} = -\frac{c D_{AB}}{R} \ln \frac{1 - x_{A0}}{1 - x_{A\infty}} \tag{B.18}$$

Note that if $x_{A0} < x_{A\infty}$, $N_{A_R} < 0$ and the flux of $A$ is toward the sphere; and that if $x_{A0} > x_{A\infty}$, $N_{A_R} > 0$ and the flux of $A$ is away from the sphere.

The mass average velocity at any point is related to the fluxes of $A$ and $B$ by $\mathbf{n}_A + \mathbf{n}_B = \rho \mathbf{u}$. Since $n_{B_r} = 0$ everywhere, $n_{A_r} = \rho u_r$. At constant $T$ and $p$, $\rho$ is constant, so

$$\begin{aligned} u_r &= \frac{N_{A_r} M_A}{\rho} \\ &= -\frac{M_A}{\rho} \frac{R}{r^2} c D_{AB} \ln \frac{1 - x_{A0}}{1 - x_{A\infty}} \end{aligned} \tag{B.19}$$

If $x_{A0} < x_{A\infty}$, $u_r < 0$ and the net flux is toward the particle, and vice versa. We note that $u_r$ satisfies the overall continuity equation for a fluid at constant density,

$$\frac{d}{dr}(r^2 u_r) = 0$$

A frequently used approximation to the foregoing development is that the mole fraction of $A$ is so small (i.e., $x_A << 1$) that the flux (B.11) may be approximated by the pure diffusive contribution

$$N_{A_r} = -c D_{AB} \frac{dx_A}{dr} \tag{B.20}$$

From

$$4\pi r^2 c D_{AB} \frac{dx_A}{dr} = C_1$$

integrating and using the two boundary conditions gives

$$\frac{x_A - x_{A\infty}}{x_{A0} - x_{A\infty}} = \frac{R}{r} \tag{B.21}$$

with the flux at the sphere surface

$$N_{A_R} = \frac{cD_{AB}}{R}(x_{A0} - x_{A\infty}) \tag{B.22}$$

We can compute the rate of change of the size of the sphere due to the flux of $A$. If the molar density of the sphere is $c_s$, then

$$-4\pi R^2 N_{A_R} = c_s \frac{d}{dt}\left(\frac{4}{3}\pi R^3\right) \tag{B.23}$$

or

$$-c_s \frac{dR}{dt} = N_{A_R} \tag{B.24}$$

Inherent in this result, if we use (B.18) or (B.22) for $N_{A_R}$, is the assumption that even though the size of the sphere is changing due to mass transfer of $A$ between the sphere and the ambient gas, the size change occurs slowly enough that the flux of $A$ can be computed from its steady-state value. In other words, the characteristic time to achieve the steady-state concentration profile of $A$ in the gas is short compared to the characteristic time for the particle to grow or shrink. We will return to this point shortly.

## B.3 Heat Transfer

The equation for conservation of energy for a pure fluid, neglecting viscous dissipation, is (Bird et al., 1960, Table 10.4-1)

$$\rho\bar{c}_p\left(\frac{\partial T}{\partial t} + \boldsymbol{u}\cdot\nabla T\right) = k\nabla^2 T + \frac{\partial \ln(1/\rho)}{\partial \ln T}\left(\frac{\partial p}{\partial t} + \boldsymbol{u}\cdot\nabla p\right) \tag{B.25}$$

where $\bar{c}_p$ is the heat capacity at constant pressure per unit mass and $k$ is the thermal conductivity. At constant pressure the second term on the right-hand side of (B.25) vanishes.

Let us continue with our spherical example. At steady state and constant pressure, the temperature distribution around the sphere satisfies

$$\rho\bar{c}_p u_r \frac{dT}{dr} = k\frac{1}{r^2}\frac{d}{dr}\left(r^2\frac{dT}{dr}\right) \tag{B.26}$$

The left-hand side of (B.26) is the contribution from the bulk, convective motion of the fluid, and the right-hand side represents that from conduction.

We can determine the temperature profile by solving (B.26) with $u_r$ from (B.19)

$$-\rho \bar{c}_p \frac{M_A}{\rho} \frac{R}{r^2} cD_{AB} \left( \ln \frac{1 - x_{A0}}{1 - x_{A\infty}} \right) \frac{dT}{dr} = k \frac{1}{r^2} \frac{d}{dr} \left( r^2 \frac{dT}{dr} \right)$$

Letting $\beta = \bar{c}_p M_A R c D_{AB} \ln [(1 - x_{A0})/(1 - x_{A\infty})]/k$, this equation becomes

$$-\beta \frac{dT}{dr} = \frac{d}{dr} \left( r^2 \frac{dT}{dr} \right)$$

To solve this equation, we let $y = dT/dr$, giving

$$\frac{d}{dr} (r^2 y) = -\beta y$$

which, upon integration, becomes

$$r^2 y = D_1 e^{\beta/r}$$

or

$$r^2 \frac{dT}{dr} = D_1 e^{\beta/r}$$

Integrating again gives

$$T = -D_1 \left( \frac{1}{\beta} e^{\beta/r} \right) + D_2$$

The two constants of integration are determined from the boundary conditions $T = T_\infty$ as $r \to \infty$ and $T = T_0$ at $r = R$. The final result is

$$\frac{T - T_\infty}{T_0 - T_\infty} = \frac{1 - e^{\beta/r}}{1 - e^{\beta/R}} \tag{B.27}$$

The heat flow at the surface of the sphere is

$$4\pi R^2 \left[ \rho \bar{c}_p u_r T - k \frac{dT}{dr} \right]_{r=R} \tag{B.28}$$

In the case in which we can neglect the velocity $u_r$, which corresponds to assuming a very dilute system [e.g., (B.20)], the original shell energy balance gives

$$\frac{d}{dr} (4\pi r^2 q_r) = 0 \tag{B.29}$$

which, with the use of Fourier's law, $q_r = -k\partial T/\partial r$, gives (B.26) with $u_r = 0$. Solving that equation, namely

$$\frac{d}{dr}\left(r^2\frac{dT}{dr}\right) = 0 \tag{B.30}$$

subject to $T = T_\infty$ as $r \to \infty$ and $T = T_0$ at $r = R$ produces the steady-state temperature profile,

$$\frac{T - T_\infty}{T_0 - T_\infty} = \frac{R}{r} \tag{B.31}$$

by direct analogy to (B.21). The heat flux at the sphere's surface is just

$$q_R = \frac{k}{R}(T_0 - T_\infty) \tag{B.32}$$

Clearly, if $T_0 > T_\infty$, $q_R > 0$, and vice versa.

Let us say that the transfer of species $A$ to or from the sphere is accompanied by heat generation or consumption in the particle, such as if species $A$ is condensing or evaporating. Let $\Delta h$ be the enthalpy change per mole of $A$. Then if the steady-state heat flux to or from the sphere is that due to the heat consumed or generated by species $A$, and if the concentration of $A$ is sufficiently small that the convective contribution to the heat flux at the sphere surface can be neglected,

$$-4\pi R^2 N_{A_R}\Delta h = 4\pi R^2 q_R \tag{B.33}$$

## B.4 Characteristic Times

We have just been obtaining expressions for steady-state profiles of mole fraction and temperature around a sphere in an infinite fluid. If conditions are changing in time, the appropriate unsteady-state equations are needed. In such a case it will always be necessary to evaluate first the characteristic times of the processes occurring to see if phenomena are taking place on vastly different time scales. For example, we noted above that if the characteristic time to achieve a steady-state profile is very short compared to other times in the system, steady-state profiles may be assumed to exist at any instant of time.

For a spherical particle of radius $R$, the characteristic times for relaxation of the temperature and concentration profiles in the gas phase to their steady-state values are

$$\tau_h = \frac{R^2}{\alpha} \tag{B.34}$$

$$\tau_m = \frac{R^2}{D_{AB}} \tag{B.35}$$

where $\alpha = k/\rho\bar{c}_p$, the thermal diffusivity. The characteristic time for heat conduction *within* the sphere is

$$\tau_{hp} = \frac{R^2}{\alpha_p} \tag{B.36}$$

**TABLE 1.11** CHARACTERISTIC TIMES FOR HEAT AND MASS TRANSFER INVOLVING A SPHERICAL PARTICLE IN AIR[a]

| Material | $\frac{\tau_{hp}}{\tau_h} = \frac{\alpha}{\alpha_p}$ | $\frac{\tau_m}{\tau_h} = \frac{\alpha}{D_{AB}}$ |
|---|---|---|
| Organics | 200 | 2–4 |
| Water | 90 | 0.86 |
| Metals | 5 | 1.5 |

[a] $\tau_h = R^2/\alpha$, heat conduction in gas phase; $\tau_m = R^2/D_{AB}$, diffusion in gas phase; $\tau_{hp} = R^2/\alpha_p$, heat conduction in particle.

where $\alpha_p = k_p/\rho_p\bar{c}_{pp}$. Table 1.11 compares the characteristic times for spherical particles having the properties of organics, water, and metals. We see that the characteristic time for equilibration of the temperature profile in the particle is generally considerably longer than that for the gas phase. The characteristic times for relaxation of the temperature and concentration profiles in the gas phase are the same order of magnitude. The large value of $\tau_{hp}/\tau_h$ suggests that any heat released at the particle's surface is conducted primarily outward. As combustion, for example, begins, the particle surface temperature rises until the rate of outward heat conduction balances the rate of heat release. The formation of the external temperature and vapor concentration profiles occurs simultaneously on approximately the same time scales. Consequently, the steady-state fluxes of energy and mass may be related by a steady-state energy balance, such as in (B.33), to determine the surface temperature during the combustion of the particle.

Although we do not yet introduce characteristic times for the rate of change of the particle size, due, for example, to consumption by combustion reactions or growth by vapor condensation, we will see later that such times are generally considerably longer than those for heat and mass transfer in the vapor phase. Thus the gas-phase temperature and concentration profiles may be assumed to be in a pseudo-steady state at any instant of time.

## APPENDIX C ELEMENTS OF PROBABILITY THEORY

For processes that occur under turbulent flow conditions it is impossible to predict the exact values of variables at any time or location; consequently, it is necessary to analyze such situations within the language of random variables. The main process of interest to us in this book that falls in this category is turbulent combustion.

### C.1 The Concept of a Random Variable

Think of a laboratory experiment the result of which is unknown until the experiment has been completed, as opposed to an experiment the result of which can be predicted precisely knowing all the conditions beforehand (of course, in the latter case we probably would not want to perform the experiment in the first place if we knew what the outcome would be). Fortunately (or maybe unfortunately), we usually do not know the result of an experiment until it is finished. Let us call the result of such an experiment an *event*. There may be a discrete number or a continuous spectrum of possible events, or outcomes, but the most we might be able to say about which event will result from a particular trial is the *probability* of observing each possible event.

We have now introduced a concept, *probability*, that we must define in order to be more precise about the outcomes of our experiment. Let us write Prob $\{X\} = P(X)$ to signify the probability that an event $X$ occurs. Even though $X$ denotes an outcome rather than a number, $P(X)$ is a number (i.e., the *probability* that $X$ occurs). Now, how do we define *probability*? We can only define probability on the basis of our physical reasoning about the process the outcome of which may be $X$. Nevertheless, we can set down some general rules to help us determine the outcome probabilities for a particular process.

Intuitively, we would say that the probability of a certain event is 1, while the probability of an event that cannot occur is zero. Also, if there are two *mutually exclusive* events $X_1$ and $X_2$, the probability of obtaining either $X_1$ or $X_2$ is the sum of the individual probabilities, $P(X_1)$ and $P(X_2)$. We can state the following axioms of probability theory formally:

1. $P(X) \geq 0$ (probability is nonnegative).
2. If $X$ is certain, $P(X) = 1$.
3. If two events $X_1$ and $X_2$ are mutually exclusive,

$$\text{Prob}\left\{X_1 \text{ or } X_2\right\} = P(X_1) + P(X_2)$$

Consider for a moment the roll of a single die. Each of the outcomes corresponds to a number, the number being one of the integers 1, 2, . . . , 6. We can let $X$ denote the number that will result from the roll of the die; that is, $X$ is a random variable that can assume the integer values 1, 2, . . . , 6. If we throw the die $n$ times and the value $j$ occurs $n_j$ times, it seems reasonable to define the probability of outcome $j$ as

$$\text{Prob}\left\{X = j\right\} = P(X = j) = \lim_{n \to \infty} \frac{n_j}{n}$$

The random variable, which we have denoted by $X$, is a symbol for the outcome of the event in which we are interested. In the die example, the random variable $X$ assumes discrete values, the integers from 1 to 6. However, $X$ can also denote the instantaneous value of a continuous variable, say the concentration in a flame at a given location and time. Even if the outcome of an event has no obvious numerical value, we can assign numerical values to the random variable $X$ corresponding to the possible outcomes. For example, in the tossing of a coin, we can let $X = 1$ denote a head and $X = 2$ denote a tail.

It is customary to denote random variables by capital letters and to denote the values they may assume by the corresponding lowercase letters. Therefore, $x$ denotes one of the values that $X$ may assume. In the case of a throw of the die we could have written $P(X = x)$, where $x$ assumes the values 1 to 6.

The random variable $X$ is called a *discrete random variable* if it may assume a finite or a denumerably infinite set of values, denoted as $x_j$, $j = 1, 2, \ldots, N$. The probability that $X$ assumes the value $x_j$ is denoted by $p(j)$. $p(j)$ is called the *probability mass function* of the random variable $X$, because the values of $X$ may be thought of as being confined to the mass points $x_1, \ldots, x_N$. The probability mass function $p(j)$ has the following properties:

$$0 \le p(j) \le 1$$

$$\sum_{j=1}^{N} p(j) = 1$$

If $X$ is a *continuous random variable*, we define the *probability density function* $p(x)$ such that

$$p(x)\,dx = \text{Prob}\,\{x < X \le x + dx\}$$

where $p(x)\,dx$ may be thought of as the fraction of the total mass of $X$ that lies in the range $x$ to $x + dx$; that is, $p(x)$ is the *density* of $X$ in this region. The probability density function has the following properties:

$$0 \le p(x) \le 1$$

$$\int_{-\infty}^{\infty} p(x)\,dx = 1$$

We stress that $p(x)$ itself is not a probability; rather, $p(x)\,dx$ is a probability. Whatever the units of the random variable, the probability density function is measured in $(\text{units})^{-1}$.

We shall now introduce the *distribution function*. Consider the probability that a random variable $X$ will not exceed a given value $x$. Clearly, this probability is a function of the threshold value $x$. We call this function the distribution function and denote it by $F(x)$. For a discrete random variable $F(x_j)$ or $F(j)$ is the probability that $X \le x_j$, where $x_1 < \cdots < x_N$. In terms of the probability mass function

$$F(j) = \sum_{i=1}^{j} p(i) \tag{C.1}$$

Let us return to the die-throwing experiment. For $x_1 = 1$, $p(1) = \frac{1}{6}$, so $F(1)$, the probability that $X \le 1$ (clearly, $X$ cannot be less than 1, but only exactly equal to 1), is $\frac{1}{6}$. At $x = 1$, $F(j)$ jumps from 0 to $\frac{1}{6}$, just as at $x = 2$, $F(j)$ jumps from $\frac{1}{6}$ to $\frac{1}{3}$. Finally, $F(6) = 1$, since no numbers higher than 6 can be obtained. Thus the distribution function for a discrete random variable is a staircase function of discrete jumps.

For a continuous random variable, $F(x)$ is related to the probability density function by

$$F(x) = \int_{-\infty}^{x} p(\zeta)\, d\zeta \tag{C.2}$$

A distribution function $F(x)$ has the following general properties:

1. $0 \le F(x) \le 1$.
2. $\lim_{x \to -\infty} F(x) = 0$ and $\lim_{x \to \infty} F(x) = 1$.
3. As $x$ increases, $F(x)$ must not decrease.
4. Prob $\{x_1 < X \le x_2\} = F(x_2) - F(x_1)$.

Properties 1 and 2 follow since $F(x)$ is itself a probability and since the value of $X$ must lie somewhere on the $x$-axis. Clearly, as $x$ increases, the probability that $X \le x$ must not decrease, since new intervals are continually being added to $x$. If these new intervals have probability zero, then $F(x)$ must at least remain constant. Property 4 follows from axiom 3, namely that

$$\text{Prob}\,\{X \le x_2\} = \text{Prob}\,\{X \le x_1\} + \text{Prob}\,\{x_1 < X \le x_2\}$$

For a continuous random variable, this becomes

$$F(x_2) - F(x_1) = \int_{x_1}^{x_2} p(x)\, dx \tag{C.3}$$

At point $x$ where the derivative of $F(x)$ exists, the derivative is equal to the density $p(x)$ at that point:

$$p(x) = \frac{dF(x)}{dx} \tag{C.4}$$

In Figure 1.2 we illustrate probability mass and density functions and their associated distribution functions. We also show the region corresponding to the probability that $x_1 < X \le x_2$.

## C.2 Properties of Random Variables

We shall now consider special properties of random variables, namely, their *expectations*. The expected value of the random variable $X$ is denoted by $E(X)$ and is also called the *mean* of $X$. The expected value is computed by

$$E(X) = \begin{cases} \sum_{j=1}^{N} x_j p(j) & \text{(discrete)} \quad \text{(C.5)} \\ \int_{-\infty}^{\infty} x p(x)\, dx & \text{(continuous)} \quad \text{(C.6)} \end{cases}$$

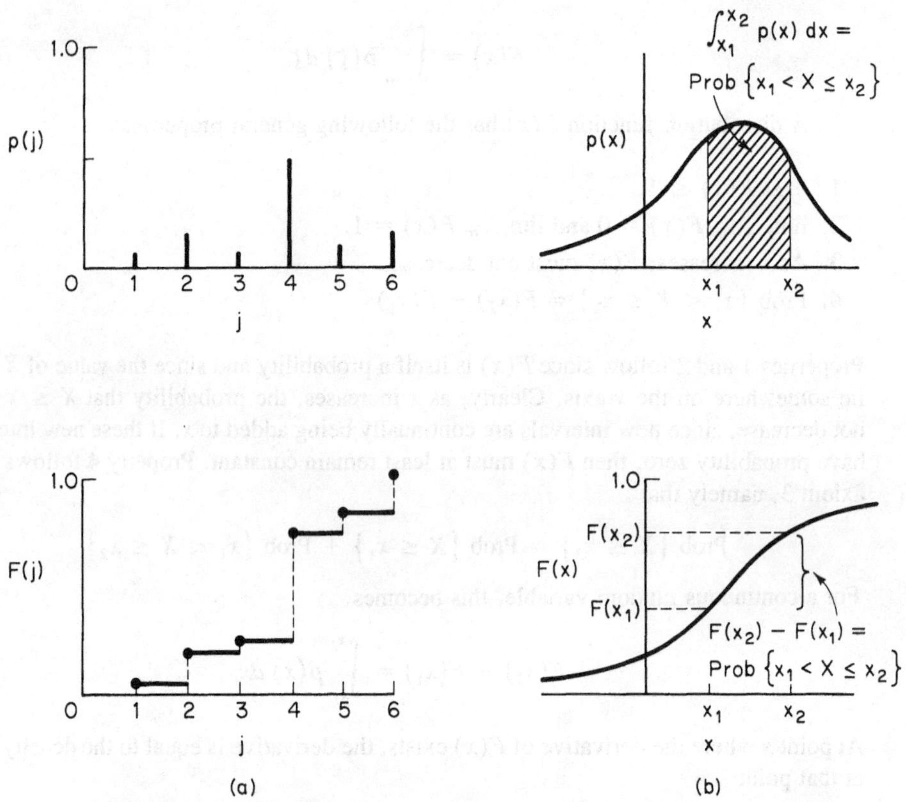


**Figure 1.2** Typical probability mass, density, and distribution functions: (a) typical probability mass function $p(j)$ and distribution function $F(j)$ for a discrete random variable $X$; (b) typical probability density function $p(x)$ and distribution function $F(x)$ for a continuous random variable $X$.

The expected value of a function of $X$, $f(X)$, is found from

$$E[f(X)] = \begin{cases} \sum_{j=1}^{N} f(x_j)\, p(j) & \text{(discrete)} \quad \text{(C.7)} \\ \int_{-\infty}^{\infty} f(x)\, p(x)\, dx & \text{(continuous)} \quad \text{(C.8)} \end{cases}$$

**Example 1.5** ***The Poisson Distribution***

Let the discrete random variable $X$ assume the values $j = 0, 1, 2, \ldots$ with probability mass function (the Poisson distribution)

$$p(j) = \frac{\lambda^j}{j!} e^{-\lambda} \qquad \lambda > 0$$

Compute the expected value of $X$ using (C.5):

$$E(X) = \sum_{j=0}^{\infty} j \frac{\lambda^j}{j!} e^{-\lambda} = \lambda e^{-\lambda} \sum_{j=1}^{\infty} \frac{\lambda^{j-1}}{(j-1)!}$$

$$= \lambda e^{-\lambda} \sum_{k=0}^{\infty} \frac{\lambda^k}{k!} = \lambda e^{-\lambda} e^{\lambda} = \lambda$$

**Example 1.6** ***The Binomial Distribution***

Let the discrete random variable $X$ assume the values $j = 0, 1, 2, \ldots, N$ with probability mass function (the binomial distribution)

$$p(j) = \frac{N!}{j!(N-j)!} p^j (1-p)^{N-j} \qquad 0 < p < 1$$

Compute the expected value of $X$.

$$E(X) = \sum_{j=0}^{N} j \frac{N!}{j!(N-j)!} p^j (1-p)^{N-j}$$

$$= \sum_{j=1}^{N} j \frac{N!}{j!(N-j)!} p^j (1-p)^{N-j}$$

$$= Np \sum_{j=1}^{N} \frac{(N-1)!}{(j-1)!\,(N-j)!} p^{j-1} (1-p)^{N-j}$$

$$= Np \sum_{j=0}^{N-1} \frac{(N-1)!}{j!(N-1-j)!} p^j (1-p)^{N-1-j}$$

$$= Np[p + (1-p)]^{N-1} = Np$$

The general expectations of probability distributions are the *moments* of the distribution, of which the mean is a special case. We shall define the expectation $E(X^r)$ as the $r$th *noncentral moment of* $X$. We shall use the notation $\mu_r'$ for this moment. Thus for discrete and continuous random variables,

$$\mu_r' = \sum_{j=0}^{N} x_j^r p(j) \tag{C.9}$$

$$= \int_{-\infty}^{\infty} x^r p(x)\, dx \tag{C.10}$$

By definition, $\mu_0' = 1$, and $\mu_1'$ is the mean of $X$. We shall define the expectation $E[(X - \mu_1')^r]$ as the $r$th *central moment of* $X$, that is, the $r$th moment about the mean $\mu_1'$. Thus, denoting the central moments by $\mu_r$,

$$\mu_r = \sum_{j=0}^{N} (x_j - \mu_1')^r\, p(j) \tag{C.11}$$

$$= \int_{-\infty}^{\infty} (x - \mu_1')^r\, p(x)\, dx \tag{C.12}$$

From the definition of the central moments, $\mu_0 = 1$ and $\mu_1 = 0$. The second central moment $\mu_2$ is called the *variance* of the distribution and is often denoted by $\sigma^2$. $\mu_3$ is a measure of the skewness of the distribution about the mean, and $\mu_3 = 0$ if the distribution is symmetric. $\mu_4$ is a measure of the distortion from a normal (Gaussian) distribution resulting in a low center and high ends of the distribution or a high center and low ends, that is, a measure of the flatness of the distribution.

The two most important moments of a distribution are the mean $\mu_1'$ and the variance $\mu_2$. The nonnegative square root of the variance, usually denoted $\sigma$, is called the *standard deviation* of the distribution. The standard deviation $\sigma$ has the same units as the random variable $X$ and is often used as a measure of the dispersion of a distribution about its mean value.

Two other parameters often used to characterize distributions are the *coefficient of variation* $\nu$, defined as the ratio of the standard deviation to the expected value,

$$\nu = \frac{\sigma}{\mu_1'} \tag{C.13}$$

and the *coefficient of skewness* $\gamma$, given by

$$\gamma = \frac{\mu_3}{\sigma^3} \tag{C.14}$$

The coefficient of skewness $\gamma$ measures the extent to which a distribution departs from a symmetric distribution [one for which $F(a - x) = 1 + F(a + x) - \text{Prob}\{X = a + x\}$]. In a symmetric distribution all the central moments of odd order equal zero. Thus each central moment of odd order serves as a measure of asymmetry of the distribution. The coefficient of skewness expresses the third central moment relative to the standard deviation. $\gamma$ may be positive or negative, giving rise to what is called positive or negative asymmetry.

### C.3 Common Probability Distributions

We shall now present several of the more common discrete and continuous probability distributions that arise in the analysis of physical system. The distributions are summarized in Tables 1.12 and 1.13.

**Example 1.7 *The Binomial Distribution***

Let us derive the mean and variance of the binomial distribution shown in Table 1.12. We can describe a sequence of $n$ Bernoulli trials in terms of $n$ independent random variables $X_1, X_2, \ldots, X_n$, where $X_i = 1$ if the $i$th trial is successful and $X_i = 0$ if unsuccessful. Each variable $X_i$ has the probability mass function

$$\text{Prob}\{X_i = 1\} = p(1) = p$$

$$\text{Prob}\{X_i = 0\} = p(0) = q = 1 - p$$

**TABLE 1.12 SOME COMMON PROBABILITY MASS FUNCTIONS**

| Name | Probability mass function | Moments | Remarks |
|---|---|---|---|
| Binomial | $p(m) = \binom{n}{m} p^m q^{n-m}$, $m = 0, 1 \ldots, n$ $\binom{n}{m} = \frac{n!}{m!(n-m)!}$ | $\mu_1' = np$ $\mu_2' = npq + n^2p^2$ $\mu_2 = npq$ | The binomial distribution arises when an event has two possible outcomes, the probability of the first outcome (success) being $p$ and the second (failure) being $q = 1 - p$, and the event is repeated $n$ times. $p(m)$ represents the probability that in $n$ trials there will be $m$ successes. Each event is assumed independent of the others, and the probability of success and failure are the same for every trial. An experiment of this type is called a sequence of Bernoulli trials. |
| Poisson | $p(j) = \frac{\lambda^j}{j!} e^{-\lambda}, \lambda > 0,$ $j = 0, 1, 2, \ldots$ | $\mu_1' = \lambda$ $\mu_2' = \lambda(\lambda + 1)$ $\mu_2 = \lambda$ | Consider random events occurring in time, such as radioactive disintegrations or incoming telephone calls to an exchange. Each event is represented by a point on the time axis, and we are interested in the probability that we shall discover exactly $j$ points in an interval of fixed length if the average rate of occurrence of points in that interval is $\lambda$. Events occur independently, obtained as the limit of the binomial distribution as $n$ becomes large with $np = \lambda$. |

**TABLE 1.13 SOME COMMON PROBABILITY DENSITY FUNCTIONS**

| Name | Probability density function | Moments | Remarks |
|---|---|---|---|
| Uniform | $p(x) = \begin{cases} \dfrac{1}{b-a}, & a \le x \le b \\ 0, & x < a, x > b \end{cases}$ | $\mu'_r = \dfrac{b^{r+1} - a^{r+1}}{(r+1)(b-a)}$<br>$\mu_2 = (b-a)^2/12$<br>$\mu_3 = 0$<br>$\mu_4 = (b-a)^4/80$ | $X$ has a uniform likelihood of being found anywhere on the segment of the real line between $x = a$ and $x = b$. The simplest case of a continuous random variable. |
| Normal (Gaussian) | $p(x) = \dfrac{1}{\sqrt{2\pi}\sigma} \exp\left[-\dfrac{1}{2}\left(\dfrac{x-\mu}{\sigma}\right)^2\right],$<br>$-\infty < x < \infty$ | $\mu'_1 = \mu$<br>$\mu'_2 = \sigma^2 + \mu^2$<br>$\mu'_3 = 3\sigma^2\mu + \mu^3$<br>$\mu'_4 = 3\sigma^4 + 6\mu^2\sigma^2 + \mu^4$<br>$\mu_{2k+1} = 0$, all $k$<br>$\mu_{2k} = 1 \cdot 3 \cdots (2k-1)\,\sigma^{2k}$ | The distribution approached by the sum of a large number of independent random variables under most conditions, and usually describing experimental errors. Approached as the limit of the binomial and Poisson distributions. |
| Exponential | $p(x) = ae^{-ax}, x \ge 0$ | $\mu'_r = \dfrac{r!}{a^r}$ | Consider events occurring randomly in time, such as radioactive disintegration or incoming telephone calls to an exchange. Each event is represented by a point on the time axis, and we are interested in the density function for $X$, the time elapsed between occurrences, where the |

| | | | |
|---|---|---|---|
| | | | events occur at an average rate of $a$ per time. Let $p(x)\,dx$ = probability that the time elapsed between two events is between $x$ and $x + dx$. This distribution also describes the residence time density of a particle in an ideally mixed vessel. |
| Gamma | $p_n(z) = \dfrac{a^n z^{n-1} e^{-az}}{(n-1)!}$, $z \geq 0$ | $\mu_1' = \dfrac{n}{a}$<br>$\mu_2' = \dfrac{n(n+1)}{a^2}$<br>$\mu_2 = \dfrac{n}{a^2}$ | Consider the sum $Z$ of $n$ independent random variables, $Z = X_1 + X_2 + \cdots + X_n$, each exponentially distributed with average rate $a$. Then $p_n(z)$, the distribution of $Z$, is given by the gamma distribution. For example, $Z$ can represent the residence time of a particle in a series of $n$ ideally mixed vessels. |
| Poisson | $p_n(x) = \dfrac{(ax)^n}{n!} e^{-ax}$<br>$x \geq 0$ | $\mu_1' = ax$<br>$\mu_2' = (ax)^2 + ax$<br>$\mu_3' = (ax)^3 + 3(ax)^2 + ax$<br>$\mu_4' = (ax)^4 + 6(ax)^3 + 7(ax)^2 + ax$<br>$\mu_2 = ax$<br>$\mu_3 = ax$ | Consider the probability that exactly $n$ events will occur in a time interval $x$ if the time between individual events is exponentially distributed. In terms of the gamma distribution, if $Z_n$ is the total time taken for $n$ events to occur, then $Z_n > x$ and $Z_{n+1} \leq x$. The probability that exactly $n$ events have occurred in time $x$ is $F_{Z_n}(x) - F_{Z_{n+1}}(x)$, where $F_Z(x)$ is the distribution function for the gamma density. Then $p_n(x)$ for the Poisson distribution is $p_n(x) = F_{Z_n}(x) - F_{Z_{n+1}}(x)$. |

The total number of successes in $n$ trials is simply the sum of the $X_i$s,

$$Z = X_1 + X_2 + \cdots + X_n$$

since $X_i = 0$ if the $i$th trial is a failure.

We can determine the mean and variance of the number of successes $Z$. The mean of $Z$ is simply the sum of the means, which are all identical to that of $X_1$:

$$E(Z) = nE(X_1)$$

Since the $X_i$s are independent, the variance of $Z$ is the sum of the individual variances which are all equal to that of $X_1$. Since $E(X_1) = 0 \cdot q + 1 \cdot p = p$,

$$E(Z) = np$$

Also, $E[(X_1 - p)^2] = p^2q + q^2p = pq$. Thus

$$E[(Z - np)^2] = npq$$

We are often interested in obtaining the probability density of a random variable $Y$ which is some given function of a random variable $X$ that has a known probability density $p_X(x)$. Let us say that we want to compute

$$Y = f(X) \tag{C.15}$$

The unknown density is written as $p_Y(y)$. If $X$ changes its value by $dx$ and the corresponding change in $Y$ is $dy$, the probability of finding $X$ between $x$ and $x + dx$ is the same as observing $Y$ between the corresponding $y$ and $y + dy$. Therefore,

$$p_X(x)\, dx = p_Y(y)\, dy \tag{C.16}$$

which really should be written as

$$p_X(x)\, |dx| = p_Y(y)\, |dy| \tag{C.17}$$

since the probabilities are equal to the magnitudes of the areas under $dx$ and $dy$. Thus

$$p_Y(y) = \frac{p_X(x)}{|dy/dx|} = \frac{p_X(x)}{|df/dx|} \tag{C.18}$$

Since $p_Y(y)$ is a function of $y$, the right-hand side of (C.18) must be expressed in terms of $y$. Assuming that (C.15) can be inverted, we can express $x$ in terms of $y$ by $x = f^{-1}(y)$ in (C.18).

## APPENDIX D TURBULENT MIXING

Mixing plays an important role in combustion. The fuel and air must be brought into rapid and intimate contact for the combustion reactions to occur. Since most combustion systems are operated in the turbulent mode, to understand the role of mixing in combustion, we must devote some attention to turbulent mixing. Turbulence cannot maintain itself without a source of energy, and we will find that energy must be supplied to achieve the high mixing rates desired in combustion systems.

## D.1 Scales of Turbulence

Turbulence can be viewed as a flow characterized by chaotic, random motions on length scales that vary from those of the container in which the flow exists down to scales at which the viscosity of the fluid prevents the generation of even smaller scales by dissipating small-scale energy into heat. The largest scales, or eddies, are responsible for most of the transport of momentum and species. A classic picture of turbulence, due to Kolmogorov, is one in which the energy is transferred from larger eddies to smaller ones, where it is ultimately dissipated. Since the small-scale motions have correspondingly small time scales, it can be assumed that these motions are statistically independent of the relatively large scale, slower motions. In such a case, the nature of the small-scale motion should depend only on the rate at which it is supplied with energy by the large-scale motion and on the viscosity of the fluid. Since the small-scale motions adjust rapidly to changes in this cascade of energy, the instantaneous rate of dissipation is equal to the rate of energy supplied from the large-scale motion.

If the properties of the small-scale eddies are independent of those of the large scale, these properties can be dependent only on viscosity (since dissipation depends on viscosity) and the rate of energy transfer $\epsilon\,(\mathrm{cm}^2\,\mathrm{s}^{-3})$ into the small eddies. This is Kolmogorov's first hypothesis. Between the large energy-containing eddies and the small dissipation eddies there is a region of eddies containing little energy and dissipating little energy whose properties must depend solely on $\epsilon$. This region is called the *inertial subrange* and constitutes Kolmogorov's second hypothesis.

In the smallest eddies, the parameters governing the small-scale motion are the dissipation rate per unit mass $\epsilon\,(\mathrm{cm}^2\,\mathrm{s}^{-3})$ and the kinematic viscosity $\nu\,(\mathrm{cm}^2\,\mathrm{s}^{-1})$. With these two parameters, one can define characteristic length, time, and velocity scales as follows:

$$\eta = \left(\frac{\nu^3}{\epsilon}\right)^{1/4} \qquad \tau = \left(\frac{\nu}{\epsilon}\right)^{1/2} \qquad v = (\nu\epsilon)^{1/4} \tag{D.1}$$

These scales are referred to as the Kolmogorov microscales of length, time, and velocity. Note that the Reynolds number formed from $\eta$ and $v$, $\eta v/\nu = 1$, indicating that the small-scale motion is indeed viscous and that the rate of dissipation $\epsilon$ adjusts itself to the energy supply through adjustment of the length and velocity scales.

The largest scale of turbulent motion, $L$, corresponds approximately to the length scale of the container (e.g., the diameter of a pipe in turbulent pipe flow). The rate of generation of energy in the large-scale motion, and hence the rate of dissipation by the small-scale motion, $\epsilon$, can be estimated as the product of the kinetic energy per unit mass of the large-scale turbulence, proportional to $u^2$, and the rate of transfer of that energy, $u/L$ (i.e., $u^3/L$), where $u$ is a characteristic large-scale velocity. Thus $\epsilon \sim u^3/L$. This estimate implies that the large-scale eddies lose a significant fraction of their kinetic energy, $u^2/2$, within one "turnover" time $L/u$. The Reynolds number of the turbulent flow is $\mathrm{Re} = uL/\nu$. On this basis, with $\epsilon \sim u^3/L$, we have that

$$\frac{\eta}{L} \sim \mathrm{Re}^{-3/4} \qquad \frac{\tau u}{L} \sim \mathrm{Re}^{-1/2} \qquad \frac{v}{u} \sim \mathrm{Re}^{-1/4}$$

and since Re >> 1, we see that the length, time, and velocity scales of the smallest eddies are very much smaller than those of the largest eddies, and that the separation increases as the Reynolds number increases.

## D.2 Statistical Properties of Turbulence

Turbulence is *stationary* if all its statistical properties are independent of time. The variables themselves, such as the velocity components, are indeed functions of time, but if the turbulence is stationary, their statistical properties, such as the mean, are independent of time. Turbulence is *homogeneous* if all statistical properties are indepenent of location in the field. Turbulence is *isotropic* if all statistical properties are independent of the orientation of the coordinate axes. Kolmogorov's hypothesis was that the small-scale eddies are isotropic.

The velocity component in direction $i$ at location $x$, $u_i(x, t)$, is a random function of time $t$. The mean of such a random function is computed in theory by an *ensemble average* in which the function is averaged over an infinite ensemble of identical experiments. Unfortunately, in the real world we cannot repeat an experiment indefinitely to obtain an ensemble average. In some cases we can replace the ensemble average with a time average over a single experiment, for example,

$$\overline{u}_i(x) = \lim_{T\to\infty} \frac{1}{T}\int_{t_0}^{t_0+T} u_i(x, t)\, dt \tag{D.2}$$

A direct correspondence between the ensemble average, denoted by $\langle u_i(x, t)\rangle$, and $\overline{u}_i(x)$, exists only when $u_i$ is a stationary function, because then $\langle u_i \rangle$ is independent of $t$ and $\overline{u}_i$ is independent of $t_0$.

It is common practice to decompose a turbulent quantity such as the velocity component $u_i$ into its mean $\overline{u}_i$, and a random component, $u_i'$. The mean kinetic energy per unit mass is then $\frac{1}{2}(\overline{u_1^2} + \overline{u_2^2} + \overline{u_3^2}) = \frac{1}{2}(\overline{u}_1^2 + \overline{u}_2^2 + \overline{u}_3^2) + \frac{1}{2}(\overline{u_1'^2} + \overline{u_2'^2} + \overline{u_3'^2})$. The *intensity* of turbulence, say in direction $i$, is defined by $(\overline{u_i'^2})^{1/2}/\overline{u}_i$, a measure of the size of the fluctuations relative to the mean velocity component at the same location.

The properties $\overline{u}_i$ and $\overline{u_i'^2}$ are based on a single velocity component at a single location and time. If we want to describe the time evolution of a random function like $u_i(x, t)$ at point $x$, we can define the Eulerian autocorrelation, $\overline{u_i(x, t)\, u_i(x, t + \tau)}$, between the values of $u_i$ at point $x$ at times $t$ and $t + \tau$. Because we are generally interested in stationary turbulence, this correlation does not depend on the time origin $t$, only on the time separation $\tau$. We thus define the Eulerian temporal velocity autocorrelation coefficient by

$$R_{ii}(\tau) = \frac{\overline{u_i(x, t)\, u_i(x, t + \tau)}}{\overline{u_i^2}} \tag{D.3}$$

where $R_{ii}(0) = 1$. Based on $R_{ii}(\tau)$ we can define the Eulerian integral time scale $T_E$ by

$$T_E = \max_i \int_0^\infty R_{ii}(\tau)\, d\tau \tag{D.4}$$

$T_E$ is a measure of the length of time over which the velocity at a point is correlated with itself.

In a similar vein, we can define the autocorrelation $\overline{u_i(x, t)\, u_i(x + r, t)}$ between the values of $u_i$ at time $t$ separated by a distance $r$, the corresponding spatial velocity autocorrelation coefficient,

$$S_{ii}(r) = \frac{\overline{u_i(x, t)\, u_i(x + r, t)}}{\overline{u_i^2}} \tag{D.5}$$

in homogeneous turbulence and the integral length scale (for scalar separation $r$)

$$L = \max_i \int_0^\infty S_{ii}(r)\, dr \tag{D.6}$$

## D.3 The Microscale

These velocity autocorrelation coefficients have the general shape shown in Figure 1.3. Note that the integral time or length scale, the value of the integral, can be represented as shown. Also, the curvature of the autocorrelation coefficient at the origin can be used to define a characteristic time or length scale $\lambda$ by, for example (Tennekes and Lumley, 1972),

$$\left(\frac{d^2R}{d\tau^2}\right)_{\tau=0} = -\frac{2}{\lambda^2} \tag{D.7}$$

Expanding $R(\tau)$ in a Taylor series about the origin, we can write for small $\tau$,

$$R(\tau) \simeq 1 - \frac{\tau^2}{\lambda^2} \tag{D.8}$$

This scale, called the *microscale*, is thus the intercept of the parabola that matches $R(\tau)$

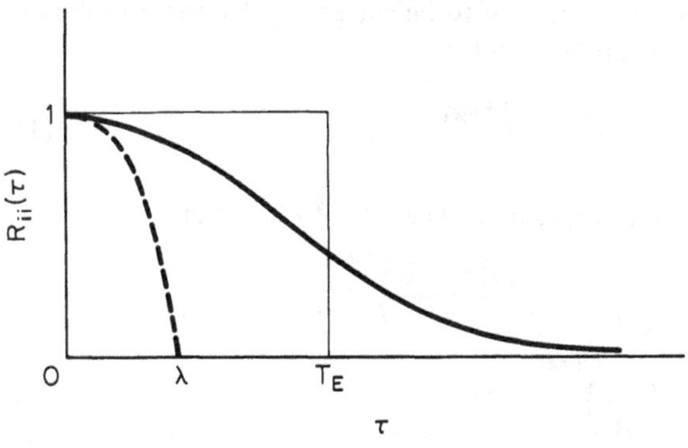


**Figure 1.3** Velocity autocorrelation coefficients showing the integral time scale $T_E$ and the Taylor microscale $\lambda$.

at the origin. Since $u_i(t)$ is stationary, we can write

$$0 = \frac{d^2}{dt^2}\overline{(u_i^2)} = 2\overline{u_i \frac{d^2 u_i}{dt^2}} + 2\overline{\left(\frac{du_i}{dt}\right)^2} \tag{D.9}$$

Using (D.7)–(D.9), we find that

$$\overline{\left(\frac{du_i}{dt}\right)^2} = \frac{2\overline{u_i^2}}{\lambda^2} \tag{D.10}$$

a relation that will turn out to be useful shortly.

Let us return to the discussion of the large and small scales of motion. The dissipation of energy at the small-scale end of the spectrum depends on several terms like $(\partial u_i/\partial x_j)^2$, most of which cannot be measured conveniently. Since the small-scale motions tend to be isotropic, it can be shown that the dissipation rate $\epsilon$ in isotropic turbulence is given by (Hinze, 1959)

$$\epsilon = 15\nu\overline{\left(\frac{\partial u_1}{\partial x_1}\right)^2} \tag{D.11}$$

The coefficient 15 arises because of summing a number of terms like $\overline{(\partial u_i/\partial x_j)^2}$. By analogy to (D.10) we define a length scale $\lambda$ by (omitting the factor of 2)

$$\overline{\left(\frac{\partial u_1}{\partial x_1}\right)^2} = \frac{\overline{u_1^2}}{\lambda^2} \tag{D.12}$$

The length scale in (D.12) is called the *Taylor microscale* in honor of G. I. Taylor who first defined (D.12). Note that in isotropic turbulence $\overline{u_1^2} = \overline{u_2^2} = \overline{u_3^2}$. Thus, from (D.11) and (D.12),

$$\epsilon = \frac{15\nu\overline{u_1^2}}{\lambda^2} \tag{D.13}$$

is a convenient estimate of $\epsilon$.

A relation between $\lambda$ and $L$, the macroscopic length scale of the system, can be obtained if we equate the rate of production of turbulent energy by shear to the rate of viscous dissipation (Tennekes and Lumley, 1972),

$$\frac{A\overline{u_1^3}}{L} = \frac{15\nu\overline{u_1^2}}{\lambda^2} \tag{D.14}$$

where $A$ is an undetermined constant order of 1. The ratio $\lambda/L$ is then

$$\frac{\lambda}{L} = \left(\frac{15}{A}\right)^{1/2}\left(\frac{(\overline{u_1^2})^{1/2}L}{\nu}\right)^{-1/2}$$

$$= \left(\frac{15}{A}\right)^{1/2} \mathrm{Re}^{-1/2} \tag{D.15}$$

Since Re >> 1, the Taylor microscale $\lambda$ is always much smaller than the macroscopic length scale $L$ of the flow.

The Taylor microscale is not the smallest length scale in turbulence; that is the Kolmogorov microscale $\eta$. The Taylor microscale does not represent the eddy sizes at which dissipation effects dominate, which is $\eta$. Rather, $\lambda$ is defined on the basis of the characteristic length scale of the velocity gradients.

## D.4 Chemical Reactions

Of primary interest to us is the assessment of chemical reaction rates when the reactants are embedded in a turbulent fluid and are inhomogeneously mixed. For a second-order reaction at constant temperature, if the chemical reaction rate is slow compared with the molecular diffusion rate (which determines how quickly local inhomogeneities are smoothed out), no effect of inhomogeneous mixing is observed and the reaction rate is that predicted on the basis of the spatial average concentrations. On the other hand, if the chemical reaction rate is fast compared with the molecular diffusion rate, the rate of reaction is limited by the rate at which the reactants can be brought together by molecular diffusion. Combustion reactions fall within the latter category.

We now derive the basic equations for predicting the mean concentrations of two reacting species, call them $A$ and $B$, in a turbulent flow. For a second-order isothermal reaction with rate constant $k$, the local, instantaneous rates of consumption of $A$ and $B$ are

$$\frac{\partial c_A}{\partial t} = -kc_A c_B \tag{D.16}$$

$$\frac{\partial c_B}{\partial t} = -kc_A c_B \tag{D.17}$$

If we let $c_A = \overline{c}_A + c'_A$ and $c_B = \overline{c}_B + c'_B$, and then substitute into (D.16) and (D.17) and average, we obtain the mean rates of consumption as

$$\frac{\partial \overline{c}_A}{\partial t} = -k(\overline{c}_A\overline{c}_B + \overline{c'_A c'_B}) \tag{D.18}$$

$$\frac{\partial \overline{c}_B}{\partial t} = -k(\overline{c}_A\overline{c}_B + \overline{c'_A c'_B}) \tag{D.19}$$

We see that the local mean rate of disappearance of $A$ and $B$ is $-k\overline{c}_A\overline{c}_B$ only if $\overline{c}_A\overline{c}_B >> \overline{c'_A c'_B}$ for the particular conditions in question. Let us see if we can determine the conditions for this approximation to hold. To do so we need to examine the continuity equation that governs the mean of the product of the concentration fluctuations of $A$ and $B$, $\overline{c'_A c'_B}$. Although the general equation governing $\overline{c'_A c'_B}$ is rather complicated, we can consider the behavior of $\overline{c'_A c'_B}$ in the absence of any appreciably large gradients. In that case the contributions to the local rate of change of the correlation $\overline{c'_A c'_B}$ arise only from the chemical reaction and molecular diffusion (Donaldson and Hilst, 1972),

$$\frac{\partial \overline{c'_A c'_B}}{\partial t} = \left(\frac{\partial \overline{c'_A c'_B}}{\partial t}\right)_{\text{chem}} - 2D\left(\overline{\frac{\partial c'_A}{\partial x_i}\frac{\partial c'_B}{\partial x_i}}\right) \tag{D.20}$$

where we have assumed for simplicity that both $A$ and $B$ have molecular diffusivity $D$. The first term on the right-hand side of (D.20) represents the generation of the correlation $\overline{c'_A c'_B}$ by the chemical reaction, and the second term is the decay of the correlation by the action of molecular diffusion.

To derive the second term on the right-hand side of (D.20), begin with the species conservation equations, (B.7), expressed here as (assuming $D_A = D_B = D$)

$$\frac{\partial c_A}{\partial t} + u_i \frac{\partial c_A}{\partial x_i} = D \frac{\partial^2 c_A}{\partial x_i \partial x_i} \tag{D.21a}$$

$$\frac{\partial c_B}{\partial t} + u_i \frac{\partial c_B}{\partial x_i} = D \frac{\partial^2 c_B}{\partial x_i \partial x_i} \tag{D.21b}$$

where $u_i\, \partial c/\partial x_i$ is a shorthand notation for $\mathbf{u} \cdot \nabla c$ and $\partial^2 c/\partial x_i \partial x_i$ for $\nabla^2 c$. Thus the repeated index $i$ denotes the summation over the three components of the term. If the concentration field is homogeneous, the spatial gradient of any mean quantity is zero. Using $c_A = \bar{c}_A + c'_A$ and $c_B = \bar{c}_B + c'_B$ in (D.21) yields

$$\frac{\partial}{\partial t}(\bar{c}_A + c'_A) + u_i \frac{\partial}{\partial x_i}(\bar{c}_A + c'_A) = D \frac{\partial^2}{\partial x_i \partial x_i}(\bar{c}_A + c'_A) \tag{D.22a}$$

$$\frac{\partial}{\partial t}(\bar{c}_B + c'_B) + u_i \frac{\partial}{\partial x_i}(\bar{c}_B + c'_B) = D \frac{\partial^2}{\partial x_i \partial x_i}(\bar{c}_B + c'_B) \tag{D.22b}$$

Multiply (D.22a) by $c'_B$ and (D.22b) by $c'_A$ and add the resulting equations,

$$c'_B \frac{\partial}{\partial t}(\bar{c}_A + c'_A) + c'_A \frac{\partial}{\partial t}(\bar{c}_B + c'_B) + u_i c'_B \frac{\partial}{\partial x_i}(\bar{c}_A + c'_A) + u_i c'_A \frac{\partial}{\partial x_i}(\bar{c}_B + c'_B)$$

$$= D c'_B \frac{\partial^2}{\partial x_i \partial x_i}(\bar{c}_A + c'_A) + D c'_A \frac{\partial^2}{\partial x_i \partial x_i}(\bar{c}_B + c'_B) \tag{D.23}$$

Now take the mean of this equation. Note that the means of $c'_A$ and $c'_B$, by themselves, are by definition zero. The terms in (D.23) become, upon averaging,

$$\overline{c'_B \frac{\partial}{\partial t}(\bar{c}_A + c'_A)} + \overline{c'_A \frac{\partial}{\partial t}(\bar{c}_B + c'_B)} = \frac{\partial}{\partial t}\overline{c'_A c'_B}$$

$$\overline{u_i c'_B \frac{\partial}{\partial x_i}(\bar{c}_A + c'_A)} = \overline{u_i c'_A \frac{\partial}{\partial x_i}(\bar{c}_B + c'_B)} = 0$$

$$\overline{D c'_B \frac{\partial^2}{\partial x_i \partial x_i}(\bar{c}_A + c'_A)} + \overline{D c'_A \frac{\partial^2}{\partial x_i \partial x_i}(\bar{c}_B + c'_B)} = -2D \overline{\frac{\partial c'_A}{\partial x_i}\frac{\partial c'_B}{\partial x_i}}$$

Thus (D.23) reduces to

$$\frac{\partial}{\partial t}\overline{c_A' c_B'} = -2D\,\overline{\frac{\partial c_A'}{\partial x_i}\frac{\partial c_B'}{\partial x_i}} \tag{D.24}$$

By analogy to (D.10), we can represent the right-hand side of (D.24) by means of a microscale $\lambda_c$ (Corrsin, 1957)

$$\overline{\frac{\partial c_A'}{\partial x_i}\frac{\partial c_B'}{\partial x_i}} = 6\,\frac{\overline{c_A' c_B'}}{\lambda_c^2} \tag{D.25}$$

where the additional factor of 3 results from the summation over the 3 coordinate components. Thus the characteristic time for the destruction of the correlation $\overline{c_A' c_B'}$ by molecular diffusion is

$$\tau_d = \frac{\lambda_c^2}{12D} \tag{D.26}$$

The effect of the first term on the right-hand side of (D.20) can be seen by examining the local equation governing $\bar{c}_A\bar{c}_B$ due only to chemistry.

$$\frac{\partial}{\partial t}(\bar{c}_A\bar{c}_B) = -k(\bar{c}_A + \bar{c}_B)(\bar{c}_A\bar{c}_B + \overline{c_A' c_B'}) \tag{D.27}$$

From this equation we see that the effect of the chemical reaction is to drive $\bar{c}_A\bar{c}_B$ to the negative of $\overline{c_A' c_B'}$, or vice versa, with a characteristic time

$$\tau_c = \frac{1}{k(\bar{c}_A + \bar{c}_B)} \tag{D.28}$$

Equation (D.27) states that the reaction between $A$ and $B$ will always stop (i.e., $\bar{c}_A\bar{c}_B + \overline{c_A' c_B'} = 0$) short of the exhaustion of $A$ or $B$ unless $A$ and $B$ are perfectly mixed in the turbulent field. The reason for this is that in the absence of diffusion, if $A$ and $B$ are not perfectly mixed to start with, there is no mechanism to replenish the reactants once they have been consumed in a local volume element.

From (D.25) if the concentration microscale $\lambda_c$ is small enough and the reaction rate is slow enough, the dissipation term in (D.20) will dominate and $\overline{c_A' c_B'} \simeq 0$. Physically, this means that molecular diffusion is rapid enough to keep $A$ and $B$ well mixed locally. On the other hand, if $\lambda_c$ is large and the reaction rate is fast, the generation of $\overline{c_A' c_B'}$ by the chemistry will be dominant and $\overline{c_A' c_B'}$ will tend toward $-\bar{c}_A\bar{c}_B$, indicating that the two species are poorly mixed. In this case the overall rate of reaction in the flow is governed not by the rate of reaction but by the rate at which $A$ and $B$ are brought together locally by molecular diffusion.

For gases at sufficiently high Reynolds number, the concentration microscale $\lambda_c$ is related to the Taylor microscale by (Corrsin, 1957)

$$\frac{\lambda_c}{\lambda} \approx 2\,\frac{D}{\nu} \tag{D.29}$$

Using (D.26) and (D.29), the decay time for concentration fluctuations is related to the Taylor microscale by

$$\tau_d \approx \frac{\lambda^2}{6\nu} \tag{D.30}$$

## APPENDIX E UNITS

The units used in this book are largely those of the International System of Units (SI). The SI system of units consists of a set of basic units, prefixes to extend the range of values that can be handled conveniently with the units, and a set of derived units. All the derived units can be expressed in terms of the basic units, although each derived unit has its own symbol that can be used for conciseness. The prefixes can be used with either the basic units or the derived units.

TABLE 1.14 DERIVED SI UNITS

| Quantity | Name of unit | Unit symbol or abbreviation | Relation to basic units |
|---|---|---|---|
| Frequency | hertz | Hz | $s^{-1}$ |
| Angular velocity | radian per second | | rad $s^{-1}$ |
| Force | newton | N | kg m $s^{-2}$ |
| Surface tension | newton per meter | N $m^{-1}$ | kg $s^{-2}$ |
| Pressure | pascal, newton per square meter | Pa, N $m^{-2}$ | kg $m^{-1}$ $s^{-2}$ |
| Viscosity | newton-second per square meter | N s $m^{-2}$ | kg $m^{-1}$ $s^{-1}$ |
| Kinematic viscosity, Diffusivity, Thermal diffusivity | meter squared per second | $m^2$ $s^{-1}$ | $m^2$ $s^{-1}$ |
| Energy | joule | J | kg $m^2$ $s^{-2}$ |
| Power | watt, joule per second | W, J $s^{-1}$ | kg $m^2$ $s^{-3}$ |
| Heat capacity, Gas constant | joule per kilogram degree | J $kg^{-1}$ $K^{-1}$ | $m^2$ $s^{-2}$ $K^{-1}$ |
| Enthalpy per unit mass | joules per kilogram | J $kg^{-1}$ | $m^2$ $s^{-2}$ |
| Entropy per unit mass | joules per kilogram degree | J $kg^{-1}$ $K^{-1}$ | $m^2$ $s^{-2}$ $K^{-1}$ |
| Thermal conductivity | joules per meter second degree | J $m^{-1}$ $s^{-1}$ $K^{-1}$ | kg m $s^{-3}$ $K^{-1}$ |
| Mass transfer coefficient | meter per second | m $s^{-1}$ | m $s^{-1}$ |
| Electric charge | coulomb | C | A s |
| Electromotive force | volt | V | kg $m^2$ $A^{-1}$ $s^{-3}$ |
| Electric field strength | volt per meter | V $m^{-1}$ | kg m $A^{-1}$ $s^{-3}$ |
| Electric resistance | ohm | Ω | kg $m^2$ $A^{-2}$ $s^{-2}$ |
| Electric conductivity | ampere per volt meter | A $V^{-1}$ $m^{-1}$ | $A^2$ $s^3$ $kg^{-1}$ $m^{-3}$ |
| Electric capacitance | farad | F | $A^2$ $s^4$ $kg^{-1}$ $m^{-2}$ |

**TABLE 1.15 CONVERSION FACTORS TO SI UNITS**

| Quantity | Conversion factors | | |
|---|---|---|---|
| Length | $10^{-6}\ \frac{\text{m}}{\mu\text{m}}$ | $10^{-10}\ \frac{\text{m}}{\text{Å}}$ | $0.3048\ \frac{\text{m}}{\text{ft}}$ |
| Mass | $0.4536\ \frac{\text{kg}}{\text{lbm}}$ | | |
| Time | $3600\ \frac{\text{s}}{\text{h}}$ | $86{,}400\ \frac{\text{s}}{\text{day}}$ | |
| Temperature | $0.5555\ \frac{\text{K}}{°\text{R}}$ | $0.5555\ \frac{\text{K}}{°\text{F}^a}$ | $1.0\ \frac{\text{K}}{°\text{C}^a}$ |
| Volume | $3.785 \times 10^{-3}\ \frac{\text{m}^3}{\text{gal (U.S.)}}$ | | |
| Density | $10^3\ \frac{\text{kg m}^{-3}}{\text{g cm}^{-3}}$ | | |
| Force | $10^{-5}\ \frac{\text{N}}{\text{dyn}}$ | | |
| Pressure | $0.1\ \frac{\text{Pa}}{\text{dyn cm}^{-2}}$ | $10^5\ \frac{\text{Pa}}{\text{bar}}$ | |
| | $1.0133 \times 10^5\ \frac{\text{Pa}}{\text{std atm}}$ | $133.3\ \frac{\text{Pa}}{\text{mmHg}}$ | |
| Energy | $4.187\ \frac{\text{J}}{\text{cal}}$ | $4187\ \frac{\text{J}}{\text{kcal}}$ | $3.6 \times 10^6\ \frac{\text{J}}{\text{kWh}}$ |
| | $10^{-7}\ \frac{\text{J}}{\text{erg}}$ | $1055\ \frac{\text{J}}{\text{Btu}}$ | $2.685 \times 10^6\ \frac{\text{J}}{\text{hp h}}$ |
| Power | $4.187\ \frac{\text{W}}{\text{cal s}^{-1}}$ | $10^{-7}\ \frac{\text{W}}{\text{erg s}^{-1}}$ | $0.293\ \frac{\text{W}}{\text{Btu h}^{-1}}$ |
| | $745.8\ \frac{\text{W}}{\text{hp}}$ | | |
| Specific energy (energy per unit mass) | $4187\ \frac{\text{J kg}^{-1}}{\text{cal g}^{-1}}$ | $4.187 \times 10^6\ \frac{\text{J kg}^{-1}}{\text{kcal g}^{-1}}$ | |
| | $2326\ \frac{\text{J kg}^{-1}}{\text{Btu lbm}^{-1}}$ | | |
| Specific heat and gas constant[b] | $4187\ \frac{\text{J kg}^{-1}\ \text{K}^{-1}}{\text{cal g}^{-1}\,°\text{C}^{-1}}$ | | |
| Thermal conductivity | $418.7\ \frac{\text{W m}^{-1}\ \text{K}^{-1}}{\text{cal s}^{-1}\ \text{cm}^{-1}\ °\text{C}^{-1}}$ | | |

[a] Temperature difference
[b] Universal gas constant

$$R = 8.314 \times 10^3 \text{ J kg-mol}^{-1}\ \text{K}^{-1}$$
$$= 1.987 \text{ cal g-mol}^{-1}\ \text{K}^{-1}$$
$$= 82.05 \text{ cm}^3 \text{ atm g-mol}^{-1}\ \text{K}^{-1}$$
$$= 8.314 \text{ Pa m}^3 \text{ g-mol}^{-1}\ \text{K}^{-1}$$

The basic SI units are:

| | | |
|---|---|---|
| Length: | meter | m |
| Mass: | kilogram | kg |
| Time: | second | s |
| Electric current: | ampere | A |
| Temperature: | kelvin | K |

Some of the standard prefixes are:

| | | |
|---|---|---|
| $10^{-12}$: | pico | p |
| $10^{-9}$: | nano | n |
| $10^{-6}$: | micro | $\mu$ |
| $10^{-3}$: | milli | m |
| $10^{-2}$: | centi | c |
| $10^{3}$: | kilo | k |
| $10^{6}$: | mega | M |
| $10^{9}$: | giga | G |
| $10^{12}$: | tera | T |

The symbol for the prefix is attached to the symbol for the unit. For example, $10^{-6}$ m = 1 $\mu$m. This particular unit, the micrometer, is often referred to simply as a micron.

From time to time it will be more convenient to employ the centimeter (cm) as the basic unit of length and the gram (g) as the basic unit of mass. For example, mass density is usually expressed in terms of g cm$^{-3}$. The derived units that are used in this book are listed in Table 1.14, and Table 1.15 lists several common conversion factors.

## PROBLEMS

**1.1.** The 1-hour NAAQS for carbon monoxide is 40 mg m$^{-3}$. Convert this to ppm at standard temperature and pressure.

**1.2.** A solid-waste incinerator emits particulate matter at the NSPS rate of 0.18 g/dscm. The incinerator burns 50 metric tons per day (1 metric ton = $10^3$ kg) and exhausts gases at a ratio of 20 kg of gases per kilogram of feed at atmospheric pressure and 453 K. Assume that the average molecular weight of the emitted gases is 30 and that they contain 12% $CO_2$ and 10% $H_2O$. What is the daily emission rate of particulate matter?

**1.3.** Consider the following sequence for the thermal decomposition of the hydrocarbon molecule $M_1$:

$$M_1 \xrightarrow{1} 2R_1 \qquad (E = 80)$$

$$R_1 + M_1 \xrightarrow{2} R_1H + R_2 \qquad (E = 15)$$

$$R_2 \xrightarrow{3} R_1 + M_2 \qquad (E = 38)$$

$$R_1 + R_2 \xrightarrow{4} M_3 \qquad (E = 8)$$

If the chains (reactions 2 and 3) are long and $k_1$ is relatively small, show that the overall rate of disappearance of $M_1$ is appproximately

$$\frac{d[M_1]}{dt} = -\left(\frac{k_1k_2k_3}{k_4}\right)^{1/2} [M_1]$$

and that the overall activation energy is 62.5.

**1.4.** For the reaction mechanism of Problem 1.3, show that the overall order of the reaction depends on the termination reaction as indicated in the following table.

| Termination reaction | Overall order |
|---|---|
| $R_2 + R_2$ | $\frac{1}{2}$ |
| $R_1 + R_2$ | 1 |
| $R_1 + R_1$ | $\frac{3}{2}$ |

**1.5.** If the random variable $X$ obeys the normal distribution

$$p_X(x) = \frac{1}{\sqrt{2\pi}\sigma} e^{-(x-\mu)^2/2\sigma^2}$$

determine $p_Y(y)$ if $Y = X^2$.

## REFERENCES


ARIS, R. *Introduction to the Analysis of Chemical Reactors*, Prentice-Hall, Englewood Cliffs, NJ (1965).

BENSON, S. W. *Foundations of Chemical Kinetics*, McGraw-Hill, New York (1960).

BIRD, R. B., STEWART, W. E., and LIGHTFOOT, E. N. *Transport Phenomena*, Wiley, New York (1960).

CORRSIN, S. "Simple Theory of an Idealized Turbulent Mixer," *A.I.Ch.E.J.*, 3, 329–330 (1957).

DONALDSON, C. DU P., and HILST, G. R. "Chemical Reactions in Inhomogeneous Mixtures: The Effect of the Scale of Turbulent Mixing," in *Proceedings of the 1972 Heat Transfer and Fluid Mechanics Institute*, R. B. Landis and G. J. Hordemann, Eds., Stanford University Press, Stanford, CA (1972).

HINZE, J. O. *Turbulence*, McGraw-Hill, New York (1959).

SINGH, H. B., SALAS, L. J., SMITH, A., STILES, R., and SHIGEISHI, H. "Atmospheric Measurements of Selected Hazardous Organic Chemicals," Project Summary, U.S. Environmental Protection Agency Report No. EPA-600/53-81-032 (1981).

TENNEKES, H., and LUMLEY, J. L. *A First Course in Turbulence*, MIT Press, Cambridge, MA (1972).

U.S. Environmental Protection Agency. "Compilation of Air Pollutant Emission Factors," AP-42 (1977).

U.S. Environmental Protection Agency. "Air Quality Criteria for Ozone and Other Photochemical Oxidants," Report No. EPA-600/8-78-004 (1978).

U.S. Environmental Protection Agency. "Air Quality Criteria for Oxides of Nitrogen," Report No. EPA-600/8-82-026 (1982a).

U.S. Environmental Protection Agency. "Air Quality Criteria for Particulate Matter and Sulfur Oxides," Report No. EPA-600/8-82-029 (1982b).

# 2

# Combustion Fundamentals

To understand the formation of pollutants in combustion systems, we must first understand the nature of the fuels being burned, the thermodynamics of the combustion process, and some aspects of flame structure. In this chapter we discuss fundamental aspects of hydrocarbon fuel combustion that relate directly to the formation of pollutants or to the control of emissions. Questions of flame stability, detonations, and several other important aspects of combustion science are beyond the scope of the present discussion and will not be treated. Specific pollution control problems will be addressed in detail in later chapters.

## 2.1 FUELS

Of the spectrum of fuels currently in widespread use, the simplest in composition is natural gas, which consists primarily of methane but includes a number of other constituents as well. The compositions of other gaseous fuels are generally more complex, but they are, at least, readily determined. Table 2.1 illustrates the range of compositions encountered in gaseous fuels, both natural and synthetic.

Information on the composition of liquid or solid fuels is generally much more limited than that for gaseous fuels. Rarely is the molecular composition known since liquid fuels are usually complex mixtures of a large number of hydrocarbon species. The most commonly reported composition data are derived from the *ultimate analysis*, which consists of measurements of the elemental composition of the fuel, generally presented as mass fractions of carbon, hydrogen, sulfur, oxygen, nitrogen, and ash, where appro-

**TABLE 2.1** PROPERTIES OF GASEOUS FUELS

| | $CH_4$ | $C_2H_6$ | $C_3H_8$ | Other hydrocarbons | CO | $H_2$ | $H_2S$ | $N_2$ | $CO_2$ | Heating value[a] ($10^6$ J $m^{-3}$) |
|---|---|---|---|---|---|---|---|---|---|---|
| Natural gas | | | | | | | | | | |
| No. 1 | 77.7 | 5.6 | 2.4 | 1.8 | — | — | 7.0 | — | — | — |
| No. 2[b] | 88.8 | 6.4 | 2.7 | 2.0 | — | — | 0.0004 | — | 0 | 41.9 |
| No. 3 | 59.2 | 12.9 | — | — | — | — | — | 0.7 | 26.2 | 30.7 |
| No. 4 | 99.2 | — | — | — | — | — | — | 0.6 | 0.2 | 36.3 |
| Refinery gas | | | | | | | | | | |
| No. 1 | 41.6 | 20.9 | 19.7 | 15.6 | — | — | 2.2 | — | — | 68.6 |
| No. 2 | 4.3 | 82.7 | 13.0 | — | — | — | — | — | — | 67.1 |
| No. 3 | 15.9 | 5.0 | — | 2.4 | 14.3 | 50.9 | — | 8.4 | 2.2 | 18.7 |
| Coke oven gas | — | — | — | 35.3 | 6.3 | 53.0 | — | 3.4 | 1.8 | 21.5 |
| Blast furnace gas | — | — | — | — | 26.2 | 3.2 | — | 57.6 | 13 | 3.4 |

[a] $p$, 101 kPa; $T$, 25°C.

[b] "Sweetened," $H_2S$ removed.

**TABLE 2.2 PROPERTIES OF TYPICAL LIQUID FUELS**

| Gasoline | Percent by weight C | H | N | O | S | Ash | Specific gravity | Heating value ($10^6$ J $kg^{-1}$) |
|---|---|---|---|---|---|---|---|---|
| Kerosene (No. 1) | 86.5 | 13.2 | 0.1 | 0.1 | 0.1 | Trace | 0.825 | 46.4 |
| Fuel oil | | | | | | | | |
| No. 2 | 86.4 | 12.7 | 0.1 | 0.1 | 0.4–0.7 | Trace | 0.865 | 45.5 |
| No. 4 | 85.6 | 11.7 | 0.3 | 0.4 | <2 | 0.05 | 0.953 | 43.4 |
| No. 6 | 85.7 | 10.5 | 0.5 | 0.4 | <2.8 | 0.08 | 0.986 | 42.5 |

priate. The heating value, a measure of the heat release during complete combustion, is also reported with the ultimate analysis. Ultimate analyses of a number of liquid fuels are presented in Table 2.2.

In addition to the limited composition data given in Tables 2.1 and 2.2, physical properties that influence the handling and use of a particular fuel are frequently measured. For liquid fuels, the specific gravity or API gravity,* viscosity (possibly at several temperatures), flash point (a measure of the temperature at which the fuel is sufficiently volatile to ignite readily), and distillation profiles (fraction vaporized as a function of temperature) may be reported.

The properties of solid fuels vary even more widely than those of liquid fuels. The most common solid fuel is coal. Formed by biological decomposition and geological transformation of plant debris, coals are classified by rank, a measure of the degree to which the organic matter has been transformed from cellulose. Low-rank fuels such as peat or lignite have undergone relatively little change, whereas high-rank anthracite is nearly graphitic in structure. Low-rank fuels contain large amounts of volatile matter that are released upon heating. High-rank fuels contain much more *fixed carbon*, which remains after the volatiles are released.

Solid fuels are characterized by the ultimate analysis and by the so-called *proximate analysis*, which identifies the degree of coalifaction of a solid fuel (Table 2.3). Coal samples that have been air dried are subjected to a number of standardized tests to determine the amount of *moisture* inherent to the coal structure, the quantity of *volatile matter* released by the coal upon heating to 1200 K for several minutes, and the mass of *ash* or noncombustible inorganic (mineral) impurities that remains after low temperature (700 to 1050 K) oxidation. The difference between the initial mass of coal and the sum of masses of moisture, volatile matter, and ash is called *fixed carbon*. The conditions of these standardized tests differ markedly from typical combustion environments, and the values reported in the proximate analysis do not necessarily represent yields actually encountered in practical combustors. This point is discussed in more detail in the section on solid fuel combustion.

*Degrees API = [141.5/(specific gravity 16°C/water at 16°C) − 131.5].

**TABLE 2.3 PROPERTIES OF SELECTED SOLID FUELS**

| | Percent by weight | | | | | | | | | |
|---|---|---|---|---|---|---|---|---|---|---|
| | Proximate analysis | | | | Ultimate analysis | | | | | Heating value |
| Fuel (state) | Carbon | Volatile matter | Moisture | Ash | C | H | N | O | S | ($10^6$ J kg$^{-1}$) |
| Meta-anthracite (RI) | 65.3 | 2.5 | 13.3 | 18.9 | 64.2 | 0.4 | 0.2 | 2.7 | 0.3 | 21.7 |
| Anthracite (PA) | 77.1 | 3.8 | 5.4 | 13.7 | 76.1 | 1.8 | 0.6 | 1.8 | 0.6 | 27.8 |
| Semianthracite (PA) | 78.9 | 8.4 | 3.0 | 9.7 | 80.2 | 3.3 | 1.1 | 2.0 | 0.7 | 31.3 |
| Bituminous (PA) | 70.0 | 20.5 | 3.3 | 6.2 | 80.7 | 4.5 | 1.1 | 2.4 | 1.8 | 33.3 |
| High-volatile bituminous | | | | | | | | | | |
| (PA) | 58.3 | 30.3 | 2.6 | 9.1 | 76.6 | 4.9 | 1.6 | 3.9 | 1.3 | 31.7 |
| (CO) | 54.3 | 32.6 | 1.4 | 11.7 | 73.4 | 5.1 | 1.3 | 6.5 | 0.6 | 30.7 |
| (KY) | 45.3 | 37.7 | 7.5 | 9.5 | 66.9 | 4.8 | 1.4 | 6.4 | 3.5 | 28.1 |
| (IL) | 39.1 | 40.2 | 12.1 | 8.6 | 62.8 | 4.6 | 1.0 | 6.6 | 4.3 | 26.7 |
| Subbituminous (CO) | 45.9 | 30.5 | 19.6 | 4.0 | 58.8 | 3.8 | 1.3 | 12.2 | 0.3 | 23.6 |
| Lignite (ND) | 30.8 | 28.2 | 34.8 | 6.2 | 42.4 | 2.8 | 0.7 | 12.4 | 0.7 | 16.8 |
| Brown coal (Australia) | 15.3 | 17.7 | 66.3 | 0.7 | | | | | 0.1 | 8.6 |
| Wood (Douglas fir, as received) | 17.2 | 82.0 | 35.9 | 0.8 | 52.3 | 6.3 | 0.1 | 40.5 | 0 | 21.0 |

## 2.2 COMBUSTION STOICHIOMETRY

Complete oxidation of simple hydrocarbon fuels forms carbon dioxide ($CO_2$) from all of the carbon and water ($H_2O$) from the hydrogen, that is, for a hydrocarbon fuel with the general composition $C_nH_m$,

$$C_nH_m + \left(n + \frac{m}{4}\right)O_2 \longrightarrow nCO_2 + \frac{m}{2} H_2O$$

Even in the idealized case of complete combustion, the accounting of all species present in combustion exhaust involves more than simply measuring the $CO_2$ and $H_2O$. Since fuels are burned in air rather than in pure oxygen, the nitrogen in the air may participate in the combustion process to produce nitrogen oxides. Also, many fuels contain elements other than carbon, and these elements may be transformed during combustion. Finally, combustion is not always complete, and the effluent gases contain unburned and partially burned products in addition to $CO_2$ and $H_2O$.

Air is composed of oxygen, nitrogen, and small amounts of carbon dioxide, argon, and other trace species. Since the vast majority of the diluent in air is nitrogen, for our purposes it is perfectly reasonable to consider air as a mixture of 20.9% (mole basis) $O_2$ and 79.1% (mole basis) $N_2$. Thus for every mole of oxygen required for combustion, 3.78 mol of nitrogen must be introduced as well. Although nitrogen may not significantly alter the oxygen balance, it does have a major impact on the thermodynamics, chemical kinetics, and formation of pollutants in combustion systems. For this reason it is useful to carry the "inert" species along in the combustion calculations. The stoichiometric relation for complete oxidation of a hydrocarbon fuel, $C_nH_m$, becomes

$$C_nH_m + \left(n + \frac{m}{4}\right)(O_2 + 3.78N_2) \longrightarrow nCO_2 + \frac{m}{2} H_2O + 3.78\left(n + \frac{m}{4}\right)N_2$$

Thus for every mole of fuel burned, $4.78(n + m/4)$ mol of air are required and $4.78(n + m/4) + m/4$ mol of combustion products are generated. The molar *fuel/air* ratio for stoichiometric combustion is $1/[4.78(n + m/4)]$.

Gas compositions are generally reported in terms of mole fractions since the mole fraction does not vary with temperature or pressure as does the concentration (moles/unit volume). The product mole fractions for complete combustion of this hydrocarbon fuel are

$$y_{CO_2} = \frac{n}{4.78(n + m/4) + m/4}$$

$$y_{H_2O} = \frac{m/2}{4.78(n + m/4) + m/4}$$

$$y_{N_2} = \frac{3.78(n + m/4)}{4.78(n + m/4) + m/4}$$

The large quantity of nitrogen diluent substantially reduces the mole fractions of the combustion products from the values they would have in its absence.

**Example 2.1** ***Combustion of Octane in Air***

Determine the stoichiometric fuel/air mass ratio and product gas composition for combustion of octane ($C_8H_{18}$) in air.

The overall stoichiometry is

$$C_8H_{18} + 12.5(O_2 + 3.78N_2) \longrightarrow 8CO_2 + 9H_2O + 47.25N_2$$

For each mole of fuel burned, 59.75 mol of air is required. The molecular weight of octane is 114. The fuel/air mass ratio for stoichiometric combustion is, therefore,

$$\left(\frac{m_f}{m_a}\right)_s = \frac{114}{12.5(32 + 3.78 \times 28)} = \frac{114}{1723} = 0.0662$$

The total number of moles of combustion products generated is

$$8 + 9 + 47.25 = 64.25$$

Finally, the product gas composition is, on a mole fraction basis,

$$y_{CO_2} = \frac{8}{64.25} = 0.125 = 12.5\%$$

$$y_{H_2O} = \frac{9}{64.25} = 0.140 = 14.0\%$$

$$y_{N_2} = \frac{47.25}{64.25} = 0.735 = 73.5\%$$

Minor components and impurities in the fuel complicate our analysis of combustion products somewhat. Fuel sulfur is usually oxidized to form sulfur dioxide ($SO_2$). (Even though there are cases where sulfur compounds involving higher oxidation states of sulfur or reduced sulfur compounds are produced, it is a reasonable first approximation to assume that all of the fuel sulfur forms $SO_2$.) Upon combustion, organically bound fuel-nitrogen is converted to both $N_2$ and NO, with molecular nitrogen generally dominating. For the moment we shall assume that all of the fuel-nitrogen forms $N_2$. Ash, the noncombustible inorganic (mineral) impurities in the fuel, undergoes a number of transformations at combustion temperatures, which will also be neglected for the time being, so that the ash will be assumed to be inert.

For most common fuels, the only chemical information available is its elemental composition on a mass basis, as determined in the ultimate analysis. Before we can proceed with combustion calculations it is necessary to convert these data to an effective molar composition.

**Example 2.2** ***Coal Composition***

Consider a Pittsburgh seam coal that contains 77.2% C, 5.2% H, 1.2% N, 2.6% S, 5.9% O, and 7.9% ash by weight. The ultimate analysis is generally reported on an "as received" basis, including the moisture in the chemical analysis. The molar composition may be determined by dividing each of the mass percentages by the atomic weight of the constituent. For convenience in stoichiometric calculations, the composition is then normalized with respect to carbon:

| Element | wt % | mol/100 g | mol/mol C |
|---|---|---|---|
| C | 77.2 ÷ 12 | = 6.43 | ÷ 6.43 = 1.00 |
| H | 5.2 ÷ 1 | = 5.20 | ÷ 6.43 = 0.808 |
| N | 1.2 ÷ 14 | = 0.0857 | ÷ 6.43 = 0.013 |
| S | 2.6 ÷ 32 | = 0.0812 | ÷ 6.43 = 0.013 |
| O | 5.9 ÷ 16 | = 0.369 | ÷ 6.43 = 0.057 |
| Ash | 7.9 | | ÷ 6.43 = 1.23 g/mol C |

The chemical formula that can be used to describe this particular coal is, thus,

$$CH_{0.808}N_{0.013}S_{0.013}O_{0.057}$$

The formula weight of the fuel, or, as written here, the mass per mole of carbon, including ash, is

$$M_f = \frac{100}{6.43}\frac{g}{\text{mol C}} = 15.55\,\frac{\text{g}}{\text{mol C}}$$

The combustion stoichiometry of this fuel must include the minor species, ash, and oxygen in the fuel. Making the simplifying assumptions described above, we may write the stoichiometry as

$$CH_{0.808}N_{0.013}S_{0.013}O_{0.057} + \alpha(O_2 + 3.78N_2) \longrightarrow CO_2 + 0.404H_2O + 0.013SO_2 + (3.78\alpha + 0.0065)N_2$$

where

$$\alpha = 1 + \frac{0.808}{4} + 0.013 - \frac{0.057}{2} = 1.19$$

The fuel/air mass ratio for stoichiometric combustion is

$$\left(\frac{m_f}{m_a}\right)_s = \frac{15.55\ \text{g/mol C}}{1.19(32 + 3.78 \times 28)\ \text{g/mol C}} = 0.0948$$

The total number of moles of gaseous combustion products per mole of C is

$$N_T = 1 + 0.404 + 0.013 + 4.504 = 5.921$$

The species mole fractions in the combustion products are, therefore,

$$y_{CO_2} = \frac{1}{5.921} = 0.169 = 16.9\%$$

$$y_{H_2O} = \frac{0.404}{5.921} = 0.068 = 6.82\%$$

$$y_{SO_2} = \frac{0.013}{5.921} = 0.00220 = 2200 \text{ ppm}$$

$$y_{N_2} = \frac{4.504}{5.921} = 0.761 = 76.1\%$$

where the $SO_2$ mole fraction has been expressed as parts per million (ppm) on a mole (or volume) basis, a common form for presenting data on minor species in the gas (recall Section 1.3).

Few combustion systems are operated precisely at the stoichiometric condition because of the difficulty of achieving such intimate mixing between fuel and air that perfect conversion is attained. More commonly, combustors are operated with a margin for error using more than the stoichiometric amount of air. The *fuel/air ratio* is used to define the operating conditions of a combustor. Comparison of the two examples presented above shows that the fuel/air ratio required for complete combustion varies with fuel composition. Values of the fuel/air and air/fuel mass ratios for stoichiometric combustion of a variety of fuels are presented in Table 2.4. Because the mass ratios vary widely with fuel composition, they are not a convenient base for comparison of systems burning different fuels.

The stoichiometric condition is a logical reference point for comparison of systems operating on different fuels. Several normalized ratios are used in the combustion literature to overcome the ambiguity of the mass ratio. The *equivalence ratio*, $\phi$, is defined as the fuel/air ratio normalized with respect to the stoichiometric fuel/air ratio,

$$\phi = \frac{m_f/m_a}{(m_f/m_a)_s} \tag{2.1}$$

Alternatively, the stoichiometric ratio, $\lambda$, is the air/fuel ratio normalized with respect to stoichiometric, that is,

$$\lambda = \frac{m_a/m_f}{(m_a/m_f)_s} = \frac{1}{\phi} \tag{2.2}$$

Other ratios that appear in the literature include the percent excess air $[EA = (\lambda - 1) \times 100\%]$ and the percent theoretical air $(TA = \lambda \times 100\%)$. In reading the combustion

**TABLE 2.4** MASS RATIOS FOR STOICHIOMETRIC COMBUSTION

| Fuel | Molar H/C ratio | $(m_f/m_a)_s$ | $(m_a/m_f)_s$ |
|---|---|---|---|
| $H_2$ | $\infty$ | 0.029 | 34 |
| $CH_4$ | 4 | 0.058 | 17 |
| Kerosene $C_nH_{2n}$ | 2 | 0.068 | 15 |
| Benzene (coke) | 1 | 0.076 | 13 |
| Char | 0.5 | 0.081 | 12 |
| Carbon | 0 | 0.087 | 11 |
| Methanol $CH_3OH$ | 4 | 0.155 | 6.46 |

literature, one should be careful to ascertain which of the various terms is being used since neither names nor symbols have been fully standardized. The fuel/air equivalence ratio, $\phi$, will be used in this book unless otherwise stated.

The mix of combustion products varies with the equivalence ratio. Combustion may be complete under fuel-lean conditions (excess air, $\phi < 1$) with some oxygen remaining unreacted in the combustion products. The composition of the products of fuel-lean combustion is, to a good approximation, determined by atom balances alone. Consider, for example, the combustion of methane at $\phi = 0.85$,

$$CH_4 + \frac{2}{0.85}(O_2 + 3.78N_2) \longrightarrow CO_2 + 2H_2O + \left(\frac{2}{0.85} - 2\right)O_2 + \frac{2 \times 3.78}{0.85} N_2$$

$$\longrightarrow CO_2 + 2H_2O + 0.353O_2 + 8.89N_2$$

The composition of the combustion products now includes $O_2$:

$$y_{CO_2} = \frac{1}{12.24} = 0.0817 = 8.17\%$$

$$y_{H_2O} = \frac{2}{12.24} = 0.163 = 16.3\%$$

$$y_{O_2} = \frac{0.353}{12.24} = 0.0288 = 2.88\%$$

$$y_{N_2} = \frac{8.89}{12.24} = 0.726 = 72.6\%$$

In some references, the combustion condition is not stated in terms of a fuel/air ratio but, rather, in terms of the amount of oxygen in the combustion products (i.e., 2.9% $O_2$ in this case).

The problem of specifying the products of combustion is more complicated for fuel-rich combustion, $\phi > 1$, than for fuel-lean combustion. Since there is insufficient oxygen for complete combustion under fuel-rich conditions, some carbon monoxide, hydrogen, and possibly, unburned hydrocarbons remain in the combustion products. Thus there are at least five products present ($CO$, $CO_2$, $H_2$, $H_2O$, $N_2$), but only four elemental balances are possible. An auxiliary condition based on thermodynamics or kinetics is needed to determine the exhaust composition. We now turn our attention to combustion thermodynamics before returning to the question of product gas composition in fuel-rich combustion.

## 2.3 COMBUSTION THERMODYNAMICS

Substantial energy is released in a very short time when a fuel is burned, leading to a dramatic temperature increase of the combustion gases. Temperatures in excess of 2000 K are common in flames. It is the high temperature that allows rapid oxidation of hydrocarbons and carbon monoxide to carbon dioxide and water but also makes possible the oxidation of $N_2$ to form nitric oxide. The temperature in the flame must be known to consider the formation and control of pollutants.

Thermodynamics provides us with good estimates of the flame temperature that are needed not only to assess the combustion process itself but also to calculate the concentrations of the many chemical species that play a role in the formation and destruction of pollutants. We begin our study of the combustion process with a brief review of the relevant thermodynamics.

### 2.3.1 First Law of Thermodynamics

The first law of thermodynamics states that the change in the total energy of a closed system of fixed mass and identity is equal to the heat transfer to the system from its surroundings minus the work done by the system on its surroundings; that is, for an infinitesimal change of state,

$$dE = \delta Q - \delta W \tag{2.3}$$

The total energy of the system, $E$, includes the internal energy, $U$, the kinetic energy, and the potential energy. The energy is a property of the system that is independent of the path taken in going from one state to another. In contrast, the heat transfer, $\delta Q$, and the work transfer, $\delta W$, for any change in the state of the system depend on the manner in which the state of the system is changed. The change in the system energy is described by a total differential, $dE$. Since the work and heat transfer depend on the path followed by the system, the $\delta$ is used to indicate that these increments are not total differentials. For most systems of concern here, the kinetic and potential energy terms can be neglected, so we may express the system energy in terms of the internal energy, that is,

$$dU = \delta Q - \delta W \tag{2.4}$$

Integrating over a finite change of state from state 1 to state 2, the first law for a closed system becomes

$$U_2 - U_1 = Q_{12} - W_{12} \tag{2.5}$$

Only rarely in the consideration of combustion processes can we limit ourselves to a fixed mass in a closed system. More generally, the fuel and air enter the combustion zone across certain boundaries, and combustion products are exhausted across other boundaries. It is convenient, therefore, to derive an expression for the change in state of a fixed volume in space, called a *control volume*, rather than a fixed mass.

A control volume may be defined in terms of any volume in space in which one has interest for a particular analysis. Figure 2.1 illustrates a control volume that is prescribed by a surface, $S$. We would like to derive an equation that describes the change in the properties of the control volume when a small increment of mass, $\delta m$, crosses $S$ and enters the control volume. To do this, we first define a closed system that includes both the material initially in the control volume, mass $m$, energy $E_1$, and the increment of mass to be added, $\delta m$. The initial state of the combined system consists of the control

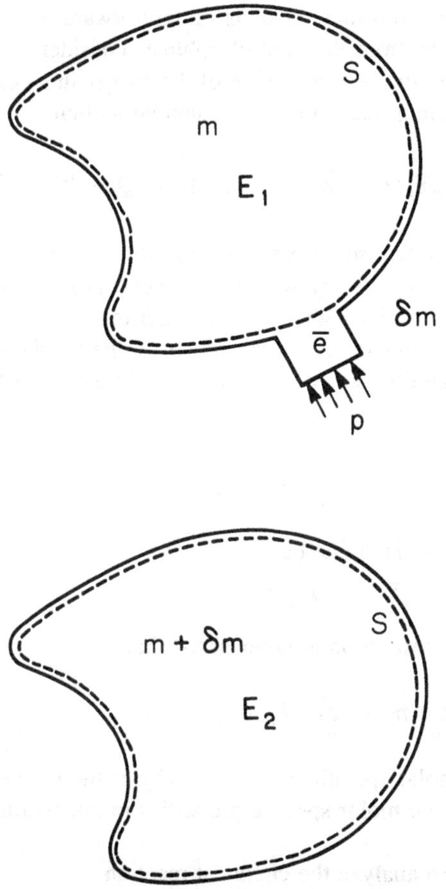


**Figure 2.1** Schematic of mass addition to a control volume and related thermodynamic system. The control volume is enclosed by the dashed curve. The solid curve denotes the closed system.

volume with its initial mass and energy and the incremental mass. After $\delta m$ is added, the mass in the control volume is $m + \delta m$, and the energy in the control volume is $E_2$. The first law for the change of state of the combined closed system may be written as

$$E_2 - (E_1 + \bar{e}\,\delta m) = Q_{12} + p\bar{v}\,\delta m - W_{x12}$$

where $\bar{e}$ denotes the energy/unit mass (called the mass specific energy) of $\delta m$, $\bar{v} = 1/\rho$ is the mass specific volume, $p\bar{v}\,\delta m$ is the work done on the combined system by the environment as the small volume is moved across the control volume surface, and $W_x$ is any work other than that associated with that volume displacement. Overbars are used to denote mass specific properties. Rearranging, we find

$$E_2 - E_1 = \bar{e}\,\delta m + p\bar{v}\,\delta m + Q_{12} - W_{x12}$$

For a small increment of change of state, this becomes

$$dE = (\bar{e} + p\bar{v})\,\delta m + \delta Q - \delta W_x \tag{2.6}$$

The extension to a number of mass increments is straightforward: simply sum over all mass flows entering and leaving from the control volume, considering the relevant properties of each increment. The time rate of change of the energy in a control volume with a number of entering and exiting mass flows may then be written

$$\frac{dE}{dt} + \sum_{j,\text{out}} (\bar{e}_j + p\bar{v}_j)\bar{f}_j - \sum_{i,\text{in}} (\bar{e}_i + p\bar{v}_i)\bar{f}_i = Q - W_x \tag{2.7}$$

where $\bar{f}_j$ and $\bar{f}_i$ are the mass flow rates (mass per time) leaving or entering the control volume, $Q$ is the rate of heat transfer to the system (energy per time), and $W_x$ is the rate at which work is done by the system on its surroundings other than that associated with flows across the control volume boundary. As noted above, in the combustion applications of interest here we can generally neglect the kinetic and potential energy contributions to the total energy, giving

$$\frac{dU}{dt} = \sum_{i,\text{in}} \bar{f}_i\bar{h}_i - \sum_{j,\text{out}} \bar{f}_j\bar{h}_j + Q - W_x \tag{2.8}$$

where the mass specific enthalpy, $\bar{h}$, is defined as

$$\bar{h} = \bar{u} + p\bar{v} \tag{2.9}$$

The energy equation may also be written on a molar basis, that is,

$$\frac{dU}{dt} = \sum_{i,\text{in}} f_i h_i - \sum_{j,\text{out}} f_j h_j + Q - W_x \tag{2.10}$$

where $h = u + pv$ denotes the molar specific enthalpy, and $f_i$ is the molar flow rate of species $i$. We shall generally use the molar specific properties in our treatment of combustion systems.

Let us apply the foregoing to analyze the chemical reaction

$$aA + bB \longrightarrow cC + dD$$

occurring at steady state and constant pressure in the isothermal flow reactor illustrated in Figure 2.2. The feed and effluent flows are at their stoichiometric values.

Applying the steady-state form of (2.10) to this system gives

$$cfh_C(T_1) + dfh_D(T_1) - afh_A(T_1) - bfh_B(T_1) = Q$$

where no work is done by the combustion gases except that due to flows across the boundary, so $W_x = 0$. (The expansion work is already accounted for in the enthalpy.) The molar flow of $A$ into the control volume is $af$, that of $C$ is $cf$, and so on, and the temperature is $T_1$. Dividing through by $f$ yields

$$ch_C(T_1) + dh_D(T_1) - ah_A(T_1) - bh_B(T_1) = \frac{Q}{f}$$

The heat transfer per mole that is required to maintain the process at a constant temperature, $T = T_1$, is called the *enthalpy of reaction*, and is given the symbol $\Delta h_r(T_1)$, that

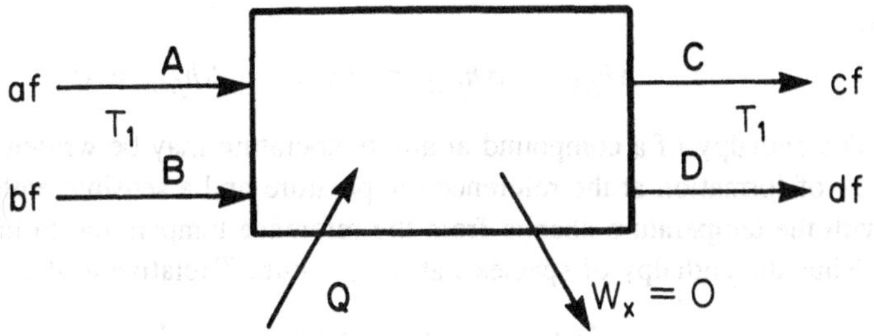


**Figure 2.2** Isothermal steady flow reactor.

is,

$$\Delta h_r(T_1) = \frac{Q}{f} = ch_C(T_1) + dh_D(T_1) - ah_A(T_1) - bh_B(T_1) \tag{2.11}$$

We see that the enthalpy of reaction is just the difference between the molar specific enthalpies of the products and reactants taking into account the stoichiometry of the reaction. To define the enthalpy of a species requires a reference state at which the enthalpy is taken to be zero. That reference state is arbitrary as long as a consistent reference state is used throughout the calculations. Generally, the reference temperature and pressure are taken to be $T_0 = 298$ K and $p_0 = 1$ atm $= 101$ kPa, respectively. It should be noted, however, that some sources report thermodynamic data relative to other reference temperatures or pressures. The chemical reference state is usually based on the pure elements in their predominant forms at $T_0$ and $p_0$, that is,

- C as solid graphite
- H as $H_2$ gas
- N as $N_2$ gas
- O as $O_2$ gas
- S as solid sulfur etc.

The enthalpy of a compound relative to the reference states of its constituent elements is the enthalpy of the reaction of these elemental species that form 1 mole of the compound. When evaluated for reactants and products at the same temperature, $T$, this quantity is called the *enthalpy of formation*. Thus the enthalpy of formation of water is the enthalpy of the reaction

$$H_2 + \tfrac{1}{2}O_2 \longrightarrow H_2O$$

namely

$$\Delta h^\circ_{f,H_2O}(T) = h_{H_2O}(T) - h_{H_2}(T) - \tfrac{1}{2}h_{O_2}(T)$$

The superscript ° denotes evaluation with respect to the chemical reference state. By definition, the enthalpies of formation of the elemental reference compounds are zero, that is,

$$\Delta h^{\circ}_{f,C_s} = \Delta h^{\circ}_{f,H_2} = \Delta h^{\circ}_{f,N_2} = \Delta h^{\circ}_{f,O_2} = 0$$

The enthalpy of a compound at any temperature may be written as the sum of the enthalpy of formation at the reference temperature and a sensible enthalpy term associated with the temperature change from the reference temperature to the desired temperature. Thus the enthalpy of species $i$ at temperature $T$ relative to the reference state is

$$h^{\circ}_i(T) = h_i(T) - h_i(T_0) + \Delta h^{\circ}_{fi}(T_0) \tag{2.12}$$

The sensible enthalpy term may be evaluated as an integral over temperature of the specific heat at constant pressure, $c_p = (\partial h/\partial T)_p$, that is,

$$h_i(T) - h_i(T_0) = \int_{T_0}^{T} c_{p,i}(T')\, dT' \tag{2.13}$$

The specific heat generally varies with temperature. If the range of temperature variation is large, as is commonly the case in combustion applications, one must account for the dependence of $c_{p,i}$ on temperature. For the present purposes, it is sufficient to approximate the specific heat as a linear function of temperature,

$$c_{p,i} \approx a_i + b_i T \tag{2.14}$$

This approximate form allows calculation of the sensible enthalpy over the range of temperatures commonly encountered in combustion calculations (i.e., 300 to 3000 K) within about 10%. Table 2.5 presents specific heats, enthalpies of formation, and additional data to which we shall refer later for species encountered in combustion problems. While the linear approximation to $c_p$ is sufficient for present purposes, tabulations of thermodynamic data such as the *JANAF Thermochemical Tables* (Stull and Prophet, 1971) should be used, in general, for more precise calculations.

The first law of thermodynamics for a chemically reacting open system may now be written as

$$\frac{dU}{dt} + \sum_{j,\text{out}} f_j[h_j(T) - h_j(T_0) + \Delta h^{\circ}_{fj}(T_0)] - \sum_{i,\text{in}} f_i[h_i(T) - h_i(T_0) + \Delta h^{\circ}_{fi}(T_0)] = Q - W_x \tag{2.15}$$

If the chemical composition and thermodynamic properties of the fuel are known, (2.15) allows us to calculate temperature changes, heat transfer, or work performed in combustion systems.

As an example, consider a steady-flow furnace burning a stoichiometric methane-air mixture. The combustion reaction is

$$CH_4 + 2(O_2 + 3.78N_2) \longrightarrow CO_2 + 2H_2O + 7.56N_2$$

**TABLE 2.5** APPROXIMATE THERMODYNAMIC DATA FOR SPECIES OF COMBUSTION INTEREST

| Species | Name | $\Delta h_f^\circ$(298 K) (J mol$^{-1}$) | $s^\circ$(298 K) (J mol$^{-1}$ K$^{-1}$) | $c_p = a + bT$ (J mol$^{-1}$ K$^{-1}$) $a$ | $b$ |
|---|---|---|---|---|---|
| C | Carbon, monatomic | 716,033 | 158.215 | 20.5994 | 0.00026 |
| C($s$) | Graphite (ref.) | 0 | 5.694 | 14.926 | 0.00437 |
| CH | Methylidine | 594,983 | 183.187 | 27.6451 | 0.00521 |
| $CH_2$ | Methylene | 385,775 | 181.302 | 35.5238 | 0.01000 |
| $CH_3$ | Methyl | 145,896 | 194.337 | 42.8955 | 0.01388 |
| $CH_4$ | Methane | −74,980 | 186.413 | 44.2539 | 0.02273 |
| CN | Cyano | 435,762 | 202.838 | 28.2979 | 0.00469 |
| CO | Carbon monoxide | −110,700 | 197.810 | 29.6127 | 0.00301 |
| COS | Carbonyl sulfide | −138,605 | 231.804 | 47.6042 | 0.00659 |
| $CO_2$ | Carbon dioxide | −394,088 | 213.984 | 44.3191 | 0.00730 |
| $C_2H$ | CCH radical | 447,662 | 207.615 | 40.4732 | 0.00880 |
| $C_2H_2$ | Acetylene | 227,057 | 201.137 | 51.7853 | 0.01383 |
| $C_2H_4$ | Ethylene | 52,543 | 219.540 | 60.2440 | 0.02637 |
| $C_2H_4O$ | Ethylene oxide | −52,710 | 243.272 | 70.1093 | 0.03319 |
| $C_2N_2$ | Cyanogen | 309,517 | 241.810 | 63.7996 | 0.00913 |
| H | Hydrogen, monatomic | 218,300 | 114.773 | 20.7859 | 0 |
| HCHO | Formaldehyde | −116,063 | 218.970 | 43.3037 | 0.01465 |
| HCN | Hydrogen cyanide | 135,338 | 202.000 | 38.9985 | 0.00885 |
| HCO | Formyl | −12,151 | 245.882 | 37.3667 | 0.00766 |
| HNO | Nitroxyl hydride | 99,722 | 220.935 | 38.2143 | 0.00750 |
| $HNO_2$ | Nitrous acid, *cis-* | −76,845 | 249.666 | 54.0762 | 0.01100 |
| $HNO_2$ | Nitrous acid, *trans-* | −78,940 | 249.498 | 54.5058 | 0.01075 |
| $HNO_3$ | Nitric acid vapor | −134,499 | 266.749 | 68.1195 | 0.01549 |
| $HO_2$ | Hydroperoxyl | 20,950 | 227.865 | 38.3843 | 0.00719 |
| $H_2$ | Hydrogen (ref.) | 0 | 130.770 | 27.3198 | 0.00335 |
| $H_2O$ | Water vapor | −242,174 | 188.995 | 32.4766 | 0.00862 |
| $H_2O_2$ | Hydrogen peroxide | −136,301 | 232.965 | 41.6720 | 0.01952 |
| $H_2S$ | Hydrogen sulfide | −20,447 | 205.939 | 35.5142 | 0.00883 |
| $H_2SO_4$ | Sulfuric acid vapor | −741,633 | 289.530 | 101.7400 | 0.02143 |
| $H_2SO_4$ | Sulfuric acid liquid | −815,160 | 157.129 | 144.0230 | 0.02749 |
| N | Nitrogen, monatomic | 473,326 | 153.413 | 20.7440 | 0.00004 |
| NH | Imidogen | 339,392 | 181.427 | 28.0171 | 0.00349 |
| $NH_2$ | Amidogen | 167,894 | 194.785 | 33.5349 | 0.00837 |
| $NH_3$ | Ammonia | −45,965 | 192.866 | 38.0331 | 0.01593 |
| NO | Nitric oxide | 90,421 | 210.954 | 30.5843 | 0.00278 |
| $NO_2$ | Nitrogen dioxide | 33,143 | 240.255 | 43.7014 | 0.00575 |
| $NO_3$ | Nitrogen trioxide | 71,230 | 253.077 | 61.1847 | 0.00932 |
| $N_2$ | Nitrogen (ref.) | 0 | 191.777 | 29.2313 | 0.00307 |
| $N_2H$ | Diimide | 213,272 | 218.719 | 43.2755 | 0.01466 |
| $N_2O$ | Nitrous oxide | 82,166 | 220.185 | 44.9249 | 0.00693 |
| $N_2O_5$ | Dinitrogen pentoxide | 11,313 | 346.933 | 122.4940 | 0.01018 |
| O | Oxygen, monatomic | 249,553 | 161.181 | 21.2424 | −0.0002 |
| OH | Hydroxyl | 39,520 | 183.858 | 28.0743 | 0.00309 |
| $O_2$ | Oxygen (ref.) | 0 | 205.310 | 30.5041 | 0.00349 |

**TABLE 2.5** *(Continued)*

| Species | Name | $\Delta h_f^\circ(298\ \text{K})$ ($\text{J mol}^{-1}$) | $s^\circ(298\ \text{K})$ ($\text{J mol}^{-1}\ \text{K}^{-1}$) | $c_p = a + bT$ ($\text{J mol}^{-1}\ \text{K}^{-1}$) $a$ | $b$ |
|---|---|---|---|---|---|
| $O_3$ | Ozone | 142,880 | 239.166 | 46.3802 | 0.00553 |
| S($g$) | Sulfur, gas | 279,391 | 168.019 | 22.4619 | −0.0004 |
| S($l$) | Sulfur, liquid | 1,425 | 35.364 | 28.5005 | 0.00976 |
| S($s$) | Sulfur, solid (ref.) | 0 | 31.970 | 13.9890 | 0.02191 |
| $SO_2$ | Sulfur dioxide | −297,269 | 248.468 | 45.8869 | 0.00574 |
| $SO_3$ | Sulfur trioxide | −396,333 | 256.990 | 62.1135 | 0.00877 |

The energy equation becomes

$$f_{CH_4}\Big\{[h(T_2) - h(T_0) + \Delta h_f^\circ(T_0)]_{CO_2} + 2[h(T_2) - h(T_0) + \Delta h_f^\circ(T_0)]_{H_2O}$$
$$+ 7.56[h(T_2) - h(T_0) + \Delta h_f^\circ(T_0)]_{N_2} - [h(T_1) - h(T_0) + \Delta h_f^\circ(T_0)]_{CH_4}$$
$$- 2[h(T_1) - h(T_0) + \Delta h_f^\circ(T_0)]_{O_2} - 7.56[h(T_1) - h(T_0) + \Delta h_f^\circ(T_0)]_{N_2}\Big\}$$
$$= Q - W_x = Q$$

where $T_1$ and $T_2$ are the temperatures of the reactants entering and the products leaving the furnace, respectively. $W_x$ has been set equal to zero since we are dealing with a heat transfer system in which no work is performed. Using thermodynamic data for all of the chemical species involved, the heat transfer rate can readily be computed.

When the chemical composition of a fuel is not known, instead of using fundamental thermochemical data on the constituents, we must rely on the empirical characterization provided by the ultimate analysis. The enthalpy of the combustion reaction is readily measured using a calorimeter, such as the flow calorimeter illustrated schematically in Figure 2.3.

Fuel and air are introduced to the calorimeter at $T_1$ and $p_1$. The fuel is burned completely, and the products are cooled to $T_1$. The heat transfer required for this cooling is measured. Applying the first law, (2.8), at steady-state conditions in the absence of any work performed yields

$$\bar{f}_{\text{products}}\bar{h}^\circ_{\text{products}}(T_1) - \bar{f}_{\text{fuel}}\bar{h}^\circ_{\text{fuel}}(T_1) - \bar{f}_{\text{air}}\bar{h}^\circ_{\text{air}}(T_1) = Q_c - 0$$

We have used the first law on a mass rather than a molar basis to be consistent with the way enthalpies of combustion are commonly measured and reported, since if the molecular structure of the fuel is not known, we cannot uniquely define the enthalpy of reaction on a molar basis. The heat released per unit mass of fuel burned is, however, readily determined, so enthalpy of combustion data are commonly reported on a mass specific

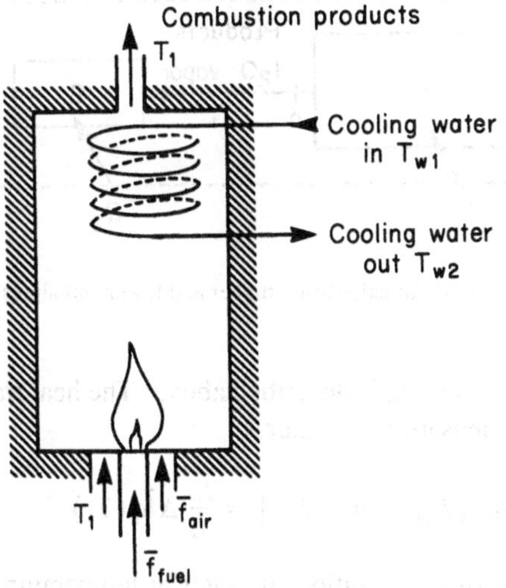


**Figure 2.3** Flow calorimeter.

basis. We find the enthalpy of combustion of a unit mass of fuel

$$\Delta \bar{h}_c(T_1) = \frac{Q_c}{\bar{f}_{\text{fuel}}}$$

$$= \frac{\bar{f}_{\text{products}}}{\bar{f}_{\text{fuel}}} \bar{h}^{\circ}_{\text{products}}(T_1) - \bar{h}^{\circ}_{\text{fuel}}(T_1) - \frac{\bar{f}_{\text{air}}}{\bar{f}_{\text{fuel}}} \bar{h}^{\circ}_{\text{air}}(T_1) \tag{2.16}$$

Since the combustion process is exothermic (releases heat), $\Delta \bar{h}_c(T_1)$ is negative. For combustion chemistry calculations, it is convenient to convert the mass specific enthalpy of combustion to a mole specific value using the formula weight, that is,

$$\Delta h_c(T_1) = M_f \Delta \bar{h}_c(T_1) \tag{2.17}$$

Flow calorimeter measurements of the heating value are usually performed at temperatures in the range 288 to 298 K, introducing a problem in the interpretation of the enthalpy of combustion. The measurement requires complete combustion; that is, all carbon and hydrogen must be oxidized to form $CO_2$ and $H_2O$, respectively. If the calorimeter is operated near stoichiometric, the product gases may contain several percent $H_2O$, considerably more than the saturation vapor pressure of water at that temperature. Hence water will condense in the calorimeter, increasing the apparent heat release due to the latent heat of vaporization. The measured heating value thus depends on the phase of the product water.

The effect of this phase transition may be seen by examining the heat transfer required to generate vapor-phase water in the products and that needed for the condensation of that vapor. This analysis requires introduction of a second control volume in the thermodynamic model, as illustrated in Figure 2.4. The enthalpy of combustion as

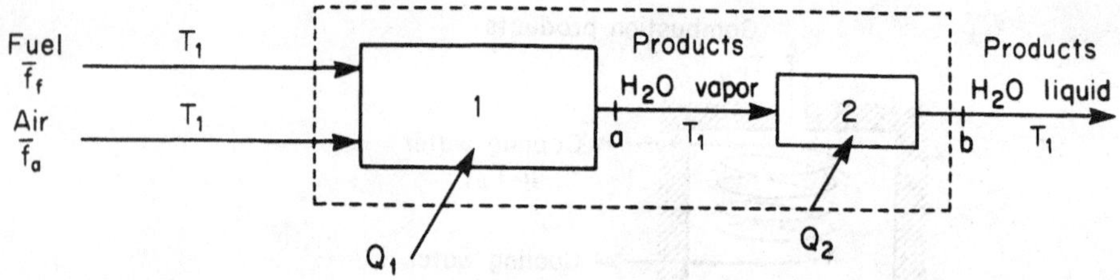


**Figure 2.4** Thermodynamic model for calculating higher and lower enthalpies of combustion.

measured by reactor 1, $\Delta \bar{h}_c(T_1) = Q_1/\bar{f}_f$, is described above. The heat transfer to reactor 2 is that associated with the condensation of water.

$$\frac{Q_2}{\bar{f}_f} = \frac{\bar{f}_2}{\bar{f}_f}\left[\bar{h}^\circ_{w2}(T_1) - \bar{h}^\circ_{w1}(T_1)\right] = \frac{\bar{f}_w}{\bar{f}_f}\Delta \bar{h}_v(T_1)$$

where $\Delta \bar{h}_v(T_1)$ is the latent heat of vaporization of water at temperature $T_1$. At 298 K, $\Delta \bar{h}_v$ (298 K) $= -2442$ J g$^{-1}$, or $\Delta h_v$ (298 K) $= -44{,}000$ J mol$^{-1}$.* The enthalpy of combustion measured with $H_2O$ present as liquid (reactors 1 and 2 combined) is, therefore,

$$\Delta \bar{h}_{c(1+2)}(T_1) = \Delta \bar{h}_{c1}(T_1) + \frac{\bar{f}_w}{\bar{f}_f}\Delta \bar{h}_v(T_1) \tag{2.18}$$

The term *heating value* is used to denote heat release due to combustion, $-\Delta \bar{h}_c(T_1)$. The two measures of the enthalpy of combustion are generally specified in terms of the heating value. The higher heating value, HHV, corresponds to the heat of reaction when the latent heat of condensation of water is recovered:

$$\text{HHV} = -\Delta \bar{h}_{c(1+2)}(T_1)$$

The lower heating value, LHV, corresponds to the case when the water is present as vapor:

$$\text{LHV} = -\Delta \bar{h}_{c1}(T_1)$$

While heating values in the U.S.A. are usually reported as higher heating values, lower heating values are often given in other parts of the world. Exhaust temperatures for most combustors are sufficiently high that the water is exhausted as vapor. At the temperatures of a flame, water is present only as vapor. Thus the lower heating value is more relevant. It is frequently necessary to compute lower heating values from the commonly reported higher heating value data.

The subscripts $L$ and $H$ will be used to indicate the enthalpies of combustion corresponding to the lower and higher heating values, respectively, that is,

$$\Delta \bar{h}_{cL}(T_1) = -\text{LHV}$$

$$\Delta \bar{h}_{cH}(T_1) = -\text{HHV}$$

*The units J mol$^{-1}$ will, throughout this book, mean J g-mol$^{-1}$

Given the heating value of a fuel, an effective enthalpy of formation can readily be calculated. Working in terms of molar quantities, we may write

$$\Delta h_c(T_1) = \sum_{i,\text{prod}} \frac{f_i}{f_{\text{fuel}}} \Delta h_{fi}^{\circ}(T_1) - \frac{f_{O_2}}{f_{\text{fuel}}} \Delta h_{f,O_2}^{\circ}(T_1) - \Delta h_{f,\text{fuel}}^{\circ}(T_1)$$

where, if the molecular form of the fuel is not known, the molar flow rate of fuel may be expressed in terms of moles of carbon per second. Rearranging, we find the enthalpy of formation of the fuel

$$\Delta h_{f,\text{fuel}}^{\circ}(T_1) = \sum_{i,\text{prod}} \frac{f_i}{f_{\text{fuel}}} \Delta h_{fi}^{\circ}(T_1) - \frac{f_{O_2}}{f_{\text{fuel}}} \Delta h_{f,O_2}(T_1) - \Delta h_{cL}^{\circ}(T_1) \qquad (2.19)$$

With this information, an estimate for the temperature in the flame can be calculated.

**Example 2.3** ***Higher and Lower Heating Values and Enthalpy of Formation***

A fuel oil contains 86.96% carbon and 13.04% hydrogen by weight. Its heating value is reported to be 44 kJ $g^{-1}$. Determine the higher and lower heating values and enthalpy of formation on a mole basis.

The fuel composition is

| Element | wt % | Normalize with respect to C |
|---|---|---|
| C | 86.96 ÷ 12 = | 7.25 ÷ 7.25 = 1 |
| H | 13.04 ÷ 1 = | 13.04 ÷ 7.25 = 1.8 |

The fuel composition is $CH_{1.8}$, and its formula weight is

$$M_f = 12 + (1.8)(1) = 13.8$$

Heating values are most commonly reported as the higher heating value; thus

$$\Delta \bar{h}_{cH}(T_1) = -\text{HHV} = -44 \text{ kJ g}^{-1}$$

The molar enthalpy of combustion is

$$\Delta h_{cH} = M_f \Delta \bar{h}_{cH} = (13.8)(-44000 \text{ J g}^{-1}) = -607{,}200 \text{ J (mol C)}^{-1}$$

Combustion of this fuel proceeds according to

$$CH_{1.8} + 1.45O_2 \longrightarrow CO_2 + 0.9H_2O$$

Thus, 0.9 mol of water is generated for each mole of fuel (carbon) burned. The latent heat of vaporization of water at 298 K is $\Delta h_v(298 \text{ K}) = -44{,}000 \text{ J mol}^{-1}$. Thus

$$\begin{aligned} \Delta h_{cL} &= \Delta h_{cH} - 0.90\, \Delta h_v \\ &= -607{,}200 + 0.90 \times 44{,}000 \\ &= -567{,}600 \text{ J (mol C)}^{1} \end{aligned}$$

The enthalpy of formation of this fuel may be determined using the lower heating value and the enthalpy of formation data from Table 2.5.

$$\begin{aligned}\Delta h_{f,CH_{1.8}}(T_1) &= \Delta h_{f,CO_2}(T_1) + 0.90\,\Delta h_{f,H_2O}(T_1) - 1.45\,\Delta h_{f,O_2}(T_1) - \Delta h_{cL}(T_1)\\ &= -394{,}088 + 0.90 \times (-242{,}174) - 1.45(0) - (-567{,}600)\\ &= -44{,}440\ \mathrm{J\ (mol\ C)^{-1}}\end{aligned}$$

### 2.3.2 Adiabatic Flame Temperature

Combustion reactions generally occur very fast, on the order of 1 ms and little heat or work transfer takes place on the time scale of combustion. For this reason the maximum temperature achieved in the combustion process is often near that for adiabatic combustion. This so-called *adiabatic flame temperature* may readily be calculated by applying the first law of thermodynamics to an adiabatic combustor. Consider a steady-flow combustor, illustrated in Figure 2.5, burning a fuel with composition $CH_m$.

The combustion stoichiometry for fuel-lean combustion is

$$CH_m + \frac{\alpha_s}{\phi}(O_2 + 3.78N_2) \longrightarrow CO_2 + \frac{m}{2}H_2O + \alpha_s\left(\frac{1}{\phi} - 1\right)O_2 + \frac{3.78\alpha_s}{\phi}N_2 \tag{2.20}$$

where $\alpha_s = 1 + m/4$. The first law of thermodynamics becomes

$$\begin{aligned}f\Bigg[&[h(T) - h(T_0) + \Delta h_f^\circ(T_0)]_{CO_2} + \frac{m}{2}[h(T) - h(T_0) + \Delta h_f^\circ(T_0)]_{H_2O}\\ &+ \alpha_s\left(\frac{1}{\phi} - 1\right)[h(T) - h(T_0) + \Delta h_f^\circ(T_0)]_{O_2} + \frac{3.78}{\phi}\alpha_s[h(T)\\ &- h(T_0) + \Delta h_f^\circ(T_0)]_{N_2}\\ &- [h(T_f) - h(T_0) + \Delta h_f^\circ(T_0)]_f - \alpha_s\frac{1}{\phi}[h(T_a) - h(T_0) + \Delta h_f^\circ(T_0)]_{O_2}\\ &- \frac{3.78}{\phi}\alpha_s[h(T_a) - h(T_0) + \Delta h_f^\circ(T_0)]_{N_2}\Bigg] = Q - W_x = 0\end{aligned} \tag{2.21}$$

Sensible enthalpy and enthalpy of formation data for each of the species are used to solve for the adiabatic flame temperature, $T$. Using the linear approximation for the temperature dependence of the specific heats, $c_{pi} = a_i + b_i T$, we have

$$h_i(T) - h_i(T_0) = a_i(T - T_0) + \frac{b_i}{2}(T^2 - T_0^2) \tag{2.22}$$

Thus, with this approximate representation of the temperature dependence of the specific heat, the problem of determining the adiabatic flame temperature is reduced to solving a quadratic equation.

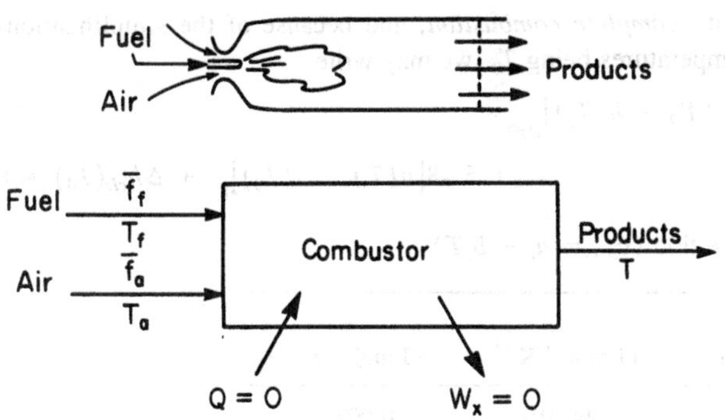


**Figure 2.5** Steady flow combustor.

**Example 2.4 *Adiabatic Flame Temperature***

A heavy fuel oil with composition $CH_{1.8}$ and a higher heating value of 44 kJ g$^{-1}$ is burned in stoichiometric air. The initial fuel and air temperatures, denoted by subscripts $f$ and $a$, respectively, are $T_f = T_a = T_0 = 298$ K. The pressure is 101 kPa (1 atm). Calculate the temperature of the products of adiabatic combustion.

1. We are given the higher heating value that includes the latent heat of condensation of water vapor. The lower heating value is given by (2.18). Converting the higher heating value to the mole-based enthalpy of combustion, we have

$$\Delta h_{cL}(T_0) = -(44 \times 10^3 \text{ J g}^{-1})(12 + 1.8 \times 1) \text{ g mol}^{-1} + 0.9(44 \times 10^3) = -568 \times 10^3 \text{ J mol}^{-1}$$

2. Combustion stoichiometry yields from (2.20):

$$CH_{1.8} + 1.45(O_2 + 3.78N_2) \longrightarrow CO_2 + 0.9H_2O + 5.48N_2$$

3. First law of thermodynamics:

$$1\left[h(T) - h(T_0) + \Delta h_f(T_0)\right]_{CO_2} + 0.9\left[h(T) - h(T_0) + \Delta h_f(T_0)\right]_{H_2O}$$
$$+ 5.48\left[h(T) - h(T_0) + \Delta h_f(T_0)\right]_{N_2} - \left[h(T_f) - h(T_0) + \Delta h_f(T_0)\right]_{CH_{1.8}}$$
$$- 1.45\left[h(T_a) - h(T_0) + \Delta h_f(T_0)\right]_{O_2} - 5.48\left[h(T_a) - h(T_0) + \Delta h_f(T_0)\right]_{N_2} = \frac{Q}{f_f} - \frac{W_x}{f_f}$$

Grouping enthalpy of formation terms and noting that $T_f = T_a = T_0$ yields

$$\left[h(T) - h(T_0)\right]_{CO_2} + 0.9\left[h(T) - h(T_0)\right]_{H_2O} + 5.48\left[h(T) - h(T_0)\right]_{N_2}$$
$$+ \Delta h_{f,CO_2}(T_0) + 0.9\,\Delta h_{f,H_2O}(T_0) - \Delta h_{f,CH_{1.8}}(T_0)$$
$$- 1.45\,\Delta h_{f,O_2}(T_0) + 5.48\left[\Delta h_{f,N_2}(T_0) - \Delta h_{f,N_2}(T_0)\right] = 0$$

But

$$\Delta h_{cL}(T_0) = \Delta h_{f,CO_2}(T_0) + 0.9\,\Delta h_{f,H_2O}(T_0) - \Delta h_{f,CH_{1.8}}(T_0) - 1.45\,\Delta h_{f,O_2}(T_0)$$

So, since we are dealing with *complete combustion*, and because of the simplifications associated with the initial temperatures being $T_0$, we may write

$$[h(T) - h(T_0)]_{CO_2} + 0.9[h(T) - h(T_0)]_{H_2O} + 5.48[h(T) - h(T_0)]_{N_2} + \Delta h_{cL}(T_0) = 0$$

4. From Table 2.5, we find ($c_{p,i} = a_i + b_i T$)

| Species | $a_i$ (J mol$^{-1}$ K$^{-1}$) | $b_i$ (J mol$^{-1}$ K$^{-2}$) |
|---|---|---|
| $CO_2$ | 44.319 | 0.00730 |
| $H_2O_{(v)}$ | 32.477 | 0.00862 |
| $N_2$ | 29.231 | 0.00307 |

$$h_i(T) - h_i(T_0) = \int_{T_0}^{T} c_{p,i}(T')\, dT' = a_i(T - T_0) + \frac{b_i}{2}(T^2 - T_0^2)$$

Substituting into the energy equation gives us

$$44.319(T - T_0) + \frac{0.00730}{2}(T^2 - T_0^2) + 0.9\left[32.477(T - T_0) + \frac{0.00862}{2}(T^2 - T_0^2)\right] + 5.48\left[29.231(T - T_0) + \frac{0.00307}{2}(T^2 - T_0^2)\right] + (-568{,}000) = 0$$

Grouping terms, we find

$$233.734(T - T_0) + 0.01594(T^2 - T_0^2) - 568{,}000 = 0$$

Solving this quadratic equation for $T$ yields

$$T = 2356 \text{ K}$$

(*Note:* A solution based on linear interpolation on the more precise JANAF Tables data yields $T = 2338$ K, so the error associated with using $c_p = a + bT$ is, in this case, about 18 K or 0.8%.)

### 2.3.3 Chemical Equilibrium

We have, so far, assumed that the fuel reacts completely, forming only $CO_2$, $H_2O$, and other fully oxidized products. For fuel-lean combustion with product temperatures below about 1250 K, the stable species, $CO_2$, $H_2O$, $O_2$, and $N_2$, are the usual products and this is a good assumption (Glassman, 1977). Element balances are sufficient to determine the composition of the combustion products under these conditions. Most combustion systems, however, reach temperatures much higher than 1250 K. We have seen that adiabatic flame temperatures can reach 2300 K for stoichiometric combustion. At such high temperatures, species that are stable at ambient temperatures can dissociate by reactions such as

$$CO_2 \rightleftharpoons CO + \tfrac{1}{2} O_2$$

$$H_2O \rightleftharpoons H_2 + \tfrac{1}{2} O_2$$

so carbon monoxide, hydrogen, and other reduced species may be present even though sufficient oxygen is available for complete combustion. In fact, these species are oxidized rapidly, but they are continually replenished by dissociation and other reactions that occur in the hot gases. The concentrations of these species are determined by the balance between those reactions that lead to their formation and those that consume them.

Chemical equilibrium provides a reasonable first approximation to the composition of the combustion products at high temperatures since the equilibrium state is that which would be achieved given a time sufficiently long for the chemical reactions to proceed. We will see that chemical equilibrium calculations also provide insight into pollutant formation.

The conditions for thermodynamic equilibrium are derived from the second law of thermodynamics. These conditions may be concisely stated in terms of the Gibbs free energy, $G = H - TS$ (Denbigh, 1971). For a closed system at a constant temperature and pressure, the Gibbs free energy is a minimum at thermodynamic equilibrium. Thus, for any change *away* from an equilibrium state at constant $T$ and $p$, $dG > 0$. The Gibbs free energy is a function of the temperature, pressure, and composition [i.e., $G = G(T, p, n_1, n_2 \ldots)$]. Thus we may write

$$dG = \left(\frac{\partial G}{\partial T}\right)_{p,n_j} dT + \left(\frac{\partial G}{\partial p}\right)_{T,n_j} dp + \left(\frac{\partial G}{\partial n_1}\right)_{T,p,n_{j\neq 1}} dn_1 + \left(\frac{\partial G}{\partial n_2}\right)_{T,p,n_{j\neq 2}} dn_2 + \cdots \tag{2.23}$$

The partial derivative of the Gibbs free energy with respect to the number of moles of a species, $i$, is the chemical potential

$$\mu_i \equiv \left(\frac{\partial G}{\partial n_i}\right)_{T,p,n_{j\neq i}} \tag{2.24}$$

Recalling the definition of $G$, we may write

$$dG = dU + p\,dV - T\,dS + V\,dp - S\,dT + \sum_i \mu_i\,dn_i$$

Using the first law of thermodynamics, it can be shown that

$$dU + p\,dV - T\,dS = 0$$

Hence

$$dG = V\,dp - S\,dT + \sum_i \mu_i\,dn_i \tag{2.25}$$

The partial molar Gibbs free energy may be written

$$\mu_i = \frac{\partial}{\partial n_i}(H - TS)_{T,p,n_{j \neq i}} = h_i - Ts_i \tag{2.26}$$

where $s_i$ *is the partial molar entropy of species i*. For the purposes of examining most combustion equilibria, we may focus on ideal gases and simple condensed phases since the pressures of combustion are generally near atmospheric. The enthalpy of an ideal gas is independent of pressure. The entropy is

$$s_i(T, p) = s_i^\circ(T_0) + \int_{T_0}^{T} \frac{c_{p,i}(T')}{T'}\,dT' + R \ln \frac{p_i}{p_0} \tag{2.27}$$

where $s_i^\circ(T_0)$ is the entropy at the reference state. Since the partial pressure is usually expressed in units of atmospheres, the partial pressure term of (2.27) is commonly expressed as $\ln p_i$. Since the heat capacity of an ideal gas is not a function of pressure, the pressure dependence of the partial molar Gibbs free energy for an ideal gas is simply that associated with the entropy change from the reference state, and we may write

$$\mu_i = \mu_i^\circ(T) + RT \ln p_i \tag{2.28}$$

where $\mu_i^\circ(T)$, the standard chemical potential of species $i$, is the chemical potential of $i$ at the reference pressure, $p_0 = 1$ atm. Values of $s_i^\circ(T_0)$ are included with the thermodynamic data in Table 2.5.

For a pure condensed phase at modest pressures, the entropy depends only on temperature,

$$s(T) = s^\circ(T_0) + \int_{T_0}^{T} \frac{c_p(T')}{T'}\,dT'$$

Since the enthalpy is also independent of pressure, the partial molar Gibbs free energy is a function only of the temperature, that is,

$$\mu_i = \mu_i^\circ(T) \tag{2.29}$$

The condition for thermodynamic equilibrium may now be written as

$$(dG)_{T,p} = \sum_i \mu_i\,dn_i \geq 0 \tag{2.30}$$

for any change away from the equilibrium state. Consider a chemical reaction

$$\sum_j \nu_j A_j = 0 \tag{2.31}$$

We may express the progress of the reaction in terms of the number of moles of a product species generated divided by the stoichiometric coefficient, the extent of reaction [recall

(A.5)],

$$d\xi = \frac{dn_j}{\nu_j} \tag{2.32}$$

The condition of chemical equilibrium at constant $T$ and $p$ is then

$$\sum_j \nu_j \mu_j = 0 \tag{2.33}$$

This condition must be satisfied at equilibrium for any $d\xi$, regardless of sign. Using (2.28) we obtain

$$\sum_j \nu_j \mu_j^\circ + \sum_{j,\text{gas}} RT \ln p_j^{\nu_j} = 0 \tag{2.34}$$

at equilibrium. This expression now defines the equilibrium composition of the gas.

Separating the pressure-dependent terms from the temperature-dependent terms yields a relation between the partial pressures of the gaseous species and temperature, that is,

$$\prod_{j,\text{gas only}} p_j^{\nu_j} = \exp\left(-\sum_j \nu_j \frac{\mu_j^\circ}{RT}\right) \equiv K_p(T) \tag{2.35}$$

The function $K_p(T)$ is the equilibrium constant in terms of partial pressures. Note that the quantities of the pure condensed phases do not enter explicitly into this relation.

It is often convenient to work in terms of mole fractions or concentrations instead of partial pressures. The partial pressure is, according to Dalton's law,

$$p_i = y_i p \tag{2.36}$$

where $y_i$ is the mole fraction of species $i$, calculated considering gas-phase species only. Substituting into the expression for $K_p$ yields

$$\prod_{j,\text{gas only}} (y_j p)^{\nu_j} = K_p(T) \tag{2.37}$$

Similarly, using the ideal gas relation, $p_i = c_i RT$, the equilibrium constant in terms of concentrations is found to be

$$K_c(T) = K_p(T)\,(RT)^{-\sum_{i,\text{gas only}} \nu_i} = \prod_{j,\text{gas only}} c_j^{\nu_j} \tag{2.38}$$

The composition of a system at equilibrium is determined by solving a set of the equilibrium relations [(2.34), (2.35), (2.37), or (2.38)] subject to element conservation constraints.

When reactions involving condensed-phase species are considered, equilibria involving the condensed-phase species do not explicitly indicate the amounts of each of those species present. For example, the reaction

$$C_{(s)} + O_2 \rightleftharpoons CO_2$$

yields the equilibrium relation

$$K_p(T) = \frac{p_{CO_2}}{p_{O_2}}$$

Only if the quantity of carbon in the system is sufficiently large relative to the amount of oxygen can the ratio $p_{CO_2}/p_{O_2}$ equal $K_p(T)$, bringing this equilibrium into play. For smaller amounts of carbon, no solid carbon will be present at equilibrium.

**Example 2.5** ***Carbon Oxidation***

Carbon is oxidized in stoichiometric air at $T = 3000$ K and atmospheric pressure. What are the mole fractions of carbon monoxide, carbon dioxide, and oxygen at chemical equilibrium? How much solid carbon remains?

From Table 2.5 we find

| Species | $\Delta h_f^\circ(T_0)$ (J mol$^{-1}$) | $s^\circ(T_0)$ (J mol$^{-1}$) | $c_p = a + bT$ (J mol$^{-1}$ K$^{-1}$) $a$ | $b$ |
|---|---|---|---|---|
| $C_{(s)}$ | 0 | 5.694 | 14.926 | 0.00437 |
| CO | −110,700 | 197.810 | 29.613 | 0.00301 |
| $CO_2$ | −394,088 | 213.984 | 44.319 | 0.00730 |
| $N_2$ | 0 | 191.777 | 29.231 | 0.00307 |
| $O_2$ | 0 | 205.310 | 30.504 | 0.00349 |

The general expression for the chemical potential of species $i$ is

$$\mu_i^\circ(T) = \int_{T_0}^{T} c_{p,i}\, dT' + \Delta h_{fi}^\circ(T_0) - T\left[s_i^\circ(T_0) + \int_{T_0}^{T} \frac{c_{p,i}}{T'}\, dT'\right]$$

$$= a_i\left(T - T_0 - T\ln\frac{T}{T_0}\right) - \frac{b_i}{2}(T - T_0)^2 + \Delta h_{fi}^\circ(T_0) - Ts_i^\circ(T_0)$$

At 3000 K and 1 atm:

| Species | $\mu_i^\circ$ (J mol$^{-1}$) |
|---|---|
| $C_{(s)}$ | −96,088 |
| CO | −840,216 |
| $CO_2$ | −1,249,897 |
| $N_2$ | −710,007 |
| $O_2$ | −757,525 |

Neglecting any solid carbon in the products, the stoichiometry under consideration is

$$C + O_2 + 3.78N_2 \longrightarrow xCO + (1 - x)CO_2 + \frac{x}{2}O_2 + 3.78N_2$$

The species mole fractions are

$$y_{CO} = \frac{x}{4.78 + x/2}$$

$$y_{CO_2} = \frac{1 - x}{4.78 + x/2}$$

$$y_{O_2} = \frac{x/2}{4.78 + x/2}$$

$$y_{N_2} = \frac{3.78}{4.78 + x/2}$$

The problem of determining the equilibrium composition is now reduced to that of evaluating the parameter, $x$. We assume that CO and $CO_2$ are in equilibrium

$$CO_2 \overset{1}{\rightleftarrows} CO + \tfrac{1}{2} O_2$$

The change in the chemical potential associated with a mole of CO formation by this reaction is

$$\begin{aligned}\Delta G_1 &= \sum_j \nu_{j1}\mu_j^\circ = \mu_{CO}^\circ + \tfrac{1}{2}\mu_{O_2}^\circ - \mu_{CO_2}^\circ \\ &= -840{,}216 + \tfrac{1}{2}(-757{,}525) - (-1{,}249{,}897) \\ &= +30{,}918 \text{ J mol}^{-1}\end{aligned}$$

where $\nu_{j1}$ denotes the stoichiometric coefficient for species $j$ in reaction 1. Thus the equilibrium constant for this reaction is

$$\begin{aligned}K_{p1} &= \exp\left(-\frac{\sum_j \nu_{j1}\mu_j^\circ}{RT}\right) \\ &= \exp\left[-\frac{30{,}918 \text{ J mol}^{-1}}{(8.3144 \text{ J mol}^{-1}\text{ K})(3000 \text{ K})}\right] \\ &= 0.2895 \text{ atm}^{1/2}\end{aligned}$$

We may now solve for the equilibrium mole fractions. Since

$$K_{p1} = \frac{y_{CO}\, y_{O_2}^{1/2}}{y_{CO_2}} p^{1/2}$$

we may write

$$p^{-1/2}K_{p1} = \frac{x}{1 - x}\left(\frac{x/2}{4.78 + x/2}\right)^{1/2}$$

which leads to a cubic equation for $x$,

$$f(x) = \left(1 - \frac{K_{p1}^2}{p}\right)x^3 - 7.56\frac{K_{p1}^2}{p}x^2 + 18.12\frac{K_{p1}^2}{p}x - 9.56\frac{K_{p1}^2}{p} = 0$$

This equation may be solved iteratively using Newton's method. Beginning with a guess, $x'$, an improved estimate is generated by

$$x = x' - \frac{f(x')}{df(x')/dx}$$

This new estimate is then used to obtain a better estimate until the desired degree of precision is achieved. Guessing initially that $x' = 0.9$, successive iterations yield

| *Estimate number* | $x$ |
|---|---|
| 1 | 0.9 |
| 2 | 0.623 |
| 3 | 0.556 |
| 4 | 0.553 |
| 5 | 0.553 |

Thus the equilibrium composition is

$$y_{CO} = 0.109$$

$$y_{CO_2} = 0.0884$$

$$y_{O_2} = 0.0547$$

$$y_{N_2} = 0.748$$

We must now test to see whether there will be any residual carbon at equilibrium. Consider the reaction

$$CO_2 \overset{2}{\rightleftharpoons} C_{(s)} + O_2$$

for which

$$\Delta G_2 = 396{,}284 \text{ J mol}^{-1}$$

Thus

$$K_{p2} = \exp\left(\frac{-396{,}284}{8.3144 \times 3000}\right)$$

$$= 1.26 \times 10^{-7}$$

In terms of mole fractions at equilibrium,

$$\frac{y_{O_2}}{y_{CO_2}} = 0.619 > K_{p2} = 1.26 \times 10^{-7}$$

Thus there is too much oxygen in the system to allow any carbon to remain unreacted at chemical equilibrium.

The temperature dependence of the equilibrium constant can readily be expressed in terms of the enthalpy of reaction (Denbigh, 1971). Equation (2.35) may be written

$$\ln K_p = -\frac{1}{R}\sum_j \nu_j \frac{\mu_j^\circ}{T}$$

Differentiation yields

$$\frac{d \ln K_p}{dT} = -\frac{1}{R} \sum_j \nu_j \frac{d}{dT}\left(\frac{\mu_j^\circ}{T}\right) \tag{2.39}$$

To evaluate the derivative on the right-hand side, we observe from (2.25) that

$$\mu_i = \left(\frac{\partial G}{\partial n_i}\right)_{p,T,nj}$$

$$S = -\left(\frac{\partial G}{\partial T}\right)_{p,ni,nj}$$

Since $G$ is a state function, $dG$ is an exact differential. Thus, from (2.25) we may obtain the reciprocity relations

$$\left(\frac{\partial \mu_i}{\partial T}\right)_{p,ni,nj} = -\left(\frac{\partial S}{\partial n_i}\right)_{p,T,nj} = -s_i$$

Equation (2.26) may now be written

$$\mu_i = h_i + T\left(\frac{\partial \mu_i}{\partial T}\right)_{p,ni,nj}$$

which may be rearranged in the form we seek:

$$\left(\frac{\partial(\mu_i/T)}{\partial T}\right)_{p,ni,nj} = -\frac{h_i}{T^2}$$

Finally, recalling (2.28), this becomes

$$\left(\frac{\partial(\mu_i^\circ/T)}{\partial T}\right)_{p,ni,nj} = -\frac{h_i}{T^2} \tag{2.40}$$

Substituting (2.40) into (2.39) gives

$$\frac{d \ln K_p}{dT} = \frac{\sum_i \nu_i h_i}{RT^2}$$

The term $\Sigma_i \nu_i h_i$ is just the enthalpy of reaction $\Delta h_r(T)$. The resulting relation is called *van't Hoff's equation*,

$$\frac{d \ln K_p}{dT} = \frac{\Delta h_r}{RT^2} \tag{2.41}$$

Over small temperature ranges the enthalpy of reaction may be assumed to be approximately constant. Although either exact numerical evaluation of $K_p$ from polynomial fits to the specific heat (e.g., Table 2.5) or the use of thermodynamic data tabulations is preferred for calculations of compositions of mixtures at chemical equilibrium, the assumption of constant $\Delta h_r$ and use of (2.41) will greatly simplify kinetic expressions we shall develop later using equilibrium constants.

The conditions for thermodynamic equilibrium have been derived for a system maintained at a prescribed temperature and pressure. The energy, enthalpy, entropy, and specific volume of a system may be calculated using the composition of the system, as determined from the equilibrium condition, and the thermodynamic properties of the constituents of the system. The equilibrium state of the system is, however, independent of the manner in which it was specified. Any two independent properties could be used in place of the pressure and temperature.

The temperature of a combustion system is rarely known a priori. The adiabatic flame temperature is often a good estimate of the peak temperature reached during combustion, provided that the reaction equilibria are taken into account. This requires solving a chemical equilibrium problem subject to constraints on the pressure and enthalpy (for a flow system) rather than temperature and pressure. Iterations on temperature to satisfy the first law of thermodynamics are now needed in addition to iterations on the composition variables. This procedure is best shown with an example.

**Example 2.6** ***Adiabatic Combustion Equilibrium***

Example 2.4 considered stoichiometric combustion of a heavy fuel oil, $CH_{1.8}$, in stoichiometric air at atmospheric pressure. Initial fuel and air temperatures were 298 K. The adiabatic flame temperature calculated assuming complete combustion was 2356 K. How do reaction equilibria influence the temperature and composition of the reaction products?

Allowing for incomplete combustion, the combustion stoichiometry may be written

$$CH_{1.8} + 1.45(O_2 + 3.78N_2) \longrightarrow (1 - x)CO_2 + xCO$$
$$+ (0.9 - y)H_2O + yH_2 + \left(\frac{x}{2} + \frac{y}{2}\right)O_2 + 5.48N_2$$

The total number of moles of reaction products is

$$N_T = (1 - x) + x + (0.9 - y) + y + \left(\frac{x}{2} + \frac{y}{2}\right) + 5.48$$
$$= 7.38 + \frac{x}{2} + \frac{y}{2}$$

Two linearly independent equilibrium relations are needed to compute $x$ and $y$. The reactions we choose to represent the equilibrium are arbitrary, as long as they are linearly independent. Possible reactions include

$$CO + H_2O \overset{1}{\rightleftharpoons} CO_2 + H_2 \quad \text{(the so-called water-gas shift reaction)}$$

$$H_2O \overset{2}{\rightleftharpoons} H_2 + \tfrac{1}{2}O_2$$

$$CO_2 \overset{3}{\rightleftharpoons} CO + \tfrac{1}{2}O_2$$

We see by inspection that the first reaction can be obtained by subtracting reaction 3 from reaction 2, but any two of these reactions are linearly independent. The choice is dictated by computational expediency. We may choose, for example,

$$CO + H_2O \overset{1}{\rightleftarrows} CO_2 + H_2$$

$$H_2O \overset{2}{\rightleftarrows} H_2 + \tfrac{1}{2} O_2$$

The corresponding equilibrium relations are

$$K_{p1} = \frac{1-x}{x}\frac{y}{0.9-y}$$

$$p^{-1/2}K_{p2} = \frac{y}{0.9-y}\left(\frac{x/2 + y/2}{7.38 + x/2 + y/2}\right)^{1/2}$$

$$= \frac{y}{0.9-y}\left(\frac{x + y}{14.76 + x + y}\right)^{1/2}$$

If we had replaced reaction 1 with 3, the first equilibrium relation would be replaced with

$$p^{-1/2}K_{p3} = \frac{x}{1-x}\left(\frac{x + y}{14.76 + x + y}\right)^{1/2}$$

By selecting reaction 1 rather than 3 we have a somewhat simpler equilibrium expression to solve. In either case, the equilibrium composition corresponding to a specified temperature (and, therefore, specified $K_p$s) may now be calculated by simultaneous solution of the two nonlinear equilibrium relations. The same solution will be obtained regardless of the choice of equilibrium relations.

A number of methods are available for solving simultaneous nonlinear equations. Newton's method may be applied readily in this case. Suppose that we want the solution to two simultaneous equations:

$$f(x, y) = 0,$$
$$g(x, y) = 0$$

From an initial approximation $(x_0, y_0)$ we attempt to determine corrections, $\Delta x$ and $\Delta y$, such that

$$f(x_0 + \Delta x, y_0 + \Delta y) = 0 \qquad g(x_0 + \Delta x, y_0 + \Delta y) = 0$$

are simultaneously satisfied. If the functions are approximated by a Taylor series and only the linear terms are retained, the equations become

$$f_0 + f_{x0}\,\Delta x + f_{y0}\,\Delta y = 0$$
$$g_0 + g_{x0}\,\Delta x + g_{y0}\,\Delta y = 0$$

where the 0 subscripts indicate that the functions have been evaluated at $(x_0, y_0)$ and the subscripts $x$ and $y$ denote $\partial/\partial x$ and $\partial/\partial y$, respectively. These linear equations are readily solved for the correction terms, $\Delta x$ and $\Delta y$. Improved estimates are then computed by

$$x = x_0 + \Delta x$$
$$y = y_0 + \Delta y$$

By iterating until $\Delta x$ and $\Delta y$ become sufficiently small, the solution of the equations can be found.

We may define the functions to be solved in the present problem as

$$f(x, y) = \frac{1 - x}{x} \frac{y}{0.9 - y} - K_{p1} = 0$$

$$g(x, y) = \frac{y}{0.9 - y} \left( \frac{x + y}{14.76 + x + y} \right)^{1/2} - p^{-1/2} K_{p2} = 0$$

The partial derivatives are

$$f_x = \frac{\partial f}{\partial x} = -\frac{y}{x^2(0.9 - y)}$$

$$f_y = \frac{0.9(1 - x)}{x(0.9 - y)^2}$$

$$g_x = \frac{y}{0.9 - y} \left( \frac{1}{2} \right) \left[ \frac{14.76 + x + y}{x + y} \right]^{1/2} \left[ \frac{14.76}{(14.76 + x + y)^2} \right]$$

$$g_y = \left[ \frac{x + y}{14.76 + x + y} \right]^{1/2} \frac{0.9}{(0.9 - y)^2} + g_x$$

and the correction terms are

$$\Delta x = \frac{g_0 f_{y0} - f_0 g_{y0}}{f_{x0} g_{y0} - f_{y0} g_{x0}}$$

$$\Delta y = \frac{f_0 g_{x0} - g_0 f_{x0}}{f_{x0} g_{y0} - f_{y0} g_{x0}}$$

Thus, for specified equilibrium constants, we may readily iterate on $x$ and $y$ to find the corresponding equilibrium composition. Poor initial guesses $x_0$ and $y_0$ may lead to estimates of $x$ and $y$ outside the domain of solution,

$$0 \leq x \leq 1$$
$$0 \leq y \leq 0.9$$

If this occurs, one may still use the information regarding the *direction* of the solution by letting

$$x = x_0 + \beta \, \Delta x$$

$$y = y_0 + \beta \, \Delta y$$

where $\beta (0 < \beta \leq 1)$ is chosen to step toward the solution but not beyond the limits of feasible solutions.

Since the temperature of the equilibrium mixture is not known a priori, we must guess the temperature before the equilibrium constants can be evaluated and any calculations can be performed. We may note this temperature estimate as $T'$. Once the equilibrium composition is determined, we can see how good our guess was by applying the first law of thermodynamics,

$$F(T') = \sum_i^{\text{products}} \nu_i[h_i(T') - h_i(T_0) + \Delta h_{fi}^\circ(T_0)] - \sum_i^{\text{reactants}} \nu_i[h_i(T_i) - h_i(T_0) + \Delta h_{fi}^\circ(T_0)]$$

For adiabatic combustion, we should have $F(T) = 0$, but we are unlikely to find this on our first guess. If $F(T') > 0$, the initial temperature guess was too high. It is as if heat were transferred to the control volume. The temperature for adiabatic combustion must be lower than that estimate. If, on the other hand, $F(T') < 0$, the temperature is that of a system that has rejected heat to the environment. The temperature estimate must be increased in this case. We may use the first law, assuming constant composition, to give an improved estimate of the gas composition. The composition corresponding to this new temperature estimate can then be evaluated as was done for our initial guess. The whole process is then repeated until satisfactory convergence is achieved.

Returning to our example, the first law becomes

$$\begin{aligned}
(1 - x)&[h(T) - h(T_0) + \Delta h_f^\circ(T_0)]_{CO_2} + x[h(T) - h(T_0) + \Delta h_f^\circ(T_0)]_{CO} \\
&+ (0.9 - y)[h(T) - h(T_0) + \Delta h_f^\circ(T_0)]_{H_2O} + y[h(T) - h(T_0) + \Delta h_f^\circ(T_0)]_{H_2} \\
&+\left(\frac{x}{2} + \frac{y}{2}\right) [h(T) - h(T_0) + \Delta h_f^\circ(T_0)]_{O_2} + 5.48[h(T) - h(T_0) + \Delta h_f^\circ(T_0)]_{N_2} \\
&- [h(T_f) - h(T_0) + \Delta h_f^\circ(T_0)]_{\text{fuel}, CH_{1.8}} - 1.45[h(T_a) - h(T_0) + \Delta h_f^\circ(T_0)]_{O_2} \\
&- 5.48[h(T_a) - h(T_0) + \Delta h_f^\circ(T_0)]_{N_2} = Q - W = 0
\end{aligned}$$

where $T_f$ and $T_a$ are the temperatures of the fuel and air, respectively. Grouping terms and noting that, for this problem, $T_f = T_a = T_0$, we have

$$\begin{aligned}
&[h(T) - h(T_0)]_{CO_2} + 0.9[h(T) - h(T_0)]_{H_2O} + 5.48[h(T) - h(T_0)]_{N_2} \\
&+ \Delta h_{f,CO_2}^\circ(T_0) + 0.9\,\Delta h_{f,H_2O}^\circ(T_0) - \Delta h_{f,CH_{1.8}}^\circ(T_0) - 1.45\,\Delta h_{f,O_2}^\circ(T_0) \\
&- x\left[[h(T) - h(T_0)]_{CO_2} - [h(T) - h(T_0)]_{CO} - \tfrac{1}{2}[h(T) - h(T_0)]_{O_2}\right] \\
&- x[\Delta h_{f,CO_2}^\circ(T_0) - \Delta h_{f,CO}^\circ(T_0) - \tfrac{1}{2}\,\Delta h_{f,O_2}^\circ(T_0)] \\
&- y\left[[h(T) - h(T_0)]_{H_2O} - [h(T) - h(T_0)]_{H_2} - \tfrac{1}{2}[h(T) - h(T_0)]_{O_2}\right] \\
&- y[\Delta h_{f,H_2O}^\circ(T_0) - \Delta h_{f,H_2}^\circ(T_0) - \tfrac{1}{2}\,\Delta h_{f,O_2}^\circ(T_0)] = 0
\end{aligned}$$

The first group of enthalpies of formation is seen to be the enthalpy of the complete combustion reaction at $T = T_0$. The enthalpy of formation terms that are multiplied by $x$ equal the enthalpy of the dissociation reaction

$$CO_2 \rightleftharpoons CO + \tfrac{1}{2}O_2$$

at temperature $T$. We have already seen that this reaction is simply the difference between reactions 2 and 1. Similarly, the last group of enthalpy of formation terms equals the enthalpy of reaction 2:

$$H_2O \rightleftharpoons H_2 + \tfrac{1}{2}O_2$$

Thus we see that the heat release of the combustion process is reduced by the amount consumed by the dissociation reactions.

The thermodynamic data necessary for these calculations, from Table 2.5, are summarized below:

| Species | $\Delta h_f^\circ(T_0)$ (J mol$^{-1}$) | $s^\circ(T_0)$ (J mol$^{-1}$ K$^{-1}$) | $c_p = a + bT$ (J mol$^{-1}$ K$^{-1}$) $a$ | $b$ |
|---|---|---|---|---|
| CO | −110,700 | 197.81 | 29.613 | 0.00301 |
| $CO_2$ | −394,088 | 213.98 | 44.319 | 0.00730 |
| $H_2$ | 0 | 130.77 | 27.320 | 0.00335 |
| $H_2O$ | −242,174 | 188.99 | 32.477 | 0.00862 |
| $N_2$ | 0 | 191.78 | 29.231 | 0.00307 |
| $O_2$ | 0 | 205.31 | 30.504 | 0.00349 |

In terms of these thermodynamic data the chemical potentials become

$$\mu_i^\circ = a_i\left(T - T_0 - T\ln\frac{T}{T_0}\right) - \frac{b_i}{2}(T - T_0)^2 + \Delta h_{fi}^\circ(T) - T\, s_i^\circ(T)$$

In preparation for determinations of the equilibrium constants, it is convenient to compute the following sums:

Reaction 1

$$\Delta a_1 = a_{CO_2} + a_{H_2} - a_{CO} - a_{H_2O} = 9.549 \text{ J mol}^{-1}\text{ K}^{-1}$$

$$\Delta b_1 = b_{CO_2} + b_{H_2} - b_{CO} - b_{H_2O} = -0.00098 \text{ J mol}^{-1}\text{ K}^{-2}$$

$$\Delta h_1^\circ = \Delta h_{f,CO_2}^\circ + \Delta h_{f,H_2}^\circ - \Delta h_{f,CO}^\circ - \Delta h_{f,H_2O}^\circ = -41{,}214 \text{ J mol}^{-1}$$

$$\Delta s_1^\circ = s_{CO_2}^\circ + s_{H_2}^\circ - s_{CO}^\circ - s_{H_2O}^\circ = -42.05 \text{ J mol}^{-1}\text{ K}^{-1}$$

Reaction 2

$$\Delta a_2 = a_{H_2} + \tfrac{1}{2}a_{O_2} - a_{H_2O} = 10.095 \text{ J mol}^{-1}\text{ K}^{-1}$$

$$\Delta b_2 = b_{H_2} + \tfrac{1}{2}b_{O_2} - b_{H_2O} = -0.003525 \text{ J mol}^{-1}\text{ K}^{-2}$$

$$\Delta h_2^\circ = \Delta h_{f,H_2}^\circ + \tfrac{1}{2}\Delta h_{f,O_2}^\circ - \Delta h_{f,H_2O}^\circ = 242{,}174 \text{ J mol}^{-1}$$

$$\Delta s_2^\circ = s_{H_2}^\circ + \tfrac{1}{2}s_{O_2}^\circ - s_{H_2O}^\circ = 44.435 \text{ J mol}^{-1}\text{ K}^{-1}$$

Thus we have

$$K_{p1}(T) = \exp\left[-\frac{9.549\,(T - T_0 - T\ln(T/T_0)) + (0.00098/2)\,(T - T_0)^2 - 41{,}214 + 42.05T}{8.3144T}\right]$$

$$K_{p2}(T) = \exp\left[-\frac{10.095\,(T - T_0 - T\ln(T/T_0)) + (0.003525/2)\,(T - T_0)^2 + 242{,}174 - 44.435T}{8.3144T}\right]$$

Since the complete combustion calculation using these approximate thermodynamic data (Example 2.4) yielded a flame temperature estimate of 2356 K, we begin with a guess of 2300 K. At $T = 2300$ K,

$$K_{p1} = 0.1904$$

$$K_{p2} = 0.001900$$

Guessing initially that $x = y = 0.01$, our iterations yield the following successive estimates:

| | | |
|---|---|---|
| 1 | $x = 0.01$ | $y = 0.01$ |
| 2 | $x = 0.0407$ | $y = 0.0325$ |
| 3 | $x = 0.0585$ | $y = 0.0222$ |
| 4 | $x = 0.0818$ | $y = 0.0198$ |
| 5 | $x = 0.0967$ | $y = 0.0189$ |
| 6 | $x = 0.1002$ | $y = 0.0187$ |
| 7 | $x = 0.1003$ | $y = 0.0187$ |

The energy equation becomes

$$234.213(T - T_0) + \frac{0.03188}{2}(T^2 - T_0^2) - 567{,}605$$

$$- x\left[-0.5456(T - T_0) + \frac{0.002545}{2}(T^2 - T_0^2) - 283{,}388\right]$$

$$- y\left[10.095(T - T_0) - \frac{0.003525}{2}(T^2 - T_0^2) - 242{,}174\right] = 0$$

which simplifies to

$$[0.01592 - 0.001273x + 0.001763y]T^2 + [234.213 + 0.5456x - 10.095y]T$$

$$+ [-638{,}853 + 283{,}338x + 244{,}870y] = 0$$

Substituting in the values for $x$ and $y$, the temperature that satisfies the first law for this composition can be evaluated explicitly. We find

$$T = 2252 \text{ K}$$

The equilibrium constants at this temperature are

$$K_{p1} = 0.1960$$

$$K_{p2} = 0.001422 \text{ atm}^{1/2}$$

We may continue to iterate on $x$, $y$, and $T$ until the results converge. We find

| $T$ | $x$ | $y$ |
|---|---|---|
| 2300 | 0.1003 | 0.0187 |
| 2245 | 0.0802 | 0.0152 |
| 2266 | 0.0875 | 0.0165 |
| 2259 | 0.0875 | 0.0165 |
| 2261 | 0.0850 | 0.0160 |
| 2261 | 0.0857 | 0.0161 |
| 2261 | 0.0857 | 0.0161 |

Thus $T = 2261$ K. The mole fractions of the equilibrium reaction products for adiabatic combustion are

$$y_{CO_2} = 0.123$$

$$y_{CO} = 0.0115$$

$$y_{H_2O} = 0.119$$

$$y_{H_2} = 0.00217 = 2170 \text{ ppm}$$

$$y_{O_2} = 0.00685 = 6850 \text{ ppm}$$

$$y_{N_2} = 0.737$$

Comparing the present results with those for complete combustion, Example 2.4, we see that the dissociation reactions reduce the adiabatic flame temperature by about 95 K.

**Example 2.7** ***Detailed Balancing***

The primary reaction leading to NO formation in flames is

$$N_2 + O \underset{k_-}{\overset{k_+}{\rightleftharpoons}} NO + N$$

The forward rate constant is

$$k_+ = 1.8 \times 10^8 \exp\left(-\frac{38{,}370}{T}\right) \text{ m}^3 \text{ mol}^{-1} \text{ s}^{-1}$$

Let us derive an expression for $k_-$ using detailed balancing. From detailed balancing we may write

$$k_- = \frac{k_+}{K_c} = \frac{k_+}{K_p}$$

where we can use either $K_c$ or $K_p$ since the number of moles of reactants and products are equal. The thermodynamic data necessary to evaluate $K_p$ are obtained from Table 2.5.

| | | | $c_p = a + bT$ | |
|---|---|---|---|---|
| Species | $\Delta h_f^\circ(T_0)$ | $s^\circ(T_0)$ | $a$ | $b$ |
| $N_2$ | 0 | 191.777 | 29.2313 | 0.00307 |
| O | 249,553 | 161.181 | 21.2424 | −0.0002 |
| NO | 90,420 | 210.954 | 30.5843 | 0.00278 |
| N | 473,326 | 153.413 | 20.7440 | 0.00004 |

The standard chemical potentials may be written

$$\mu_i^\circ = a_i\left(T - T_0 - T\ln\frac{T}{T_0}\right) - \frac{b_i}{2}(T - T_0)^2 + \Delta h_{fi}^\circ(T_0) - Ts_i^\circ(T_0)$$

The equilibrium constant thus becomes

$$K_p = \exp\left[-\frac{\begin{array}{c}0.8546(T - T_0 - T\ln(T/T_0)) \\ -\ 7.84\times10^{-5}(T - T_0)^2 + 314{,}193 - 11.409T\end{array}}{8.3144T}\right]$$

Direct use of this form of the equilibrium constant will give a complicated expression for the rate constant. Van't Hoffs' equation, (2.41),

$$\frac{d\ln K_p}{dT} = \frac{\Delta h_r}{RT^2}$$

provides a method for estimating the variation of $K_p$ over a temperature range that is sufficiently narrow that the enthalpy of reaction, $\Delta h_r$, can be assumed to be constant. Integrating (2.41) from $T_1$ to $T$ yields

$$\ln K_p(T) - \ln K_p(T_1) = -\frac{\Delta h_r(T_1)}{RT} + \frac{\Delta h_r(T_1)}{RT_1}$$

Rearranging, we find

$$K_p = K_p(T_1)\exp\left(\frac{\Delta h_r}{RT_1}\right)\exp\left(-\frac{\Delta h_r}{RT}\right) = B\exp\left(-\frac{\Delta h_r(T_1)}{RT}\right) \qquad (2.42)$$

where $B = K_p(T_1)\exp(\Delta h_r(T_1)/RT_1)$. Since NO formation occurs primarily at flame temperatures, we evaluate $K_p$ at $T_1 = 2300$ K,

$$K_p(2300\text{ K}) = 3.311\times10^{-7}$$

The enthalpy of reaction is

$$\Delta h_r(2300\text{ K}) = 316{,}312\text{ J mol}^{-1}$$

Thus we find

$$K_p = 5.05\exp\left(-\frac{38{,}044}{T}\right)$$

The rate constant for the reverse reaction becomes

$$k_- = 3.6 \times 10^7 \exp\left(-\frac{330}{T}\right) \text{ m}^3 \text{ mol}^{-1} \text{ s}^{-1}$$

The rate of the exothermic reverse reaction is found to be essentially independent of temperature.

We have, so far, limited our attention to the major products of combustion. Many of the pollutants with which we shall be concerned and the chemical species that influence their formation are present only in small concentrations. Calculations of the chemical equilibria governing trace species can be performed in the manner described above; however, care must be exercised to ensure that the equilibrium reactions used in the calculations are all linearly independent.

The calculation of the equilibrium concentrations can be simplified for species that are present only in such low concentrations that they do not significantly influence either the energy balance or the mole balances. The equilibrium distribution of the more abundant species can, in such cases, be calculated ignoring the minor species. The minor species can then be calculated using equilibrium reactions involving the major species. For example, the equilibrium concentration of nitric oxide, NO, in fuel-lean combustion products, generally can be calculated using the equilibrium between $N_2$ and $O_2$,

$$\tfrac{1}{2}\, N_2 + \tfrac{1}{2}\, O_2 \rightleftarrows NO$$

$$y_{NO} = K_p(y_{N_2} y_{O_2})^{1/2}$$

If such equilibrium calculations indicate that the concentration of the species in question is large enough to influence the energy or element balances (i.e., larger than a few thousand parts per million), a more exact calculation taking the influence on element and energy balances into account is in order.

While the conditions for chemical equilibrium have been stated in terms of equilibrium constants and reactions, these reactions are only stoichiometric relationships between the species present in the system. The number of equilibrium relations required is equal to the number of species to be considered less the number of element balances available for the system. The reactions must be linearly independent but are otherwise arbitrary; that is, they have no relationship to the mechanism by which the reactions actually occur.

An alternative to the specification of a set of reactions for the equilibrium calculations is to minimize the Gibbs free energy directly, subject to constraints on the total number of moles of each of the elements in the system (White et al., 1958). Let $b_i^\circ$ be the number of moles of element $i$ in the system and $a_{ij}$ be the number of moles of element $i$ in a mole of species $j$. If $n_j$ is the number of moles of species $j$ in the system, the elemental conservation constraint that must be satisfied takes the form

$$b_i^\circ - \sum_{j=1}^{n} a_{ij} n_j = 0, \qquad i = 1, 2, \ldots, l \tag{2.43}$$

where $n$ is the total number of species in the system and $l$ is the number of elements. The method of Lagrange multipliers can be used to solve this constrained minimization problem. We define $\Gamma$ to be

$$\Gamma = G - \sum_{i=1}^{l} \lambda_i (b_i - b_i^\circ)$$

where

$$b_i = \sum_{j=1}^{n} a_{ij} n_j \tag{2.44}$$

and $\lambda_i$ are Lagrange multipliers. The condition for equilibrium becomes

$$\delta\Gamma = 0 = \sum_{j=1}^{n} \left( \mu_j - \sum_{i=1}^{l} \lambda_i a_{ij} \right) \delta n_j - \sum_{i=1}^{l} (b_i - b_i^\circ) \delta\lambda_i$$

This must hold for all $\delta n_j$ and $\delta\lambda_i$, so we must have

$$\mu_j - \sum_{i=1}^{l} \lambda_i a_{ij} = 0, \qquad j = 1, 2, \ldots, n \tag{2.45}$$

and the elemental constraints as $l + n$ equations in $l + n$ unknowns.

For ideal gases,

$$\mu_j = \mu_j^\circ + RT \ln \frac{n_j}{n_{\text{gas}}} + RT \ln \frac{p}{p_0}$$

where

$$n_{\text{gas}} = \sum_{j=1}^{\text{gas only}} n_j \tag{2.46}$$

is the total number of moles of gaseous species. For simple condensed phases,

$$\mu_j = \mu_j^\circ$$

Thus for gaseous species, the condition for equilibrium becomes

$$\frac{\mu_j^\circ}{RT} + \ln \frac{n_j}{n_g} + \ln \frac{p}{p_0} - \sum_{i=1}^{l} \pi_i a_{ij} = 0, \qquad j = 1, \ldots, n_g \tag{2.47}$$

where $\pi_i = \lambda_i / RT$, and for condensed-phase species,

$$\frac{\mu_j^\circ}{RT} - \sum_{i=1}^{l} \pi_i a_{ij}, \qquad j = n_{g+1}, \ldots, n \tag{2.48}$$

To determine the equilibrium composition, $n + l + 1$ simultaneous equations, (2.43), (2.46)–(2.48), must be solved. The number of moles of gaseous species $j$ can be found by rearranging (2.47):

$$n_j = n_{\text{gas}} \frac{p_0}{p} \exp\left(-\frac{\mu_j^\circ}{RT} - \sum_{i=1}^{l} \pi_i a_{ij}\right), \qquad j = 1, 2, \ldots, n_g$$

eliminating $n_g$ of the equations, so only $n - n_g + l + 1$ equations must be solved. The exponential is similar to that obtained in deriving the equilibrium constant for a reaction leading to the formation of a mole of the gaseous species from the elements. The Lagrange multipliers, called *elemental potentials* because of this structure (Reynolds, 1986), thus are the key to determining the equilibrium composition by this route. The details of the procedures for determining the element potentials are beyond the scope of this book. Powerful, general-purpose equilibrium codes that use this method are available, however, and should be considered for complex equilibrium calculations [e.g., Gordon and McBride (1971) and Reynolds (1981)].

### 2.3.4 Combustion Equilibria

We have seen that at chemical equilibrium for stoichiometric combustion, substantial quantities of carbon monoxide and hydrogen remain unreacted, and that this incomplete combustion reduces the adiabatic flame temperature by nearly 100 K. Figure 2.6 shows how the equilibrium composition and temperature for adiabatic combustion of kerosene, $CH_{1.8}$, vary with equivalence ratio. The results determined using stoichiometry alone for fuel-lean combustion are shown with dashed lines. It is apparent that the major species concentrations and the adiabatic flame temperature for complete combustion are very good approximations for equivalence ratios less than about 0.8. As the equivalence ratio approaches unity, this simple model breaks down due to the increasing importance of the dissociation reactions. For fuel-rich combustion, the number of chemical species that are present in significant quantities exceeds the number of elements in the system, so we must rely on equilibrium to determine the adiabatic flame temperature and composition.

Chemical equilibrium provides our first insight into the conditions that favor the formation of pollutants. Carbon monoxide is a significant component of the combustion products at the adiabatic flame temperature for equivalence ratios greater than about 0.8. Nitric oxide formation from gaseous $N_2$ and $O_2$,

$$\tfrac{1}{2} N_2 + \tfrac{1}{2} O_2 \rightleftharpoons NO$$

is highly endothermic, $\Delta h_r(298\ \text{K}) = 90{,}420\ \text{J mol}^{-1}$. Because of the large heat of reaction, NO formation is favored only at the highest temperatures. Hence, as we will see in the next chapter, the equilibrium NO concentration peaks at equivalence ratios near unity and decreases rapidly with decreasing equivalence ratio due to the decrease in temperature. The equilibrium NO level decreases for fuel-rich combustion due to the combined effects of decreasing temperature and decreasing oxygen concentration.

The equilibrium composition of combustion gases is a strong function of temperature. The reason for this case can readily be seen by examining the equilibrium con-

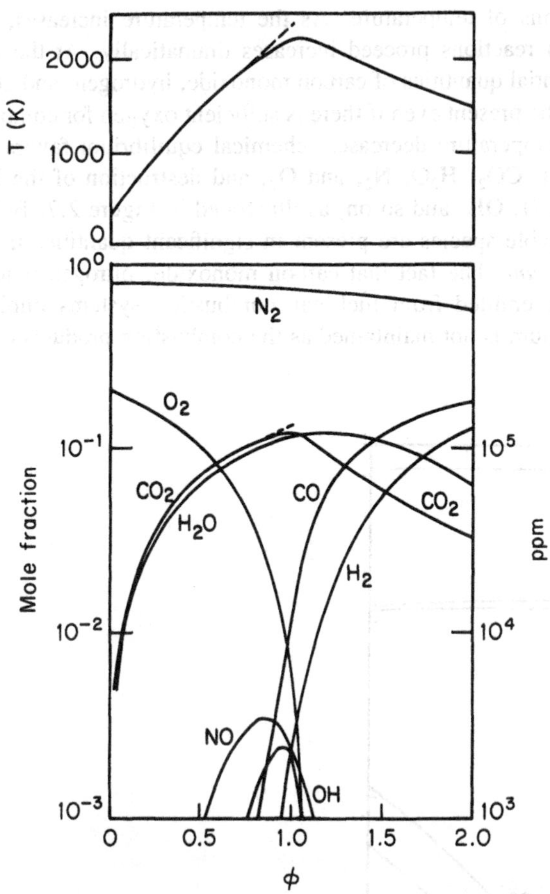


**Figure 2.6** Equilibrium composition and temperature for adiabatic combustion of kerosene, $CH_{1.8}$, as a function of equivalence ratio.

stants for combustion reactions using the integrated form of van't Hoff's relation,

$$K_p = B \exp\left(-\frac{\Delta h_r(T_1)}{RT}\right)$$

where $T_1$ is a reference temperature at which the preexponential factor $B$, is evaluated. The dissociation reactions, for example,

$$CO_2 \overset{1}{\rightleftharpoons} CO + \tfrac{1}{2}O_2$$

$$H_2O \overset{2}{\rightleftharpoons} H_2 + \tfrac{1}{2}O_2$$

have large positive heats of reaction,

$$\Delta h_{r1} = 283{,}388 \text{ J mol}^{-1}$$

$$\Delta h_{r2} = 242{,}174 \text{ J mol}^{-1}$$

and are therefore strong functions of temperature. As the temperature increases, the extent to which the dissociation reactions proceed increases dramatically. At the adiabatic flame temperature, substantial quantities of carbon monoxide, hydrogen, and other partially oxidized products may be present even if there is sufficient oxygen for complete combustion available. As the temperature decreases, chemical equilibrium favors the formation of the stable products, $CO_2$, $H_2O$, $N_2$, and $O_2$, and destruction of the less stable species, CO, $H_2$, NO, O, H, OH, and so on, as illustrated in Figure 2.7. Below about 1300 K, only the most stable species are present in significant quantities in the combustion products *at equilibrium*. The fact that carbon monoxide, nitrogen oxides, and unburned hydrocarbons are emitted from fuel-lean combustion systems implies, therefore, that chemical equilibrium is not maintained as the combustion products cool.

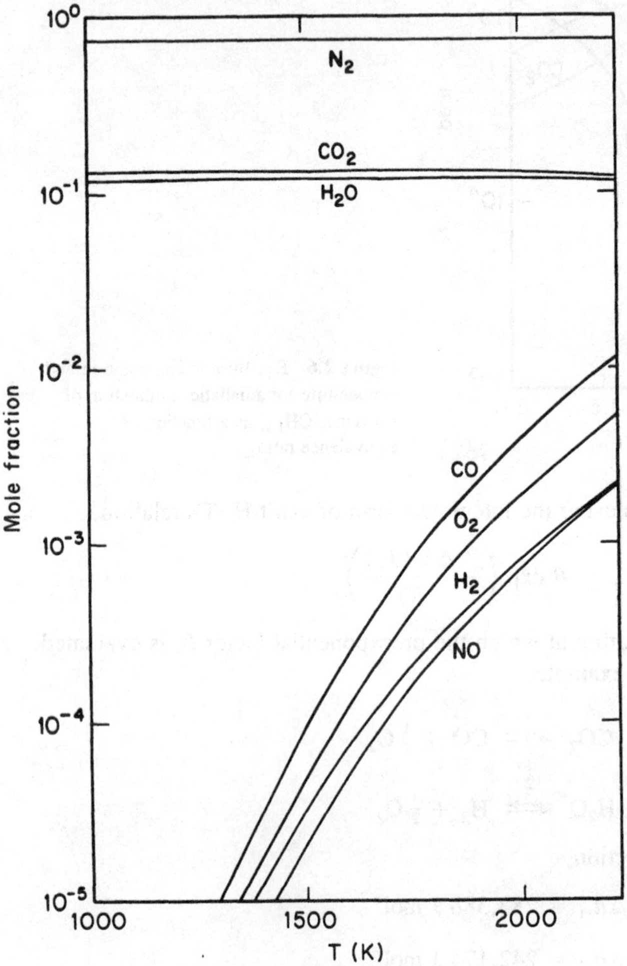


**Figure 2.7** Variation of equilibrium composition with temperature for stoichiometric combustion of kerosene, $CH_{1.8}$.

## 2.4 COMBUSTION KINETICS

Chemical equilibrium describes the composition of the reaction products that would ultimately be reached if the system were maintained at constant temperature and pressure for a sufficiently long time. Chemical reactions proceed at finite rates, however, so equilibrium is not established instantaneously. We have seen that at equilibrium there would only be very small amounts of pollutants such as CO, NO, or unburned hydrocarbons in the gases emitted from combustors operated at equivalence ratios less than unity. Slow reactions allow the concentrations of these pollutants to be orders of magnitude greater than the equilibrium values when gases are finally emitted into the atmosphere. The sharp peak in the equilibrium NO concentration near $\phi = 1$ suggests that the amount of NO in the flame could be reduced significantly by reducing the equivalence ratio below about 0.5. Unfortunately, the combustion reactions also proceed at finite rates. Reducing the equivalence ratio lowers the temperature in the flame, thereby slowing the hydrocarbon oxidation reactions and the initial approach to equilibrium within the flame. The residence time in combustion systems is limited, so reducing the combustion rate eventually results in the escape of partially reacted hydrocarbons and carbon monoxide.

To understand the chemical factors that control pollutant emissions, therefore, it is necessary to examine the rate at which a chemical system approaches its final equilibrium state. The study of these rate processes is called *chemical kinetics*. The *reaction mechanism*, or the sequence of reactions involved in the overall process, and the rates of the individual reactions must be known to describe the rate at which chemical equilibrium is approached. In this section we examine the chemical kinetics of hydrocarbon fuel combustion, beginning with an overview of the detailed kinetics. Several approximate descriptions of combustion kinetics will then be examined. The kinetics that directly govern pollutant emissions will be treated in Chapter 3.

### 2.4.1 Detailed Combustion Kinetics

Combustion mechanisms involve large numbers of reactions even for simple hydrocarbon fuels. Consider propane combustion for which the overall stoichiometry for complete combustion is

$$C_3H_8 + 5O_2 \longrightarrow 3CO_2 + 4H_2O$$

The combustion reactions must break 15 chemical bonds (C—C, C—H, O—O) and form 14 new ones (C—O, H—O). As described in Chapter 1, hydrocarbon oxidation involves a large number of elementary bimolecular reaction steps. The many elementary reactions that comprise the combustion process generate intermediate species that undergo rapid reaction and, therefore, are not present in significant quantities in either the reactants or the products. A detailed description of combustion must include the intermediate species.

Detailed simulation of the chemical kinetics of combustion becomes quite formidable, even for simple, low-molecular-weight hydrocarbons such as $CH_4$, $C_2H_2$, $C_2H_4$, $C_2H_6$, $C_3H_8$, $CH_3OH$, and so on. Numerous studies of combustion mechanisms of such simple fuels have been presented (Westbrook and Dryer, 1981a; Miller et al., 1982; Vandooren and Van Tiggelen, 1981; Westbrook, 1982; Venkat et al., 1982; Warnatz, 1984). Rate constants have been measured for many, but not all, of the 100 or so reactions in these mechanisms.

The description of the combustion kinetics for practical fuels is complicated by our incomplete knowledge of the fuel composition. Only rarely is the fuel composition sufficiently well known that detailed mechanisms could be applied directly, even if they were available for all the components of the fuel.

Our ultimate goal here is to develop an understanding of the processes that govern the formation and destruction of pollutants in practical combustion systems. Once combustion is initiated (as described below), the combustion reactions generally proceed rapidly. Such pollutant formation processes involve slow reaction steps or physical processes that restrain the approach to equilibrium, either during combustion or as the combustion products cool, and lead to unoxidized or partially oxidized fuel or intermediate species in the exhaust gases. Let us first examine the important features common to hydrocarbon combustion reaction mechanisms.

A mixture of a hydrocarbon (RH) fuel with air at normal ambient temperature will not react unless an ignition source is present. When the mixture is heated, the fuel eventually begins to react with oxygen. *Initiation* of the combustion reactions is generally thought to occur via the abstraction of a hydrogen atom from the hydrocarbon molecule by an oxygen molecule.

$$RH + O_2 \xrightarrow{1} R\cdot + HO_2\cdot$$

An alternative initiation reaction for large hydrocarbon molecules is thermally induced dissociation to produce hydrocarbon radicals, that is,

$$RR' + M \xrightarrow{2} R\cdot + R'\cdot + M$$

This reaction involves breaking a carbon–carbon or carbon–hydrogen bond. The energy required for bond breakage can be estimated using the bond strengths summarized in Table 2.6. Hydrogen abstraction reactions (reaction 1) involve breaking a carbon–hydrogen bond with a strength ranging from 385 to 453 kJ mol$^{-1}$ and forming $HO_2$, leading to a net energy of reaction of 190 to 250 kJ mol$^{-1}$. Reaction 2 involves breaking a carbon–carbon bond. The single bond requires 369 kJ mol$^{-1}$, with double and triple bonds requiring considerably more energy. Thus both reactions are endothermic, with reaction 2 having a significantly larger enthalpy of reaction since no new bonds are formed as the initial bond is broken.

The large enthalpy of reaction makes the reaction rate in the endothermic direction a strong function of temperature. Detailed balancing provided us with a relationship between the forward and reverse rate constants for elementary reactions, that is,

$$\frac{k_f(T)}{k_r(T)} = K_c(T)$$

**TABLE 2.6 TYPICAL BOND STRENGTHS**

| Bond | kJ mol$^{-1}$ |
|---|---|
| Diatomic Molecules | |
| H—H | 437 |
| H—O | 429 |
| H—N | 360 |
| C—N | 729 |
| C—O | 1076 |
| N—N | 950 |
| N—O | 627 |
| O—O | 498 |
| Polyatomic Molecules | |
| H—CH | 453 |
| $H—CH_2$ | 436 |
| $H—CH_3$ | 436 |
| $H—C_2H_3$ | 436 |
| $H—C_2H_5$ | 411 |
| $H—C_3H_5$ | 356 |
| $H—C_6H_5$ | 432 |
| H—CHO | 385 |
| $H—NH_2$ | 432 |
| H—OH | 499 |
| HC≡CH | 964 |
| $H_2C=CH_2$ | 699 |
| $H_3C—CH_3$ | 369 |
| O=CO | 536 |

The temperature dependence of the equilibrium constant can be expressed approximately using van't Hoff's relation (2.42),

$$K_p(T) \approx \left\{ K_p(T_1) \exp\left[\frac{\Delta h_r(T_1)}{RT_1}\right] \right\} \exp\left[-\frac{\Delta h_r(T)}{RT}\right]$$

$$\approx B(T_1) \exp\left[-\frac{\Delta h_r(T)}{RT}\right]$$

and the definition of $K_p(T)$, (2.38). Thus the rate constant in the forward direction is

$$k_f(T) = k_r(T)\, B(T_1) \exp\left[-\frac{\Delta h_r(T_1)}{RT}\right] [RT]^{-1}$$

Consider, for example, the dissociation of methane

$$CH_4 + M \underset{k_r}{\overset{k_f}{\rightleftharpoons}} CH_3\cdot + H\cdot + M$$

for which $k_r = 0.282T \exp[-9835/T]$ m$^6$ mol$^{-2}$ s$^{-1}$ (Westbrook, 1982). From the thermochemical property data of Table 2.5 and application of van't Hoff's relation, we find

$$K_p(T) = 4.11 \times 10^7 \exp\left[\frac{-34{,}700}{T}\right] \text{ atm}$$

from which we find

$$k_f(T) = k_r(T)\, K_p(T)\, [RT]^{-1}$$

$$= 1.41 \times 10^{11} \exp\left[\frac{-44{,}535}{T}\right] \text{ m}^3 \text{ mol}^{-1} \text{ s}^{-1}$$

While the rate of the exothermic recombination reaction is, in this case, a strong function of temperature, the endothermic dissociation reaction is even more strongly dependent on temperature. In cases where the temperature dependence of rate coefficients results entirely from the exponential factor, that is, the rates are of the Arrheneus form, $k = A \exp(-E/RT)$, a plot of log $k$ versus $T^{-1}$, known as an Arrheneus plot, clearly illustrates the influence of the large positive enthalpy of reaction on the temperature dependence of the rate of this reaction. The slope of the rate curve, shown in Figure 2.8, is equal to $-(\ln 10)^{-1}(E/R)$ and thus indicates the activation energy. The rates of exothermic or mildly endothermic reactions may be fast or slow for a variety of reasons as discussed in Chapter 1, but in general, highly endothermic reactions are slow except at very high temperatures.

Because of the relatively low rates of the highly endothermic initiation reactions, radicals are generated very slowly. After the radicals have accumulated for a period of time, their concentrations become high enough for the faster radical chemistry to become important. This delay between the onset of the initiation reactions and rapid combustion is called an induction period or ignition delay. After this delay, other reactions dominate the oxidation of the fuel and the initiation reactions are no longer important.

Hydrocarbon radicals react rapidly (due to low activation energies) with the abundant oxygen molecules to produce peroxy radicals

$$R\cdot + O_2 + M \xrightarrow{3} RO_2\cdot + M$$

or olefins (alkenes, R = R') and the hydroperoxyl radical

$$R\cdot + O_2 \xrightarrow{4} \text{olefin} + HO_2\cdot$$

The olefin is then oxidized in a manner similar to the original hydrocarbon. Peroxy radicals undergo dissociation at high temperatures:

$$RO_2\cdot + M \xrightarrow{5} R'CHO + R''\cdot + M$$

These are called *chain carrying* reactions since the number of radicals produced equals the number consumed. The aldehydes (RCHO) may react with $O_2$:

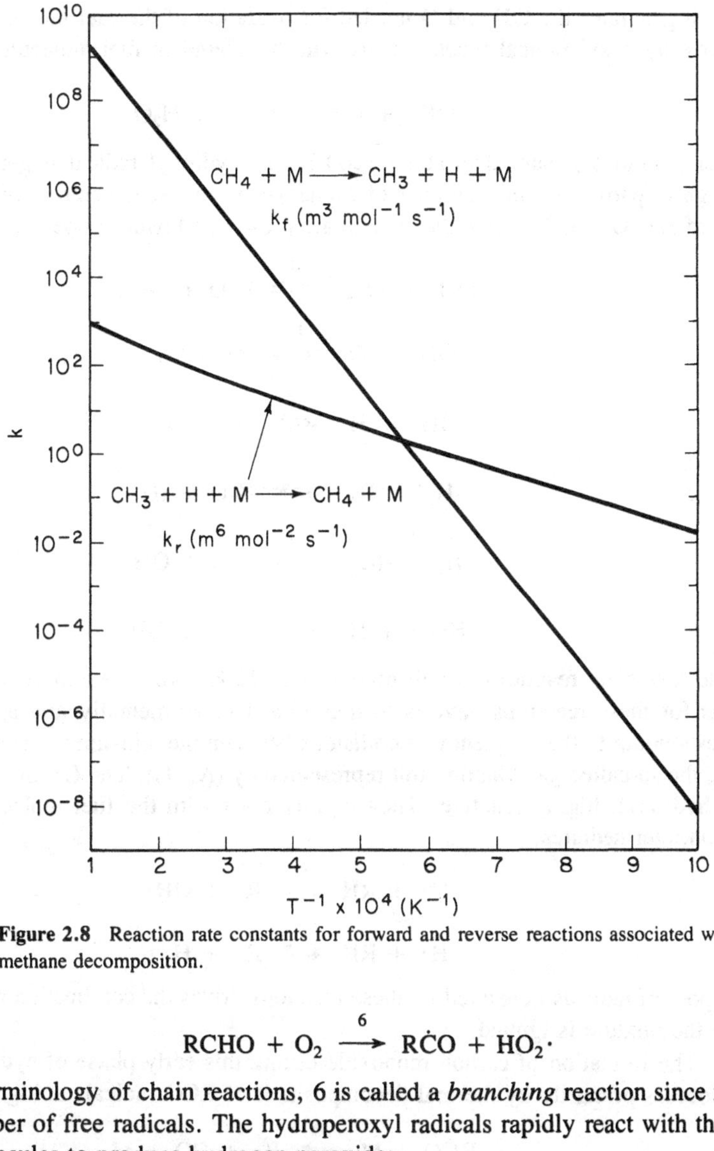


**Figure 2.8** Reaction rate constants for forward and reverse reactions associated with methane decomposition.

$$\mathrm{RCHO} + \mathrm{O_2} \xrightarrow{6} \mathrm{R\dot{C}O} + \mathrm{HO_2}\cdot$$

In the terminology of chain reactions, 6 is called a *branching* reaction since it increases the number of free radicals. The hydroperoxyl radicals rapidly react with the abundant fuel molecules to produce hydrogen peroxide:

$$\mathrm{HO_2}\cdot + \mathrm{RH} \xrightarrow{7} \mathrm{HOOH} + \mathrm{R}\cdot$$

Actually the single most important reaction in combustion is the chain-branching step:

$$\mathrm{H}\cdot + \mathrm{O_2} \xrightarrow{8} \mathrm{OH}\cdot + \mathrm{O}\cdot$$

since it generates the OH and O needed for oxidation of the fuel molecules. The highly reactive hydroxyl radical reacts readily with the abundant fuel molecules:

$$OH\cdot + RH \xrightarrow{9} R\cdot + H_2O$$

At temperatures greater than about 1200 K, the hydroxyl radical is generally abundant enough to participate in a number of exchange reactions, generating much larger numbers of $H\cdot$, $O\cdot$, and $OH\cdot$ radicals than are present at lower temperatures:

$$OH\cdot + OH\cdot \overset{10}{\rightleftharpoons} H_2O + O\cdot$$

$$OH\cdot + O\cdot \overset{11}{\rightleftharpoons} H\cdot + O_2$$

$$OH\cdot + H\cdot \overset{12}{\rightleftharpoons} O\cdot + H_2$$

$$H_2 + OH\cdot \overset{13}{\rightleftharpoons} H_2O + H\cdot$$

$$H_2 + HO_2\cdot \overset{14}{\rightleftharpoons} H_2O + OH\cdot$$

$$HO_2\cdot + H\cdot \overset{15}{\rightleftharpoons} OH\cdot + OH\cdot$$

These reversible reactions are all mildly (8 to 72 kJ $mol^{-1}$) exothermic. The rate constants for these reactions have been determined experimentally and approach the rate corresponding to the frequency of collisions between the relevant radicals and molecules [i.e., the so-called gas kinetic limit represented by (A.11)]. The $O\cdot$ and $H\cdot$ radicals are, like hydroxyl, highly reactive. They rapidly react with the fuel molecules and hydrocarbon intermediates,

$$O\cdot + RH \overset{16}{\rightleftharpoons} R\cdot + OH\cdot$$

$$H\cdot + RH \overset{17}{\rightleftharpoons} R\cdot + H_2$$

The pool of radicals generated by these reactions drives the combustion reactions rapidly once the mixture is ignited.

The formation of carbon monoxide during this early phase of hydrocarbon oxidation occurs primarily by thermal decomposition of RCO radicals at high temperatures,

$$R\dot{C}O + M \overset{18}{\rightleftharpoons} R\cdot + CO + M$$

The dominant carbon monoxide oxidation process is the reaction with hydroxyl,

$$CO + OH\cdot \overset{19}{\rightleftharpoons} CO_2 + H\cdot$$

Three-body recombination reactions,

$$H\cdot + H\cdot + M \overset{20}{\rightleftharpoons} H_2 + M$$

$$H\cdot + OH\cdot + M \overset{21}{\rightleftharpoons} H_2O + M$$

$$O\cdot + O\cdot + M \overset{22}{\rightleftharpoons} O_2 + M$$

$$H\cdot + O_2 + M \overset{23}{\rightleftharpoons} HO_2\cdot + M$$

$$OH\cdot + OH\cdot + M \overset{24}{\rightleftharpoons} HOOH + M$$

reduce the total number of moles in the system. These reactions are exothermic but relatively slow since they require the intervention of a third molecule to stabilize the product. As combustion products cool, the slow recombination steps may allow radical concentrations to persist long after the equilibrium concentrations have dropped to extremely low levels.

Even though we have not attempted to list all the free-radical reactions involved in the combustion of hydrocarbons, we have already identified a large number of reactions. Detailed mechanisms for specific hydrocarbon molecules typically involve more than 100 reactions. It is noteworthy that the most important reactions in combustion, the chain branching reactions, do not involve the fuel molecules. This fact permits prediction of gross combustion features without full knowledge of the detailed reaction mechanism.

The mechanisms for different fuels involve common submechanisms (Westbrook and Dryer, 1981b). Combustion of carbon monoxide in the presence of hydrogen or water vapor involves the reactions of the hydrogen-oxygen mechanism. The combined $CO$–$H_2$–$O_2$ mechanism is, in turn, part of the mechanism for formaldehyde oxidation, which is a subset of the methane mechanism. In combustion of methane under fuel-lean conditions the carbon atom follows the sequence: $CH_4 \rightarrow CH_3 \rightarrow HCHO \rightarrow HCO \rightarrow CO \rightarrow CO_2$. Westbrook and Dryer (1981b) develop this hierarchical approach for fuels through $C_2$ and $C_3$ hydrocarbons, providing a framework for understanding the detailed combustion kinetics for a range of hydrocarbon fuels using as a starting point for each successive fuel the knowledge of the mechanisms of the simpler fuels. More complicated molecules, such as aromatic hydrocarbons (Venkat et al., 1982), will introduce additional reactions into this hierarchy, but the reactions already identified in studies of simpler molecules still contribute to the expanded overall mechanisms.

A detailed description of the dynamics of so many simultaneous reactions requires solution of a large number of simultaneous ordinary differential equations. The large enthalpies of combustion reaction and relatively slow heat transfer from a flame lead to large temperature changes during combustion. The first law of thermodynamics must be applied to evaluate the temperatures continuously throughout the combustion process. The large temperature changes result in very large changes in the many reaction rate constants. The integration of these rate equations is difficult since the equations contain several very different time scales, from the very short times of the free-radical reactions to the longer times of the initiation reactions. Such sets of equations are called *stiff*.

Since much of the chemistry with which we shall be concerned in our study of the formation and destruction of pollutants takes place late in the combustion process, a complete description of the combustion process is not generally required for our purposes. Hydrocarbon oxidation in combustion is generally fast, leading to a rapid approach to equilibrium. This is fortunate since detailed combustion mechanisms are simply not known for many practical fuels such as coal or heavy fuel oils. Simplified models of the combustion process will, for these reasons, be used extensively in the discussion to follow.

### 2.4.2 Simplified Combustion Kinetics

One way to overcome the difficulties in modeling the combustion reactions is to represent the process by a small number of artificial reactions, each of which describes the results of a number of fundamental reaction steps. These so-called *global mechanisms* are stoichiometric relationships for which approximate kinetic expressions may be developed. Global reaction rate expressions may be derived from detailed kinetic mechanisms by making appropriate simplifying assumptions (e.g., steady-state or partial-equilibrium assumptions, which will be discussed later). Alternatively, correlations of observed species concentration profiles, flame velocity measurements, or other experimental data may be used to estimate global rate parameters.

Global mechanisms greatly reduce the complexity of kinetic calculations since a small number of steps are used to describe the behavior of a large number of reactions. Moreover, the simplified reactions generally involve the major stable species, greatly reducing the number of chemical species to be followed. This reduction may be either quite useful or an oversimplification, depending on the use to which the mechanism is to be put. If a combustion mechanism is to be used to describe the net rate of heat release during combustion, minor species are of little concern and a global mechanism can be quite effective. The minor species, on the other hand, strongly influence the formation of pollutants, and the simplified global mechanisms therefore may not contain sufficient chemical detail to describe the pollutant formation steps.

The simplest model of hydrocarbon combustion kinetics is the one-step, global model given at the beginning of Section 2.2,

$$C_nH_m + \left(n + \frac{m}{4}\right) O_2 \xrightarrow{k_{ov}} n\, CO_2 + \frac{m}{2} H_2O$$

where the subscript ov refers to "overall" model. The rate of this reaction can be expressed empirically by

$$R_{ov} = AT^n \exp\left(\frac{-E_a}{RT}\right) [C_nH_m]^a [O_2]^b \tag{2.49}$$

where the parameters $A$, $n$, $E_a$, $a$, and $b$ are generally determined by matching $R_{ov}$ to the observed oxidation rate inferred from flame speed or the rich and lean limits of stable laminar flames. The obvious advantage of the single-step model is its simplicity. It is very useful for calculating heat release rates and for examining flame stability. Unfortunately, the single-step model does not include intermediate hydrocarbon species or carbon monoxide.

The hydrocarbons are rapidly consumed during combustion, forming CO, $H_2$, and $H_2O$. The oxidation of CO to $CO_2$ proceeds somewhat more slowly. The difference in reaction rates can be taken into account using two-step models that are only slightly more complicated than the single-step model but can separate the relatively slow oxidation of CO to $CO_2$ from the more rapid oxidation of the hydrocarbon to CO and $H_2O$ (Hautman et al., 1981), that is,

$$C_nH_m + \left(\frac{n}{2} + \frac{m}{4}\right) O_2 \xrightarrow{k_A} nCO + \frac{m}{2} H_2O$$

$$CO + \tfrac{1}{2} O_2 \xrightarrow{k_B} CO_2$$

This description lumps together reactions 1–18 and 20–24 from the detailed mechanism of Section 2.4.1, with reaction 19 being treated separately. The rate for reaction $A$ is generally expressed in the same empirically derived form as the hydrocarbon oxidation in the single-step model

$$R_A = A_A T^{n_A} \exp\left[\frac{-E_A}{RT}\right] [C_nH_m]^a [O_2]^b \tag{2.50}$$

Carbon monoxide oxidation is described empirically by

$$R_B = A_B T^{n_B} \exp\left[\frac{-E_B}{RT}\right] [H_2O]^c [O_2]^d [CO] \tag{2.51}$$

where the dependence on $[H_2O]$ may be determined empirically or estimated based on kinetic arguments as noted below. The inclusion of $H_2O$ in the rate expression can be explained because most CO is consumed by reaction with OH that, to a first approximation, may be assumed to be in equilibrium with water.

Westbrook and Dryer (1981b) have used flammability limit data (the minimum and maximum equivalence ratios for sustained combustion) and flame speed data (which we will discuss shortly) for a variety of hydrocarbon fuels to determine the rate parameters for the various approximate combustion models. These parameters are summarized in Table 2.7. For each mechanism, the rate of the hydrocarbon consumption has been fitted to the form

$$r = A \exp\left(\frac{-E_a}{RT}\right) [\text{fuel}]^a [O_2]^b \tag{2.52}$$

For the two-step model, the oxidation of CO,

$$CO + \tfrac{1}{2} O_2 \longrightarrow CO_2$$

might, to a first approximation, be described using the global rate from Dryer and Glassman (1973):

$$r_f = 1.3 \times 10^{10} \exp\left(\frac{-20{,}130}{T}\right) [CO] [H_2O]^{0.5} [O_2]^{0.25} \text{ mol m}^{-3} \text{ s}^{-1} \tag{2.53}$$

**TABLE 2.7** RATE PARAMETERS FOR QUASI-GLOBAL REACTION MECHANISMS GIVING BEST AGREEMENT BETWEEN EXPERIMENTAL AND COMPUTED FLAMMABILITY LIMITS[a]

| | Single-step mechanism $C_nH_m + \left(n + \frac{m}{4}\right) O_2 \rightarrow nCO_2 + \frac{m}{2} H_2O$ | | | | Two-step mechanism $C_nH_m + \left(\frac{n}{2} + \frac{m}{4}\right) O_2 \rightarrow nCO + \frac{m}{2} H_2O$ | | | | Quasi-global mechanism $C_nH_m + \frac{n}{2} O_2 \rightarrow nCO + \frac{m}{2} H_2$ | | | |
|---|---|---|---|---|---|---|---|---|---|---|---|---|
| Fuel | $A \times 10^{-6}$ | $(E_a/R) \times 10^{-3}$ | $a$ | $b$ | $A \times 10^{-6}$ | $(E_a/R) \times 10^{-3}$ | $a$ | $b$ | $A \times 10^{-6}$ | $(E_a/R) \times 10^{-3}$ | $a$ | $b$ |
| $CH_4$ | 130 | 24.4 | −0.3 | 1.3 | 2800 | 24.4 | −0.3 | 1.3 | 4000 | 24.4 | −0.3 | 1.3 |
| $C_2H_6$ | 34 | 15.0 | 0.1 | 1.65 | 41 | 15.0 | 0.1 | 1.65 | 63 | 15.0 | 0.3 | 1.3 |
| $C_3H_8$ | 27 | 15.0 | 0.1 | 1.65 | 31 | 15.0 | 0.1 | 1.65 | 47 | 15.0 | 0.1 | 1.65 |
| $C_4H_{10}$ | 23 | 15.0 | 0.15 | 1.6 | 27 | 15.0 | 0.15 | 1.6 | 41 | 15.0 | 0.15 | 1.6 |
| $C_5H_{12}$ | 20 | 15.0 | 0.25 | 1.5 | 24 | 15.0 | 0.25 | 1.5 | 37 | 15.0 | 0.25 | 1.5 |
| $C_6H_{14}$ | 18 | 15.0 | 0.25 | 1.5 | 22 | 15.0 | 0.25 | 1.5 | 34 | 15.0 | 0.25 | 1.5 |
| $C_7H_{16}$ | 16 | 15.0 | 0.25 | 1.5 | 19 | 15.0 | 0.25 | 1.5 | 31 | 15.0 | 0.25 | 1.5 |
| $C_8H_{18}$ | 14 | 15.0 | 0.25 | 1.5 | 18 | 15.0 | 0.25 | 1.5 | 29 | 15.0 | 0.25 | 1.5 |
| $C_9H_{20}$ | 13 | 15.0 | 0.25 | 1.5 | 16 | 15.0 | 0.25 | 1.5 | 27 | 15.0 | 0.25 | 1.5 |
| $C_{10}H_{22}$ | 12 | 15.0 | 0.25 | 1.5 | 14 | 15.0 | 0.25 | 1.5 | 25 | 15.0 | 0.25 | 1.5 |
| $CH_3OH$ | 101 | 15.0 | 0.25 | 1.5 | 117 | 15.0 | 0.25 | 1.5 | 230 | 15.0 | 0.25 | 1.5 |
| $C_2H_5OH$ | 47 | 15.0 | 0.15 | 1.6 | 56 | 15.0 | 0.15 | 1.6 | 113 | 15.0 | 0.15 | 1.6 |
| $C_6H_6$ | 6 | 15.0 | −0.1 | 1.85 | 7 | 15.0 | −0.1 | 1.85 | 13 | 15.0 | −0.1 | 1.85 |
| $C_7H_8$ | 5 | 15.0 | −0.1 | 1.85 | 6 | 15.0 | −0.1 | 1.85 | 10 | 15.0 | −0.1 | 1.85 |
| $C_2H_4$ | 63 | 15.0 | 0.1 | 1.65 | 75 | 15.0 | 0.1 | 1.65 | 136 | 15.0 | 0.1 | 1.65 |
| $C_3H_6$ | 13 | 15.0 | −0.1 | 1.85 | 15 | 15.0 | −0.1 | 1.85 | 25 | 15.0 | −0.1 | 1.85 |
| $C_2H_2$ | 205 | 15.0 | 0.5 | 1.25 | 246 | 15.0 | 0.5 | 1.25 | 379 | 15.0 | 0.5 | 1.25 |

[a]Units: m, s, mol, K.

*Source:* Westbrook and Dryer, 1981b.

The rate of the reverse of the CO oxidation reaction was estimated by Westbrook and Dryer (1981b) to be

$$r_r = 1.6 \times 10^7 \exp\left(\frac{-20,130}{T}\right)[CO_2][H_2O]^{0.5}[O_2]^{-0.25} \text{ mol m}^{-3}\text{ s}^{-1} \quad (2.54)$$

One must be cautious in using such rate expressions. Since (2.53) and (2.54) were obtained from flame observations, they may not be appropriate to postflame burnout of CO. This issue will be addressed in Chapter 3 when we discuss the CO emission problem.

Lumping all the reactions that lead to CO formation into a single step means that the dynamics of these reactions can only be described approximately. The endothermic initiation reactions proceed slowly for some time before the radical population becomes large enough for rapid consumption of fuel and $O_2$. Little CO is produced during this ignition delay, so efforts to model CO formation frequently overlook the initiation process. Assuming direct production of $H_2O$ means the transients in the production and equilibration of H, OH, O, $HO_2$, and so on, are not described. Thus the two-step model does not accurately describe the processes occurring early in combustion. It is, however, a marked improvement over the single-step model in that it allows CO oxidation to proceed more slowly than fuel consumption. Although the two-step model does not adequately describe processes occurring early in combustion, the omission of the radical chemistry is not serious if one is primarily interested in processes that take place after the main combustion reactions are complete (e.g., the highly endothermic oxidation of $N_2$ to form NO).

Additional reactions can be incorporated to develop quasi-global reaction mechanisms with improved agreement between calculations and experimental observations while avoiding the complications and uncertainties in describing detailed hydrocarbon oxidation kinetics. Edelman and Fortune (1969) pushed this process toward its logical limit, describing the oxidation of the fuel to form CO and $H_2$ by a single reaction and then using the detailed reaction mechanisms for CO and $H_2$ oxidation. Because all the elementary reactions and species in the CO–$H_2$–$O_2$ system are included, this approach can provide an accurate description of the approach to equilibrium and of postflame processes such as nitric oxide formation from $N_2$ and CO burnout as the combustion products are cooled.

The quasi-global model requires oxidation rates for both CO and $H_2$. Although lumped reaction models can be used, the major advantage of the quasi-global model is that it can be used in conjunction with a detailed description of the final stages of combustion. Westbrook and Dryer (1981b) compared the flame structure predictions of the quasi-global model with those of a detailed mechanism for methanol-air flames. The reactions and corresponding rate coefficients for the CO–$H_2$–$O_2$ system that were needed for the quasi-global model are summarized in Table 2.8. Predictions of temperature profiles, fuel concentrations, and general flame structure are in close agreement for the two models. The predicted concentrations of CO and radical species (O, H, and OH) showed qualitatively different behavior for the two models because reactions of the radicals with unburned fuel are not taken into account in the quasi-global model.

TABLE 2.8 C–H–O KINETIC MECHANISM

| Reaction | $k_f$ (units: $m^3$, mol, K, s) | Reference |
|---|---|---|
| | CO oxidation | |
| $CO + OH \rightleftharpoons CO_2 + H$ | $4.4\ T^{1.5} \exp(+373/T)$ | Warnatz (1984) |
| $CO + O_2 \rightleftharpoons CO_2 + O$ | $2.5 \times 10^6 \exp(-24{,}060/T)$ | Warnatz (1984) |
| $CO + O + M \rightleftharpoons CO_2 + M$ | $5.3 \times 10^1 \exp(+2285/T)$ | Warnatz (1984) |
| $CO + HO_2 \rightleftharpoons CO_2 + OH$ | $1.5 \times 10^8 \exp(-11{,}900/T)$ | Westbrook and Dryer (1981a) |
| | Exchange reactions | |
| $H + O_2 \rightleftharpoons O + OH$ | $1.2 \times 10^{11}\ T^{-0.91} \exp(-8310/T)$ | Warnatz (1984) |
| $H_2 + O \rightleftharpoons H + OH$ | $1.5 \times 10^1\ T^2 \exp(-3800/T)$ | Warnatz (1984) |
| $O + H_2O \rightleftharpoons OH + OH$ | $1.5 \times 10^4\ T^{1.14} \exp(-8680/T)$ | Warnatz (1984) |
| $OH + H_2 \rightleftharpoons H + H_2O$ | $1.0 \times 10^2\ T^{1.6} \exp(-1660/T)$ | Warnatz (1984) |
| $O + HO_2 \rightleftharpoons O_2 + OH$ | $2.0 \times 10^7$ | Warnatz (1984) |
| $H + HO_2 \rightleftharpoons OH + OH$ | $1.5 \times 10^8 \exp(-500/T)$ | Warnatz (1984) |
| $H + HO_2 \rightleftharpoons H_2 + O_2$ | $2.5 \times 10^7 \exp(-350/T)$ | Warnatz (1984) |
| $OH + HO_2 \rightleftharpoons H_2O + O_2$ | $2.0 \times 10^7$ | Warnatz (1984) |
| $HO_2 + HO_2 \rightleftharpoons H_2O_2 + O_2$ | $2.0 \times 10^7$ | Warnatz (1984) |
| | Recombination reactions | |
| $H + O_2 + M \rightleftharpoons HO_2 + M$ | $1.5 \times 10^3 \exp(-500/T)$ | Westbrook and Dryer (1984) |
| $OH + OH + M \rightleftharpoons H_2O_2 + M$ | $1.3 \times 10^{10}\ T^{-2}$ | Warnatz (1984) |
| $O + H + M \rightleftharpoons OH + M$ | $1.0 \times 10^4$ | Westbrook and Dryer (1984) |
| $O + O + M \rightleftharpoons O_2 + M$ | $1.0 \times 10^5\ T^{-1}$ | Warnatz (1984) |
| $H + H + M \rightleftharpoons H_2 + M$ | $6.4 \times 10^5\ T^{-1}$ | Warnatz (1984) |
| $H + OH + M \rightleftharpoons H_2O + M$ | $1.41 \times 10^{11}\ T^{-2}$ | Warnatz (1984) |

Quasi-global rate models may be suitable for the systems from which they were derived, but caution must be exercised in their use. Assumptions made in their derivation or the conditions of the particular experiment used in the estimation of the rate parameters strongly influence the predicted rates. For example, different preexponential factors must be used for flow systems and stirred reactors (Edelman and Fortune, 1969) and for flames (Westbrook and Dryer, 1981b). Nevertheless, quasi-global models often represent a practical compromise between comprehensive kinetic mechanisms based entirely on elementary reaction steps and simple one-step models of the combustion process.

While the chemical kinetics of combustion describe much of what happens when a fuel burns, chemical kinetics alone cannot describe combustion in practical systems. Calculations of the rate of combustion reactions, using either detailed combustion mechanisms or global models, reveal that the reactions proceed extremely slowly unless the temperature exceeds a critical value. To understand combustion, therefore, we need to examine the physical processes that heat the reactants to this temperature so that reaction can take place.

## 2.5 FLAME PROPAGATION AND STRUCTURE

We now turn our attention from combustion thermochemistry to the physical processes that govern the way fuels burn. One of the striking features of most combustion is the existence of a flame, a luminous region in the gas that is associated with the major heat release. Some flames, such as that of a candle, are relatively steady, whereas others fluctuate wildly due to turbulent motions of the gas. The flame is a reaction front created by diffusion of energy or free radicals from the hot burned gases into the cooler unreacted gas or by the mixing of fuel and air.

A flame that is stabilized at a fixed location is actually propagating into a flow of fuel and/or air. In this case the propagation velocity must match the gas velocity for the flame itself to remain fixed in space. This is illustrated schematically in Figure 2.9(a) for the bunsen burner flame. Here a mixture of fuel and air is introduced through a pipe at a velocity, $v$. The flame appears as a conical region of luminosity above the pipe outlet. The height of the conical flame depends on the gas velocity; that is, low velocities produce short flames while higher velocities produce longer flames. The shape of the flame is determined by the way the reaction propagates from hot burned gases into cooler unburned gases. We shall see that the reaction front moves into the unburned gases at a velocity that is determined by the combined effects of molecular diffusion and chemical kinetics. This propagation velocity is the laminar flame speed, $S_L$. If the gas velocity, $v$, is greater than $S_L$, the flame assumes a shape such that the component of the gas velocity normal to the flame exactly balances the flame speed, as illustrated in Figure 2.9(a).

The gas velocity at the wall is zero, allowing the flame to propagate close to the burner outlet. Heat transfer to the pipe prevents the flow from propagating into the pipe in normal operation, so the flame is stabilized at the pipe outlet by the combined effects of diffusion and heat transfer.

When, as in the bunsen burner flame, a gaseous fuel and air are uniformly mixed prior to combustion, a *premixed flame* results. Such a combustible mixture can easily explode, so premixed combustion is used in relatively few systems of practical importance; for example, laboratory bunsen and Meeker burners mix fuel and air prior to combustion, the carburetor on an automobile engine atomizes liquid gasoline into the combustion air in order to achieve premixed combustion, and some premixing of fuel and air takes place in gas cooking stoves.

Within the automobile engine immediately prior to combustion, there is no flow. Combustion is initiated by a spark and then propagates through the mixture as illustrated in Figure 2.9(b). In spite of the different appearances of these two flames, the physics and chemistry that govern the structure of the propagating flame are the same as those of stabilized flames, although geometry and fluid motions can vary greatly from situation to situation.

More commonly, fuel and air enter the combustion zone separately and must mix before reaction is possible. The chemistry of the so-called *diffusion flame* that results cannot be described by a single equivalence ratio since, as illustrated in Figure 2.9(c), gases with the entire range of compositions, from pure air to pure fuel, are present in

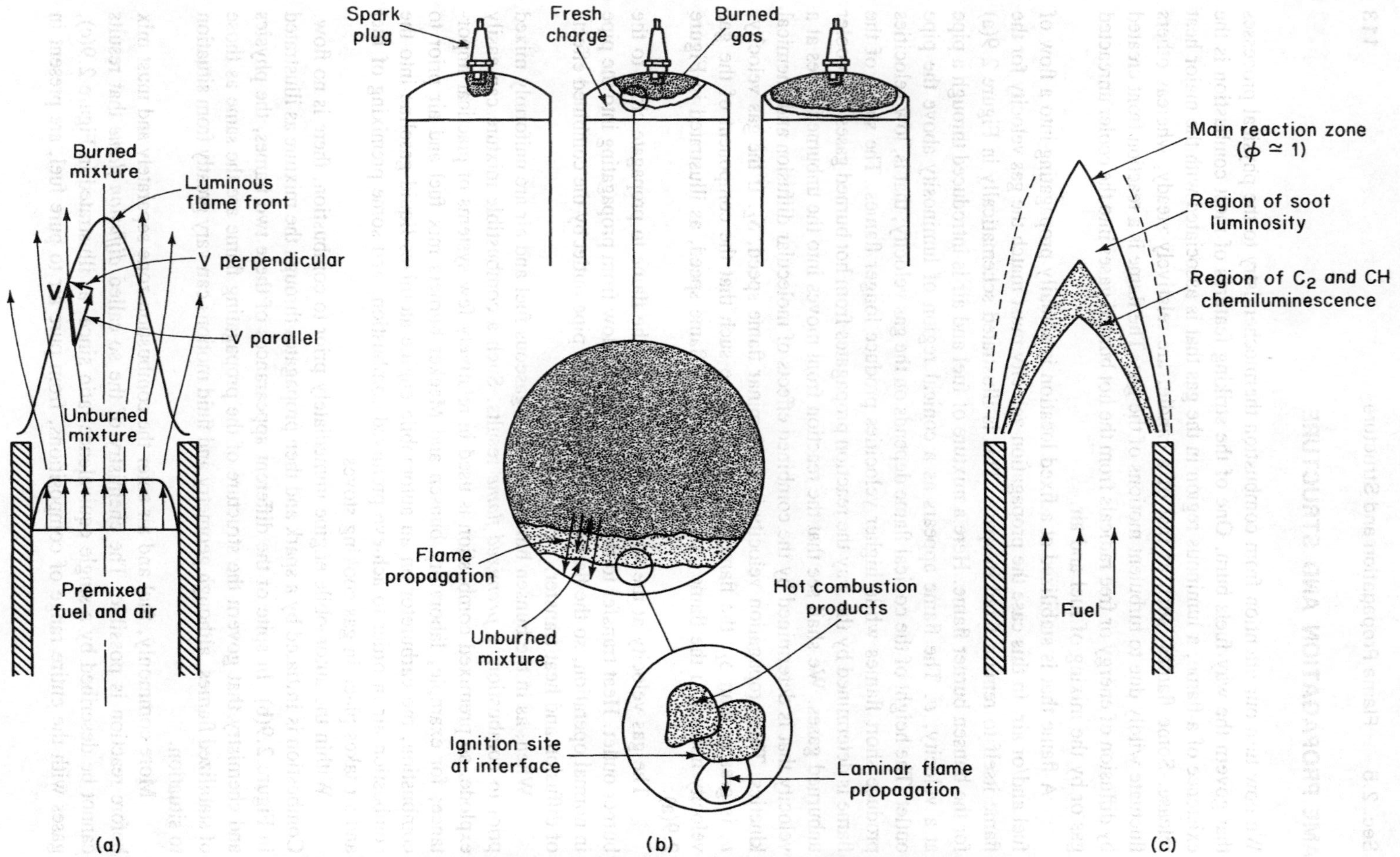


**Figure 2.9** Flame propagation and structure: (a) bunsen burner flame; (b) spark ignition engine; (c) diffusion flame.

the combustion zone. An overall equivalence ratio may be useful in describing the net flow of fuel and air into the combustor, but that value does not correspond to the local composition (on a molecular scale) that governs combustion chemistry. Within the diffusion flame illustrated in Figure 2.9(c), there is a central core that contains pure gaseous fuel. This core is surrounded by a zone in which air diffuses inward and fuel diffuses outward. The visible flame front sits approximately at the location of stoichiometric composition in this zone. The different luminous zones correspond to regions in which chemiluminescent (or light-emitting) chemical reactions take place and in which soot (small carbonaceous particles) emits thermal radiation.

The shape of this diffusion flame is, as in the stabilized premixed flame, determined by the competition between flow and diffusion. The flame length increases as the fuel velocity increases. At sufficiently high fuel velocities, the flame ceases to be uniform in shape due to the onset of turbulence. The turbulent velocity fluctuations increase the rate at which fuel and air come into contact and, therefore, cause the flame to shorten as the velocity is increased further. Ultimately, as illustrated in Figure 2.10, the flame length approaches an asymptotic value.

Each of these flame types, the premixed flame and the diffusion flame, can be further subdivided into *laminar* and *turbulent* flames. Heat and mass transfer in laminar flames occur by molecular conduction and diffusion. In many systems, the existence of either laminar or turbulent flow is determined by the value of the Reynolds number

$$\mathrm{Re} \equiv \frac{\rho v L}{\mu}$$

a dimensionless ratio of inertial to viscous forces, where $\rho$ is the gas density, $v$ and $L$ are characteristic velocity and length scales for the flow, and $\mu$ is the viscosity of the fluid. For example, the Reynolds number must be less than about 2200 to assure laminar

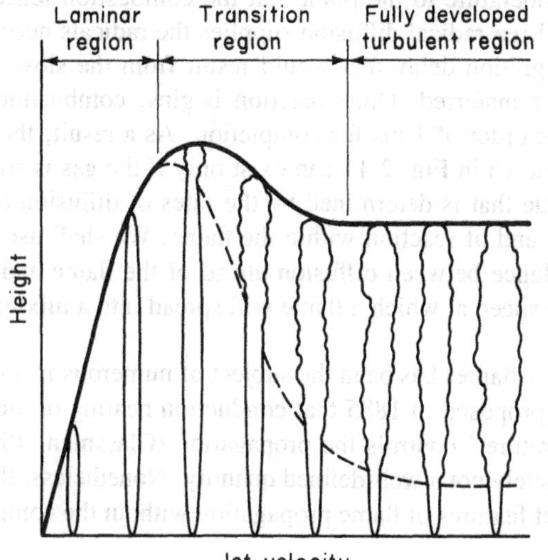


**Figure 2.10** Diffusion flame length variation with fuel jet velocity.

flow in a pipe. Turbulence can, however, be promoted at Reynolds numbers below this value by flow obstructions.

Most flames of practical significance are turbulent. Even in turbulent flames, molecular diffusion plays an important role, albeit on a scale much smaller than that of the flow system. For this reason, we shall examine the structure of laminar flames before addressing turbulent flames.

Our primary objective is to understand those aspects of flame structure that directly influence the production of pollutants. The rate of flame spread and consumption of fuel determines whether combustion will be complete and how long the combustion products will reside at high temperature. Flame stability is also important since a flame that nearly extinguishes and reignites due to instabilities may allow combustion products or reaction intermediates to escape.

### 2.5.1 Laminar Premixed Flames

The simplest type of flame is the laminar premixed flame. From the study of these flames, we can explore important aspects of flame propagation. The propagation velocity, or *laminar flame speed*, is particularly important to our discussion since it determines how rapidly a fuel–air mixture is burned.

A conceptually simple laminar, premixed flame can be produced by flowing a fuel–air mixture through a porous plug, as illustrated in Figure 2.11. A luminous flame appears as a thin planar front that remains at a fixed distance from the porous plug. The fuel–air mixture passes through the plug with a velocity, $v$. For the flame to remain stationary, it must propagate at an equal and opposite velocity toward the fuel–air flow. The laminar flame speed, $S_L$, is the speed at which the flame propagates into the cold fuel–air mixture (i.e., $S_L = v$ ). The flame speed is determined by the rates of the combustion reactions and by diffusion of energy and species into the cold unreacted mixture. Heat transfer raises the gas temperature to the point that the combustion reactions can proceed at an appreciable rate. Free-radical diffusion supplies the radicals necessary for rapid combustion without the ignition delay that would result from the slow initiation reactions if only energy were transferred. Once reaction begins, combustion is very rapid, typically requiring on the order of 1 ms for completion. As a result, the flame is generally thin. The flat flame shown in Fig. 2.11 can exist only if the gas is supplied at a velocity below a limiting value that is determined by the rates of diffusion of energy and radicals ahead of the flame and of reaction within the flame. We shall use a simple model to examine how this balance between diffusion ahead of the flame and reaction within the flame determines the speed at which a flame will spread into a mixture of fuel and air.

The propagation of laminar flames has been the subject of numerous investigations since Mallard and LeChâtelier proposed in 1885 that conduction heating of the fuel–air mixture to an "ignition temperature" controls the propagation (Glassman, 1977). The ignition temperature is unfortunately not a well-defined quantity. Nonetheless, this model illustrates many of the important features of flame propagation without the complications of the more elaborate theories.

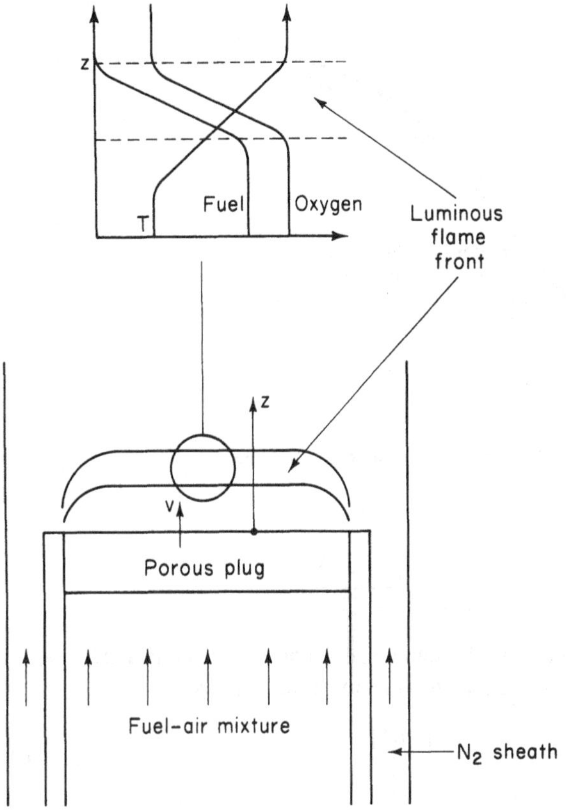


**Figure 2.11** Porous plug burner with a flat flame.

Mallard and LeChâtelier stated that the heat conducted ahead* of the flame must, for steady propagation, be equal to that required to heat the unburned gases from their initial temperature, $T_0$, to the ignition temperature, $T_i$. The flame is then divided into two zones, as illustrated in Figure 2.12. The enthalpy rise of the fuel-air mixture in the preheat zone is $\bar{f}\bar{c}_p(T_i - T_0)$, where $\bar{f}$ is the mass flux through the flame front. This enthalpy must be supplied by conduction from the reaction zone. Thus we have

$$\bar{f}\bar{c}_p(T_i - T_0) = k\frac{dT}{dz} \tag{2.55}$$

where $k$ is the thermal conductivity of the gas. The mass flux is directly related to the speed, $S_L$, at which the laminar flame propagates into the cold fuel-air mixture,

$$\bar{f} = \rho_0 S_L \tag{2.56}$$

Approximating the temperature profile with a constant slope, $dT/dz = (T_f - T_i)/\delta$, where $\delta$ is the flame thickness and $T_f$ is the adiabatic flame temperature, the laminar

*The terminology "ahead of the flame" customarily refers to the cold fuel/air mixture into which the flame is propagating.

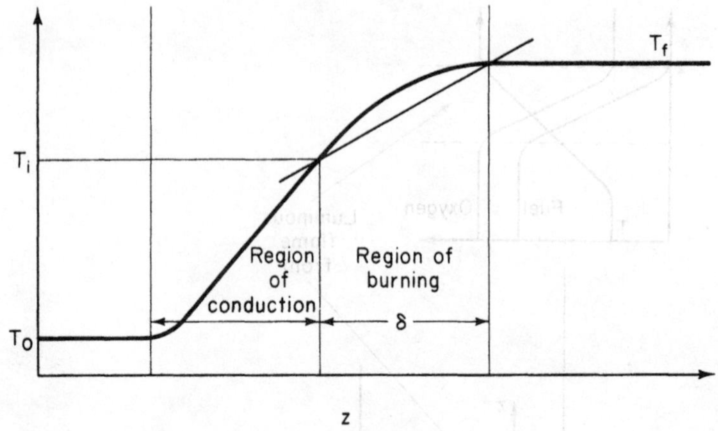


**Figure 2.12** Two zones of a premixed flame.

flame speed becomes

$$S_L = \frac{k}{\rho_0 \bar{c}_p} \frac{T_f - T_i}{T_i - T_0} \frac{1}{\delta} \tag{2.57}$$

The flame thickness is related to the flame speed and the characteristic time for the combustion reactions. Defining the characteristic reaction time as

$$\tau_c = \frac{[\text{fuel}]_0}{r_f} \tag{2.58}$$

where $r_f$ is the overall fuel oxidation rate, the flame thickness becomes

$$\delta = S_L \tau_c = \frac{S_L [\text{fuel}]_0}{r_f} \tag{2.59}$$

Substituting into (2.57) and rearranging yields

$$S_L = \left[ \frac{k}{\rho_0 \bar{c}_p} \frac{T_f - T_i}{T_i - T_0} \frac{r_f}{[\text{fuel}]_0} \right]^{1/2} \tag{2.60}$$

Global oxidation rates such as (2.49) can be used to explore factors that influence the flame speed. It is apparent from (2.49) and (2.60) and the ideal gas law that

$$S_L \approx p^{(a+b-2)/2}$$

Since the overall reaction order $(a + b)$ for most hydrocarbon fuels is approximately two (see Table 2.7), the flame speed is seen to be only weakly dependent on pressure. The reaction rate is a highly nonlinear function of temperature due to the exponential term. Although the reaction may begin at lower temperature, most of the reaction takes place after the gases have been heated very nearly to the final temperature (i.e., near the adiabatic flame temperature). The activation energies for the combustion of most hydrocarbon fuels are similar, as are the adiabatic flame temperatures and the thermal conductivities of the fuel–air mixtures. Thus one would expect flame velocities of different hydrocarbon fuels to be similar. The flame temperature is highest near stoichiometric combustion and drops significantly at lower or higher equivalence ratios.

**Example 2.8** ***Laminar Flame Speed***

Use the single-step global rate expression for methane combustion to estimate the laminar flame speed for stoichiometric combustion in air with $T_0 = 298$ K and at $p = 1$ atm.

To estimate the laminar flame speed, we need to know the flame temperature, $T_f$, and the ignition temperature, $T_i$. $T_f$ may be approximated by the adiabatic flame temperature. The combustion stoichiometry is

$$CH_4 + 2(O_2 + 3.78N_2) \longrightarrow CO_2 + 2H_2O + 7.56N_2$$

and the energy equation for adiabatic combustion becomes

$$[h(T) - h(T_0)]_{CO_2} + 2[h(T) - h(T_0)]_{H_2O} + 7.56[h(T) - h(T_0)]_{N_2}$$
$$+ \Delta h^\circ_{f,CO_2}(T_0) + 2\,\Delta h^\circ_{f,H_2O}(T_0) - \Delta h^\circ_{f,CH_4}(T_0) - 2\,\Delta h^\circ_{f,O_2}(T_0) = 0$$

Using the approximate thermodynamic data of Table 2.5, we recall that

$$[h(T) - h(T_0)]_i = \int_{T_0}^{T} c_{pi}\, dT' = \int_{T_0}^{T} (a_i + b_i T')\, dT' = a_i(T - T_0) + \frac{b_i}{2}(T^2 - T_0^2)$$

The necessary data are:

| Species | $\Delta h^\circ_f\,(T_0)$ (J mol$^{-1}$) | $a$ | $b$ |
|---|---|---|---|
| $CH_4$ | −74,980 | 44.2539 | 0.02273 |
| $O_2$ | 0 | 30.5041 | 0.00349 |
| $CO_2$ | −394,088 | 44.3191 | 0.00730 |
| $H_2O$ | −242,174 | 32.4766 | 0.00862 |
| $N_2$ | 0 | 29.2313 | 0.00307 |

Substituting in the coefficient values, the energy equation become

$$330.26(T - T_0) + 0.02387(T^2 - T_0^2) - 803{,}456 = 0$$

or

$$0.02387T^2 - 330.26T - 904{,}045 = 0$$

which yields

$$T = 2341 \text{ K}$$

We now have an estimate for $T_f = 2341$ K. The global rate expression for methane combustion is (from Table 2.7)

$$r_{CH_4} = 1.3 \times 10^8 \exp\left(-\frac{24{,}400}{T}\right) [CH_4]^{-0.3} [O_2]^{1.3} \text{ mol m}^{-3}\text{ s}^{-1}$$

Since the reaction rate is a strong function of temperature, the characteristic time for the reaction should be evaluated near the peak temperature. At the adibatic flame temperature,

$$\tau_c = \frac{[CH_4]_0}{1.3 \times 10^8 \exp(-24{,}400/T)[CH_4]^{-0.3}[O_2]^{1.3}}$$

In the reactants, the species mole fractions are

$$y_{CH_4} = \frac{1}{10.56} = 0.0947$$

$$y_{O_2} = \frac{2}{10.56} = 0.189$$

Using the ideal gas law to calculate the concentrations, we find

$$\tau_c = 1.05 \times 10^{-4} \text{ s}$$

The heat transfer occurs at lower temperature. Evaluating the gas mixture properties at, say, 835 K (the geometric mean of the extreme values), we find for the mixture

$$\rho_0 = 0.403 \text{ kg m}^{-3}$$

$$\bar{c}_p = 1119 \text{ J kg}^{-1} \text{ K}^{-1}$$

As an approximation, we use the thermal conductivity for air,

$$k = 0.0595 \text{ J m}^{-1} \text{ K}^{-1} \text{ s}^{-1}$$

The ignition temperature should be near the flame temperature due to the exponential dependence of the reaction rate (Glassman, 1977). Substituting into (2.60) assuming that $T_i$ = 2100 K yields

$$S_L = \left(\frac{0.0595 \text{ J m}^{-1} \text{ K}^{-1} \text{s}^{-1}}{0.403 \text{ kg m}^{-3} \times 1119 \text{ J kg}^{-1} \text{ K}^{-1}} \times \frac{2341 - 2100}{2100 - 298} \times \frac{1}{1.05 \times 10^{-4} \text{ s}}\right)^{1/2}$$
$$= 0.41 \text{ m s}^{-1}$$

We may also examine the flame thickness using (2.59):

$$\delta = S_L \tau_c \approx 0.41 \text{ m s}^{-1} \times 1.05 \times 10^{-4} \text{ s}$$
$$\approx 4 \times 10^{-5} \text{ m}$$
$$\approx 0.04 \text{ mm}$$

Because the flame is so thin, studies of the structure of premixed flames are frequently conducted at reduced pressures to expand the flame.

### 2.5.2 Turbulent Premixed Flames

The automobile is the major practical system in which fuel and air are thoroughly mixed prior to burning. In the automobile engine, combustion takes place in a confined volume. Combustion is initiated in a small fraction of this volume by a spark. A flame spreads from the ignition site throughout the volume. The fluid motion in the cylinder is chaotic due to the turbulence generated by the high-velocity flows through the intake valves and by motions induced as the piston compresses the gas.

The velocities of the random turbulent motions may far exceed the laminar flame velocity, leading to wild distortions of the flame front as it propagates. Figure 2.13 is a simplified schematic of the way that turbulent velocity fluctuations may influence the propagation of a premixed flame. Here we consider a flame front that is initially flat. If this flame were to propagate at the laminar flame speed, it would move a distance $S_L\,\delta t$ in a time $\delta t$. This motion is limited to propagation from the burned gas into the unburned gas. On the other hand, velocity fluctuations with a root-mean-square value of $u'$ would distort the front between the burned and unburned gases about the initial flame-front location. Without bringing molecular diffusion into play, no molecular scale mixing of burned and unburned gases would take place, and the quantity of burned gas would not increase. The rate of diffusive propagation of the flame from the burned gases into the unburned gases is governed by a balance between molecular diffusion and the kinetics of the combustion reactions (i.e., the same factors that were considered in the original analysis of laminar flame propagation). Thus the propagation of the flame from the distorted front into the unburned gases is characterized by the laminar flame speed, and the position of the flame front after a small time is the combination of these two effects.

The microscales for the composition and velocity fluctuations in a turbulent flow, $\lambda_c$ and the Taylor microscale, $\lambda$, respectively, discussed in Appendix D of Chapter 1,

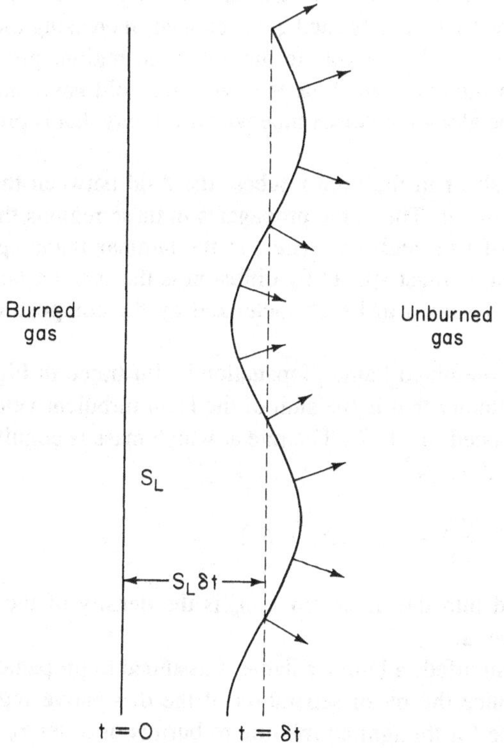


**Figure 2.13** Enhancement of flame speed by turbulent motion.

are the scales that are characteristic of the fluctuations in the position of the flame front and, therefore, of the distance over which the flame must propagate diffusively. A time scale that characterizes the burning of the gas in these regions of entrained unburned gas is

$$\tau_b = \frac{\lambda_c}{S_L} \tag{2.61}$$

This time scale differs from that for dissipation of concentration fluctuations in nonreacting flows, (D.30),

$$\tau_d = \frac{\lambda^2}{6\nu}$$

since the rapid combustion reactions lead to large gradients in the flame front, enhancing the rate of diffusion of energy and radicals.

Observations of the small-scale structure of turbulent flow (Tennekes, 1968) provide important insights into the mechanism of turbulent flame propagation and the basis for a quantitative model of combustion rates and flame spread (Tabaczynski et al., 1977). Within the turbulent fluid motion, turbulent dissipation occurs in small so-called vortex tubes with length scales on the order of the Kolmogorov microscale, $\eta$ (D.1). Chomiak (1970, 1972) postulated that the vortex tubes play an essential role in the flame propagation. When the combustion front reaches the vortex tube, the high shear rapidly propagates the combustion across the tube. The burned gases expand, increasing the pressure in the burned region of the vortex tube relative to the unburned region, providing the driving force for the motion of the hot burned gases toward the cold gases and leading to rapid propagation of the flame along the vortex tube with a velocity that is proportional to $u'$, the turbulent intensity.

In contrast to the vigorous shear in the vortex tubes, the fluid between the tubes is envisioned to be relatively quiescent. The flame propagates in these regions through the action of molecular diffusion of heat and mass (i.e., at the laminar flame speed, $S_L$). The distance over which the flame must spread by diffusion is the spacing between the vortex tubes. This distance is assumed to be characterized by the composition microscale, $\lambda_c$.

This model for turbulent premixed flame propagation is illustrated in Figure 2.14. Ignition sites propagate at a velocity that is the sum of the local turbulent velocity fluctuation and the laminar flame speed, $u' + S_L$. The rate at which mass is engulfed within the flame front can be expressed as

$$\frac{dm_e}{dt} = \rho_u A_e (u' + S_L) \tag{2.62}$$

where $m_e$ is the mass engulfed into the flame front, $\rho_u$ is the density of the unburned gas, and $A_e$ is the flame front area.

Once unburned fluid is engulfed, a laminar flame is assumed to propagate through it from the burned regions. Since the mean separation of the dissipative regions is of order $\lambda_c$, the characteristic time for the ignited mixture to burn is of order $\tau_b = \lambda_c/S_L$.

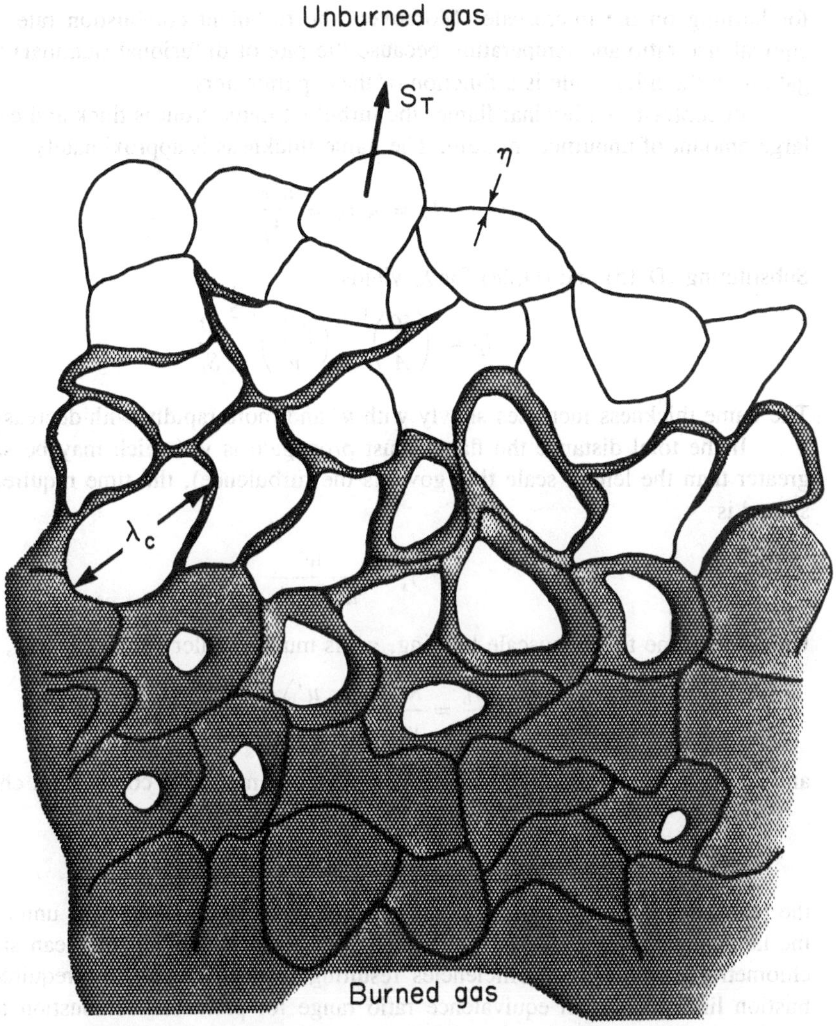


**Figure 2.14** Turbulent premixed flame.

The mass of unburned mixture behind the flame is $(m_e - m_b)$. The rate at which the entrained mixture is burned may be approximated by

$$\frac{dm_b}{dt} = \frac{m_e - m_b}{\tau_b} \tag{2.63}$$

In the limit of instantaneous burning, $\tau_b \to 0$, of the engulfed gas (i.e., $m_b = m_e$), this degenerates to Damkohler's (1940) model for the turbulent flame in which $S_T = S_L + u'$ and all the gas behind the flame front is assumed to be burned. The rate of burning is generally slower than the rate of engulfment, however, because of the time required for burning on the microscale. Moreover, the turbulent combustion rate depends on equivalence ratio and temperature because the rate of diffusional (laminar) flow propagation on the microscale is a function of these parameters.

In contrast to a laminar flame, the turbulent flame front is thick and can contain a large amount of unburned mixture. The flame thickness is approximately

$$l_F \approx u'\tau_b \approx \frac{u'\lambda_c}{S_L} \tag{2.64}$$

Substituting (D.15) and (D.28) for $\lambda_c$ yields

$$l_F \approx \left(\frac{60}{A}\right)^{1/2} \left(\frac{u'L}{\nu}\right)^{1/2} \frac{D}{S_L} \tag{2.65}$$

The flame thickness increases slowly with $u'$ and more rapidly with decreasing $S_L$.

If the total distance the flame must propagate is $w$ (which may be substantially greater than the length scale that governs the turbulence), the time required for flame spread is

$$\tau_s \approx \frac{w}{u' + S_L} \tag{2.66}$$

Unless the time for microscale burning, $\tau_b$, is much smaller than $\tau_s$, that is,

$$\frac{\tau_b}{\tau_s} = \frac{\lambda_c}{w}\left(1 + \frac{u'}{S_L}\right) << 1$$

and is also much less than the available residence time in the combustion chamber, $\tau_R$,

$$\frac{\tau_b}{\tau_R} = \frac{\lambda_c}{S_L \tau_R} << 1$$

the possibility exists that some of the mixture will leave the chamber unreacted. Since the laminar flame speed drops sharply on both the fuel-rich and fuel-lean sides of stoichiometric, combustion inefficiencies resulting from the finite time required for combustion limit the useful equivalence ratio range for premixed combustion to a narrow band about stoichiometric. Automobiles are thus generally restricted to operating in the range $0.8 < \phi < 1.2$.

**Example 2.9** ***Flame Propagation in a Pipe Flow***

Estimate the flame propagation velocity and flame thickness for stoichiometric combustion of premixed methane in air flowing in a 0.1-m-diameter pipe with a cold gas velocity of 10 m s$^{-1}$. The initial pressure and temperature are 1 atm and 298 K, respectively.

The Reynolds number of the cold flowing gas is

$$\text{Re} = \frac{Ud}{\nu} = \frac{10 \text{ m s}^{-1} \times 0.1 \text{ m}}{1.5 \times 10^{-5} \text{ m}^2 \text{ s}^{-1}} = 66{,}700$$

which is greater than that required for turbulent flow (Re = 2200), so the flow may be assumed to be turbulent. To estimate the turbulent flame speed, we need to know the turbulent dissipation rate. The dissipation rate can be estimated by considering the work done due to the pressure drop in the pipe flow since the work done by the fluid is dissipated through the action of turbulence. From thermodynamics, we estimate the work per unit mass due to a pressure drop, $\Delta p$, to be

$$\overline{w} = -\frac{1}{\rho}\Delta p$$

The pressure drop in a turbulent pipe flow can be calculated using the Fanning friction factor, $f_F$:

$$\Delta p = -f_F \frac{L}{d}\frac{\rho U^2}{2}$$

where $L$ is the length of the segment of pipe being considered. The mass flow rate through the pipe is $\rho U(\pi/4)\,d^2$, so the total power dissipated in the length, $L$, is

$$\begin{aligned} P &= -\frac{1}{\rho}\Delta p \frac{\pi}{4} d^2 \rho U \\ &= -\frac{1}{\rho}\left(-f_F \frac{L}{d}\frac{\rho U^2}{2}\right)\frac{\pi}{4} d^2 \rho U \\ &= f_F \frac{\pi}{8}\rho\, dLU^3 \end{aligned}$$

The turbulent dissipation rate is the rate of energy dissipation per unit mass, which we find by dividing by the total mass contained in the length $L$,

$$\epsilon = \frac{P}{m} = \frac{f_F(\pi/8)\rho\, dLU^3}{(\pi/4)\rho\, d^2L} = f_F\frac{U^3}{2d}$$

From Bird et al. (1960), we find

$$f_F = \frac{0.0791}{\mathrm{Re}^{1/4}} \qquad 2100 < \mathrm{Re} < 10^5$$

so for the present problem, $f_F = 0.00492$ and

$$\epsilon = (0.00492)\frac{(10\ \mathrm{m\ s^{-1}})^3}{2 \times 0.1\ \mathrm{m}} = 24.6\ \mathrm{m^2\ s^{-3}}$$

$\epsilon$ is related to the characteristic velocity fluctuation by (D.14):

$$\epsilon \approx \frac{Au'^3}{d}$$

where $A$ is a constant of order unity. Assuming that $A = 1$, we estimate

$$u' \approx (\epsilon d)^{1/3} = (24.6\ \mathrm{m^2\ s^{-3}} \times 0.1\ \mathrm{m})^{1/3}$$

or

$$u' \approx 1.35\ \mathrm{m\ s^{-1}}$$

From (D.15) we find the Taylor microscale:

$$\lambda = d\left(\frac{15}{A\ \mathrm{Re}}\right)^{1/2} = 0.1\left(\frac{15}{66{,}700}\right)^{1/2} = 0.00150 \text{ m}$$

This is significantly larger than the smallest scale of the turbulent motion, the Kolmogorov microscale (D.1)

$$\eta = \left(\frac{\nu^3}{\epsilon}\right)^{1/4} = \left[\frac{(1.5 \times 10^{-5} \text{ m}^2 \text{ s}^{-1})^3}{24.6 \text{ m}^2 \text{ s}^{-3}}\right]^{1/4}$$

$$= 0.00011 \text{ m}$$

According to our model, the flame spreads at a velocity of

$$S_T = u' + S_L$$

The measured laminar flame speed for stoichiometric combustion of methane in air is $S_L = 0.38 \text{ m s}^{-1}$, so

$$S_T \approx 1.35 + 0.38 \approx 1.73 \text{ m s}^{-1}$$

$$l_F \approx \frac{u'\lambda_c}{S_L}$$

Assuming that $\lambda_c \approx \lambda$, we find

$$l_F \approx \frac{1.35 \times 0.00150}{0.38}$$

$$\approx 0.00533 \text{ m} = 5.33 \text{ mm}$$

which is considerably larger than the laminar flame thickness calculated in Example 2.8.

### 2.5.3 Laminar Diffusion Flames

When fuel and air enter a combustion system separately, they must mix on a molecular level before reaction can take place. The extent of reaction is strongly influenced by the extent to which that mixing has occurred prior to combustion. This mixing may be achieved solely by molecular diffusion, as in a candle flame, or may be enhanced by turbulence. We shall again begin our discussion with the laminar flame because of the simplicity it affords.

A laminar diffusion flame was illustrated in Figure 2.9(c). Fuel and air enter in separate streams, diffuse together, and react in a narrow region. While a single value of the equivalence ratio could be used to characterize a premixed flame, the equivalence ratio in the diffusion flame varies locally from zero for pure air to infinity for pure fuel. Combustion in a confined flow may be characterized by an overall equivalence ratio based on the flow rates of fuel and air, but that value may differ dramatically from the value in the flame region. A hydrocarbon diffusion flame may have two distinct zones: (1) the primary reaction zone, which is generally blue, and (2) a region of yellow luminosity. Most of the combustion reactions take place in the primary reaction zone where the characteristic blue emission results from the production of electronically excited molecules that spontaneously emit light, so-called chemiluminescence. Small particles composed primarily of carbon, known as *soot*, are formed in extremely fuel-rich (C/O ratio of order 1), hot regions of the flame and emit the brighter yellow radiation. The soot particles generally burn as they pass through the primary reaction zone, but may escape unburned if the flame is disturbed.

If the combustion reactions were infinitely fast, the combustion would take place entirely on a surface where the local equivalence ratio is equal to 1. This "thin flame sheet" approximation is the basis of an early model developed by Burke and Schumann (1928) and has been used in much more recent work [e.g., Mitchell and Sarofim (1975)]. Assuming that fuel and oxygen cannot coexist at any point greatly simplifies the calculations by replacing the chemical kinetics with stoichiometry or, at worst, chemical equilibrium calculations. The simplified calculations yield remarkably good results for adiabatic laminar diffusion flames larger than several millimeters in size since the reaction times at the adiabatic flame temperature near stoichiometric combustion are short compared to typical diffusion times. Only when heat is transferred from the flame at a high rate, as when the flame impinges on a cold surface, or when the scale of the flame is very small, as in the combustion of a small droplet, does the reaction time approach the diffusion time.

### 2.5.4 Turbulent Diffusion Flames

The small-scale structures of the turbulent flow fields in premixed and diffusion flames are similar. Many of the features of the flow in diffusion flames are made apparent by the distribution of composition in the flame. Large-scale eddies, shown in Figure 2.15 persist for long times in turbulent flows (Brown and Roshko, 1974). The development of a turbulent flow is controlled by such structures. Entrainment of one fluid stream into another takes place when fluid is engulfed between large coherent vortices.

Fuel and air are introduced separately into turbulent diffusion flames. Since the reactants must be mixed on a molecular scale to burn, this entrainment and the subsequent mixing control the combustion rate. As in the laminar diffusion flame, the gas composition in the flame is distributed continuously from pure fuel to pure air. The

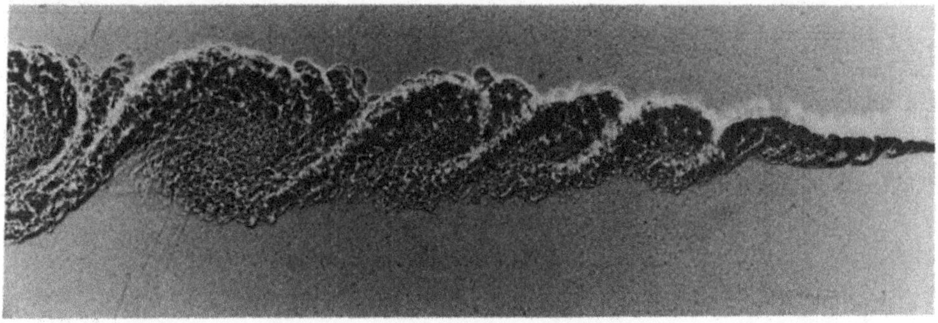

**Figure 2.15** Coherent structures in turbulent shear flow (Brown and Roshko, 1974). Reprinted by permission of Cambridge University Press.

structure of a turbulent diffusion flame that results when a fuel jet is released into air was illustrated in Figure 2.9(c). For some distance, the central core of the jet contains unreacted fuel. Combustion takes place at the interface between the fuel and air flows. The flame front is distorted by the turbulent motion but is, as in the laminar diffusion flame, relatively thin. Whether combustion will be complete depends on both the combustion kinetics and the mixing processes in the flame.

Simple jet flames are used in relatively few combustion systems because they are easily extinguished. A continuous ignition source must be supplied to achieve stable combustion. This is commonly accomplished by inducing flow recirculation, either with a bluff body or with a swirling flow, as illustrated in Figure 2.16. The low-pressure region in the near wake of the bluff body or in the center of the swirling flow causes a reverse flow bringing hot combustion products into the vicinity of the incoming fuel and air. Generally, only a small fraction of the combustion takes place within the recirculation zone. The remaining fuel burns as it mixes with air and hot products downstream of the recirculation zone. The flame in this downstream region may be a clearly defined jet that entrains gases from its surroundings, as in large industrial boilers, or may fill the entire combustor.

The extent of mixing in the flame can be characterized in terms of a *segregation* factor, originally proposed by Hawthorne et al. (1951). Arguing that in a high-temperature hydrogen-oxygen flame, hydrogen and oxygen would not be present together at any time, the time-average hydrogen and oxygen concentrations were used as a measure of the fraction of the fluid in the sample that is locally fuel-rich or fuel-lean.

Pompei and Heywood (1972) used similar arguments to infer the distribution of composition in a turbulent flow combustor burning a hydrocarbon fuel. Their combustor consisted of a refractory tube into which kerosene fuel was injected using an air-blast atomizer in which a small, high-velocity airflow disperses the fuel, as illustrated in Figure 2.17. Swirl, induced using stationary vanes, was used to stabilize the flame. The turbulence level in the combustor was controlled by the input of mechanical power introduced by the flow of high-pressure air used to atomize the fuel. Mixing in this ap-

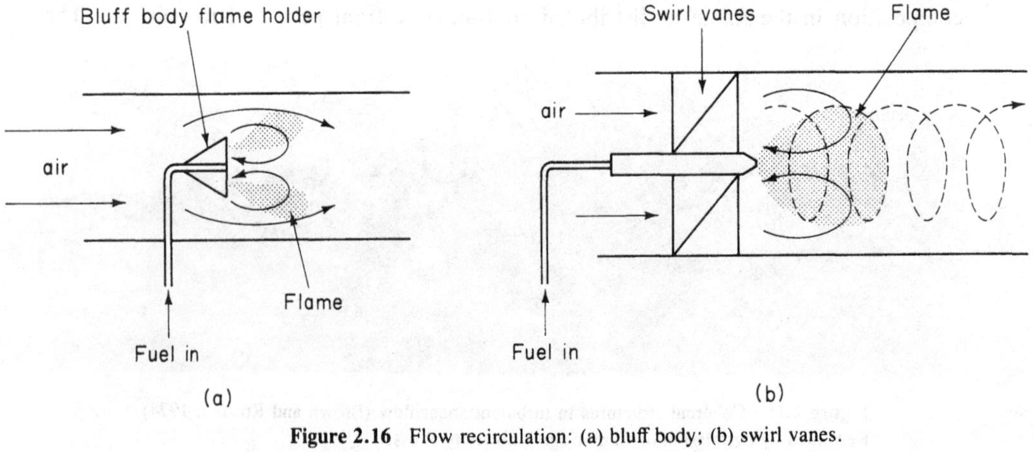


**Figure 2.16** Flow recirculation: (a) bluff body; (b) swirl vanes.

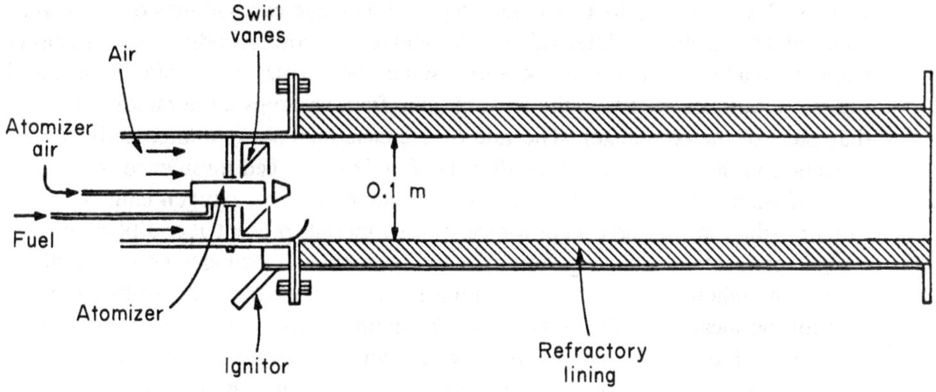


**Figure 2.17** Turbulent flow combustor used by Pompei and Heywood (1972).

paratus is readily characterized since the mean composition at any axial location is uniform over the entire combustor cross section. For this reason and because of the volume of pollutant formation data obtained with this system, we shall make extensive use of this system to illustrate the influence of turbulence on combustion and emissions.

Figure 2.18 shows the measured mean oxygen concentration for stoichiometric combustion as a function of position along the length of the combustor. Several profiles are shown, each corresponding to a different pressure for the atomizing air, which, as noted above, controls the initial turbulence level in the combustor. At a low atomizing pressure and correspondingly low turbulence level, the oxygen mole fraction decreases

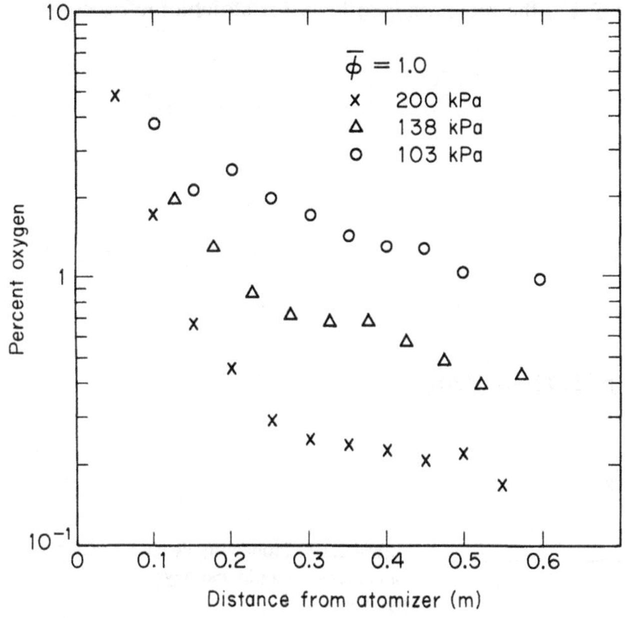


**Figure 2.18** Measured mean oxygen concentration for stoichiometric combustion as a function of length along the combustor (Pompei and Heywood, 1972). Reprinted by permission of The Combustion Institute.

to about 3% within the first two diameters and then decreases more slowly to an ultimate value of 1%. Even this final value is far above that corresponding to chemical equilibrium at the adiabatic flame temperature. When the atomizing pressure is increased, raising the turbulence intensity, the oxygen mole fraction drops more rapidly in the first two diameters of the combustor. The rate of decrease then slows dramatically, indicating a reduction in the turbulence level after the atomizer-induced turbulence is dissipated.

If we assume that combustion is instantaneous (i.e., oxygen cannot coexist with fuel or carbon monoxide except for the minor amounts present at equilibrium), the mean oxygen concentration during stoichiometric combustion provides us with a direct measure of the inhomogeneity or segregation in the combustor. A probability density function for the local equivalence ratio may be defined such that the fraction of the fluid at an axial position, $z$, in the combustor with equivalence ratio between $\phi$ and $\phi + d\phi$ is $p(\phi, z)\, d\phi$. If the number of moles of $O_2$ per unit mass is $w_{O_2}(\phi) = [O_2]_\phi/\rho$, the mean amount of oxygen in the combustor at $z$ is

$$\overline{w_{O_2}(z)} = \int_0^\infty w_{O_2}(\phi)\, p(\phi, z)\, d\phi \tag{2.67}$$

We have used moles per unit mass since mass is conserved in combustion but moles are not. The mean mole fraction of oxygen is

$$\overline{y_{O_2}(z)} = \overline{M} \int_0^\infty w_{O_2}(\phi)\, p(\phi, z)\, d\phi \tag{2.68}$$

where $\overline{M}$ is the mean molecular weight. Since the oxygen level decreases with increasing equivalence ratio and is insignificant (for present purposes) in the fuel-rich portions of the flame (as illustrated in Figure 2.19) the mean oxygen level for stoichiometric com-

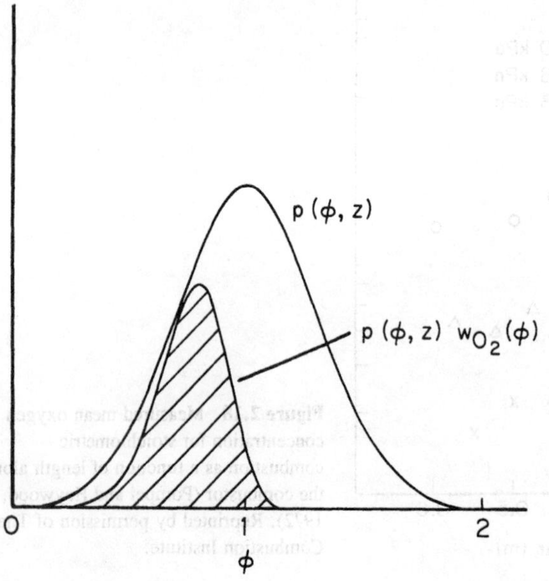


**Figure 2.19** Probability density function for equivalence ratio and oxygen distribution.

bustion gives a direct indicator of the breadth of the probability density function, $p(\phi, z)$.

The form of the probability density function, $p(\phi, z)$, is not known a priori. In order to examine the effects of composition fluctuations on emissions, Pompei and Heywood (1972) assumed that the distribution of local equivalence ratios would be Gaussian, that is,

$$p(\phi, z) = \frac{1}{\sqrt{2\pi}\sigma} \exp\left[-\frac{(\phi - \overline{\phi})^2}{2\sigma^2}\right] \tag{2.69}$$

where $\sigma(z)$ is the standard deviation of the distribution and $\overline{\phi}$ is the mean equivalence ratio. Since the mean equivalence ratio is controlled by the rates at which fuel and air are fed to the combustor, it is known. Only one parameter is required to fit the distribution to the data, namely $\sigma$. This fit is readily accomplished by calculating and plotting the mean oxygen concentration as a function of $\sigma$. The value of $\sigma$ at any position in the combustor is then determined by matching the observed oxygen level with that calculated using the assumed distribution function, shown in Figure 2.20 as mean concentration as a function of the segregration parameter $S$.

To describe the extent of mixing in nonstoichiometric combustion, Pompei and

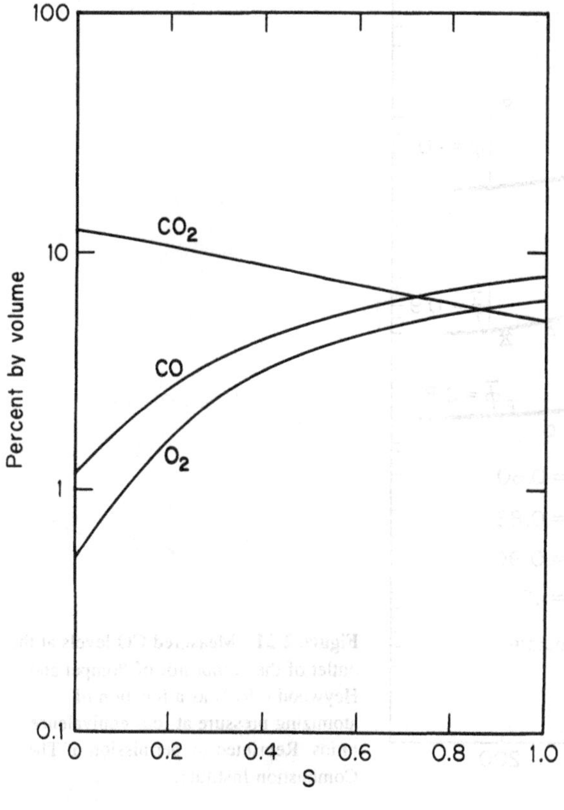


**Figure 2.20** Mean composition of products of stoichiometric combustion as a function of the segregation parameter $S$.

Heywood (1972) assumed that turbulent mixing in their combustor is not affected significantly by changes in the equivalence ratio as long as the flow rate is maintained constant. Under these conditions, the coefficient of variation of the composition probability density function, or segregation.

$$S \equiv \frac{\sigma}{\bar{\phi}} \tag{2.70}$$

determined from the stoichiometric combustion experiments can be applied to other mean equivalence ratios.

The fact that oxygen remains in the products of stoichiometric combustion means that carbon monoxide and other products of incomplete combustion must also be present. Figure 2.21 shows the measured CO levels at the outlet of the combustor as a function of atomizing pressure for several equivalence ratios (Pompei and Heywood, 1972). Assuming that chemical equilibrium is established locally, the mean CO concentration may be calculated using the values of $S$ inferred from the oxygen data, that is,

$$w_{CO} = \int_0^\infty \frac{[CO]_e}{\rho} p(\phi, S)\, d\phi \tag{2.71}$$

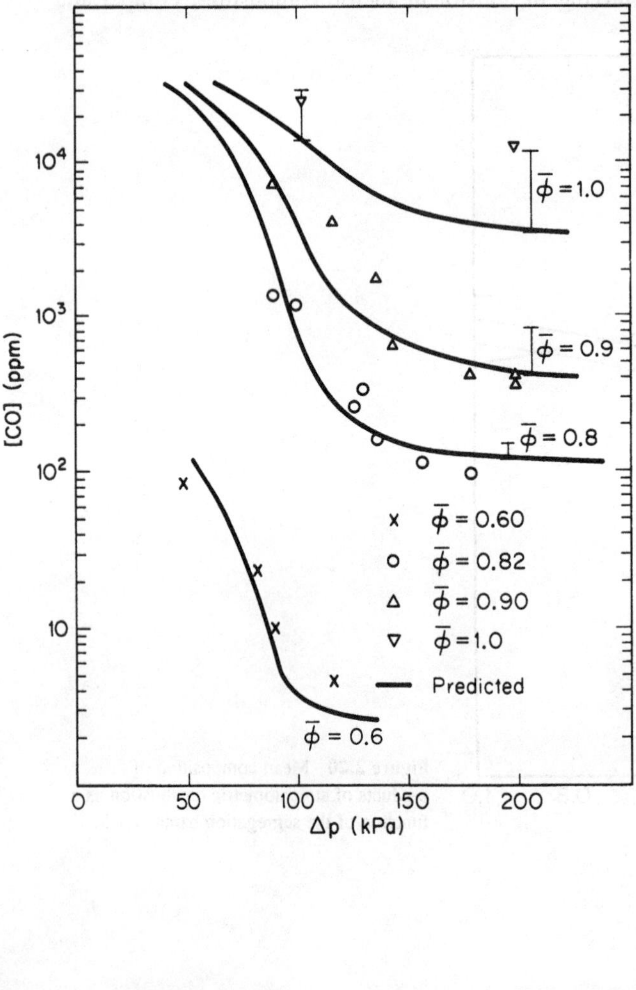


**Figure 2.21** Measured CO levels at the outlet of the combustor of Pompei and Heywood (1972) as a function of atomizing pressure at four equivalence ratios. Reprinted by permission of The Combustion Institute.

The results from these calculations are also shown in Figure 2.21. Heat losses to the combustor wall have been taken into account in computing the local equilibrium composition (Pompei and Heywood, 1972). At high atomizing pressures, the combustor approaches the well-mixed condition. The higher CO levels at low atomizing pressures clearly result from incomplete mixing. It is interesting to note that CO emissions from a single piece of combustion equipment can vary by more than two orders of magnitude at fixed equivalence ratio and total fuel and air flow rates due to relatively minor changes in the combustor operating parameters.

## 2.6 TURBULENT MIXING

We have seen that good mixing is required to achieve high combustion efficiency and corresponding low emission of partially oxidized products like CO. It would seem that, as in the laboratory studies, all combustors should be operated with the turbulence levels necessary to achieve good mixing. In this section we examine the extent to which this can be achieved in practical combustors.

Following Appendix D of Chapter 1 it can be shown that the rate of change in the concentration of a nonreactive tracer due to turbulent mixing can be described by

$$\frac{d\langle c^2 \rangle}{dt} = -\frac{\langle c^2 \rangle}{\tau_d(t)} \tag{2.72}$$

where the characteristic time for turbulent mixing is a function of the correlation scale for the composition fluctuations, $\lambda_c$ (D.26):

$$\tau_d = \frac{\lambda_c^2(t)}{12D} \tag{2.73}$$

The concentration microscale may vary with time due to variations in the turbulent dissipation rate, $\epsilon$. The variance of the composition of a nonreactive tracer becomes

$$\langle c^2(t) \rangle = \langle c^2(0) \rangle \exp\left[-\int_0^t \frac{dt'}{\tau_d(t')}\right] \tag{2.74}$$

A convenient tracer for characterization of mixing in a turbulent flame is total carbon per unit mass, that is, the sum of contributions of fuel, hydrocarbon intermediates, CO, and $CO_2$. This is directly related to the equivalence ratio, so $\langle c^2 \rangle$ is related to the variance, $\sigma^2$, in the equivalence ratio distribution. Thus $\tau_d$ is the characteristic time describing the approach of the equivalence ratio distribution to uniformity.

The mixing time can be related to turbulence quantities through application of (D.15) and (D.30):

$$\tau_d = \frac{\lambda_c^2}{12D} \approx \frac{\lambda^2}{6\nu} = A'\left(\frac{\epsilon}{L^2}\right)^{-1/3} \tag{2.75}$$

where $A'$ is an undetermined constant that is presumably of order unity. Here we see a major problem in achieving efficient mixing: the time scale decreases only as the one-third power of increasing turbulent energy.

To maintain a steady turbulence level in a burner, turbulent kinetic energy must be supplied to the system at a rate equal to the dissipation rate. The rate at which kinetic energy, $E_k$, is supplied to the system is the sum of the contributions of all the flows entering the system:

$$E_k = \sum_i (\rho_i u_i A_i) \frac{u_i^2}{2} \tag{2.76}$$

The air blast atomizer used by Pompei and Heywood (1972) uses a sonic velocity air jet to atomize a liquid fuel and to generate turbulence. For a choked (sonic) flow through the atomizer orifice, the air mass flow rate is directly proportional to the absolute pressure on the upstream side of the orifice. The total flow rate through the combustor was about 0.016 kg s$^{-1}$, with a maximum atomizer airflow rate of 0.0012 kg s$^{-1}$. The power input by the atomizer jet was

$$E_k \approx 0.0012 \text{ kg s}^{-1} \tfrac{1}{2} (330 \text{ m s}^{-1})^2 \approx 65 \text{ W}$$

Flame structure observations suggest that this energy was dissipated in the first two diameters of the 0.1-m-diameter combustor, which contained a mass of approximately

$$m \approx \frac{\pi}{4} (0.1 \text{ m})^2 (0.2 \text{ m})(0.18 \text{ kg m}^{-3})$$
$$\approx 0.00028 \text{ kg}$$

assuming a mean temperature of 2000 K. Thus the minimum mixing time is of order

$$\tau_d \approx \left( \frac{0.00028 \text{ kg } (0.1 \text{ m})^2}{65 \text{ W}} \right)^{1/3} \approx 0.0035 \text{ s}$$

The minimum atomizer airflow rate was about 0.00047 kg s$^{-1}$, leading to a power input of 25 W and a mixing time of about 0.0048 s. These times may be compared with the residence time in the mixing zone,

$$\tau_R = \frac{(\pi/4)(0.1 \text{ m})^2 (0.2 \text{ m})(0.18 \text{ kg m}^{-3})}{0.016 \text{ kg s}^{-1}} \approx 0.018 \text{ s}$$

Thus we see that the mixing time is comparable to the residence time. When the time scales are similar, small changes in the mixing time significantly affect the combustion efficiency.

What would happen if no effort were made to enhance the turbulence in the combustor? If we assume that the turbulence would correspond to that in a pipe, $u' \approx 0.1U$, the dissipation rate is

$$\epsilon \approx \frac{0.001 U^3}{L}$$

and the mixing time becomes

$$\tau_d \approx \left[\frac{L^3}{0.001U^3}\right]^{1/3} \approx \frac{0.1 \text{ m}}{0.001^{1/3}(11 \text{ m s}^{-1})} \approx 0.09 \text{ s}$$

Without the air-assist atomization the turbulence would not be sufficient to mix fuel and air, even within the 0.05-s residence time in the combustor. The slow mixing downstream of the atomizer influenced zone is indicative of this low dissipation rate.

Equation (2.75) provides a simple scaling criterion for geometrically similar burners (Corrsin, 1957). Consider the power required to maintain a constant mixing time when the size of burner is increased by a factor, $\kappa$. The integral scale of turbulence is proportional to the flow-system dimensions; hence

$$L' = \kappa L \quad \text{and} \quad m' = \kappa^3 m$$

$\tau' = \tau$ is achieved when

$$E_k' = \kappa^5 E_k$$

Thus we see that the power required to achieve constant mixing time increases as the fifth power of the burner size. The power per unit mass increases as the square of the scale factor. The rate at which kinetic energy can be supplied to a burner is limited, so mixing rates for large burners tend to be lower than for small burners.

For very large combustors, such as utility boilers, a number of relatively small burners, typically 1 m in diameter, are generally used instead of one larger burner to achieve good mixing. Air velocities through these burners are generally limited to about 30 m s$^{-1}$, leading to mixing times in the range 0.03 to 0.3 s, depending on the efficiency of conversion of the input kinetic energy (about 10 kW) to turbulence. These times are long compared with the laboratory experiments described above, but are short enough to assure good mixing within typical residence times of several seconds in large boilers. The initial combustion will, however, take place under poorly mixed conditions, a fact that strongly influences the formation of NO and other pollutants.

A typical utility boiler has 10 to 20 burners of this size. If they were combined into one large burner using the same air velocity, the mixing time would increase to 0.2 to 2 s, large enough that good mixing is unlikely.

## 2.7 COMBUSTION OF LIQUID FUELS

Liquid fuels are generally sprayed into a combustor as relatively fine droplets. Volatile fuels, such as kerosene, vaporize completely prior to combustion. Heavy fuel oils may partially vaporize, leaving behind a carbonaceous solid or liquid residue that then undergoes surface oxidation. The nature of the combustion process and pollutant emissions depends strongly on the behavior of the condensed-phase fuel during combustion.

The combustion of a fuel spray is governed by the size and volatility of the fuel droplets. Fuel droplets take a finite amount of time to vaporize, so not all of the fuel is immediately available for reaction. In order to vaporize, the droplet temperature must first be raised from the temperature at which it is introduced, $T_i$, to its vaporization temperature, $T_s$. The latent heat of vaporization, $\overline{L}$, must then be supplied. The energy required to vaporize a unit mass of fuel is

$$\overline{q} = \overline{c}_{pf}(T_s - T_i) + \overline{L} \tag{2.77}$$

where $\overline{c}_{pf}$ is the specific heat of the liquid. For a small particle of radius $a$ moving at a low velocity relative to the gas ($\text{Re} = 2\rho ua/\mu < 1$), the convective heat transfer rate to the particle is

$$Q = 4\pi a^2 k \left(\frac{dT}{dr}\right)_s \tag{2.78}$$

where $k$ is the thermal conductivity of the gas. Although it may be important in some flames, radiative heat transfer will be neglected in the present analysis. Once the droplet temperature has been raised to $T_s$, only the energy corresponding to the latent heat of vaporization, $\overline{L}$, must be supplied. An energy balance at the particle surface of the vaporizing droplet then yields

$$\overline{L}\overline{R}_v = 4\pi a^2 k \left(\frac{dT}{dr}\right)_s \tag{2.79}$$

where $\overline{R}_v$ is the rate of mass loss from the droplet by vaporization.

The vapor is transported away from the surface by convection and diffusion [see (B.3)]. Since only vapor leaves the surface, we may write

$$\overline{R}_v = 4\pi a^2 \rho u_s x_{vs} - 4\pi a^2 \rho D \left(\frac{dx_v}{dr}\right)_s \tag{2.80}$$

where $x_{vs}$ is the vapor mass fraction at the droplet surface. Noting that $4\pi a^2 \rho u_s = \overline{R}_v$, this becomes

$$\overline{R}_v(1 - x_{vs}) = -4\pi a^2 \rho D \left(\frac{dx_v}{dr}\right)_s \tag{2.81}$$

The mass, energy, and species conservation equations are (B.1), (B.25), and (B.5). The time required to establish the temperature and composition profiles around the evaporating droplet is generally short compared to that for the droplet to evaporate, so we may assume that the radial profiles of temperature and composition achieve a quasi-steady state. For the case of pure evaporation, the chemical reaction source term can also be eliminated. The conservation equations thus reduce to*

$$\frac{1}{r^2}\frac{d}{dr}(\rho u r^2) = 0 \tag{2.82}$$

$$\rho u \overline{c}_p \frac{dT}{dr} = \frac{1}{r^2}\frac{d}{dr}\left(r^2 k \frac{dT}{dr}\right) \tag{2.83}$$

*The transport properties $k$ and $\rho D$ are generally functions of temperature.

$$\rho u \frac{dx_v}{dr} = \frac{1}{r^2}\frac{d}{dr}\left(r^2 \rho D \frac{dx_v}{dr}\right) \tag{2.84}$$

From (2.82), we see that the velocity at any radial position can be related to that at the droplet surface, i.e.,

$$4\pi \rho u r^2 = 4\pi \rho u_s a^2 = \overline{R}_v \tag{2.85}$$

Substituting into the energy and species conservation equations yields

$$\overline{R}_v \overline{c}_p \frac{dT}{dr} = 4\pi \frac{d}{dr}\left(r^2 k \frac{dT}{dr}\right) \tag{2.86}$$

$$\overline{R}_v \frac{dx_v}{dr} = 4\pi \frac{d}{dr}\left(r^2 \rho D \frac{dx_v}{dr}\right) \tag{2.87}$$

Integrating, we find

$$\overline{R}_v \overline{c}_p (T + C_1) = 4\pi r^2 k \frac{dT}{dr}$$

$$\overline{R}_v (x_v + C_2) = 4\pi r^2 \rho D \frac{dx_v}{dr}$$

Applying the boundary conditions at the particle surface ($r = a$) yields

$$C_1 = \frac{4\pi a^2 k (dT/dr)_s}{\overline{R}_v \overline{c}_p} - T_s$$

$$= \frac{\overline{L}}{\overline{c}_p} - T_s$$

and

$$C_2 = \frac{4\pi a^2 \rho D (dx_v/dr)_s}{\overline{R}_v} - x_{vs}$$

$$= -(1 - x_{vs}) - x_{vs}$$

$$= -1$$

Thus we have

$$\overline{R}_v \overline{c}_p \left(T - T_s + \frac{\overline{L}}{\overline{c}_p}\right) = 4\pi r^2 k \frac{dT}{dr}$$

$$\overline{R}_v (x_v - 1) = 4\pi r^2 \rho D \frac{dx_v}{dr}$$

Integrating again, assuming $k$ and $\rho D$ are independent of $T$, and evaluating the constants in terms of the values as $r \to \infty$, we find

$$\frac{\bar{R}_v \bar{c}_p}{4\pi r k} = \ln \frac{T - T_s + \bar{L}/\bar{c}_p}{T_\infty - T_s + \bar{L}/\bar{c}_p}$$

$$\frac{\bar{R}_v}{4\pi r \rho D} = \ln \frac{x_v - 1}{x_{v\infty} - 1}$$

from which we obtain two expressions for the vaporization rate in terms of the values at the droplet surface:

$$\bar{R}_v = 4\pi a (k/\bar{c}_p) \ln \left[1 + \frac{\bar{c}_p(T_\infty - T_s)}{\bar{L}}\right] \tag{2.88}$$

$$\bar{R}_v = 4\pi a \rho D \ln \left(1 + \frac{x_{v\infty} - x_{vs}}{x_{vs} - 1}\right) \tag{2.89}$$

To evaluate $T_s$ and $x_{vs}$, we can equate these two rates:

$$\frac{k}{\bar{c}_p} \ln \left[1 + \frac{\bar{c}_p(T_\infty - T_s)}{\bar{L}}\right] = \rho D \ln \left(1 + \frac{x_{v\infty} - x_{vs}}{x_{vs} - 1}\right) \tag{2.90}$$

Using thermodynamic data to relate the equilibrium vapor mass fraction at the droplet surface to temperature, one iterates on $T$ until (2.90) is satisfied. The temperature dependence of the vapor pressure can be described with the Clapeyron equation, that is,

$$p_v(T) = [p_v(T_1) \exp (L/RT_1)] \exp (-L/RT) \tag{2.91}$$

The equilibrium vapor mass fraction is obtained from

$$x_{vs} = \frac{\rho_{vs}}{\rho} = \frac{c_{vs}}{c}\frac{M_v}{M} = \frac{p_{vs}}{p}\frac{M_v}{M} \tag{2.92}$$

Since the particle radius does not appear in (2.90), the surface temperature and vapor mass fraction at the surface are seen to be independent of droplet size.

Once the surface temperature is known, we may calculate the time required to vaporize the droplet. Define the transfer number

$$B_T = \frac{\bar{c}_p(T_\infty - T_s)}{\bar{L}} \tag{2.93}$$

The time rate of change of the droplet mass is

$$\frac{dm}{dt} = -4\pi a \frac{k}{\bar{c}_p} \ln (1 + B_T) \tag{2.94}$$

Since the droplet mass is $m = (4\pi/3)\rho_d a^3$, (2.94) may be integrated to find the droplet radius as a function of time:

$$a^2 - a_0^2 = 2(k/\bar{c}_p) \ln (1 + B_T) t \tag{2.95}$$

where $a_0$ is the initial radius. The time required for the droplet to vaporize completely is found by setting $a = 0$,

$$\tau_e = \frac{a_0^2 \bar{c}_p}{2k \ln (1 + B_T)} \tag{2.96}$$

This droplet lifetime is an important parameter in spray combustion. If $\tau_e$ approaches the residence time in the combustor, $\tau_R$, liquid fuel may be exhausted from the combustor, or vapors that are released late may not have time to mix with fuel-lean gases and burn completely. Thus a long droplet lifetime may lead to low combustion efficiencies and emissions of unreacted fuel vapor or products of incomplete combustion. Since the droplet lifetime varies as the square of its initial radius, it is imperative that the maximum droplet size be carefully limited.

As a droplet vaporizes in hot, fuel-lean gases, the fuel vapors may burn in a thin diffusion flame surrounding the droplet as illustrated in Figure 2.22. The approach outlined above can be applied to determine the rate of droplet burning by considering the rate of diffusion of oxygen (or other oxidant) toward the droplet and the energy release by the combustion reactions. In this analysis it is common to make the thin flame approximation, assuming that fuel vapor diffusing from the droplet is instantaneously consumed by the counter diffusing oxygen at the flame front. The heat release due to combustion substantially increases the gas temperature at this surface. This increases the rate of heat transfer to the droplet surface and, therefore, accelerates the evaporation. The rates of transport of energy, fuel vapor, and oxygen must be balanced at the flame front.

The steady-state conservation equations may be written

$$\bar{R}_v \bar{c}_p \frac{dT}{dr} = \frac{d}{dr}\left(4\pi r^2 k \frac{dT}{dr}\right) + 4\pi r^2 q \tag{2.97}$$

$$\bar{R}_v \frac{dx_v}{dr} = \frac{d}{dr}\left(4\pi r^2 \rho D_v \frac{dx_v}{dr}\right) + 4\pi r^2 r_v \tag{2.98}$$

$$\bar{R}_v \frac{dx_o}{dr} = \frac{d}{dr}\left(4\pi r^2 \rho D_o \frac{dx_o}{dr}\right) + 4\pi r^2 r_o \tag{2.99}$$

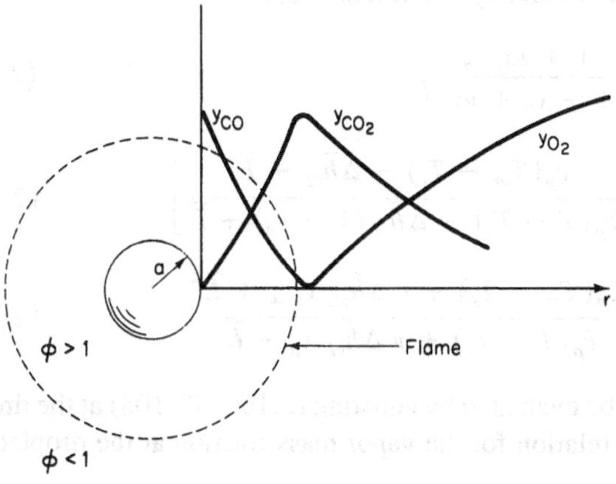


**Figure 2.22** Vaporization of a fuel droplet and associated diffusion flame.

where $-r_v$ and $-r_o$ denote the rates of consumption of fuel vapor and oxidant by gas-phase reaction, respectively, and $q$ is the heat release due to the combustion reactions. The boundary conditions at $r = a$ are

$$4\pi a^2 k\left(\frac{dT}{dr}\right)_s = \bar{R}_v \bar{L}$$

$$4\pi a^2 \rho D_v\left(\frac{dx_v}{dr}\right)_s = -\bar{R}_v(1 - x_{vs}) \tag{2.100}$$

$$4\pi a^2 \rho D_o\left(\frac{dx_o}{dr}\right)_s = 0, \qquad x_{os} = 0$$

From combustion stoichiometry we may write

$$r_v = \nu r_o \tag{2.101}$$

where $\nu$ is the mass-based stoichiometric coefficient for complete combustion. The heat release, $q$, due to the combustion reactions is

$$q = -\Delta \bar{h}_{cL}\, r_v = -\nu\, \Delta \bar{h}_{cL}\, r_o \tag{2.102}$$

If $D_o = D_v = k/\rho \bar{c}_p$ (equal molecular and thermal diffusivities), (2.97)–(2.99) can be combined to eliminate the reaction rate terms, i.e.,

$$\bar{R}_v \frac{d}{dr}(x_v - \nu x_o) = \frac{d}{dr}\left(4\pi r^2 \rho D \frac{d}{dr}(x_v - \nu x_o)\right) \tag{2.103}$$

$$\bar{R}_v \frac{d}{dr}(\bar{c}_p T + \Delta \bar{h}_{cL}\, x_v) = \frac{d}{dr}\left(4\pi r^2 \rho D \frac{d}{dr}(\bar{c}_p T + \Delta \bar{h}_{cL}\, x_v)\right) \tag{2.104}$$

$$\bar{R}_v \frac{d}{dr}(\bar{c}_p T + \nu\, \Delta \bar{h}_{cL}\, x_o) = \frac{d}{dr}\left(4\pi r^2 \rho D \frac{d}{dr}(\bar{c}_p T + \nu\, \Delta \bar{h}_{cL}\, x_o)\right) \tag{2.105}$$

Integrating twice and imposing the boundary conditions yield

$$\frac{\bar{R}_v}{4\pi r \rho D} = \ln\left(\frac{1 + \nu x_{o,\infty}}{1 - x_v + \nu x_o}\right) \tag{2.106}$$

$$\frac{\bar{R}_v}{4\pi r \rho D} = \ln\left[\frac{\bar{c}_p(T_\infty - T_s) - \Delta \bar{h}_{cL} + \bar{L}}{\bar{c}_p(T - T_s) - \Delta \bar{h}_{cL}(1 - x_v) + \bar{L}}\right] \tag{2.107}$$

$$\frac{\bar{R}_v}{4\pi r \rho D} = \ln\left[\frac{\bar{c}_p(T_\infty - T_s) + \nu\, \Delta \bar{h}_{cL}\, x_{o,\infty} + \bar{L}}{\bar{c}_p(T - T_s) + \nu\, \Delta \bar{h}_{cL}\, x_o + \bar{L}}\right] \tag{2.108}$$

Finally, the evaporation rate may be evaluated by equating (2.106)–(2.108) at the droplet surface and using an equilibrium relation for the vapor mass fraction at the droplet surface. In terms of the surface conditions, the vaporization rate becomes

$$\overline{R}_v = 4\pi a\rho D \ln\left[1 + \frac{x_{vs} + \nu x_{o,\infty}}{1 - x_{vs}}\right] \tag{2.109}$$

$$\overline{R}_v = 4\pi a\rho D \ln\left[1 + \frac{\overline{c}_p(T_\infty - T_s) - \Delta\overline{h}_{cL}\, x_{vs}}{\overline{L} - \Delta\overline{h}_{cL}(1 - x_{vs})}\right] \tag{2.110}$$

$$\overline{R}_v = 4\pi a\rho D \ln\left[1 + \frac{\overline{c}_p(T_\infty - T_s) + \nu\, \Delta\overline{h}_{cL}\, x_{o,\infty}}{\overline{L}}\right] \tag{2.111}$$

Since we have assumed equal diffusivities and unit Lewis number ($\text{Le} = k/\rho\overline{c}_p D$), the surface conditions in steady-state evaporation are obtained from

$$\frac{x_{vs} + x_{o,\infty}}{1 - x_{vs}} = \frac{\overline{c}_p(T_\infty - T_s) - \Delta\overline{h}_{cL}\, x_{vs}}{\overline{L} - \Delta\overline{h}_{cL}(1 - x_{vs})}$$

$$= \frac{\overline{c}_p(T_\infty - T_s) + \nu\, \Delta\overline{h}_{cL}\, x_{o,\infty}}{\overline{L}} \tag{2.112}$$

Equation (2.111) quickly yields a reasonable estimate of the burn time since the sensible enthalpy term is generally small,

$$\overline{c}_p(T_\infty - T_s) << \left|\Delta\overline{h}_{cL}\, x_{vs}\right|$$

The vaporization rate can thus be approximated by

$$\overline{R}_v \approx 4\pi a\rho D \ln\left[1 + \frac{\nu\, \Delta\overline{h}_{cL}\, x_{o,\infty}}{\overline{L}}\right] \tag{2.113}$$

Once the burning rate is known, the position of the flame front can readily be calculated from (2.106) by setting $x_v = x_o = 0$. We find

$$r_{\text{flame}} = \frac{\overline{R}_v}{4\pi\rho D \ln(1 + \nu x_{o,\infty})} \tag{2.114}$$

For small oxygen concentrations the flame radius rapidly becomes large. In dense fuel sprays, the oxygen is quickly consumed. The predicted flame radius may then exceed the mean droplet separation, $\delta$, as illustrated in Figure 2.23. Although some reaction may take place in the fuel-rich regions between particles in such a dense spray, the flame front will ultimately surround the cloud of droplets (Labowsky and Rosner, 1978). The cloud of droplets then acts as a distributed source of fuel vapor.

The way a fuel is atomized can profoundly influence the combustion process through (1) the droplet lifetime, which varies as the square of the droplet radius; (2) the uniformity of the spatial distribution of droplets; and (3) the generation of turbulence due to the kinetic energy delivered to the flow by the spray. The drop size is determined by a balance between the fluid mechanical forces, which tend to pull the liquid apart, and the surface tension, $\sigma$, which tends to hold it together. The classical model for drop

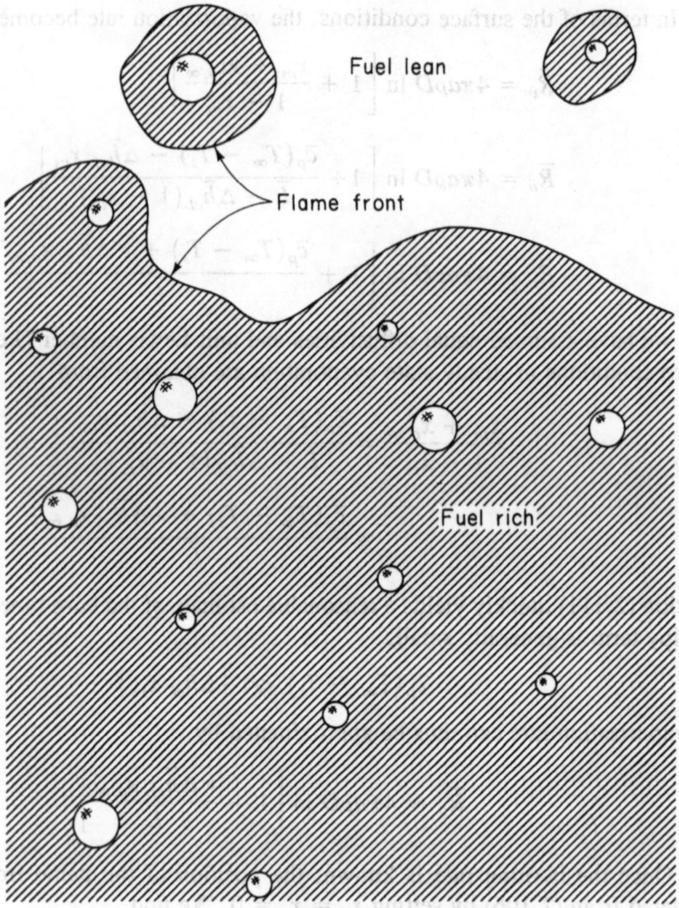


**Figure 2.23** Droplet combustion in a dense spray.

stability was developed by Prandtl (1949). An estimate of the maximum stable diameter of droplets is obtained by equating the dynamic pressure of the airflow past the drop and the surface tension force,

$$d_{\max} \approx C\sigma/\rho v^2 \tag{2.115}$$

where $v$ is the relative velocity between the liquid and the gas, $\rho$ is the gas density, and the proportionality constant, $C$, is equal to 15.4. A high relative velocity will produce the smallest droplets.

The motion of the liquid relative to the gas can be created by forcing the liquid through an orifice with high pressure (pressure atomization) or by using a high-velocity gas flow (air-assist atomization). These two types of atomizers are illustrated in Figure 2.24. The more common pressure atomizer is somewhat limited in the maximum velocity due to pressure constraints and the fouling of very small orifices by contaminant

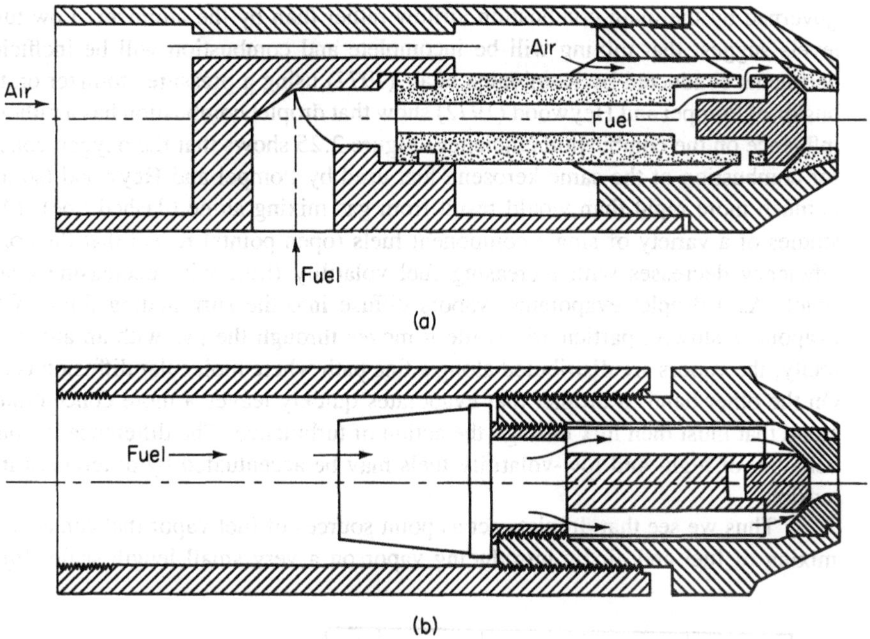


**Figure 2.24** Liquid fuel atomizers: (a) air-assist atomizer; (b) pressure atomizer. Komiyama et al. (1977). Reprinted by permission of The Combustion Institute.

particles in the fuel. The fluid velocity through the orifice of a pressure atomizer is

$$v = (2\,\Delta p\, C_D/\rho_f)^{1/2} \tag{2.116}$$

where $\Delta p$ is the pressure drop across the orifice and $C_D$ is the discharge coefficient (typically, $0.6 < C_D < 1$). High atomizer pressures, say 50 atm ($5 \times 10^6$ Pa), can result in velocities as high as 100 m s$^{-1}$. Typical surface tensions for hydrocarbon liquids are 0.2 to 0.8 N m$^{-1}$. As the liquid stream breaks up, it quickly decelerates, so the maximum droplet size can be much larger than the approximately 40 $\mu$m that this velocity would suggest. In diesel engines, where the fuel must be injected very quickly, much higher atomization pressures are generally used, and smaller droplets may be generated.

In the air-assist atomizer, a high-velocity gas flow is used to atomize the liquid and disperse the droplets. The airflow velocity can be as high as the local speed of sound, about 330 m s$^{-1}$ in ambient-temperature air. Drop formation takes place within the atomizer, where the velocity is high, so small drops can be generated [i.e., (2.115) suggests an upper bound on the droplet size of about 4 $\mu$m].

The air-assist atomizer also introduces a large amount of kinetic energy that is ultimately dissipated through turbulence. The pressure atomizer does not have this impact due to the lower velocity and the lower mass flow entering through the atomizer. Thus, when the pressure atomizer is used, the turbulence levels in the flame region are governed by the main combustion airflow rather than by the atomizer. Low turbulence levels suggest that mixing will be incomplete and combustion will be inefficient. Experiments conducted by Komiyama et al. (1977) using a pressure atomizer on the combustor of Pompei and Heywood (1972) show that droplet combustion has a rather striking influence on the combustion efficiency. Figure 2.25 shows that the oxygen consumption for combustion of the same kerosene fuel used by Pompei and Heywood (solid points) is much more rapid than would result from gas mixing alone (dashed line). Moreover, studies of a variety of single component fuels (open points) reveal that the combustion efficiency decreases with increasing fuel volatility (i.e., with decreasing evaporation time). As a droplet evaporates, vapors diffuse into the surrounding fluid. If the drop evaporates slowly, particularly while it moves through the gas with an appreciable velocity, the vapors are distributed along a fine path where molecular diffusion is effective. On the other hand, a droplet that evaporates quickly leaves a more concentrated vapor cloud that must then mix through the action of turbulence. The difference in combustion behavior of high- and low-volatility fuels may be accentuated by differences in surface tension.

Thus we see that droplets act as point sources of fuel vapor that can accelerate the mixing of fuel and air by introducing vapor on a very small length scale. Injection at

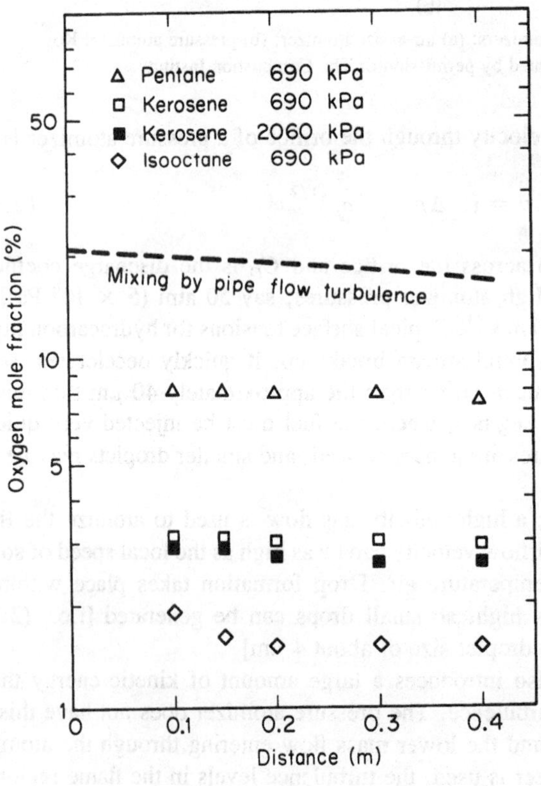


**Figure 2.25** Oxygen concentrations for combustion of four fuels as a function of distance along the combustor length. Komiyama et al. (1977). Reprinted by permission of The Combustion Institute.

high velocities can distribute the fuel throughout the combustion gases. As droplets enter the combustor at high velocity, their Reynolds numbers may be large enough that convective transport enhances drop evaporation. The correlation proposed by Ranz and Marshall (1952) can be used to take forced convection into account, that is,

$$\overline{R}_v = \overline{R}_{v0}(1 + 0.36\ \mathrm{Re}^{1/2}\ \mathrm{Sc}^{1/3}) \tag{2.117}$$

where $\overline{R}_{v0}$ is the vaporization rate for purely diffusive vaporization, $\mathrm{Re} = \rho v d/\mu$ is the droplet Reynolds number, and $\mathrm{Sc} = \mu/\rho D$ is the Schmidt number of the gas.

We have examined the evaporation and combustion of a single-component fuel. Practical fuels are complex mixtures of hydrocarbons. As droplets are heated, the more volatile components evaporate more rapidly than the less volatile components, so the fuel volatility gradually decreases. Diffusional resistance within the droplet can become a significant hindrance to vaporization (Hanson et al., 1982; Law and Law, 1981). Components that vaporize slowly may be heated to high temperatures, possibly leading to the formation of solid carbonaceous particles of coke. These solid particles can be very difficult to burn and are often emitted with the exhaust gases.

## 2.8 COMBUSTION OF SOLID FUELS

Solid fuels are burned in a variety of systems, some of which are similar to those fired by liquid fuels. In large industrial furnaces, particularly boilers for electric power generation, coal is pulverized to a fine powder (typically, 50 $\mu$m mass mean diameter and 95% smaller by mass than about 200 $\mu$m) which is sprayed into the combustion chamber and burned in suspension as illustrated in Figure 2.26. The combustion in the pulverized coal system has many similarities to the combustion of heavy fuel oils. Smaller systems generally utilize fixed- or fluidized-bed combustors that burn larger particles. The latter technologies are also applied to the combustion of wood, refuse, and other solid fuels. Air is fed into a fluidized bed at a sufficiently high velocity to levitate the particles, producing a dense suspension that appears fluidlike. Heat transfer in the bed must be high enough and heat release rates low enough to keep the bed relatively cool and prevent the ash particles from fusing together to form large ash agglomerates known as clinkers. Noncombustible solids are often used to dilute the fuel and keep the temperature low; most commonly, limestone is used in order to retain the sulfur in the bed at the same time. In contrast to the rapid mixing in a fluidized bed, only a fraction of the air comes in contact with the fuel in a fixed bed, or stoker, combustion system, with the remainder being introduced above the bed of burning fuel. Large amounts of excess air are required to achieve reasonable combustion efficiency, and even with the large airflows, hydrocarbon and carbon monoxide emissions can be quite high, due to poor mixing above the bed. The increased air requirements lower the thermal efficiency of stoker units, so pulverized coal or fluidized-bed combustion is favored for large systems. Most large systems currently in use burn pulverized coal. We shall, for this reason, focus on these systems.

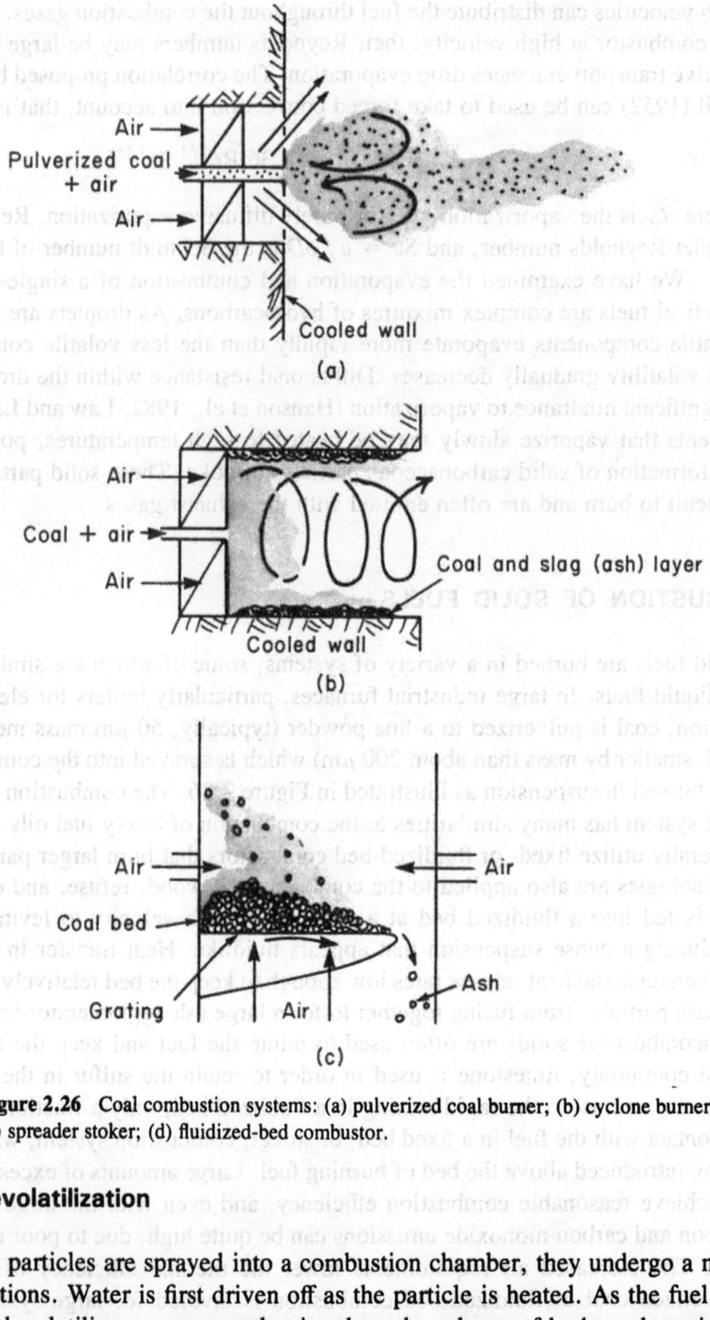


**Figure 2.26** Coal combustion systems: (a) pulverized coal burner; (b) cyclone burner; (c) spreader stoker; (d) fluidized-bed combustor.

## 2.8.1 Devolatilization

When coal particles are sprayed into a combustion chamber, they undergo a number of transformations. Water is first driven off as the particle is heated. As the fuel is heated further, it devolatilizes, a process that involves the release of hydrocarbons in the coal and the cracking of the molecular structure of the coal. This complex chemical process has received considerable attention (Gavalas, 1982), but we shall examine here only one

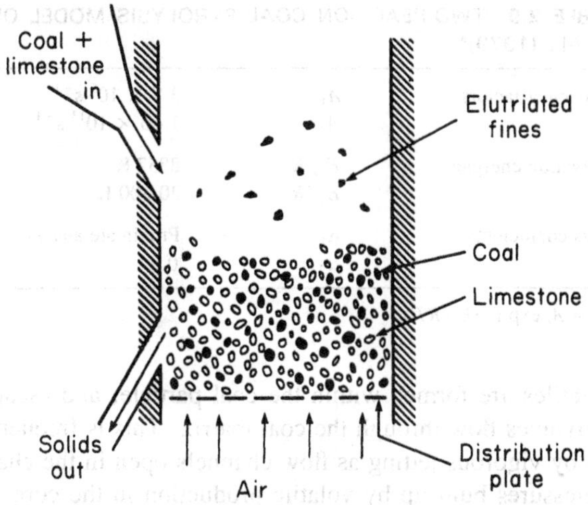


(d)

**Figure 2.26** *(Continued)*

of the simpler models. The devolatilization has been modeled as competing chemical reactions (Kobayashi et al., 1977), that is,

$$\text{coal} \begin{cases} \xrightarrow{k_1} (1 - \alpha_1) R_1 + \alpha_1 V_1 \\ \xrightarrow{k_2} (1 - \alpha_2) R_2 + \alpha_2 V_2 \end{cases}$$

Each of the two reactions produces volatile matter ($V_i$) and residual char ($R_i$), which does not undergo additional pyrolysis. The fraction of the mass of coal undergoing reaction $i$ that is released as volatile matter is $\alpha_i$. Reaction 1 is assumed to be a low-temperature reaction that produces an asymptotic volatile yield $\alpha_1$. Reaction 2 is assumed to have a higher activation energy and therefore contributes significantly only at high temperature. Rapid heating brings reaction 2 into play, while substantial coal remains unreacted, leading to higher volatile yields than at low temperature.

The parameters in this simplified model must be empirically determined. $\alpha_1$ is generally chosen to equal the volatile yield measured in the proximate analysis, a low-temperature pyrolysis test. The remaining parameters of this model, as estimated by Stickler et al. (1979), are summarized in Table 2.9. With this simple model, the release of volatile matter and the quantity of char residue can be estimated. The fractional conversion of the char is described by the rate equation

$$\frac{dx_{\text{coal}}}{dt} = -k_1 x_{\text{coal}} - k_2 x_{\text{coal}} \tag{2.118}$$

where $x_{\text{coal}}$ is the mass fraction of the coal that has not undergone reaction.

**TABLE 2.9** TWO-REACTION COAL PYROLYSIS MODEL OF STICKLER ET AL. (1979)[a]

| | | |
|---|---|---|
| Frequency factors | $A_1$ | $3.7 \times 10^5\ s^{-1}$ |
| | $A_2$ | $1.46 \times 10^{13}\ s^{-1}$ |
| Activation energies | $E_1/R$ | 8857 K |
| | $E_2/R$ | 30,200 K |
| Mass coefficients | $\alpha_1$ | Proximate analysis volatile matter |
| | $\alpha_2$ | 0.8 |

[a] $k_i = A_i \exp(-E_i/RT)$.

The volatiles are formed within the coal particle, and escape to the surrounding atmosphere involves flow through the coal matrix. This is frequently a violent process, characterized by vigorous jetting as flow channels open in the char to allow the release of the high pressures built up by volatile production in the core of the particle. These complications preclude the application of the drop combustion models derived in the preceding section to the combustion of coal volatiles. Nonetheless, the volatiles play a very important role in coal combustion, particularly in ignition and stabilization of coal flames. Knowledge of volatile release is also essential to specifying the initial condition for the next phase of coal combustion, the surface oxidation of char residue.

**Example 2.10** ***Coal Devolatilization***

A dry coal particle intially 50 $\mu$m in diameter is suddenly heated to 2000 K in air. The proximate analysis of the coal is

| | |
|---|---|
| Volatile matter | 43.69% |
| Fixed carbon | 46.38% |
| Ash | 9.94% |

The carbon and ash densities are 1.3 and 2.3 g $cm^{-3}$, respectively. Using the Kobayashi model and assuming that the particle temperature remains constant throughout devolatilization, estimate the mass of char remaining after devolatilization.

Consider first devolatilization. Equation (2.118) expresses a relation for the fraction of unreacted coal as a function of time. Integrating (2.118) and noting that $x_{coal} = 1$ at $t = 0$, we find

$$x_{coal} = e^{-t/\tau_D}$$

where the characteristic time for devolatilization is

$$\tau_D = (k_1 + k_2)^{-1}$$

The fraction of the coal that is converted to char by reaction $i$ is $1 - \alpha_i$, so

$$\frac{dx_{char}}{dt} = (1 - \alpha_1)\, k_1 x_{coal} + (1 - \alpha_2)\, k_2 x_{coal}$$

$$= \left[(1 - \alpha_1)\, k_1 + (1 - \alpha_2)\, k_2\right] e^{-t/\tau_D}$$

Integrating and letting $x_{\text{char}} = 0$ at $t = 0$ yields

$$x_{\text{char}} = \frac{(1 - \alpha_1) k_1 + (1 - \alpha_2) k_2}{k_1 + k_2} (1 - e^{-t/\tau_D})$$

The limit as $t \to \infty$ is

$$x_{\text{char}}(t \longrightarrow \infty) = \frac{(1 - \alpha_1) k_1 + (1 - \alpha_2) k_2}{k_1 + k_2}$$

From Table 2.9,

$$k_i = A_i e^{-E_i/RT}$$

$$\alpha_1 = \frac{0.4369}{1 - 0.0994} = 0.485 \qquad \alpha_2 = 0.8$$

$$A_1 = 3.7 \times 10^5 \text{ s}^{-1} \qquad A_2 = 1.46 \times 10^{13} \text{ s}^{-1}$$

$$E_1/R = 8857 \text{ K} \qquad E_2/R = 30{,}200 \text{ K}$$

Evaluating at 2000 K, we find that

$$k_1 = 4400 \text{ s}^{-1} \qquad k_2 = 4.06 \times 10^6 \text{ s}^{-1}$$

Thus

$$\tau_D = 2.46 \times 10^{-7} \text{ s} = 0.246\ \mu\text{s}$$

and

$$x_{\text{char}} = \frac{(1 - 0.485)\, 4.4 \times 10^3 + (1 - 0.8) \times 4.06 \times 10^6}{4.4 \times 10^3 + 4.06 \times 10^6} = 0.200$$

Only 20% of the original carbonaceous material remains in the char.

### 2.8.2 Char Oxidation

The combustion of coal char or other entrained carbonaceous particles (such as coke produced in combustion of heavy fuel oils) is governed by the diffusion of oxidizer ($O_2$, OH, O, $CO_2$, $H_2O$, etc.) to the carbon surface and by surface reaction kinetics. Coal char is highly porous and presents a surface area for oxidation that is much larger than the external surface. Mulcahy and Smith (1969) identified several modes of char combustion: regime 1, in which the rate is controlled strictly by surface reactions, allowing the reaction to take place uniformly throughout the char volume; regime 2, in which pore diffusion and surface reactions both influence the rate; and regime 3, in which external diffusion controls the oxidation rate. Pulverized coal combustion generally falls in regime 2.

We begin our discussion of char oxidation with an examination of the role of the external diffusional resistance in the char oxidation kinetics. For this purpose we may use detailed models of the intrinsic surface reaction kinetics in combination with a model of the porous structure of the char and the diffusional resistance within these pores. Alternatively, we may use a global rate expression that describes the total rate of reaction in terms of the apparent external surface area. We shall use the latter approach at present and examine the processes taking place inside the char particle later.

The rate of char oxidation is the sum of several reactions that convert carbon to CO, primarily

$$C_{(s)} + \tfrac{1}{2} O_2 \xrightarrow{k_1} CO$$

$$C_{(s)} + CO_2 \xrightarrow{k_2} 2CO$$

The apparent rates of these reactions may be expressed in the form

$$\bar{r}_i = A_i e^{E_i/RT} p_i^n \qquad (\text{kg C m}^{-2}\ \text{s}^{-1}) \tag{2.119}$$

where $p_i$ is the partial pressure of the oxidizer. Numerous measurements of the rate parameters for various chars have been reported. Table 2.10 presents selected rate coefficients.

**TABLE 2.10 CHAR OXIDATION RATE PARAMETERS**[a]

| Parent coal | $A$ | $E/R$ (K) | $n$ |
|---|---|---|---|
| Petroleum coke | 70 | 9,910 | 0.5 |
| Pittsburgh seam (swelling bituminous coal, USA) | 41,870 | 17,100 | 0.17 |
| Illinois No. 6 (swelling bituminous coal, USA) | 63,370 | 17,200 | 0.17 |
| Brodsworth (swelling bituminous coal, UK) | 1,113 | 121,300 | 1.0 |
| East Hetton (swelling bituminous coal, UK) | 6,358 | 17,100 | 1.0 |
| Anthracite and semianthracite (UK and Western Europe) | 204 | 9,560 | 1.0 |
| Millmerran (nonswelling subbituminous coal, Australia) | 156 | 8,810 | 0.5 |
| Ferrymoor (nonswelling subbituminous coal, UK) | 703 | 10,820 | 1.0 |
| Whitwick (nonswelling bituminous coal, UK) | 504 | 8,910 | 1.0 |
| Yallourn brown coal (Australia) | 93 | 8,150 | 0.5 |

[a] $\bar{r} = Ae^{-E/RT} p_{O_2}^n$ kg m$^{-2}$ s$^{-1}$, $p$ in atm.

*Source:* Data from Smith (1982).

The char oxidation rate is the net result of the rate of oxidizer diffusion to the char surface and the rate of surface reaction. Since the activation energies of the oxidation reactions are large, the particle temperature is very important in determining the rate of oxidation. Because of the high temperatures reached and the large emissivity of carbon, radiation can be a major mechanism for heat transfer from the burning char particle. If the temperature of the surrounding surfaces is $T_w$, the radiative flux to the particle is

$$q_r = \sigma\epsilon(T_w^4 - T_s^4) \qquad \text{W m}^{-2} \tag{2.120}$$

where $\sigma = 5.67 \times 10^{-8}$ W m$^{-2}$ K$^{-4}$ is the Stefan-Boltzmann constant, and $\epsilon$ is the particle emissivity. Conduction heat transfer

$$q_c = -k\left(\frac{\partial T}{\partial r}\right)_s \tag{2.121}$$

is also very important for small particles. Large particles encountered in fluidized-bed combustors and small particles injected at high velocities may have large enough Reynolds numbers (Re$= \rho v d/\mu$) that convection must be taken into account, and the rate of heat transfer is expressed in terms of a heat transfer coefficient, $h$,

$$q_c = h(T_\infty - T_s) \tag{2.122}$$

where $h$ is generally obtained from correlations of the Nusselt number (Nu $= hd/k$), for example (Bird et al., 1960),

$$\text{Nu} = 2 + 0.60\ \text{Re}^{1/2}\ \text{Pr}^{1/3} \tag{2.123}$$

where the Prandtl number is defined by Pr $= c_p\mu/k$. Species transport to and from the particle surface also influences the energy balance. The net enthalpy flux to the particle due to species transport is given by

$$q_s = \sum_j \bar{f}_j \bar{h}_j^\circ(T_s) \tag{2.124}$$

where $\bar{h}_j^\circ = [(\bar{h}(T_s) - \bar{h}(T_0) + \Delta\bar{h}_f^\circ(T_0)]_j$ is the total enthalpy of species $j$ at the temperature of the particle surface and $\bar{f}_j$ is the species mass flux toward the particle surface.

The net species transport from the particle surface is directly related to the rate of reaction:

$$\bar{R} = 4\pi a^2(\bar{r}_1 + \bar{r}_2) = 4\pi a^2 \bar{f}_{CO,s}/\nu_{CO} = -4\pi a^2 \bar{f}_{O_2,s}/\nu_{O_2} - 4\pi a^2 \bar{f}_{CO_2,s}/\nu_{CO_2} \tag{2.125}$$

where $\nu_j$ denotes the mass-based stoichiometric coefficients for reactions 1 and 2, respectively:

$$\nu_{CO} = \frac{12 + 16}{12} = 2.33$$

$$\nu_{O_2} = \frac{32}{2 \times 12} = 1.33$$

$$\nu_{CO_2} = \frac{44}{12} = 3.67$$

Combining (2.120), (2.121), and (2.124), the time rate of change of the particle energy becomes

$$\frac{dU}{dt} = 4\pi a^2 \{-\nu_{CO,1}\bar{r}_1\bar{h}^\circ_{CO}(T_s) + \nu_{O_2}\bar{r}_1\bar{h}^\circ_{O_2}(T_s) - \nu_{CO,2}\bar{r}_2\bar{h}^\circ_{CO}(T_s) + \nu_{CO_2}\bar{r}_2\bar{h}^\circ_{CO_2}(T_s)\} \tag{2.126}$$

The rate of change of the particle energy is

$$\frac{dU}{dt} = \frac{d}{dt}(m\bar{u}^\circ) = \bar{u}^\circ \frac{dm}{dt} + m\frac{d\bar{u}^\circ}{dt}$$

Assuming quasi-steady combustion ($d\bar{u}^\circ/dt \approx 0$) and noting that, for solid carbon, $\bar{h}^\circ_C \approx \bar{u}^\circ_C$, we find

$$\frac{dU}{dt} \approx \bar{h}^\circ_C(\bar{r}_1 + \bar{r}_2)$$

Combining these terms, the quasi-steady energy balance on the surface of the burning particle becomes

$$\sigma\epsilon(T_v^4 - T_s^4) - k\left(\frac{dT}{dr}\right)_s + \bar{r}_1\,\Delta\bar{h}_{r1}(T_s) + \bar{r}_2\,\Delta\bar{h}_{r2}(T_s) = 0 \tag{2.127}$$

Uniform particle temperature has been assumed in this analysis.

To evaluate the reaction rates, $\bar{r}_1$ and $\bar{r}_2$, we need to know both the temperature and the concentrations of the oxidizing species at the particle surface. The species fluxes at the particle surface are obtained from the condition

$$\bar{f}_{js} = x_{js}\sum_k \bar{f}_{ks} - 4\pi\rho D_j\left(\frac{dx_j}{dz}\right)_s \tag{2.128}$$

The net mass flux away from the particle surface equals the rate of carbon consumption

$$\sum_k \bar{f}_{ks} = \bar{r}_1 + \bar{r}_2 \tag{2.129}$$

The fluxes of the oxidizing species are

$$\bar{f}_{js} = -\bar{r}_j\nu_j \tag{2.130}$$

where $\nu_j$ is the mass of oxidizer $j$ consumed per unit mass of carbon consumed by reaction $j$. The surface boundary conditions for the oxidizing species become

$$\nu_j\bar{r}_j + (\bar{r}_1 + \bar{r}_2) = 4\pi\rho D_j\left(\frac{dx_j}{dz}\right)_s \tag{2.131}$$

Assuming that the particle is spherical and that CO is only oxidized far from the particle surface, the combustion rate determination follows the approach of the liquid fuel evaporation problem. We begin with the energy and species conservation equations:

$$(\overline{R}_1 + \overline{R}_2)\overline{c}_p \frac{dT}{dr} = \frac{d}{dr}\left(4\pi r^2 k \frac{dT}{dr}\right) \tag{2.132}$$

$$(\overline{R}_1 + \overline{R}_2)\frac{dx_j}{dr} = \frac{d}{dr}\left(4\pi r^2 \rho D_j \frac{dx_j}{dr}\right) \tag{2.133}$$

where the total rate of reaction $j$ is $\overline{R}_j = 4\pi a^2 \overline{r}_j$. The solutions are

$$\frac{\overline{R}_1 + \overline{R}_2}{4\pi a(k/\overline{c}_p)} = \ln\left[1 + \frac{\overline{c}_p(T_s - T_\infty)(\overline{R}_1 + \overline{R}_2)}{-\overline{R}_1\,\Delta\overline{h}_{R1}(T_s) - \overline{R}_2\,\Delta\overline{h}_{R2}(T_s) + 4\pi a^2\sigma\epsilon(T_w^4 - T_s^4)}\right] \tag{2.134}$$

$$\frac{\overline{R}_1 + \overline{R}_2}{4\pi a\rho D_j} = \ln\left[1 + \frac{(x_{j\infty} - x_{js})(\overline{R}_1 + \overline{R}_2)}{(\overline{R}_1 + \overline{R}_2)x_{js} + \overline{R}_j\nu_j}\right] \tag{2.135}$$

Equating these rates and requiring that the reaction rate expressions, (2.119), be satisfied, yields, after iterative solutions, the values of the temperature and mass fractions of the oxidizers at the particle surface. The terms on the right-hand side of (2.134) and (2.135) are analogous to the transfer number for the evaporation of a liquid fuel, $B_T$ (2.93). If the thermal and molecular diffusivities are equal [i.e., Lewis number = Le $= k/(\rho\overline{c}_p D) = 1$], the transfer numbers derived from (2.134) and (2.135) are equal:

$$\begin{aligned} B_T &= \frac{\overline{c}_p(T_s - T_\infty)(\overline{R}_1 + \overline{R}_2)}{-\overline{R}_1\,\Delta\overline{h}_{R1}(T_s) - \overline{R}_2\,\Delta\overline{h}_{R2}(T_s) + 4\pi a^2\sigma\epsilon(T_w^4 - T_s^4)} \\ &= \frac{(x_{j\infty} - x_{js})(\overline{R}_1 + \overline{R}_2)}{(\overline{R}_1 + \overline{R}_2)x_{js} + \overline{R}_j\nu_j} \end{aligned} \tag{2.136}$$

The special case of very rapid surface reaction corresponds to diffusion-limited combustion (i.e., Mulcahy and Smith's regime 3 combustion) and allows significant simplification. Assume that only the oxygen reaction 1 is important. If oxygen is consumed as fast as it reaches the particle surface, $x_{O_2,s} = 0$. Thus (2.135) becomes

$$\overline{R}_1 = 4\pi a\rho D \ln\left(1 + \frac{x_{O_2,\infty}}{\nu_{O_2}}\right) \tag{2.137}$$

This is an upper bound on the char combustion rate. Diffusion-limited combustion is a reasonable assumption for combustion of large particles at high temperatures. As either particle size or temperature decreases, reaction kinetics become increasingly important in controlling char oxidative kinetics. Combustion of small particles of pulverized coal is generally in regime 2 (i.e., both diffusional and kinetic resistances become important).

The prediction of the combustion rate requires knowledge of the reaction rate as a function of external surface area and oxidizer concentration. So far, we have relied on global rate expressions, which, as shown by the data in Table 2.10, vary widely from one char to another. Much of this variability can be attributed to differences in the porous structure of the char and the resistance to diffusion to the large surface area contained in that structure. The pore structure varies from one coal to another. Since the quantity of char residue depends on the heating rate, it stands to reason that the char structure will also vary with the devolatilization history. A priori prediction of the char structure is not possible at this time, but the role of the porous structure can be understood.

A number of models of combustion of porous particles have been developed (Simons, 1982; Gavalas, 1981; Smith, 1982). We limit our attention to one of the simpler diffusional resistance models.

The pore structure presents a large surface area for surface oxidation within the volume of the char particle. The pores are small enough that they present a substantial resistance to diffusion. The pore structure can be crudely characterized in terms of the total surface area per unit mass, $S$, most commonly measured by the BET gas adsorption method (Hill, 1977), and the total pore volume fraction in the char, $\epsilon_p$. If we assume the pores to be uniformly sized and cylindrical, the pore volume fraction and surface area per unit mass of char are

$$\epsilon_p = \pi \overline{\xi}^2 L_{tot}$$

$$S = \frac{2\pi \overline{\xi} L_{tot}}{\rho_a}$$

where $L_{tot}$ is the total length of pores per unit volume of char, $\overline{\xi}$ is the mean pore radius, and $\rho_a$ is the apparent density of the char. Combining these expressions to eliminate $L_{tot}$ and solving for $\overline{\xi}$, we find

$$\overline{\xi} = \frac{2\epsilon_p}{\rho_a S} \tag{2.138}$$

Consider, for example, a char with a BET surface area of 100 $m^2\ g^{-1}$ and a porosity of $\epsilon_p = 0.4$. The mean pore radius is 0.012 $\mu$m (assuming a char density of 1.5 g $cm^{-3}$).

If the pore radius is large compared to the mean free path $\lambda$ of the gas molecules, then the mechanism of diffusion through the pore is the usual continuum transport. (We will discuss the mean free path of gas molecules in Section 5.2.) If the pore radius is small compared to $\lambda$, then diffusion of the molecules through the pore occurs by collisions with the walls of the pore. For air at ambient temperature and pressure, $\lambda \approx 0.065$ $\mu$m. At combustion temperatures, the mean free path increases to 0.2 to 0.5 $\mu$m. The ratio of the mean free path to the pore radius, known as the Knudsen number,

$$\text{Kn} = \frac{\lambda}{\overline{\xi}} \tag{2.139}$$

indicates whether it is reasonable to apply continuum transport models. The continuum models are valid for Kn << 1 (see also Chapter 5). At very large Knudsen numbers, the kinetic theory of gases gives the following result for diffusivity of molecules in cylindrical pores,

$$D_k = \frac{2}{3}\bar{\xi}\sqrt{\frac{8RT}{\pi M}} \tag{2.140}$$

At intermediate values of Kn, the pore diffusivity is approximately (best for equimolar counter diffusion, $N_A = -N_B$)

$$D_p \approx \frac{1}{1/D_k + 1/D_{AB}} \tag{2.141}$$

The effective diffusivity within the porous particle is reduced by the fraction of voids in the particle, $\epsilon_p$, and by the tortuous path through which the gas must diffuse in the particle, characterized by a tortuosity factor, $\tau$, that is typically about 2, that is,

$$D_e \approx \frac{\epsilon_p D_p}{\tau} \tag{2.142}$$

The diffusion of oxidant within the pores of the char and the reactions on the pore surfaces can now be calculated. Consider the reaction of oxygen with the char,

$$2C_{(s)} + O_2 \longrightarrow 2CO$$

for which we shall assume first-order intrinsic reaction kinetics,

$$\bar{r}_i = k_1(T)p_{O_2} \quad \text{kg C m}^{-2}\,\text{s}^{-1} \tag{2.143}$$

The net local rate of carbon oxidation per unit of char volume is $\bar{r}_i S\rho_a$ (kg m$^{-3}$ s$^{-1}$), where $\rho_a$ is the apparent density of the char (density of carbon, including pores). The quasi-steady transport and reaction of oxygen within the porous char can be expressed as

$$4\pi r^2 \rho u \frac{dx_{O_2}}{dr} = \frac{d}{dr}\left(4\pi r^2 \rho D_e \frac{dx_{O_2}}{dr}\right) - 4\pi r^2 S\rho_a k' x_{O_2} \nu_{O_2} \tag{2.144}$$

where the reaction rate has been expressed in terms of species mass fraction (i.e., $\bar{r}_i = k_1 p_{O_2} = k' x_{O_2}$). The mass flux at any position in the char can be evaluated from the mass continuity equation,

$$4\pi \frac{d}{dr}(r^2 \rho u) = 4\pi r^2 S\rho_a k' x_{O_2} \tag{2.145}$$

The solution to (2.144) is greatly simplified if the convective transport term is small compared to diffusive transport, whence

$$\frac{d}{dr}\left(4\pi r^2 \rho D_e \frac{dx_{O_2}}{dr}\right) = 4\pi r^2 S\rho_a k' x_{O_2} \nu_{O_2} \tag{2.146}$$

The boundary conditions for this differential equation are

$$r = 0: \quad x_{O_2} = \text{finite}$$
$$r = a: \quad x_{O_2} = x_{O_2,s} \tag{2.147}$$

Solution to equations of the form of (2.146) is facilitated by substituting $x = f(r)/r$. We find

$$x = \frac{A \sinh(\phi r/a)}{r/a} + \frac{B \cosh(\phi r/a)}{r/a} \tag{2.148}$$

where

$$\phi^2 = \frac{\rho_a a^2 Sk'}{\rho D_e} \tag{2.149}$$

The parameter $\phi$ is known as the Thiele modulus (Hill, 1977). Applying the boundary conditions, we find

$$x_{O_2} = x_{O_2,s} \frac{a \sinh(\phi r/a)}{r \sinh \phi} \tag{2.150}$$

The net diffusive flux of oxygen into the porous particle exactly equals the rate of oxygen consumption by reaction. Thus the net rate of carbon consumption is

$$\overline{R}_p = 4\pi a^2 \rho D_e \left(\frac{dx}{dr}\right)_a$$
$$= 4\pi a \rho D_e x_{O_2,s} (\phi \coth \phi - 1) \tag{2.151}$$

If access to the interior surface were not limited by diffusion, the entire area would be exposed to $x_{O_2,s}$, and the net reaction rate would be

$$\overline{R}_{p,\text{ideal}} = \tfrac{4}{3} \pi a^3 \rho_a Sk' x_{O_2,s} \tag{2.152}$$

The ratio of the actual reaction rate to the ideal rate gives a measure of the effectiveness of the pores in the combustion chemistry,

$$\eta \equiv \frac{\overline{R}_p}{\overline{R}_{p,\text{ideal}}} = \frac{4\pi a \rho D_e x_{O_2,s} (\phi \coth \phi - 1)}{\tfrac{4}{3} \pi a^3 \rho_a Sk' x_{O_2,s}}$$

which with (2.149) yields

$$\eta = \frac{3}{\phi^2} (\phi \coth \phi - 1) \tag{2.153}$$

For fast reaction (i.e., large $\phi$), (2.153) approaches

$$\eta = \frac{3}{\phi} \tag{2.154}$$

so only a small fraction of the char surface area is available for reaction, that area near the char surface. In this limit the particle will shrink, but its density will remain constant once the pores near the char surface establish a steady-state profile of pore radius with depth. This opening of the pore mouth has been neglected in this derivation. On the other hand, in the limit of small $\phi$, $\eta$ tends to unity:

$$\lim_{\phi \to 0} \eta = 1 - \frac{\phi^2}{15}$$

and all of the interior surface contributes to the char oxidation. In this limit, regime 1 combustion, the pores must enlarge and the apparent density of the char must decrease during oxidation. Pulverized coal combustion corresponds most closely to the former case. Lower-temperature combustion in fluidized beds or in stokers may, however, result in a low Thiele modulus.

Coal particles are not, generally, spherical as illustrated by the scanning electron microscope photograph shown in Figure 2.27. Bird et al. (1960) suggest that a non-spherical particle be approximated as a sphere with the same ratio of apparent external surface area to volume,

$$a_{\text{nonsphere}} = \frac{3V_p}{A_p} \tag{2.155}$$

in calculating $\eta$ for a variety of shapes (spheres, cylindrical, rods, flat plates, etc.). The deviations of exact results from that obtained using (2.154) are small over the range of Thiele moduli important in char combustion.

The char oxidation kinetics are not necessarily first order. The analysis can readily be carried out for reactions of arbitrary (but constant) order. The variation of the pore diffusional resistance with reaction order is relatively weak.

With this analysis it is possible to estimate the intrinsic reaction kinetics from observations of char consumption rates. Smoot et al. (1984) have shown that a single intrinsic rate expression can correlate data on a broad spectrum of coal char. Moreover,

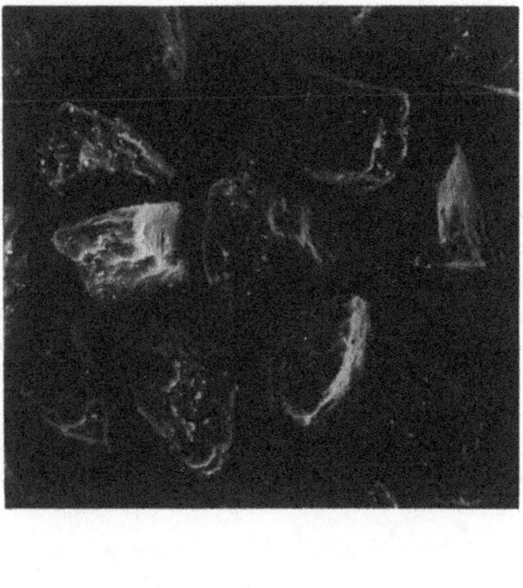

**Figure 2.27** Scanning electron microscope photograph of pulverized coal particles.

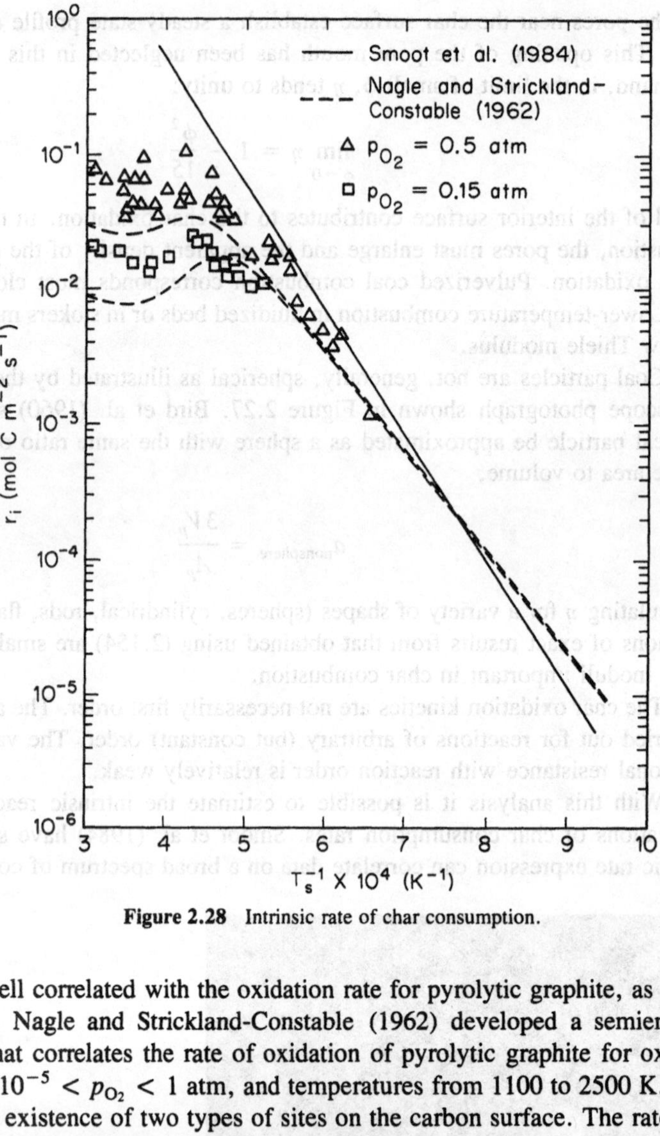


**Figure 2.28** Intrinsic rate of char consumption.

this rate is well correlated with the oxidation rate for pyrolytic graphite, as illustrated in Figure 2.28. Nagle and Strickland-Constable (1962) developed a semiempirical rate expression that correlates the rate of oxidation of pyrolytic graphite for oxygen partial pressures of $10^{-5} < p_{O_2} < 1$ atm, and temperatures from 1100 to 2500 K. This rate is based on the existence of two types of sites on the carbon surface. The rate of reaction at the more reactive type $A$ sites is governed by the fraction of sites not covered by surface oxides, so the reaction order varies between 0 and 1. Desorption from the less reactive type $B$ sites is rapid, so the rate of reaction is first order in the oxygen concentration. Thermal rearrangement of type $A$ sites into type $B$ is allowed. A steady-state analysis of this mechanism yields

$$\frac{r_i}{12} = \frac{k_A p_{O_2}}{1 + k_z p_{O_2}} + k_B p_{O_2}(1 - \chi) \qquad \text{mol C m}^{-2}\ \text{s}^{-1} \tag{2.156}$$

**TABLE 2.11 RATE CONSTANTS FOR THE NAGLE AND STRICKLAND-CONSTABLE MODEL**

| | |
|---|---|
| $k_A = 200 \exp(-15{,}100/T)$ | $\text{kg m}^{-2}\ \text{s}^{-1}\ \text{atm}^{-1}$ |
| $k_B = 4.45 \times 10^{-2} \exp(-7640/T)$ | $\text{kg m}^{-2}\ \text{s}^{-1}\ \text{atm}^{-1}$ |
| $k_T = 1.51 \times 10^{4} \exp(-48{,}800/T)$ | $\text{kg m}^{-2}\ \text{s}^{-1}$ |
| $k_z = 21.3 \exp(2060/T)$ | $\text{atm}^{-1}$ |

where $\chi$, the fraction of the surface covered by type $A$ sites, is

$$\chi = \left(1 + \frac{k_T}{k_B p_{O_2}}\right)^{-1}$$

The empirically determined rate constants for this model are given in Table 2.11.

According to this mechanism, the reaction is first order at low oxygen partial pressures but approaches zero order at higher $p_{O_2}$. At low temperatures and at fixed oxygen partial pressure, the rate increases with temperature with an activation energy corresponding to $E/R = 17{,}160$ K. Above a certain temperature, the rate begins to decrease due to the formation of unreactive type $B$ sites by thermal rearrangement. At very high temperatures, the surface is entirely covered with type $B$ sites and the rate becomes first order in $p_{O_2}$.

From the close correspondence of the char oxidation kinetics and the Nagle and Strickland-Constable rate for pyrolytic graphite, one may surmise that after processing at high temperature, carbons from a variety of sources may exhibit similar kinetics for the reaction of $O_2$ and, very likely, for other oxidants. Indeed, data for carbon black, soot, and some petroleum cokes are also in reasonable agreement with those from coal chars.

The model we have used to describe the porous structure is simplistic. The pores in char are not uniform in size, nor are they cylindrical in shape. Furthermore, as the char burns, the pores change shape and, in the regions near the char surface, will enlarge significantly. At high temperatures, where the surface reaction rates are high and oxygen is consumed quickly as it diffuses into the char pores, the pore mouths will enlarge until neighboring pores merge while the interior pores remain unchanged. More detailed models of the porous structure have been developed to take some of these variations into account (Gavalas, 1981; Simons, 1979, 1980). These models require more data on the pore size distribution than is commonly available at present, but may ultimately eliminate much of the remaining uncertainty in char oxidation rates.

## PROBLEMS

**2.1.** Methanol ($CH_3OH$) is burned in dry air at an equivalence ratio of 0.75.

**(a)** Determine the fuel/air mass ratio.

**(b)** Determine the composition of the combustion products.

**2.2.** A high-volatile bituminous coal has the following characteristics:

| | |
|---|---|
| Proximate analysis | |
| Fixed carbon | 54.3% |
| Volatile matter | 32.6% |
| Moisture | 1.4% |
| Ash | 11.7% |
| Ultimate analysis | |
| C | 74.4% |
| H | 5.1% |
| N | 1.4% |
| O | 6.7% |
| S | 0.7% |
| Heating value | $30.7 \times 10^6$ J kg$^{-1}$ |

It is burned in air at an equivalence ratio of 0.85. $500 \times 10^6$ W of electric power is produced with an overall process efficiency (based on the input heating value of the fuel) of 37%.

(a) Determine the fuel and air feed rates in kg s$^{-1}$.

(b) Determine the product gas composition.

(c) Sulfur dioxide is removed from the flue gases with a mean efficiency of 80% and the average output of the plant is 75% of its rated capacity. What is the $SO_2$ emission rate in metric tonnes ($10^3$ kg) per year?

**2.3.** A liquid fuel has the composition:

| | |
|---|---|
| C | 86.5% |
| H | 13.0% |
| O | 0.2% |
| S | 0.3% |

Its higher heating value is HHV $= 45 \times 10^6$ J kg$^{-1}$. Determine an effective chemical formula and enthalpy of formation for this fuel.

**2.4.** Methane is burned in air at $\phi = 1$. Using the thermodynamic data of Table 2.5 and assuming complete combustion, compute the adiabatic flame temperature. The initial temperatures of the fuel and air are both 298 K.

**2.5.** Methanol shows promise as an alternate fuel that could reduce nitrogen oxide emissions. The reduction is attributed to lower flame temperatures. Compare the adiabatic flame temperature for combustion of pure methanol at $\phi = 1$ with that of methane (Problem 2.4). Initial fuel and air temperatures are 298 K. The enthalpy of formation of liquid methanol is $\Delta h_f^\circ$ (298 K) $= -239{,}000$ J mol$^{-1}$.

**2.6.** The bituminous coal of Problem 2.2 is burned in air that has been heated to 590 K. To estimate the maximum temperature in combustion, compute the adiabatic flame temperature for stoichiometric combustion assuming complete combustion. The specific heats of the coal carbon and ash may be taken as $\bar{c}_{pc} = 1810$ and $\bar{c}_{pa} = 1100$ J kg$^{-1}$ K$^{-1}$, respectively. The ash melts at 1500 K with a latent heat of melting of $\Delta \bar{h}_m = 140$ kJ kg$^{-1}$

**2.7.** Kerosene (88% C, 12% H) is burned in air at an equivalence ratio of 0.8. Determine the equilibrium mole fractions of carbon monoxide and nitric oxide at $T = 2000$ K and $p = 1$ atm.

**2.8.** Graphite (C) is burned in dry air at $\phi = 2$ and $p = 1$ atm. Determine the equilibrium composition (mole fractions of CO, $CO_2$, $O_2$, and amount of solid carbon) of the combustion products at $T = 2500$ K.

**2.9.** A fuel oil containing 87% C and 13% H has a specific gravity of 0.825 and a higher heating value of $3.82 \times 10^{10}$ J m$^{-3}$. It is injected into a combustor at 298 K and burned at atmospheric pressure in stoichiometric air at 298 K. Determine the adiabatic flame temperature and the equilibrium mole fractions of CO, $CO_2$, $H_2$, $H_2O$, $O_2$, and $N_2$.

**2.10.** For Problem 2.9, determine the equilibrium mole fractions of NO, OH, H, and O. How much is the flame temperature reduced in producing these species?

**2.11.** Carbon monoxide is oxidized by the following reactions:

$$CO + OH \overset{1}{\rightleftharpoons} CO_2 + H$$

$$CO + O_2 \overset{2}{\rightleftharpoons} CO_2 + O$$

$$CO + O + M \overset{3}{\rightleftharpoons} CO_2 + M$$

$$CO + HO_2 \overset{4}{\rightleftharpoons} CO_2 + OH$$

Rate coefficients for these reactions are given in Table 2.8.

**(a)** Write the full rate equation for carbon monoxide consumption.

**(b)** Assuming chemical equilibrium for the combustion of methane at atmospheric pressure and $\phi = 0.85$, compare the effectiveness of these reactions in terms of characteristic times for CO destruction

$$\tau_i = \frac{[CO]_e}{R^e_{+i}}$$

where $R^e_{+i}$ is the rate of reaction in the forward direction only based on equilibrium concentrations of all species. Plot the $\tau_i$s from $T = 1200$ K to $T = 2000$ K.

**(c)** Considering only the dominant reaction and assuming equilibrium for the minor species, derive a global rate expression for CO oxidation and $CO_2$ reduction in terms of CO, $CO_2$, $O_2$, and $H_2O$ concentrations and temperature.

**(d)** Compare your oxidation rate with that obtained by Dryer and Glassman (1973), (2.53). Plot the two rates as a function of temperature.

**2.12.** An industrial process releases 500 ppm of ethane into an atmospheric pressure gas stream containing 2% oxygen at $T = 1000$ K. Use the single-step global combustion model for ethane to estimate how long the gases must be maintained at this temperature to reduce the ethane concentration below 50 ppm.

**2.13.** A combustor burning the fuel oil of Example 2.3 at $\phi = 1$ contains 1.5% $O_2$ in the combustion products. Using the data in Figure 2.6, estimate the CO level in the combustion products assuming local equilibrium.

**2.14.** Natural gas (assumed to be methane) is burned in atmospheric pressure air ($T_f = T_a = 300$ K) at an equivalence ratio of 0.9. For a characteristic mixing time of $\tau_m = 0.01$ s, and assuming that $B = 1$ in

$$\tau_m^{-1} = B\left(\frac{\epsilon}{l^2}\right)^{1/3}$$

compute and plot as a function of burner diameter, $d$, the ratio of the rate of kinetic energy dissipation in turbulence to the heat released by combustion. Assume that intense recirculation limits the volume in which the kinetic energy is dissipated to $2d^3$ and that the mean gas temperature in this volume is 2000 K. What is a reasonable maximum burner size? How many burners would be required to generate 100 MW ($100 \times 10^6$ J s$^{-1}$) of electric power at an overall process efficiency of 40%? If the maximum burner gas velocity is limited to 30 m s$^{-1}$, what is the maximum burner diameter and how many burners will be needed? Suppose that we allow the mixing time to be 0.05 s. How would this influence the results?

**2.15.** Carbon ($\rho = 2000$ kg m$^{-3}$) is injected into atmospheric pressure air in a furnace that maintains both gas and walls at the same temperature. Compute and plot the particle temperature and time for complete combustion of a 50-$\mu$m diameter char particle as a function of furnace temperature over the range from 1300 to 2000 K, assuming diffusion-limited combustion. The thermodynamic properties of the carbon may be taken to be those of pure graphite. Use the following physical properties:

$$D = 1.5 \times 10^{-9} T^{1.68} \text{ m}^2 \text{ s}^{-1}$$
$$k = 3.4 \times 10^{-4} T^{0.77} \text{ W m}^{-1} \text{ K}^{-1}$$
$$\epsilon = 1$$

**2.16.** For the system of Problem 2.15 and a fixed wall and gas temperature of $T = 1700$ K, compute and plot the particle temperature and characteristic time for combustion of a 50-$\mu$m diameter particle as the function of the oxygen content of an $O_2$–$N_2$ mixture over the range of oxygen contents from 1 to 20.9%, assuming

(a) diffusion limited combustion.

(b) combustion according to the global rate expression of Smith (1982) for anthracite (Table 2.10).

What are the implications of these results for char burnout in pulverized coal combustion?

**2.17.** For the conditions of Problem 2.15 and combustion in air at 1700 K, compute and plot the particle temperature and characteristic time as a function of the char particle size over the range 1 to 200 $\mu$m.

**2.18.** A Millmerran subbituminous coal has the following composition:

| | |
|---|---|
| C | 72.2% |
| H | 5.8% |
| N | 1.4% |
| O | 10.0% |
| S | 1.2% |
| Ash | 9.4% |
| Volatile matter | 41.9% |

Fixed carbon 44.5%
Moisture 4.2%

$$\rho_{\text{ash-free coal}} = 1300 \text{ kg m}^{-3}$$

$$\rho_{\text{ash}} = 2300 \text{ kg m}^{-3}$$

A 150-$\mu$m-diameter particle is injected into atmospheric pressure air in a furnace that maintains both the gas and wall temperatures at 1700 K.

**(a)** Assuming that the particle is instantaneously heated to the gas temperature and maintained at that temperature throughout the devolatilization process, determine the amount of volatile matter released and the quantity of char remaining. Assuming that the physical dimension of the particle has not changed, what is the final density of the particle?

**(b)** Oxidation begins immediately following devolatilization. Assuming quasi-steady combustion and that the particle density remains constant, and using the apparent reaction kinetics of Table 2.10, calculate the particle temperature and size throughout the combustion of the particle. Assume that the enthalpy of formation of the char is the same as that of graphite. How long does it take for the particle to burn out?

**2.19.** A fuel oil with composition

C 86%
H 14%

is burned in dry air. Analysis of the combustion products indicates, on a dry basis (after condensing and removing all water),

$O_2$ 1.5%
CO 600 ppm

What is the equivalence ratio of combustion?

## REFERENCES


BIRD, R. B., STEWART, W. E., and LIGHTFOOT, E. N. *Transport Phenomena*, Wiley, New York (1960).

BROWN, G. L., and ROSHKO, A. "On Density Effects and Large Structure in Turbulent Mixing Layers," *J. Fluid Mech.*, *64*, 775–816 (1974).

BURKE, S. P., and SCHUMANN, T. E. W. "Diffusion Flames," *Ind. Eng. Chem.*, *20*, 998–1004 (1928).

CHOMIAK, J. "A Possible Propagation Mechanism of Turbulent Flames at High Reynolds Numbers," *Combust. Flame*, *15*, 319–321 (1970).

CHOMIAK, J. "Application of Chemiluminescence Measurement to the Study of Turbulent Flame Structure," *Combust. Flame*, *18*, 429–433 (1972).

CORRSIN, S. "Simple Theory of an Idealized Turbulent Mixer," *AIChE J.*, *3*, 329–330 (1957).

DAMKOHLER, G. "The Effect of Turbulence on the Flame Velocities in Gas Mixtures," NACA TM 1112, *Der Einfluss der Turbolenz auf die Flammengeschwindigkeit und angewandte Gasgemischen, Z. Elektrochem. Angew. Phys. Chem.*, *46*, 601–626 (1940).

DENBIGH, K. *The Principles of Chemical Equilibrium*, Cambridge University Press, London (1971).

DRYER, F. L., and GLASSMAN, I. "High Temperature Oxidation of CO and $CH_4$," in *Fourteenth Symposium (International) on Combustion*, The Combustion Institute, Pittsburgh, PA, 987–1003 (1973).

EDELMAN, R. B., and FORTUNE, O. F. "A Quasi-global Chemical Kinetic Model for the Finite Rate Combustion of Hydrocarbon Fuels with Application to Turbulent Burning and Mixing in Hypersonic Engines and Nozzles," AIAA Paper No. 69-86, Amer. Inst. of Aero. Astro. (1969).

GAVALAS, G. R. "Analysis of Char Combustion Including the Effect of Pore Enlargement," *Combust. Sci. Technol.*, *24*, 197–210 (1981).

GAVALAS, G. R. *Coal Pyrolysis*, Elsevier, New York (1982).

GLASSMAN, I. *Combustion*, Academic Press, New York (1977).

GORDON, S., and MCBRIDE, B. J. "Computer Program for Calculation of Complex Chemical Equilibrium Compositions, Rocket Performance, Incident and Reflected Shocks, and Chapman-Jonquet Detonations," NASA Report No. SP-273 (1971).

HANSON, S. P., BEER, J. M., and SAROFIM, A. F. "Non-equilibrium Effects in the Vaporization of Multicomponent Fuel Droplets," in *Nineteenth Symposium (International) on Combustion*, The Combustion Institute, Pittsburgh, PA, 1029–1036 (1982).

HAUTMAN, D. J., DRYER, F. L., SCHUG, K. P., and GLASSMAN, I. "A Multiple-Step Overall Kinetic Mechanism for the Oxidation of Hydrocarbons," *Combust. Sci. Technol.*, *25*, 219–235 (1981).

HAWTHORNE, W. R., WEDDELL, D. S., and HOTTEL, H. C. "Mixing and Combustion in Turbulent Gas Flames," in *Third Symposium (International) on Combustion*, Combustion Institute, Pittsburgh, PA, 266–288 (1951).

HILL, C. G., JR. *An Introduction to Chemical Engineering Kinetics and Reactor Design*, Wiley, New York (1977).

KOBAYASHI, H., HOWARD, J. B., and SAROFIM, A. F. "Coal Devolatilization at High Temperatures," in *Sixteenth Symposium (International) on Combustion*, The Combustion Institute, Pittsburgh, PA, 411–425 (1977).

KOMIYAMA, K., FLAGAN, R. C., and HEYWOOD, J. B. "The Influence of Droplet Evaporation on Fuel-Air Mixing Rate in a Burner," in *Sixteenth Symposium (International) on Combustion*, The Combustion Institute, Pittsburgh, PA, 549–560 (1977).

LABOWSKY, M., and ROSNER, D. E. "Group Combustion of Droplets in Fuel Clouds: I. Quasi-steady Predictions," in *Advances in Chemistry Series No. 166: Evaporation-Combustion of Fuels*, J. T. Zung, Ed., American Chemical Society, Washington, DC, 64–79 (1978).

LAW, C. K., and LAW, H. K. "A $d^2$-Law for Multicomponent Droplet Vaporization and Combustion," *AIAA J.*, *20*, 522–527 (1981).

MILLER, J. A., MITCHELL, R. E., SMOOKE, M. D., and KEE, R. J. "Toward a Comprehensive Chemical Kinetic Mechanism for the Oxidation of Acetylene: Comparison of Model Predictions with Results from Flame and Shock Tube Experiments," in *Nineteenth Symposium (International) on Combustion*, The Combustion Institute, Pittsburgh, PA, 181–196 (1982).

MITCHELL, R. E., SAROFIM, A. F., and CLOMBERG, L. A. "Partial Equilibrium in the Reaction Zone of Methane-Air Diffusion Flames," *Combust. Flame*, *37*, 201–206 (1980).

MULCAHY, M. F. R., and SMITH, W. "Kinetics of Combustion of Pulverized Fuel: A Review of Theory and Experiment," *Rev. Pure Appl. Chem.*, *18*, 81–108 (1969).

NAGLE, J., and STRICKLAND-CONSTABLE, R. F. "Oxidation of Carbon between 1000–2000°C," in *Proceedings of the Fifth Carbon Conference*, *1*, 154–164 (1962).

POMPEI, F., and HEYWOOD, J. B. "The Role of Mixing in Burner-Generated Carbon Monoxide and Nitric Oxide," *Combust. Flame*, *19*, 407–418 (1972).

PRANDTL, L. *Essentials of Fluid Dynamics with Applications to Hydraulics, Aeronautics, Meteorology and Other Subjects*, Hafner (1952).

RANZ, W. E., and MARSHALL, W. R., JR. "Evaporation from Drops," *Chem. Eng. Progr.*, *48*, 141–146 and 173–180 (1952).

REYNOLDS, W. C. "STANJAN-Interactive Computer Programs for Chemical Equilibrium Analysis," Dept. of Mechanical Engineering, Stanford Univ. (1981).

SIMONS, G. A. "Char Gasification: Part I. Transport Model," *Comb. Sci. Tech.*, *20*, 107–116 (1979a).

SIMONS, G. A. "Char Gasification: Part II. Oxidation Results," *Comb. Sci. Tech.*, *20*, 117–124 (1979b).

SIMONS, G. A. "The Pore Tree Structure of Porous Char," in *Nineteenth Symposium (International) on Combustion*, The Combustion Institute, Pittsburgh, PA, 1067–1076 (1982).

SMITH, I. W. "The Combustion Rates of Coal Chars: A Review," in *Nineteenth Symposium (International) on Combustion*, The Combustion Institute, Pittsburgh, PA, 1045–1065 (1982).

SMOOT, L. D., HEDMAN, P. O., and SMITH, P. J. "Pulverized-Coal Combustion Research at Brigham Young University," *Prog. Energy Combust. Sci.*, *10*, 359–441 (1984).

STICKLER, D. B., BECKER, F. E., and UBHAYAKAR, S. K. "Combustion of Pulverized Coal at High Temperature," AIAA Paper No. 79-0298, Amer. Inst. of Aero. Astro. (1979).

STULL, D. R., and PROPHET, H. *JANAF Thermochemical Tables*, 2nd Ed., National Bureau of Standards NSRDS-NBS37 (1971).

TABACZYNSKI, R. J., FERGUSON, C. R., and RADHAKRISHNAN, K. "A Turbulent Entrainment Model for Spark-Ignition Engine Combustion," SAE Paper No. 770647, Society of Automotive Engineers, Warrendale, PA (1977).

TENNEKES, H. "Simple Model for the Small-Scale Structure of Turbulence," *Phys. Fluids*, *11*, 669–671 (1968).

VANDOOREN, J., and VAN TIGGELEN, P. J. "Experimental Investigation of Methanol Oxidation in Flames: Mechanisms and Rate Constants of Elementary Steps," in *Eighteenth Symposium (International) on Combustion*, The Combustion Institute, Pittsburgh, PA, 473–483 (1981).

VENKAT, C., BREZINSKY, K., and GLASSMAN, I. "High Temperature Oxidation of Aromatic Hydrocarbons," in *Nineteenth Symposium (International) on Combustion*, The Combustion Institute, Pittsburgh, PA, 143–152 (1982).

WARNATZ, J. "Rate Coefficients in the C/H/O System," in *Combustion Chemistry*, W. C. Gardiner, Jr., Ed., Springer-Verlag, New York, 197–360 (1984).

WESTBROOK, C. K. "Chemical Kinetics of Hydrocarbon Oxidation in Gaseous Detonations," *Combust. Flame*, *46*, 191–210 (1982).

WESTBROOK, C. K., and DRYER, F. L. "Chemical Kinetics of Modeling of Combustion Processes," in *Eighteenth Symposium (International) on Combustion*, The Combustion Institute, Pittsburgh, PA, 749–767 (1981a).

WESTBROOK, C. K., and DRYER, F. L. "Simplified Reaction Mechanisms for the Oxidation of Hydrocarbon Fuels in Flames," *Combust. Sci. Technol.*, *27*, 31–43 (1981b).

WESTBROOK, C. K., and DRYER, F. L. "Chemical Kinetic Modeling of Hydrocarbon Combustion," *Prog. Energy Combust. Sci.*, *10*, 1–57 (1984).

WHITE, W. B., JOHNSON, W. M., and DANTZIG, G. B. "Chemical Equilibrium in Complex Mixtures," *J. Chem. Phys.*, *28*, 751–755 (1958).

# 3

# Pollutant Formation and Control in Combustion

In this chapter we study the formation of pollutants in combustion systems. Our ultimate objectives are to be able to predict the levels of pollutants in the combustion products, leaving practical combustion systems, and to use such predictions to suggest combustion modifications to achieve lower emission levels. We focus on the basic mechanisms of pollutant formation in continuous-flow combustors. Internal combustion engines and gas-cleaning systems will be treated in subsequent chapters.

## 3.1 NITROGEN OXIDES

Nitrogen oxides are important air pollutants, the primary anthropogenic source of which is combustion. Motor vehicles account for a large fraction of the nitrogen oxide emissions, but stationary combustion sources ranging from electric power generating stations to gas-fired cooking stoves also release nitrogen oxides.

Both nitric oxide, NO, and nitrogen dioxide, $NO_2$, are produced in combustion, but the vast majority of nitrogen oxides are emitted as NO (recall Table 1.4). Because NO is converted to $NO_2$ in the atmosphere, emissions of both species frequently are lumped together with the designation $NO_x$. When $NO_x$ emissions are presented in mass units, the mass of $NO_x$ is calculated as if all the NO had been converted to $NO_2$. Because NO is the predominant $NO_x$ species formed in combustion, we shall concentrate on it in this section.

Nitric oxide is formed both from atmospheric nitrogen, $N_2$, and from nitrogen contained in some fuels. The latter source depends on the fuel composition and is not important for fuels with low nitrogen contents but is a major source of $NO_x$ in coal combustion. Nitric oxide can be formed, however, when any fuel is burned in air because of the high-temperature oxidation of $N_2$. We begin our discussion with the fixation of atmospheric nitrogen.

### 3.1.1 Thermal Fixation of Atmospheric Nitrogen

The formation of NO by oxidation of atmospheric nitrogen can be expressed in terms of the overall reaction

$$\tfrac{1}{2}\,N_2 + \tfrac{1}{2}\,O_2 \rightleftharpoons NO$$

which is highly endothermic [i.e., $\Delta h_f^\circ$ (298 K) = 90.4 kJ mol$^{-1}$]. As a result, the equilibrium concentration of NO is high at the very high temperatures encountered near stoichiometric combustion and decreases rapidly away from that point.

Even though we express the overall reaction as above, the direct reaction of $N_2$ with $O_2$ is too slow to account for significant NO formation. Free oxygen atoms, produced in flames by dissociation of $O_2$ or by radical attack on $O_2$, attack nitrogen molecules and begin a simple chain mechanism that was first postulated by Zeldovich et al. (1947), that is,

$$N_2 + O \underset{-1}{\overset{+1}{\rightleftharpoons}} NO + N$$

$$N + O_2 \underset{-2}{\overset{+2}{\rightleftharpoons}} NO + O$$

The concentration of $O_2$ is low in fuel-rich combustion, so reaction 2 is less important than in fuel-lean combustion. Reaction with the hydroxyl radical eventually becomes the major sink for N:

$$N + OH \underset{-3}{\overset{+3}{\rightleftharpoons}} NO + H$$

The rate constants for the so-called extended Zeldovich mechanism are (Hanson and Salimian, 1984)

$$k_{+1} = 1.8 \times 10^8\, e^{-38,370/T} \quad m^3\, mol^{-1}\, s^{-1}$$

$$k_{-1} = 3.8 \times 10^7\, e^{-425/T} \quad m^3\, mol^{-1}\, s^{-1}$$

$$k_{+2} = 1.8 \times 10^4\, T\, e^{-4680/T} \quad m^3\, mol^{-1}\, s^{-1}$$

$$k_{-2} = 3.8 \times 10^3\, T\, e^{-20,820/T} \quad m^3\, mol^{-1}\, s^{-1}$$

$$k_{+3} = 7.1 \times 10^7\, e^{-450/T} \quad m^3\, mol^{-1}\, s^{-1}$$

$$k_{-3} = 1.7 \times 10^8\, e^{-24,560/T} \quad m^3\, mol^{-1}\, s^{-1}$$

The high activation energy of reaction 1, resulting from its essential function of breaking the strong $N_2$ triple bond, makes this the rate-limiting step of the Zeldovich mechanism.

Due to the high activation energy, NO production by this mechanism proceeds at a slower rate than the oxidation of the fuel constituents and is extremely temperature sensitive. The production of atomic oxygen required for the first reaction is also highly temperature sensitive.

To understand the rate of NO formation, let us examine the rate equations corresponding to the mechanism of reactions 1–3. For example, the net rates of formation of NO and N are

$$R_{NO} = k_{+1}[N_2][O] - k_{-1}[N][NO] + k_{+2}[N][O_2] - k_{-2}[NO][O] + k_{+3}[N][OH] - k_{-3}[NO][H] \quad (3.1)$$

$$R_{N} = k_{+1}[N_2][O] - k_{-1}[N][NO] - k_{+2}[N][O_2] + k_{-2}[NO][O] - k_{+3}[N][OH] + k_{-3}[NO][H] \quad (3.2)$$

The concentrations of O, H, and OH are required for calculation of the N and NO formation rates. The high activation energy of the initial $N_2$ attack allows us to make an important simplification. Since the reaction rate is fast only at the highest temperatures, most of the reaction takes place after the combustion reactions are complete and before significant heat is transferred from the flame. It is a reasonable first approximation, therefore, to assume that the O, H, and OH radicals are present in their equilibrium concentrations.

This suggests a simplification proposed by Lavoie et al. (1970). At thermodynamic equilibrium, we may write

$$k_{+1}[N_2]_e[O]_e = k_{-1}[N]_e[NO]_e \quad (3.3)$$

We may define the equilibrium, one-way rate of reaction as

$$R_1 = k_{+1}[N_2]_e[O]_e = k_{-1}[N]_e[NO]_e \quad (3.4)$$

Similarly, at equilibrium

$$R_2 = k_{+2}[N]_e[O_2]_e = k_{-2}[NO]_e[O]_e \quad (3.5)$$

$$R_3 = k_{+3}[N]_e[OH]_e = k_{-3}[NO]_e[H]_e \quad (3.6)$$

Further defining the quantities,

$$\alpha = \frac{[NO]}{[NO]_e}$$

$$\beta = \frac{[N]}{[N]_e}$$

the rate equations may now be expressed in the abbreviated form

$$R_{NO} = R_1 - R_1\alpha\beta + R_2\beta - R_2\alpha + R_3\beta - R_3\alpha \quad (3.7)$$

$$R_{N} = R_1 - R_1\alpha\beta - R_2\beta + R_2\alpha - R_3\beta + R_3\alpha \quad (3.8)$$

We must determine the N atom concentration if we are to calculate the rate of NO formation. Since the activation energy for oxidation of the nitrogen atom is small and, for fuel-lean conditions, the reaction involves $O_2$, a major component of the gas, the free nitrogen atoms are consumed as rapidly as they are generated, establishing a quasi-steady state. Setting the left-hand side of (3.8) equal to zero and solving for the steady-state nitrogen atom concentration, we find

$$\beta_{ss} = \frac{R_1 + R_2\alpha + R_3\alpha}{R_1\alpha + R_2 + R_3} = \frac{\kappa + \alpha}{\kappa\alpha + 1} \tag{3.9}$$

where

$$\kappa = \frac{R_1}{R_2 + R_3} \tag{3.10}$$

Substituting (3.9) into (3.7) yields a rate equation for NO formation in terms of $\alpha$ and known quantities,

$$R_{NO} = \frac{2R_1(1 - \alpha^2)}{1 + \kappa\alpha} \tag{3.11}$$

In general, the NO formation rate is expressed based on the total mass in the system [see (A.7)–(A.10)]

$$R_{NO} = \rho \frac{d}{dt}\left(\frac{[NO]}{\rho}\right)$$

For constant temperature and pressure, this may be written as a differential equation for $\alpha$:

$$\frac{d\alpha}{dt} = \frac{1}{[NO]_e} \frac{2R_1(1 - \alpha^2)}{1 + \kappa\alpha} \tag{3.12}$$

The initial NO formation rate (at $\alpha = 0$) is twice the rate of reaction 1. Figure 3.1 shows how this initial rate varies with equivalence ratio for adiabatic combustion. The conditions for these calculations are the same as for the equilibrium calculations previously shown in Figures 2.6 and 2.7. The sharp peak near stoichiometric is due to the high flame temperatures.

Equation (3.12) can be integrated analytically to describe NO formation in a constant-temperature system. Assuming that there is no NO present initially (i.e., $\alpha = 0$ at $t = 0$), the result is

$$(1 - \kappa) \ln (1 + \alpha) - (1 + \kappa) \ln (1 - \alpha) = \frac{t}{\tau_{NO}} \tag{3.13}$$

where the characteristic time for NO formation is

$$\tau_{NO} = \frac{[NO]_e}{4R_1} \tag{3.14}$$

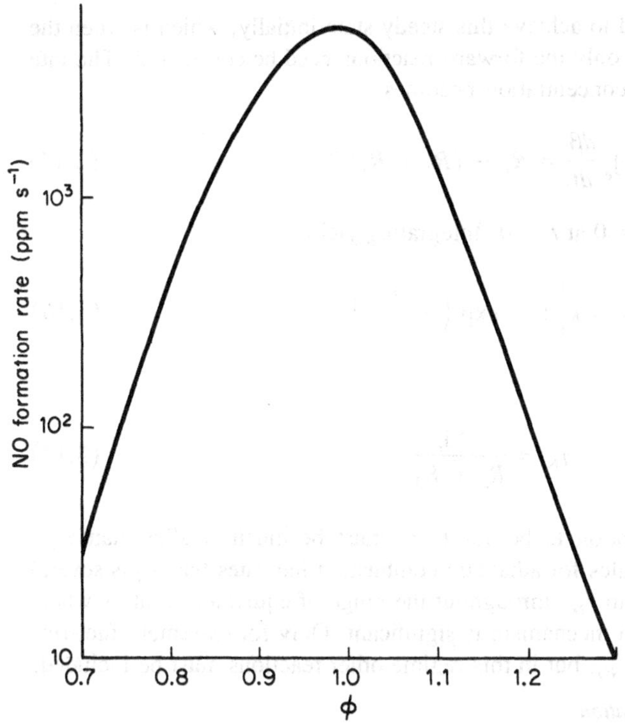


**Figure 3.1** Variation of the initial NO formation rate with equivalence ratio for adiabatic combustion of kerosene with composition $CH_{1.8}$.

This approach to equilibrium is illustrated in Figure 3.2. It is apparent that $\tau_{NO}$ corresponds to the time that would be required for NO to reach the equilibrium level if the reaction continued at its initial rate and were not slowed by the reverse reactions.

Two major assumptions have been made in the derivation of (3.11): (1) a quasi-steady state for the nitrogen atom concentration and (2) equilibrium concentrations for the O, H, and OH radicals. The validity of the first assumption can readily be examined.

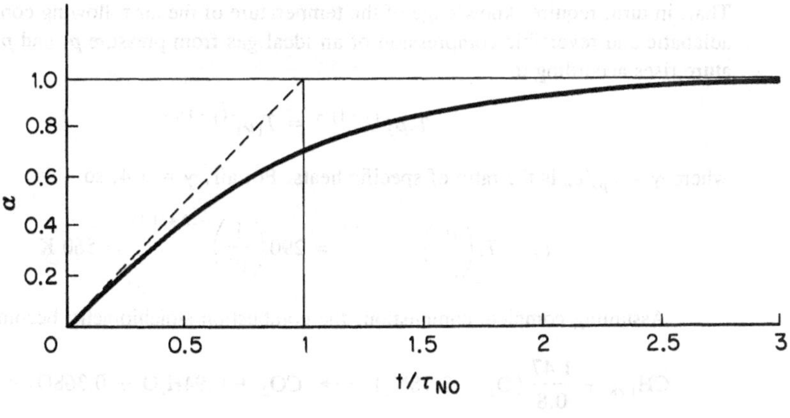


**Figure 3.2** Approach of the dimensionless NO concentration to equilibrium.

If we consider the time required to achieve this steady state initially, which is when the NO concentration is small, and only the forward reactions need be considered. The rate equation for the nitrogen atom concentration becomes

$$[\mathrm{N}]_e \frac{d\beta}{dt} = R_1 - (R_2 + R_3)\beta \tag{3.15}$$

with the initial condition of $\beta = 0$ at $t = 0$. Integrating yields

$$\beta = \kappa\left[1 - \exp\left(-\frac{t}{\tau_{\mathrm{N}}}\right)\right] \tag{3.16}$$

where

$$\tau_{\mathrm{N}} = \frac{[\mathrm{N}]_e}{R_2 + R_3} \tag{3.17}$$

For the quasi-steady N assumption to be valid, $\tau_{\mathrm{N}}$ must be much smaller than $\tau_{\mathrm{NO}}$. Comparison of the two time scales for adiabatic combustion indicates that $\tau_{\mathrm{N}}$ is several orders of magnitude smaller than $\tau_{\mathrm{NO}}$ throughout the range of equivalence ratios where NO formation by the Zeldovich mechanism is significant. Only for extremely fuel-rich combustion does $\tau_{\mathrm{N}}$ approach $\tau_{\mathrm{NO}}$, but in this regime other reactions must be included.

**Example 3.1** ***Thermal-NO$_x$ Formation***

In a gas turbine, air is compressed adiabatically from atmospheric pressure and 290 K to a pressure of 10 atm. Fuel (aviation kerosene, $CH_{1.88}$, LHV = 600 kJ (mol C)$^{-1}$ is injected into the hot air, mixed rapidly, and burned. After a brief residence in the primary combustion zone, typically 0.005 s, the combustion products are diluted with additional compressed air to lower the temperature below the limiting turbine inlet temperature. Assuming perfect mixing and adiabatic combustion at $\phi = 0.8$ in the primary zone, estimate the mole fraction of $NO_x$ formed in the primary combustion zone.

To calculate the NO formation rate, we first need to know the flame temperature. That, in turn, requires knowledge of the temperature of the air following compression. For adiabatic and reversible compression of an ideal gas from pressure $p_1$ and $p_2$, the temperature rises according to

$$T_2 p_2^{-(\gamma-1)/\gamma} = T_1 p_1^{-(\gamma-1)/\gamma}$$

where $\gamma = c_p/c_v$ is the ratio of specific heats. For air, $\gamma = 1.4$, so

$$T_2 = T_1\left(\frac{p_1}{p_2}\right)^{-(\gamma-1)/\gamma} = 290\left(\frac{1}{10}\right)^{-(0.4/1.4)} = 560\ \mathrm{K}$$

Assuming complete combustion, the combustion stoichiometry becomes

$$CH_{1.88} + \frac{1.47}{0.8}(O_2 + 3.78N_2) \longrightarrow CO_2 + 0.94H_2O + 0.368O_2 + 6.95N_2$$

The energy equation may be written as

$$[h(T) - h(T_0)]_{CO_2} + 0.94[h(T) - h(T_0)]_{H_2O} + 0.368[h(T) - h(T_0)]_{O_2}$$
$$+ 6.95[h(T) - h(T_0)]_{N_2} - 1.84[h(T_a) - h(T_0)]_{O_2} - 6.95[h(T_a) - h(T_0)]_{N_2}$$
$$- [h(T_f) - h(T_0)]_{CH_{1.88}} + \Delta h_{cL}(T_0) = 0$$

Using the data from Table 2.5, the sensible enthalpy terms become

$$[h(T) - h(T_0)]_i = a_i(T - T_0) + \frac{b_i}{2}(T_i^2 - T_0^2)$$

Neglecting the small sensible enthalpy term for the fuel, we find

$$T = 2304 \text{ K}$$

To calculate the NO formation rate, we need the equilibrium concentrations of NO, N, O, OH, and H. Use the following reactions and equilibrium constants:

$$\tfrac{1}{2} N_2 + \tfrac{1}{2} O_2 \rightleftharpoons NO \qquad K_{pNO} = 0.0416$$
$$\tfrac{1}{2} N_2 \rightleftharpoons N \qquad K_{pN} = 3.80 \times 10^{-8} \text{ atm}^{1/2}$$
$$\tfrac{1}{2} O_2 \rightleftharpoons O \qquad K_{pO} = 4.77 \times 10^{-3} \text{ atm}^{1/2}$$
$$\tfrac{1}{2} H_2O + \tfrac{1}{4} O_2 \rightleftharpoons OH \qquad K_{pOH} = 0.0322 \text{ atm}^{1/4}$$
$$\tfrac{1}{2} H_2O \rightleftharpoons H + \tfrac{1}{4} O_2 \qquad K_{pH} = 4.30 \times 10^{-4} \text{ atm}^{3/4}$$

From the combustion stoichiometry and these equilibrium relationships, we calculate

$$y_{N_2} = 0.751 \qquad y_{N_e} = 1.04 \times 10^{-8}$$
$$y_{O_2} = 0.0397 \qquad y_{O_e} = 3.01 \times 10^{-4}$$
$$y_{H_2O} = 0.102 \qquad y_{OH_e} = 2.58 \times 10^{-3}$$
$$y_{NO_e} = 7.18 \times 10^{-3} \qquad y_{H_e} = 5.47 \times 10^{-5}$$

The equilibrium, one-way reaction rates are

$$R_1 = k_{+1}c^2 y_{N_2} y_{O_e} = 6.61 \text{ mol m}^{-3} \text{ s}^{-1}$$
$$R_2 = k_{+2}c^2 y_{N_e} y_{O_2} = 6.29 \text{ mol m}^{-3} \text{ s}^{-1}$$
$$R_3 = k_{+3}c^2 y_{N_e} y_{OH_e} = 4.39 \text{ mol m}^{-3} \text{ s}^{-1}$$

leading to a characteristic time for NO formation of

$$\tau_{NO} = \frac{[NO]_e}{4R_1} = \frac{7.18 \times 10^{-3} \times 52.9 \text{ mol m}^{-3}}{4 \times 6.61 \text{ mol m}^{-3} \text{ s}^{-1}}$$
$$= 0.0143 \text{ s}$$

which is more than the residence time of 0.005 s, so we do not expect NO to reach equilibrium. To find the actual conversion, we can solve (3.13) by iteration using, from (3.10),

$\kappa = 0.619$. The result is

$$\alpha = \frac{y_{NO}}{y_{NO_e}} = 0.165$$

or

$$y_{NO} = \alpha y_{NO_e} = 1.18 \times 10^{-3}$$

$$= 1180 \text{ ppm}$$

It is also worthwhile to examine the validity of the assumptions regarding the nitrogen atoms. The steady-state nitrogen atom level is given by (3.9)

$$\beta_{ss} = \frac{\kappa + \alpha}{1 + \kappa\alpha} = 0.711$$

The time required to reach steady state is estimated by (3.17):

$$\tau_N = \frac{[N]_e}{R_2 + R_3} = \frac{c y_{N_e}}{R_1/\kappa} = 5.15 \times 10^{-8} \text{ s}$$

Clearly, the steady state is achieved on a time scale that is very short compared to that for NO formation.

### 3.1.2 Prompt NO

Nitric oxide can be formed from $N_2$ in the air through a mechanism distinct from the thermal mechanism. This other route, leading to what is termed *prompt* NO, occurs at low temperature, fuel-rich conditions and short residence times. This mechanism was first identified by C.P. Fenimore in 1971. In studying NO formation in fuel-rich hydrocarbon flames, Fenimore found that NO concentration profiles in the postflame gases did not extrapolate to zero at the burner surface. He did not find such behavior in either CO or $H_2$ flames, which are not hydrocarbons. Fenimore concluded that the NO formed early in the flame was the result of the attack of a hydrocarbon free radical on $N_2$, in particular by

$$CH + N_2 \longrightarrow HCN + N$$

The rate of oxidation of the fuel is usually sufficiently rapid that fuel radicals such as CH are at such low concentrations that reactions such as $CH + N_2$ are negligible. Under certain fuel-rich conditions, however, such hydrocarbon radicals can reach high enough concentration levels that reactions with $N_2$ can become an important mode of breaking the $N_2$ bond and, in turn, be responsible for significant NO formation. Such reactions appear to have relatively low activation energy and can proceed at a rate comparable to that of the oxidation of the fuel. Because of the early (that is, within the flame rather than in the post-flame gases) formation of NO by this mechanism, relative to that formed by the Zeldovich mechanism, NO thus formed is often referred to as *prompt* NO (Bowman, 1975). Even in fuel-lean flames where the hydrocarbon radical attack of $N_2$ is unimportant, the nonequilibrium chemistry in the flame front can lead to prompt NO formation. The quantity of NO formed in the flame zone increases as the equivalence ratio increases. Miller and Fisk (1987) have found that in the combustion of methane in a well-stirred reactor at 2 ms residence time the CH + $N_2$ reaction accounts for virtually all of the fixed nitrogen at equivalence ratios greater than 1.2; in fact, the mechanism explains 25 percent of the fixed nitrogen even at stoichiometric conditions. Longer residence times, however, increasingly favor the Zeldovich mechanism. The prompt NO route adds the complication that fixed nitrogen can be emitted from combustors in forms other than $NO_x$, that is HCN or products of its oxidation.

Prediction of NO formation within the flame requires coupling the NO kinetics to an actual hydrocarbon combustion mechanism. As noted in Chapter 2, hydrocarbon combustion involves several steps, that is, attack of the hydrocarbon molecules leading to CO formation, CO oxidation, and radical consumption by three-body recombination reactions. Since the attack of $N_2$ by O is highly endothermic, most prompt NO is formed relatively late in fuel-lean flames, after CO has been formed but before the final C/H/O equilibrium is achieved. Sarofim and Pohl (1973) proposed estimating the concentrations of the radicals in the region of the flame where CO is consumed using a partial equilibrium approximation. In such an approximation, the rapid bimolecular reactions of the radicals (reactions 10–15 of the hydrocarbon oxidation mechanism presented in Section 2.4.1) are assumed to be locally equilibrated long before the CO oxidation,

$$CO + OH \rightleftharpoons CO_2 + H$$

and the three-body recombination reactions approach equilibrium. After some rearrangement, the radical exchange reactions yield overall reactions that relate the radical concentrations to those of the major species:

$$H_2 + O_2 \rightleftharpoons H_2O + O$$

$$H_2 + O_2 \rightleftharpoons 2OH$$

$$3H_2 + O_2 \rightleftharpoons 2H_2O + 2H$$

Sarofim and Pohl (1973) used such relationships to calculate the concentrations of H, O, and OH based on measurements of the major species concentrations in a laminar flame front. The partial equilibrium radical concentrations were then used to make improved estimates of the rate of NO formation within the flame front by the reactions of the Zeldovich mechanism. The calculated NO levels in the flame were in reasonable agreement with experimental observations for fuel-lean and slightly fuel-rich flames.

To predict NO formation in the flame front, the kinetics of the CO oxidation and the three-body recombination reactions must be followed along with the kinetics of NO formation. Using the partial-equilibrium approximation, overall rate expressions can be derived for these additional processes, as will be demonstrated in our discussion of CO oxidation. The intricate coupling of the NO chemistry to the hydrocarbon oxidation mechanism in fuel-rich flames precludes the development of simplified models for such flames.

Although NO formation rates in the vicinity of the flame can be large, the quantity of NO formed in the postflame region is large compared to the prompt NO in many practical combustions. The coupling between NO formation and the combustion process, to a first approximation, can be neglected in this case, and the extended Zeldovich mechanism and equilibrium properties of the postcombustion gases can be used to calculate NO emissions. As $NO_x$ emission levels are reduced to very low levels, however, the relative importance of the prompt NO can be expected to increase, possibly limiting the effectiveness of the $NO_x$ emission controls.

### 3.1.3 Thermal-$NO_x$ Formation and Control in Combustors

The inhomogeneities in composition in nonpremixed combustion strongly influence $NO_x$ emissions. The experiments and approach of Pompei and Heywood (1972) provide a convenient vehicle for exploring the role of turbulent mixing in $NO_x$ formation. In Section 2.5.4 we discussed their use of oxygen and a Gaussian composition distribution in the estimation of the degree of inhomogeneity in a nonpremixed combustor. The influence of inhomogeneity on the initial rate of NO formation can be seen by calculating the mean NO formation rate. In the discussion to follow, we limit our attention to NO formation by the Zeldovich mechanism before the NO concentration has accumulated to the point that the reverse reactions become significant. Prompt NO will not be considered.

The mean NO formation rate

$$\overline{R_{NO}} = \bar{\rho} \int_0^{\infty} \frac{R_{NO}(\phi)}{\rho(\phi)} p(\phi)\, d\phi \tag{3.18}$$

is shown in Figure 3.3 as a function of equivalence ratio for several values of the segregation parameter, $S = \sigma/\bar{\phi}$. For $S = 0$ (perfect mixing) we see the sharp peak in the NO formation rate. Poorer mixing substantially reduces the maximum NO formation rate but extends the domain of significant NO formation to lower equivalence ratios.

The strong dependence of the Zeldovich kinetics on temperature provides the major tool used in the control of NO formation in combustion systems. Any modifications of the combustion process that reduce the peak temperatures in the flame can be used to reduce $NO_x$ emissions. Because of this temperature dependence, the NO formation rate varies strongly with equivalence ratio, with a sharp peak near $\phi = 1$, as was shown in Figure 3.1. Reduction of the equivalence ratio is one possible method for $NO_x$ control, but as we have just seen, this method is substantially less effective in nonpremixed combustion than simple theory might predict. While chemical considerations would suggest that reducing $\bar{\phi}$ from 0.9 to 0.7 should reduce the $NO_x$ formation rate by two orders of magnitude, a typical combustor with $S \approx 0.5$ would show virtually no change. Fuel-lean combustion reduces the flame temperature by diluting the combustion gases with excess air. If a material that does not participate in the combustion reactions is used as a diluent instead of air, the adiabatic temperature of stoichiometric combustion can be reduced and more effective control can be achieved. One common method is flue gas recirculation (FGR), in which the most readily available nonreactive gas, cooled com-

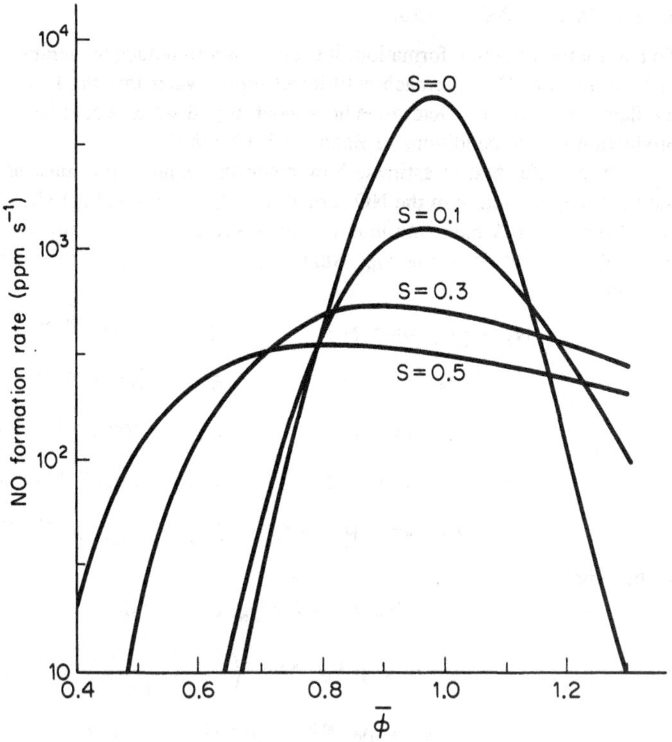


**Figure 3.3** Influence of mixing on the NO formation rate in adiabatic combustion of kerosene, $CH_{1.8}$.

bustion products, is mixed with the combustion air. This approach is used extensively in utility boilers and other large stationary combustors. Injection of other diluents such as water or steam can also be used to reduce the NO formation rates, but the penalty in reduced system efficiency may be larger.

Modifications of the combustion system design or operation are also used to control $NO_x$ emissions. One way the combustion rate is lowered is to introduce only a fraction of the air with fuel, with the remaining air being added in so-called overfire air ports above the main burners. Heat transfer from the fuel-rich region lowers the ultimate flame temperature and, therefore, the NO formation rate when the air required to complete the combination is finally added. Different types of burners yield widely different $NO_x$ emission levels depending on the time required for combustion and the amount of heat rejected during the combustion process. A long flame that slowly entrains air allows a large fraction of the heat of combustion to be transferred to the furnace walls before combustion is complete, while a highly turbulent flame with intense recirculation may be more nearly adiabatic. $NO_x$ emissions from the former type of burner are generally lower than those from the latter.

**Example 3.2** ***Thermal-NO$_x$ Control***

To control thermal-NO$_x$ formation, it is necessary to reduce the temperature to slow the rate of $N_2$ oxidation. One approach is to inject liquid water into the fuel–air mixture to reduce the flame temperature. Determine how much liquid water would be required to lower NO$_x$ production for the conditions of Example 3.1 by 90%.

It is useful first to estimate how much the temperature must be reduced to achieve the necessary reduction in the NO formation rate. Since the final NO level was well below equilibrium, a 90% reduction in emissions corresponds approximately to a 90% reduction in $R_1$. We need to determine $y_{Oe}$. Applying van't Hoff's relation, we find the equilibrium constants:

$$\tfrac{1}{2}N_2 + \tfrac{1}{2}O_2 \rightleftharpoons NO \qquad K_{pNO} = 4.71e^{-10,900/T}$$

$$\tfrac{1}{2}N_2 \rightleftharpoons N \qquad K_{pN} = 3030e^{-57,830/T}\ \text{atm}^{1/2}$$

$$\tfrac{1}{2}O_2 \rightleftharpoons O \qquad K_{pO} = 3030e^{-30,790/T}\ \text{atm}^{1/2}$$

$$\tfrac{1}{2}H_2O + \tfrac{1}{4}O_2 \rightleftharpoons OH \qquad K_{pOH} = 166e^{-19,680/T}\ \text{atm}^{1/4}$$

$$\tfrac{1}{2}H_2O \rightleftharpoons H + \tfrac{1}{4}O_2 \qquad K_{pH} = 44,100e^{-42,500/T}\ \text{atm}^{3/4}$$

$R_1$ becomes

$$R_1 = k_1c^2y_{N_2}y_{Oe} = k_1c^2y_{N_2}K_{pO}p^{-1/2}y_{O_2}^{1/2}$$

$$= 1.8 \times 10^8\, e^{-38,370/T}\left(\frac{1.013 \times 10^6}{8.314T}\right)^2 \times 0.751$$

$$\times\ 3030e^{-30,790/T}\ 10^{-1/2} \times 0.0397^{1/2}$$

$$= 3.83 \times 10^{20}T^{-2}e^{-69,160/T}\ \text{mol m}^{-3}\ \text{s}^{-1}$$

For 90% reduction in NO$_x$ formation, we want $R_1 \approx 0.661$ mol m$^{-3}$ s$^{-1}$. Solving iteratively for temperature, we find

$$T \approx 2130\ \text{K}$$

Thus it appears that $T$ must be reduced by only 175 K to achieve 90% control.

The amount of water required for this reduction is determined by a first-law analysis. The energy equation becomes

$$[h(T) - h(T_0)]_{CO_2} + (0.94 + \zeta)\,[h(T) - h(T_0)]_{H_2O} + 0.368[h(T) - h(T_0)]_{O_2}$$

$$+\ 6.95[h(T) - h(T_0)]_{N_2} - 1.84[h(T_a) - h(T_0)]_{O_2}$$

$$-\ 6.95[h(T_a) - h(T_f)]_{N_2} - [h(T_f) - h(T_0)]_{CH_{1.88}}$$

$$-\ \zeta[h(T_w) - h(T_0)] + \zeta\,\Delta h_v(T_0) + \Delta h_{cL}(T_0)$$

where $\zeta$ is the number of moles of water added per mole of carbon and $\Delta h_v$ is the molar latent heat of vaporization of the water.

Using data from Table 2.5, we find

$$(289 + 32.5\zeta)\,(T - T_0) + (0.0190 + 0.00431\zeta)\,(T^2 - T_0^2)$$

$$-\ 259.4(T_a - T_0) - 0.0139(T_a^2 - T_0^2) + 44,000\zeta - 600,000 = 0$$

Recalling from Example 3.1 that $T_a = 560$ K, and imposing $T = 2130$ K, this becomes

$$122{,}711\zeta - 57{,}126 = 0$$

or

$$\zeta = 0.47$$

Thus we require 0.60 kg of water to be added for every 1 kg of fuel burned.

Now let us confirm the amount of NO formed. The combustion product composition for $T = 2130$ K and $\zeta = 0.47$ is

$$y_{N_2} = \frac{6.95}{9.26 + 0.47} = 0.714 \qquad y_{N_e} = 1.31 \times 10^{-9}$$

$$y_{O_2} = 0.0378 \qquad y_{O_e} = 9.82 \times 10^{-5}$$

$$y_{H_2O} = 0.145 \qquad y_{OH_e} = 1.52 \times 10^{-3}$$

$$y_{CO_2} = 0.103 \qquad y_{H_e} = 4.63 \times 10^{-4}$$

$$y_{NO_e} = 4.64 \times 10^{-3}$$

from which we find

$$R_1 = 0.620 \text{ mol m}^{-3}\text{ s}^{-1}$$

$$R_2 = 0.690$$

$$R_3 = 0.369$$

and

$$\tau_{NO} = \frac{c y_{NO_e}}{4R_1} = 0.107 \text{ s}$$

$$\kappa = \frac{R_1}{R_2 + R_3} = 0.585$$

For a residence time of 0.005 s, we find by iteration

$$\alpha = 2.33 \times 10^{-2}$$

and

$$y_{NO} = 1.08 \times 10^{-4} = 108 \text{ ppm}$$

Thus, emissions would be reduced by 91% through the addition of 0.60 kg of water for each 1 kg of fuel burned. The effects of this large water addition on engine performance would also have to be considered in assessing this approach to $NO_x$ emission control.

### 3.1.4 Fuel-$NO_x$

Many fuels contain organically bound nitrogen that is readily oxidized to NO during combustion (Sarofim and Flagan, 1976). Crude oils contain 0.1 to 0.2% nitrogen on a mass basis, but levels as high as 0.5% are found in some oils. In refining the oil, this

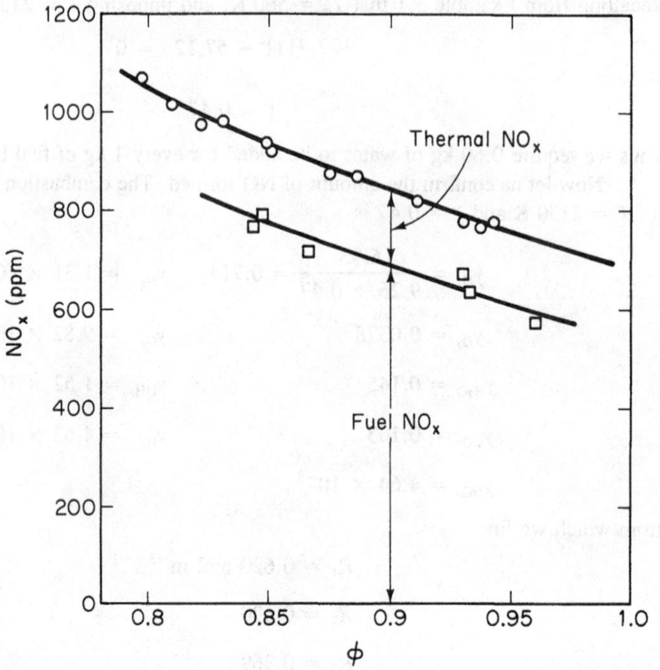


**Figure 3.4** Contributions of thermal-$NO_x$ and fuel-$NO_x$ to total $NO_x$ emissions in the laboratory pulverized coal combustion experiments of Pershing and Wendt (1977). Reprinted by permission of The Combustion Institute.

nitrogen is concentrated in the residual fractions, that is, in that portion of the oil that is most likely to be burned in large combustion systems such as power plants or industrial boilers rather than used as transportation fuels. Coal typically contains 1.2 to 1.6% nitrogen. The range of nitrogen contents of coals is much narrower than that of the sulfur contents. Thus burning a low-nitrogen coal is not a practical solution to the problems of fuel-$NO_x$ emissions from coal-fired boilers unless one dilutes the coal with a low-nitrogen fuel oil. New fuel sources may further aggravate the problems associated with fuel-nitrogen. Some of the major shale oil deposits in the United States contain 2 to 4% nitrogen.

The contribution of fuel-nitrogen to $NO_x$ emissions is most clearly shown by experiments in which the possibility of forming NO from $N_2$ is eliminated. Pershing and Wendt (1977) compared the $NO_x$ emissions from combustion of pulverized coal in air to the emissions when the air was replaced with a mixture of oxygen, argon, and carbon dioxide that was selected to achieve the same adiabatic flame temperature as combustion at the same equivalence ratio in air. The former case includes both NO formed from $N_2$ and NO from the fuel nitrogen while fuel-nitrogen contributes to NO formation in the latter case. The data of Pershing and Wendt, shown in Figure 3.4, clearly demonstrate

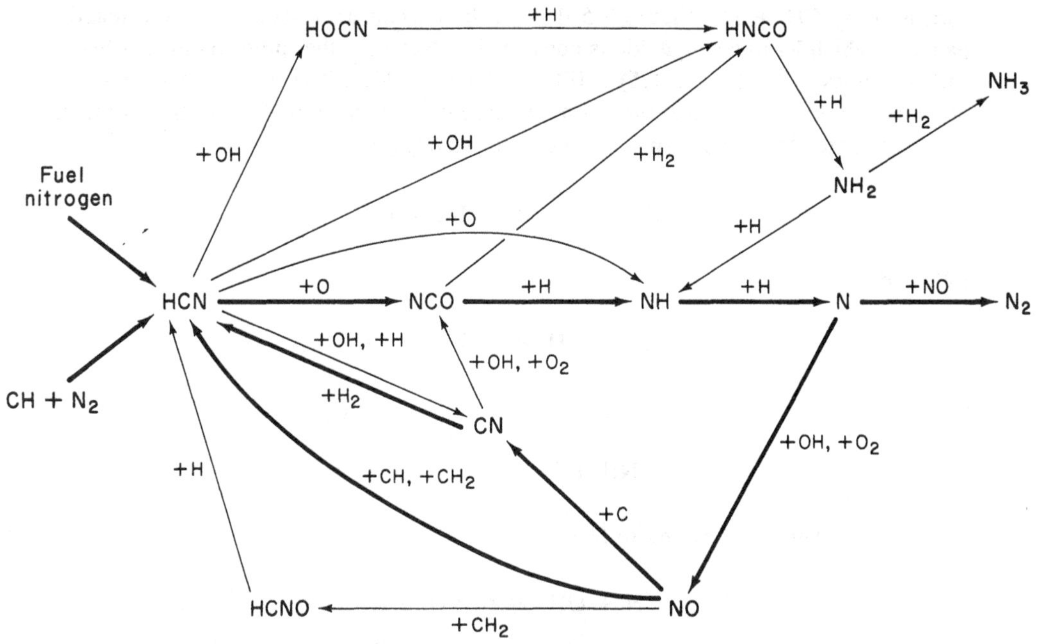


**Figure 3.5** Reaction paths in rich hydrocarbon flames ($1.0 < \phi < 1.5$). Thick arrows show the dominant steps. (Miller and Fisk, 1987).

the importance of the fuel-nitrogen. Even though fuel-nitrogen was the major source of $NO_x$, not all of the fuel-nitrogen was converted to NO. In these experiments, 20 to 30% conversion was observed.

The nitrogen in these fuels is present predominantly in pyridine and pyrrole groups:

pyridine pyrrole

In the early phases of combustion, these molecules undergo ring schism. Further attack yields molecules such as HCN or $NH_3$ or nitrogen-containing radicals such as $NH_2$ or CN. These reactions are an integral part of the combustion of the parent hydrocarbon and therefore proceed rapidly in the combustion process.

The principal paths by which fuel nitrogen species are combusted are thought to begin with the conversion of the fuel nitrogen molecule to hydrogen cyanide, HCN (Haynes et al., 1975). We recall that the key step in prompt NO formation is also HCN formation by $CH + N_2$. Figure 3.5 depicts the current understanding of the reaction paths by which hydrogen cyanide is converted to NO and other products in rich hydrocarbon flames ($1.0 < \phi < 1.5$) (Miller and Fisk, 1987). The thick arrows show the dominant steps. Under most conditions the dominant path from HCN to NO is the sequence initiated by reaction of HCN with atomic oxygen,

$$HCN + O \rightleftarrows NCO + H$$

followed by

$$NCO + H \rightleftarrows NH + CO$$

and

$$NH + H \rightleftarrows N + H_2$$

with the nitrogen atom leading to NO,

$$N + OH \rightleftarrows NO + H$$

$$N + O_2 \rightleftarrows NO + O$$

The NO produced in this way may itself react with N atoms to form $N_2$,

$$NO + N \rightleftarrows N_2 + O$$

or it can be recycled to form CN or HCN by reaction with hydrocarbon free radicals,

$$NO + C \rightleftarrows CN + O$$

$$NO + CH \rightleftarrows HCN + O$$

$$NO + CH_2 \rightleftarrows HCN + OH$$

The reaction paths through CN and HNCO can also be important, for example the sequence,

$$HCN + OH \rightleftarrows HNCO + H$$

$$HNCO + H \rightleftarrows NH_2 + CO$$

$$NH_2 + H_2 \rightleftarrows NH_3 + H$$

leads to ammonia formation.

Under leaner conditions the oxygen atom concentration may be sufficiently high that NCO and CN may react with oxygen atoms to produce NO directly or indirectly. NH may react with O or OH to produce NO directly or indirectly through HNO, for example

$$NH + O \rightleftharpoons NO + H$$

$$\rightleftharpoons N + OH$$

The nitrogen species produced by this chain of reactions can also react to form $N_2$. The formation of $N_2$ from fuel-nitrogen requires the reaction of two fixed nitrogen species, for example,

$$NO + N \underset{-1}{\overset{+1}{\rightleftharpoons}} N_2 + O$$

$$NH + N \underset{-2}{\overset{+2}{\rightleftharpoons}} N_2 + H$$

Note that the reverse of reaction 1 is the first reaction of the Zeldovich mechanism, which, as we have already seen, is the predominant reaction contributing to NO formation from $N_2$ in fuel-lean or near stoichiometric combustion. Other reactions that form the strong nitrogen–nitrogen bond may ultimately lead to $N_2$ formation. Such reactions include

$$NH + NH \underset{-3}{\overset{+3}{\rightleftharpoons}} N_2H + H$$

$$NO + NH \underset{-4}{\overset{+4}{\rightleftharpoons}} N_2O + H$$

$$NO + NH_2 \overset{5a}{\rightleftharpoons} N_2 + H_2O$$

$$\overset{5b}{\rightleftharpoons} N_2 + H + OH$$

$$\overset{5c}{\rightleftharpoons} N_2O + H_2$$

The NO + $NH_2$ reaction has three possible outcomes but may be described approximately by a single rate expression for the net formation of the nitrogen–nitrogen bond (Hanson and Salimian, 1984). The concentrations of the fixed nitrogen species are relatively low, so these reactions that are second order in fixed nitrogen (RN) proceed slowly.

The fuel-nitrogen chemistry may be described schematically by

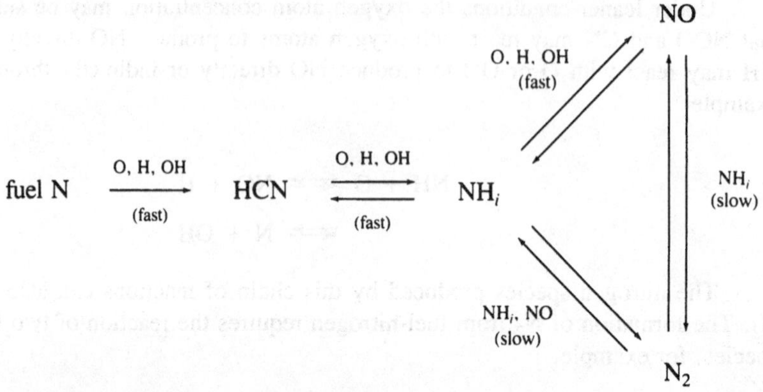


Since the reactions leading to $NH_i$ and NO are much faster than those leading to $N_2$ formation, a rate-constrained partial-equilibrium model can be developed to describe the rate at which fuel-nitrogen is converted to $N_2$ (Flagan et al., 1974).

We begin our quantitative discussion of fuel-nitrogen chemistry by examining how the fuel-nitrogen would be distributed if, during the initial hydrocarbon attack, no $N_2$ formation were to occur. Because the formation of $N_2$ requires the reaction of two fixed nitrogen species, one being present only in very small concentrations due to the rapid consumption of N, NH and $NH_2$, $N_2$ formation proceeds much more slowly than the flame chemistry. Hence the distribution of the fuel-nitrogen subject to the constraint that no $N_2$ be formed from fuel-N is a reasonable approximation of the gas composition immediately downstream of the flame. This partial-equilibrium distribution of single nitrogen species for adiabatic combustion of a fuel oil containing 1% by weight of nitrogen is illustrated in Figure 3.6. The total amount of fuel-nitrogen in the parent fuel is indicated by the dashed line labeled RN. Given the constraint on equilibrium, the sum of the concentrations of all single nitrogen (RN) species must equal this value. At equivalence ratios from 0.1 to 1.6, the major fixed nitrogen species in this partial equilibrium is NO. At higher equivalence ratios, $NH_3$ dominates. $N_2$ is the primary species in very fuel-lean gases. Nitrogen atoms and other radicals are present only in very low concentrations.

The NO concentration at full thermodynamic equilibrium ($NO_e$) is also shown by a dashed line. The NO derived from the fuel-nitrogen is below the equilibrium level for equivalence ratios ranging from 0.5 to 1.05, so additional NO formation from $N_2$ may be expected in this regime. The primary reactions leading to the fixation of $N_2$ are those of the Zeldovich mechanism, so the model developed in Section 3.1.1 describes the $N_2$ fixation in this regime. Outside this region, the conversion of fixed nitrogen to $N_2$ is favored thermodynamically. The rate at which $N_2$ is formed from fuel-nitrogen intermediates in these rich and lean regimes can be examined with the partial-equilibrium approach. The basic assumption of this model is that within the flame, the fuel-nitrogen is distributed among all the possible fixed nitrogen species according to a local thermodynamic equilibrium. The conversion of fixed nitrogen to $N_2$ can then be described using the known kinetics of reactions 1–5.

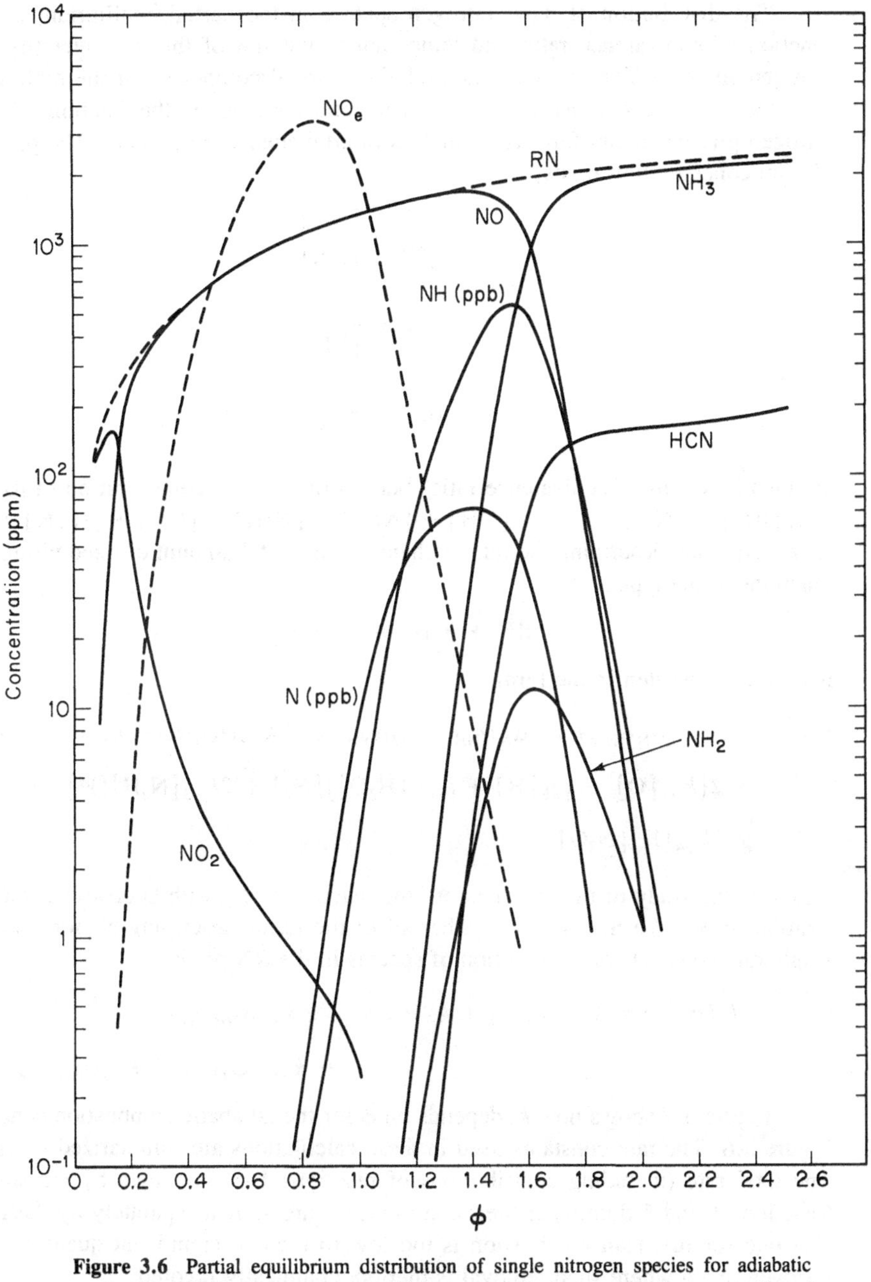


**Figure 3.6** Partial equilibrium distribution of single nitrogen species for adiabatic combustion of a fuel oil containing 1% by weight of nitrogen. Total concentration of fixed nitrogen for this fuel and the NO concentration corresponding to full thermodynamic equilibrium are shown by broken lines.

The distribution of fixed nitrogen species in the partial equilibrium is a strong function of equivalence ratio and temperature, but not of the total quantity of fixed nitrogen since the fixed nitrogen is a relatively small component of the fuel. To a first approximation we may assume that in adiabatic combustion the fraction of the fixed nitrogen present in any form is a function of equivalence ratio alone. The partial equilibrium concentration ratios,

$$\alpha_{NO} = \frac{[NO]}{[RN]}$$

$$\alpha_{N} = \frac{[N]}{[RN]}$$

$$\alpha_{NH} = \frac{[NH]}{[RN]} \tag{3.19}$$

are thus functions of equivalence ratio, but not of the total concentration of fixed nitrogen, $[RN] = [NO] + [N] + [NH] + [NH_2] + [NH_3] + [CN] + [HCN] + [NO_2]$. A rate equation describing the rate of change in the total amount of fixed nitrogen by all reactions of the type

$$RN + R'N \rightleftharpoons N_2 + \cdots$$

may now be written in the form

$$\begin{aligned} R_{RN} = & -2(k_{+1}\alpha_N\alpha_{NO} + k_{+2}\alpha_N\alpha_{NH} + k_{+3}\alpha_{NH}\alpha_{NH} + k_{+4}\alpha_{NO}\alpha_{NH} + k_{+5}\alpha_{NO}\alpha_{NH_2})[RN]^2 \\ & + 2(k_{-1}[O] + k_{-2}[H] + k_{-5a}[H_2O])[N_2] + 2k_{-3}[N_2H][H] \\ & + 2k_{-4}[H][N_2O] \end{aligned} \tag{3.20}$$

Based on the study of thermal fixation, the reaction of $N_2$ with O clearly dominates the fixation of $N_2$. We may lump together all of the terms describing $N_2$ formation into a single rate constant for the reaction of species in the RN pool:

$$\begin{aligned} k_e(\phi, T) = & \; k_{+1}\alpha_N\alpha_{NO} + k_{+2}\alpha_N\alpha_{NH} + k_{+3}\alpha_{NH}\alpha_{NH} \\ & + k_{+4}\alpha_{NO}\alpha_{NH} + k_{+5}\alpha_{NO}\alpha_{NH_2} \end{aligned} \tag{3.21}$$

Figure 3.7 shows how $k_e$ depends on $\phi$ for the adiabatic combustion conditions of Figure 3.6. The rate constants used in these calculations are summarized in Table 3.1. The total rate (including contributions of reactions 1–5) is shown by the solid lines. Reactions 1 and 5 dominate; their contributions are shown separately by dashed lines. The rate for fuel-lean combustion is too low to remove significant quantities of fixed nitrogen, even where such removal is thermodynamically favored.

In the fuel-rich region where the initial RN concentration is far in excess of equilibrium, only the RN removal term is important, so we may write

$$R_{RN} = -k_e(\phi, T)[RN]^2 \tag{3.22}$$

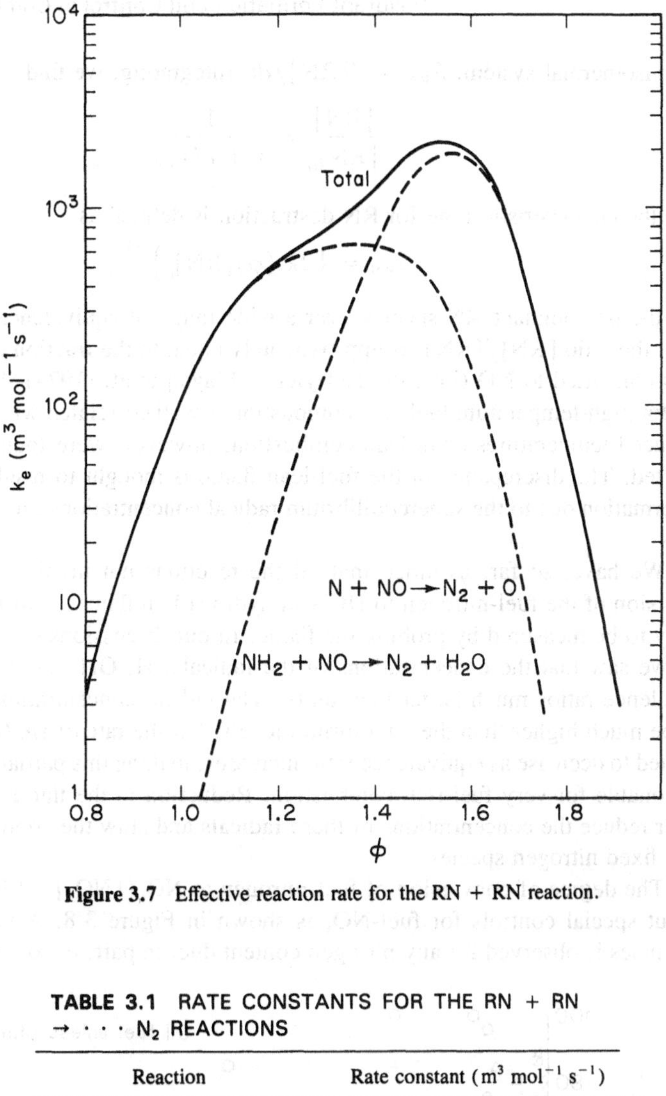


**Figure 3.7** Effective reaction rate for the RN + RN reaction.

**TABLE 3.1 RATE CONSTANTS FOR THE RN + RN → · · · $N_2$ REACTIONS**

| Reaction | Rate constant ($m^3\ mol^{-1}\ s^{-1}$) |
|---|---|
| $N + NO \underset{-1}{\overset{+1}{\rightleftharpoons}} N_2 + O$ | $k_{+1} = 3.8 \times 10^7\ e^{-425/T}$<br>$k_{-1} = 1.8 \times 10^8\ e^{-38,370/T}$ |
| $N + NH \overset{2}{\rightleftharpoons} N_2 + H$ | $k_2 = 6.3 \times 10^5\ T^{0.5}$ |
| $NH + NH \overset{3}{\rightleftharpoons} N_2H + H$ | $k_3 = 7.9 \times 10^5\ T^{0.5} e^{-500/T}$ |
| $NO + NH \overset{4}{\rightleftharpoons} N_2O + H$ | $k_4 = 1.1 \times 10^6\ e^{-230/T}$ |
| $NO + NH_2 \overset{5}{\rightleftharpoons}$ products $N_2$, $N_2O$ | $k_5 = 1.2 \times 10^{14}\ T^{-2.46} e^{-938/T}$ |

*Source:* Hanson and Salimian (1984).

For an isothermal system, $R_{RN} = d[RN]/dt$. Integrating, we find

$$\frac{[RN]}{[RN]_0} = \frac{1}{1 + t/\tau_{RN}} \tag{3.23}$$

where the characteristic time for RN destruction is defined as

$$\tau_{RN} = \{2k_e(\phi)[RN]_0\}^{-1} \tag{3.24}$$

Since the predominant RN species over a wide range of equivalence ratios ($\phi < 1.6$) is NO, the ratio $[RN]/[RN]_0$ is approximately equal to the fraction of the fuel-nitrogen that is converted to NO (i.e., the NO yield). Flagan et al. (1974) showed that the NO yield for high-temperature fuel-rich combustion is well correlated with (3.23). NO yields at reduced temperatures or in lean combustion, however, were found to be lower than predicted. The discrepancy in the fuel-lean flame is thought to result from accelerated NO formation due to the superequilibrium radical concentrations present in hydrocarbon flames.

We have, so far, assumed that all the reactions not involving $N_2$ are fast. The conversion of the fuel-nitrogen to HCN in hydrocarbon flames is usually completed too rapidly to be measured by probing the flame. In our discussions of combustion equilibrium we saw that the concentrations of the radicals, H, OH, and O, become small at equivalence ratios much larger than unity. The radical concentrations within the flame may be much higher than the equilibrium levels, but the rate of HCN attack can still be expected to decrease as equivalence ratio increases, making this partial-equilibrium model questionable for very fuel-rich combustion. Reduction in the flame temperature would further reduce the concentrations of these radicals and slow the approach to equilibrium of the fixed nitrogen species.

The degree of conversion of fuel-nitrogen to $NO_x([NO_x]/[RN]_0)$ in combustors without special controls for fuel-$NO_x$ is shown in Figure 3.8. A range of conversion efficiencies is observed for any nitrogen content due, in part, to contributions from ther-

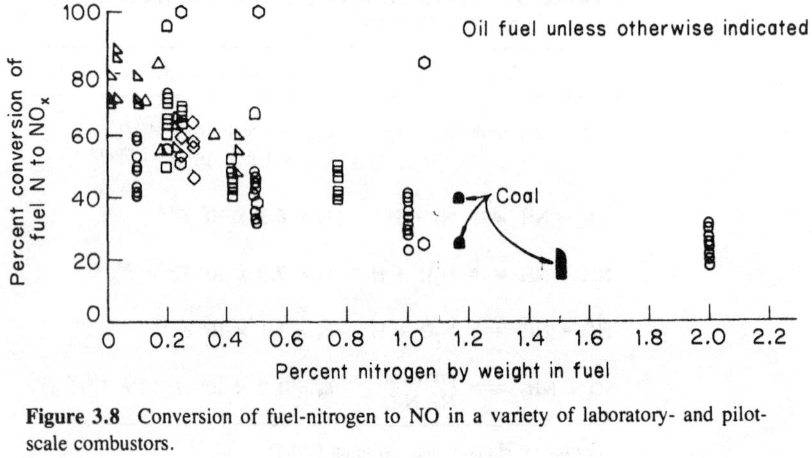


**Figure 3.8** Conversion of fuel-nitrogen to NO in a variety of laboratory- and pilot-scale combustors.

mal fixation of $N_2$. There is, however, a distinct lower bound to the degree of conversion, and that lower limit to the conversion decreases as the nitrogen content of the fuel increases.

The fuel-nitrogen conversion is lower than one would expect based on the mechanism described above and the overall equivalence ratios (generally less than 1) at which the combustion systems are operated. The formation of $N_2$ is favored by fuel-rich combustion. This discrepancy is attributed to the influence of mixing on fuel-$NO_x$ formations. Imperfect mixing allows combustion gases to remain fuel-rich even though the combustor is fuel-lean overall, thereby reducing the amount of fuel-$NO_x$ formed.

Figure 3.9 shows the effects of changing mixing rates on the formation of $NO_x$ from combustion of kerosene doped with 0.51% fuel-nitrogen in the same plug flow combustor on which we have focused in our previous discussions of mixing. Once again, high atomizing pressures yield high mixing rates and relatively uniform compositions for the combustion gases. The amount of NO corresponding to 100% conversion of the fuel-nitrogen is shown by the solid line. The dashed line adds to this the quantity of NO formed in well-mixed combustion in the absence of fuel nitrogen, i.e., that due to thermal fixation of $N_2$. Consider first the results for well-mixed combustion. NO yields for fuel-lean combustion are close to the amount of fuel-nitrogen and exceed the fuel nitrogen near stoichiometric. The excess NO is due to the thermal fixation of atmospheric nitrogen. As the equivalence ratio is increased beyond unity, the NO level drops rapidly. This we expect from the mechanism described above. As the mixing rate (atomizing pressure) is decreased, the NO yield decreases at all equivalence ratios. At the lowest mixing rates, the NO mole fraction is almost independent of the equivalence ratio.

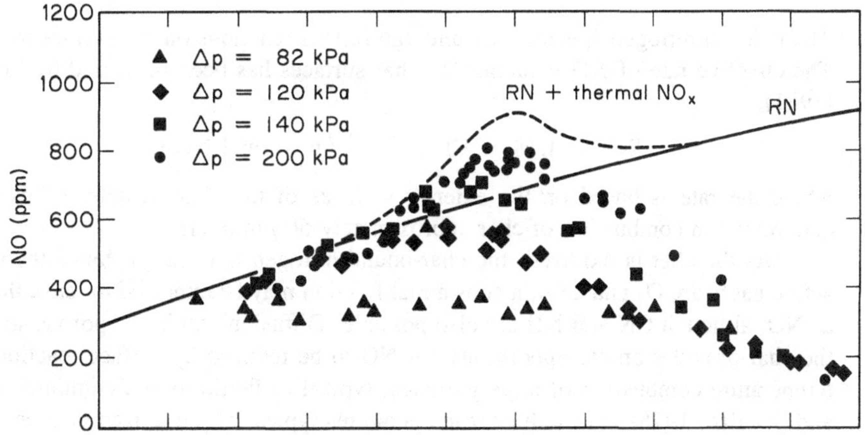


**Figure 3.9** Influence of mixing on fuel nitrogen conversion to $NO_x$. Data are from combustion of kerosene doped with 0.51% nitrogen by weight using an air-assist atomizer (Flagan and Appleton, 1974). Reprinted by permission of The Combustion Institute.

The interaction between turbulent mixing and the chemical kinetics of fuel-nitrogen conversion is too complicated to be treated with the simple probability density function approach that we have applied to thermal fixation and CO concentrations. The length of time a fluid element resides at high equivalence ratios determines the amount of fixed nitrogen that will remain when it is finally diluted with air to substoichiometric conditions. More elaborate descriptions of the evolution of the probability density function in turbulent mixing (e.g., Flagan and Appleton, 1974; Pope, 1985) are needed for calculations of fuel-nitrogen conversion in turbulent flames, but these models are beyond the scope of this book.

The nitrogen in solid or heavy fuel oils may be released with the volatiles and will behave like the volatile nitrogen compounds discussed above, or it may remain with the refractory materials, forming part of the char. In distillation of heavy oils, the nitrogen is generally concentrated in the heavy fractions. Studies of the fate of organically bound nitrogen in coal have shown the char to be slightly enriched in nitrogen (Pohl and Sarofim, 1975). The effect appears to be small, so to a first approximation the fraction of nitrogen in the char may be assumed to be in proportion with the char yield.

The char introduces two factors that we have not yet taken into account. The char particle consumes oxygen, thereby providing a locally reducing atmosphere that can promote the conversion of NO to $N_2$. Second, there is evidence of NO being reduced on carbon surfaces (Wendt et al., 1979). There are two possible paths for nitric oxide reduction on carbon:

1. Direct reduction to $N_2$
2. Transformation of NO to HCN or $NH_3$

The reduced nitrogen species may undergo further reactions on the surface to form $N_2$. The effective rate of NO reduction on char surfaces has been measured by Levy et al. (1981),

$$R_{NO} = 4.18 \times 10^4 e^{-17,500/T} p_{NO} \quad \text{mol NO m}^{-2} \text{ s}$$

where the rate is based on the exterior surfaces of the char particles. This rate was determined in combustion of char approximately 50 $\mu$m in size.

As the char is oxidized, the char-bound nitrogen is released. Since the major reactive gases are $O_2$ and $CO_2$, a substantial fraction may be expected to leave the surface as NO, although CN and NH are also possible. Diffusion within the porous structure of the char provides ample opportunity for NO to be reduced by surface reactions in low temperature combustion of large particles, typical of fluidized-bed combustion (Wendt and Schulze, 1976). At the higher temperatures typical of pulverized coal combustion, the release of char nitrogen as NO was found to decrease with increasing particle size. That is to be expected since the atmosphere at the particle surface becomes increasingly reduced as the limit of diffusion-controlled combustion is approached. In fuel-rich pulverized fuel combustion experiments, the NO level is observed to rise rapidly to a maximum and then decay slowly (Wendt et al., 1979). In char combustion, the decline was attributed to heterogeneous reduction of NO on the char surface. The hydrogen released with the volatile matter from coal increases the concentration of $NH_3$, HCN, and other fixed nitrogen species, accelerating the rate of homogeneous conversion of NO to $N_2$ beyond the heterogenous reactions. In either case, extended residence times in fuel-rich conditions promote the conversion of fuel-nitrogen to $N_2$.

### 3.1.5 Fuel-$NO_x$ Control

Since the conversion of fuel-nitrogen to NO is only weakly dependent on temperature but is a strong function of the combustion stoichiometry, temperature reduction by flue gas recycle or steam injection, which are effective methods for thermal-$NO_x$, have little influence on fuel-$NO_x$. What is required to minimize the amount of fuel-nitrogen leaving a combustor as NO is that the gases be maintained fuel-rich long enough for the $N_2$-forming reactions to proceed. Since the overall combustion process must be fuel-lean if high combustion efficiency is to be maintained, this generally requires dividing the combustion process into separate fuel-rich and fuel-lean stages.

A number of names are applied to the various implementations of staged combustion, including: overfire air, off-stoichiometric combustion, and low-$NO_x$ burners. Most commonly, only part of the air required for complete combustion is supplied with the fuel. The remaining air is supplied through separate "overfire air" ports. Staged combustion was first applied to the control of thermal-$NO_x$ because it allowed some of the heat to be rejected before completing the combustion process, but it is better suited to fuel-$NO_x$ control since it provides the time required for $N_2$ formation. This can be carried too far, however. If the primary combustion zone is operated too fuel-rich, the fixed nitrogen can be retained in a combustible form (e.g., HCN). When the secondary air is added, such compounds may act as fuel-N and form $NO_x$ rather than the $N_2$ that was sought. The $NO_x$ emissions from staged combustors, as a result, may pass through a minimum as the equivalence ratio of the primary combustion zone is increased.

The "low-$NO_x$" burners, illustrated in Figure 3.10, utilize burner aerodynamics to slow the rate at which fuel and air are mixed. Whereas most burners are designed to achieve a highly turbulent zone of intense combustion, the low-$NO_x$ burners are designed to produce a long, "lazy" flame. The degree of control that can be reached by this method is limited by the need to achieve complete combustion within the volume of the combustor. Low-$NO_x$ burners have the important advantage of being a relatively low cost technology that can be used as a retrofit on existing sources to reduce $NO_x$ emissions.

### 3.1.6 Postcombustion Destruction of $NO_x$

Reactions similar to those that convert fixed nitrogen to $N_2$ in combustion can be used to destroy nitrogen oxides in the postflame gases. Because of the similarity of these processes to combustion, we shall discuss these postcombustion treatment methods here rather than in the chapter on gas cleaning.

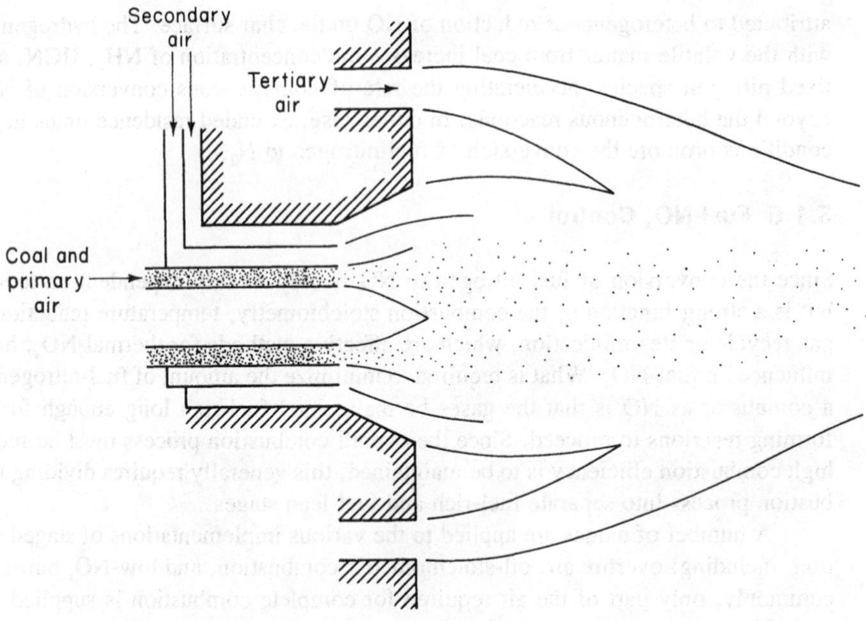


**Figure 3.10** Coal burner designed to reduce formation of $NO_x$ by spreading the mixing of fuel and air.

Wendt et al. (1973) demonstrated that nitrogen oxides could be reduced, presumably to $N_2$, by injection and oxidation of fuel in the partially cooled combustion products. This method takes advantages of the shift in the equilibrium NO concentration associated with the temperature reduction. The NO formed in the high-temperature flame region is generally not reduced as the equilibrium level decreases due to the low concentration of nitrogen atoms. When a fuel such as methane is added at sufficiently high temperature (e.g., 1800 K), it is oxidized, generating high concentrations of radicals. These radicals promote the formation of N and other reactive fixed nitrogen species from NO, for example,

$$NO + H \rightleftharpoons N + OH$$

These species then react with NO to form $N_2$.

At lower temperatures, selective reduction of NO by fixed nitrogen species (e.g., $NH_3$) may be used to destroy $NO_x$ in the products of combustion, even in the presence of a large excess of oxygen (Wendt et al., 1973; Lyon, 1976; Muzio et al., 1979). The noncatalytic process using ammonia is called the thermal de-$NO_x$ process. The temperature range in which the reaction between $NH_i$ species and NO is favored over the formation of additional NO (i.e., less than 1500 K) is much lower than typical flame temperatures.

Several detailed studies of the kinetics and mechanisms of selective reduction of NO by ammonia have been reported (Branch et al., 1982; Miller et al., 1981; Lucas and Brown, 1982; Dean et al., 1982). The partial-equilibrium assumption used in the discussion of fuel-$NO_x$ is not valid at such low temperatures since the endothermic reverse reactions are too slow to maintain the equilibrium among the single nitrogen species. The $NH_3$ does not react directly with the NO. Before any NO can be reduced to $N_2$, the ammonia must decompose by reactions such as

$$NH_3 + M \xrightarrow{1} NH_2 + H + M$$

$$NH_3 + OH \xrightarrow{2} NH_2 + H_2O$$

$$NH_3 + O \xrightarrow{3} NH_2 + OH$$

$$NH_3 + H \xrightarrow{4} NH_2 + H_2$$

The $NH_2$ may undergo further oxidation, that is,

$$NH_2 + OH \xrightarrow{5} NH + H_2O$$

$$NH_2 + O \xrightarrow{6a} NH + OH$$

$$\xrightarrow{6b} HNO + H$$

$$NH_2 + H \xrightarrow{7} NH + H_2$$

$$NH_2 + O_2 \xrightarrow{8a} HNO + OH$$

$$\xrightarrow{8b} NH + HO_2$$

Alternatively, the $NH_2$ may react with NO, leading to the ultimate formation of $N_2$,

$$NH_2 + NO \xrightarrow{9} N_2 + H_2O$$

The NH produced by reactions 5–7 is highly reactive and can be attacked by $O_2$ with a low activation energy:

$$NH + O_2 \xrightarrow{10} HNO + O$$

forming the NO bond. Thus, once the $NH_2$ is oxidized, the formation of NO or related species quickly follows.

The overall reaction sequence may be written

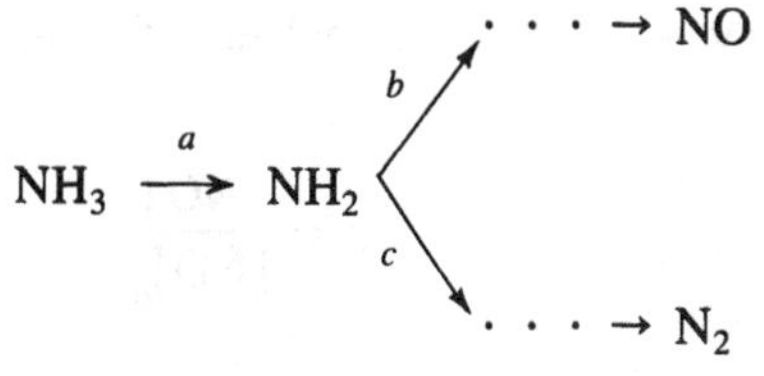

The rate equations may be written in terms of the characteristic times of each of the three types of reactions:

$$R_{NH_3} = -\frac{[NH_3]}{\tau_a} \tag{3.25}$$

$$R_{NH_2} = \frac{[NH_3]}{\tau_a} - \frac{[NH_2]}{\tau_b} - \frac{[NH_2][NO]}{\tau_c[NO]_0} \tag{3.26}$$

$$R_{NO} = \frac{[NH_2]}{\tau_b} - \frac{[NH_2][NO]}{\tau_c[NO]_0} \tag{3.27}$$

where the characteristic reaction times are defined by

$$\tau_a^{-1} = k_1[M] + k_2[OH] + k_3[O] + k_4[H] \tag{3.28}$$

$$\tau_b^{-1} = k_5[OH] + k_6[O] + k_7[H] + k_8[O_2] \tag{3.29}$$

$$\tau_c^{-1} = k_9[NO]_0 \tag{3.30}$$

Since we are dealing with combustion products that have generally had several seconds to equilibrate, a reasonable first approximation to the concentrations of the radicals OH, O, and H is that chemical equilibrium is achieved. The NO and added $NH_3$ are minor species, so their effect on the equilibrium composition should be small. Restricting our consideration to an isothermal system, the density and radical concentrations are constant, and we may integrate (3.25) to find

$$[NH_3] = [NH_3]_0\, e^{-t/\tau_a} \tag{3.31}$$

We immediately see one of the limitations to the use of ammonia injection for $NO_x$ control. If the rates of the ammonia reactions are too slow due to low temperature, the ammonia will be emitted unreacted along with the nitric oxide.

Once $NH_2$ is produced, it will react rapidly, either with radicals or with NO. If we assume that these reactions are sufficiently fast to establish a steady state, the $NH_2$ concentration may be estimated as

$$[NH_2]_{ss} = \frac{([NH_3]_0/\tau_a)\, e^{-t/\tau_a}}{(1/\tau_b) + ([NO]/[NO]_0)\,(1/\tau_c)} \tag{3.32}$$

We now have the estimates of the concentrations of $NH_2$ and $NH_3$ that we need to determine the NO levels. It is convenient to define the following dimensionless quantities:

$$z = \frac{[NO]}{[NO]_0}$$

$$\chi = \frac{[NH_3]_0}{[NO]_0}$$

$$\theta = \frac{t}{\tau_a}$$

$$\gamma = \frac{\tau_b}{\tau_c}$$

The rate equation for the NO concentration in an isothermal system becomes

$$\frac{dz}{d\theta} = \frac{1 - \gamma z}{1 + \gamma z} \chi e^{-\theta} \tag{3.33}$$

The initial condition is $z = 1$ at $\theta = 0$. Integrating, we find

$$1 - z - \frac{2}{\gamma} \ln \frac{1 - \gamma z}{1 - \gamma} = \chi(1 - e^{-\theta}) \tag{3.34}$$

Equation (3.34) may be solved iteratively to determine the level of NO control in the selective reduction system. Fig 3.11 compares the results of these calculations with measurements made by Muzio and Arand (1976) on a pilot-scale facility. The rate constants used in this calculation are summarized in Table 3.2. The calculations are based on a 0.35-s residence time in an isothermal system. In the test facility the temperature decreased by about 200 K within this time. The reported temperatures correspond approximately to the temperature at the point where the ammonia was injected. The radical concentrations and reaction rates will decrease as the gases cool, so our simple model is not strictly valid.

Nevertheless, this simple model reproduces most of the important features of the selective reduction system. NO is effectively reduced only in a narrow temperature window centered about 1200 K. As shown in Figure 3.11(c), $\theta$ is small at lower temperatures due to the slow reaction of ammonia. At higher temperatures, the $NH_2$ oxidation becomes faster than NO reduction (i.e., $\gamma > 1$), as shown in Figure 3.11(c), allowing additional NO to be formed from the ammonia. Even when excess ammonia is added near the optimal temperature, not all of the nitric oxide is reduced since some of the $NH_3$ forms NO.

It is apparent in Figure 3.11 that this model predicts a much broader temperature window than was observed experimentally. The high-temperature limit of the window is reproduced reasonably well. The model, however, predicts that $NH_3$ is oxidized more rapidly than is observed at low temperatures. The primary oxidation reaction is reaction 2:

$$NH_3 + OH \xrightarrow{2} NH_2 + H_2O$$

An overestimate of the hydroxyl concentration due to ignoring the temperature variation along the length of the experimental section could account for much of the discrepancy.

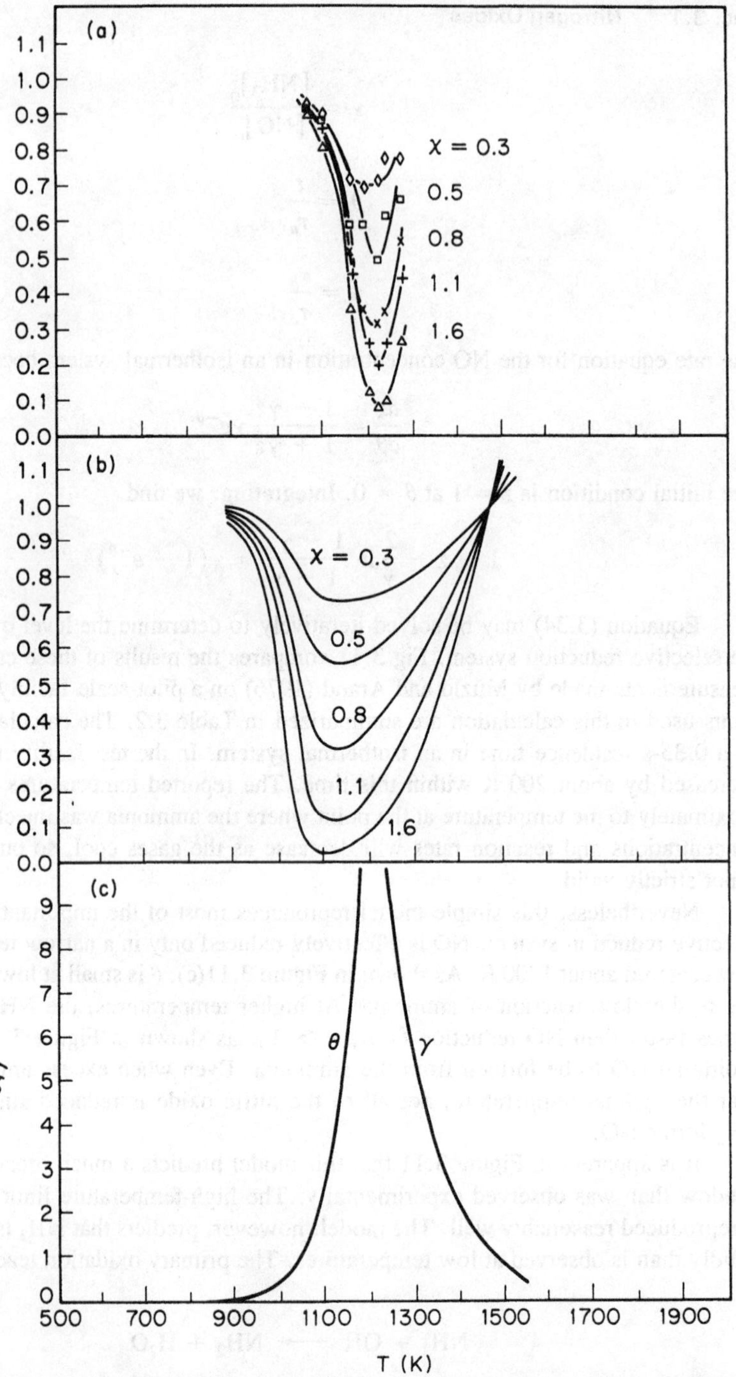


**Figure 3.11** Performance of ammonia injection in the destruction of $NO_x$ in combustion products: (a) $NO_x$ penetration data of Muzio and Arand (1976); (b) calculated NO penetration; (c) dimensionless times for $NH_3$ and $NH_2$ oxidation.

TABLE 3.2 RATE CONSTANTS FOR THE $NH_3$/NO REACTIONS

| Reaction | Rate constant ($m^3\ mol^{-1}\ s^{-1}$) |
|---|---|
| $NH_3 + M \xrightarrow{1} NH_2 + H + M$ | $k_1 = 2.5 \times 10^{10}\ e^{-47,200/T}$ |
| $NH_3 + OH \xrightarrow{2} NH_2 + H_2O$ | $k_2 = 5.8 \times 10^7\ e^{-4055/T}$ |
| $NH_3 + O \xrightarrow{3} NH_2 + OH$ | $k_3 = 2.0 \times 10^7\ e^{-4470/T}$ |
| $NH_3 + H \xrightarrow{4} NH_2 + H_2$ | $k_4 = 1.3 \times 10^8\ e^{-10,280/T}$ |
| $NH_2 + OH \xrightarrow{5} NH + H_2O$ | $k_5 = 5.0 \times 10^5\ T^{0.5}\ e^{-1000/T}$ |
| $NH_2 + O \xrightarrow{6a} NH + OH$ | $k_{6a} = 1.3 \times 10^8\ T^{-0.5}$ |
| $\xrightarrow{6b} HNO + H$ | $k_{6b} = 6.3 \times 10^8\ T^{-0.5}$ |
| $NH_2 + H \xrightarrow{7} NH + H_2$ | $k_7 = 1.9 \times 10^7$ |
| $NH_2 + O_2 \xrightarrow{8a} HNO + OH$ | $k_{8a} = 10^8\ e^{-25,000/T}$ |
| $\xrightarrow{8b} NH + HO_2$ | $k_{8b} = 1.8 \times 10^6\ e^{-7500/T}$ |
| $NH_2 + NO \xrightarrow{9} N_2 + H_2O$ | $k_9 = 1.2 \times 10^{14}\ T^{-2.46} e^{-938/T}$ |
| $NH + O_2 \xrightarrow{10} HNO + O$ | $k_{10} = 10^7\ e^{-6000/T}$ |

*Source:* Hanson and Salimian (1984).

The extreme temperature sensitivity of the $NH_3$–NO reaction mechanism makes the location of the ammonia injection extremely important. The temperature window corresponds to the gas temperature in boiler superheaters and convective heat exchangers where the temperature drops rapidly. The temperature at a given point in a boiler also changes with load. If the location of the injectors were optimized for full-load operation, the temperature would drop at reduced load. Instead of reducing NO, the ammonia would then be emitted into the atmosphere. Optimizing for a lower load condition would allow ammonia oxidation when the temperature increases at full load. One way around this problem is to use injectors at multiple locations, although the heat exchangers limit the locations of the injectors, particularly when an existing boiler is being retrofitted with the control.

The original implementation of selective reduction was based on ammonia, but the narrow-temperature window severely limits the range of systems to which it can be applied. Major efforts have been undertaken to elucidate the chemistry involved, with the ultimate objective of making the temperature restrictions less severe. One approach is to use catalysts to promote the reaction at lower temperatures, but catalyst poisoning by contaminants like ash in the combustion products presents major obstacles to this approach. Alternatively, a radical source could reduce the lower bound of the temperature window. Salimian and Hanson (1980) found that the optimal temperature for the ammonia reaction could be reduced to about 1000 K by adding hydrogen along with the ammonia. Azuhata et al. (1981) observed that $H_2O_2$ could promote the $NH_3$ reaction at temperatures as low as 800 K.

The essence of the selective reduction technique is the introduction of nitrogen compounds that will react with NO, leading to the ultimate formation of $N_2$. The temperature at which the reaction proceeds must be slow enough that NO formation from the additive is avoided. Compounds other than ammonia have been proposed. Urea, $H_2NCONH_2$, lowers the temperature window to approximately 1000 K (Salimian and Hanson, 1980). Other compounds look even more promising. Perry and Siebers (1986) have found that isocyanic acid is extremely effective at reducing NO at temperatures as low as 670 K. Isocyanic acid was produced by flowing hot combustion products over a bed of cyanic acid which undergoes thermal decomposition at temperatures in excess of 600 K,

$$(HOCN)_3 \longrightarrow 3HNCO$$

The combustion products were then passed over a bed of stainless steel pellets. Their proposed mechanism for the reactions within the packed bed is

$$HNCO \longrightarrow NH + CO$$

$$NH + NO \longrightarrow H + N_2O$$

$$H + HNCO \longrightarrow NH_2 + CO$$

$$CO + OH \longrightarrow CO_2 + H$$

$$NH_2 + NO \longrightarrow N_2H + OH$$

$$\longrightarrow N_2 + H_2O$$

$$N_2H + M \longrightarrow N_2 + H + M$$

At sufficiently low temperature, the gas-phase NO formation is avoided, but surface reactions of oxygen in the bed may lead to NO formation. Above about 670 K, very low $NO_x$ levels were observed in tests on a diesel engine. Thus promising technologies for reducing NO in the combustion products are under development.

### 3.1.7 Nitrogen Dioxide

Most of the nitrogen oxides emitted from combustion systems are in the form of nitric oxide (NO), but nitrogen dioxide ($NO_2$) is usually present as well. In the combustion zone, $NO_2$ levels are usually low, but exhaust levels can be significant at times. $NO_2$ can account for as much as 15 to 50% of the total $NO_x$ emitted by gas turbines (Diehl, 1979; Hazard, 1974; Levy, 1982). $NO_2$ levels far in excess of the NO concentration have been measured in some regions of laminar diffusion flames (Hargreaves et al., 1981).

Nitrogen dioxide is formed by the oxidation of NO. The overall reaction for the process is

$$NO + \tfrac{1}{2} O_2 \rightleftharpoons NO_2$$

which is exothermic [i.e., $\Delta h_r(298\ \text{K}) = -57{,}278\ \text{J mol}^{-1}$]. Thus the formation of $NO_2$ is thermodynamically favored at low temperatures. The fact that $NO_2$ is usually much less abundant than NO in cooled combustion products clearly indicates that the rate of NO oxidation is a relatively slow process. The overall reactions of NO with $O_2$ or O are termolecular

$$NO + NO + O_2 \xrightarrow{1} NO_2 + NO_2$$

$$NO + O + M \xrightarrow{2} NO_2 + M$$

for which the measured rate constants are (Baulch et al., 1973)

$$k_{+1} = 1.2 \times 10^{-3} \exp\left(\frac{530}{T}\right) \quad \text{m}^6\ \text{mol}^{-2}\ \text{s}^{-1}$$

$$k_{+2} = 1.5 \times 10^{3} \exp\left(\frac{940}{T}\right) \quad \text{m}^6\ \text{mol}^{-2}\ \text{s}^{-1}$$

The conversion of NO to $NO_2$ by the hydroperoxyl radical

$$NO + HO_2 \xrightarrow{3} NO_2 + OH$$

with a rate constant (Hanson and Salimian, 1984)

$$k_{+3} = 2.1 \times 10^{6} \exp\left(\frac{240}{T}\right) \quad \text{m}^3\ \text{mol}^{-1}\ \text{s}^{-1}$$

is generally slow because $HO_2$ is present only at low concentrations, at least as long as equilibrium of the C-H-O system is maintained. Within the flame front, superequilibrium radical concentrations can lead to some $NO_2$ formation, but $NO_2$ dissociation by the reverse of reaction 2 is likely if it passes through the hot region of the flame,

$$k_{-2} = 1.1 \times 10^{10} \exp\left(-\frac{33{,}000}{T}\right) \quad \text{m}^3\ \text{mol}^{-1}\text{s}^{-1}$$

High $NO_2$ levels are observed following rapid cooling of combustion products (Hargreaves et al., 1981). While $HO_2$ is a relatively minor species in the flame, relatively high concentrations can be formed by the three-body recombination reaction

$$H + O_2 + M \overset{4}{\rightleftharpoons} HO_2 + M$$

in regions where fuel-lean combustion products are rapidly cooled, accelerating $NO_2$ formation in the cool gases. Reactions of $HO_2$ with major species (CO, $H_2$, $O_2$, etc.) are slow at low temperatures. It does, however, react with NO via reaction 3 and with itself.

$$HO_2 + HO_2 \overset{5}{\rightleftharpoons} H_2O_2 + O_2$$

The $H_2O_2$ produced by reaction 5 is stable at low temperature, but at higher temperatures may react further,

$$H_2O_2 + M \overset{6}{\rightleftharpoons} OH + OH + M$$

$$H_2O_2 + OH \overset{7}{\rightleftharpoons} HO_2 + H_2O$$

regenerating $HO_2$ and minimizing the effect of reaction 5. Bimolecular exchange reactions of OH, for example,

$$OH + O \overset{8}{\rightleftharpoons} O_2 + H$$

generate the H necessary for reaction 4.

Reactions 3–8 can explain substantial conversion of NO to $NO_2$ that is observed when combustion products are rapidly cooled to intermediate temperatures, on the order of 1000 K (Hargreaves et al., 1981). A detailed treatment of $NO_2$ formation requires an understanding of the rates of destruction of the radicals in combustion products. The radicals are ultimately consumed by three-body recombination reactions. Since these reactions are relatively slow, rapid cooling can allow substantial radical concentrations to remain at the low temperatures where $NO_2$ is stable. Slower cooling would allow the radicals to remain equilibrated to lower temperature, thereby eliminating the conditions necessary for fast NO oxidation. A detailed treatment of this radical chemistry in cooled combustion products will be presented as part of our discussion of CO oxidation.

The foregoing analysis clearly indicates that $NO_2$ formation is favored by rapid cooling of combustion products in the presence of substantial $O_2$ concentrations. Gas turbine engines provide such conditions due to the need to limit the temperatures of the combustion products entering the power turbine. Stable combustion requires near stoichiometric operation that yields temperatures much higher than the tolerable ($<1400$ K) turbine inlet temperatures. To cool the combustion products to this temperature, additional air is injected downstream of the primary combustion zone. This reduces the overall equivalence ratio to 0.2 to 0.4 and provides abundant oxygen for $HO_2$ formation. Since engine size is a premium in aircraft gas turbines, the residence time is kept very short, of order 10 ms, providing the rapid quenching needed to favor $NO_2$ formation.

To measure the composition of gas samples extracted from flames, it is necessary to cool the gases rapidly to quench the oxidation of CO and other species. Rapid cooling in sample probes also provides the conditions that favor $NO_2$ formation (Cernansky and Sawyer, 1975; Allen, 1975; Levy, 1982). A rapid quench sample probe also provides a large surface that can catalyze the recombination reaction

$$NO + O \xrightarrow{\text{probe wall}} NO_2$$

Some of the early reports of high $NO_2$ levels in flames were plagued by probe-induced sampling biases. A better understanding of the kinetics of $NO_2$ formation has made it possible to minimize probe sampling biases in in-flame studies. Estimates of $NO_2$ levels in flames have decreased accordingly.

## 3.2 CARBON MONOXIDE

In Chapter 2 we saw that carbon monoxide is an intermediate species in the oxidation of hydrocarbon fuels to $CO_2$ and $H_2O$. In fuel-rich regions of a flame, the CO levels are necessarily high since there is insufficient oxygen for complete combustion. Only if sufficient air is mixed with such gases at sufficiently high temperature can the CO be oxidized. Thus, imperfect mixing can allow carbon monoxide to escape from combustors that are operated fuel-lean overall. Even in premixed combustion systems, carbon monoxide levels can be relatively high due to the high equilibrium concentrations at the flame temperature, particularly in internal combustion engines where the gases are hot prior to ignition due to compression. As the combustion products are cooled by heat or work transfer, the equilibrium CO level decreases. If equilibrium were maintained as the temperature decreased, carbon monoxide emissions from automobiles and other well-mixed combustors would be very low in fuel-lean operation. The extent to which CO is actually oxidized, however, depends on the kinetics of the oxidation reactions and the manner of cooling. In this section we explore the kinetics of CO oxidation and the mechanisms that allow CO to escape oxidation in locally fuel-lean combustion.

The predominant reaction leading to carbon monoxide oxidation in hydrocarbon combustion is

$$\mathrm{CO + OH \underset{-1}{\overset{+1}{\rightleftharpoons}} CO_2 + H}$$

where

$$k_{+1} = 4.4T^{1.5}e^{372/T} \qquad \mathrm{m^3\ mol^{-1}\ s^{-1}}$$

The rate of carbon dioxide production by reaction 1 is

$$R_{+1} = k_{+1}[\mathrm{CO}][\mathrm{OH}]$$

The rate equation describing the total change in the CO level must include the reverse reactions:

$$\begin{aligned} R_{\mathrm{CO}} &= -R_{+1} + R_{-1} \\ &= -k_{+1}[\mathrm{CO}][\mathrm{OH}] + k_{-1}[\mathrm{CO_2}][\mathrm{H}] \end{aligned}$$

where, by detailed balancing,

$$k_{-1} = \frac{k_{+1}}{K_{c1}}$$

Thus, to describe the CO oxidation kinetics, we must know the concentrations of OH and H.

In this discussion, our primary concern is the oxidation of CO in the postflame gases as they cool. The speed of the reactions in responding to a perturbation from the equilibrium state may be expressed in terms of the characteristic reaction time,

$$\tau_{CO} = \frac{[CO]}{R_{+1}} = \frac{1}{k_{+1}[OH]}$$

As a first approximation, the OH may be assumed to be present at its equilibrium concentration. The dotted line in Figure 3.12 shows the variation of $\tau_{CO}$ with equivalence ratio as calculated using the results of the combustion equilibrium calculations from Figure 2.6 for adiabatic combustion of a fuel oil ($CH_{1.8}$). At equivalence ratios greater than about 0.6, corresponding to temperatures greater than 1650 K, the reaction time is less than 1 ms, indicating that the CO level can quickly respond to changes in the equilibrium state of the system. At equivalence ratios below about 0.4 ($T < 1250$ K), the reaction time exceeds 1 s, so chemical equilibrium will be very difficult to maintain in

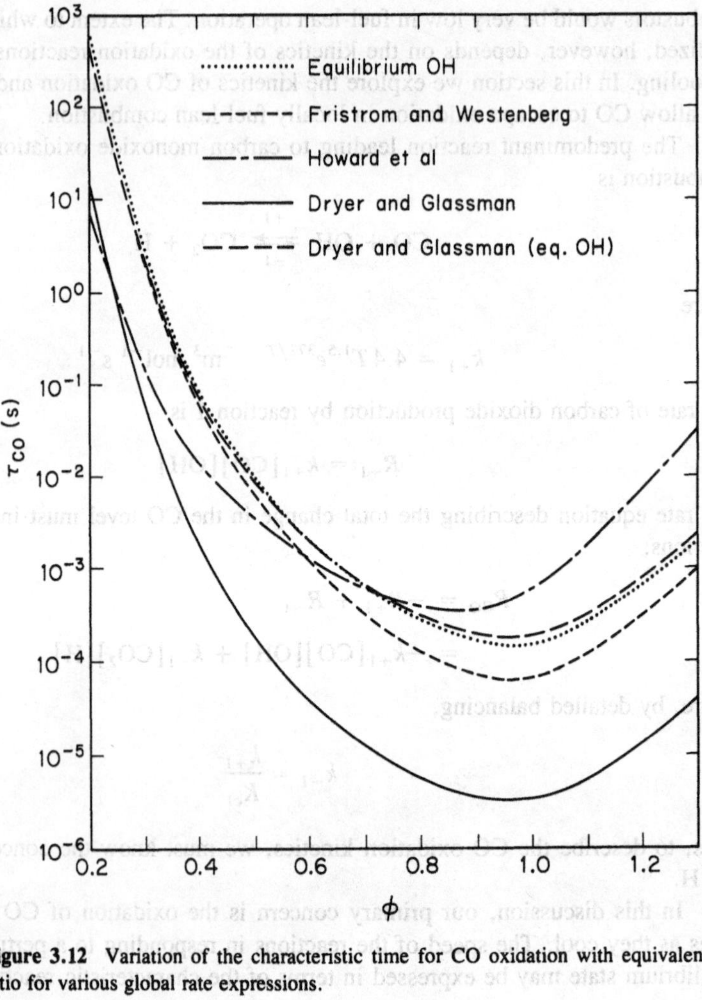


**Figure 3.12** Variation of the characteristic time for CO oxidation with equivalence ratio for various global rate expressions.

any combustion system. In the discussion to follow, we show that temperature has the predominant influence on the oxidation rate.

This simple approach to describing CO oxidation can readily be used to derive a global rate expression that will make it possible to estimate oxidation rates without resorting to elaborate chemical equilibrium calculations. The equilibration of OH with $H_2O$ and $O_2$ can be described by the reaction

$$\tfrac{1}{2}\,H_2O + \tfrac{1}{4}\,O_2 \rightleftharpoons OH$$

giving the equilibrium concentration

$$[OH]_e = K_{cOH}[H_2O]^{1/2}[O_2]^{1/4} \tag{3.35}$$

Similarly, the H concentration may be estimated by assuming equilibrium of the reaction

$$\tfrac{1}{2}\,H_2O \rightleftharpoons H + \tfrac{1}{4}\,O_2$$

leading to

$$[H]_e = K_{cH}[H_2O]^{1/2}[O_2]^{-1/4} \tag{3.36}$$

The rate equation for CO thus becomes

$$R_{CO} = -k_f[CO][H_2O]^{1/2}[O_2]^{1/4} + k_r[CO_2][H_2O]^{1/2}[O_2]^{-1/4} \tag{3.37}$$

where $k_f = k_{+1}K_{c\mathrm{OH}}$ and $k_r = k_{-1}K_{c\mathrm{H}}$ are the global rate constants. Several investigators have reported global CO oxidation rates in this form (Table 3.3). The first rate expression was fitted to measured CO oxidation rates in postflame gases. The second was derived using the measured rate for reaction 1 and the equilibrium assumption for OH. These two rates agree closely with the calculations made using an equilibrium code to determine $[OH]_e$, as illustrated by the reaction times shown in Figure 3.12. The third rate expression was derived by fitting CO oxidation data obtained from measurements made in flames. The resulting reaction times are two orders of magnitude shorter than those for equilibrium OH, as illustrated in Figure 3.12. This discrepancy results from superequilibrium OH concentrations within the flame front and clearly indicates the need for caution in applying global reaction rates. Although they may be very useful, the

TABLE 3.3 GLOBAL CO OXIDATION RATE EXPRESSIONS $-R_{CO,ox}$ (mol m$^{-3}$ s$^{-1}$)

| | Temperature range (K) | Reference |
|---|---|---|
| (1) $1.3 \times 10^9\,[CO][H_2O]^{1/2}[O_2]^{1/4}\exp(-22{,}660/T)$ | 1750–2000 | Fristrom and Westenberg (1965) |
| (2) $1.3 \times 10^7\,[CO][H_2O]^{1/2}[O_2]^{1/2}\exp(-15{,}100/T)$ | 840–2360 | Howard et al. (1973) |
| (3) $1.3 \times 10^{10}\,[CO][H_2O]^{1/2}[O_2]^{1/4}\exp(-20{,}140/T)$ | 1030–1230 | Dryer and Glassman (1973) |
| (4) $1.3 \times 10^8\,[CO][H_2O]^{1/2}[O_2]^{1/4}\exp(-19{,}870/T)$ | Equilibrium OH | Dryer and Glassman (1973) |

assumptions implicit in global rate expressions sometimes severely limit the range of conditions to which they are applicable.

From these global rate expressions, we can see that the dependence of reaction times on the equivalence ratio is primarily a temperature effect. Since the rate coefficient for the CO + OH reaction is not strongly dependent on temperature at low temperatures, this temperature dependence arises primarily from the influence of temperature on the equilibrium OH concentration. While the influence of temperature on the CO + OH reaction rate is minor, other reactions with larger activation energies are more strongly affected. The reactions involved in the equilibration of hydroxyl and other minor species with the major species include a number with large activation energies. We expect, therefore, that at some point as the gases are cooled the radical concentrations will begin to deviate from chemical equilibrium. This reaction quenching strongly influences the CO oxidation rate and must be taken into consideration.

### 3.2.1 Carbon Monoxide Oxidation Quenching

While it is beyond the scope of this book to undertake detailed kinetic modeling of the postflame reactions, it is useful to examine the process of reaction quenching qualitatively. Fenimore and Moore (1974) analyzed the problem of CO oxidation quenching in a constant-pressure system. Their analysis allows one to derive an expression for the maximum cooling rate beyond which CO oxidation reactions will be frozen.

As we have noted previously, the radicals O, OH, H, and $HO_2$ and other reaction intermediates (e.g., $H_2$ and $H_2O_2$) undergo a number of rapid exchange reactions, for example,

$$H + O_2 \underset{-2}{\overset{+2}{\rightleftharpoons}} OH + O$$

$$O + H_2 \underset{-3}{\overset{+3}{\rightleftharpoons}} OH + H$$

$$OH + H_2 \underset{-4}{\overset{+4}{\rightleftharpoons}} H + H_2O$$

$$O + H_2O \underset{-5}{\overset{+5}{\rightleftharpoons}} OH + OH$$

$$HO_2 + OH \underset{-6}{\overset{+6}{\rightleftharpoons}} H_2O + O_2$$

$$H_2O_2 + O_2 \underset{-7}{\overset{+7}{\rightleftharpoons}} HO_2 + HO_2$$

We must describe the dynamics of these reaction intermediates to determine the instantaneous OH concentration and calculate the rate of CO oxidation.

The exchange reactions are fast compared to the three-body recombination reactions that ultimately eliminate the radicals from the system. To a first approximation the reaction intermediates H, O, OH, $HO_2$, and $H_2$ may be assumed to be equilibrated with one another by the action of reactions 2–7. These partial-equilibrium concentrations may be expressed in terms of the hydroxyl concentration as follows:

$$
\begin{aligned}
[O]_{pe} &= \frac{[OH]^2}{K_5[H_2O]} \\
[H]_{pe} &= \frac{[OH]^3}{K_2K_5[O_2][H_2O]} \\
[HO_2]_{pe} &= \frac{[H_2O][O_2]}{K_6[OH]} \\
[H_2]_{pe} &= \frac{[OH]^2}{K_2K_4K_5[O_2]} \\
[H_2O_2]_{pe} &= \frac{[H_2O]^2[O_2]}{K_6^2K_7[OH]^2}
\end{aligned}
\tag{3.38}
$$

where all the equilibrium constants are in concentration units ($K_c$). Using the expression for $[H]_{pe}$, the rate of CO oxidation may be written

$$R_{CO} = -k_{+1}[OH]([CO] - [CO]_{pe}) \tag{3.39}$$

where

$$[CO]_{pe} = \frac{[CO_2][OH]^2}{K_1K_2K_5[O_2][H_2O]} \tag{3.40}$$

is the CO concentration corresponding to a partial equilibrium with this pool of reaction intermediates.

The instantaneous rate of CO oxidation is proportional to the hydroxyl concentration, which, in turn, depends on the rate at which OH is consumed. The variation of the ratio, $[CO]/[CO]_{pe}$, with $[OH]$ can be used to identify conditions that lead to deviations from the partial equilibrium.

The ratio, $[CO]/[CO]_{pe}$, depends on both the hydroxyl concentration and time. Taking the total derivative with respect to $[OH]$,

$$\frac{d([CO]/[CO]_{pe})}{d[OH]} = -\frac{[CO]}{[CO]_{pe}^2}\frac{d[CO]_{pe}}{d[OH]} + \frac{1}{[CO]_{pe}}\frac{d[CO]/dt}{d[OH]/dt}$$

and applying (3.39) and (3.40), we find

$$\alpha[OH]\frac{d([CO]/[CO]_{pe})}{d[OH]} = (1 - 2\alpha)\frac{[CO]}{[CO]_{pe}} - 1 \tag{3.41}$$

where

$$\alpha = \frac{-d[OH]/dt}{k_{+1}[OH]^2} \tag{3.42}$$

The value of $\alpha$ determines whether CO will remain equilibrated with the other trace species. If $\alpha = 0$, $[CO]/[CO]_{pe} = 1$ and CO is equilibrated with the pool of reaction intermediates. For $\alpha \to \infty$,

$$\frac{d([CO]/[CO]_{pe})}{d[OH]} = -\frac{2}{[OH]}\frac{[CO]}{[CO]_{pe}}$$

or

$$\frac{[CO]}{[CO]_{pe}} \propto [OH]^{-2}$$

Since $[CO]_{pe} \propto [OH]^2$, CO is independent of OH and, therefore, strictly frozen.

For $[CO]/[CO]_{pe}$ to be greater than but decreasing toward unity as [OH] decreases, $\alpha$ must be between 0 and $\frac{1}{2}$. Thus, only if $\alpha < \frac{1}{2}$ can the CO partial equilibrium be continuously maintained. When CO oxidation is quenched, $[CO]/[CO]_{pe} \gg 1$; so (3.41) becomes, approximately,

$$\frac{\alpha[OH]d([CO]/[CO]_{pe})}{d[OH]} = \frac{[CO]}{[CO]_{pe}}(1 - 2\alpha)$$

Noting that

$$\frac{d[CO]_{pe}}{[CO]_{pe}} = 2\frac{d[OH]}{[OH]}$$

we find

$$d \ln [CO] = \frac{1}{2\alpha} d \ln [CO]_{pe} \tag{3.43}$$

If $\alpha = 2$, a 10-fold decrease in [OH] and therefore a 100-fold decrease in $[CO]_{pe}$ yield only a factor of 3 decrease in [CO]. Thus the CO oxidation reactions may be considered to be effectively quenched for $\alpha > 2$.

We now need to evaluate $\alpha$ to determine whether or not the CO partial equilibrium is maintained. The rate of decay of the hydroxyl concentration is tied to the other reaction intermediates through the fast exchange reactions 2–7. The partial equilibrium assumption greatly simplifies the analysis of an otherwise very complex kinetics problem. There are several ways to evaluate the partial equilibrium. We have identified a number of reactions (2–7) that maintain the partial equilibrium among the reaction intermediates. We now need to develop a description of how the entire pool of intermediates evolves due to reactions other than 2–7. This can be done by following a weighted sum of the concentrations of the species in that pool, i.e.,

$$P = [OH] + a[O] + b[H_2] + c[H] + d[HO_2] + e[H_2O_2]$$

$P$ is defined such that it is not affected by reactions 2–7. Other reactions are needed for it to change. If, for example, the temperature were changed, the distribution of reaction intermediates would change as described by (3.38); but as long as the partial equilibrium is maintained, the value $P$ will remain unchanged unless other reactions take place. If we write

$$R_2 = k_{+2}[H][O_2] - k_{-2}[OH][O]$$

$$R_3 = k_{+3}[O][H_2] - k_{-3}[OH][H] \quad \text{etc.}$$

The time rate of change of $P$ may be written

$$\rho \frac{d(P/\rho)}{dt} = (R_2 + R_3 - R_4 + 2R_5 - R_6) + a(R_2 - R_3 - R_5) + b(-R_3 - R_4) + c(-R_2 + R_3 + R_4) + d(-R_6 + 2R_7) + e(-R_7) + R_p$$

where $R_p$ is the total contribution of other reactions to the time rate of change of $P$. Rearranging, we have

$$\rho \frac{d(P/\rho)}{dt} = R_2(1 + a - c) + R_3(1 - a - b + c) + R_4(-1 - b + c) + R_5(2 - a) + R_6(-1 - d) + R_7(2d - e) + R_p \tag{3.44}$$

At the partial equilibrium, the net contribution of reactions 2–7 to changing $P$ must be zero regardless of the rates of the individual reactions. This condition is satisfied by setting the coefficients of each of the rates equal to zero. This yields

$$P = [OH] + 2[O] + 2[H_2] + 3[H] - [HO_2] - 2[H_2O_2] \tag{3.45}$$

Three-body recombination reactions are responsible for the decrease in $P$ as the combustion products cool. These reactions include

$$H + O_2 + M \underset{-8}{\overset{+8}{\rightleftharpoons}} HO_2 + M$$

$$H + H + M \underset{-9}{\overset{+9}{\rightleftharpoons}} H_2 + M$$

$$H + O + M \underset{-10}{\overset{+10}{\rightleftharpoons}} OH + M$$

$$O + O + M \underset{-11}{\overset{+11}{\rightleftharpoons}} O_2 + M$$

$$H + OH + M \underset{-12}{\overset{+12}{\rightleftharpoons}} H_2O + M$$

$$OH + OH + M \underset{-13}{\overset{+13}{\rightleftharpoons}} H_2O_2 + M$$

**TABLE 3.4 CARBON MONOXIDE OXIDATION AND RECOMBINATION REACTION RATES**

| Reaction | Rate coefficient |
|---|---|
| $CO + OH \xrightarrow{+1} CO_2 + H$ | $k_{+1} = 4.4\ T^{1.5} \exp(+372/T)\ \text{m}^3\ \text{mol}^{-1}\ \text{s}^{-1}$ |
| $H + O_2 + M \xrightarrow{+8} HO_2 + M$ | $k_{+8} = 1.5 \times 10^3 \exp(-500/T)\ \text{m}^6\ \text{mol}^{-2}\ \text{s}^{-1}$ |
| $H + H + M \xrightarrow{+9} H_2 + M$ | $k_{+9} = 6.4 \times 10^5\ T^{-1}\ \text{m}^6\ \text{mol}^{-2}\ \text{s}^{-1}$ |
| $H + O + M \xrightarrow{+10} OH + M$ | $k_{+10} = 1.0 \times 10^4\ \text{m}^6\ \text{mol}^{-2}\ \text{s}^{-1}$ |
| $O + O + M \xrightarrow{+11} O_2 + M$ | $k_{+11} = 1.0 \times 10^5\ T^{-1}\ \text{m}^6\ \text{mol}^{-2}\ \text{s}^{-1}$ |
| $H + OH + M \xrightarrow{+12} H_2O + M$ | $k_{+12} = 1.4 \times 10^{11}\ T^{-2}\ \text{m}^6\ \text{mol}^{-2}\ \text{s}^{-1}$ |
| $OH + OH + M \xrightarrow{+13} H_2O_2 + M$ | $k_{+13} = 1.3 \times 10^{10}\ T^{-2}\ \text{m}^6\ \text{mol}^{-2}\ \text{s}^{-1}$ |

The forward rate constants for these reactions are summarized in Table 3.4. In fuel-lean combustion, reaction 8 is the predominant recombination reaction at equilibrium, but other reactions may become important at nonequilibrium states.

The rate of change in $P$ may be written

$$\begin{aligned}\rho \frac{d(P/\rho)}{dt} = &-4k_{+8}[\text{H}][\text{O}_2][\text{M}] + 4k_{-8}[\text{HO}_2][\text{M}] - 4k_{+9}[\text{H}][\text{H}][\text{M}] \\ &+4k_{-9}[\text{H}_2][\text{M}] - 4k_{+10}[\text{H}][\text{O}][\text{M}] + 4k_{-10}[\text{OH}][\text{M}] \\ &-4k_{+11}[\text{O}][\text{O}][\text{M}] + 4k_{-11}[\text{O}_2][\text{M}] - 4k_{+12}[\text{H}][\text{OH}][\text{M}] \\ &+4k_{-12}[\text{H}_2\text{O}][\text{M}] - 4k_{+13}[\text{OH}][\text{OH}][\text{M}] + 4k_{-13}[\text{H}_2\text{O}_2][\text{M}]\end{aligned}$$

where the factors of 4 result from the net change of $P$ as one mole recombines. Using the partial equilibrium concentrations of the reaction intermediates, we find

$$\begin{aligned}\rho \frac{d(P/\rho)}{dt} = &-(4k_{+8}[\text{H}][\text{O}_2][\text{M}] + 4k_{+9}[\text{H}][\text{H}][\text{M}] + 4k_{+10}[\text{H}][\text{O}][\text{M}] \\ &+4k_{+11}[\text{O}][\text{O}][\text{M}] + 4k_{+12}[\text{H}][\text{OH}][\text{M}] \\ &+4k_{+13}[\text{OH}][\text{OH}][\text{M}])\left(1 - \frac{[\text{OH}]_e^4}{[\text{OH}]^4}\right)\end{aligned} \tag{3.46}$$

where the subscript $e$ denotes the concentration at full thermodynamic equilibrium.

$P$ may be expressed in terms of any of the reaction intermediates. In terms of hydroxyl, (3.46) becomes

$$\rho \frac{d[\mathrm{OH}]/\rho}{dt} = -\left[\frac{4k_{+8}[\mathrm{OH}]_e^3[\mathrm{M}]}{K_2K_5[\mathrm{H_2O}]} y^3 + \frac{4k_{+9}[\mathrm{OH}]_e^6[\mathrm{M}]}{K_2^2K_5^2[\mathrm{H_2O}]^2[\mathrm{O_2}]^2} y^6\right.$$
$$+ \frac{4k_{+10}[\mathrm{OH}]_e^5[\mathrm{M}]}{K_2K_5^2[\mathrm{O_2}][\mathrm{H_2O}]^2} y^5 + \frac{4k_{+11}[\mathrm{OH}]_e^4[\mathrm{M}]}{K_5^2[\mathrm{H_2O}]^2} y^4 \tag{3.47}$$
$$\left. + \frac{4k_{+12}[\mathrm{OH}]_e^4[\mathrm{M}]}{K_2K_5[\mathrm{H_2O}][\mathrm{O_2}]} y^4 + 4k_{+13}[\mathrm{OH}]_e^2[\mathrm{M}]\, y^2\right]\left(1 - \frac{1}{y^4}\right)$$
$$\div \left[1 + Ay + By^2 + Cy^{-2} + Dy^{-3}\right]$$

where

$$y = \frac{[\mathrm{OH}]}{[\mathrm{OH}]_e}$$

and

$$A = 4\left(\frac{[\mathrm{O}]_e}{[\mathrm{OH}]_e} + \frac{[\mathrm{H_2}]_e}{[\mathrm{OH}]_e}\right)$$

$$B = 9\frac{[\mathrm{H}]_e}{[\mathrm{OH}]_e}$$

$$C = \frac{[\mathrm{HO_2}]_e}{[\mathrm{OH}]_e}$$

$$D = \frac{4[\mathrm{H_2O_2}]_e}{[\mathrm{OH}]_e}$$

In terms of $y$, $P$ becomes

$$P = [\mathrm{OH}]_e y + 2[\mathrm{O}]_e y^2 + 2[\mathrm{H_2}]_e y^2 + 3[\mathrm{H}]_e y^3 - [\mathrm{HO_2}]_e y^{-1} - 2[\mathrm{H_2O_2}]_e y^{-2} \tag{3.48}$$

The coefficients depend on fuel composition (C/H ratio), temperature, and equivalence ratio. Only for extreme deviations from equilibrium or for near stoichiometric or fuel-rich combustion will the reaction intermediates be present in high enough concentrations to alter the concentrations of the major species appreciably and thus to modify the coefficients. Limiting our attention to fuel-lean combustion, (3.47) can now be used to evaluate $\alpha$ and to determine the conditions for which the CO partial equilibrium can be maintained:

$$\alpha = \frac{E_8 y + E_9 y^4 + E_{10} y^3 + E_{11} y^2 + E_{12} y^2 + E_{13}}{1 + Ay + By^2 + Cy^{-2} + Dy^{-3}}(1 - y^{-4}) \tag{3.49}$$

where

$$E_8 = \frac{4k_{+8}[OH]_e[M]}{k_{+1}K_2K_5[H_2O]}$$

$$E_9 = \frac{4k_9[OH]_e^4[M]}{k_{+1}K_2^2K_5^2[H_2O]^2[O_2]^2}$$

$$E_{10} = \frac{4k_{+10}[OH]_e^3[M]}{k_{+1}K_2K_5^2[H_2O]^2[O_2]}$$

$$E_{11} = \frac{4k_{+11}[OH]_e^2[M]}{k_{+1}K_5^2[H_2O]^2}$$

$$E_{12} = \frac{4k_{+12}[OH]_e^2[M]}{k_{+1}K_2K_5[H_2O][O_2]}$$

$$E_{13} = \frac{4k_{+13}[M]}{k_{+1}}$$

Figure 3.13 shows the variation of $\alpha$ with $y$ for combustion of aviation kerosene in air at an equivalence ratio of 0.91 and atmospheric pressure. To examine the quenching of CO oxidation, we limit our attention to the decrease of $P$ from an initial value

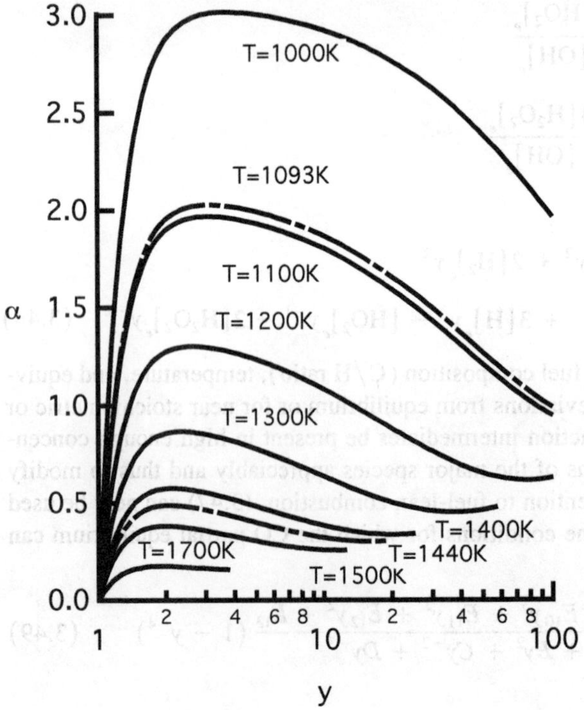


**Figure 3.13** Variation of $\alpha$ with $y$ for various temperatures. The maximum value of $y$ for each temperature corresponds to the value at which $P$ equals the value at 2000 K.

corresponding to thermodynamic equilibrium in the flame region. At temperatures above 1440 K, $\alpha$ is less than $\frac{1}{2}$ for all $y$, so the carbon monoxide partial equilibrium will be maintained regardless of cooling rate. Since

$$[\mathrm{CO}]_{pe} = [\mathrm{CO}]_e y^2 \tag{3.50}$$

the CO levels can still be much larger than the equilibrium value. The very rapid initial rise of $\alpha$ with $y$ means that it is very difficult to maintain the partial equilibrium at low temperatures. At 1000 K, $y$ must be smaller than 1.04 for the partial equilibrium to hold.

Complete quenching of CO oxidation requires $\alpha$ to be larger than about 2. This occurs only at temperatures below 1255 K. At 1000 K, $\alpha$ is less than 2 for $y < 1.26$. If cooling is sufficiently rapid such that the recombination reactions are unable to maintain the radical concentrations very close to equilibrium, CO will begin to deviate from the partial equilibrium below 1440 K and will be fully frozen between 1000 and 1100 K. The level at which the CO concentration will freeze depends strongly on the deviation from equilibrium of the concentration of reaction intermediates.

The cooling rate for which $y$ equals a specified value can be used to estimate the maximum cooling rate that will lead to acceptable CO emissions. Our analysis of Figure 3.13 has provided guidelines on the values of $y$ for which CO will be oxidized or frozen. To eliminate the temperature dependence of the concentrations in our calculations, it is convenient to express $y$ in terms of mole fractions, that is,

$$y = \frac{x_{\mathrm{OH}}}{x_{\mathrm{OH},e}}$$

noting that this formulation also requires that the total number of moles in the combustion products (or the mean molecular weight) not change significantly due to the recombination reactions. This limits the analysis to fuel-lean combustion products.

$y$ is a function of time through $x_{\mathrm{OH}}$ and of temperature through $x_{\mathrm{OH},e}$. The time rate of change of $y$ is thus

$$\frac{dy}{dt} = \frac{1}{x_{\mathrm{OH},e}}\frac{dx_{\mathrm{OH}}}{dt} - \frac{x_{\mathrm{OH}}}{x_{\mathrm{OH},e}^2}\frac{dx_{\mathrm{OH},e}}{dT}\frac{dT}{dt}$$

To cool at constant $y$, $dy/dt = 0$, so

$$\frac{dT}{dt} = \frac{(1/x_{\mathrm{OH}})(dx_{\mathrm{OH}}/dt)}{d \ln x_{\mathrm{OH},e}/dT} \tag{3.51}$$

Substituting (3.42) yields

$$\frac{dT}{dt} = \frac{\alpha k_{+1} c x_{\mathrm{OH},e} y}{d \ln x_{\mathrm{OH},e}/dT} \tag{3.52}$$

where $c$ is the molar concentration of the gas.

The equilibrium OH concentration can be expressed in terms of species whose concentrations will not vary significantly with $T$ by the reaction

$$\tfrac{1}{2}\,H_2O + \tfrac{1}{4}\,O_2 \rightleftarrows OH$$

that is,

$$x_{OH,e} = K_{p,OH}\, x_{H_2O}^{1/2}\, x_{O_2}^{1/4}\, p^{-1/4}$$

Thus

$$\frac{d \ln x_{OH,e}}{dT} = \frac{d \ln K_{p,OH}}{dT}$$

Van't Hoff's equation, (2.41), can be used to develop expressions for the equilibrium constants:

$$K_{p2} = 13.2e^{-8220/T} \qquad K_{p4} = 0.241e^{7670/T}$$

$$K_{p5} = 9.64e^{-8690/T} \qquad K_{p6} = 0.0822e^{36,600/T}$$

$$K_{p,OH} = 159e^{-19,620/T} \qquad K_{p7} = 6.50e^{-21,155/T}$$

Using these equilibrium constants and the rate coefficients from Table 3.4, the cooling rate for which $y$ will remain constant may be evaluated. Cooling rates corresponding to several values of $y$ are shown in Figure 3.14. For any value of $y$, the allowable cooling rate decreases rapidly as $T$ decreases. The limits of maintenance of the CO partial equilibrium ($\alpha = 0.5$) and freezing of CO oxidation ($\alpha = 2$) for constant $y$ cooling are also shown. It is clear that very slow cooling rates are required to maintain the partial equilibrium below 1300 K. Considering that combustion products are generally cooled by 1000 to 2000 K before being exhausted, this slow cooling can be achieved only in systems with residence times in excess of 1 s. Thus the CO partial equilibrium might be maintained to 1300 K in utility boilers, but not in many smaller systems and certainly not in engines where the total residence time may be tens of milliseconds or less. As we have noted previously, complete quenching of the CO oxidation only occurs below 1100 K. At high cooling rates, deviations from the CO partial equilibrium begin at the limiting temperature of 1450 K. As the cooling rate is increased, the value of $y$ and, therefore, the CO partial equilibrium level also increase.

Prediction of the actual CO emission levels from a combustion system requires integration of the chemical rate equations through the cooling process. The CO equilibrium is rapidly established in the high-temperature region of most flames, so equilibrium provides a reasonable initial condition for the examination of oxidation quenching in flames.

We can now see the reasons for the variation in the global rate expressions of Table 3.3. Rates 1 and 2 corresponded to carbon monoxide oxidation in an equilibrated pool of reaction intermediates. This corresponds to

$$R_{CO} = -k_1[OH]_e\{[CO] - [CO]_{pe}\}$$

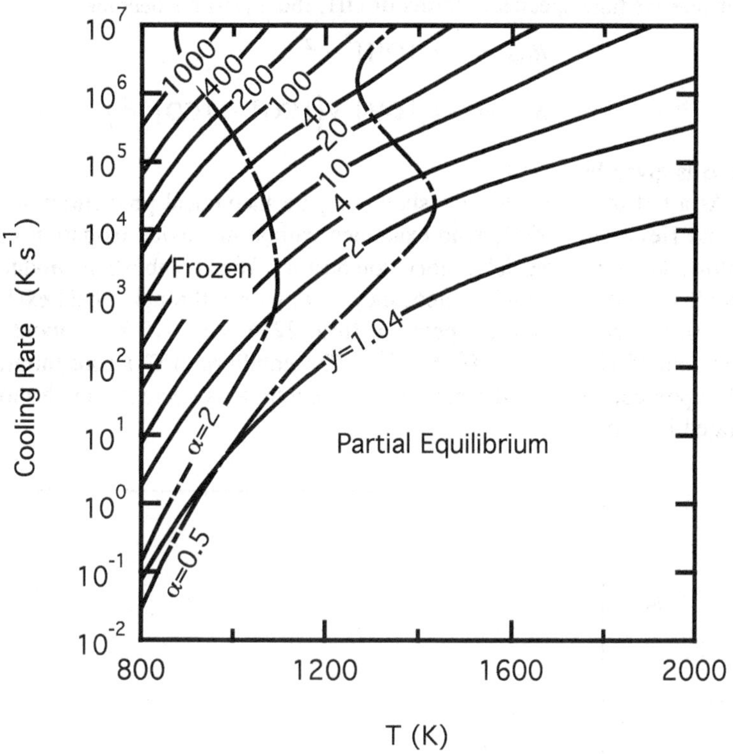


**Figure 3.14** Cooling rates corresponding to constant values of $y$ as a function of temperature. The cooling rates corresponding to the onset of deviations from partial equilibrium ($\alpha = 0.5$) and complete quenching of CO oxidation ($\alpha = 2$) are indicated by the dashed curves. The region of partial quenching is shaded.

Rate 3 was determined within the flame where radicals are present in superequilibrium concentrations.

$$R_{CO} = -k_1[OH]_e y\{[CO] - [CO]_{pe}\}$$

The excess of radicals over equilibrium (value for $y$) for which this rate expression was derived is specific to the flame in which the rate was measured. It might correspond to a particular nonflame situation, but different cooling rates will clearly lead to different radical concentrations. Thus no one global expression is universally applicable. In contrast to the global models, the partial-equilibrium approach provides an estimate of the radical concentration corresponding to a particular cooling history.

To apply the partial-equilibrium model, rate equations for both the pool of reaction intermediates and carbon monoxide must be integrated. Again expressing the concentration of intermediate species in terms of OH, the equations become

$$R_{OH} = -k_1[OH]_e^2 \alpha y^2$$

$$R_{CO} = -k_1[OH]_e y\{[CO] - [CO]_e y^2\}$$

where $\alpha$ is given by (3.49).

As a test of this model, we shall compare theoretical predictions with the data of Morr and Heywood (1974), who examined carbon monoxide oxidation quenching in a plug-flow, kerosene-fired laboratory combustor. After combustion products were fully mixed ($S < 0.05$) and equilibrated, they were passed through a heat exchanger which, at $\phi = 0.91$, dropped the temperature from 2216 K to 975 K in about 5 ms (i.e., a cooling rate of about $2 \times 10^5$ K $s^{-1}$). The equilibrium CO concentration decreases rapidly upon cooling, but the measured concentrations decrease much more slowly, as illustrated in Figure 3.15.

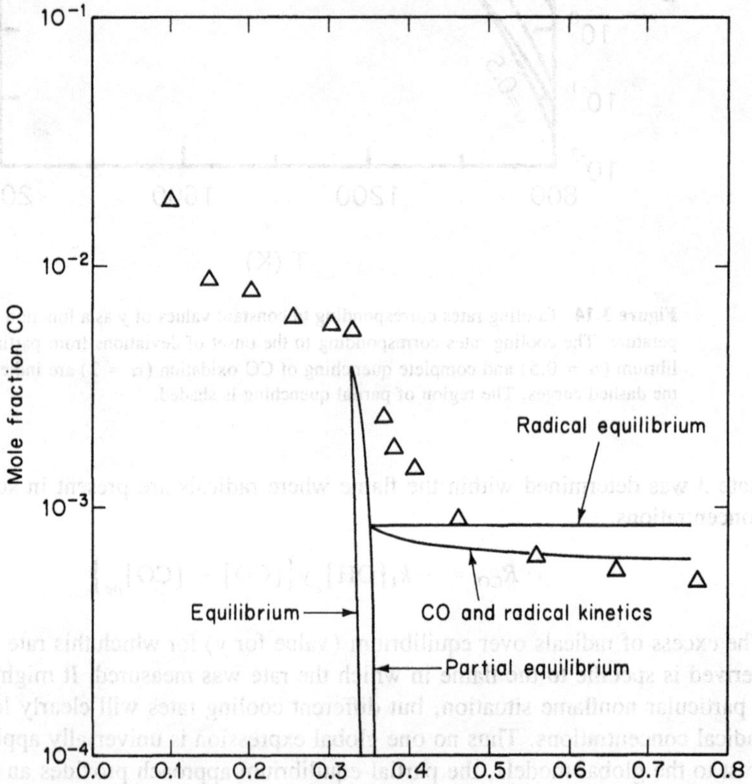


**Figure 3.15** Comparison of measured and calculated carbon monoxide profiles as a function of axial position in a plug flow combustor with a heat exchange (data from Morr and Heywood, 1974).

Three predictions of the CO level, obtained by numerical integration of the rate equations, are also shown in Figure 3.15. If CO is assumed to be equilibrated with OH, its concentration decreases rapidly to levels far below the measurements. The prediction for equilibrium OH shows a rapid initial rate of CO oxidation, with the reactions being completely frozen within the heat exchanger due to the very low equilibrium OH concentration at low temperature. This contrasts markedly with the slower but continuous decrease in CO that was observed. The partial equilibrium model, including both the recombination reactions and CO oxidation, yields a slightly more gradual decline in the CO level. The ultimate predictions are in close agreement with the experimental data. Thus we see that both the recombination reactions and the carbon monoxide oxidation reactions are important to CO oxidation.

The treatment presented here is subject to a number of limiting assumptions, notably (1) the members of the radical pool must all be minor species so that the concentrations of $H_2O$ and $O_2$ are constant, and (2) the mean molecular weight must be constant. Detailed modeling of CO formation and destruction in flames or in stoichiometric or fuel-rich combustion products requires that these constraints be eliminated. Calculation of the partial equilibrium by direct minimization of the Gibbs free energy as described by Keck and Gillespie (1971) and Morr and Heywood (1974) allows the model developed here to be applied in general without such restrictions. Furthermore, the Gibbs free-energy minimization code can be applied without the tedious algebra since chemical reaction constraints can be treated in exactly the same way as element conservation constraints to the minimization problem.

## 3.3 HYDROCARBONS

Like CO, hydrocarbon emissions from combustion systems result from incomplete combustion. Equilibrium levels of hydrocarbons are low at the equivalence ratios at which practical combustion systems are normally operated, and the oxidation reactions are fast. Hydrocarbons can escape destruction if poor mixing allows very fuel-rich gases to persist to the combustor exhaust or if the oxidation reactions are quenched early in the combustion process.

The composition and quantity of hydrocarbons in exhaust gases depend on the nature of the fuel and of the process that limits oxidation. The range of hydrocarbon species that are emitted from combustion systems is too broad to present a detailed accounting here. Polycyclic aromatic hydrocarbons (PAH) are a particularly significant class of combustion-generated hydrocarbons since they include a number of known carcinogens or mutagens. These compounds form in extremely fuel-rich regions of the flame, where hydrocarbon polymerization reactions are favored over oxidation. Figure 3.16 illustrates pathways that are thought to lead to a number of PAH species. The complexities and uncertainties in hydrocarbon oxidation mechanisms make a detailed analysis beyond the scope of this book. Rough estimates of hydrocarbon emission rates can be made using global oxidation models such as those discussed in Chapter 2, but quantitative models have yet to be developed. Combustion conditions that result in low

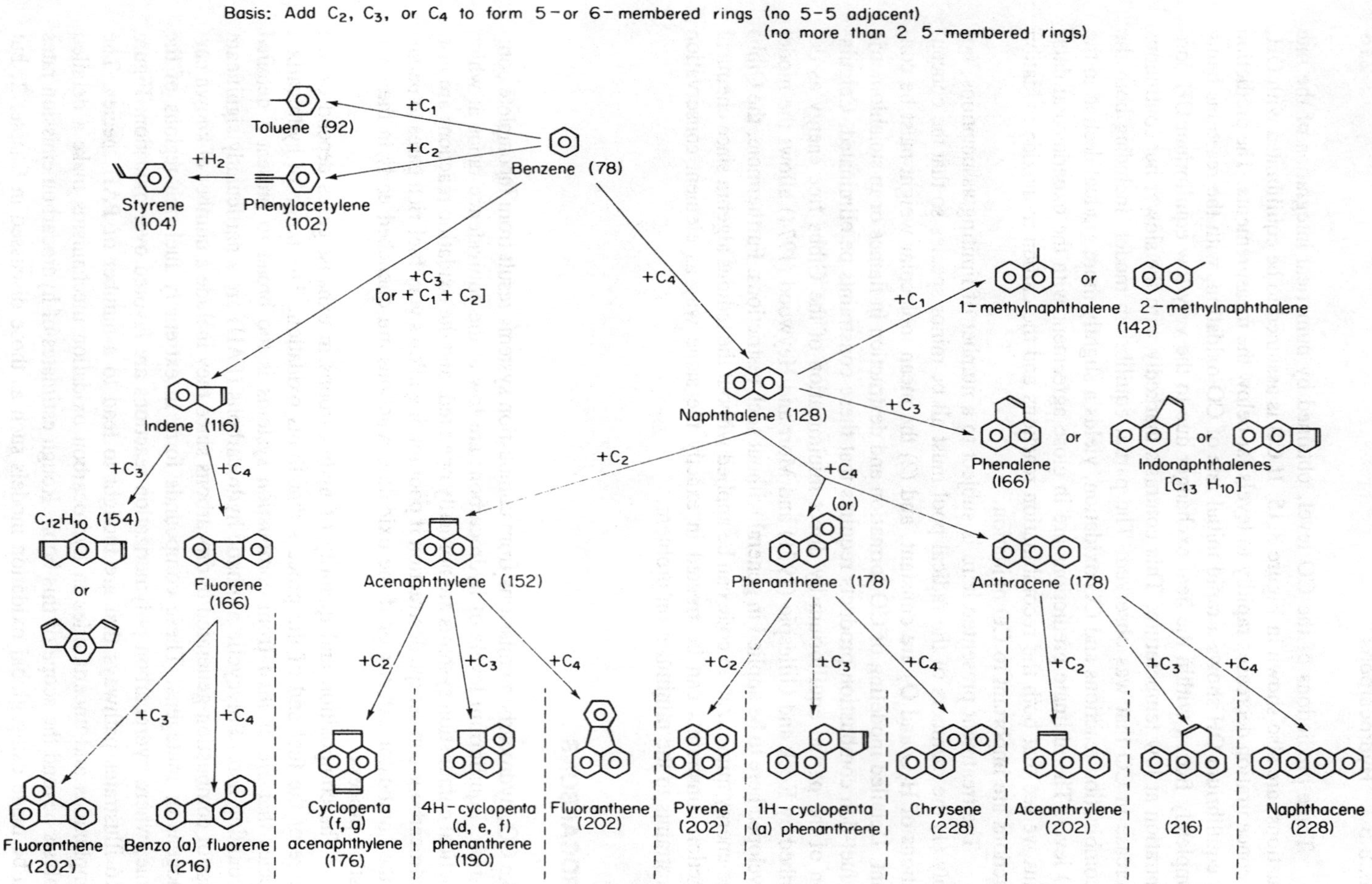


**Figure 3.16** Mechanisms of polycyclic aromatic hydrocarbon formation and growth (courtesy of A. Sarofim). The numbers in parenthesis are the molecular weights.

carbon monoxide emission will generally yield low exhaust concentrations of unburned hydrocarbons. Noncombustion sources of hydrocarbons, such as fuel or lubricant evaporation, are frequently major contributors to hydrocarbon emissions.

## 3.4 SULFUR OXIDES

Coal and heavy fuel oils contain appreciable amounts of sulfur (Tables 2.2 and 2.3). The dominant inorganic sulfur species in coal is pyrite ($FeS_2$). Organic sulfur forms include thiophene, sulfides, and thiols (Attar, 1978):

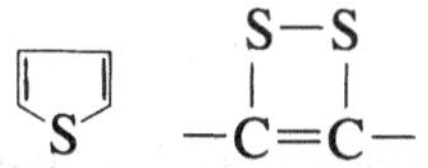

thiophene thiol

In fuel-lean combustion, the vast majority of this sulfur is oxidized to form $SO_2$. Sulfides, predominantly $H_2S$ and COS, may survive in fuel-rich flames (Kramlich et al., 1981). All of these compounds are considered air pollutants. The sulfides are odoriferous as well. Sulfur removal, either from the fuel or from the combustion products, is required for emission control.

A fraction of the sulfur is further oxidized beyond $SO_2$ to form $SO_3$ (Barrett et al., 1966). Sulfur trioxide is a serious concern to boiler operators since it corrodes combustion equipment. When combustion products containing $SO_3$ are cooled, the $SO_3$ may react with water to form sulfuric acid,

$$SO_3 + H_2O \longrightarrow H_2SO_4$$

which condenses rapidly at ambient temperatures to form a fine sulfuric acid aerosol that frequently appears as a blue fume near the stack outlet.

The equilibrium level of $SO_3$ in fuel-lean combustion products is determined by the overall reaction

$$SO_2 + \tfrac{1}{2}\,O_2 \rightleftharpoons SO_3$$

Applying van't Hoff's equation to the data of Table 2.5, we find

$$K_p = 1.53 \times 10^{-5}\, e^{11,760/T}\ \text{atm}^{-1/2}$$

We see that the equilibrium yield of $SO_3$ increases with decreasing temperature. In combustion products below about 900 K, $SO_3$ would be the dominant species at chemical equilibrium. Actual conversion of $SO_2$ to $SO_3$ is too slow to reach the equilibrium level.

$SO_3$ is formed primarily by the reaction

$$SO_2 + O + M \xrightarrow{1} SO_3 + M$$

for which Westenberg and de Haas (1975) recommended a rate coefficient of

$$k_1 = 8.0 \times 10^4\, e^{-1400/T} \quad \text{m}^6\ \text{mol}^{-2}\ \text{s}^{-1}$$

$SO_3$ can also react with O,

$$SO_3 + O + M \xrightarrow{2} SO_2 + O_2 + M$$

where the rate constant is

$$k_2 = 7.04 \times 10^4 \, e^{785/T} \qquad m^6 \, mol^{-2} \, s^{-1}$$

Another possible sink for $SO_3$ is

$$SO_3 + H \xrightarrow{3} SO_2 + OH$$

Muller et al. (1979) have estimated

$$k_3 = 1.5 \times 10^7 \, m^3 \, mol^{-1} \, s^{-1}$$

The destruction of $SO_3$ is dominated by these radical reactions rather than by the thermal decomposition of $SO_3$ (Fenimore and Jones, 1965; Cullis and Mulcahy, 1972). At the low temperatures where $SO_3$ is formed, the endothermic reverse reactions are not important, so the rate equation for $SO_3$ may be written

$$R_{SO_3} = k_1[SO_2][O][M] - k_2[SO_3][O][M] - k_3[SO_3][H]$$

If we assume constant-temperature, fuel-lean combustion, this becomes

$$\frac{d[SO_3]}{dt} = k_1[O][M]\{[S_T] - [SO_3]\} - \{k_2[O][M] + k_3[H]\}[SO_3]$$

where $[S_T]$ is the total concentration of sulfur oxides. Further assuming that the concentrations of the radicals are constant, this may be integrated to determine $[SO_3]$:

$$[SO_3] = \frac{k_1[S_T][O][M]}{k_1[O][M] + k_2[O][M] + k_3[H]}(1 - e^{-t/\tau}) + [SO_3]_0 e^{-t/\tau}$$

where

$$\tau = \{k_1[O][M] + k_2[O][M] + k_3[H]\}^{-1}$$

and $[SO_3]_0$ denotes the amount of $SO_3$ formed in the flame where the H and O concentrations are not equilibrated. At large time, $SO_3$ approaches a steady-state level

$$[SO_3]_{ss} = \frac{k_1[S_T][O][M]}{k_1[O][M] + k_2[O][M] + k_3[H]}$$

Assuming equilibrium O and H, the characteristic time for $SO_3$ oxidation is about 0.009 s for $\phi = 0.9$ at 1700 K, but increases to 0.13 s at 1500 K. Thus the reactions become quite slow as the combustion products are cooled below 1500 K, where the equilibrium yield becomes appreciable.

Steady-state $SO_3$ levels are on the order of 10% of the total sulfur in fuel-lean combustion products at low temperatures. Actual conversions are frequently lower, on the order of 3%. Clearly, the steady state is not maintained throughout the cooling process. The quenching of $SO_2$ oxidation is directly coupled to the radical chemistry that was discussed in conjunction with CO oxidation. Rapid cooling leads to superequilibrium radical concentrations and the maintenance of the $SO_3$ steady state at lower temperatures than the foregoing estimates of chemical relaxation times would suggest.

Sulfur dioxide oxidation is catalyzed by metals such as vanadium, so $SO_3$ production in some systems is much greater than the homogeneous gas-phase chemistry would suggest. $SO_3$ yields as high as 30% have been reported for combustion of some high-vanadium fuel oils. These high levels produce an acidic aerosol that has been called "acid smut." In extreme cases, plumes become opaque due to the condensation of water on the hygroscopic acid aerosol. Fuel additives such as MgO have been used to control such $SO_3$ formation.

Sulfur reactions with inorganic species can be used, under special circumstances, to retain sulfur in the solid phase so that it can be removed from combustion products with particulate emission control equipment. We consider processes in Chapter 8 that use such species for gas cleaning. A commonly used sorbent is lime, CaO, which reacts with sulfur oxides to form calcium sulfate, or gypsum,

$$CaO_{(s)} + SO_2 + \tfrac{1}{2}\,O_2 = CaSO_{4(s)}$$

The equilibrium constant for this $SO_2$ capture reaction is

$$K_p = 1.17 \times 10^{-14}\, e^{58,840/T}\ \text{atm}^{-3/2}$$

The equilibrium mole fraction of $SO_2$ is

$$y_{SO_2} = p^{-3/2} K_p^{-1} y_{O_2}^{-1/2}$$

Thus low temperatures and fuel-lean combustion favor the retention of $SO_2$ as calcium sulfate. For combustion products containing 3% oxygen ($\phi = 0.85$) the equilibrium $SO_2$ level is below 100 ppm for temperatures below about 1360 K. Sulfur capture in the combustion zone is most promising for systems in which the peak temperatures are low [e.g., fluidized-bed or fixed-bed (stoker-fired) combustors]. The high temperatures typical of pulverized coal flames promote the decomposition of any $CaSO_4$ that may be formed to CaO and $SO_2$. While sulfur capture is favored thermodynamically as the products of pulverized coal combustion cool, the reactivity of lime that is processed through high-temperature flames is reduced dramatically by sintering or fusion that reduces the surface area on which the sulfur capture reactions can take place.

A major application of sorbents for in-flame sulfur capture is in fluidized-bed combustors where the temperature must be kept low to prevent ash agglomeration. In fluidized-bed combustors, relatively large (millimeter-sized) coal particles are fluidized by an airflow from below at a velocity that lifts the particles but does not entrain them and carry them out of the bed. To keep the temperature below the ash fusion temperature, the coal bed is diluted with a large excess of noncombustible particles. If limestone or another sorbent is used as the diluent, the release of fuel sulfur from the flame can be effectively minimized.

The reactions leading to sulfur capture take place on the surface of the CaO. The active surface includes large numbers of small pores that are generated as the lime is produced by calcination of limestone, $CaCO_3$. The density of $CaCO_3$ is 2710 kg m$^{-3}$, considerably smaller than that of CaO, 3320 kg m$^{-3}$, but the mass loss in calcination at temperatures below about 1400 K occurs without appreciable change in the dimensions of the particle (Borgwardt and Harvey, 1972). The resulting pore volume ($\approx 3.6 \times 10^{-4}$ m$^3$ kg$^{-1}$) accounts for more than one-half of the total lime volume. The spongy material thus provides a large surface area (0.6 to 4 m$^2$ g$^{-1}$) on which the surface reactions of $SO_2$ can take place.

Reaction of CaO to form $CaSO_4$ increases the solid volume by a factor of 3.3. If the particle does not expand upon reaction, the amount of $CaSO_4$ formed in the pores is determined by the pore volume, that is,

$$\text{volume of } CaSO_4 \text{ formed} = \text{volume of CaO reacted} + \text{pore volume}$$

This corresponds to conversion of about one-half of the CaO to $CaSO_4$. Before the pore volume is filled, the reaction rate is limited by the rate of the surface reaction. For large particles, the diffusional resistance in the pores reduces the sulfur capture rate. After the pore volume is filled, $SO_2$ must diffuse through the solid reaction products for further reaction to take place, so the reaction rate slows considerably.

The rate of sulfur capture may be expressed in the form

$$r_0 = k_s S_g [SO_2] \eta \qquad \text{mol } SO_2 \text{ (kg CaO s)}^{-1}$$

where $S_g$ is the initial surface area as measured by $N_2$ adsorption (BET method), and $\eta$ is an effectiveness factor that accounts for the resistance to diffusion in the pore structure of the sorbent. Borgwardt and Harvey (1972) measured an intrinsic rate coefficient, $k_s$, of

$$k_s = 135\, e^{-6290/T} \qquad \text{m s}^{-1}$$

For small particles, large pores, or both, the effectiveness factor may be taken as unity, and this rate may be applied directly. For larger particles, a pore transport model is required to estimate $\eta$ and sulfur capture. (See Section 2.8.2.)

Within fixed beds, in the early fuel-rich regions of pulverized coal flames, and within the porous structure of burning char, the reducing atmosphere makes sulfur capture as mineral sulfides possible. For equivalence ratios not too far from unity, $SO_2$ may react with CaO:

$$CaO_{(s)} + SO_2 + 3CO \rightleftarrows CaS_{(s)} + 3\,CO_2$$

$$K_p = 2.31 \times 10^{-10}\, e^{47{,}355/T} \text{ atm}^{-1}$$

In extremely fuel-rich gases, $H_2S$ and COS are the predominant sulfur species, so reactions such as

$$CaO_{(s)} + H_2S = CaS_{(s)} + H_2O$$

$$K_p = 19.4\, e^{6156/T}$$

also contribute to sulfur retention. The sulfur capture reactions are not strongly dependent on temperature at high equivalence ratios, so sulfur capture can be achieved at higher temperatures than would be possible in a fuel-lean environment. Freund and Lyon (1982) have reported sulfur retention as high as 80% in the ash of coal that was enriched with Ca by ion exchange prior to combustion at equivalence ratios greater than 3.

Control of $NO_x$ formation from fuel-bound nitrogen also requires combustion under fuel-rich conditions, so there has been considerable interest in combining $NO_x$ control with in-flame sulfur capture. One version of this technology involves limestone injection in multistage (low $NO_x$) burners (LIMB). The limestone is calcined early in the flame and then may react with the sulfur released from the fuel. One problem is that the addition of sufficient air to complete the combustion process provides the thermodynamic driving force for the release of any sulfur that has been captured. Sintering of the sorbent at the high temperatures in the near stoichiometric regions of the flame, however, may inhibit the release of previously captured sulfur.

## PROBLEMS

**3.1.** A furnace is fired with natural gas (assume methane). The inlet air and fuel temperatures are 290 K and the pressure is atmospheric. The residence time in the combustor is 0.1 s. Assuming the combustion to be adiabatic, calculate the NO mole fraction at the combustor outlet for combustion at $\phi = 0.85$.

**3.2.** Combustion products at 600 K are to be mixed with the incoming air to lower the NO emissions from the furnace in Problem 3.1. Assuming that the total mass flow rate through the furnace does not change, determine what fraction of the flue gases must be recycled to reduce $NO_x$ formation by 90%. What is the corresponding change in the amount of heat that can be transferred to a process requiring a temperature of 600 K?

**3.3.** For the furnace of Problem 3.1, determine the equivalence ratio that would achieve a 90% reduction in NO formation. What is the corresponding change in the amount of heat that can be transferred to a process requiring a temperature of 600 K?

**3.4.** Examine the influence of mixing on NO formation by determining the amount of NO formed in the furnace of Problem 3.1 if the segregation parameter is $S = 0.3$. Assume a Gaussian equivalence ratio distribution.

**3.5.** As a model of the low $NO_x$ burner, consider a flame in which combustion initially takes place fuel-rich, at an equivalence ratio of $\phi_1 = 1.5$, and that air is then entrained into the flame region at a constant mass entrainment rate until the flame is diluted to an equivalence ratio of $\phi_2 = 0.8$. Let the time over which this dilution takes place be $\tau$. The flame may be assumed to be adiabatic, with no further change in $\phi$ occurring after an equivalence ratio of 0.8 is achieved. Calculate the NO formation when a fuel with composition $CH_{1.8}N_{0.01}$ and lower heating value of 41 MJ $kg^{-1}$ is burned in 298 K air at atmospheric pressure. Use the effective reaction rate from Figure 3.7 and $T$ from Figure 2.6 to calculate the rate of $N_2$ formation. What is the effect of increasing the entrainment time from 0.1 s to 1 s?

**3.6.** In the derivation of the model for selective reduction of NO, it was assumed that the $NH_2$ concentration achieved a steady state. Examine the validity of that assumption.

**3.7.** The temperature is not constant in the reaction zone where the $NH_3$–NO reactions take place in practical implementations of the thermal de-$NO_x$ process. Numerically integrate the rate equations for the kinetic model for this system to determine the impact of cooling at a rate of 500 K $s^{-1}$ on the temperature window of operation. Compare your results with isothermal operation. Assume that the fuel has composition $CH_2$ and was burned at an equivalence ratio of 0.85 and that the initial NO concentration was 400 ppm, and consider $NH_3/NO$ ratios of 0.5, 1, and 2.

**3.8.** Combustion products with composition

$$y_{O_2} = 0.02$$

$$y_{H_2O} = 0.10$$

$$y_{CO_2} = 0.11$$

$$y_{N_2} = 0.77$$

$$y_{NO} = 500 \text{ ppm}$$

are cooled from 2300 K to 500 K.

(a) Determine the equilibrium NO mole fraction as a function of temperature for this cooling process.

(b) Numerically integrate the NO rate equation to calculate the amount of NO formed when the cooling takes place at constant cooling rates of 1000 and $10^5$ K $s^{-1}$.

(c) Determine the influence of radical nonequilibrium on NO formation by adapting the model developed to describe CO oxidation.

**3.9.** Use the rate constants presented in Table 2.8 to examine the validity of the partial-equilibrium assumption for the species H, OH, O, $H_2$, $HO_2$, and $H_2O_2$. For combustion of $CH_{1.88}$ in atmospheric pressure air at $\phi = 0.91$ and a cooling rate of $10^4$ K $s^{-1}$, beginning with equilibrium at 2000 K, at what temperature does the partial-equilibrium model begin to break down?

## REFERENCES

ALLEN, J. D. "Probe Sampling of Oxides of Nitrogen from Flames," *Combust. Flame, 24*, 133–136 (1975).

ATTAR, A. "Chemistry, Thermodynamics, and Kinetics of Sulfur in Coal Gas Reactions—A Review," *Fuel, 57*, 201–212 (1978).

AZUHATA, S., KAJI, R., AKIMOTO, H., and HISHINUMA, Y. "A Study of the Kinetics of the $NH_3$-NO-$O_2$-$H_2O_2$ Reaction," in *Eighteenth Symposium (International) on Combustion*, The Combustion Institute, Pittsburgh, PA, 845–852 (1981).

BARRETT, R. E., HUMMELL, J. D., and REID, W. T. "Formation of $SO_3$ in a Noncatalytic Combustor," *J. Eng. Power Trans. ASME, A88*, 165–172 (1966).

BAULCH, D. L., DRYSDALE, D. D., HORNE, D. G., and LLOYD, A. C. *Evaluated Kinetic Data for High Temperature Reactions*, Vol. 2, *Homogeneous Gas Phase Reactions of the* $H_2$-$N_2$-$O_2$ *System*, Butterworth, London (1973).

BORGWARDT, R. H., and HARVEY, R. D. "Properties of Carbonate Rocks Related to $SO_2$ Reactivity," *Environ. Sci. Technol., 6*, 350–360 (1972).

BOWMAN, C. T. "Kinetics of Pollutant Formation and Destruction in Combustion," *Prog. Energy Combust. Sci.*, *1*, 33–45 (1975).

BRANCH, M. C., KEE, R. J., and MILLER, J. A. "A Theoretical Investigation of Mixing Effects in the Selective Reduction of Nitric Oxide by Ammonia," *Combust. Sci. Technol.*, *29*, 147–165 (1982).

CERNANSKY, N. P., and SAWYER, R. F. "NO and $NO_2$ Formation in a Turbulent Hydrocarbon/Air Diffusion Flame," in *Fifteenth Symposium (International) on Combustion*, The Combustion Institute, Pittsburgh, PA, 1039–1050 (1975).

CULLIS, C. F., and MULCAHY, M. F. R. "The Kinetics of Combustion of Gaseous Sulfur Compounds," *Combust. Flame*, *18*, 225–292 (1972).

DEAN, A. M., HARDY, J. E., and LYON, R. K. "Kinetics and Mechanism of $NH_3$ Oxidation," in *Nineteenth Symposium (International) on Combustion*, The Combustion Institute, Pittsburgh, PA, 97–105 (1982).

DIEHL, L. A. "Reduction of Aircraft Gas Turbine Pollutant Emissions—A Status Report," 71st Annual Meeting, Air Pollution Control Assoc., 78-26.4 (1979).

DRYER, F. L. and GLASSMAN, I. "High Temperature Oxidation of Carbon Monoxide and Methane," in *Fourteenth Symposium (International) on Combustion*, The Combustion Institute, Pittsburgh, PA, 987–1003 (1974).

FENIMORE, C. P., and JONES, G. W. "Sulfur in the Burnt Gas of Hydrogen-Oxygen Flames," *J. Phys. Chem.*, *69*, 3593–3597 (1965).

FENIMORE, C. P., and MOORE, J. "Quenched Carbon Monoxide in Fuel-Lean Flame Gases," *Combust. Flame*, *22*, 343–351 (1974).

FLAGAN, R. C., and APPLETON, J. P. "A Stochastic Model of Turbulent Mixing with a Chemical Reaction: Nitric Oxide Formation in a Plug-Flow Burner," *Combust. Flame*, *23*, 249–267 (1974).

FLAGAN, R. C., GALANT, S., and APPLETON, J. P. "Rate Constrained Partial Equilibrium Models for the Formation of Nitric Oxide from Organic Fuel Nitrogen," *Combust. Flame*, *22*, 299–311 (1974).

FREUND, H., and LYON, R. K. "The Sulfur Retention of Calcium-Containing Coal during Fuel-Rich Combustion," *Combust. Flame*, *45*, 191–203 (1982).

FRISTROM, R. M. and WESTENBERG, A. A. *Flame Structure*, McGraw-Hill, New York (1965).

HANSON, R. K., and SALIMIAN, S. "Survey of Rate Constants in the N-H-O System," in *Combustion Chemistry*, W. C. Gardiner, Ed., Springer-Verlag, New York, 361–421 (1984).

HARGREAVES, K. J. A., HARVEY, R., ROPER, F., and SMITH, D. B. "Formation of $NO_2$ in Laminar Flames," in *Eighteenth Symposium (International) on Combustion*, The Combustion Institute, Pittsburgh, PA, 133–142 (1981).

HAYNES, B. S., IVERACH, D., and KIROV, N. Y. "The Behavior of Nitrogen Species in Fuel Rich Hydrocarbon Flames," in *Fifteenth Symposium (International) on Combustion*, The Combustion Institute, Pittsburgh, PA, 1103–1112 (1975).

HAZARD, H. R. "Conversion of Fuel Nitrogen to $NO_x$ in a Compact Combustor," *J. Eng. Power Trans. ASME, A96*, 185–188 (1974).

HOWARD, J. B., WILLIAMS, G. C., and FINE, D. H. "Kinetics of Carbon Monoxide Oxidation in Postflame Gases," in *Fourteenth Symposium (International) on Combustion*, The Combustion Institute, Pittsburgh, PA, 975–986 (1973).

KECK, J. C., and GILLESPIE, D. "Rate-Controlled Partial-Equilibrium Method for Treating Reacting Gas Mixtures," *Combust. Flame, 17,* 237–241 (1971).

KRAMLICH, J. C., MALTE, P. C., and GROSSHANDLER, W. L. "The Reaction of Fuel Sulfur in Hydrocarbon Combustion," in *Eighteenth Symposium (International) on Combustion,* The Combustion Institute, Pittsburgh, PA, 151–161 (1981).

LAVOIE, G. A., HEYWOOD, J. B., and KECK, J. C. "Experimental and Theoretical Study of Nitric Oxide Formation in Internal Combustion Engines," *Combust. Sci. Technol., 3,* 313–326 (1970).

LEVY, A. "Unresolved Problems in $SO_x$, $NO_x$, Soot Control in Combustion," in *Nineteenth Symposium (International) on Combustion,* The Combustion Institute, Pittsburgh, PA, 1223–1242 (1982).

LEVY, J. M., CHAN, L. K., SAROFIM, A. F., and BEER, J. M. "NO/Char Reactions at Pulverized Coal Flame Conditions," in *Eighteenth Symposium (International) on Combustion,* The Combustion Institute, Pittsburgh, PA, 111–120 (1981).

LUCAS, D., and BROWN, N. "Characterization of the Selective Reduction of NO by $NH_3$," *Combust. Flame, 47,* 219–234 (1982).

LYON, R. K. "The $NH_3$-NO-$O_2$ Reaction," *Int. J. Chem. Kinet., 8,* 315–318 (1976).

MILLER, J. A., BRANCH, M. C., and KEE, R. J., "A Chemical Kinetic Model for the Selective Reduction of Nitric Oxide by Ammonia," *Combust. Flame, 43,* 81–98 (1981).

MILLER, J. A., and FISK, G. A. "Combustion Chemistry," *Chem. & Eng. News,* 22–46, August 31, 1987.

MORR, A. R., and HEYWOOD, J. B. "Partial Equilibrium Model for Predicting Concentration of CO in Combustion," *Acta Astronautica, 1,* 949–966 (1974).

MULLER, C. H., SCHOFIELD, K., STEINBERG, M., and BROIDA, H. P. "Sulfur Chemistry in Flames," in *Seventeenth Symposium (International) on Combustion,* The Combustion Institute, Pittsburgh, PA, 867–879 (1979).

MUZIO, L. J., and ARAND, J. K. "Homogeneous Gas Phase Decomposition of Oxides of Nitrogen," Electric Power Research Institute Report No. EPRI FP-253, Palo Alto, CA (1976).

MUZIO, L. J., MALONEY, K. L., and ARAND, J. K. "Reactions of $NH_3$ with NO in Coal-Derived Combustion Products," in *Seventeenth Symposium (International) on Combustion,* The Combustion Institute, Pittsburgh, PA, 89–96 (1979).

PERRY, R. A., and SIEBERS, D. L. "Rapid Reduction of Nitrogen Oxides in Exhaust Gas Streams," *Nature, 324,* 657–658 (1986).

PERSHING, D. W., and WENDT, J. O. L. "Pulverized Coal Combustion: The Influence of Flame Temperature and Coal Composition on Thermal and Fuel $NO_x$," in *Sixteenth Symposium (International) on Combustion,* The Combustion Institute, Pittsburgh, PA, 389–399 (1977).

POHL, J. H., and SAROFIM, A. F. "Devolatilization and Oxidation of Coal Nitrogen," in *Sixteenth Symposium (International) on Combustion,* The Combustion Institute, Pittsburgh, PA, 491–501 (1977).

POMPEI, F., and HEYWOOD, J. B. "The Role of Mixing in Burner Generated Carbon Monoxide and Nitric Oxide," *Combust. Flame, 19,* 407–418 (1972).

POPE, S. B. "PDF Methods for Turbulent Reactive Flows," *Prog. Energy Comb. Sci., 11,* 119–192 (1985).

SALIMIAN, S., and HANSON, R. K. "A Kinetic Study of NO Removal from Combustion Gases by Injection of $NH_i$-Containing Compounds," *Combust. Sci. Technol., 23,* 225–230 (1980).

SAROFIM, A. F., and FLAGAN, R. C. "$NO_x$ Control for Stationary Combustion Sources," *Prog. Energy Combust. Sci.*, *2*, 1–25 (1976).

SAROFIM, A. F., and POHL, J. "Kinetics of Nitric Oxide Formation in Premixed Laminar Flames," in *Fourteenth Symposium (International) on Combustion*, The Combustion Institute, Pittsburgh, PA, 739–754 (1973).

WENDT, J. O. L., PERSHING, D. W., LEE, J. W. and GLASS, J. W. "Pulverized Coal Combustion: Nitrogen Oxide Formation under Fuel Rich and Staged Combustion Conditions," *Seventeenth Symposium (International) on Combustion*, The Combustion Institute, Pittsburgh, PA (1979).

WENDT, J. O. L., and SCHULZE, O. E. "On the Fate of Fuel Nitrogen during Coal Char Combustion," *A.I.Ch.E.J.*, *22*, 102–110 (1976).

WENDT, J. O. L., STERNLING, C. V., and MATOVICH, M. A. "Reduction of Sulfur Trioxide and Nitrogen Oxides by Secondary Fuel Injection," in *Fourteenth Symposium (International) on Combustion*, The Combustion Institute, Pittsburgh, PA, 897–904 (1973).

WESTENBERG, A. A., and DE HAAS, H. "Rate of the Reaction $O + SO_2 + M \rightarrow SO_3 + M^*$," *J. Chem. Phys.*, *63*, 5411-5415 (1975).

ZELDOVICH, Y. B., SADOVNIKOV, P. Y., and FRANK-KAMENETSKII, D. A. *Oxidation of Nitrogen in Combustion*, M. Shelef, Trans., Academy of Sciences of USSR, Institute of Chemical Physics, Moscow-Leningrad (1947).

# 4

# Internal Combustion Engines

Internal combustion engines are devices that generate work using the products of combustion as the working fluid rather than as a heat transfer medium. To produce work, the combustion is carried out in a manner that produces high-pressure combustion products that can be expanded through a turbine or piston. The engineering of these high-pressure systems introduces a number of features that profoundly influence the formation of pollutants.

There are three major types of internal combustion engines in use today: (1) the spark ignition engine, which is used primarily in automobiles; (2) the diesel engine, which is used in large vehicles and industrial systems where the improvements in cycle efficiency make it advantageous over the more compact and lighter-weight spark ignition engine; and (3) the gas turbine, which is used in aircraft due to its high power/weight ratio and also is used for stationary power generation.

Each of these engines is an important source of atmospheric pollutants. Automobiles are major sources of carbon monoxide, unburned hydrocarbons, and nitrogen oxides. Probably more than any other combustion system, the design of automobile engines has been guided by the requirements to reduce emissions of these pollutants. While substantial progress has been made in emission reduction, automobiles remain important sources of air pollutants. Diesel engines are notorious for the black smoke they emit. Gas turbines emit soot as well. These systems also release unburned hydrocarbons, carbon monoxide, and nitrogen oxides in large quantities.

In this chapter we examine the air pollutant emissions from engines. To understand the emissions and the special problems in emission control, it is first necessary that we understand the operating principles of each engine type. We begin our discussion with a system that has been the subject of intense study and controversy—the spark ignition engine.

## 4.1 SPARK IGNITION ENGINES

The operating cycle of a conventional spark ignition engine is illustrated in Figure 4.1. The basic principle of operation is that a piston moves up and down in a cylinder, transmitting its motion through a connecting rod to the crankshaft which drives the vehicle. The most common engine cycle involves four strokes:

1. *Intake.* The descending piston draws a mixture of fuel and air through the open intake valve.

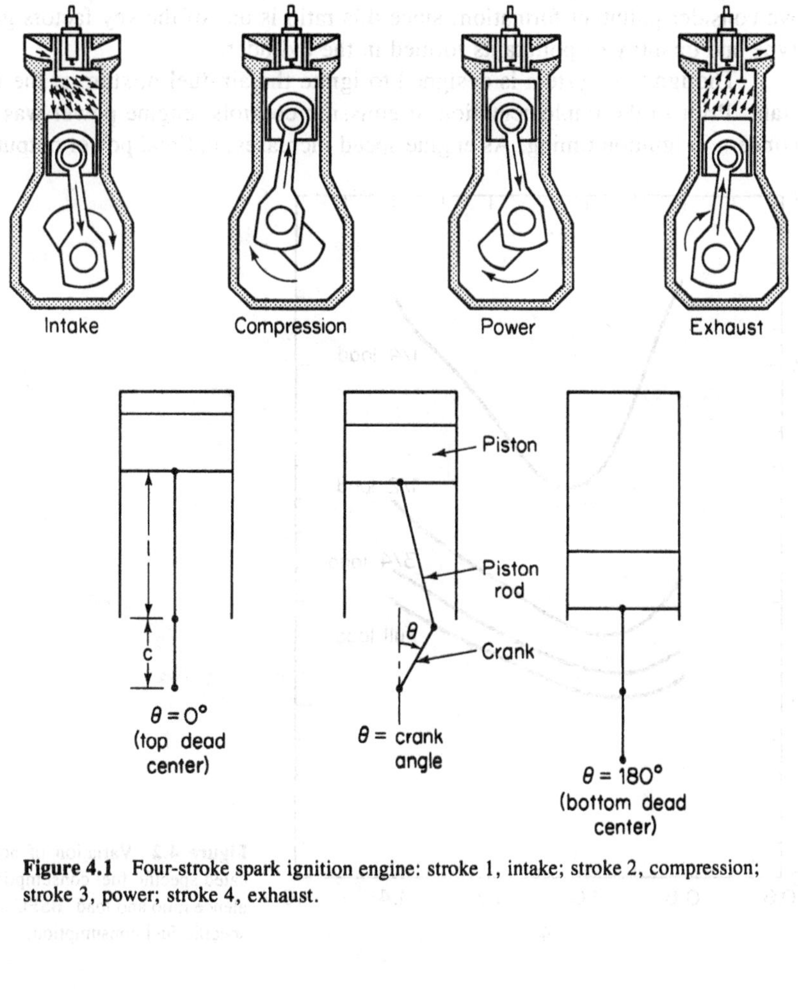


**Figure 4.1** Four-stroke spark ignition engine: stroke 1, intake; stroke 2, compression; stroke 3, power; stroke 4, exhaust.

2. *Compression.* The intake valve is closed and the rising piston compresses the fuel–air mixture. Near the top of the stroke, the spark plug is fired, igniting the mixture.
3. *Expansion.* The burning mixture expands, driving the piston down and delivering power.
4. *Exhaust.* The exhaust valve opens and the piston rises, expelling the burned gas from the cylinder.

The fuel and air mixture is commonly premixed in a carburetor. Figure 4.2 shows how engine power and fuel consumption depend on equivalence ratio over the range commonly used in internal combustion engines. Ratios below 0.7 and above 1.4 generally are not combustible on the time scales available in reciprocating engines. The maximum power is obtained at a higher ratio than is minimum fuel consumption. As a vehicle accelerates, high power is needed and a richer mixture is required than when cruising at constant speed. We shall return to the question of the equivalence ratio when we consider pollutant formation, since this ratio is one of the key factors governing the type and quantity of pollutants formed in the cylinder.

The ignition system is designed to ignite the air–fuel mixture at the optimum instant. Prior to the implementation of emission controls, engine power was the primary concern in ignition timing. As engine speed increases, optimal power output is achieved

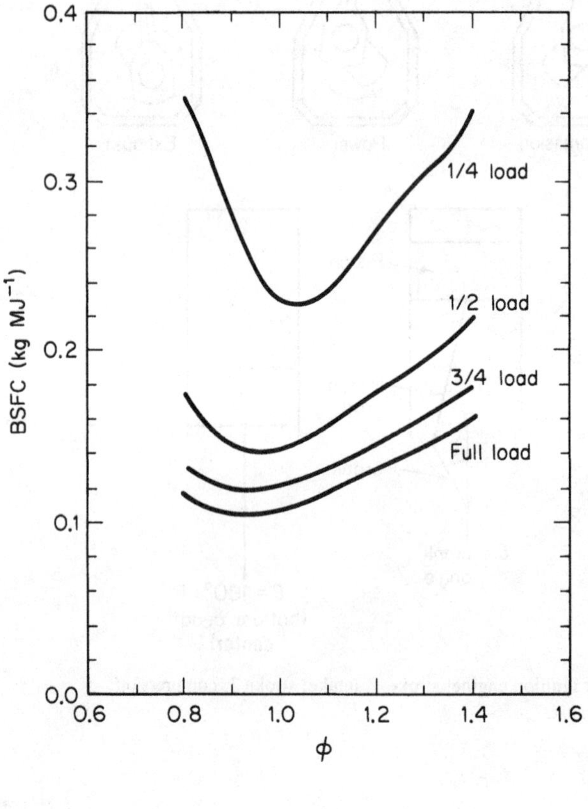


**Figure 4.2** Variation of actual and indicated specific fuel consumption with equivalence ratio and load. BSFC denotes "brake specific fuel consumption."

by advancing the time of ignition to a point on the compression stroke before the piston reaches the top of its motion where the cylinder volume is smallest. This is because the combustion of the mixture takes a certain amount of time, and optimum power is developed if the completion of the combustion coincides with the piston arriving at so-called top dead center. The spark is automatically advanced as engine speed increases. Also, a pressure diaphragm senses airflow through the carburetor and advances the spark as airflow increases.

Factors other than power output must be taken into account, however, in optimizing the engine operation. If the fuel–air mixture is compressed to an excessive pressure, the mixture temperature can become high enough that the preflame reactions can ignite the charge ahead of the propagating flame front. This is followed by very rapid combustion of the remaining charge and a correspondingly fast pressure increase in the cylinder. The resultant pressure wave reverberates in the cylinder, producing the noise referred to as *knock* (By et al., 1981). One characteristic of the fuel composition is its tendency to autoignite, expressed in terms of an octane rating.

High compression ratios and ignition spark timing that optimize engine power and efficiency lead to high octane requirements. The octane requirement can be reduced by using lower compression ratios and by delaying the spark until after the point for optimum engine performance. Emission controls require additional compromises in engine design and operation, sacrificing some of the potential engine performance to reduce emissions.

### 4.1.1 Engine Cycle Operation

The piston sweeps through a volume that is called the displacement volume, $V_d$. The minimum volume occurs when the piston is in its uppermost position. This volume is called the clearance volume, $V_c$. The maximum volume is the sum of these two. The ratio of the maximum volume to the clearance volume is called the compression ratio,

$$R_c = \frac{V_c + V_d}{V_c} \tag{4.1}$$

The efficiency of the engine is a strong function of the compression ratio. We shall see that $R_c$ also has a strong influence on the formation of pollutants. The volume in the cylinder can be expressed as a simple function of the crank angle, $\theta$, and the ratio of the length of the piston rod to that of the crank, that is,

$$V = V_c + \frac{V_d}{2}\left(1 + \frac{l}{c} - \cos\theta - \sqrt{\frac{l^2}{c^2} - \sin^2\theta}\right) \tag{4.2}$$

where $l$ is the piston rod length and $c$ is the length of the crank arm as defined in Figure 4.1. The minimum volume occurs at $\theta = 0°$, commonly referred to as *top dead center*, TDC. The maximum volume occurs at *bottom dead center*, BDC, $\theta = 180°$. These positions are illustrated in Figure 4.1.

Engine speeds range from several hundred revolutions per minute (rpm) for large industrial engines to 10,000 rpm or more for high-performance engines. Most automobiles operate with engine speeds in the vicinity of 3000 rpm. At this speed, each stroke in the cycle takes place in 20 ms. As an automobile is driven, the equivalence ratio and intake pressure vary with the engine load. Such changes in engine operation, however, are slow by comparison with the individual strokes. In discussing engine operation, we can assume that in any one cycle the engine operates at constant speed, load, and equivalence ratio.

We begin with a discussion of the thermodynamics of the spark ignition engine cycle and develop a model that has been used extensively in optimizing engine operation to minimize emissions and to maximize performance.

The spark ignition engine is one of the few combustion systems that burns premixed fuel and air. Fuel is atomized into the air as it flows through a carburetor and vaporizes before it enters the cylinder. Even though the fuel and air are premixed prior to combustion, the gas in the cylinder becomes segmented into burned and unburned portions once ignition occurs. A flame front propagates through the cylinder as illustrated in Figure 4.3. The fuel–air mixture ahead of the flame is heated somewhat by adiabatic compression as the burning gas expands. Not only are the burned and unburned gases at widely different temperatures, but also there are large variations in the properties of the burned gases. These variations must be taken into account to predict accurately the formation and destruction of $NO_x$ and CO in the engine.

Another important feature that distinguishes reciprocating engines from the systems discussed thus far is that the volume in which the combustion proceeds is tightly constrained. While the individual elements of fluid do expand as they burn, this expansion requires that other elements of fluid, both burned and unburned, be compressed. As a result, the burning element of fluid does work on the other fluid in the cylinder, $\delta W = p\,dV$, increasing its internal energy and therefore its temperature.

While the engine strokes are brief, the time is still long by comparison with that required for pressure equilibration. For an ideal gas, the propagation rate for small pressure disturbances is the speed of sound,

$$a_s = \sqrt{\gamma RT/M} \tag{4.3}$$

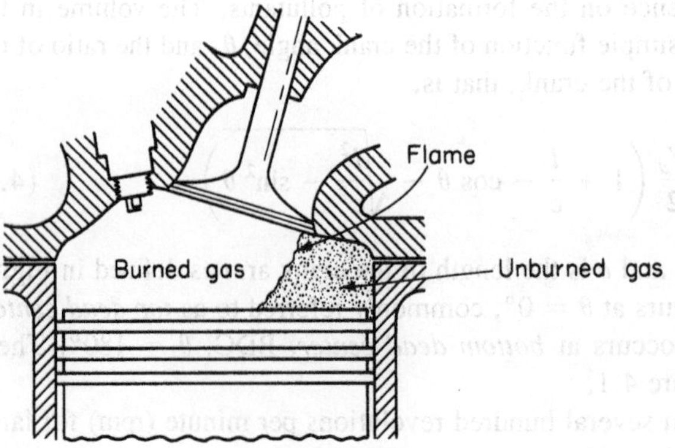


**Figure 4.3** Flame propagation in the cylinder.

where $\gamma$ is the ratio of specific heats, $c_p/c_v$, and $M$ is the molecular weight of the gas; $a_s$ is of the order of 500 to 1000 m s$^{-1}$ for typical temperatures in internal combustion engines. For a cylinder 10 cm in diameter, the time required for a pressure disturbance to propagate across the cylinder is on the order of 0.2 ms, considerably shorter than the time required for the stroke. Thus, to a first approximation, we may assume that the pressure is uniform throughout the cylinder at any instant of time, at least during normal operation.

### 4.1.2 Cycle Analysis

The essential features of internal combustion engine operation can be seen with a "zero-dimensional" thermodynamic model (Lavoie et al., 1970; Blumberg and Kummer, 1971). This model describes the thermodynamic states of the burned and unburned gases as a function of time, but does not attempt to describe the complex flow field within the cylinder.

We consider a control volume enclosing all the gases in the cylinder. Mass may enter the control volume through the intake valve at flow rate, $\bar{f}_i$. Similarly, mass may leave through the exhaust valve and possibly through leaks at a flow rate $\bar{f}_e$. The first law of thermodynamics (2.8) for this control volume may be written in the general form

$$\frac{dU}{dt} = \bar{f}_i \bar{h}_i - \bar{f}_e \bar{h}_e + \frac{dQ}{dt} - \frac{dW}{dt}$$

where $U$ is the total internal energy of the gases contained in the cylinder and $\bar{h}_i$ and $\bar{h}_e$ are the mass specific enthalpies of the incoming and exiting flows, respectively. $Q$ denotes the heat transferred to the gases. The work done by the gases, $W$, is that of a pressure acting through a change in the volume of the control volume as the piston moves. If we limit our attention to the time between closing the intake valve and opening the exhaust valve and assume that no leaks occur, no mass enters or leaves the cylinder (i.e., $\bar{f}_i = \bar{f}_e = 0$). The energy equation then simplifies to

$$\frac{d}{dt}(m\bar{u}_T) = \frac{dQ}{dt} - p\frac{dV}{dt}$$

where $\bar{u}_T$ is the total mass specific internal energy (including energies of formation of all species in the cylinder), $-Q$ is heat transferred out of the charge, and $m$ is the total mass of the charge. The only work done by the gases is due to expansion against the piston, so the work is expressed as $p\,dV/dt$. If we further limit our attention to constant engine speed, the time derivations may be expressed as

$$\frac{d}{dt} = \omega\frac{d}{d\theta}$$

where $\omega$ is the engine rotation speed (crank angle degrees per s). Thus we have

$$\frac{d}{d\theta}(m\bar{u}_T) = \frac{dQ}{d\theta} - p\frac{dV}{d\theta} \tag{4.4}$$

The total specific internal energy of the gas includes contributions of burned and unburned gases, with a mass fraction $\alpha$ of burned gas,

$$\overline{u}_T = \alpha \langle \overline{u}_b \rangle + (1 - \alpha) \langle \overline{u}_u \rangle \tag{4.5}$$

where $\langle\ \rangle$ denotes an average over the entire mass of burned or unburned gas in the cylinder. The unburned gas is quite uniform in temperature (i.e., $\langle \overline{u}_u \rangle = \overline{u}_u$) but the burned gas is not. Due to the progressive burning, a temperature gradient develops in the burned gas. As a fluid element burns, its expansion compresses both unburned and burned gases. Because the volume per unit mass of the hot burned gas is larger than that of the cooler unburned gas, the increase in the mass specific internal energy due to the compression work is higher for burned gas than for unburned gas. Therefore, we need to keep track of when individual fluid elements burn. Let $\overline{u}_b(\alpha, \alpha')$ represent the energy when the combustion has progressed to burned gas mass fraction $\alpha$ of a fluid element that burned when the burned gas mass fraction was $\alpha'$. Averaging over all burned gas, we find

$$\langle \overline{u}_b \rangle (\alpha) = \frac{1}{\alpha} \int_0^\alpha \overline{u}_b(\alpha, \alpha')\, d\alpha' \tag{4.6}$$

The internal energy of either burned or unburned gas may be expressed in terms of the specific heat,

$$\overline{u}_i = \Delta \overline{u}_{fi}^\circ (T_0) + \int_{T_0}^{T} \overline{c}_{vi}(T')\, dT' \tag{4.7}$$

While the specific heats vary with temperature, we have already seen in Chapter 2 that variation is small over a limited temperature range. We assume constant specific heats since that will greatly simplify our analysis of the engine cycle. To minimize the errors introduced by this simplification, the specific heats should be evaluated for the actual composition of the gases in the cylinder as an average over the temperature range encountered by those gases. In terms of the linear correlations of specific heats presented in Table 2.5 and evaluating over the temperature interval, $T_1 \leq T \leq T_2$, this average can be expressed in terms of the correlations for $c_p$ in Table 2.5 by

$$\overline{c}_{vi} = \frac{\displaystyle\int_{T_1}^{T_2} (a_i' + b_i T)\, dT}{M_i (T_2 - T_1)} = \frac{a_i}{M_i} + \frac{b_i}{2M_i}(T_1 + T_2) \tag{4.8}$$

The internal energies of the burned and unburned portions of the gas may be expressed in terms of the average specific heats by

$$\begin{aligned} \overline{u}_b &= a_b + \overline{c}_{vb} T_b \\ \overline{u}_u &= a_u + \overline{c}_{vu} T_u \end{aligned} \tag{4.9}$$

where $a_u$ and $a_b$ include the reference temperature terms and the energies of formation. Substituting into (4.6), the mean burned gas energy becomes

$$\langle \bar{u}_b \rangle = \frac{1}{\alpha} \int_0^{\alpha} \left[ a_b + \bar{c}_{vb} T_b(\alpha, \alpha') \right] d\alpha'$$

where $T_b(\alpha, \alpha')$ is the temperature of an element that burned at $\alpha'$ at a later time when combustion has progressed to $\alpha$. Thus the mean burned gas energy can be expressed in terms of the mean burned gas temperature,

$$\langle \bar{u}_b \rangle = a_b + \bar{c}_{vb} \langle T_b \rangle \tag{4.10}$$

where

$$\langle T_b \rangle = \frac{1}{\alpha} \int_0^{\alpha} T_b(\alpha, \alpha')\, d\alpha'$$

Substitution of (4.5), (4.9), and (4.10) into the energy equation yields

$$\frac{d}{d\theta} \left[ m(1 - \alpha)(a_u + \bar{c}_{vu} T_u) + m\alpha \left( a_b + \bar{c}_{vb} \langle T_b \rangle \right) \right] = \frac{dQ}{d\theta} - p \frac{dV}{d\theta} \tag{4.11}$$

The total volume of burned and unburned gases must, at all times, equal the volume in the cylinder:

$$V = m\alpha \langle \bar{v}_b \rangle + m(1 - \alpha) \bar{v}_u \tag{4.12}$$

Assuming ideal gases with constant composition, the mean specific volume of the burned gas is

$$\langle \bar{v}_b \rangle = \int_0^{\alpha} \frac{\bar{R}_b T_b(\alpha, \alpha')}{p} d\alpha' = \frac{\bar{R}_b \langle T_b \rangle}{p} \tag{4.13}$$

Noting that $\bar{R}_b = (\gamma_b - 1)\bar{c}_{vb}$, where $\gamma_b = \bar{c}_{pb}/\bar{c}_{vb}$ is the ratio of specific heats, (4.12) may now be simplified to

$$m\alpha \bar{c}_{vb} \langle T_b \rangle = \frac{pV}{\gamma_b - 1} - m(1 - \alpha) \frac{\gamma_u - 1}{\gamma_b - 1} \bar{c}_{vu} T_u \tag{4.14}$$

Substituting this result into (4.11) eliminates the burned gas temperature from the energy equation:

$$\frac{d}{d\theta} \left[ m(1 - \alpha) a_u + m(1 - \alpha) \left( \frac{\gamma_b - \gamma_u}{\gamma_b - 1} \right) \bar{c}_{vu} T_u + m\alpha a_b + \frac{pV}{\gamma_b - 1} \right] = \frac{dQ}{d\theta} - p \frac{dV}{d\theta} \tag{4.15}$$

A simple approach can be used to eliminate the unburned gas temperature. At the end of the intake stroke, the cylinder is assumed to be filled with a uniform mixture of fuel and air and possibly some combustion products from previous cycles. The pressure, cylinder volume, and gas temperature at the time the intake valve closes are $p_i$, $V_i$, and $T_i$, respectively. Because the temperature difference between these gases and the cylinder wall is small (at least compared to that between combustion products and the wall), compression of these gases is approximately adiabatic. Prior to firing the spark at $\theta_0$, the pressure in the cylinder can be determined from the formula for the relation between pressure and volumes in adiabatic compression,

$$p(\theta) = p_i\left[\frac{V_i}{V(\theta)}\right]^{\gamma_u} \qquad \theta_i \le \theta \le \theta_0 \tag{4.16}$$

The temperature of the unburned gas throughout the cycle is that determined by adiabatic compression

$$T_u(\theta) = T_i\left[\frac{p(\theta)}{p_i}\right]^{(\gamma_u-1)/\gamma_u} \tag{4.17}$$

Substituting (4.17) into (4.15) and differentiating yield

$$\begin{aligned} &m(1-\alpha)\frac{\gamma_b-\gamma_u}{\gamma_b-1}\bar{c}_{vu}T_i\left(\frac{p}{p_i}\right)^{(\gamma_u-1)/\gamma_u}\frac{1}{p}\frac{\gamma_u-1}{\gamma_u}\frac{dp}{d\theta} \\ &+ m\left[a_b - a_u - \frac{\gamma_b-\gamma_u}{\gamma_b-1}\bar{c}_{vu}T_i\left(\frac{p}{p_i}\right)^{(\gamma_u-1)/\gamma_u}\right]\frac{d\alpha}{d\theta} \\ &+ \frac{p}{\gamma_b-1}\frac{dV}{d\theta} + \frac{V}{\gamma_b-1}\frac{dp}{d\theta} \\ &= \frac{dQ}{d\theta} - p\frac{dV}{d\theta} \end{aligned} \tag{4.18}$$

This equation may be rearranged to express the rate of change of the cylinder pressure in terms of the conditions at the end of the intake stroke, the rate of volume change, and the combustion and heat transfer rates, that is,

$$\frac{dp}{d\theta} = \frac{\dfrac{dQ}{d\theta} - \dfrac{\gamma_b}{\gamma_b-1}p\dfrac{dV}{d\theta} - m\left[a_b - a_u - \dfrac{\gamma_b-\gamma_u}{\gamma_b-1}\bar{c}_{vu}T_i\left(\dfrac{p}{p_i}\right)^{(\gamma_u-1)/\gamma_u}\right]\dfrac{d\alpha}{d\theta}}{m(1-\alpha)\bar{c}_{vu}\dfrac{\gamma_b-\gamma_u}{\gamma_b-1}\dfrac{\gamma_u-1}{\gamma_u}\dfrac{T_i}{p}\left(\dfrac{p}{p_i}\right)^{(\gamma_u-1)/\gamma_u} + \dfrac{V}{\gamma_b-1}} \tag{4.19}$$

### 4.1.3 Cylinder Turbulence and Combustion Rate

We need to know the combustion rate, $d\alpha/d\theta$, to use the model of (4.19). To efficiently convert the heat released by combustion to work on the piston, the charge must be burned completely in the early part of the expansion stroke. The duration of the stroke in automotive engines is on the order of 20 ms, so the combustion can take at most a few milliseconds. Since typical laminar flame speeds are less than 1 m s$^{-1}$, tens of milliseconds would be required for laminar flame propagation across a cylinder several centimeters in diameter. We see, therefore, that the acceleration of flame propagation that turbulence provides is essential to efficient engine operation.

As discussed in Chapter 2, the turbulent flame speed depends on the turbulent intensity, $u'$. The turbulent intensity is governed by engine design and operation, and varies during the stroke as described below. The mixture entrained in the flame front by the turbulent motion burns at a rate that depends on combustion kinetics through the laminar flame speed, $S_L$. The laminar flame speed peaks near stoichiometric and decreases for richer or leaner mixtures, so there is also some dependence of flame speed on the equivalence ratio. To make general statements about the factors governing pollutant formation in spark ignition engines, therefore, we need to understand how turbulence varies with engine operation.

The generation of turbulence in an internal combustion engine is a complex, unsteady process. As the mixture passes through the intake valve, the flow separates, resulting in a highly unsteady motion (Hoult and Wong, 1980). The intensity of the resulting turbulent motion depends on the detailed geometry of the intake port and valve, on the geometry of the cylinder and piston, and on the speed of the piston.

As we discussed in Chapter 2, the turbulence may be characterized in terms of two quantities: (1) the turbulent kinetic energy per unit mass

$$E_k = \frac{1}{2}\left(\overline{u_1^2} + \overline{u_2^2} + \overline{u_3^2}\right) \tag{4.20}$$

which describes the large-scale behavior of the turbulence, and (2) the rate of turbulent kinetic energy dissipation

$$\epsilon = \nu \left[\overline{\frac{\partial u_i'}{\partial x_j}\,\frac{\partial u_i'}{\partial x_j}}\right] \tag{4.21}$$

which decribes the effects of the small-scale turbulent motions.

The mixture passes through the intake valve at a velocity that is proportional to the piston speed and hence to the angular rotation speed, $\omega$. The kinetic energy of this incoming flow contributes to the turbulent kinetic energy within the cylinder. How much of that kinetic energy remains at bottom dead center when the compression begins depends on the geometry of the particular engine.

The turbulent kinetic energy is not constant during the compression and power strokes. Dissipation tends to decrease $E_k$, while the distortion due to compression of the existing turbulent field tends to increase it. Turbulent kinetic energy may also be produced by shear associated with fluid motions. Shrouds on the intake valves, illustrated in Figure 4.4, are used to create a swirling motion in the cylinder. Complex piston or cylinder head shapes induce fluid motions during the final approach to top dead center, also shown in Figure 4.4. This so-called *squish* can greatly enhance the turbulent kinetic energy level immediately prior to combustion.

Neglecting diffusion of the turbulent kinetic energy, the rate of change of the turbulent kinetic energy is a balance between production and dissipation:

$$\rho \frac{dE_k}{dt} = \rho P - \rho\epsilon \tag{4.22}$$

where $P$ is the rate of turbulent kinetic energy production.

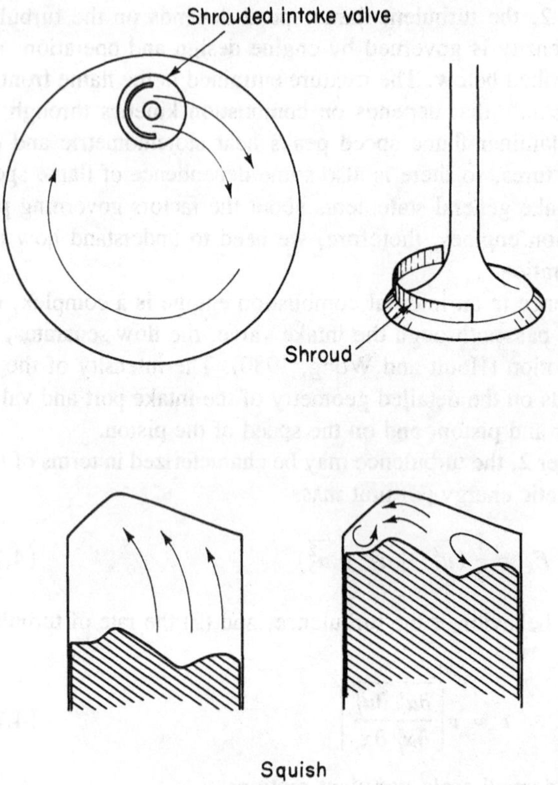


**Figure 4.4** Valve, head, and piston design features that enhance mixing.

The dissipation rate was shown in Appendix D of Chapter 1 to be related to $u'$ for homogeneous, isotropic turbulence,

$$\epsilon \approx \nu \frac{u'^2}{\lambda^2} \approx \frac{u'^3}{l}$$

where $\lambda$ and $l$ are the Taylor microscale and integral scale, respectively. Using the definition of $E_k$, we find

$$\epsilon \approx \frac{E_k^{3/2}}{l} \tag{4.23}$$

Assuming that angular momentum in the turbulent field is conserved during the rapid compression:

$$u'l \approx E_k^{1/2} l \approx \text{const.}$$

we see that $\epsilon$ is proportional to $E_k^2$,

$$\epsilon \propto E_k^2 \tag{4.24}$$

The gas density and integral scale are related by conservation of mass,

$$\rho l^3 = \text{const.}$$

or

$$l \propto \rho^{-1/3} \tag{4.25}$$

Using (4.24), this yields

$$u' \propto \rho^{1/3} \tag{4.26}$$

or

$$E_k \propto \rho^{2/3} \tag{4.27}$$

We may use these scaling arguments to simplify the rate equation for the turbulent kinetic energy. Assuming that due to the rapid distortion of the flow caused by the compression due to both piston motion and the expansion of gases upon burning, the production of turbulent kinetic energy is much more rapid than its dissipation (Borgnakke et al., 1980),

$$\rho P \approx \rho \frac{dE_k}{dt}$$

and applying (4.27), the production of turbulent kinetic energy due to the rapid distortion of the turbulent field during compression, yields

$$P \approx \frac{2}{3} \frac{E_k}{\rho} \frac{d\rho}{dt} \tag{4.28}$$

The rate equation for $E_k$ becomes

$$\frac{dE_k}{dt} = \frac{2}{3} \frac{E_k}{\rho} \frac{d\rho}{dt} - CE_k^2 \tag{4.29}$$

where $\epsilon$ has been eliminated using (4.24).

The production term generally dominates during the compression and combustion processes due to the rapid change in density, so (4.29) may be rewritten as

$$\omega \frac{d \ln (E_k/E_{k0})}{d\theta} \approx \omega \frac{2}{3} \frac{d \ln (\rho/\rho_0)}{d\theta} \tag{4.30}$$

where $E_{k0}$ and $\rho_0$ denote the initial kinetic energy and density. We see that the relative change of the turbulent kinetic energy from bottom dead center to any crank angle, $\theta$, is, to a first approximation, independent of the crank rotation speed, $\omega$. The initial turbulent kinetic energy depends on piston speed as

$$E_{k0} \approx \omega^2 \tag{4.31}$$

because the inlet flow velocity is proportional to the piston speed. Thus, for a given engine geometry, the value of $u'$ at any crank angle, $\theta$, is approximately proportional to the angular speed

$$u_0' \approx \omega \tag{4.32}$$

and the turbulent flame propagation velocity increases with the engine speed.

This dependence of flame speed on engine speed means that the number of crank angle degrees required for combustion in a given engine does not depend strongly on the engine speed. Thus, if $\alpha(\theta)$ is known for one engine speed, we may use that result as an estimate of the burn rate for other engine speeds with reasonable confidence.

Rather than attempt to develop detailed fluid mechanical models of the combustion process, therefore, we shall simply specify a functional form for $\alpha(\theta)$ that exhibits the essential features of actual combustion profiles, that is, a delay from the time the spark is fired until the pressure rise associated with combustion becomes appreciable, an accelerating combustion rate until a large fraction of the charge is burned, followed by a decreasing burn rate. A simple function with this sigmoidal behavior is the cosine function,

$$\alpha(\theta) = \frac{1}{2}\left[1 - \cos\left(\frac{\theta - \theta_0}{\Delta\theta_c}\pi\right)\right] \qquad \theta_0 \leq \theta \leq \theta_0 + \Delta\theta_c \tag{4.33}$$

where $\theta_0$ is the crank angle at which the spark is fired and $\Delta\theta_c$ is the burn duration. Other functions that allow the shape of the combustion profile to be varied have been used in the literature, but this simple function is adequate for our present purpose of exploring engine operation. We do not attempt to predict the burn duration, since it is a complex function of engine design and operation.

### 4.1.4 Cylinder Pressure and Temperature

The pressure in the cylinder can be determined by integrating (4.19) with $\alpha(\theta)$ given by (4.33) or another suitable model and with an expression for the heat transfer $dQ/d\theta$. The heat transfer is also a function of the turbulent field (Borgnakke et al., 1980). For our present purposes, it is sufficient to assume that the engine is adiabatic (i.e., $dQ/d\theta = 0$).

Once the pressure in the cylinder is known the mean burned and unburned gas temperatures can be calculated using (4.14) and (4.17), respectively. The temperatures of individual burned gas elements can be calculated if it is assumed that no mixing of the burned gases occurs and that heat transfer from a burned gas element is negligible. Under these assumptions, the burned gases can be assumed to undergo adiabatic compression and expansion from the time they burn. The temperature of an element burned when the mass fraction burned was $\alpha'$ is

$$T_b(\alpha, \alpha') = T_b(\alpha', \alpha')\left[\frac{p(\alpha)}{p(\alpha')}\right]^{(\gamma_b - 1)/\gamma_b} \tag{4.34}$$

The temperature of the element immediately following combustion, $T_b(\alpha', \alpha')$, may be evaluated by applying the first law of thermodynamics to the combustion of an infinitesimal mass of charge, $dm$. For combustion of a sufficiently small incremental mass, the pressure change during combustion is insignificant. The enthalpy of the burned gas equals that for the unburned gas, that is,

$$\overline{h}_u = \overline{u}_u + \overline{R}_u T_u = \overline{h}_b = \overline{u}_b + \overline{R}_b T_b$$

The burned gas temperature becomes

$$T_b(\alpha', \alpha') = \frac{a_u - a_b + \bar{c}_{pu} T_u}{\bar{c}_{pb}} \tag{4.35}$$

From (4.19), (4.34), and (4.35) we can determine the pressure-temperature history of each element in the charge from the beginning to the end of combustion. Figure 4.5 shows the results of calculations of Heywood (1976) for an engine with a compression ratio of 7.0. The spark is fired at 40° before top dead center. The combustion duration, $\Delta\theta_c$, is 60°. The fraction of charge burned and the cylinder pressure are shown as a function of crank angle in Figure 4.5. The temperatures of the first and last gases to burn are shown as solid lines. The dashed curves represent the temperature of the unburned gas.

The first gas to burn rises to a high temperature immediately. As additional gas burns, the pressure in the cylinder rises, compressing both burned and unburned gases.

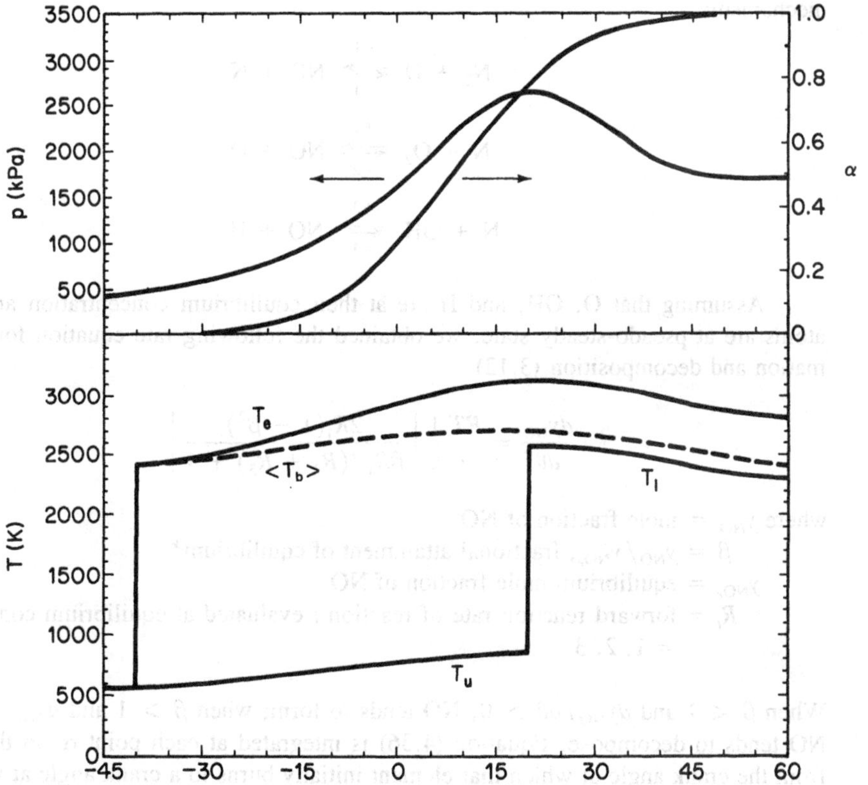


**Figure 4.5** Burned mass fraction, cylinder pressure, and temperatures of the gas that burns early, $T_e$, late, $T_l$, and the mean gas temperature inside the cylinder (after Heywood, 1976).

The work done on a gas element by this compression is $p\,dV$. Because the volume of a mass of burned gas is larger than that of an equal mass of unburned gas, more work is done on the gas that burns early in the cycle than is done on that that burns at a later time. The first gas burned, therefore, is the hottest gas in the cylinder.

### 4.1.5 Formation of Nitrogen Oxides

The foregoing model simulates the essential features of the combustion in the spark ignition engine and provides a basis for understanding the formation of pollutants in the cylinder. We first examine the rate of NO formation. In Chapter 3 we saw that NO formation is highly temperature dependent, so we expect that the NO formation rate will vary with location in the charge, depending on the temperature history of each element. Since the NO reactions require the thermal energy released by the combustion process, NO formation will take place only in the burned gases.

The dominant reactions in NO formation are those of the extended Zeldovich mechanism:

$$N_2 + O \underset{-1}{\overset{+1}{\rightleftharpoons}} NO + N$$

$$N + O_2 \underset{-2}{\overset{+2}{\rightleftharpoons}} NO + O$$

$$N + OH \underset{-3}{\overset{+3}{\rightleftharpoons}} NO + H$$

Assuming that O, OH, and H are at their equilibrium concentration and that N atoms are at pseudo-steady state, we obtained the following rate equation for NO formation and decomposition (3.12):

$$\frac{dy_{NO}}{d\theta} = \frac{RT}{p}\frac{1}{\omega}\left[\frac{2R_1(1-\beta^2)}{\beta R_1/(R_2+R_3)+1}\right] \tag{4.36}$$

where $y_{NO}$ = mole fraction of NO
$\beta = y_{NO}/y_{NO_e}$, fractional attainment of equilibrium*
$y_{NO_e}$ = equilibrium mole fraction of NO
$R_i$ = forward reaction rate of reaction $i$ evaluated at equilibrium conditions, $i = 1, 2, 3$

When $\beta < 1$ and $dy_{NO}/d\theta > 0$, NO tends to form; when $\beta > 1$ and $dy_{NO}/d\theta < 0$, NO tends to decompose. Equation (4.36) is integrated at each point $\alpha'$ in the charge from the crank angle at which that element initially burns to a crank angle at which the reaction rates are negligible. At this point the quenched value of the NO mole fraction

*We use $\beta$ here as this fraction to avoid confusion with the fraction burned $\alpha$.

$y_{NO_q}$ is achieved. The overall mole fraction of NO in the entire charge is given by

$$\bar{y}_{NO} = \int_0^1 y_{NO_q}(\alpha')\, d\alpha' \tag{4.37}$$

Nitric oxide concentrations versus crank angle, computed by Blumberg and Kummer (1971), are shown in Figure 4.6. Both rate calculated and equilibrium NO are shown at three positions in the charge, $\alpha' = 0, 0.5, 1.0$. The major contribution to the total NO formed results from the elements that burn first. They experience the highest temperatures and have the longest time in which to react. Considerable decomposition of NO occurs in the first element because of the high temperatures. However, as the first element cools during expansion, the rate of NO decomposition rapidly decreases, so that after about 40 crank angle degrees, the NO kinetics are effectively frozen.

We can now summarize the processes responsible for the production of nitric oxide

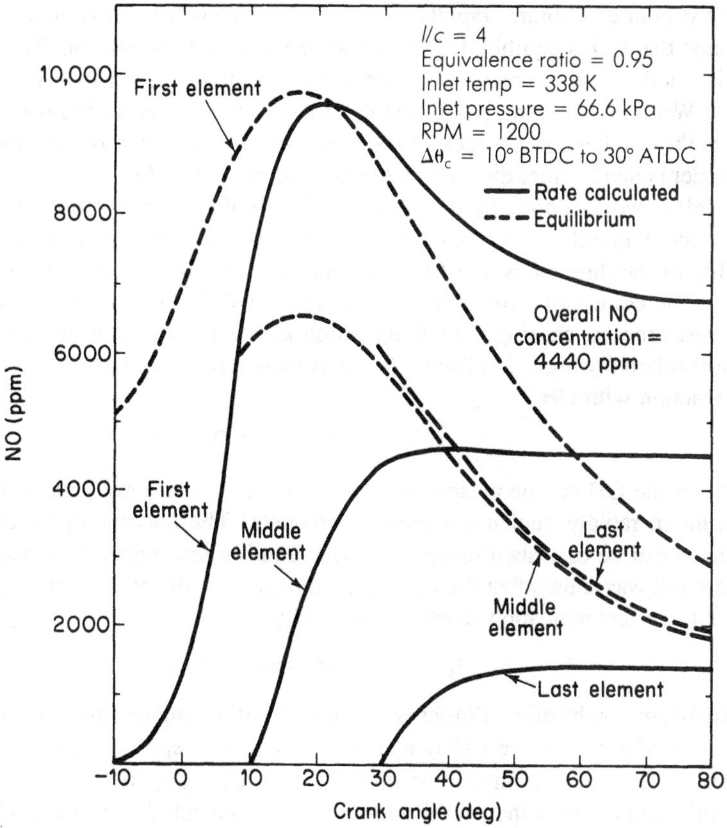


**Figure 4.6** Nitric oxide concentration in the burned gas as a function of crank angle for the first, middle, and last element to burn for $\phi = 0.95$ (Blumberg and Kummer, 1971). Reprinted by permission of Gordon and Breach Science Publishers.

in the internal combustion engine. During the flame propagation, NO is formed by chemical reactions in the hot just-burned gases. As the piston recedes, the temperatures of the different burned elements drop sharply, "freezing" the NO (i.e., the chemical reactions that would remove the NO become much slower) at the levels formed during combustion, levels well above these corresponding to equilibrium at exhaust temperatures. As the valve opens on the exhaust stroke, the bulk gases containing the NO exit. It is to the processes that occur prior to the freezing of the NO levels that we must devote our attention if we wish to reduce NO formation in the cylinder.

### 4.1.6 Carbon Monoxide

The compression due to piston motion and combustion in a confined volume leads to very high burned gas temperatures in reciprocating engines. Peak temperatures may range from 2400 to 2800 K, with pressures of 15 to 40 atm. In Chapter 3 we saw that the C–H–O system equilibrates rapidly at such high temperatures. It is therefore reasonable to assume that CO is equilibrated immediately following combustion. The equilibrium CO mole fraction at these peak temperatures is very high, greater than 1%.

Work done by the gas in the cylinder on the piston during the expansion stroke cools the combustion products. When the exhaust valve first opens, the pressure in the cylinder is much larger than that in the exhaust manifold. As the gas is forced out through the valve, work is done by the gas remaining in the cylinder, so the temperature drops even more rapidly. Ultimately, this cooling of the combustion products exceeds the ability of the three-body and CO oxidation reactions to maintain equilibrium.

The combustion products are rapidly cooled during the expansion stroke and the exhaust process, causing the CO oxidation kinetics to be quenched while the CO level is still relatively high. In Chapter 3 it was shown that CO oxidation proceeds primarily by reaction with OH,

$$CO + OH \rightleftharpoons CO_2 + H$$

and that the OH can be present at concentrations significantly greater than that at equilibrium in rapidly cooled combustion products. The concentrations of OH and other radicals can be described using the partial-equilibrium model developed in Chapter 3, wherein it was shown that the rate of CO oxidation is directly coupled to the rates of the three-body recombination reactions, primarily,

$$H + O_2 + M \rightleftharpoons HO_2 + M$$

in fuel-lean combustion. CO levels in spark ignition engines are generally high enough that the influence of the CO oxidation on the major species concentrations cannot be ignored. The direct minimization of the Gibbs free energy is better suited to incorporating this detail than is the equilibrium-constant approach developed in Chapter 3.

Heywood (1975) used the rate-constrained, partial-equilibrium model (based on direct minimization of the Gibbs free energy) to study CO behavior in spark ignition engines. His calculations confirm that at the peak temperatures and pressures the equilibration of CO is fast compared to the changes due to compression or expansion, so equilibrium may reasonably be assumed immediately following combustion. The burned gases are not uniform in temperature, however; so the equilibrium CO level depends on when the element burned. Furthermore, the blowdown of the cylinder pressure to the exhaust manifold pressure in the initial phase of the exhaust process lasts about 90 crank angle degrees. Thus the temperature–time profiles of fluid elements within the charge differ depending on the time of burning and on when they pass from the cylinder through the valve into the exhaust manifold.

These effects are illustrated by the results of an idealized calculation shown in Figure 4.7. CO mole fractions for individual fluid elements in the burned gas mixture are shown as a function of crank angle. The elements are identified in terms of the fraction of the total charge burned when the element burned, $\alpha$, and the mass fraction that has left the cylinder when the element leaves the cylinder, $z$. The partial-equilibrium calculations are close to equilibrium until about 50 crank angle degrees after top dead center, when the rapid cooling due to adiabatic expansion leads to partial quenching of the CO oxidation.

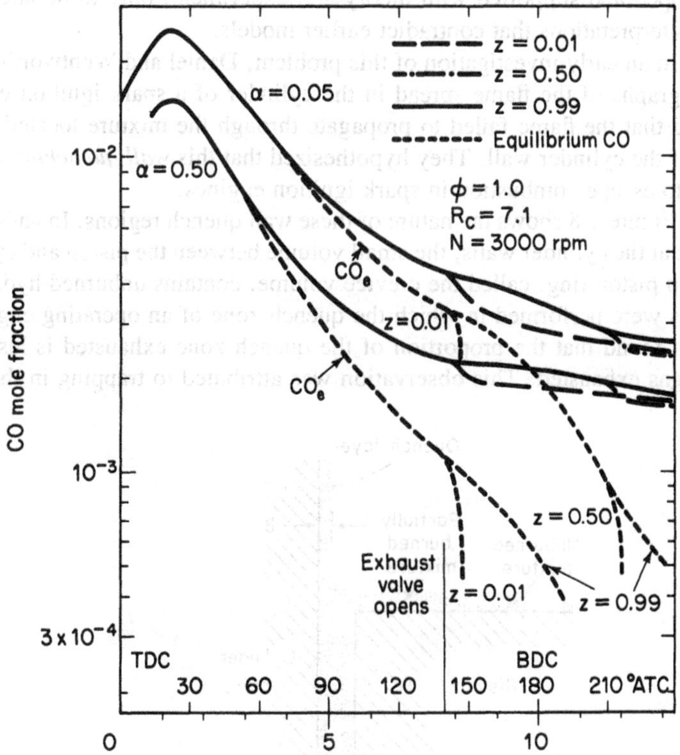


**Figure 4.7** Carbon monoxide concentration in two elements in the charge that burn at different times, during expansion and exhaust processes. $\alpha$ is the mass fraction burned and $z$ is the fraction of the gas that has left the cylinder during the exhaust process (Heywood, 1975). Reprinted by permission of The Combustion Institute.

The CO levels measured in fuel-lean combustion are substantially higher than those predicted with the partial-equilibrium model, but agreement is good near stoichiometric (Heywood, 1976). In fuel-rich combustion, the CO levels in the exhaust gases are close to the equilibrium concentrations, as predicted by the partial-equilibrium model. The reasons for the high levels in fuel-lean combustion are not fully understood, but may be coupled to the oxidation of unburned hydrocarbons in the exhaust manifold.

### 4.1.7 Unburned Hydrocarbons

The range of equivalence ratios over which spark ignition engines operate is narrow, typically $0.7 < \phi < 1.3$, the fuel and air are premixed, and the flame temperatures are high. These conditions, in steady-flow combustion systems, generally would lead to very low emissions of unburned hydrocarbons. Why, then, are relatively large quantities of hydrocarbon gases present in the combustion products of automobile engines? This question has been the subject of numerous investigations in which hypotheses have been developed and supported with theory and experiment, only to be later challenged with new interpretations that contradict earlier models.

In an early investigation of this problem, Daniel and Wentworth (1962) magnified photographs of the flame spread in the cylinder of a spark ignition engine. It was observed that the flame failed to propagate through the mixture located within 0.1 to 0.7 mm of the cylinder wall. They hypothesized that this *wall quenching* allowed hydrocarbons to escape combustion in spark ignition engines.

Figure 4.8 shows the nature of these wall quench regions. In addition to the quench layers at the cylinder walls, the small volume between the piston and cylinder wall above the top piston ring, called the crevice volume, contains unburned hydrocarbons. Experiments were performed in which the quench zone of an operating engine was sampled. It was found that the proportion of the quench zone exhausted is less than that of the total gas exhausted. This observation was attributed to trapping in the boundary layer.

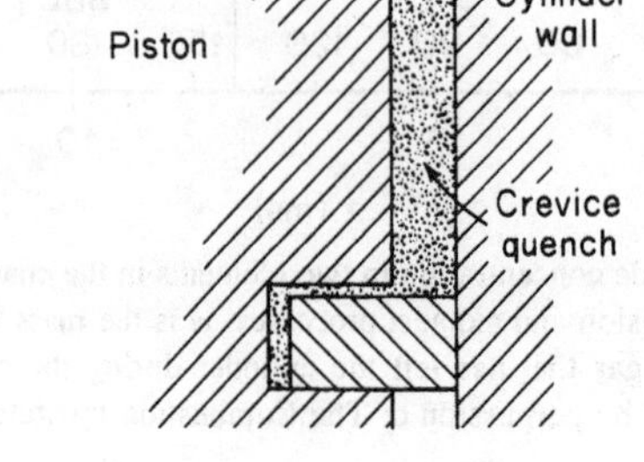


**Figure 4.8** Schematic showing the quench layer and crevice volume where heat transfer to the walls may quench the combustion (Tabaczynski et al., 1972; © SAE, Inc.).

A fraction of the gas remains in the cylinder at the end of the exhaust stroke. Although this residual gas amounts to a small fraction of the total gas in the cylinder in a normally operating engine, the residual gas hydrocarbon concentration tends to be very high. The recycled hydrocarbons may be a significant fraction of the hydrocarbons left unburned in the cylinder.

The trapping effect can be explained as follows. Gases adjacent to the wall opposite the exhaust valve are the farthest from the exit and least likely to be exhausted. Gases along the walls near the exhaust valve have a better chance to be exhausted, but viscous drag slows their movement. Some quenched gases do escape, but on the whole the more completely burned gases at the center of the chamber are preferentially exhausted first, with the result that the residual gas has a higher concentration of hydrocarbons than the exhaust gas. It is likely that the quench zone hydrocarbons that remain in the cylinder are burned in the succeeding cycle. In the experiments reported by Daniel and Wentworth (1962), about one-third of the total hydrocarbons were recycled and probably burned in succeeding cycles.

Figure 4.9 shows the measured variation in the exhaust hydrocarbon concentration and mass flow rate with crank angle. As the exhaust valve opens and the emptying of the combustion chamber starts, the hydrocarbon concentration in the exhaust manifold increases rapidly to a peak of 600 ppm. The hydrocarbon concentration then drops and remains at 100 to 300 ppm for much of the exhaust stroke. Late in the exhaust stroke, the hydrocarbon level again rises sharply. The hydrocarbon mass flow rate shows two distinct peaks corresponding to these concentration maxima. The early peak in the hydrocarbon concentration was attributed to the entrainment of the quench layer gases near the exhaust valve immediately after it opens. The low hydrocarbon concentration during the middle portion of the exhaust stroke is most probably due to the release of burned gases from the center of the cylinder.

Tabaczynski et al. (1972) further observed that, during the expansion stroke, the gases in the crevice volumes are laid along the cylinder wall. As the piston moves up during the exhaust stroke, the layer is scraped off the wall and rolled up into a vortex, as depicted in Figure 4.10. The second peak in the hydrocarbon concentration was attributed to the passage of this vortex through the exhaust valve late in the exhaust stroke.

Although the quench layer model does appear to explain many of the observations of hydrocarbons in spark ignition engines, recent studies have questioned the importance of quench layers as sources of unburned hydrocarbons (Lavoie et al., 1980). The cooling effect of the wall does, indeed, prevent the flame from propagating all the way to the cylinder wall. Hydrocarbon vapors can diffuse from this cool region, however, into the hotter gases farther from the wall. If this occurs early in the cycle when the temperature of the burned gases is high, the hydrocarbons from the quench layer will be burned.

We can gain some insight into the quench-layer problem by examining the time scales of diffusion and reaction of the hydrocarbon gases. The characteristic time for diffusion of gases from the quench layer into the bulk gases is $\tau_D \approx \delta^2/D$. Adamczyk and Lavoie (1978) report values of $\delta$ of order 50 to 75 $\mu$m and diffusion times ranging from 0.1 to 0.3 ms at atmospheric pressure. Inasmuch as this time is short compared to that of the expansion stroke and typical combustion times, a considerable amount of the

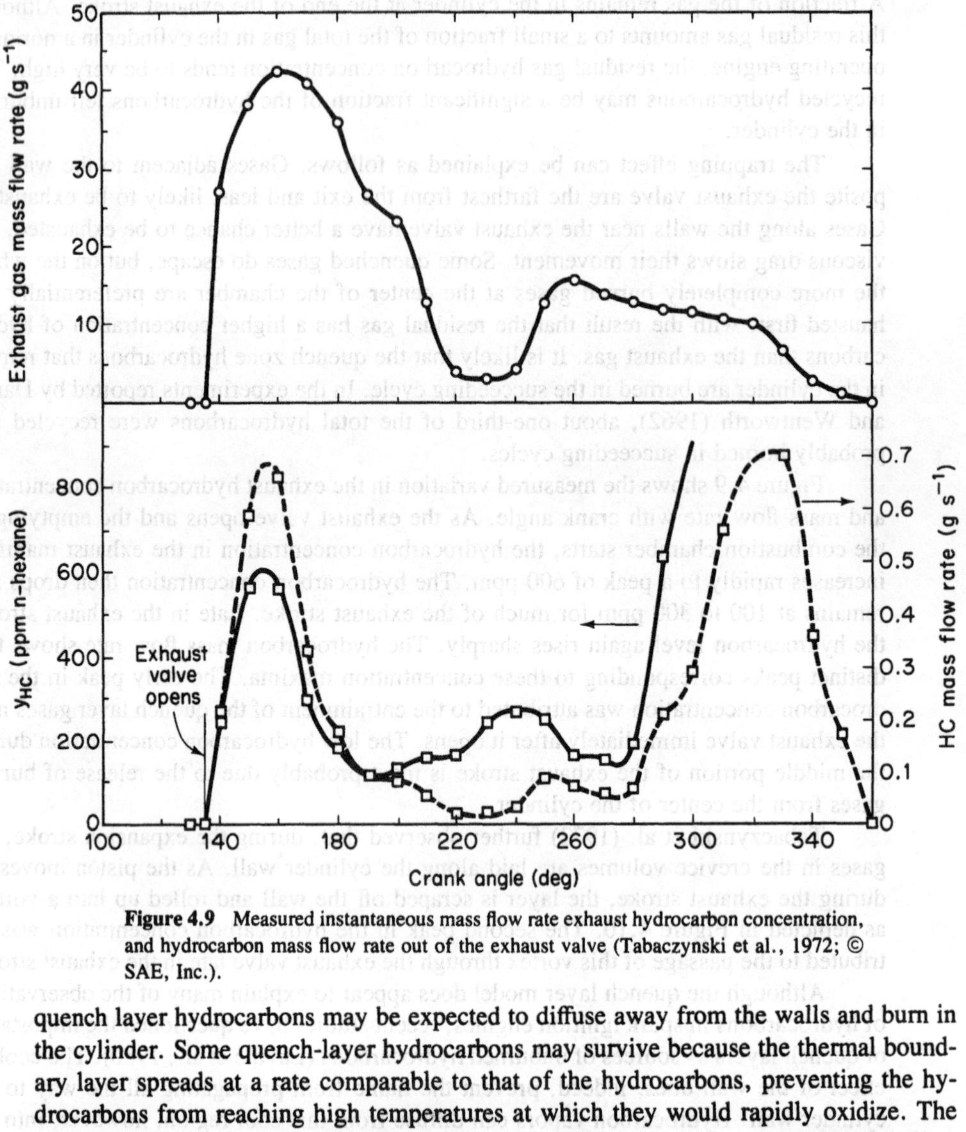


**Figure 4.9** Measured instantaneous mass flow rate exhaust hydrocarbon concentration, and hydrocarbon mass flow rate out of the exhaust valve (Tabaczynski et al., 1972; © SAE, Inc.).

quench layer hydrocarbons may be expected to diffuse away from the walls and burn in the cylinder. Some quench-layer hydrocarbons may survive because the thermal boundary layer spreads at a rate comparable to that of the hydrocarbons, preventing the hydrocarbons from reaching high temperatures at which they would rapidly oxidize. The quantities of hydrocarbons that survive by this route, however, are much too small to explain the observed hydrocarbon levels. In one study in which the quench-layer gases were sampled directly, it was estimated that the quench-layer gases could account for not more than 3 to 12% of the hydrocarbons measured in the exhaust (LoRusso et al., 1983).

Hydrocarbons contained in the crevice volume between the piston, piston ring, and cylinder wall account for much of the hydrocarbon release. These vapors expand out from the crevices late in the expansion stroke, so lower temperatures are encountered

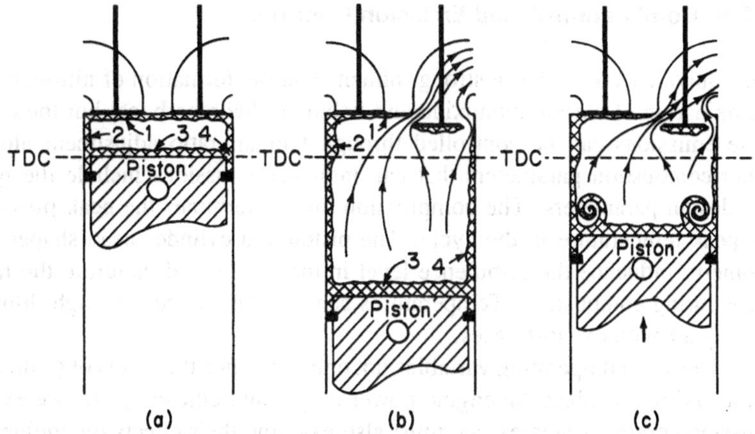


**Figure 4.10** Schematic illustrating the quench layer model for hydrocarbon emissions. (a) Quench layers are formed as heat transfer extinguishes the flame at the cool walls and in the crevice volume. (b) Gas in the crevice volume expands and is spread along the cylinder wall as the pressure falls. When the exhaust valve opens, the quench layers near the valve exit the cylinder. (c) The hydrocarbon-rich cylinder wall boundary layer rolls up into a vortex as the piston moves up the cylinder during the exhaust stroke (Tabaczynski et al., 1972; © SAE, Inc.).

by crevice gases than by the quench-layer gases (Wentworth, 1971). Adamczyk et al. (1983) examined the retention of hydrocarbons in a combustion bomb that consisted of a fixed piston in an engine cylinder. About 80% of the hydrocarbons remaining after combustion were attributed to the piston crevice, with most of the remaining hydrocarbons surviving in smaller crevices associated with the head gasket and with the threads on the spark plug. The crevice volumes contribute primarily to the peak in the hydrocarbon flux late in the exhaust process, since those gases originate far from the exhaust valve.

Other sources must therefore contribute significantly to the hydrocarbon emissions, particularly those that exit the cylinder early in the exhaust process. Haskell and Legate (1972) and Wentworth (1968) suggested that lubicating oil layers on the cylinder walls may adsorb or dissolve hydrocarbon vapors during the compression stroke. These stored hydrocarbons are protected from the flame. As the pressure in the cylinder drops during the expansion stroke and exhaust process, these hydrocarbons desorb into the combustion products. Kaiser et al. (1982) showed that fuel vapors and fuel hydrocarbon oxidation product emissions increase as the amount of oil in the cylinder increases. Carrier et al. (1981) developed a model for cyclic hydrocarbon adsorption and desorption in a liquid film, taking into account thermodynamic equilibrium at the gas–liquid interface and diffusional resistance within the liquid layer. The results from this model are qualitatively consistent with the observed reduction of hydrocarbon emission with engine speed.

### 4.1.8 Combustion-Based Emission Controls

The equivalence ratio has a strong influence on the formation of nitrogen oxides and on the oxidation of carbon monoxide and unburned hydrocarbons, but the extent to which these emissions can be controlled through fuel–air ratio adjustment alone is limited. Other combustion parameters that can influence emissions include the ignition timing and design parameters. The compression ratio determines the peak pressure and hence the peak temperature in the cycle. The piston and cylinder head shapes and the valve geometry influence the turbulence level in the engine and therefore the rate of heat release during combustion. Temperatures can also be reduced through dilution of the incoming air with exhaust gases.

Design and operating variables not only influence the levels of pollutant emissions, but also directly affect the engine power output and efficiency. As we examine various emission control strategies, we must also examine their effects on engine performance. The efficiency of an internal combustion engine is generally reported in terms of the *specific fuel consumption* (SFC), the mass of fuel consumed per unit of energy output, kg $MJ^{-1}$ or g $kW\text{-}h^{-1}$. The work output per engine cycle is presented in terms of the *mean effective pressure* (MEP), the work done per displacement volume. If the MEP is determined in terms of the net power output, $P$, the quantity is called the *brake mean effective pressure* (BMEP) and is calculated as

$$\mathrm{BMEP} = \frac{2P}{V_d\Omega} \tag{4.38}$$

where $\Omega$ is the engine rotation speed (revolutions per second). Many factors not directly involved in the combustion process influence the BMEP: friction; pumping work associated with the intake and exhaust flows; and work used to drive engine equipment such as generators, water pumps, fans, and so on. The work performed by the gas during the compression and expansion strokes,

$$W_i = \int_{0^\circ}^{360^\circ} p \frac{dV}{d\theta}\, d\theta \tag{4.39}$$

that is, that that would be *indicated* by a pressure measurement, is of more concern to us here. The mean effective pressure based on this work,

$$\mathrm{IMEP} = \frac{W_i}{V_d} \tag{4.40}$$

is called the *indicated mean effective pressure* (IMEP). It is also convenient to present the specific fuel consumption in terms of the indicated work to eliminate the influences of parasitic losses and loads. This quantity is then called the *indicated specific fuel consumption* (ISFC).

Figure 4.11 shows the influence of engine operating equivalence ratio on the indicated specific $NO_x$ emissions (g $NO_x$ $MJ^{-1}$) and fuel consumption for three different values of the combustion duration. $NO_x$ emissions are maximum at $\phi = 1$ and decrease

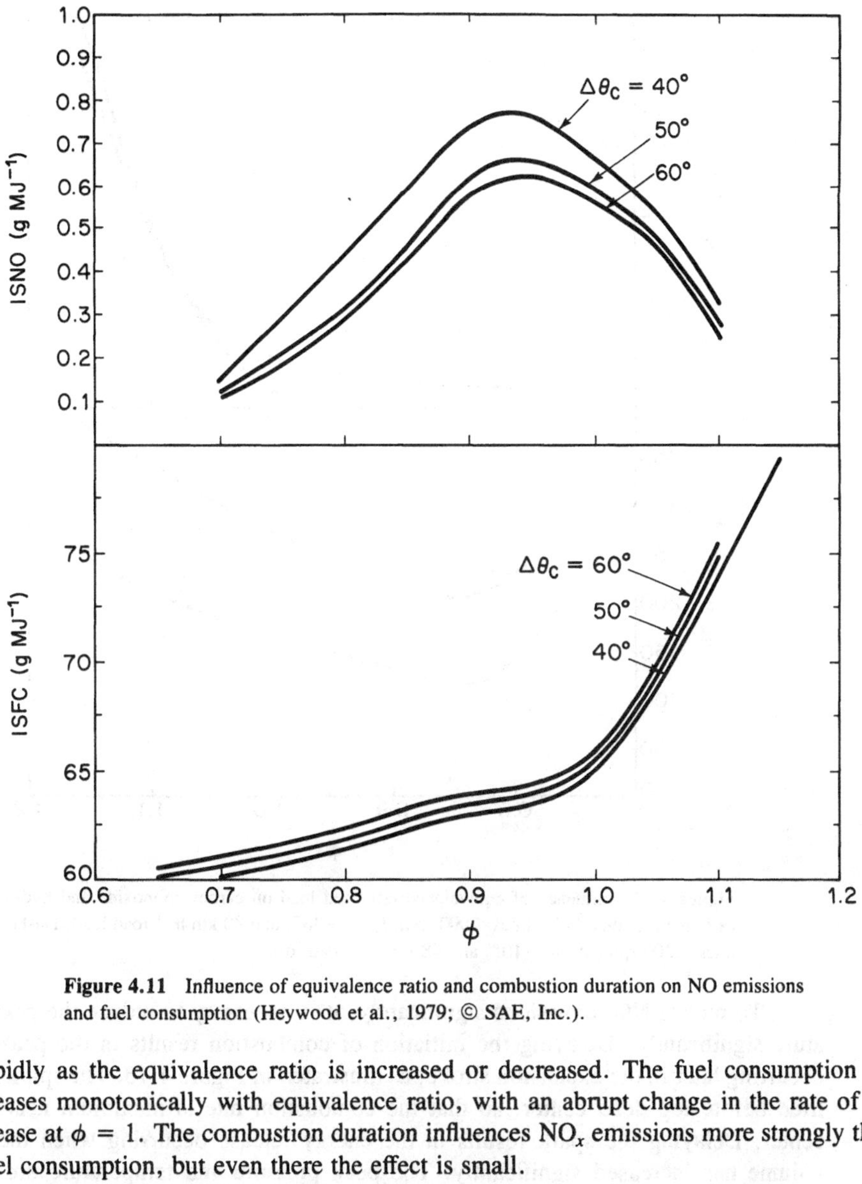


**Figure 4.11** Influence of equivalence ratio and combustion duration on NO emissions and fuel consumption (Heywood et al., 1979; © SAE, Inc.).

rapidly as the equivalence ratio is increased or decreased. The fuel consumption increases monotonically with equivalence ratio, with an abrupt change in the rate of increase at $\phi = 1$. The combustion duration influences $NO_x$ emissions more strongly than fuel consumption, but even there the effect is small.

The influence of the operating equivalence ratio on emissions of carbon monoxide and unburned hydrocarbons is illustrated in Figure 4.12. The CO level is relatively low for fuel-lean operation but rises abruptly, as expected, when the mixture becomes fuel-rich. The hydrocarbon emissions, on the other hand, exhibit a minimum and increase for very fuel-lean operation. In lean operation the temperature can be too low for hydrocarbons to burn late in the expansion stroke. Furthermore, the low laminar flame speed at low $\phi$ means that the flame may not even reach all the mixture.

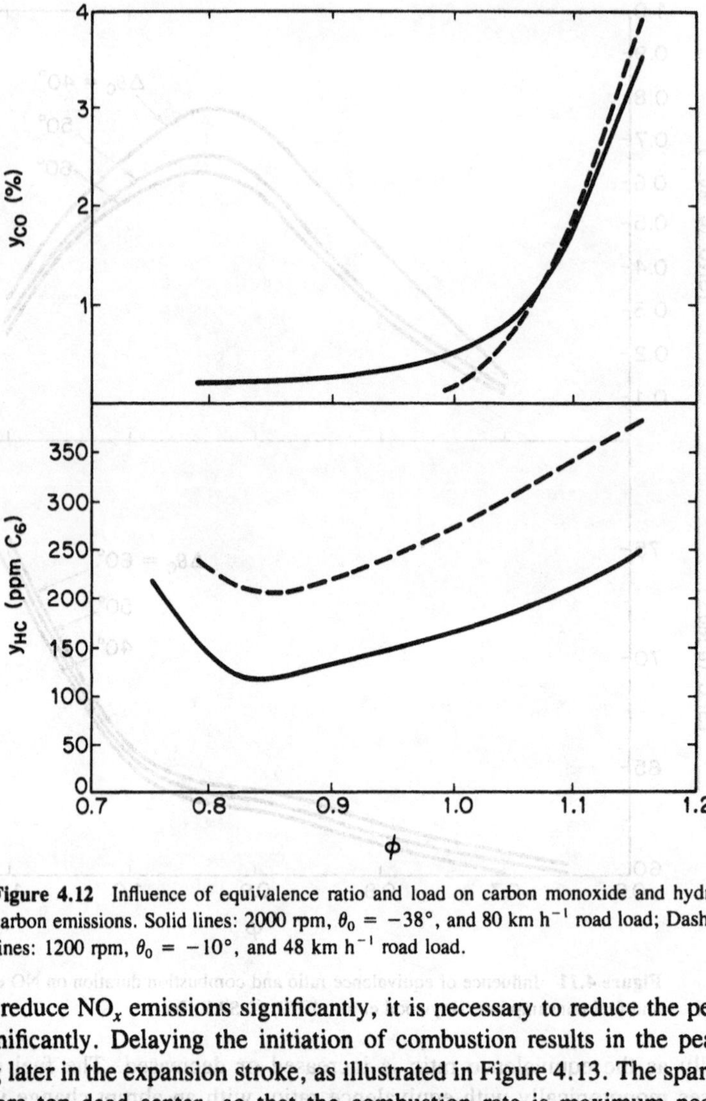


**Figure 4.12** Influence of equivalence ratio and load on carbon monoxide and hydrocarbon emissions. Solid lines: 2000 rpm, $\theta_0 = -38°$, and 80 km $h^{-1}$ road load; Dashed lines: 1200 rpm, $\theta_0 = -10°$, and 48 km $h^{-1}$ road load.

To reduce $NO_x$ emissions significantly, it is necessary to reduce the peak temperature significantly. Delaying the initiation of combustion results in the peak pressure occurring later in the expansion stroke, as illustrated in Figure 4.13. The spark is usually fired before top dead center, so that the combustion rate is maximum near top dead center. Delaying the spark results in the energy release occurring when the cylinder volume has increased significantly. The peak pressure and temperature are therefore reduced by this *spark retard*. At the most extreme level, the spark can be retarded past top dead center so that the gases begin to expand before combustion begins. The influence of equivalence ratio and ignition angle on fuel consumption and $NO_x$ emissions has been calculated by Blumberg and Kummer (1971). Their results are shown in a map of BSFC versus BSNO in Figure 4.14. Clearly, if an engine could be operated at very low equivalence ratios, $NO_x$ emissions could be reduced dramatically with only a minimal efficiency penalty. Operating at equivalence ratios more typical of premixed combustion

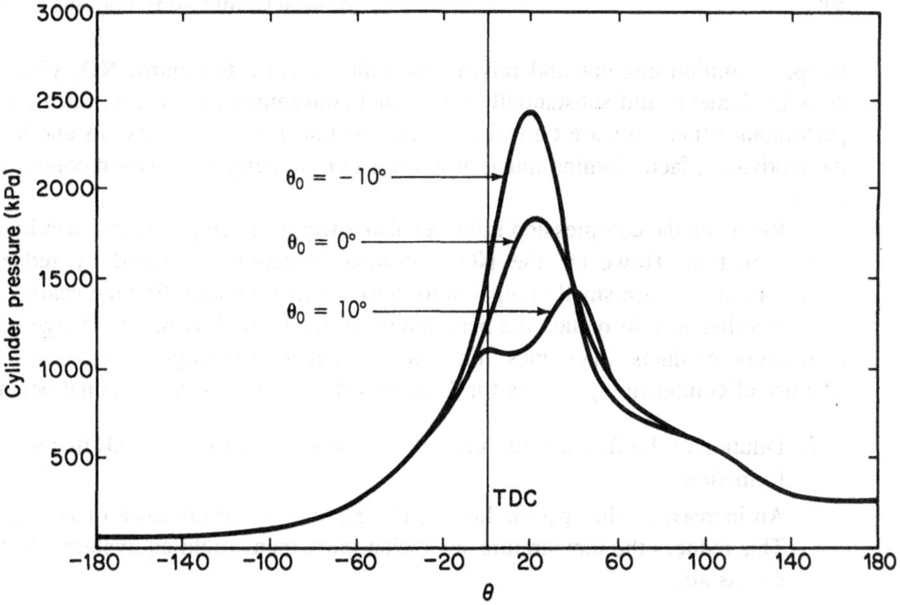


**Figure 4.13** Influence of ignition timing on cylinder pressure profiles.

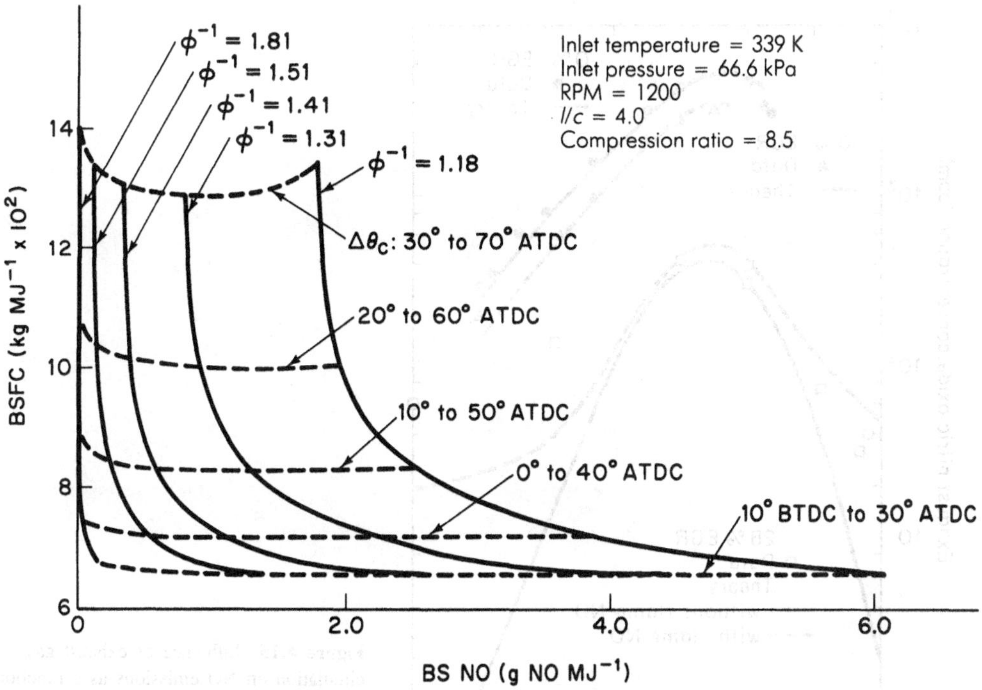


**Figure 4.14** Effect of equivalence ratio and ignition timing on efficiency and NO formation for $\Delta\theta_c = 40°$ (Blumberg and Kummer, 1971). Reprinted by permission of Gordon and Breach Science Publishers.

in spark ignition engines and relying on ignition retard to control $NO_x$ yield smaller emission benefits and substantially larger fuel consumption increases. Such emissions/performance trade-offs are typical of efforts to control engine emissions and have been the motivating factor behind much of the research into engine emission control technologies.

Reducing the compression ratio can also lower peak temperatures, thereby limiting $NO_x$ formation. However, the $NO_x$ emission reductions achieved by reducing the compression ratio are small compared to those accrued by retarding the spark.

Another way to reduce the peak temperatures is by diluting the charge with cool combustion products. In engines, this process is called *exhaust gas recirculation* (EGR). The use of combustion products for dilution instead of excess air has dual benefits:

1. Dilution of the fuel–air mixture without the addition of excess $O_2$ that aids in $NO_x$ formation.
2. An increase in the specific heat of the gas due to the presence of $H_2O$ and $CO_2$. This reduces the temperature somewhat more than would equivalent dilution with excess air.

Figure 4.15 shows how significantly EGR can reduce $NO_x$ emission levels. For small amounts of EGR, the theoretical predictions agree closely with experimental ob-

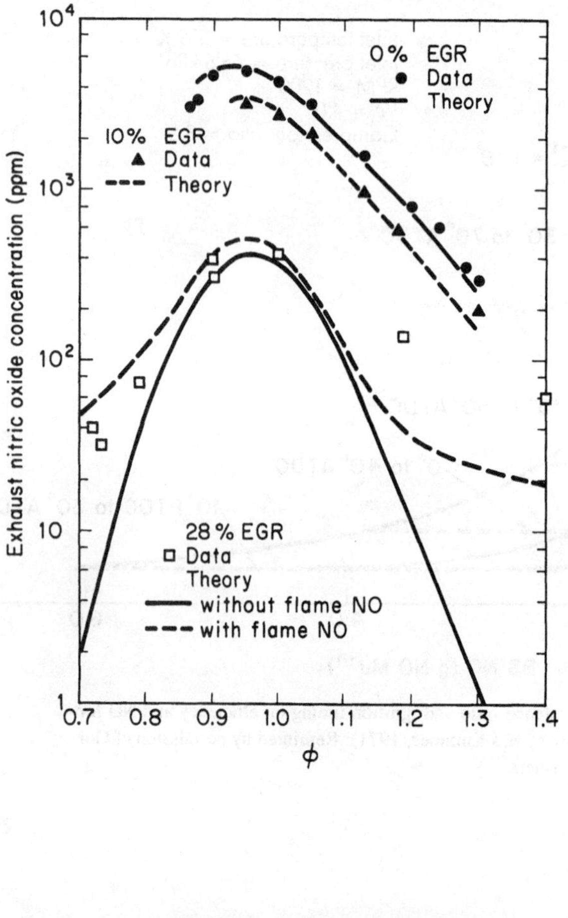


**Figure 4.15** Influence of exhaust gas recirculation on NO emissions as a function of equivalence ratio (Heywood, 1975). Reprinted by permission of The Combustion Institute.

servations; however, at 28% EGR, the measured $NO_x$ emission levels for lean or rich mixtures are significantly higher than those predicted considering only postflame chemistry. The dashed curve presents more detailed chemical mechanism calculations that take into account the nonequilibrium radical concentrations that are present within the flame front (i.e., "prompt NO"). Agreement on the fuel-lean side is very good. On the other hand, even when the flame chemistry of the O, H, and OH radicals is taken into account, the predictions of $NO_x$ formation in fuel-rich combustion are significantly lower than those observed. This discrepancy may be due to nitrogen chemistry not included in the model, particularly the reactions of $N_2$ with hydrocarbon radicals.

From these results we see the EGR can substantially reduce $NO_x$ formation in spark ignition engines, but the degree of control achievable by this method is limited. These gains are not achieved without penalties. Figure 4.16 shows calculations of the variation of fuel consumption and mean effective pressure with equivalence ratio and amount of exhaust gas recirculated. While the fuel consumption penalty is relatively small, the loss of power is significant, so the engine size must be increased to meet a particular power requirement if EGR is employed to control $NO_x$ emissions.

It is apparent that spark retard and exhaust gas recirculation are effective measures for $NO_x$ emission control. The equivalence ratio range that can be employed effectively is limited. Rich mixtures lead to high CO levels. As the mixture becomes too fuel-lean, hydrocarbon emissions rise. Hence control of emissions without the use of exhaust gas cleaning involves compromises. Spark retard and exhaust gas recirculation are usually used in combination to achieve low $NO_x$ emission levels. The introduction of strict $NO_x$ emission controls in combination with limits on CO and hydrocarbon emissions was accompanied by a substantial increase in fuel consumption of automobiles in the United

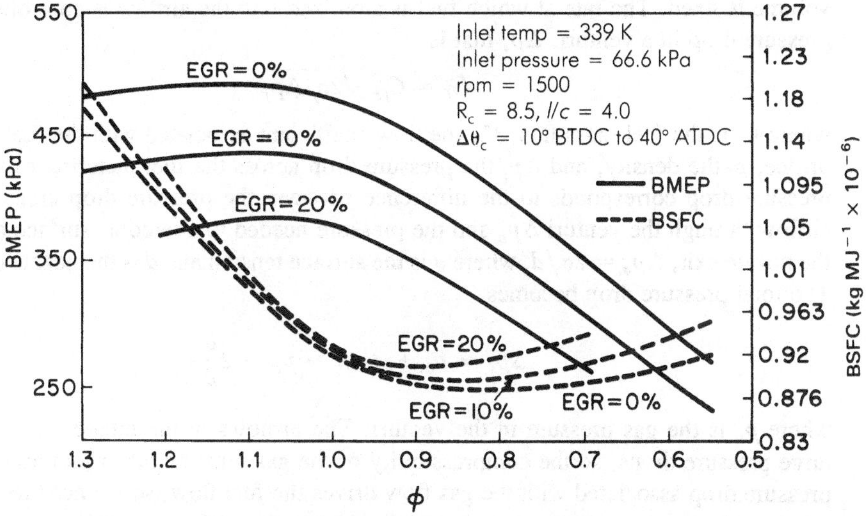


**Figure 4.16** Effect of equivalence ratio and exhaust gas recirculation on power (brake mean effective pressure) and fuel consumption (Blumberg and Kummer, 1971). Reprinted by permission of Gordon and Breach Science Publishers.

States. Ultimately, exhaust gas treatment was required to achieve acceptable emissions and performance simultaneously. Exhaust gas treatment is discussed in a subsequent section.

### 4.1.9 Mixture Preparation

The spark ignition engine burns premixed fuel and air. In conventional engines, this mixture is prepared in the *carburetor*, a complex device that controls both fuel and air flows to the engine. The mixture requirements depend on engine speed and load. A richer mixture is required at high load (such as during vehicle acceleration) than at low load. Even though combustion will be incomplete, fuel-rich mixtures have been used to increase the heat release per cycle, thereby increasing the power delivered by the engine. Carburetors have evolved as mechanically activated control systems that meet these requirements. As we have seen in the preceding discussion, emission controls place additional constraints on engine operation that are not readily met with purely mechanical control. To understand the need for and the nature of the new systems for mixture preparation that are being developed as part of integrated emission control systems, it is useful to examine the operation of a conventional carburetor.

The power output and speed of a spark ignition engine are regulated by a throttle that limits the airflow into the engine. In conventional engines, the airflow rate is used to control the fuel/air ratio. Part of the difficulty encountered in early attempts to reduce automobile emissions derived from the complex coupling of fuel and airflow rates.

A simple carburetor is illustrated in Figure 4.17. The throttle is a butterfly valve, a disk that is rotated to obstruct the airflow, producing a vacuum in the intake manifold. The low pressure reduces the mass entering the cylinders, even though the intake gas volume is fixed. The rate at which fuel is atomized into the airflow is controlled by the pressure drop in a venturi, $\Delta p$, that is,

$$G_f = C_{fF} \sqrt{2\rho_f \, \Delta p_f} \tag{4.41}$$

where $G_f$ is the fuel mass flux, $C_{fF}$ the flow coefficient associated with the fuel metering orifice, $\rho_f$ the density, and $\Delta p_f$ the pressure drop across the fuel metering orifice. This pressure drop corresponds to the difference between the pressure drop created by the airflow through the venturi $\Delta p_a$ and the pressure needed to overcome surface tension at the nozzle exit, $\Delta p_\sigma = 2\sigma / d$, where $\sigma$ is the surface tension and $d$ is the nozzle diameter. The total pressure drop becomes

$$\Delta p_f \approx p_0 + \rho_f g h - p_v - 2\frac{\sigma}{d} \tag{4.42}$$

where $p_v$ is the gas pressure in the venturi. The airflows in the intake system involve large pressure drops, so the compressibility of the gas must be taken into account. The pressure drop associated with the gas flow drives the fuel flow, so we need to know the relationship between pressure drop and flow rate. By considering the conservation of energy, we can readily derive such an expression for the adiabatic and thermodynamically reversible (i.e., isentropic) flow of an ideal gas.

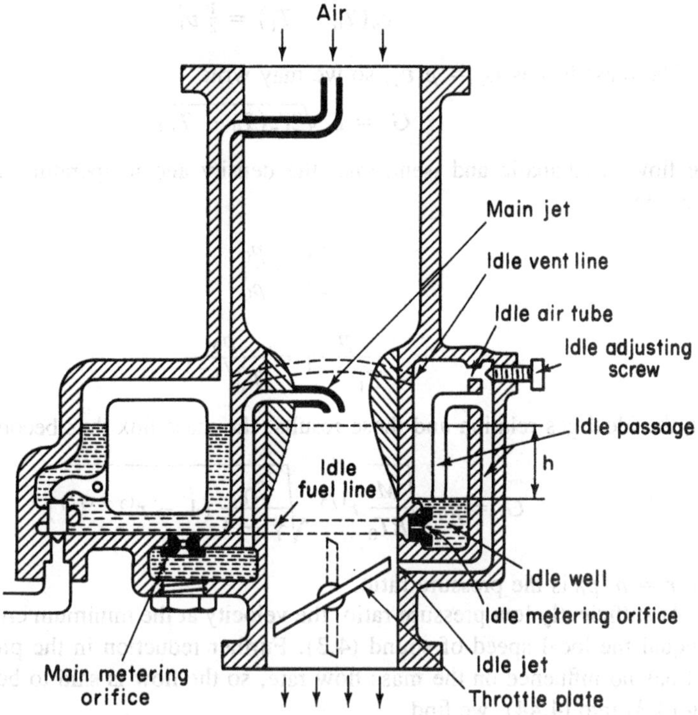


**Figure 4.17** Schematic of a simple carburetor.

The flows through real devices such as the venturi or throttle are not perfectly reversible, so the flow rate associated with a given pressure drop is lower than that for isentropic flow. The ratio of the actual flow rate to the ideal flow rate is the flow coefficient for the device, that is,

$$C_f = \frac{G}{G_s} \tag{4.43}$$

where $G$ denotes the mass flux and the subscript $s$ denotes that for isentropic flow. The flow coefficient for a sharp-edged orifice is 0.61. The venturi is designed to achieve nearly reversible flow so that $C_f$ will be closer to unity. The flow coefficient for the throttle changes as the throttle plate is rotated. It is unity when the throttle is fully open and decreases toward that for the orifice as the throttle is closed.

We consider adiabatic flow through the device in question. As the gas is accelerated, its kinetic energy must be taken into account in the fluid energy balance, that is, for the flow at velocities $v_1$ and $v_2$,

$$\bar{h}_1 + \tfrac{1}{2} v_1^2 = \bar{h}_2 + \tfrac{1}{2} v_2^2 = \bar{h}_0$$

$h_0$ is the stagnation enthalpy corresponding to $v = 0$. Assuming that the gas is ideal and the specific heats are constant, we may write

$$\bar{c}_p(T_0 - T_1) = \tfrac{1}{2} v_1^2 \tag{4.44}$$

The mass flux is $G_s = \rho_1 v_1$, so we may write

$$G_s = \rho \sqrt{2\bar{c}_p(T_0 - T_1)} \tag{4.45}$$

If the flow is adiabatic and isentropic, the density and temperature are related to the pressure by

$$\frac{p_1}{\rho_1^\gamma} = \frac{p_0}{\rho_0^\gamma} \tag{4.46}$$

$$\frac{p_1}{T_1^{\gamma/\gamma-1}} = \frac{p_0}{T_0^{\gamma/\gamma-1}} \tag{4.47}$$

Using the ideal gas relation and these results, the mass flux thus becomes

$$G_s = p_0 \sqrt{\frac{M}{RT_0}}\, r^{1/\gamma} \sqrt{\frac{2\gamma}{\gamma - 1}(1 - r^{(\gamma-1)/\gamma})} \tag{4.48}$$

where $r = p/p_0$ is the pressure ratio.

At sufficiently low pressure ratio, the velocity at the minimum cross-sectional area will equal the local speed of sound (4.3). Further reduction in the pressure below the throat has no influence on the mass flow rate, so the flow is said to be choked. Substituting (4.3) into (4.44), we find

$$\frac{T_0}{T^*} = \frac{\gamma + 1}{2} \tag{4.49}$$

where the asterisk is used to denote a property evaluated at locally sonic conditions. Using (4.47) we find the critical pressure ratio,

$$r^* = \left(\frac{2}{\gamma + 1}\right)^{\gamma/(\gamma - 1)} \tag{4.50}$$

The corresponding mass flow rate is obtained by substituting $r^*$ into (4.48),

$$G_s(r^*) = p_0 \sqrt{\frac{M\gamma}{RT_0}\left(\frac{2}{\gamma + 1}\right)^{(\gamma+1)/2(\gamma-1)}} \tag{4.51}$$

The mass flow rate for a real device becomes

$$\begin{aligned} G_f &\approx C_f p_0 \sqrt{\frac{M}{RT_0}}\, r^{1/\gamma} \sqrt{\frac{2\gamma}{\gamma - 1}(1 - r^{(\gamma-1)/\gamma})} \qquad r > r^* \\ G_f &\approx C_f p_0 \sqrt{\frac{M\gamma}{RT_0}\left(\frac{2}{\gamma + 1}\right)^{(\gamma+1)/2(\gamma-1)}} \qquad r \le r^* \end{aligned} \tag{4.52}$$

For a well-designed venturi, the flow coefficient will be nearly unity and the stagnation pressure downstream of the venturi will be close to that at the venturi inlet. Butterfly valves and other nonideal flow devices will have lower flow coefficients. If a subsonic flow separates at the minimum area, the pressure at that point will correspond approximately to the downstream stagnation pressure. Thus, closing the throttle results in the pressure in the intake manifold being substantially below atmospheric pressure.

The fuel flow rate is governed by the pressure at the throat of the venturi, so (4.41) can be expressed in terms of the pressure ratio

$$G_f = C_{fF}\sqrt{2\rho_f\left[p_0(1 - r) + gz - \frac{2\sigma}{d}\right]} \tag{4.53}$$

The fuel/air ratio becomes (for $r > r^*$)

$$\frac{G_f A_f}{G_a A_a} = \frac{A_f C_{fF}\sqrt{2\rho_f\left[p_0(1 - r) + gz - \dfrac{2\sigma}{d}\right]}}{A_v C_f p_0\sqrt{\dfrac{M}{RT_0}}\, r^{1/\gamma}\sqrt{\dfrac{2\gamma}{\gamma - 1}\left(1 - r^{(\gamma-1)/\gamma}\right)}} \tag{4.54}$$

The complex dependence of the equivalence ratio on the pressure ratio is readily apparent.

Examining (4.42) we see that, for

$$r \geq 1 - \frac{2\sigma}{p_0 d} + \frac{gz}{p_0} \tag{4.55}$$

the pressure drop in the venturi is insufficient to overcome surface tension and atomize the fuel. These high pressure ratios (low pressure drops) correspond to low engine speeds. A separate idle nozzle supplies the fuel necessary for low-speed operation. This ideal adjustment is coupled to the pressure drop at the throttle valve.

Figure 4.18 illustrates the variation of equivalence ratio with airflow that is produced by these metering devices. The pressure in the venturi throat decreases with increasing airflow. Since the difference between this pressure and that of the atmosphere provides the driving force for the main fuel flow, the fuel supplied by the main jet increases with increasing airflow. The idle jet compensates for the precipitous drop in the fuel flow supplied by the main jet. The pressure at the throttle plate provides the driving force for the idle fuel flow, so this flow is significant only when the idle plate is closed, i.e., at low airflow. As the throttle plate is opened and the airflow increases, the idle fuel flow decreases markedly. The operating equivalence ratio of the engine is determined by the sum of the two fuel flows, shown by the upper curve.

At high engine load, a richer mixture may be required than is supplied by this simple metering system. The power jet shown in Figure 4.19 is one method used to supply the additional fuel. Ideally, the throttle position at which the power jet opens would vary with engine speed. A mechanical linkage that opens gradually as the throttle

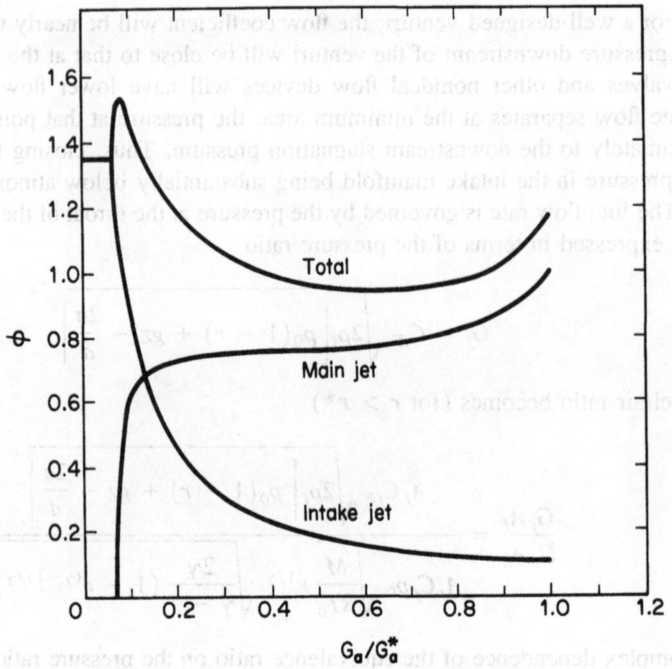


**Figure 4.18** Variation of equivalence ratio with airflow rate for a simple carburetor (Taylor, 1966). Reprinted by permission of MIT Press.

opens beyond some point is a compromise solution. When the power jet is fully open, the fuel flow is about 10% more than that supplied by the main jet.

If the throttle is rapidly opened (as when the gas pedal of a car is quickly depressed), the fuel flow does not respond instantly. To improve the engine response, an accelerator pump may be used to supply fuel at a rate that is proportional to the speed of the accelerator motion.

A very fuel-rich mixture is used to start a cold engine, on the assumption that if enough fuel is introduced into the intake manifold, some of it will surely evaporate and start the engine. A butterfly valve called a *choke* is installed between the impact tube and the venturi, as illustrated in Figure 4.19, to increase the pressure drop and therefore the fuel flow rate through the main metering orifice. The choke is frequently operated automatically, controlled by the exhaust manifold temperature and the inlet manifold pressure. Rich operation during startup leads to high CO and hydrocarbon emissions. As much as 40% of the hydrocarbons emitted during automotive test cycles may be released during the warm-up phase.

We have examined only a few of the features that have been incorporated into automotive carburetors. Since the carburetor directly controls the equivalence ratio of the mixture reaching the engine, it plays a central role in the control of automotive emissions. Much more elaborate fuel metering systems have been developed to achieve

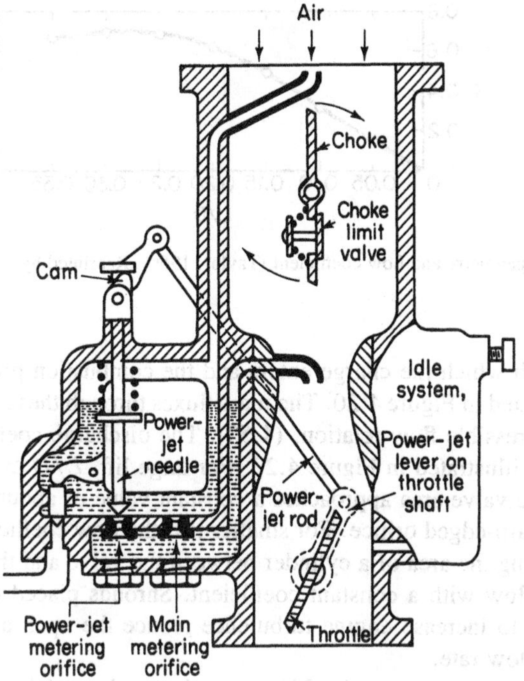


**Figure 4.19** Carburetor with power jet and choke (Taylor, 1966). Reprinted by permission of MIT Press.

the fine regulation required for emission control. Electronically manipulated valves have replaced the simple mechanically controlled fuel metering, facilitating more precise control of engine operation through the use of computers.

Fuel injection is used in place of carburetion in some spark ignition engines because the quantity of fuel introduced can be controlled independently of the airflow rate. Atomization of high-pressure fuel replaces the flow-induced fuel intake of conventional carburetors. Fuel may be injected into the intake manifold (injection carburetion) so that the mixture is controlled by an injector pump rather than being directly coupled to the airflow. Injection into the inlet ports allows cylinder-by-cylinder regulation of the equivalence ratio. Direct injection into the cylinder is also used in some engines, although this method is more sensitive to spray characteristics and may lead to imperfect mixing of fuel and air. Injection systems are becoming more common because they are so well suited to integration into feedback-controlled engine operation.

### 4.1.10 Intake and Exhaust Processes

The flows through the intake and exhaust valves also influence engine operation and emissions. We have seen that the intake flow induces turbulence that, after amplification by rapid compression, governs the flame propagation. The opening of the exhaust valve near the end of the expansion stroke causes a sudden pressure decrease and adiabatic cooling that influence carbon monoxide emissions.

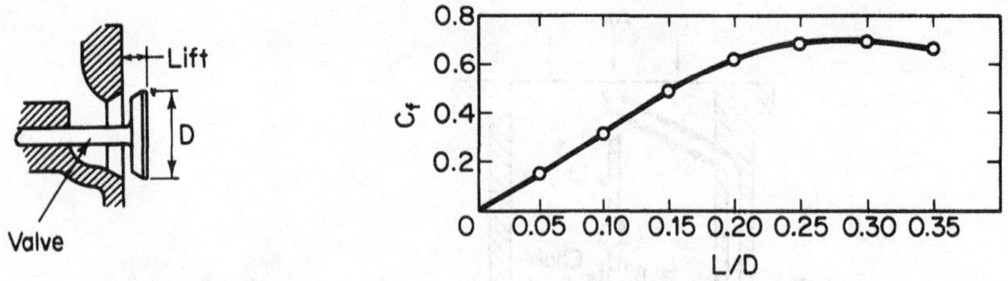


**Figure 4.20** Poppet valve geometry and flow coefficient (Taylor, 1966). Reprinted by permission of MIT Press.

The poppet valves through which the charge enters and the combustion products exit from the cylinder are illustrated in Figure 4.20. The mass fluxes through these valves are also described by the compressible flow relation, (4.53). The discharge coefficient depends on the valve lift, $L$, as illustrated in Figure 4.20. For large lift, $L/D > 0.25$, the flow coefficient based on the valve area approaches a constant value of about 0.65, slightly larger than that for a sharp-edged orifice. For smaller lift, the flow coefficient is proportional to the lift, suggesting the area of a cylinder between the valve and the port could be used to describe the flow with a constant coefficient. Shrouds placed on the intake valve to induce swirl or to increase engine turbulence reduce the open area on this cylinder and therefore the flow rate.

The intake and exhaust flows are not steady. There may be a substantial pressure difference between the cylinder and the manifold when a valve is first opened, leading to a brief period of very high flow rate. This transient flow is particularly pronounced during exhaust when the flow is initially choked. After a brief *blowdown*, the pressure drop decreases and the flow rate is governed by the piston motion. Calculated and measured flow rates from the work of Tabaczynski et al. (1972) are presented in Figure 4.9. Note that the exhaust valve opens about 50° before bottom dead center to allow the cylinder pressure to drop before the beginning of the exhaust stroke. It is also common practice to open the intake valve before the end of the exhaust stroke. This overlap reduces the amount of residual combustion products being mixed with the fresh charge. Improved scavenging achieved in this way increases the engine power output.

The exhaust system includes a length of pipe, a muffler, and gas-cleaning equipment through which the combustion products must flow before entering the atmosphere. The pressure in the exhaust manifold must therefore be greater than atmospheric pressure. The pressure of the gas entering the cylinder is lower than atmospheric pressure, due to pressure drops in the carburetor (particularly across the throttle), intake manifold, and inlet valve. The work required to draw the fuel and air into the cylinder and to pump the combustion products from the cylinder is called the *pumping work*.

The pressure in the cylinder at the end of the intake stroke only approaches atmospheric pressure for open-throttle operation at relatively low speed. From the cycle analysis, it should be apparent that the peak pressure and temperature depend on the intake pressure. Heat transfer from the hot engine block to the fuel–air mixture also influences the temperature. The variation of temperature and pressure with throttle position, engine speed, and engine temperature can be expected to be important factors in the formation of pollutants.

### 4.1.11 Crankcase Emissions

Crankcase emissions are caused by the escape of gases from the cylinder during the compression and power strokes. The gases escape between the sealing surfaces of the piston and cylinder wall into the crankcase. This leakage around the piston rings is commonly called *blowby*. Emissions increase with increasing engine airflow, that is, under heavy load conditions. The resulting gases emitted from the crankcase consist of a mixture of approximately 85% unburned fuel–air charge and 15% exhaust products. Because these gases are primarily the carbureted fuel–air mixture, hydrocarbons are the main pollutants. Hydrocarbon concentrations in blowby gases range from 6000 to 15,000 ppm. Blowby emissions increase with engine wear as the seal between the piston and cylinder wall becomes less effective. On cars without emission controls, blowby gases are vented to the atmosphere by a draft tube and account for about 25% of the hydrocarbon emissions.

Blowby was the first source of automotive emissions to be controlled. Beginning with 1963 model cars, this category of vehicular emissions has been controlled in cars made in the United States. The control is accomplished by recycling the blowby gas from the crankcase into the engine air intake to be burned in the cylinders, thereby keeping the blowby gases from escaping into the atmosphere. All control systems use essentially the same approach, which involves recycling the blowby gases from the engine oil sump to the air intake system. A typical system is shown in Figure 4.21. Ventilation air is drawn down into the crankcase and then up through a ventilator valve and hose and into the intake manifold. When airflow through the carburetor is high, additional air from the crankcase ventilation system has little effect on engine operation. However, during idling, airflow through the carburetor is so low that the returned blowby gases could alter the air–fuel ratio and cause rough idling. For this reason, the flow control valve restricts the ventilation flow at high intake manifold vacuum (low engine speed) and permits free flow at low manifold vacuum (high engine speed). Thus high ventilation rates occur in conjunction with the large volume of blowby associated with high speeds; low ventilation rates occur with low-speed operation. Generally, this principle of controlling blowby emissions is called *positive crankcase ventilation* (PCV).

### 4.1.12 Evaporative Emissions

Evaporative emissions issue from the fuel tank and the carburetor. Fuel tank losses result from the evaporation of fuel and the displacement of vapors when fuel is added to the tank. The amount of evaporation depends on the composition of the fuel and its temperature. Obviously, evaporative losses will be high if the fuel tank is exposed to high ambient temperatures for a prolonged period of time. The quantity of vapor expelled when fuel is added to the tank is equal to the volume of the fuel added.

Evaporation of fuel from the carburetor occurs primarily during the period just

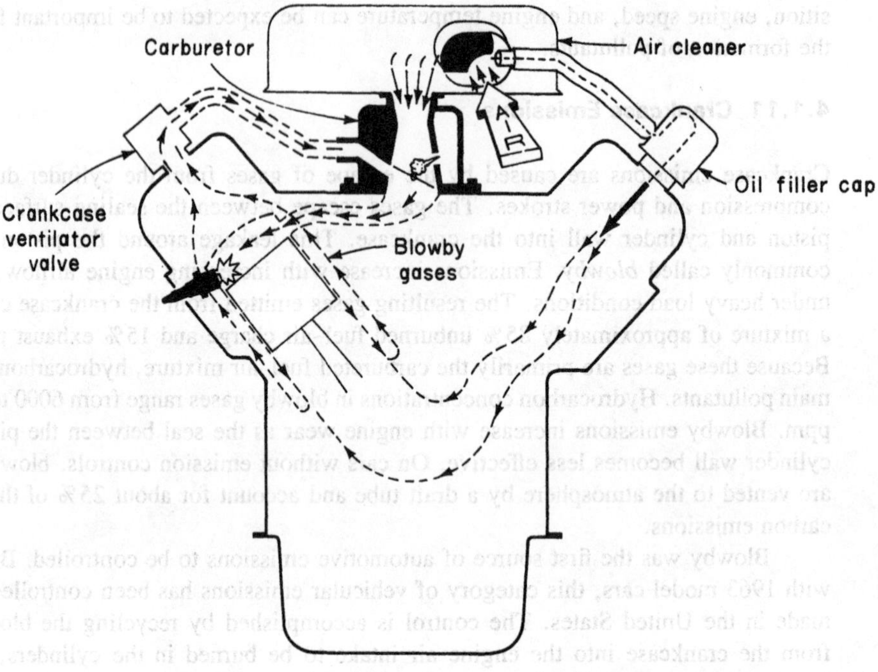


**Figure 4.21** Crankcase emission control system.

after the engine is turned off. During operation the carburetor and the fuel in the carburetor remain at about the temperature of the air under the hood. But the airflow ceases when the engine is stopped, and the carburetor bowl absorbs heat from the hot engine, causing fuel temperatures to reach 293 to 313 K above ambient and causing gasoline to vaporize. This condition is called a *hot soak*. The amount and composition of the vapors depend on the fuel volatility, volume of the bowl, and temperature of the engine prior to shutdown. On the order of 10 g of hydrocarbons may be vaporized during a hot soak. Fuel evaporation from both the fuel tank and the carburetor accounts for approximately 20% of the hydrocarbon emissions from an uncontrolled automobile.

It is clear that gasoline volatility is a primary factor in evaporative losses. The measure of fuel volatility is the empirically determined *Reid vapor pressure*, which is a composite value reflecting the cumulative effect of the individual vapor pressures of the different gasoline constituents. It provides both a measure of how readily a fuel can be vaporized to provide a combustible mixture at low temperatures and an indicator of the tendency of the fuel to vaporize. In a complex mixture of hydrocarbons, such as gasoline, the lowest-molecular-weight molecules have the greatest tendency to vaporize and thus contribute more to the overall vapor pressure than do the higher-molecular-weight constituents. As the fuel is depleted of low-molecular-weight constituents by evaporation, the fuel vapor pressure decreases. The measured vapor pressure of gasoline therefore depends on the extent of vaporization during the test. The Reid vapor-pressure determination is a standard test at 311 K in which the final ratio of vapor volume to

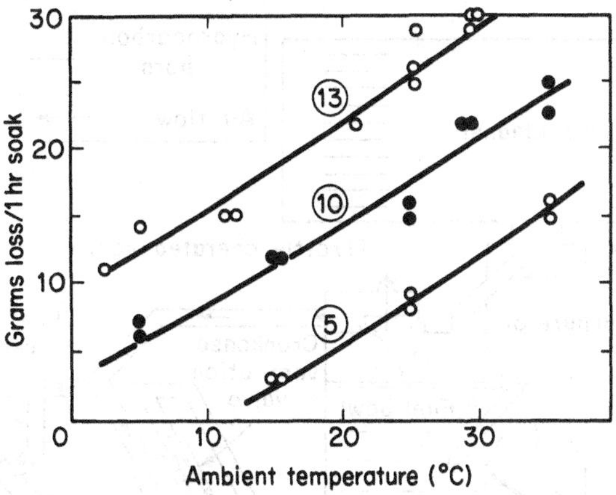


**Figure 4.22** Variation of evaporation loss from an uncontrolled carburetor with fuel vapor pressure and temperature. Numbers in large circles are Reid vapor pressure.

liquid volume is constant (4 : 1) so that the extent of vaporization is always the same. Therefore, the Reid vapor pressure for various fuels can be used as a comparative measure of fuel volatility.

Figure 4.22 shows carburetor evaporative loss as a function of temperature and Reid vapor pressure. The volatility and thus the evaporative loss increase with Reid vapor pressure. In principle, evaporative emissions can be reduced by reducing gasoline volatility. However, a decrease in fuel volatility below the 8 to 12 Reid vapor pressure range, commonly used in temperate climates, would necessitate modifications in carburetor and intake manifold design, required when low vapor pressure fuel is burned. In view of costly carburetion changes associated with reduction of fuel volatility, evaporative emission control techniques have been based on mechanical design changes. Two evaporative emission control methods are the *vapor-recovery system* and the *adsorption-regeneration system*.

In the vapor-recovery system, the crankcase is used as a storage tank for vapors from the fuel tank and carburetor. Figure 4.23(a) shows the routes of hydrocarbon vapors during shutdown and hot soak. During the hot-soak period the declining temperature in the crankcase causes a reduction in crankcase pressure sufficient to draw in vapors. During the hot soak, vapors from the carburetor are drawn into the crankcase. Vapor from the fuel tank is first carried to a condenser and vapor–liquid separator, with the vapor then being sent to the crankcase and the condensate to the fuel tank. When the engine is started, the vapors stored in the crankcase are sent to the air intake system by the positive crankcase ventilation system.

In the adsorption–regeneration system, a canister of activated charcoal collects the vapors and retains them until they can be fed back into the intake manifold to be burned. The system is shown in Figure 4.23(b). The essential elements of the system are the canister, a pressure-balancing valve, and a purge control valve. During the hot-soak period, hydrocarbon vapors from the carburetor are routed by the pressure balance valve to the canister. Vapor from the fuel tank is sent to a condenser and separator, with liquid fuel returned to the tank. When the engine is started, the pressure control valve causes

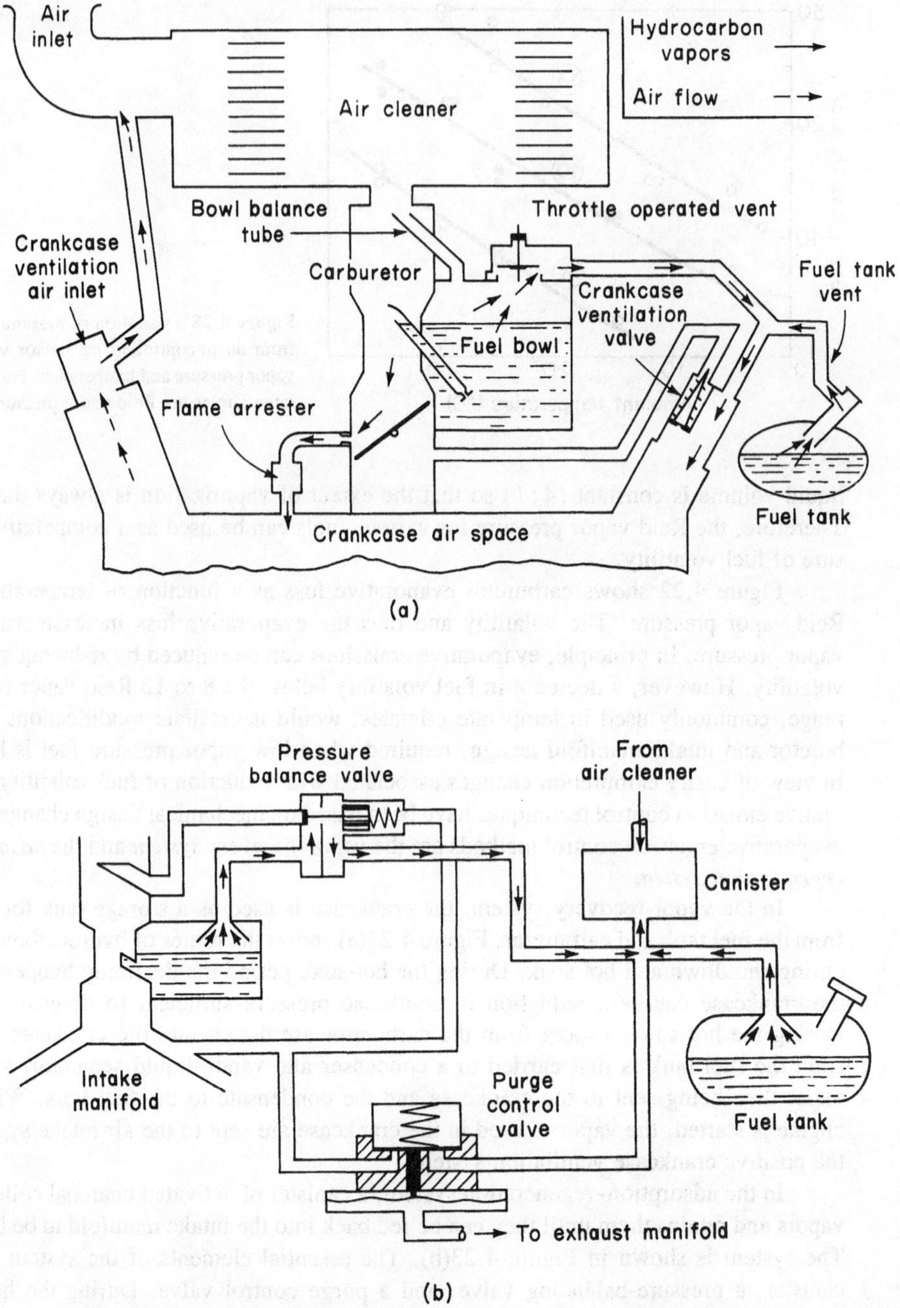


**Figure 4.23** Evaporative emission control systems: (a) use of crankcase air space; (b) adsorption-regeneration system.

air to be drawn through the canister, carrying the trapped hydrocarbons to the intake manifold to be burned.

### 4.1.13 Exhaust Gas Treatment

Modification of engine operation yields only modest emission reductions, and the penalties in engine performance and efficiency are substantial. An alternative way to control emissions involves the treatment of the exhaust gas in chemical reactors. Carbon monoxide, unburned hydrocarbons, and nitrogen oxides are all present in the exhaust gases in concentrations that are far in excess of the equilibrium values. If all the pollutants are to be controlled by exhaust gas treatment, it is necessary to *oxidize* carbon monoxide and hydrocarbons while *reducing* nitrogen oxides. Exhaust gas treatment may utilize either catalytic converters or noncatalytic thermal reactors.

**Thermal reactors.** The gas-phase oxidation of carbon monoxide slows dramatically as combustion products cool, but the reaction does not stop entirely. In fact, carbon monoxide and hydrocarbons continue to react in the exhaust manifold. To oxidize the hydrocarbons homogeneously requires a holding time of order 50 ms at temperatures in excess of 900 K. Homogeneous oxidation of carbon monoxide requires higher temperatures, in excess of 1000 K. The oxidation rate can be enhanced with a thermal reactor—an enlarged exhaust manifold that bolts directly onto the cylinder head. The thermal reactor increases the residence time of the combustion products at temperatures sufficiently high that oxidation reactions can proceed at an appreciable rate. To allow for fuel-rich operation, secondary air may be added and mixed rapidly with combustion products.

A multiple-pass arrangement is commonly used in thermal reactors to shield the hot core of the reactor from the relatively cold surroundings. This is critical since the reactions require nearly adiabatic operation to achieve significant conversion, as illustrated in Figure 4.24. Typically, only about a factor of 2 reduction in emission levels for CO and hydrocarbons can be achieved even with adiabatic operation. Higher temperatures and long residence times are typically required to achieve better conversions. The heat released in the oxidation reactions can result in a substantial temperature rise and, thereby, promote increased conversion. Removal of 1.5% CO results in a temperature rise of about 490 K (Heywood, 1976). Hence thermal reactors with fuel-rich cylinder exhaust gas and secondary air addition give greater fractional reductions in CO and hydrocarbon levels than reactors with fuel-lean cylinder exhaust. Incomplete combustion in the cylinder, however, does result in reduced fuel economy. The attainable conversion is limited by incomplete mixing of gases exhausted at various times in the cycle and any secondary air that is added.

Temperatures of the exhaust gases of automobile spark ignition engines can vary from 600 to 700 K at idle to 1200 K during high-power operation. Most of the time the exhaust temperature is between 700 and 900 K, too low for effective homogeneous oxidation. Spark retard increases the exhaust temperature, but this is accompanied by a significant loss in efficiency.

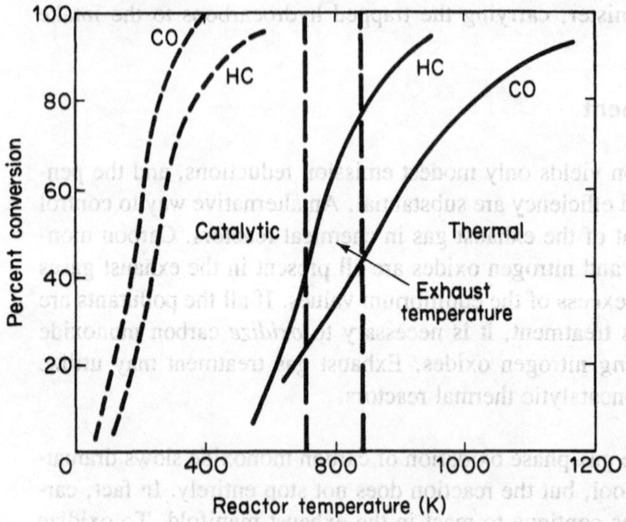


**Figure 4.24** Comparison of catalytic converter and thermal reactor for oxidation of CO and hydrocarbons.

Noncatalytic processes for vehicular emission control can yield significant improvements in carbon monoxide and hydrocarbon emissions. The problem of $NO_x$ emission control is not easily alleviated with such systems. Control of $NO_x$ emissions through noncatalytic reduction by ammonia is feasible only in a very narrow window of temperature, toward the upper limit of the normal exhaust temperature range, making joint control of products of incomplete combustion and $NO_x$ a severe technological challenge. Furthermore, the need to ensure a proper flow of ammonia presents a formidable logistical problem in the implementation of such technologies for control of vehicular emissions.

**Catalytic converters.** By the use of oxidation catalysts, the oxidation of carbon monoxide and hydrocarbon vapors can be promoted at much lower temperatures than is possible in the gas phase, as shown in Figure 4.24. The reduction of NO is also possible in catalytic converters, provided that the oxygen content of the combustion products is kept sufficiently low. In the catalytic converter, the exhaust gases are passed through a bed that contains a small amount of an active material such as a noble metal or a base metal oxide deposited on a thermally stable support material such as alumina. Alumina, by virtue of its porous structure, has a tremendous surface area per unit volume. Small pellets, typically a few millimeters in diameter, or thin-walled, honeycomb, monolithic structures, illustrated in Figure 4.25, are most commonly used as the support. Pellet supports are inexpensive, but when packed closely in a reactor, they produce large pressure drops across the device, increasing the back-pressure in the exhaust system. They may suffer from attrition of the catalyst pellets due to motion during use. This problem can be reduced, but not entirely overcome, through the use of hard, relatively high density pellets. The mass of the catalyst bed, however, increases the time required for the bed to heat to the temperature at which it becomes catalytically active, thereby allowing substantial CO and hydrocarbon emissions when the engine is first started. Mon-

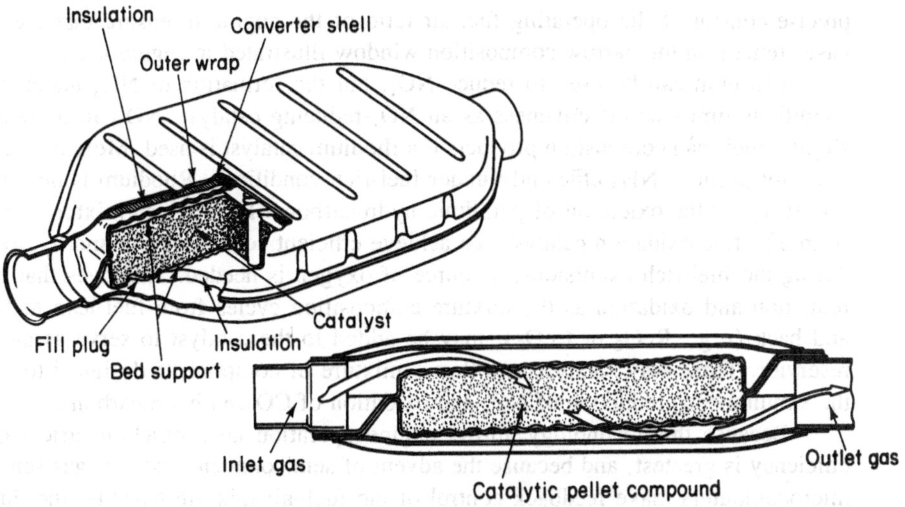


**Figure 4.25** Schematic of pellet-type catalytic converter.

olithic supports allow a freer exhaust gas flow, but are expensive and less resistant to mechanical and thermal damage. In particular, the rapid temperature changes to which a vehicular catalytic converter is exposed make thermal shock a very serious problem.

Many materials will catalyze the oxidation of CO or hydrocarbons at typical exhaust gas temperatures. The oxidation activities per unit surface area for noble metals, such as platinum, are high for both CO and hydrocarbons. Base metal oxide catalysts, notably CuO and $Co_3O_4$, exhibit similar activities for CO oxidation but are significantly less active for hydrocarbon oxidation (Kummer, 1980). Base metal catalysts degrade more rapidly at high temperature than do the noble metal catalysts. They are also more susceptible to poisoning by trace contaminants in fuels, such as sulfur, lead, or phosphorus. Hence most automotive emission catalysts employ noble metals.

NO reduction can be achieved catalytically if the concentrations of reducing species are present in sufficient excess over oxidizing species. CO levels in the exhaust gases of 1.5 to 3% are generally sufficient.

Two schemes are employed to achieve catalytic control of both $NO_x$ and products of incomplete combustion: (1) dual-bed catalytic converters and (2) three-way catalysts. The dual-bed system involves operation of the engine fuel-rich, to promote the reduction of $NO_x$. Secondary air is then added to facilitate the oxidation of CO and hydrocarbons in a second catalyst. Rich operation, while necessary for the NO reduction, results in reduced engine efficiency. Furthermore, it imposes severe restrictions on engine operation. If the exhaust gases are too rich, some of the NO may be converted to $NH_3$ or HCN. The oxidation catalyst used to eliminate CO and hydrocarbons readily oxidizes these species back to NO, particularly if the catalyst temperature exceeds 700 K.

If the engine is operated at all times at equivalence ratios very close to unity, it is possible to reduce NO and oxidize CO and hydrocarbons on a single catalyst bed known as a *three-way catalyst* (Kummer, 1980). The three-way catalytic converter requires very precise control of the operating fuel/air ratio of the engine to ensure that the exhaust gases remain in the narrow composition window illustrated in Figure 4.26.

Platinum can be used to reduce $NO_x$, but the formation of $NH_3$ under fuel-rich conditions limits its effectiveness as an $NO_x$-reducing catalyst. NO can be reduced in slightly fuel-lean combustion products if a rhodium catalyst is used. Moreover, rhodium does not produce $NH_3$ efficiently under fuel-rich conditions. Rhodium is not effective, however, for the oxidation of paraffinic hydrocarbons. In fuel-lean mixtures, platinum is an effective oxidation catalyst. To achieve efficient control of CO and hydrocarbons during the fuel-rich excursions, a source of oxygen is needed. Additives that undergo reduction and oxidation as the mixture composition cycles from fuel-lean to fuel-rich and back (e.g., $ReO_2$ or $CeO_2$), may be added to the catalyst to serve as an oxygen reservior. Three-way catalysts are thus a mixture of components designed to facilitate the simultaneous reduction of $NO_x$ and oxidation of CO and hydrocarbons.

Because this technology allows engine operation near stoichiometric where the efficiency is greatest, and because the advent of semiconductor exhaust gas sensors and microcomputers make feedback control of the fuel–air mixture feasible, the three-way converter has rapidly become the dominant form of exhaust gas treatment in the United States. The extremely narrow operating window has been a major driving force behind

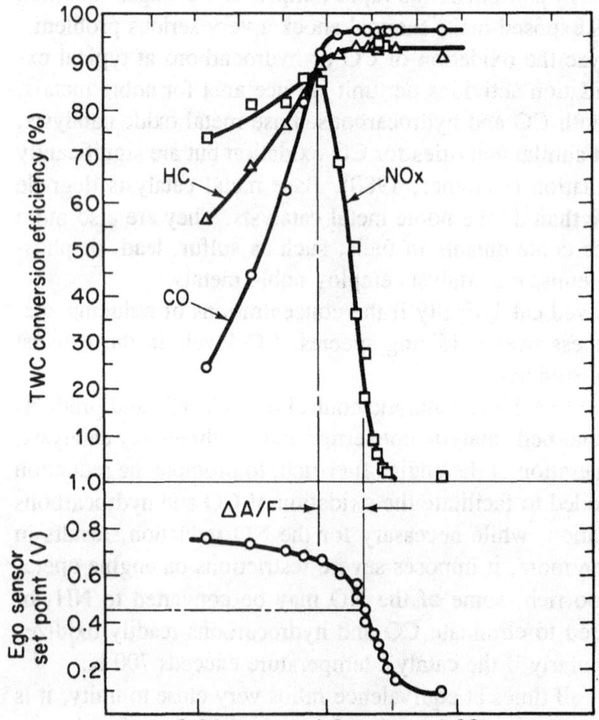


**Figure 4.26** Three-way catalyst conversion efficiency and exhaust gas oxygen sensor signal as a function of equivalence ratio (Hamburg et al., 1983; © SAE, Inc.).

the replacement of mechanically coupled carburetors with systems better suited to electronic control.

Feedback control of engines using exhaust gas sensors results in operation that oscillates about the stoichiometric condition in a somewhat periodic manner (Kummer, 1980). The frequency of these oscillations is typically on the order of 0.5 to 4 Hz, with excursions in equivalence ratio on the order of $\pm 0.01$ equivalence ratio units.

In addition to NO, CO, and unoxidized hydrocarbons, catalyst-equipped spark ignition engines can emit sulfuric acid aerosol, aldehydes, and under rich conditions, $H_2S$. Unleaded gasoline typically contains 150 to 600 ppm by weight of sulfur. The sulfur leaves the cylinder as $SO_2$, but the catalyst can promote further oxidation to $SO_3$. As the combustion products cool, the $SO_3$ combines with water to form an $H_2SO_4$ aerosol. $H_2S$ formation requires high catalyst temperatures ($>875$ K) and a reducing atmosphere. This may occur, for example, when an engine is operated steadily at high speed for some time under fuel-lean conditions and is then quickly slowed to idle fuel-rich operation. HCN formation may occur under similar conditions.

During startup, when the catalyst is cold, hydrocarbons may be only partially oxidized, leading to the emission of oxygenated hydrocarbons. Aldehyde emissions, however, are generally low when the catalyst is hot.

## 4.2 DIESEL ENGINE

Like the spark ignition engine, the diesel is a reciprocating engine. There is, however, no carburetor on the diesel. Only air (and possibly recycled combustion products for $NO_x$ control by EGR) is drawn into the cylinder through the intake valve. Fuel is injected directly into the cylinder of the diesel engine, beginning toward the end of the compression stroke. As the compression heated air mixes with the fuel spray, the fuel evaporates and ignites. Relatively high pressures are required to achieve reliable ignition. Excessive peak pressures are avoided by injecting the fuel gradually, continuing far into the expansion stroke.

The rate at which the fuel is injected and mixes with the air in the cylinder determines the rate of combustion. This injection eliminates the need to throttle the airflow into the engine and contributes to the high fuel efficiency of the diesel engine. As in the steady-flow combustor, turbulent mixing profoundly influences the combustion process and pollutant formation. The unsteady nature of combustion in the diesel engine significantly complicates the process. Rather than attempt to develop quantitative models of diesel emissions, we shall explore some of the features that govern the formation of pollutants in diesel engines.

Several diesel engine configurations are in use today. Fuel is injected directly into the cylinder of the direct injection (DI) diesel, illustrated in Figure 4.27(a). In the direct injection diesel engine, most of the turbulence is generated prior to combustion by the airflow through the intake valve and the displacement of gases during the compression stroke. The fuel jet is turbulent, but the time scale for mixing is comparable to that for entrainment, so the gas composition does not approach homogeneity within the fuel jet.

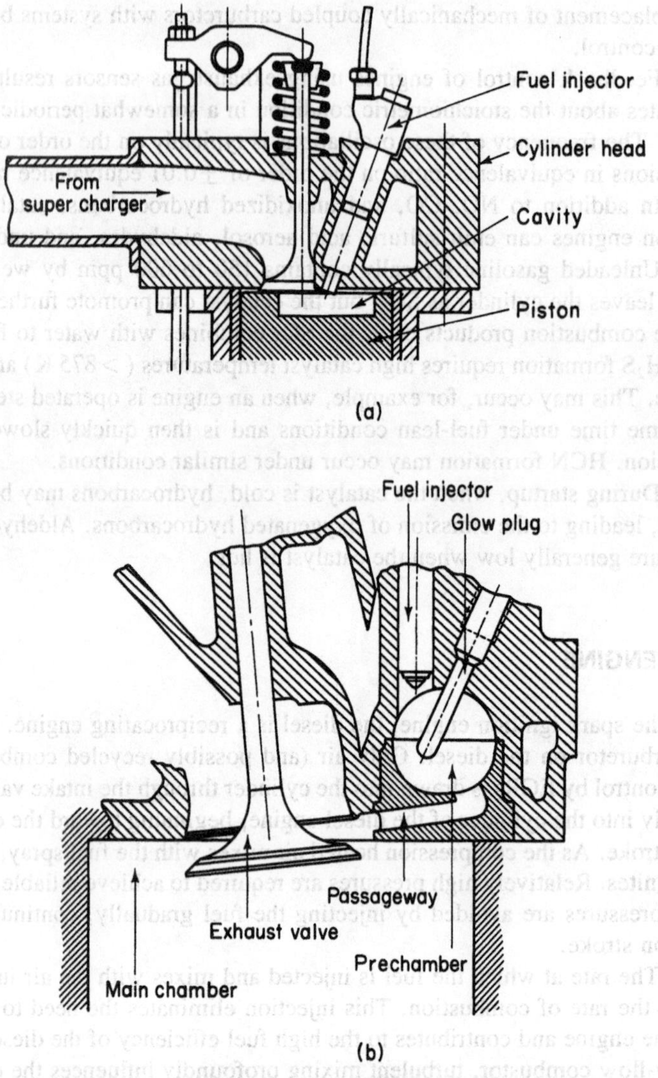


**Figure 4.27** Diesel engine types: (a) direct injection; (b) prechamber.

The use of a prechamber, as shown in Figure 4.27(b), enhances the mixing of the fuel and air in the indirect injection (IDI) or prechamber diesel engine. As the gases burn within the prechamber, they expand through an orifice into the main cylinder. The high kinetic energy of the hot gas jet is dissipated as turbulence in the jet and cylinder. This turbulence enhances mixing over that of the direct injection engine. Improved mixing limits the amount of very fuel-rich gas in the cylinder, thereby reducing soot emissions. Most light-duty diesel engines are of the indirect injection type because of the reduced particulate emissions afforded by this technology. This benefit is not without costs, however. The flow through the orifice connecting the prechamber to the cylinder results in a pressure drop, thereby reducing the efficiency of the engine.

Diesel engines may also be classified into naturally aspirated (NA), supercharged, or turbocharged types, depending on the way the air is introduced into the cylinder. In the naturally aspirated engine, the air is drawn into the cylinder by the piston motion alone. The supercharger is a mechanically driven compressor that increases the airflow into the cylinder. The turbocharger similarly enhances the intake airflow by passing the hot combustion products through a turbine to drive a centrifugal (turbine-type) compressor.

Compression of the air prior to introduction into the cylinder results in compression heating. This may be detrimental from the point of view of $NO_x$ formation because it increases the peak combustion temperature. An intercooler may be installed between the compressor and the intake valve to reduce this heating.

The fuel is sprayed into the cylinder through a number of small nozzles at very high pressure. The liquid stream issuing from the injector nozzle moves with high velocity relative to the gas. The liquid stream forms filaments that break into large droplets. The breakup of the droplets in the fuel spray is characterized by the Weber number, the ratio of the inertial body forces to surface tension forces,

$$\mathrm{We} = \frac{\rho_g D v^2}{\sigma}$$

where $\rho_g$ is the gas density, $v$ the relative velocity between the gas and the droplets, and $\sigma$ the surface tension of the liquid. As long as the Weber number exceeds a critical value of approximately 10, the droplets will continue to break into smaller droplets. Aerodynamic drag on the droplets rapidly decelerates them and accelerates the gas entrained into the fuel spray. Evaporation and combustion of the fuel can be described using the model developed in Section 2.7. In some cases, however, pressures and temperatures in the cylinder are high enough that the liquid fuel is raised above its critical point. The fuel spray then behaves like a dense gas jet.

The entrainment of air into the unsteady, two-phase, variable-density, turbulent jet has been described by a variety of empirical models, simple jet entrainment models, and detailed numerical simulations. The problem is frequently complicated further by the use of swirling air motions to enhance mixing and entrainment. The swirling air motion sweeps the fuel jet around the cylinder, spreading it and reducing impingement on the cylinder wall. Since combustion in nonpremixed systems generally occurs predominantly at equivalence ratios near unity, combustion will occur primarily on the perimeter of the jet. Mixing of hot combustion products with the fuel-rich gases in the core of the fuel spray provides the environment in which large quantities of soot can be readily generated. (We discuss soot formation in Chapter 6.) The stoichiometric combustion results in high temperatures that promote rapid $NO_x$ formation in spite of operation with large amounts of excess air in the cylinder under most operating conditions. Some of the fuel mixes with air to very low equivalence ratios before any of the mixture ignites. Temperatures in this region may be high enough for some fuel decomposition and partial oxidation to occur, accounting for the relative abundance of aldehydes and other oxygenates in the diesel emissions (Henein, 1976).

Thus we see that diesel engines exhibit all of the complications of steady-flow spray flames, in addition to being unsteady. To describe the formation of pollutants quantitatively would require the development of a probability density function description of the unsteady mixing process. While such models are being explored (Mansouri et al., 1982a,b; Kort et al., 1982; Siegla and Amann, 1984), the methods employed are beyond the scope of this book. We shall examine, instead, the general trends as seen in both experimental and theoretical studies of diesel engine emissions and emission control.

### 4.2.1 Diesel Engine Emissions and Emission Control

Relatively low levels of gaseous exhaust emissions are achieved by light-duty (automobile) diesel engines without the use of exhaust gas treatment usually applied to gasoline engines to achieve similar emission levels (Wade, 1982).

The species mole fractions in diesel exhaust are somewhat misleading because of the low and variable equivalence ratios at which diesel engines typically operate. At low-load conditions, the operating equivalence ratio may be as low as 0.2, so the pollutants are diluted significantly with excess air. Because the equivalence ratio is continually varying in normal use of diesel engines, and to facilitate comparison to other engines, it is more appropriate to report emission levels in terms of emissions per unit of output: g $MJ^{-1}$ for stationary engines or heavy-duty vehicles or g $km^{-1}$ for light-duty vehicular diesel engines.

Injection of the liquid fuel directly into the combustion chamber of the diesel engine avoids the crevice and wall quench that allows hydrocarbons to escape oxidation in carbureted engines, so hydrocarbon emissions from diesel engines are relatively low. Furthermore, diesel engines typically operate fuel-lean, so there is abundant oxygen to burn some of the hydrocarbons and carbon monoxide formed in midair in the cylinder. $NO_x$ emissions from prechamber diesel engines are lower than the uncontrolled $NO_x$ emissions from homogeneous charge gasoline engines (Wade, 1982). The low $NO_x$ emissions result from the staged combustion in the prechamber diesel and the inhomogeneous gas composition. Particulate emissions from diesel engines tend to be considerably higher than those of gasoline engines and represent a major emission control challenge.

Factors that influence the emissions from diesel engines include the timing and rate of fuel injection, equivalence ratio, compression ratio, engine speed, piston and cylinder design, including the use of prechambers, and other design factors. The influence of the overall equivalence ratio on engine performance is shown in Figure 4.28(a). The brake mean effective pressure increases with equivalence ratio, so higher equivalence ratios correspond to higher engine power output or load. The exhaust gas temperature also increases with equivalence ratio. Fuel consumption is high at low equivalence ratio, but decreases sharply as the equivalence ratio is increased. As the equivalence

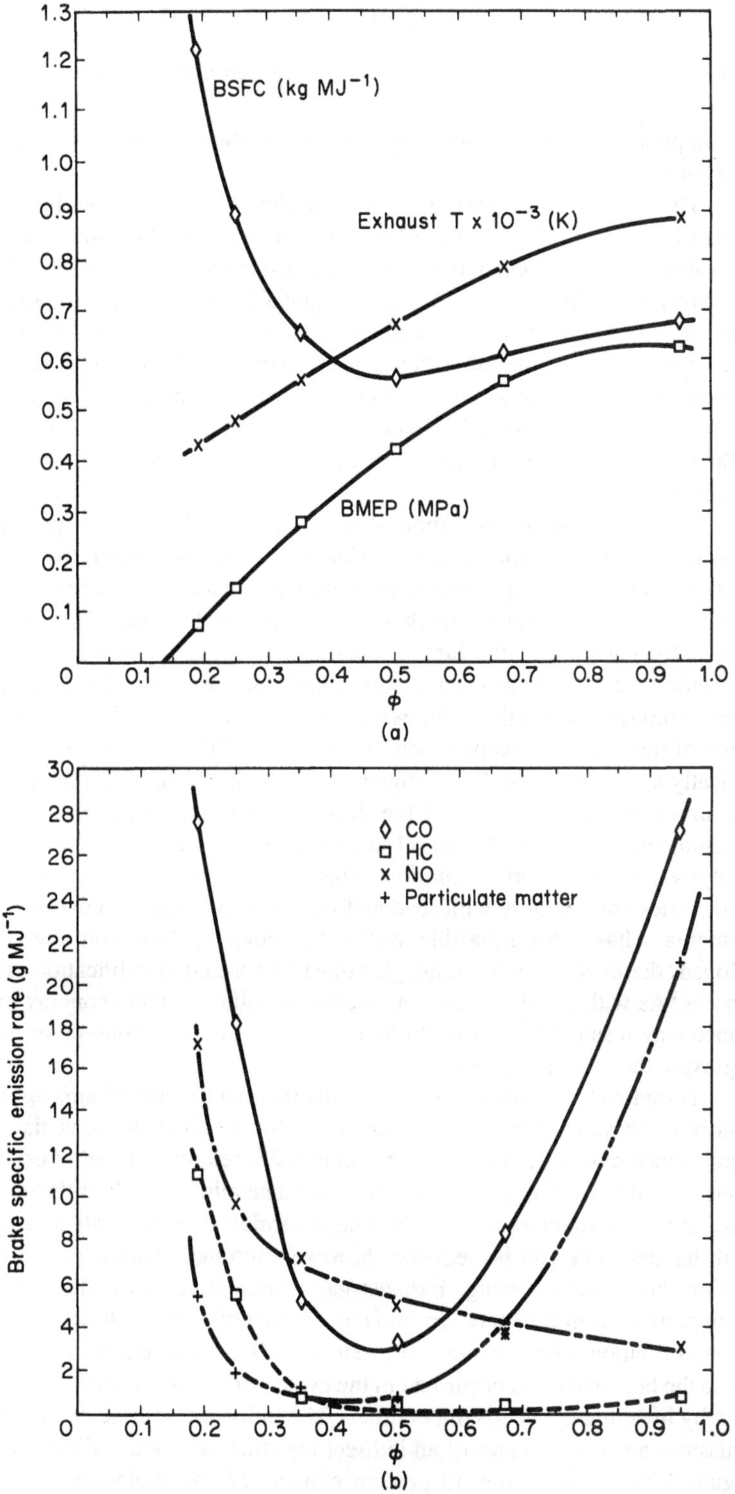


**Figure 4.28** Influence of equivalence ratio on diesel engine performance and emissions (Wade, 1982; © SAE, Inc.).

ratio approaches unity, the fuel consumption increases slightly above a minimum value at about $\phi = 0.5$.

The variation of emissions with equivalence ratio is shown in Figure 4.28(b). In terms of grams emitted per MJ of engine output, all of the emissions are high at low equivalence ratios for which the engine output is small. Carbon monoxide and particulate emissions drop sharply with increasing equivalence ratio, pass through a minimum at about $\phi = 0.5$, and then rise sharply as $\phi$ approaches unity. Hydrocarbon and nitrogen oxide emissions, on the other hand, drop sharply as $\phi$ is increased above about 0.2, reaching relatively low levels at an equivalence ratio of about 0.4, and changing only slightly thereafter. It should be noted that, while the brake specific emissions of the latter pollutants decrease with increasing load, the absolute emission rate may increase at high power output.

While CO and hydrocarbon levels in diesel exhausts are quite low, particulate emissions from diesel engines are considerably higher than those from comparable spark ignition engines. The high particulate emissions are a consequence of the direct injection of fuel into the cylinder or prechamber of the diesel engine. The mixing is relatively slow, allowing some of the fuel to remain in hot, fuel rich gases long enough for polymerization reactions to produce the high-molecular-weight hydrocarbons that ultimately form carbonaceous particles known as soot. Subsequent mixing is slow enough that many of the particles escape oxidation in spite of the large amount of excess air that is typically available in the diesel engine. $NO_x$ levels are also high because combustion in the turbulent diffusion flame of the diesel engine takes place in regions that are near stoichiometric. Because the diesel engine generally operates very fuel lean, reduction catalysts are not a practical solution to the $NO_x$ emission problem. The exhaust temperature varies considerably with load and is on the low side for noncatalytic reduction by ammonia. Thus, until a feasible system for removing $NO_x$ from diesel exhausts is developed, diesel $NO_x$ control strategies must be based on modification of the combustion process. As with the spark ignition engine, penalties in fuel economy or engine performance may result. Moreover, efforts to improve the fuel economy or performance may aggravate the emission problem.

Diesel $NO_x$ emissions result from the thermal fixation of atmospheric nitrogen, so control of these emissions can be achieved by reducing the peak flame temperatures. Equivalence ratio has, as we saw in Figure 4.28, relatively little influence on $NO_x$ emissions from the diesel engine. The diffusion flame allows much of the combustion to take place at locally stoichiometric conditions regardless of the overall equivalence ratio. The peak temperatures can be reduced, however, through exhaust gas recirculation or retarding the injection timing. Exhaust gas recirculation in the diesel engine serves the same purpose as in the spark ignition engine: that of reducing the peak flame temperature through dilution with combustion products. Like spark retard, injection timing delays cause the heat release to occur late in the cycle, after some expansion work has occurred, thereby lowering the peak temperatures. The influence of these two control strategies on emissions and performance of an indirect injection automotive diesel engine is shown in Figure 4.29. Below about 30 percent exhaust gas recirculation, hydrocarbon, carbon monoxide, and particulate emissions are not significantly influenced by exhaust gas re-

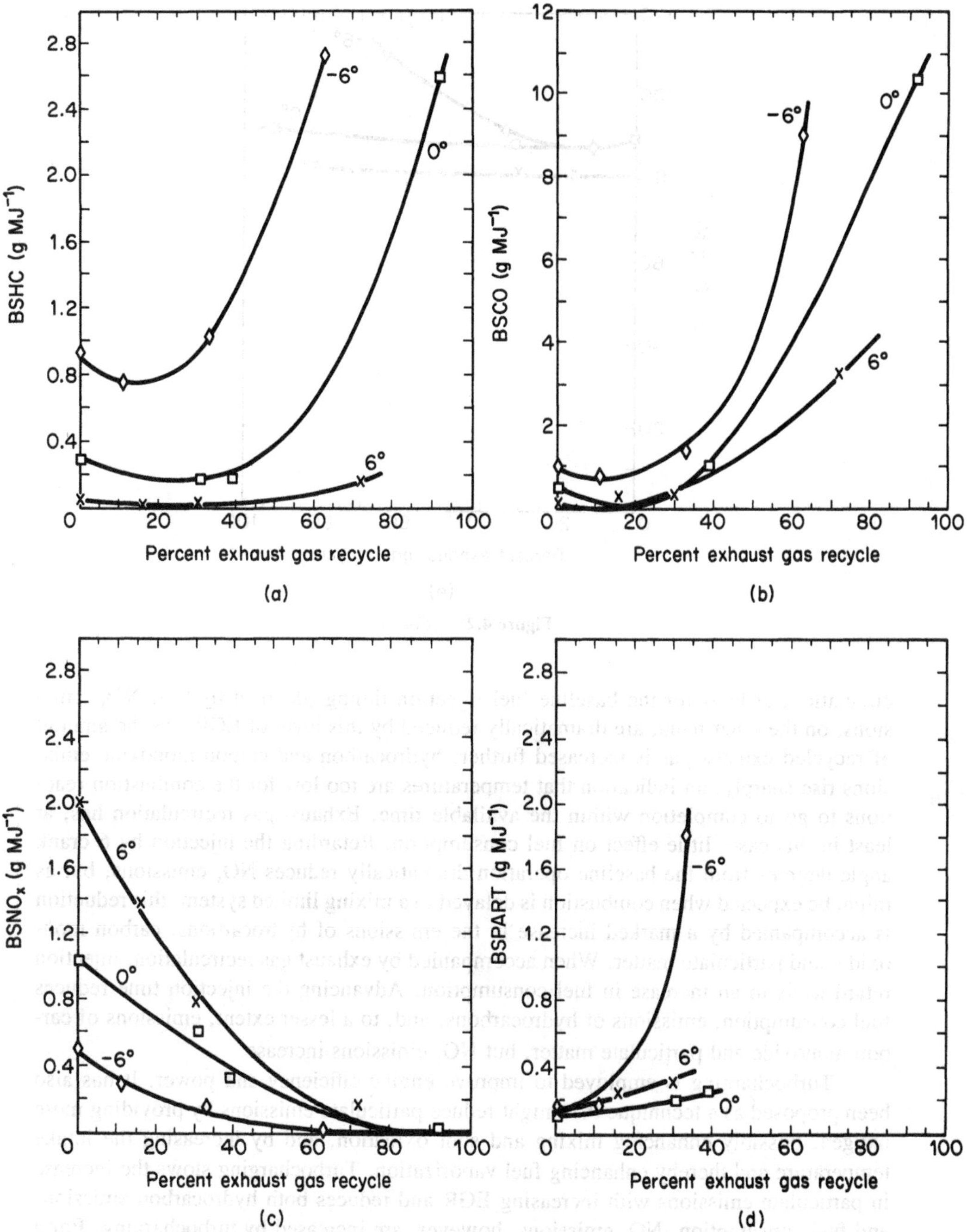


**Figure 4.29** Influence of exhaust gas recirculation and injection timing on diesel engine performance and emissions (Wade, 1982; © SAE, Inc.).

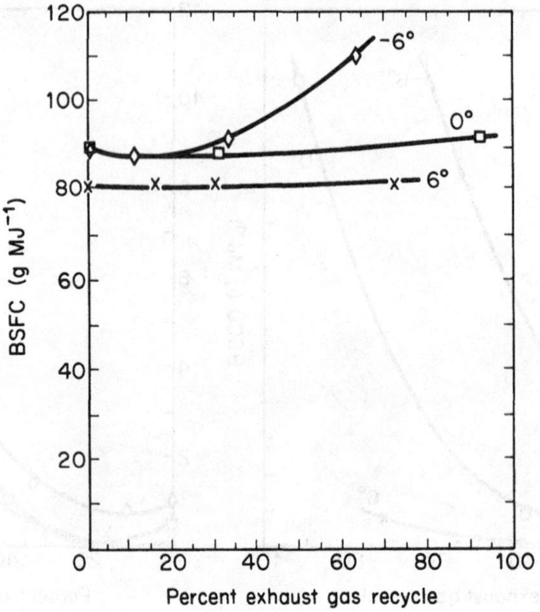


(e)

**Figure 4.29** (*Cont.*)

circulation, at least for the baseline fuel injection timing (denoted by 0°). $NO_x$ emissions, on the other hand, are dramatically reduced by this level of EGR. As the amount of recycled exhaust gas is increased further, hydrocarbon and carbon monoxide emissions rise sharply, an indication that temperatures are too low for the combustion reactions to go to completion within the available time. Exhaust gas recirculation has, at least in this case, little effect on fuel consumption. Retarding the injection by 6 crank angle degrees from the baseline operation dramatically reduces $NO_x$ emissions, but as might be expected when combustion is delayed in a mixing limited system, this reduction is accompanied by a marked increase in the emissions of hydrocarbons, carbon monoxide, and particulate matter. When accompanied by exhaust gas recirculation, injection retard leads to an increase in fuel consumption. Advancing the injection time reduces fuel consumption, emissions of hydrocarbons, and, to a lesser extent, emissions of carbon monoxide and particulate matter, but $NO_x$ emissions increase.

Turbocharging is employed to improve engine efficiency and power. It has also been proposed as a technique that might reduce particulate emissions by providing more oxygen, possibly enhancing mixing and soot oxidation, and by increasing the intake temperature and thereby enhancing fuel vaporization. Turbocharging slows the increase in particulate emissions with increasing EGR and reduces both hydrocarbon emissions and fuel consumption. $NO_x$ emissions, however, are increased by turbocharging. For a fixed level of $NO_x$ emissions, the particulate emissions, unfortunately, are not significantly reduced by turbocharging. Moreover, at light loads, hydrocarbon emissions may actually increase with turbocharging.

Due to the relatively low nitrogen content of light diesel fuels, the formation of nitric oxide in the diesel engine is primarily by the thermal fixation of atmospheric nitrogen. Diesel engines can be operated, however, on heavier liquid fuels, or even on coal. The possible contributions of fuel-nitrogen to $NO_x$ emissions must be considered for such operations.

### 4.2.2 Exhaust Gas Treatment

The diesel engine typically operates fuel-lean. The presence of excess oxygen in the combustion products suggests that the oxidation of carbon monoxide, hydrocarbon vapors, and carbonaceous particles in the exhaust gases may be possible. Wade (1980) examined the possibility of oxidizing the particles in the exhaust gases using a long exhaust reactor. At temperatures below 1000 K, the mass lost in 10 s due to oxidation was insignificant. At 1100 K, some mass was consumed by oxidation, but 3 s was required for an 80% reduction in the exhaust mass loading. Temperatures on the order of 1370 K would probably be required to achieve significant oxidation in a thermal reactor with a volume the same as the engine displacement. Thus oxidation of the particulate matter as it flows through an exhaust system does not appear to be a practical solution to the particulate emissions from vehicular diesel engines. The large temperature swings of the diesel exhaust temperature—in particular, the low temperatures encountered during low-load operation—make the thermal reactor impractical even for CO and hydrocarbon vapor emission control unless some additional fuel is added to raise the temperature.

An oxidation catalyst might be useful for the oxidation of carbon monoxide and hydrocarbons, particularly if it were effective in removing the condensible organics that contribute to the particulate emissions and the partially oxygenated hydrocarbons that are responsible for the diesel exhaust odor. The particulate matter deposition on the catalyst may affect the performance.

Filters or electrostatic precipitators can be used to remove particles from the exhaust gas stream, but disposal becomes a logistical problem in vehicular applications. These devices can be used, however, to concentrate the particulate matter for subsequent on-board incineration (Sawyer et al., 1982). After a sufficient quantity of particulate matter has been collected to sustain combustion, the collected carbonaceous material can be ignited by an auxiliary heat source and burned. This cleans the particle trap and allows extended operation without the buildup of such heavy deposits that engine back-pressure becomes a problem. A filter element may be impregnated with a catalyst to promote oxidation of the collected material at lower temperatures. Maintaining the trap at low temperature during the collection phase would facilitate the collection of vapors that would otherwise condense or adsorb on the soot particles upon release into the atmosphere. Periodic regeneration could be performed automatically using microprocessor control and sensors to monitor the status of the trap during the collection and regeneration phases of operation.

Filter elements may take the form of wire meshes, ceramic monoliths, or ceramic foams. Due to the wide range of temperatures and rapid changes to which the trap is exposed, the durability of these elements is a critical issue in the development of this technology. If thermal shock leads to the development of cracks or fissures through which the exhaust can flow, the collection efficiency may be seriously degraded.

Reduction of $NO_x$ in diesel exhaust gases represents a formidable challenge. Three-way catalysts are not effective because of the large amount of excess oxygen. The exhaust temperature varies over too wide a range for the noncatalytic ammonia injection technique to be useful, and the soot particles in the diesel exhaust are likely to foul or poison catalysts in the catalytic ammonia injection unless very effective particle removal is achieved. Alternative reducing agents that work at lower temperature, such as isocyanic acid (Perry and Siebers, 1986), may ultimately prove effective in diesel engine $NO_x$ emission control.

## 4.3 STRATIFIED CHARGE ENGINES

The emphasis in our discussion of spark ignition engines has been on the homogeneous charge engines. The range of equivalence ratios over which such engines can operate has been seen to be very narrow due to the low laminar flame speed in rich or lean mixtures. The diesel engine, on the other hand, can operate at very low equivalence ratios. Soot formation in the compression ignition engine remains a serious drawback to its use.

An alternative approach that is intermediate between these two types of reciprocating engines is a stratified charge engine. Stratified charge engines rely on a spark for ignition as in the homogeneous charge engine, but utilize a nonuniform fuel distribution to facilitate operation at low equivalence ratios. The concept of the stratified charge engine is not new. Since the 1930s attempts have been made to develop hybrid engines that incorporate the best features of both the spark ignition and diesel engines (Heywood, 1981). Some, like the diesel, involve direct injection of the fuel into the cylinder. In others, two carbureted fuel–air mixtures are introduced into the cylinder, a rich mixture into a small prechamber and a lean mixture into the main cylinder.

Figure 4.30 illustrates the latter type of engine. The idea behind the prechamber stratified charge engine is that the fuel-rich mixture is easier to ignite than is the lean

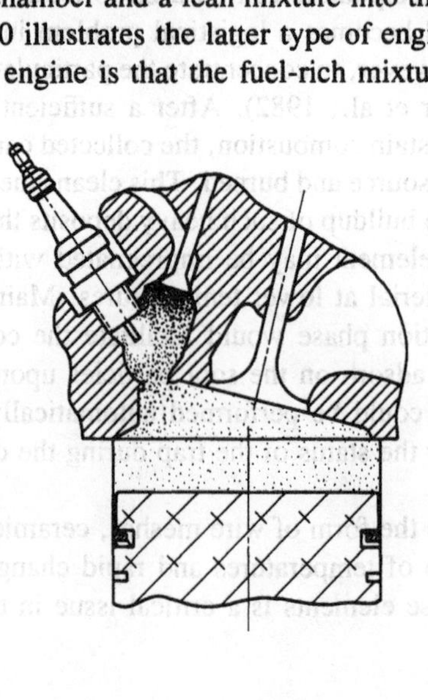

**Figure 4.30** Prechamber stratified charge engine.

mixture in the main cylinder. The spark ignites the mixture in the prechamber. As it burns, the rich mixture expands, forming a jet through the orifice connecting the two chambers. The high-velocity flow of hot combustion products rapidly mixes with and ignites the lean mixture in the main cylinder. By this route, only a small fraction of the combustion takes place near stoichiometric, with most occurring well into the fuel-lean region. Thus $NO_x$ formation is substantially slower than in a homogeneous charge engine.

The simplest conceptual model for the prechamber stratified charge engine is to assume that the mixture in the prechamber is uniform at a high equivalence ratio and that the mixture in the main cylinder is uniform at a lower value of $\phi$. This assumption, however, is rather tenuous. During the intake stroke, the two mixtures flow through the two intake valves. Because of the large displacement volume, some of the prechamber mixture may flow out through the connecting orifice into the main chamber during intake. In the compression stroke, gas from the main chamber is forced back into the prechamber, so the prechamber will contain a mix of the rich and lean charges. Some of the prechamber mixture will remain in the main chamber. Thus, in spite of using carburetors to prepare the two charges, the two mixtures may not be uniform at ignition. During combustion, further mixing between the two gases takes place.

Analysis of gas samples collected at the exhaust port suggests that no significant stratification of the mixture remains at the end of the expansion stroke. This does not mean, however, that the gas is uniformly mixed on the microscale. As in the diesel engine, the turbulent mixing process plays a critical role in determining the pollutant emission levels. The dependence of the emissions from the prechamber stratified charge engine on equivalence ratio, as a result, is somewhat weaker than that of the conventional homogeneous charge engine. The reduction in the peak $NO_x$ emissions with this technology is modest, but the ability to operate at very low equivalence ratios makes significant emission reductions possible.

Other types of stratified-charge engines utilize direct injection of the fuel into the cylinder to create local variations in composition. The fuels used in such engines generally still have the high-volatility characteristics of gasoline, and a spark is used for ignition. Two types of direct-injection stratified-charge engines are shown in Figure 4.31. The early injection version of the direct-injection stratified-charge engine typically uses a broad conical spray to distribute the fuel in the central region of the cylinder, where the piston has a cup. The spray starts early, about halfway through the compression stroke, to give the large fuel droplets produced by the low-pressure injector time to evaporate prior to ignition. In contrast, a late injection engine more closely resembles a diesel engine. A high-pressure injector introduces a narrow stream of fuel, beginning just before combustion. A swirling air motion carries the spray toward the spark plug.

The details of combustion in these engines are not well understood. Mixing and combustion occur simultaneously, so the dependence of emissions on equivalence ratio more closely resembles the weak dependence of poorly mixed steady-flow combustors than that of the conventional gasoline engine. Because pure fuel is present in the cylinder during combustion, the direct-injection stratified-charge engines suffer from one of the major difficulties of the diesel engine: soot is formed in significant quantities.

These are but a few of the possible configurations of reciprocating engines. The

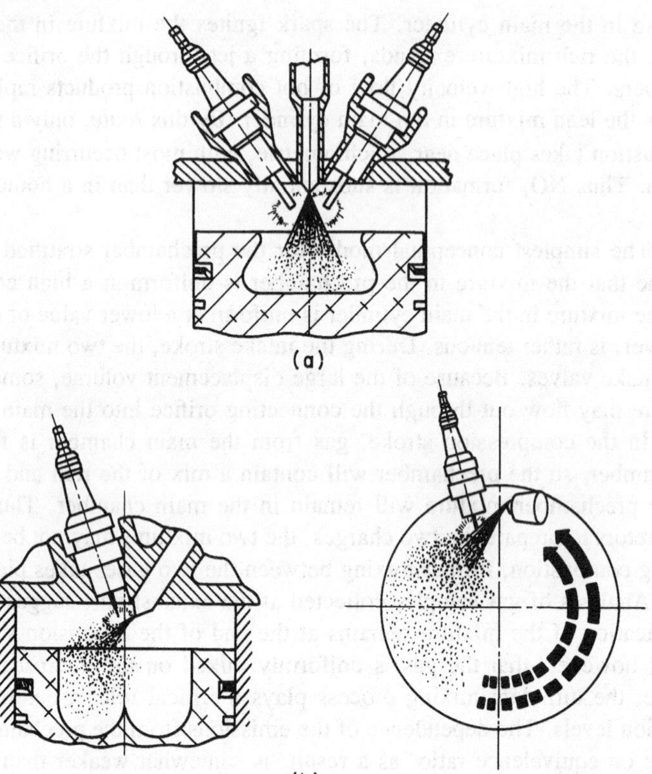
(a)
(b)

**Figure 4.31** Direct-injection stratified-charge engine: (a) early-injection; (b) late-injection.

introduction of emission limitations on automobile engines has led to major new technological developments, some successful, many that failed to meet their developers' expectations. Renewed concern over engine efficiency has once again shifted the emphasis. To satisfy both environmental and fuel utilization constraints, the use of computer control, catalytic converters, and exhaust gas sensors has been introduced. These have diminished the level of interest in stratified charge engines, since $NO_x$ reduction catalysts are needed to meet strict emission limits and require operation near stoichiometric. The problems of emission control for heavy-duty engines remain unresolved. Yet, as automobile emissions are reduced, large engines in trucks, railway engines, compressors, and so on, are becoming increasingly important sources of atmospheric pollutants, notably $NO_x$ and soot.

## 4.4 GAS TURBINES

The fourth major class of internal combustion engines is the gas turbine. The power output of the gas turbine engine can be very high, but the engine volume and weight are generally much smaller than those of reciprocating engines with comparable output. The original application of the gas turbine engine was in aircraft, where both weight and volume must be minimized. The high power output also makes the gas turbine attractive for electric power generation. Like reciprocating engine exhaust, the exhaust gases from the gas turbine are quite hot. The hot combustion products can be used to generate steam to drive a steam turbine. High efficiencies of conversion of fuel energy to electric power can be achieved by such combined-cycle power generation systems. Clearly, the constraints on these applications differ greatly, so the technologies that can be applied to control emissions for one type of gas turbine engine are not always applicable to the other.

Unlike the reciprocating engines, the gas turbine operates in steady flow. Figure 4.32 illustrates the main features of a gas turbine engine. Combustion air enters the turbine through a centrifugal compressor, where the pressure is raised to 5 to 30 atm, depending on load and the design of the engine. Part of the air is then introduced into the primary combustion zone, into which fuel is sprayed and burns in an intense flame. The fuel used in gas turbine engines is similar in volatility to diesel fuel, producing droplets that penetrate sufficiently far into the combustion chamber to ensure efficient combustion even when a pressure atomizer is used.

The gas volume increases with combustion, so as the gases pass at high velocity through the turbine, they generate more work than is required to drive the compressor. This additional work can be delivered by the turbine to a shaft, to drive an electric power generator or other machinery, or can be released at high velocity to provide thrust in aircraft applications.

The need to pass hot combustion products continuously through the turbine imposes severe limits on the temperature of those gases. Turbine inlet temperatures are limited to 1500 K or less, depending on the blade material. The development of turbine blade materials that can withstand higher temperatures is an area of considerable interest because of the efficiency gains that could result. In conventional gas turbine engines, the cooling is accomplished by dilution with additional air. To keep the wall of the combustion chamber, known as the combustor can, cool, additional air is introduced through wall jets, as shown in Figure 4.33. The distribution of the airflow along the length of the combustion chamber, as estimated by Morr (1973), is also shown.

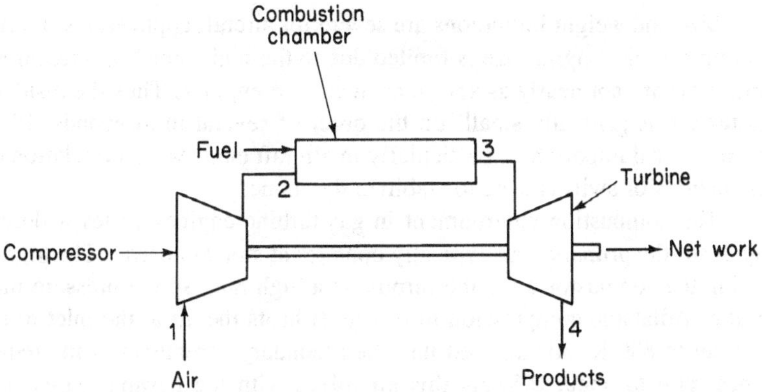


**Figure 4.32** Gas turbine engine configuration.

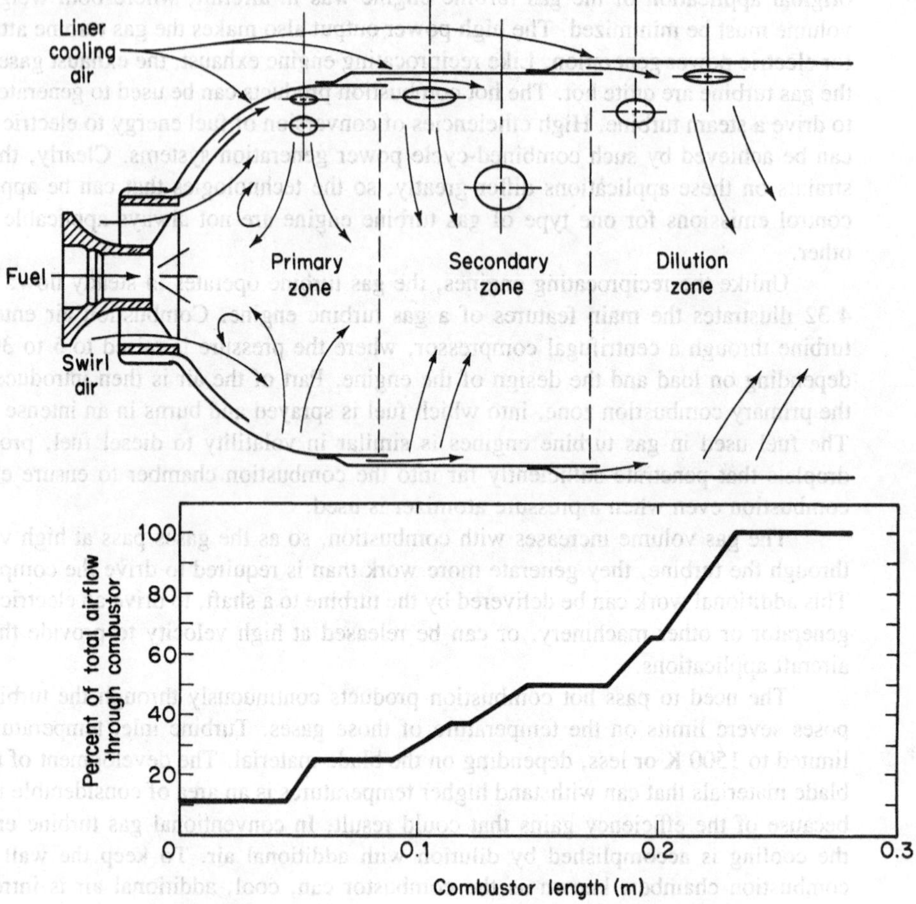


**Figure 4.33** Gas turbine combustion, airflow patterns, and air distribution. Morr (1973).

Size and weight limitations are severe for aircraft applications. Even for stationary gas turbines, the engine size is limited due to the high operating pressures, although the constraints are not nearly as severe as in aircraft engines. Thus the residence time in the gas turbine is generally small, on the order of several milliseconds. Flame stability is also of critical importance, particularly in aircraft engines. Recirculation caused by flow obstructions or swirl is used to stabilize the flame.

The combustion environment in gas turbine engines varies widely with load. At high load, the primary zone typically operates at stoichiometric. The high-velocity gases leaving the combustor drive the turbine at a high rate, so the pressure may be raised to 25 atm. Adiabatic compression to this level heats the air at the inlet to the combustion chamber to 800 K. Air injected into the secondary combustion zone reduces the equivalence ratio to about 0.5. As this air mixes with the burning gases, combustion approaches completion. Finally, dilution air reduces the equivalence ratio to below 0.3, lowering the mean gas temperature to the turbine inlet temperature of less than 1500 K.

At idle, the pressure may only increase to 1.5 atm, so the inlet air temperature is not significantly greater than ambient. The equivalence ratio in the primary combustion zone may be as low as 0.5, resulting in a mean gas temperature of only 1100 K. The distribution of air in the different regions of an aircraft engine is determined by the combustion can geometry and therefore does not change with load. The combustion products at idle may be diluted to an equivalence ratio below 0.15 at the turbine inlet, lowering the mean gas temperature to below 700 K.

Because of the very short residence time in the combustion chamber, droplet evaporation, mixing, and chemical reaction must occur very rapidly. The short time available profoundly influences the pollutant emissions from gas turbine engines. As illustrated in Figure 4.34, at low load, when the overall equivalence ratio is well below unity and the gases are rapidly quenched, $NO_x$ emissions are low. These same factors cause the CO and hydrocarbon emissions to be high in low-load operation. At high load, on the other hand, the gases remain hot longer and the equivalence ratio is higher, so more $NO_x$ is formed; while CO and hydrocarbon oxidation is more effective at high loads, causing emissions of these species to decrease with load. Soot, however, is also formed in the gas turbine engine at higher loads due to the imperfect mixing, which allows some very fuel-rich mixture to remain at high temperatures for a relatively long time.

Gas turbine engines used for electric power generation and other applications where weight is not so critical introduce additional factors into the emissions problem. While aircraft engines generally burn distillate fuels with low sulfur and nitrogen contents, stationary gas turbines frequently burn heavier fuels. Fuel-bound nitrogen may contrib-

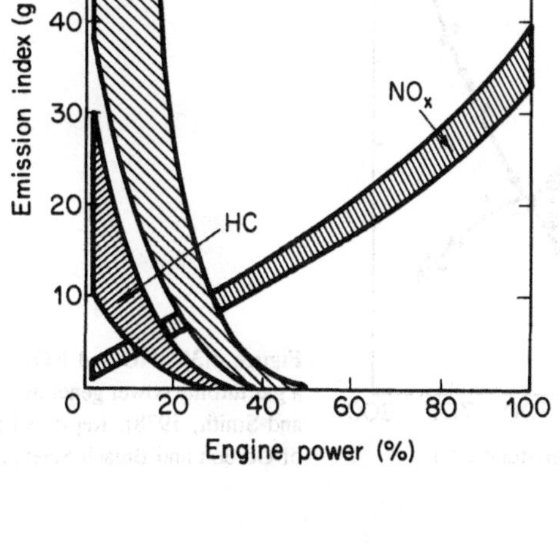


**Figure 4.34** Variation of gas turbine engine emissions with engine power.

ute significantly to $NO_x$ emissions, and sulfur oxides may be emitted as well. The use of coal directly or in combination with a gasifier could aggravate these problems and also introduce the possibility of fine ash particles being emitted. (Coarse ash particles are less likely to be a problem. Because their presence would rapidly erode the turbine and significantly degrade its performance, large particles must be efficiently removed from the combustion products upstream of the turbine inlet.)

In contrast to most other steady-flow combustors, $NO_2$ levels can be comparable to or even exceed NO levels in the exhaust gases, as shown in Figure 4.35. Rapid quenching of the combustion products has another effect: promoting the formation of large concentrations of $NO_2$.

The control of emissions from gas turbine engines involves a number of factors:

1. *Atomization and mixing.* Poor atomization and imperfect mixing in the primary combustion zone allow CO and unburned hydrocarbons to persist into the secondary zone, where the quenching occurs. It also allows stoichiometric combustion to take place even when the overall equivalence ratio of the primary zone is less than unity, thereby accelerating the formation of $NO_x$.
2. *Primary zone equivalence ratio.* To lower $NO_x$ emissions at high load, it is necessary to reduce the peak flame temperatures. Since the overall equivalence ratio of the gas turbine engine operation must be so low, reducing the primary zone equivalence ratio is a promising approach. Of course, effective mixing is required at the same time.
3. *Scheduling of dilution air introduction.* By first diluting the combustion products to an equivalence ratio at which CO and hydrocarbon oxidation are rapid and then allowing sufficient time for the reaction before completing the dilution, it may be possible to complete the oxidation, even at low loads. The total combustor volume and the need to cool the combustion can walls limit the extent to which this can be accomplished.

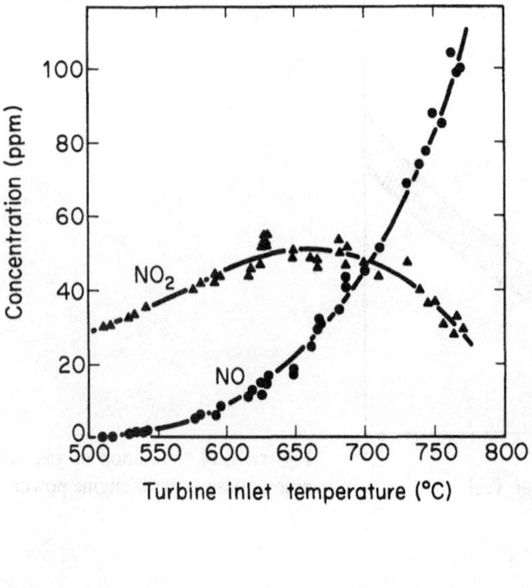


**Figure 4.35** NO and $NO_2$ emissions from a gas turbine power generating unit (Johnson and Smith, 1978). Reprinted by permission of Gordon and Breach Science Publishers.

A variety of approaches are being taken to control gas turbine engine emissions. To optimize performance at both high and low loads, many designs employ a number of small burners that are operated at near-optimum conditions all the time. An extreme example is the NASA Swirl Can Combustor, which contains 100 to 300 small (100-mm diameter) swirl can combustor modules assembled in a large annular array (Jones, 1978). To vary the load, the number of burner modules that are supplied with fuel is changed. At low loads, the flame extends only a small distance downstream of the individual modules, and most of the excess air does not mix directly with the combustion products. At high loads, so many modules are supplied with fuel that the flames merge and fill the entire combustion can. For sufficiently high combustor inlet temperatures, this design has the effect of delaying the rise in CO and hydrocarbons to quite low equivalence ratios. $NO_x$ emissions are low at low load and rise dramatically with equivalence ratio (load). The general trends in $NO_x$ emissions are well represented by a segregation parameter, (2.70), of $S = 0.3$ to 0.4, indicating the degree of inhomogeneity in the gas turbine combustor. This suggests that much of the $NO_x$ is formed in locally stoichiometric regions of the flame.

Further improvements can be realized by changing the way the fuel is introduced into the combustion chamber to minimize the residence time at stoichiometric conditions. Air-assist atomizers generate turbulence in the primary combustion zone, thereby accelerating the mixing. Some new turbine designs even involve premixing of fuel and air. Flame stability becomes a serious issue in premixed combustion as the equivalence ratio is reduced, so mechanical redistribution of the airflow is sometimes introduced to achieve stable operation at low loads (Aoyama and Mandai, 1984).

Multiple combustion stages are also incorporated into low $NO_x$ gas turbine designs (see Figure 4.36), allowing efficient combustion in one or more small burners at low load and introducing additional fuel into the secondary zone to facilitate high-load fuel-lean combustion.

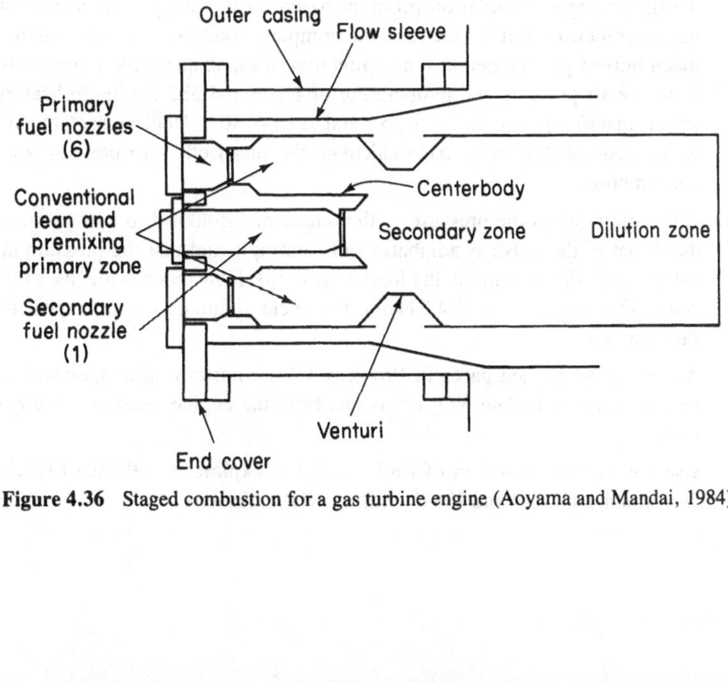


**Figure 4.36** Staged combustion for a gas turbine engine (Aoyama and Mandai, 1984).

## PROBLEMS

**4.1.** An idealized thermodynamic representation of the spark ignition engine is the ideal-air Otto cycle. This closed cycle consists of four processes:

1. Adiabatic and reversible compression from volume $V = V_c + V_d$ to $V = V_c$
2. Heat addition at constant volume, $V = V_c$
3. Adiabatic and reversible expansion from $V = V_c + V_d$ to $V = V_c$
4. Constant-volume ($V = V_c + V_d$) heat rejection

**(a)** Assuming that the ratio of specific heats, $\gamma = c_p/c_v$, is constant, derive an equation for the cycle efficiency ($\eta$ = net work out/heat in during process 2) in terms of the compression ratio, $R_c$.

**(b)** Derive an expression for the indicated mean effective pressure (IMEP) in terms of the compression ratio and the heat transfer per unit mass of gas during process 2.

**(c)** Discuss the implications of these results for $NO_x$ control by fuel-lean combustion and EGR.

**4.2.** A chamber with volume, $V$, contains a mixture of octane and air at $\phi = 0.9$. The initial temperature and pressure are 298 K and 1 atm, respectively. The vessel may be assumed to be perfectly insulated.

**(a)** The mixture is ignited and burned. Determine the final temperature and pressure.

**(b)** The pressure is measured as a function of time during the combustion process. Show how you would use the pressure measurements to determine the combustion rate. Plot the burned mass fraction as a function of the measured pressure.

**4.3.** The vessel of Problem 4.2 is a 100-mm-diameter sphere. After complete combustion, a valve is suddenly opened to create a 10-mm-diameter sharp-edged orifice. The flow coefficient for the orifice is $C_f = 0.61$. Assuming the specific heats to be constant, calculate the pressure and temperature as a function of time while the vessel discharges its contents to the environment. What is the maximum cooling rate of the gas remaining in the vessel?

**4.4.** Set up an engine simulation program based on the analysis of Section 4.1. For stoichiometric combustion of octane (assume complete combustion), calculate the pressure and the mean burned gas temperature as a function of crank angle for a compression ratio of $R_c = 8$, intake air pressure and temperature of 1 atm and 333 K (due to heating of the air in the intake manifold), and $\theta_0 = -15°$ and $\Delta\theta_c = 40°$. Neglect heat transfer to the cylinder walls. Assume $l/c = 4$. Also calculate the mean effective pressure and the specific fuel consumption.

**4.5.** A throttle reduces the pressure in the intake manifold to 0.6 atm. Assume that the flow to the throat of the valve is adiabatic and isentropic and that the pressure at the throat of the valve equals the pressure in the intake manifold. How much is the mass intake of air reduced below that for Problem 4.4? Using the cycle simulation code, repeat the calculations of Problem 4.4.

**4.6.** Assuming the burned gases in Problem 4.4 are uniform in temperature, use the Zeldovich mechanism to calculate $NO_x$ emissions from the engine. Assume an engine speed of 3000 rpm.

**4.7.** Use the engine simulation of Problem 4.4 to explore the effects of spark retard on engine performance (IMEP and ISFC) and on $NO_x$ emissions.

**4.8.** A model that is commonly used to describe the recirculation zone of a combustor is the perfectly stirred reactor. The perfectly stirred reactor is a control volume into which reactants flow and from which products are continuously discharged. The reaction

$$A + B \longrightarrow \text{products}$$

proceeds with rate

$$R = -k_{+1}[A][B]$$

Assuming that the entering concentrations are $[A]_0$ and $[B]_0$, that the volume of the control volume is $V$, and that the volumetric flow rates into and out from the reactor are equal, $Q_{in} = Q_{out} = Q$, derive expressions for the steady-state concentrations of $A$ and $B$.

**4.9.** Consider a gas turbine engine in which the incoming air is compressed adiabatically to a pressure of 6 atm. The airflow through the combustor at the design load is 3.5 kg s$^{-1}$. The fuel is aviation kerosene with composition $CH_{1.76}$ and a heating value of 46 MJ kg$^{-1}$. The combustor can is 150 mm in diameter and 300 mm long. The primary combustion zone can be considered to be a perfectly stirred reactor (see Problem 4.8) operated at $\phi = 0.7$. Downstream of the primary reaction zone, dilution air is added to reduce the gas temperature by lowering the overall equivalence ratio to 0.2. Assume that the dilution air is added at a constant amount of dilution per unit of combustion can length. Further assume that the composition at any axial position in the combustor is uniform. The entire combustor may be assumed to be adiabatic.

**(a)** What is the residence time in the primary combustion zone?

**(b)** Use the Zeldovich mechanism to calculate the NO concentration in the gases leaving the primary combustion zone.

**(c)** Calculate and plot the temperature and the flow time as functions of axial position in the dilution region.

**(d)** What is the NO concentration at the end of the combustor? What is the $NO_x$ emission index, g/kg fuel burned?

## REFERENCES

ADAMCZYK, A. A., and LAVOIE, G. A. "Laminar Head-On Flame Quenching—A Theoretical Study," SAE Paper No. 780969, Society of Automotive Engineers, Warrendale, PA (1978).

ADAMCZYK, A. A., KAISER, E. W., and LAVOIE, G. A. "A Combustion Bomb Study of the Hydrocarbon Emissions from Engine Crevices," *Combust. Sci. Technol.*, *33*, 261–277 (1983).

AOYAMA, K., and MANDAI, S. "Development of a Dry Low $NO_x$ Combustor for a 120-MW Gas Turbine," *J. Eng. Gas Turbines Power*, *106*, 795–800 (1984).

BLUMBERG, P., and KUMMER, J. T. "Prediction of NO Formation in Spark-Ignition Engines—An Analysis of Methods of Control," *Combust. Sci. Technol.*, *4*, 73–95 (1971).

BORGNAKKE, C., ARPACI, V. S., and TABACZYNSKI, R. J. "A Model for the Instantaneous Heat Transfer and Turbulence in a Spark-Ignited Engine," SAE Paper No. 800247, Society of Automotive Engineers, Warrendale, PA (1980).

BY, A., KEMPINSKI, B., and RIFE, J. M. "Knock in Spark Ignition Engines," SAE Paper No. 810147, Society of Automotive Engineers, Warrendale, PA (1981).

CARRIER, G. F., FENDELL, F., and FELDMAN, P. "Cyclic Adsorption/Desorption of Gas in a Liquid Wall Film," *Combust. Sci. Technol.*, *25*, 9–19 (1981).

DANIEL, W. A., and WENTWORTH, J. T. "Exhaust Gas Hydrocarbons—Genesis and Exodus," SAE Paper No. 486B, Society of Automotive Engineers, Warrendale, PA (1962).

HAMBURG, D. R., COOK, J. A., KAISER, W. J., and LOGOTHETIS, E. M. "An Engine Dynamometer Study of the A/F Compatibility between a Three-Way Catalyst and an Exhaust Gas Oxygen Sensor," SAE Paper No. 830986, Society of Automotive Engineers, Warrendale, PA (1983).

HASKELL, W. W., and LEGATE, C. E. "Exhaust Hydrocarbon Emissions from Gasoline Engines—Surface Phenomena," SAE Paper No. 720255, Society of Automotive Engineers, Warrendale, PA (1972).

HENEIN, N. A. "Analysis of Pollutant Formation and Control and Fuel Economy in Diesel Engines," *Prog. Energy Combust. Sci.*, *1*, 165–207 (1976).

HEYWOOD, J. B. "Pollutant Formation and Control in Spark-Ignition Engines," in *Fifteenth Symposium (International) on Combustion*, The Combustion Institute, Pittsburgh, PA 1191–1211 (1975).

HEYWOOD, J. B. "Pollutant Formation and Control in Spark-Ignition Engines," *Prog. Energy Combust. Sci.*, *1*, 135–164 (1976).

HEYWOOD, J. B. "Automotive Engines and Fuels: A Review of Future Options," *Prog. Energy Combust. Sci.*, *7*, 155–184 (1981).

HEYWOOD, J. B., HIGGINS, J. M., WATTS, P. A., and TABACZYNSKI, R. J. "Development and Use of a Cycle Simulation to Predict SI Engine Efficiency and Nitric Oxide Emissions," SAE Paper No. 790291, Society of Automotive Engineers, Warrendale, PA (1979).

HOULT, D. P., and WONG, V. W. "The Generation of Turbulence in an Internal Combustion Engine," in *Combustion Modeling in Reciprocating Engines*, J. N. Mattavi and C. A. Amann, Eds., Plenum Press, New York, 131–160 (1980).

JOHNSON, G. M., and SMITH, M. Y. "Emissions of Nitrogen Dioxide from a Large Gas-Turbine Power Station," *Combust. Sci. Technol.*, *19*, 67–70 (1978).

JONES, R. E. "Gas Turbine Emissions—Problems, Progress, and Future," *Prog. Energy Combust. Sci.*, *4*, 73–113 (1978).

KAISER, E. W., LORUSSO, J. A., LAVOIE, G. A., and ADAMCZYK, A. A. "The Effect of Oil Layers on the Hydrocarbon Emissions from Spark-Ignited Engines," *Combust. Sci. Technol.*, *28*, 69–73 (1982).

KOMIYAMA, K., and HEYWOOD, J. B. "Predicting $NO_x$ and Effects of Exhaust Gas Recirculation in Spark-Ignition Engines," SAE Paper No. 730475, Society of Automotive Engineers, Warrendale, PA (1973).

KORT, R. T., MANSOURI, S. H., HEYWOOD, J. B., and EKCHIAN, J. A. "Divided-Chamber Diesel Engine: Part II. Experimental Validation of a Predictive Cycle-Simulation and Heat Release Analysis," SAE Paper No. 820274, Society of Automotive Engineers, Warrendale, PA (1982).

KUMMER, J. T. "Catalysts for Automobile Emission Control," *Prog. Energy Combust. Sci.*, *6*, 177–199 (1980).

LAVOIE, G. A., HEYWOOD, J. B., and KECK, J. C. "Experimental and Theoretical Study of Nitric Oxide Formation in Internal Combustion Engines," *Combust. Sci. Technol.*, *1*, 313–326 (1970).

LAVOIE, G. A., LORUSSO, J. A., and ADAMCZYK, A. A. "Hydrocarbon Emissions Modeling for Spark Ignition Engines," in *Combustion Modeling in Reciprocating Engines*, J. N. Mattavi and C. A. Amann, Eds., Plenum Press, New York, 409–445 (1980).

LoRusso, J. A., Kaiser, E. W., and Lavoie, G. A. "In-Cylinder Measurements of Wall Layer Hydrocarbons in a Spark Ignited Engine," *Combust. Sci. Technol.*, *33*, 75–112 (1983).

Mansouri, S. H., Heywood, J. B., and Radhakrishnan, K. "Divided-Chamber Diesel Engine: Part I. A Cycle-Simulation which Predicts Performance and Emissions," SAE Paper No. 820273, Society of Automotive Engineers, Warrendale, PA (1982a).

Mansouri, S. H., Heywood, J. B., and Ekchian, J. A. "Studies of $NO_x$ and Soot Emissions from an IDI Diesel Using an Engine Cycle Simulation," IMechE Paper No. C120/82 (1982b).

Morr, A. R. "A Model for Carbon Monoxide Emissions from an Industrial Gas Turbine Engine," Ph.D. thesis in mechanical engineering, MIT (1973).

Perry, R. A., and Siebers, D. L. "Rapid Reduction of Nitrogen Oxides in Exhaust Gas Streams," *Nature*, *324*, 657–658 (1986).

Sawyer, R. F., Dryer, F. L., Johnson, J. R., Kliegl, J. R., Kotin, P., Lux, W. L., Myers, P. S., and Wei, J. *Impacts of Diesel-Powered Light-Duty Vehicles: Diesel Technology*, National Academy Press, Washington, DC (1982).

Siegla, D. C., and Amann, C. A. "Exploratory Study of the Low-Heat-Rejection Diesel for Passenger-Car Application," SAE Paper No. 840435, Society of Automotive Engineers, Warrendale, PA (1984).

Tabaczynski, R. J., Heywood, J. B., and Keck, J. C. "Time-Resolved Measurements of Hydrocarbon Mass Flowrate in the Exhaust of a Spark-Ignited Engine," SAE Paper No. 720112, Society of Automotive Engineers, Warrendale, PA (1972).

Taylor, C. F. *The Internal Combustion Engine in Theory and Practice*, MIT Press, Cambridge, MA (1966).

Wade, W. R. "Light-Duty Diesel $NO_x$-HC-Particulate Trade-Off Studies," in *The Measurement and Control of Diesel Particulate Emissions: Part 2*, J. H. Johnson, F. Black, and J. S. MacDonald, Eds., Society of Automotive Engineers, Warrendale, PA, 203–222 (1982).

Wentworth, J. T. "Piston and Piston Ring Variables Affect Exhaust Hydrocarbon Emissions," SAE Paper No. 680109, Society of Automotive Engineers, Warrendale, PA (1968).

Wentworth, J. T. "The Piston Crevice Volume Effect on Exhaust Hydrocarbon Emissions," *Combust. Sci. Technol.*, *4*, 97–100 (1971).

# 5

# Aerosols

An aerosol is a suspension of small particles in air or another gas. From the point of view of air pollution engineering, aerosols are important because unwanted particles are produced in combustion and other industrial processes. In addition, primary gaseous emissions may react in the atmosphere to produce secondary species that nucleate to form particles or condense on preexisting particles. An important class of industrial gas-cleaning processes is designed to remove particles from exhaust gas streams, and such processes are the subject of Chapter 7. The present chapter is devoted to fundamental aspects of aerosols that are needed to understand how particles are formed in combustion, how they change once released to the atmosphere, and how to design particulate gas-cleaning processes. In Chapter 6 we consider the mechanisms of particle formation in combustion processes.

We begin this chapter with an analysis of the dynamics of a single particle moving in a fluid. Several of the particulate gas-cleaning processes that we will study in Chapter 7 are based on the principle of imposing an external force on particles in a flowing gas stream so that they will migrate to a collector surface. Thus we will need to calculate how a particle in a flowing gas stream responds to an external force, such as gravity or an electrical force (if the particle is charged). We then proceed to an analysis of the Brownian diffusion of particles in a gas, a physical mechanism that is important in removing particles from a gas stream when an obstacle is placed in the stream to which particles may diffuse and adhere. The final portion of this chapter is devoted to the description of a population of particles through the size distribution function.

The subject of aerosols is indeed a broad one, arising in areas ranging from air pollution, where the particles are unwanted, to industrial processes, where the particles are produced intentionally. We endeavor to cover here those aspects of aerosol behavior critical to air pollution engineering.

## 5.1 THE DRAG ON A SINGLE PARTICLE: STOKES' LAW

A good place to start to study the dynamical behavior of aerosol particles in a fluid is to consider the drag force exerted on a particle as it moves in a fluid. This drag force will always be present as long as the particle is not moving in a vacuum. To calculate the drag force exerted by a fluid on a particle moving in that fluid one must solve the equations of fluid motion to determine the velocity and pressure fields around the particle. The particle will be considered to be spherical.

The velocity and pressure in an incompressible Newtonian fluid are governed by the equation of continuity,

$$\frac{\partial u_x}{\partial x} + \frac{\partial u_y}{\partial y} + \frac{\partial u_z}{\partial z} = 0 \tag{5.1}$$

and the Navier-Stokes equations (Bird et al., 1960), the $x$-component of which is

$$\rho\left(\frac{\partial u_x}{\partial t} + u_x \frac{\partial u_x}{\partial x} + u_y \frac{\partial u_x}{\partial y} + u_z \frac{\partial u_x}{\partial z}\right)$$
$$= -\frac{\partial p}{\partial x} + \mu\left(\frac{\partial^2 u_x}{\partial x^2} + \frac{\partial^2 u_x}{\partial y^2} + \frac{\partial^2 u_x}{\partial z^2}\right) + \rho g_x \tag{5.2}$$

where $g_x$ is the component of the gravity force in the $x$-direction.

By introducing a characteristic velocity $u_0$ and length $L$, the continuity and Navier-Stokes equations can be made dimensionless. The Reynolds number $\mathrm{Re} = u_0 L \rho / \mu$, the ratio of inertial to viscous forces in the flow, is the characteristic dimensionless group that emerges from the dimensionless equations. For flow around a submerged body, $L$ can be chosen as a characteristic dimension of the body, say its diameter, and $u_0$ can be chosen as the speed of the undisturbed fluid upstream of the body. For flow past a sphere the Reynolds number can be defined based on the sphere's radius or its diameter. In the first case, the Reynolds number is conventionally given the symbol $\mathrm{R} = u_0 R_p / \nu$, whereas in the second it is called $\mathrm{Re} = u_0 D_p / \nu$. Clearly, $\mathrm{Re} = 2\mathrm{R}$. We will use the Reynolds number Re defined on the basis of the sphere diameter. When viscous forces dominate inertial forces, $\mathrm{Re} << 1$, the type of flow that results is called a *creeping flow* or low-Reynolds number flow.

The solution of the equations for the velocity and pressure distribution around a sphere in creeping flow was first obtained by Stokes. The assumptions invoked to obtain that solution are: (1) an infinite medium, (2) a rigid sphere, and (3) no slip at the surface of the sphere. For the solution, we refer to the reader to Bird et al. (1960, p. 132).

The object is to calculate the net force exerted by the fluid on the sphere in the direction of flow. The force consists of two contributions. At each point on the surface of the sphere there is a pressure on the solid acting perpendicularly to the surface. This is the *normal force*. At each point there is also a *tangential force* exerted by the fluid due to the shear stress caused by the velocity gradients in the vicinity of the surface.

To obtain the normal force on the sphere, one integrates the component of the pressure acting perpendicularly to the surface. The normal and tangential forces can be shown to be

$$F_n = 2\pi\mu R_p u_\infty$$
$$F_t = 4\pi\mu R_p u_\infty \tag{5.3}$$

where $u_\infty$ is the undisturbed fluid velocity far upstream of the sphere. Note that the case of a stationary sphere in a fluid with velocity $u_\infty$ is entirely equivalent to that of a sphere moving at velocity $u_\infty$ through a stagnant fluid.

The total drag force exerted by the fluid on the sphere is

$$F_{\text{drag}} = F_n + F_t = 6\pi\mu R_p u_\infty \tag{5.4}$$

which is called *Stokes' law*. If we include gravity, the total force on the sphere is the sum of the drag force and the buoyant force. When the direction of flow and the direction of gravity coincide, the buoyant force to be added to the drag force is the force equal to the weight of the fluid displaced by the sphere,

$$F_{\text{buoyant}} = \tfrac{4}{3}\pi R_p^3 \rho g \tag{5.5}$$

At Re = 1, the drag force predicted by Stokes' law is 13% low due to the neglect of the inertial terms in the equation of motion. The correction to account for higher Reynolds numbers is (Re $\lesssim$ 2)

$$F_{\text{drag}} = 6\pi\mu R_p u_\infty \left(1 + \tfrac{3}{16}\,\text{Re} + \tfrac{9}{160}\,\text{Re}^2 \ln 2\text{Re}\right) \tag{5.6}$$

To account for the drag force over the entire range of Reynolds number, we can express the drag force in terms of an empirical *drag coefficient* $C_D$ as

$$F_{\text{drag}} = C_D A_p \rho \frac{u_\infty^2}{2} \tag{5.7}$$

where $A_p$ is the projected area of the body normal to the flow. Thus, for a spherical particle of diameter $D_p$,

$$F_{\text{drag}} = \frac{\pi}{8} C_D \rho D_p^2 u_\infty^2 \tag{5.8}$$

where the following correlations are available for the drag coefficient as a function of Reynolds number:

$$C_D = \begin{cases} \dfrac{24}{\text{Re}} & \text{Re} < 0.1 \text{ (Stokes' law)} \\ \dfrac{24}{\text{Re}}\left(1 + \frac{3}{16}\text{Re} + \frac{9}{160}\text{Re}^2 \ln 2\text{Re}\right) & 0.1 < \text{Re} < 2 \\ \dfrac{24}{\text{Re}}\left(1 + 0.15\,\text{Re}^{0.687}\right) & 2 < \text{Re} < 500 \\ 0.44 & 500 < \text{Re} < 2 \times 10^5 \end{cases} \tag{5.9}$$

The Reynolds numbers of spherical particles falling at their terminal velocities in air at 298 K and 1 atm are:

| $D_p$ ($\mu$m) | Re |
|---|---|
| 20 | 0.02 |
| 60 | 0.4 |
| 100 | 2 |
| 300 | 20 |

Thus, for particles smaller than about 20 $\mu$m in diameter, Stokes' law is an accurate formula for the drag force exerted by the air.

## 5.2 NONCONTINUUM EFFECTS

Aerosol particles are small. The particle diameter is often comparable to the distances that gas molecules travel between collisions with other gas molecules. Consequently, the basic continuum transport equations must be modified to account for the nature of the fluid/particle interaction.

### 5.2.1 The Knudsen Number

The key dimensionless group that defines the nature of the suspending fluid relative to the particle is the *Knudsen number* $\text{Kn} = 2\lambda/D_p$, where $\lambda$ is the mean free path of the fluid. Thus the Knudsen number is the ratio of two length scales, one characterizing the "graininess" of the fluid with respect to the transport of momentum, mass, or heat, and the other characterizing the particle, its radius.

Before we discuss the physical interpretation of the Knudsen number, we must consider the definition of the mean free path. The mean free path of a gas molecule can be defined as the average distance traveled between collisions with other gas molecules. If the gas consists entirely of molecules of a single type, call them $B$, the mean free path can be denoted as $\lambda_{BB}$. Even though air consists of molecules of $N_2$ and $O_2$, it is customary to talk about the mean free path of air as if air were a single chemical species.

We will denote air by $B$ in what follows. If the gas consists of molecules of two types, $A$ and $B$, several mean free paths can be defined.

$\lambda_{AB}$ is the average distance traveled by a molecule of $A$ before it encounters a molecule of $B$, with a similar interpretation for $\lambda_{BA}$. Although the idea of a mean free path can be extended to aerosol particles, it is less clear how to define the mean free path of an aerosol particle than that of a gas molecule. Aerosol particles collide only very infrequently with other particles, and, when they do, it is usually assumed that the two particles adhere. Thus aerosol-aerosol collisions are not an appropriate basis on which to define a particle mean free path. On the other hand, because of their large size and mass relative to that of gas molecules, an aerosol particle experiences a large number of collisions per unit time with the surrounding gas molecules and is not influenced significantly by any single collision. Consequently, the motion of an aerosol particle can be viewed as more or less continuous in nature, a view, however, that does not provide a convenient length to be identified as a mean free path. Fortunately, it will not be necessary for us to compute a particle mean free path in order to calculate the transport properties of aerosols.

If we are interested in characterizing the nature of the suspending gas relative to a particle, the mean free path that appears in the definition of the Knudsen number is $\lambda_g$. If the particle radius greatly exceeds $\lambda_g$, the gas appears to the particle as a continuum. Thus, when Kn $\ll 1$, we say that the particle is in the *continuum regime*, and the usual equations of continuum mechanics apply. When the mean free path of gas molecules substantially exceeds the particle radius, the particle exists in a more or less rarified medium, it appears to the surrounding gas molecules like just another gas molecule, and its transport properties can be obtained from kinetic theory of gases. This Kn $\gg 1$ limit is called the ***free molecule*** or ***kinetic regime***. The particle size range intermediate between these two extremes is called the ***transition regime***. These three regimes are depicted schematically in Figure 5.1.

The mean free path of a gas can be calculated from kinetic theory of gases from

$$\lambda_g = \frac{\mu}{0.499p(8M/\pi RT)^{1/2}} \tag{5.10}$$

where $\lambda_g$ is measured in meters and $p$ in pascal, and where $M$ is the molecular weight and $\mu$ is the gas viscosity. The mean free path of air molecules at $T = 298$ K and $p = 1$ atm is found from (5.10) to be ($\mu = 1.8 \times 10^{-5}$ kg m$^{-1}$ s$^{-1}$)

$$\lambda_g = 6.51 \times 10^{-8} \text{ m} = 0.0651\ \mu\text{m}$$

Thus, in air at standard conditions, particles with radii exceeding 0.2 $\mu$m or so can be considered, from the point of view of transport properties, to be in the continuum regime. Similarly, the transport properties of those having radii about 0.01 $\mu$m or smaller can be computed from free molecule kinetic theory. The transition regime would span the range of particle radii approximately from 0.01 to 0.2 $\mu$m.

If we are interested in the diffusion of a vapor molecule $A$ at low concentration toward a particle, both of which are contained in a background gas $B$ (e.g., air), the

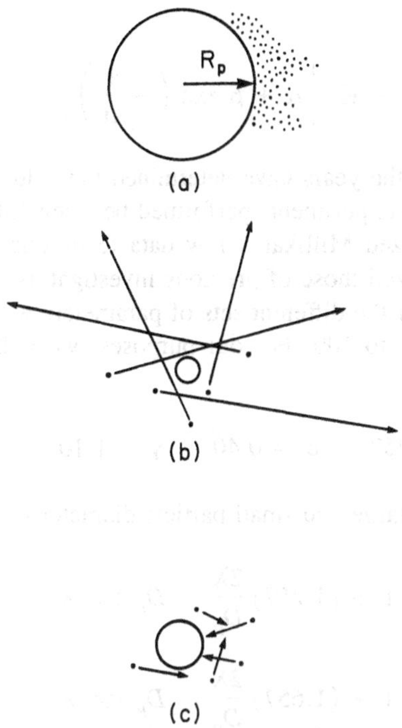


**Figure 5.1** Knudsen number ($Kn = \lambda/R_p$) and the three regimes of fluid–particle behavior: (a) continuum regime ($Kn \to 0$); (b) free molecule (kinetic) regime ($Kn \to \infty$); (c), transition regime [$Kn = O(1)$].

description of that diffusion process depends on the value of the Knudsen number defined on the basis of $\lambda_{AB}$, $Kn = 2\lambda_{AB}/D_p$. The same definitions of regimes of behavior, continuum, free molecule, and transition, can be made in that case as when determining how the particle interacts with air molecules.

The mean free path of molecules of $A$, $\lambda_{AB}$, in a binary mixture of $A$ and $B$ can be related to the binary diffusivity $D_{AB}$ and the mean molecular speed of species $A$, $\bar{c}_A = (8RT/\pi M_A)^{1/2}$, from the first-order Chapman-Enskog hard sphere solution of the Boltzmann equation by

$$\frac{D_{AB}}{\lambda_{AB}\bar{c}_A} = \frac{3\pi}{32}\left(1 + \frac{M_A}{M_B}\right) \tag{5.11}$$

### 5.2.2 Slip Correction Factor

Stokes' law is derived from the equations of continuum fluid mechanics. When the particle diameter $D_p$ approaches the same order as the mean free path $\lambda$ of the suspending fluid (e.g., air), the resisting force offered by the fluid is smaller than that predicted by Stokes' law. To account for noncontinuum effects that become important as $D_p$ becomes smaller and smaller, a slip correction factor, $C_c$, is introduced into Stokes' law written now in terms of particle diameter $D_p$,

$$F_{\text{drag}} = \frac{3\pi\mu u_\infty D_p}{C_c} \tag{5.12}$$

where $C_c$ has the general form

$$C_c = 1 + \text{Kn}\left[\alpha + \beta \exp\left(-\frac{\gamma}{\text{Kn}}\right)\right] \tag{5.13}$$

A number of investigators over the years have determined the values of the parameters $\alpha$, $\beta$, and $\gamma$ based on Millikan's experiments performed between 1909 and 1923. Allen and Raabe (1982) have reanalyzed Millikan's raw data to produce an updated set of parameters, as well as summarized those of previous investigators. The maximum difference between values of $C_c$ for the different sets of parameters was found to be about 2% over the Kn range of 0.001 to 100. For our purposes, we will use the following parameter values:

$$\alpha = 1.257 \qquad \beta = 0.40 \qquad \gamma = 1.10$$

The limiting behavior of $C_c$ for large and small particle diameter is

$$C_c = \begin{cases} 1 + (1.257)\dfrac{2\lambda}{D_p} & D_p \gg \lambda \\[2ex] 1 + (1.657)\dfrac{2\lambda}{D_p} & D_p \ll \lambda \end{cases}$$

The value of the slip correction factor for air at 1 atm and 298 K as a function of particle diameter is

| $D_p$ ($\mu$m) | $C_c$ |
|---|---|
| 0.01 | 22.7 |
| 0.05 | 5.06 |
| 0.1 | 2.91 |
| 0.5 | 1.337 |
| 1.0 | 1.168 |
| 5.0 | 1.034 |
| 10.0 | 1.017 |

At $D_p = 0.01$ $\mu$m, the limiting formula predicts $C_c = 22.54$; and at the large particle extreme, for $D_p = 10$ $\mu$m, the limiting formula predicts $C_c = 1.016$. The slip correction factor is generally neglected for particles exceeding 10 $\mu$m in diameter.

It is sometimes convenient to express the drag force on a particle in terms of a *friction coefficient* $f$ by $F_{\text{drag}} = fu_\infty$. In the Stokes' law regime, $f = 3\pi\mu D_p$, and in general

$$f = \frac{3\pi\mu D_p}{C_c} \tag{5.14}$$

## 5.3 MOTION OF AN AEROSOL PARTICLE IN AN EXTERNAL FORCE FIELD

Up to this point we have considered the drag force on a particle moving at a steady velocity $u_\infty$ through a quiescent fluid. (Recall that this case is equivalent to the flow of a fluid at velocity $u_\infty$ past a stationary sphere.) The motion of a particle arises in the first place due to the action of some external force on the particle, such as gravity or electrical forces. The drag force arises as soon as there is a difference between the velocity of the particle and that of the fluid. For a variety of applications to the design of particle collection devices it is necessary to be able to describe the motion of a particle in either a quiescent or flowing fluid subject to external forces on the particle. The basis of this description is an equation of motion for a particle. To derive an equation of motion for a particle, let us begin with a force balance on a moving particle of mass $m_p$ which we write in vector form as

$$m_p \frac{d\boldsymbol{v}}{dt} = \sum_i \boldsymbol{F}_i \tag{5.15}$$

where $\boldsymbol{v}$ is the velocity of the particle and $\boldsymbol{F}_i$ is the force acting on the particle by mechanism $i$. As long as the particle is not moving in a vacuum, the drag force will always be present, so let us remove the drag force from the summation of forces and place it permanently in (5.15)

$$m_p \frac{d\boldsymbol{v}}{dt} = \frac{3\pi\mu D_p}{C_c}(\boldsymbol{u} - \boldsymbol{v}) + \sum_i \boldsymbol{F}_{ei} \tag{5.16}$$

where the Stokes drag force on a particle moving at velocity $\boldsymbol{v}$ in a fluid having velocity $\boldsymbol{u}$ is the first term on the right-hand side of (5.16) and $\boldsymbol{F}_{ei}$ denotes external force $i$ (those forces arising from external potential fields, such as gravity force and electrical force). The fluid and particle velocities are depicted schematically in Figure 5.2. In writing (5.16) we have assumed that even though the particle motion is unsteady, its acceleration is slow enough so that at any instant, Stokes' law is still valid. Moreover, implicit in the use of the Stokes' law expression is the assumption that the particle is small enough so that the creeping flow approximation holds. Actually, (5.16) is only an approximate

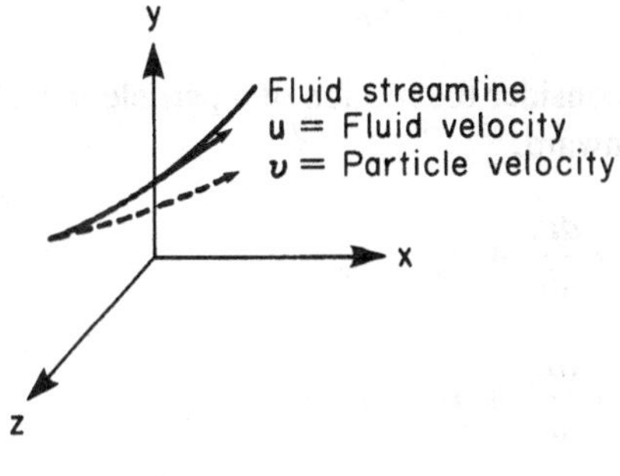


**Figure 5.2** Fluid and particle velocities.

form of the equation of motion of a particle in a fluid. The full equation is (Hinze, 1959)

$$m_p \frac{dv}{dt} = \frac{3\pi\mu D_p}{C_c}(u - v) + v_p\rho \frac{du}{dt} + \frac{v_p}{2}\rho\left(\frac{du}{dt} - \frac{dv}{dt}\right) + \frac{3D_p^2}{2}(\pi\rho\mu)^{1/2}\int_0^t \frac{(du/dt') - (dv/dt')}{(t - t')^{1/2}}\,dt' + \sum F_{ei} \tag{5.17}$$

where $v_p$ is the volume of the particle. The second term on the right-hand side is due to the pressure gradient in the fluid surrounding the particle, caused by acceleration of the gas by the particle. The third term on the right-hand side is the force required to accelerate the apparent mass of the particle relative to the fluid. The fourth term on the right-hand side, called the Basset history integral, accounts for the force arising due to the deviation of the fluid velocity from steady state. In most situations of interest for aerosol particles in air, the second, third, and fourth terms on the right-hand side of (5.17) may be neglected.

Consider for the moment the motion of a particle in a fluid in the presence only of gravity,

$$m_p \frac{dv}{dt} = \frac{3\pi\mu D_p}{C_c}(u - v) + m_p g \tag{5.18}$$

If we divide this equation by $3\pi\mu D_p/C_c$, we obtain

$$\tau \frac{dv}{dt} + v = u + \tau g \tag{5.19}$$

where $\tau = m_p C_c / 3\pi\mu D_p$ is evidently a characteristic time associated with the motion of the particle. For a spherical particle of density $\rho_p$ in a fluid of density $\rho$, $m_p = (\pi/6)D_p^3(\rho_p - \rho)$, where the factor $(\rho_p - \rho)$ is needed because $(\pi/6)D_p^3\rho g$ is the buoyancy force on the particle [recall (5.5)]. However, since for typical aerosol particles in air $\rho_p >> \rho$, $m_p \simeq (\pi/6)D_p^3\rho_p$, and

$$\tau = \frac{D_p^2 \rho_p C_c}{18\mu} \tag{5.20}$$

As an example of (5.19), consider the motion of a particle in the $x$, $z$-plane, where the $z$-axis is taken positive downward,

$$\tau \frac{dv_x}{dt} + v_x = u_x \tag{5.21}$$

$$\tau \frac{dv_z}{dt} + v_z = u_z + \tau g \tag{5.22}$$

Since $v_x = dx/dt$ and $v_z = dz/dt$, these two equations can be written in terms of particle position, $x(t)$, $z(t)$,

$$\tau \frac{d^2x}{dt^2} + \frac{dx}{dt} = u_x \tag{5.23}$$

$$\tau \frac{d^2z}{dt^2} + \frac{dz}{dt} = u_z + \tau g \tag{5.24}$$

If at time zero the initial velocity components of the particle in the two directions are $v_x(0) = v_{x0}$ and $v_z(0) = v_{z0}$, and if the fluid velocity components are constant, we can integrate (5.21) and (5.22) to obtain

$$v_x(t) = u_x + (v_{x0} - u_x)e^{-t/\tau} \tag{5.25}$$

$$v_z(t) = (u_z + \tau g) + (v_{z0} - u_z - \tau g)e^{-t/\tau} \tag{5.26}$$

If the particle starts, for example, at the origin, its position at any time is found by integrating (5.25) and (5.26) once more or by solving (5.23) and (5.24) directly with initial conditions $x(0) = z(0) = 0$ and $(dx/dt)_0 = v_{x0}$, $(dz/dt)_0 = v_{z0}$, to obtain

$$x(t) = u_x t + \tau(v_{x0} - u_x)\,(1 - e^{-t/\tau}) \tag{5.27}$$

$$z(t) = (u_z + \tau g)t + (v_{z0} - u_z - \tau g)\,(1 - e^{-t/\tau}) \tag{5.28}$$

### 5.3.1 Terminal Settling Velocity

If the particle is at rest at $t = 0$ and the air is still, the only velocity component is in the $z$-direction

$$v_z(t) = \tau g(1 - e^{-t/\tau}) \tag{5.29}$$

For $t >> \tau$, the particle attains a constant velocity, called its terminal settling velocity $v_t$,

$$v_t = \tau g = \frac{D_p^2 \rho_p g C_c}{18\mu} \tag{5.30}$$

We see that $\tau$ is the characteristic time for the particle to approach steady motion. Similarly, if a particle enters a moving airstream, it approaches the velocity of the stream with a characteristic time $\tau$. Values of $\tau$ for unit density spheres at 298 K in air are:

| $D_p$ ($\mu$m) | $\tau$ (s) |
|---|---|
| 0.05 | $4 \times 10^{-8}$ |
| 0.1 | $9 \times 10^{-8}$ |
| 0.5 | $1 \times 10^{-6}$ |
| 1.0 | $3.6 \times 10^{-6}$ |
| 5.0 | $8 \times 10^{-5}$ |
| 10.0 | $3.1 \times 10^{-4}$ |
| 50.0 | $7.7 \times 10^{-3}$ |

Thus the characteristic time for most particles of interest to achieve steady motion in air is extremely short.

Settling velocities of unit density spheres in air at 1 atm and 298 K as computed from (5.30) are:

| $D_p$ (μm) | $v_t$ (cm s$^{-1}$) |
|---|---|
| 0.1 | $8.8 \times 10^{-5}$ |
| 0.5 | $1.0 \times 10^{-3}$ |
| 1.0 | $3.5 \times 10^{-3}$ |
| 5.0 | $7.8 \times 10^{-2}$ |
| 10.0 | 0.31 |

For particles larger than about 20 μm diameter settling at their terminal velocity, the Reynolds number is too high for Stokes' law analysis to be valid. Shortly we consider how to calculate the settling velocity of such larger particles.

The velocity and position of a particle that has an initial velocity $v_{x0}$ in still air are $v_x(t) = v_{x0} \exp(-t/\tau)$ and $x(t) = \tau v_{x0}[1 - \exp(-t/\tau)]$. As $t \to \infty$, the distance a particle travels is $x(t) = \tau v_{x0}$ before it decelerates to a stop. This distance is called the *stop distance*. For a 1-μm-diameter particle, with an initial speed of 10 m s$^{-1}$, for example, the stop distance is $3.6 \times 10^{-3}$ cm.

Note that the terminal velocity $v_t$ is that velocity for which the gravity force is just balanced by the drag force,

$$\frac{3\pi\mu D_p v_t}{C_c} = \left(\frac{\pi}{6}\right) D_p^3 \rho_p g \tag{5.31}$$

which is, of course, consistent with the steady-state form of (5.22). Because of the characteristically small value of $\tau$ relative to the time scales over which other effects are changing, the velocity of a particle in a fluid very quickly adjusts to a steady state at which the drag force is balanced by the sum of the other forces acting on the particle.

**Example 5.1** ***Motion of a Particle in an Idealized Flow***

Let us consider an idealized flow, shown in Figure 5.3, in which an airflow makes an abrupt 90° turn in a corner with no change in velocity (Crawford, 1976). We wish to determine

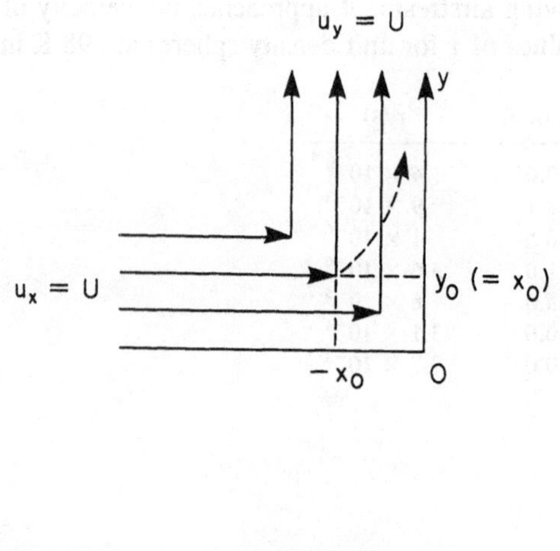


**Figure 5.3** Motion of a particle in a flow making a 90° turn with no change in velocity.

the trajectory of a particle on the streamline that makes its turn at the point $(x_0, y_0) = (-x_0, x_0)$. The trajectory of the particle is governed by

$$\tau \frac{d^2x}{dt^2} + \frac{dx}{dt} = 0 \tag{5.32}$$

$$\tau \frac{d^2y}{dt^2} + \frac{dy}{dt} = U \tag{5.33}$$

subject to

$$x(0) = -x_0 \quad y(0) = x_0 \tag{5.34}$$

$$\left(\frac{dx}{dt}\right)_{t=0} = U \quad \left(\frac{dy}{dt}\right)_{t=0} = 0 \tag{5.35}$$

Solving (5.32) and (5.33) subject to (5.34) and (5.35) gives the particle coordinates as a function of time,

$$x(t) = -x_0 + U\tau(1 - e^{-t/\tau}) \tag{5.36}$$

$$y(t) = x_0 - U\tau(1 - e^{-t/\tau}) + Ut \tag{5.37}$$

To obtain an expression for the particle trajectory in the form $y = f(x)$, we solve (5.36) for $t$ in terms of $x$ and then substitute that result in (5.37). The result is

$$y = -x - U\tau \ln\left[1 - U\tau(x + x_0)\right] \tag{5.38}$$

We see that as $t \to \infty$, $x(t) = -x_0 + U\tau$ and $y(t) = x_0 - U\tau + Ut$. Thus the particle eventually ends up at a distance $-x_0 + U\tau$ from the $y$-axis.

The flow depicted in Figure 5.3 is the most idealized one that represents the stagnation flow of a fluid toward a flat plate. If we imagine that the $x = 0$ line is a flat plate and that the $y = 0$ line is a line of symmetry, a very idealized version of the flow field that results as the fluid approaches the plate at velocity $U$ is that the fluid makes an abrupt 90° turn at points along the $y = -x$ line as shown by the streamlines drawn in Figure 5.3. If the purpose of the flat plate is to provide a surface for collecting particles in the flow, then one is interested in knowing if particles on a particular streamline will ultimately intercept the $y$-axis. The streamlines are defined by their distance from the $y = 0$ line of symmetry, that is, by the value of $x_0$ at which the streamline makes its 90° turn. In order for a particle to intercept the $y$-axis as $t \to \infty$, it is necessary that

$$x_0 = U\tau \tag{5.39}$$

Thus, for a given particle diameter $D_p$, fluid velocity $U$, and particle density $\rho_p$, from (5.39) one can determine the value of $x_0$ for which these particles just intercept the $y$-axis. For example, for $D_p = 20$ μm, $U = 20$ m s$^{-1}$, and $\rho_p = 2$ g cm$^{-3}$, we find that $x_0 = 4.83$ cm. Therefore, all 20-μm-diameter particles on streamlines from the line of symmetry ($y = 0$) out to that at $y = x_0 = 4.83$ cm will intercept the surface; all those on streamlines beyond that at $y = 4.83$ cm will escape. For 2-μm-diameter particles under the same conditions, the so-called critical streamline for collection is at $x_0 = 0.483$ mm.

**Example 5.2** ***Simplified Theory of the Cascade Impactor***

A common instrument used to measure particle size is the cascade impactor (see, e.g., Newton et al., 1977). The cascade impactor typically has several stages: the larger particles

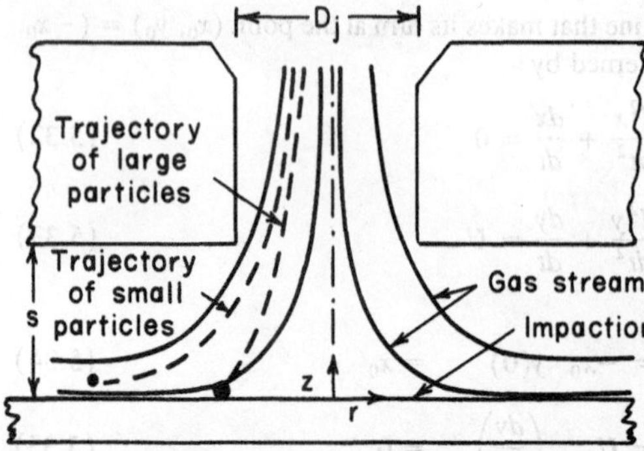


**Figure 5.4** Flow and particle trajectories on the stage of a cascade impactor.

are collected on the upper stages, with progressively smaller particles being collected on lower and lower stages. A typical impaction stage is shown in Figure 5.4. The gas flow containing the particles is accelerated through a cylindrical opening of diameter $D_j$ above the cylindrical collection plate. The gas flow streamlines must bend and flow around the impactor plate. The larger particles will not be able to follow the streamlines and will collide with the plate and be collected, whereas the smaller particles are carried around the plate to the next stage. We wish to analyze the particle trajectories using a simplified model of a cascade impactor stage. Although, as noted above, many impactors are cylindrical, let us consider a rectangular version; the cylindrical analog will be addressed in Problem 5.6.

Figure 5.5 shows an idealization of the flow in an impactor stage as the two-dimensional, inviscid flow in a corner,

$$u_x = bx \qquad u_y = -by$$

Thus the $y$-axis corresponds to the line of symmetry of the flow, and the $x$-axis corresponds to the surface of the plate. For a particle of diameter $D_p$ that starts out at $(x_0, y_0)$, we wish to derive an equation for the particle trajectory and determine the value of $x_0$ for which all particles that start at $x < x_0$ strike the plate.

We seek to solve

$$\tau \frac{d^2x}{dt^2} + \frac{dx}{dt} - bx = 0$$

$$\tau \frac{d^2y}{dt^2} + \frac{dy}{dt} + by = 0$$

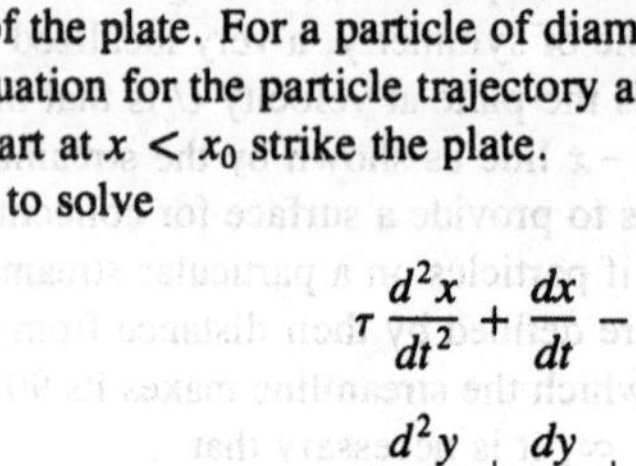


**Figure 5.5** Steady, two-dimensional, inviscid flow in a corner, $u_x = bx$, $u_y = -by$.

subject to

$$x(0) = x_0 \qquad y(0) = y_0$$

$$\frac{dx}{dt}\bigg|_{t=0} = bx_0 \qquad \frac{dy}{dt}\bigg|_{t=0} = -by_0$$

where we assume that the particle velocity is identical to the gas velocity at point $(x_0, y_0)$.

We need to look for a solution to the general, second-order ODE

$$\frac{d^2p}{dt^2} + a\frac{dp}{dt} \pm abp = 0$$

The solutions for $x(t)$ and $y(t)$ are found to be

$$\frac{x(t)}{x_0} = \frac{e^{-at/2}}{2(a^2+4ab)^{1/2}}\left\{\left[2b + a + (a^2+4ab)^{1/2}\right]\exp\left[(a^2+4ab)^{1/2}\frac{t}{2}\right]\right.$$
$$\left. + \left[-2b - a + (a^2+4ab)^{1/2}\right]\exp\left[-(a^2+4ab)^{1/2}\frac{t}{2}\right]\right\}$$

$$\frac{y(t)}{y_0} = \frac{e^{-at/2}}{2(a^2+4ab)^{1/2}}\left\{\left[-2b + a + (a^2-4ab)^{1/2}\right]\exp\left[(a^2-4ab)^{1/2}\frac{t}{2}\right]\right.$$
$$\left. + \left[2b - a + (a^2-4ab)^{1/2}\right]\exp\left[-(a^2-4ab)^{1/2}\frac{t}{2}\right]\right\}$$

where $a = 1/\tau$. These solutions can be rewritten as

$$\frac{x(t)}{x_0} = e^{-at/2}\left\{\frac{\alpha+2}{(\alpha^2+4\alpha)^{1/2}}\sinh\left[\frac{bt}{2}(\alpha^2+4\alpha)^{1/2}\right] + \cosh\left[\frac{bt}{2}(\alpha^2+4\alpha)^{1/2}\right]\right\}$$

$$\frac{y(t)}{y_0} = e^{-at/2}\left\{\frac{\alpha-2}{(\alpha^2+4\alpha)^{1/2}}\sinh\left[\frac{bt}{2}(\alpha^2-4\alpha)^{1/2}\right] + \cosh\left[\frac{bt}{2}(\alpha^2-4\alpha)^{1/2}\right]\right\}$$

where $\alpha = a/b$.

There are two ways to approach the question of whether or not the particle hits the plate. First, we can determine the time $t$ at which $y(t) \le D_p/2$, or $y \simeq 0$, from which we can then determine $x(t)$. The alternative is to assume that if the particle hits the plate, $y \to 0$ as $t \to \infty$. So we can examine the behavior of the trajectories as $t \to \infty$. The latter is the approach we will illustrate.

Let us consider $x(t)$ as $t \to \infty$. The group $(a^2 + 4ab)^{1/2} > 0$. Then as $t \to \infty$,

$$\lim_{t\to\infty}\frac{x(t)}{x_0} = \frac{e^{-at/2}}{2(a^2+4ab)^{1/2}}\left[2b + a + (a^2+4ab)^{1/2}\right]\exp\left[(a^2+4ab)^{1/2}\frac{t}{2}\right]$$

Now, what happens to $y(t)$ as $t \to \infty$? If $y \to 0$ as $t \to \infty$, the particle hits the plate. In this case we need to examine $(1 - 4b/a)^{1/2}$. If

$$\left(1 - \frac{4b}{a}\right)^{1/2} = \begin{cases} \text{real} & \dfrac{4b}{a} < 1 \\ \text{imaginary} & \dfrac{4b}{a} > 1 \end{cases}$$

Consider first the case of $4b/a < 1$. We see that

$$\frac{y(t)}{y_0} \simeq e^{-bt} + \frac{b}{a} e^{-at}$$

and, that for very large $t$, the particle will be a small distance from the plate. Small $b$ means that the fluid is moving slowly and the approach of the particle to the plate is slow.

Now consider the case of $4b/a > 1$. In this case

$$\exp\left[\frac{at}{2}\left(1 - \frac{4b}{a}\right)^{1/2}\right] = \exp\left[\frac{iat}{2}\left(\frac{4b}{a} - 1\right)^{1/2}\right]$$

$$= \cos\left[\frac{at}{2}\left(\frac{4b}{a} - 1\right)^{1/2}\right] + i \sin\left[\frac{at}{2}\left(\frac{4b}{a} - 1\right)^{1/2}\right]$$

and we find that

$$\frac{y(t)}{y_0} = e^{-at/2}\left\{\frac{1 - 2b/a}{[(4b/a) - 1]^{1/2}} \sin\left[\frac{at}{2}\left(\frac{4b}{a} - 1\right)^{1/2}\right] + \cos\left[\frac{at}{2}\left(\frac{4b}{a} - 1\right)^{1/2}\right]\right\}$$

The $y$ trajectory in this case is thus a periodic function multiplied by a decaying exponential. We see that $y(t) = 0$ occurs at finite time, so the particle hits the plate. Thus we have found that if we let the plate (i.e., the $x$-axis) extend to infinity, the particles will sooner or later hit the plate. Real impactor stages, of course, have finite size.

### 5.3.2 The Stokes Number

It will prove to be useful to be able to write the equation of motion of a particle in dimensionless form. To do so we introduce a characteristic fluid velocity $u_0$ and a characteristic length $L$ both associated with the flow field of interest. Let us consider (5.23). We define a dimensionless time $t^* = tu_0/L$, a dimensionless distance $x^* = x/L$, and a dimensionless fluid velocity $u_x^* = u_x/u_0$. Placing (5.23) in dimensionless form gives

$$\frac{\tau u_0}{L}\frac{d^2x^*}{dt^{*2}} + \frac{dx^*}{dt^*} = u_x^* \tag{5.40}$$

We call the dimensionless group $\tau u_0/L$ the *Stokes number*,

$$\text{St} = \frac{\tau u_0}{L} = \frac{D_p^2 \rho_p C_c u_0}{18\mu L} \tag{5.41}$$

The Stokes number is the ratio of the stop distance to a characteristic length scale of the flow. As particle mass decreases, the Stokes number decreases. A small Stokes number (i.e., a small stop distance relative to the chosen macroscopic length scale of the flow) implies that a particle adopts the fluid velocity very quickly. In a sense, the Stokes number can be considered as a measure of the inertia of the particle. Since the dimensionless equation of motion depends only on the Stokes number, equality of St between two geometrically similar flows indicates similitude of the particle trajectories.

### 5.3.3 Motion of a Charged Particle in an Electric Field

Situations in which a charged particle is moving in an electric field are important in several gas-cleaning and aerosol measurement methods. The electrostatic force acting on a particle having charge $q$ in an electric field of strength $\boldsymbol{E}$ is $\boldsymbol{F}_{\text{el}} = q\boldsymbol{E}$. The equation of motion for a particle of charge $q$ moving at velocity $\boldsymbol{v}$ in a fluid with velocity $\boldsymbol{u}$ in the presence of an electric field of strength $\boldsymbol{E}$ is

$$m_p \frac{d\boldsymbol{v}}{dt} = \frac{3\pi\mu D_p}{C_c}(\boldsymbol{u} - \boldsymbol{v}) + q\boldsymbol{E} \tag{5.42}$$

At steady state in the absence of a background fluid velocity, the particle velocity is such that the electrical force is balanced by the drag force and

$$\frac{3\pi\mu D_p}{C_c}\boldsymbol{v}_e = q\boldsymbol{E} \tag{5.43}$$

where $\boldsymbol{v}_e$ is termed the electrical migration velocity. Note that the characteristic time for relaxation of the particle velocity to its steady-state value is still given by $\tau = m_p C_c / 3\pi\mu D_p$, regardless of the external force influencing the particle. Thus, as long as $\tau$ is small compared to the characteristic time of changes in the electrical force, the particle velocity is given by the steady state of (5.42). The *electrical mobility* $\boldsymbol{B}_e$ is defined by $\boldsymbol{v}_e = \boldsymbol{B}_e\boldsymbol{E}$, so

$$\boldsymbol{B}_e = \frac{qC_c}{3\pi\mu D_p} \tag{5.44}$$

### 5.3.4 Motion of a Particle Using the Drag Coefficient

Earlier we introduced the drag coefficient $C_D$ as an empirical means of representing the drag force on a particle over the entire range of Reynolds number, (5.7). We did so in the case of a body moving at velocity $u_\infty$ in a quiescent fluid (or, equivalently, a fluid moving past a stationary body at speed $u_\infty$). The drag force on a particle moving with velocity $\boldsymbol{v}$ in a fluid having velocity $\boldsymbol{u}$ can be represented in a general manner by the drag coefficient $C_D$ as

$$\boldsymbol{F}_{\text{drag}} = -\tfrac{1}{2} C_D \rho A_p |\boldsymbol{v} - \boldsymbol{u}|(\boldsymbol{v} - \boldsymbol{u}) \tag{5.45}$$

where $|\boldsymbol{v} - \boldsymbol{u}|$ is the magnitude of the velocity difference. The general equation of motion then is

$$m_p \frac{d\boldsymbol{v}}{dt} = -\tfrac{1}{2} C_D \rho A_p |\boldsymbol{v} - \boldsymbol{u}|(\boldsymbol{v} - \boldsymbol{u}) + \sum_i \boldsymbol{F}_{ei} \tag{5.46}$$

We can write this equation along the direction of motion of the particle in scalar form as (assuming a single external force and no gas velocity)

$$m_p \frac{dv}{dt} = F_e - \tfrac{1}{2} C_D \rho A_p v^2 \tag{5.47}$$

If $F_e$ is constant, the motion approaches a constant velocity $v_t$ at which the external force is exactly balanced by the drag force,

$$C_D v_t^2 = \frac{2F_e}{\rho A_p} \tag{5.48}$$

For a spherical particle

$$C_D v_t^2 = \frac{8F_e}{\pi \rho D_p^2} \tag{5.49}$$

In the Stokes' law regime $C_D = 24/\text{Re}$, and

$$v_t = \frac{F_e C_c}{3\pi \mu D_p} \tag{5.50}$$

Note that (5.50) is just the generalization of (5.31), in the case of gravity force, and of (5.43), in the case of electrical force.

For terminal settling due to gravity in the non-Stokes regime, (5.49) becomes

$$\left(\frac{\pi}{6}\right) D_p^3(\rho_p - \rho)g = \left(\frac{\pi}{8}\right) C_D \rho D_p^2 v_t^2 \tag{5.51}$$

or

$$v_t = \left[\frac{4(\rho_p - \rho)gD_p}{3C_D\rho}\right]^{1/2} \tag{5.52}$$

However, since $C_D$ depends on $v_t$ through Re, this equation cannot be solved easily for $v_t$. If we form

$$C_D \text{Re}^2 = C_D \frac{v_t^2 D_p^2 \rho^2}{\mu^2} \tag{5.53}$$

and then substitute for $v_t$, we obtain

$$C_D \text{Re}^2 = C_D \frac{D_p^2 \rho^2}{\mu^2}\left[\frac{4(\rho_p - \rho)gD_p}{3C_D\rho}\right] = \frac{4D_p^3 \rho(\rho_p - \rho)g}{3\mu^2} \tag{5.54}$$

$C_D \text{Re}^2$ can be computed from this equation and from a plot of $C_D \text{Re}^2$ versus Re, the value of Re corresponding to this $C_D \text{Re}^2$ can be found, yielding $v_t$. Figure 5.6 shows $C_D \text{Re}^2$ versus Re for a spherical particle. The group on the right-hand side of (5.54) is sometimes called the Galileo number, Ga.

As an alternative to Figure 5.6, we can form the equation

$$\frac{\text{Re}}{C_D} = \frac{\text{Re}^3}{\text{Ga}} = \frac{3\rho^2 v_t^3}{4g(\rho_p - \rho)\mu} \tag{5.55}$$

Then the right-hand side of (5.55) is independent of $D_p$. The following correlation due to Koch (Licht, 1980, p. 140)

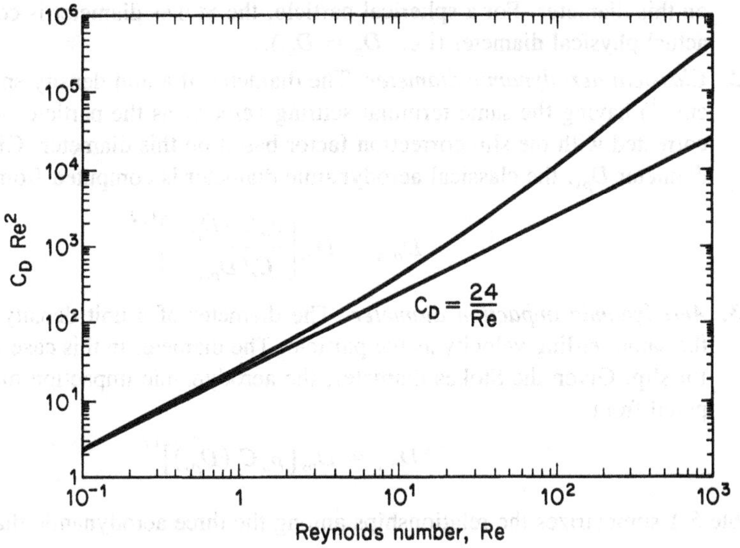


**Figure 5.6** $C_D\,\mathrm{Re}^2$ as a function of Re for a sphere.

$$\log\left(\frac{\mathrm{Re}}{C_D}\right)^{1/3} = -1.387 + 2.153 \log \mathrm{Ga}^{1/3} - 0.548(\log \mathrm{Ga}^{1/3})^2 + 0.05665(\log \mathrm{Ga}^{1/3})^3 \quad (5.56)$$

can be used. To calculate $v_t$ for a particle of any diameter, first calculate the value of Ga. Then calculate $\mathrm{Re}/C_D$ from (5.56) and finally obtain $v_t$ from (5.55).

### 5.3.5 Aerodynamic Diameter

Up to this point we have been considering spherical particles with known diameter $D_p$ and known density $\rho_p$. In dealing with actual aerosols, the particles may not be spheres and the density of the particles may not be known. Nonetheless, we want to be able to describe the motion of such particles in flow fields and under imposed external forces. To do so, we introduce the concept of an aerodynamic diameter, which is the diameter of a hypothetical spherical particle with the same motion characteristics as the particle in question.

The *aerodynamic diameter* is a measure of the reaction of a particle to inertial forces. To determine the aerodynamic diameter of a particle, one need not know the physical size, shape, or density of the particle. In fact, particles with different sizes, shapes, and densities can have the same aerodynamic size. There are several aerodynamic diameters, the three most common being:

1. *Stokes diameter.* The diameter of a sphere having the same terminal settling velocity and density as the particle and corrected with the slip correction factor based

on this diameter. For a spherical particle, the Stokes diameter is equivalent to the actual physical diameter (i.e., $D_p = D_{ps}$).

2. *Classical aerodynamic diameter*. The diameter of a unit density sphere ($\rho_p = 1$ g cm$^{-3}$) having the same terminal settling velocity as the particle, which has been corrected with the slip correction factor based on this diameter. Given the Stokes diameter $D_{ps}$, the classical aerodynamic diameter is computed from

$$D_{pca} = D_{ps}\left[\frac{\rho_p C_c(D_{ps})}{C_c(D_{pca})}\right]^{1/2} \tag{5.57}$$

3. *Aerodynamic impaction diameter*. The diameter of a unit density sphere having the same settling velocity as the particle. The diameter in this case is not corrected for slip. Given the Stokes diameter, the aerodynamic impaction diameter is computed from

$$D_{pai} = D_{ps}\left[\rho_p C_c(D_{ps})\right]^{1/2} \tag{5.58}$$

Table 5.1 summarizes the relationships among the three aerodynamic diameters.

## 5.4 BROWNIAN MOTION OF AEROSOL PARTICLES

Particles suspended in a fluid undergo irregular random motion due to bombardment by surrounding fluid molecules. Brownian motion is the name given to such motion. To describe the Brownian motion process we need not consider the details of the particle-fluid interaction, but only assume that the particle motion consists of statistically independent successive displacements.

The basis for analyzing Brownian motion is the equation of motion of a single particle (5.16). Assuming that the only forces acting on a particle are the random acceleration $\alpha(t)$ caused by the bombardment by the bath molecules and Stokes drag, we write the equation of motion as

$$m_p \frac{dv}{dt} = -\frac{3\pi\mu D_p}{C_c} v + m_p \alpha(t) \tag{5.59}$$

Dividing through by $m_p$, we write (5.59) as

$$\frac{dv}{dt} = -av + \alpha(t) \tag{5.60}$$

where $a = 1/\tau$, the inverse of the relaxation time of the particle. The random acceleration $\alpha(t)$ is a discontinuous term since it is intended to represent the force exerted by the suspending fluid molecules, which imparts an irregular, jerky motion to the particle. That the equation of motion can be decomposed into continuous and discontinuous pieces is an ad hoc assumption that is intuitively appealing and, moreover, leads to successful predictions of observed behavior. Equation (5.60) is referred to as a *Langevin equation*.

TABLE 5.1 CONVERSIONS AMONG THE THREE AERODYNAMIC DIAMETERS

| Given | Equation to convert to: Stokes diameter, $D_{ps}$ | Classical aerodynamic diameter, $D_{pca}$ | Aerodynamic impaction diameter,[a] $D_{pai}$ |
|---|---|---|---|
| Stokes diameter, $D_{ps}$ | 1.0 | $D_{pca} = D_{ps}\left[\frac{\rho_p C_c(D_{ps})}{C_c(D_{pca})}\right]^{1/2}$ | $D_{pai} = D_{ps}\left[\rho_p C_c(D_{ps})\right]^{1/2}$ |
| Classical aerodynamic diameter, $D_{pca}$ | $D_{ps} = D_{pca}\left[\frac{C_c(D_{pca})}{\rho_p C_c(D_{ps})}\right]^{1/2}$ | 1.0 | $D_{pai} = D_{pca}\left[C_c(D_{pca})\right]^{1/2}$ |
| Aerodynamic impaction diameter, $D_{pai}$ | $D_{ps} = D_{pai}\left[\rho_p C_c(D_{ps})\right]^{-1/2}$ | $D_{pca} = D_{pai} C_c(D_{pca})^{-1/2}$ | 1.0 |

[a]The units of this diameter are $\mu$m g$^{1/2}$ cm$^{-3/2}$.

Since the motion, and therefore the trajectory, of any one particle is a random process due to $\alpha(t)$, in order to study the Brownian motion phenomenon, it is necessary to consider the behavior of an entire population, or *ensemble,* of particles. Consider the trajectory of one particle along the $y$-direction released at the origin at $t = 0$. The displacement from $y = 0$ at time $t$ for this particle can be called $y(t)$. If a large number of particles are released from the origin and we average all their $y$-displacements at time $t$, we expect that average, the ensemble average, denoted by $\langle y(t) \rangle$, to be zero since there is no preferred direction inherent in $\alpha(t)$. On the other hand, $\langle y^2(t) \rangle$, the mean-square displacement of all the particles, is nonzero and is, in fact, a measure of the intensity of the Brownian motion. The mean-square displacement of the Brownian motion process can be shown to be given by (Seinfeld, 1986, p. 320 et seq.)*

$$\langle x^2 \rangle = \langle y^2 \rangle = \langle z^2 \rangle = \frac{2kTC_c t}{3\pi\mu D_p} \tag{5.61}$$

This result, first derived by Einstein in 1905, has been confirmed experimentally in numerous ways.

The movement of particles due to Brownian motion can be described as a diffusion process. The number concentration $N$ of particles undergoing Brownian motion can be assumed to be governed by the diffusion equation

$$\frac{\partial N}{\partial t} = D\,\nabla^2 N \tag{5.62}$$

where $D$ is the Brownian diffusivity. To obtain an expression for $D$, we want to relate the properties of the diffusion equation to (5.61). To do so, let us calculate the mean-square displacement of $N_0$ particles all of which start from the origin at $t = 0$. Considering $N_0$ particles all starting at the origin at the same time is conceptually the same as releasing $N_0$ single particles one at a time and averaging the results of the $N_0$ separate trajectories, which is, of course, the same idea employed to obtain (5.61). Multiply the one-dimensional version of (5.62) by $x^2$ and integrate the resulting equation over $x$ from $-\infty$ to $\infty$. The result for the left-hand side is

$$\int_{-\infty}^{\infty} x^2 \frac{\partial N}{\partial t}\,dx = N_0 \frac{\partial \langle x^2 \rangle}{\partial t} \tag{5.63}$$

and for the right-hand side is

$$\int_{-\infty}^{\infty} x^2 D \frac{\partial^2 N}{\partial x^2}\,dx = 2DN_0 \tag{5.64}$$

Equating (5.63) and (5.64) gives

$$\frac{\partial \langle x^2 \rangle}{\partial t} = 2D \tag{5.65}$$

*Where there is no possibility of confusion, the Boltzmann constant is denoted by $k$.

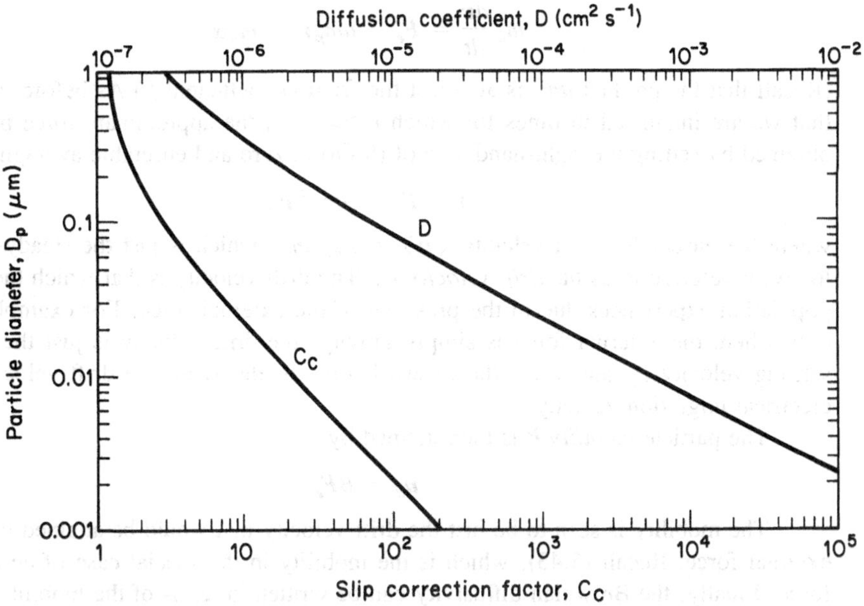


**Figure 5.7** Diffusion coefficient and slip correction factor as a function of particle size (J. H. Seinfeld, *Atmospheric Chemistry and Physics of Air Pollution*, © 1986; reprinted by permission of John Wiley & Sons).

and integrating gives the mean-square displacement in the $x$-direction,

$$\langle x^2 \rangle = 2Dt \tag{5.66}$$

We now equate this result for $\langle x^2 \rangle$ with that from (5.61) to obtain an explicit relation for $D$,

$$D = \frac{kTC_c}{3\pi\mu D_p} \tag{5.67}$$

which is known as the Stokes-Einstein relation. When $D_p >> \lambda$, $C_c \simeq 1$, and $D$ varies as $D_p^{-1}$. When $D_p << \lambda$, $C_c \simeq 1 + (1.657)(2\lambda/D_p)$, and $D$ can be approximated by $2(1.657)\lambda kT/3\pi\mu D_p^2$, so that in the free molecule regime $D$ varies as $D_p^{-2}$.

Diffusion coefficients for particles ranging from 0.001 to 1.0 $\mu$m in diameter are shown, together with the slip correction factor $C_c$, in air at 293 K and 1 atm, in Figure 5.7. The change from the $D_p^{-2}$ dependence to the $D_p^{-1}$ dependence is evident in the change of slope of the line of $D$ versus $D_p$.

### 5.4.1 Mobility and Drift Velocity

In the development of the particle equation of motion in Section 5.3, we did not include the fluctuating acceleration of $\alpha(t)$. If we generalize (5.16) to include an external force $F_e$ and assume that the fluid velocity is zero, we get

$$m_p \frac{dv}{dt} = F_e - am_p v + m_p \alpha \tag{5.68}$$

(Recall that the product $am_p$ is also just the friction coefficient $f$.) As before, assuming that we are interested in times for which $t \gg a^{-1}$, the approximate force balance is obtained by setting the right-hand side of (5.68) to zero and ensemble averaging,

$$0 = F_e - am_p \langle v \rangle \tag{5.69}$$

where the ensemble mean velocity $\langle v \rangle = F_e/am_p$, which is just the steady-state velocity, is referred to as the *drift velocity* $v_d$. The drift velocity is that which the particle population experiences due to the presence of the external force. For example, in the case where the external force is simply gravity, the drift velocity is just the terminal settling velocity $v_t$, and when the external force is electrical, the drift velocity is the electrical migration velocity.

The particle *mobility B* is then defined by

$$v_d = BF_e \tag{5.70}$$

The mobility is seen to be just the drift velocity that would be attained under unit external force. Recall (5.43), which is the mobility in the special case of an electrical force. Finally, the Brownian diffusivity can be written in terms of the mobility by

$$D = BkT \tag{5.71}$$

a result known as the Einstein relation.

### 5.4.2 Solution of Diffusion Problems for Aerosol Particles

The concentration distribution of aerosol particles in a stagnant fluid in which the particles are subject to Brownian motion and in which there is a drift velocity $v_d$ in the positive $x$-direction is governed by the convection/diffusion equation

$$\frac{\partial N}{\partial t} + v_d \frac{\partial N}{\partial x} = D \nabla^2 N \tag{5.72}$$

**Example 5.3** ***Simultaneous Diffusion and Settling of Particles above a Surface***

As an example, let us consider the simultaneous diffusion and settling of particles above a surface at $z = 0$. At $t = 0$ a uniform concentration $N_0$ of particles is assumed to exist for $z > 0$, and at all times the concentration of particles right at the surface is zero due to their "removal" at the surface. The appropriate boundary value problem describing this situation is

$$\frac{\partial N}{\partial t} - v_t \frac{\partial N}{\partial z} = D \frac{\partial^2 N}{\partial z^2}$$

$$\begin{aligned} N(z, 0) &= N_0 \\ N(0, t) &= 0 \\ N(z, t) &= N_0 \qquad z \to \infty \end{aligned} \tag{5.73}$$

where the $z$-coordinate is taken as vertically upward.

The solution of (5.73) is

$$N(z, t) = \frac{N_0}{2}\left[1 + \text{erf}\left(\frac{z + v_t t}{2(Dt)^{1/2}}\right) - \exp\left(-\frac{v_t z}{D}\right)\text{erfc}\left(\frac{z - v_t t}{2(Dt)^{1/2}}\right)\right] \tag{5.74}$$

We can calculate the deposition rate of particles on the $z = 0$ surface from the expression for the flux of particles at $z = 0$,

$$J = D\left(\frac{\partial N}{\partial z}\right)_{z=0} + v_t N(0, t) \tag{5.75}$$

which gives

$$J = N_0\left\{\frac{v_t}{2}\left[1 + \text{erf}\left(\frac{v_t t}{2(Dt)^{1/2}}\right)\right] + \left(\frac{D}{\pi t}\right)^{1/2}\exp\left(-\frac{v_t^2 t}{4D}\right)\right\} \tag{5.76}$$

Note that the flux (5.75) normally includes the term $v_t N(0, t)$ on the right-hand side, although in this case $N(0, t) = 0$.

We can identify a characteristic time $\tau_{ds} = 4D/v_t^2$ associated with this system. A large value of $\tau_{ds}$ is associated with either large $D$ or small $v_t$, that is, small particles; and a small value of $\tau_{ds}$ is associated with large particles. Thus, if there is a population of particles of different sizes, each particle size obeys (5.74) and (5.76) individually. The limiting behavior of the flux to the surface for small and large values of $t$ is

$$J(t) = \begin{cases} N_0\left[\left(\dfrac{D}{\pi t}\right)^{1/2} + \dfrac{v_t}{2}\right] & t \ll \tau_{ds} \\ N_0 v_t & t \gg \tau_{ds} \end{cases} \tag{5.77}$$

We see that at very short times the deposition flux is that due to diffusion plus one-half that due to settling, whereas for long times the deposition flux becomes solely the settling flux.

### 5.4.3 Phoretic Effects

Phoretic effects produce a directional preference in the Brownian diffusion of aerosol particles due to a difference in momentum imparted to a particle by molecules coming from different directions.

*Thermophoresis* is the migration of a particle away from the higher-temperature region and toward the lower-temperature region of the gas in which it is suspended (Talbot et al., 1980). The motion is caused by the greater force imparted by the higher energy molecules on the "hot" side of the particle than by those on the "cold" side. This phenomenon is responsible for the soiling of walls in the vicinity of hot pipes. In the absence of other forces, the thermophoretic force $F_T$ will be balanced only by the drag force, leading to a steady thermophoretic velocity $v_T$. The thermophoretic velocity may be correlated in terms of the thermal dimensionless group

$$\text{Th} = -\frac{v_T \rho T}{\mu \, dT/dx} \tag{5.78}$$

in which $dT/dx$ is the temperature gradient (always $< 0$) in the direction of motion. For particles obeying Stokes' law, the thermophoretic force is given by

$$F_T = \frac{3\pi\mu^2 D_p}{C_c \rho T}\left(-\text{Th}\,\frac{dT}{dx}\right) \tag{5.79}$$

The dimensionless group Th is essentially constant for those particles for which Kn >> 1, indicating that $v_T$ is independent of particle size, directly proportional to the temperature gradient, and only weakly dependent on temperature. The value of Th is predicted theoretically to lie between 0.42 and 1.5, with experimental measurements in the range 0.5.

For larger particles, Kn << 1, the elementary picture of gas molecules impinging on the particle from different sides with different mean velocities is oversimplified. Actually, there is a creeping flow of gas from the colder to the warmer regions along the surface of the particle which experiences a force in the cold direction. In that case, Th becomes a function of Kn as well as of the ratio of the thermal conductivities of gas and particle. Th decreases as Kn increases and may fall as low as 0.02 at Kn = 0.01 for $k_g/k_p \ll 1$. A correlation of Th is presented by Gieseke (1972).

We can estimate the importance of thermophoresis by calculating the thermophoretic velocity $v_T$. Consider a 0.1-μm-diameter particle in air at 573 K. Using Th = 0.5, (5.78) gives $v_T = 4.17 \times 10^{-4}(dT/dx)$, where $dT/dx$ is measured in °C cm$^{-1}$ and $v_T$ is in cm s$^{-1}$. At $dT/dx = 10^3$ °C cm$^{-1}$, $v_T = 0.417$ cm s$^{-1}$; whereas at $dT/dx$ = 10 °C cm$^{-1}$, $v_T = 0.00417$ cm s$^{-1}$. Unless the temperature gradient is very steep, thermophoresis can generally be neglected compared with the effects of other forces on aerosol particles.

Aerosol particles have been found to experience an external force when subjected to an intense light beam. This phenomenon, called *photophoresis*, results from the nonuniform heating of the particles. The force on the particle arises when uneven heating of the particle due to absorption of the incident light leads to uneven heating of the gas molecules in the vicinity of the particle. The net force on a spherical particle can be directed either toward (negative photophoresis) or away from (positive photophoresis) the light source. If the particle has optical characteristics such that incident light energy is absorbed and dissipated at its front side, the more energetic gas molecules are on the incident side of the particle and positive photophoresis results. Conversely, if the absorption characteristics of the particle are such that its back side absorbs more of the incident light, negative photophoresis results. Calculating the photophoretic force on a particle involves computing the internal distribution of absorbed energy in the particle and then calculating the interaction between the unevenly heated particle and the surrounding gas. The force ultimately arises because of the temperature gradient that develops in the gas around the particle, and therefore it can be analyzed in a manner similar to that of thermophoresis. For an analysis of photophoresis in the continuum regime, we refer to the reader to Reed (1977).

*Diffusiophoresis* occurs in the presence of a gradient of vapor molecules that are either lighter or heavier than the surrounding gas molecules (Derjaguin and Yalamov, 1972). For example, consider an evaporating surface above which a gradient of water vapor concentration exists. Water molecules are less massive than air molecules, so a decrease in water vapor concentration with distance above the evaporating surface will lead to a net downward force on aerosol particles that is the result of the downward flux of air molecules needed to balance the upward flux of water molecules.

A diffusiophoretic migration velocity can be defined that depends on the gradient in partial pressure of the diffusing gas $A$ and the partial pressures and molecular weights of the diffusing gas and the background gas $B$. The migration velocity is

$$v_D = -\left(\frac{M_A^{1/2}}{p_A M_A^{1/2} + p_B M_B^{1/2}}\right)\frac{pD_{AB}}{p_B}\nabla p_A \tag{5.80}$$

which can be expressed as

$$v_D = -C_1 D_{AB}\left(\frac{1}{1 - y_A}\right)\nabla y_A \tag{5.81}$$

where $y_A$ is the mole fraction of species $A$ and $C_1$ is the molecular weight and composition function that describes the effect of the molecular mass gradient due to the net motion of the gas. For water vapor in air at mole fractions between 0.1 and 0.5, $C_1$ varies from 0.8 to 0.88.

We will not consider phoretic effects further in this book.

## 5.5 DIFFUSION TO SINGLE PARTICLES

### 5.5.1 Continuum Regime

When the mean free path of vapor molecules leaving or approaching the surface of a particle is small compared with the radius of the particle, that is, when the Knudsen number defined by $\mathrm{Kn} = \lambda_{AB}/R_p$ is small, when the mixture of $A$ and $B$ is very dilute, and if the particle can be considered to exist in a stagnant medium, the transient concentration distribution of species $A$ in the background gas species $B$ around the particle obeys the diffusion equation (B.7),

$$\frac{\partial c_A}{\partial t} = \frac{D_{AB}}{r^2}\frac{\partial}{\partial r}\left(r^2\frac{\partial c_A}{\partial r}\right) \tag{5.82}$$

The characteristic time for the gas-phase concentration of species $A$ to attain its steady-state distribution around the particle is evident from (5.82) as $\tau_m = R_p^2/D_{AB}$ [recall (B.35)]. For $t \gg \tau_m$, steady state is attained and the steady-state concentration profile satisfies

$$\frac{d^2 c_A}{dr^2} + \frac{2}{r}\frac{dc_A}{dr} = 0 \tag{5.83}$$

If the concentration of $A$ just above the particle surface is $c_{As}$ and that far away from the particle is $c_{A\infty}$, the solution of (5.83) is

$$\frac{c_{A\infty} - c_A(r)}{c_{A\infty} - c_{As}} = \frac{R_p}{r} \tag{5.84}$$

The total flow of $A$ toward the particle (moles $A$ $s^{-1}$) is then given by the so-called Maxwell equation,

$$J_{Ac} = 4\pi R_p D_{AB}(c_{A\infty} - c_{As}) \tag{5.85}$$

where the subscript $c$ on $J$ denotes the continuum regime. If $c_{As} < c_{A\infty}$, the flow is toward the particle (note that we have taken $J_A$ as positive in this case), and vice versa.

### 5.5.2 Free Molecule Regime

As the Knudsen number approaches order unity, indicating that the mean free path of species $A$ approaches the particle radius, Fick's law becomes invalid. For $\mathrm{Kn} >> 1$, the mean free path of species $A$ is much greater than the particle radius, and the particle appears to the vapor to be just a large molecule. The particle does not disturb the background vapor concentration, and in this limit the flux of vapor molecules to the particle can be predicted from kinetic theory of gases.

The number of gas molecules hitting a unit area in unit time is the so-called *effusion flux* and is given by (Moore, 1962, pp. 217–219)

$$\tfrac{1}{4} N\bar{c} \tag{5.86}$$

where $N$ is the number concentration of molecules in the gas and $\bar{c}$ is the mean molecular speed of the species. We have been expressing the vapor concentration in molar units, so $c_{A\infty}$ would replace $N$ in (5.86) and the units of the effusion flux are then moles $A$ $cm^{-2}$ $s^{-1}$. The mean molecular speed of species $A$, $\bar{c}_A$, is (Moore, 1962, p. 238)*

$$\bar{c}_A = \left(\frac{8kT}{\pi m_A}\right)^{1/2} \tag{5.87}$$

and where $m_A$ is the molecular mass of species $A$.

The rate of impingement of molecules of $A$ on the particle, in units of mol $A$ $s^{-1}$, in the kinetic (or free molecule) regime is just the effusion flux multiplied by the surface area of the particle,

$$J_{Ak} = \pi R_p^2 \bar{c}_A (c_{A\infty} - c_{As}) \tag{5.88}$$

where we have used $(c_{A\infty} - c_{As})$ for consistency with (5.85).

### 5.5.3 Transition Regime

In the intermediate or transition regime of Knudsen numbers, neither continuum diffusion theory nor elementary kinetic theory of gases applies. In the continuum limit, the concentration profile of species $A$ around the particle obeys (5.84); whereas in the kinetic limit, the concentration of $A$ in the background gas is unaffected by the presence of the particle, and the flow of molecules to the particle is just the rate at which they impinge on the surface of the particle in their normal motion. The concentration profile in the

*The mean molecular speed $\bar{c}_A$ is not to be confused with the concentration $c_A$.

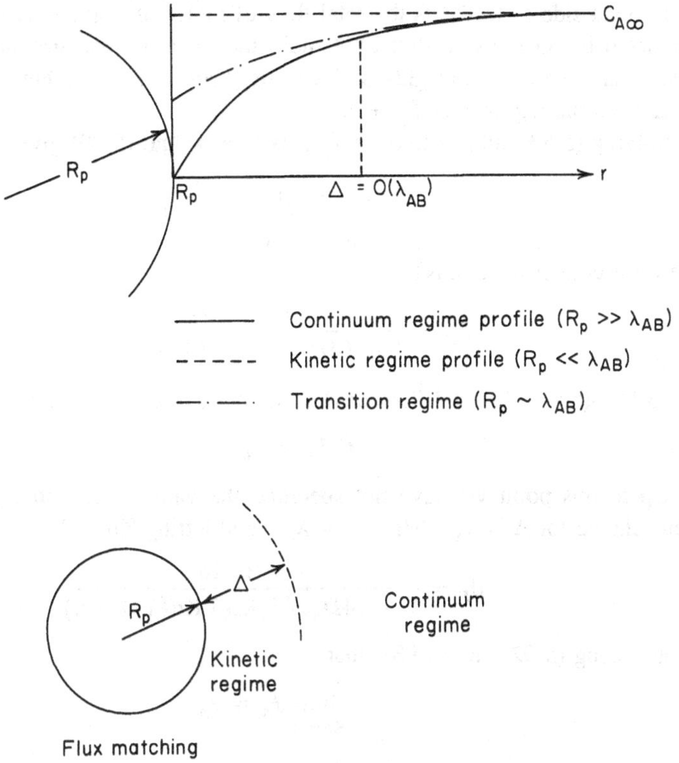


**Figure 5.8** Steady-state concentration profiles of a diffusing species $A$ around a spherical particle in different regimes of particle radius $R_p$ and diffusing mean free path $\lambda_{AB}$. Lower sketch shows idea of flux-matching at a surface at $r = R_p + \Delta$.

transition regime is expected to lie somewhere intermediate between these two extremes. The situation is depicted in Figure 5.8.

We expect in the transition regime that in the neighborhood of the particle, say out to some distance $\Delta$ from the surface, where $\Delta$ is the order of $\lambda_{AB}$, the molecules of $A$ do not experience many collisions with molecules of $B$ and, therefore, that the flux of molecules of $A$ to the surface can be approximated by the effusion flux. Far from the particle surface, on the other hand, the concentration profile of $A$ should be quite close to the continuum values. This physical picture was utilized by Fuchs in suggesting that to determine the flux of molecules to a particle in the transition regime one solve the continuum diffusion equation (5.83) up to a distance $\Delta$ of the particle surface and match the flux at $r = R_p + \Delta$ to the kinetic theory flux resulting from $R_p + \Delta$ to $R_p$ due to the concentration value at $R_p + \Delta$. The concentration profile of $A$ then satisfies (5.83) for $r \geq R_p + \Delta$ subject to $c_A = c_{A\infty}$ as $r \to \infty$ and the flux-matching condition,

$$4\pi R_p^2 (\tfrac{1}{4}\bar{c}_A)\left[c_A(R_p + \Delta) - c_{As}\right] = 4\pi (R_p + \Delta)^2 D_{AB} \left(\frac{dc_A}{dr}\right)_{r = R_p + \Delta} \tag{5.89}$$

The left-hand side of (5.89) is the total flow of moles of $A$ at the surface of the particle as predicted by an effusion flux of $A$ molecules at a net background concentration of $[c_A(R_p + \Delta) - c_{As}]$. The right-hand side of (5.89) is the continuum flow of $A$ at $r = R_p + \Delta$ from the region $r > R_p + \Delta$.

Solving (5.83) subject to $c_A = c_{A\infty}$ as $r \to \infty$ and (5.89) gives

$$\frac{c_{A\infty} - c_A(r)}{c_{A\infty} - c_{As}} = \frac{R_p}{r}\beta_F \tag{5.90}$$

where (subscript $F$ for Fuchs)

$$\beta_F = \frac{1 + \Delta/R_p}{1 + (4D_{AB}/\bar{c}_A R_p)(1 + \Delta/R_p)} \tag{5.91}$$

From (5.89)–(5.91) the total flow of $A$ at the particle surface is found to be

$$J_A = J_{Ac}\beta_F \tag{5.92}$$

Up to this point we have not specified the value of $\Delta$. On a physical basis the obvious choice for $\Delta$ is $\lambda_{AB}$. With $\Delta = \lambda_{AB}$, and letting $\mathrm{Kn} = \lambda_{AB}/R_p$, (5.91) becomes

$$\beta_F = \frac{1 + \mathrm{Kn}}{1 + (4D_{AB}/\bar{c}_A\lambda_{AB})\,\mathrm{Kn}(1 + \mathrm{Kn})} \tag{5.93}$$

We note, using (5.92) and (5.93), that

$$\lim_{\mathrm{Kn}\to 0} J_A = J_{Ac}$$

$$\lim_{\mathrm{Kn}\to \infty} J_A = J_{Ak}$$

as must of course be the case.

To use the Fuchs interpolation formula, (5.92) and (5.93), it is necessary to specify the group $D_{AB}/\bar{c}_A\lambda_{AB}$. An expression for this group is given in (5.11).

An alternative interpolation factor to $\beta_F$ is that due to Fuchs and Sutugin (1971),

$$\beta_{FS} = \frac{1 + \mathrm{Kn}}{1 + 1.71\,\mathrm{Kn} + 1.33\,\mathrm{Kn}^2} \tag{5.94}$$

In order for $J_A = J_{Ac}\beta_{FS}$ to obey the kinetic limit as $\mathrm{Kn} \to \infty$ it is necessary that $\lambda_{AB}$ be defined in accordance with

$$\frac{D_{AB}}{\bar{c}_A\lambda_{AB}} = \frac{1}{3} \tag{5.95}$$

This expression corresponds to the Chapman-Enskog exact solution for the case of $M_A \ll M_B$.

**Example 5.4** ***Evaporation of a Droplet***

The steady-state continuum molar flow of species $A$ from a particle that is evaporating is given by (5.85) with $c_{As} > c_{A\infty}$. The mass flow is just $J_{Ac}M_A$, where $M_A$ is the molecular weight of $A$. Expressed in terms of partial pressures, the mass flow from an evaporating drop of pure $A$ is

$$\frac{2\pi D_p D_{AB} M_A}{RT}(p_{As} - p_{A\infty})$$

where $p_{As}$ and $p_{A\infty}$ are the partial pressures just over the particle surface and in the background gas, respectively. The mass flow can be related to the change in size of the particle by

$$M_A J_{Ac} = -\frac{\pi}{2}\rho_p D_p^2 \frac{dD_p}{dt} \tag{5.96}$$

The vapor pressure over the particle surface is the product of the vapor pressure of the species over a flat surface, $p_A^\circ$, and a factor greater than one that expresses the increase in vapor pressure over a curved as opposed to a flat surface, the so-called Kelvin effect (see Section 5.10). That factor is

$$\exp\left(\frac{4\sigma M_A}{\rho_p RT D_p}\right)$$

where $\sigma$ is the surface tension of the material. Thus the vapor pressure of species $A$ over a droplet of pure $A$ of diameter $D_p$ is

$$p_{As} = p_A^\circ \exp\left(\frac{4\sigma M_A}{\rho_p RT D_p}\right) \tag{5.97}$$

To include particles small enough to lie in the transition or kinetic regimes we employ $J_A = J_{Ac}\beta_{FS}$.

Thus the evaporation or growth equation is

$$\frac{dD_p}{dt} = \frac{4 D_{AB} M_A}{\rho_p RT D_p}\left[p_{A\infty} - p_A^\circ \exp\left(\frac{4\sigma M_A}{\rho_p RT D_p}\right)\right]\beta_{FS}(\mathrm{Kn}) \tag{5.98}$$

It is useful to introduce the following dimensionless variables:

$$\hat{D}_p = \frac{D_p}{D_{p0}}$$

$$\Gamma = \frac{4\sigma M_A}{\rho_p RT D_{p0}}$$

$$\mathrm{Kn}_0 = \frac{2\lambda_{AB}}{D_{p0}}$$

$$\hat{t} = \frac{t D_{AB}}{D_{p0}^2}$$

$$\hat{p}_{A\infty} = \frac{4 M_A p_{A\infty}}{\rho_p RT}$$

$$\hat{p}_A = \frac{4 M_A p_A^\circ}{\rho_p RT}$$

in which case (5.98) becomes

$$\frac{d\hat{D}_p}{dt} = \frac{1}{\hat{D}_p}\left[\hat{p}_{A\infty} - \hat{p}_A \exp\left(\frac{\Gamma}{\hat{D}_p}\right)\right] \beta_{FS}(\mathrm{Kn}_0/\hat{D}_p) \tag{5.99}$$

In the special case of evaporation of the particle into a background free of the species $A$, $\hat{p}_{A\infty} = 0$, and (5.99) can be integrated to give

$$\hat{p}_A \hat{t} = \int_{\hat{D}_p}^{1} \hat{D}'_p \exp(-\Gamma/\hat{D}'_p)\, \beta_{FS}(\mathrm{Kn}_0/\hat{D}'_p)\, d\hat{D}'_p \tag{5.100}$$

The right-hand side of (5.100) can be numerically integrated for values of the final diameter $\hat{D}_p$, $0 \le \hat{D}_p < 1$, for choices of $\Gamma$ and $\mathrm{Kn}_0$. The values of the integral correspond to

$$\hat{p}_A \hat{t} = \frac{4M_A p_A^\circ D_{AB}}{\rho_p RTD_{p0}^2}\, t \tag{5.101}$$

If measurements of $D_{p0}$ and $D_p(t)$ are made, and if $M_A$, $D_{AB}$, $\rho_p$, and $T$ are known, the value of the vapor pressure $p_A^\circ$ may be inferred from such data from the slope of a plot of $\hat{p}_A \hat{t}$ versus time.

Let us apply these results to a particle of dibutylphthalate, $C_6H_4[C(O)OC_4H_9]_2$, evaporating into vapor-free air at 298 K. The characteristic time to achieve a steady-state concentration profile in the vapor phase, $R_{p0}^2/D_{AB}$, is of order $10^{-9}$ s for the initial sizes we will consider. It will be seen that particle diameter changes due to evaporation occur on a time scale of order 1 s, thus validating the use of a steady-state vapor concentration profile in the evaporation equation.

The right-hand side of (5.100) can be evaluated for various final diameters $\hat{D}_p$ in the interval $0 < \hat{D}_p \le 1$, and the actual diameter changes for three initial sizes are shown in Figure 5.9. Parameter values for dibutylphthalate are $D_{AB} = 0.0367$ cm$^2$ s$^{-1}$, $p_A^\circ = 0.0345$ dyn cm$^{-2}$, $\sigma = 33.14$ dyn cm$^{-1}$, $M_A = 278.35$, $\rho_p = 1.042$ g cm$^{-3}$, $\lambda_{AB} = 0.0504$ $\mu$m. For $D_{p0} = 0.1$ $\mu$m, $\Gamma = 0.14287$, and $\mathrm{Kn}_0 = 1.008$. We see that a 0.1-$\mu$m dibutylphthalate particle evaporates to one-half its original diameter in about 1.4 s and evaporates essentially completely in approximately 2.05 s. A particle of initial diameter 0.2 $\mu$m requires slightly more than 6 s to evaporate. An interesting feature of the process is the apparently instantaneous disappearance (evaporation) of the particle, for example, for the initially 0.1-$\mu$m-diameter particle upon reaching a diameter of 0.005 $\mu$m (50 Å), 0.125% of its original

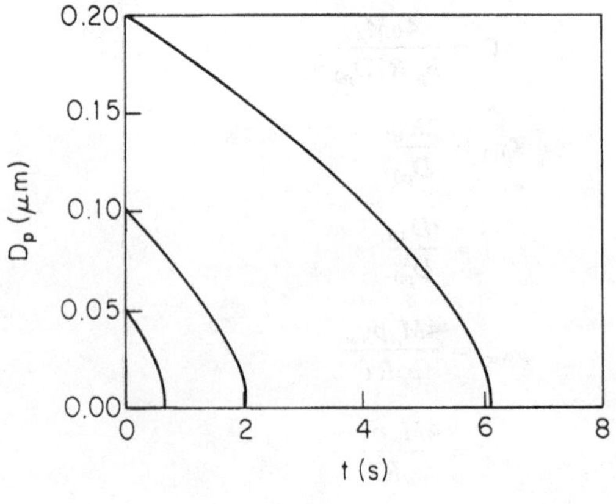


**Figure 5.9** Diameter of evaporating dibutylphthalate particles as a function of time.

volume. As the particle diameter gets smaller and smaller, the vapor pressure over the curved particle surface increases exponentially until a point where the vapor pressure is so large as to cause essentially instantaneous evaporation. Physically, increasing surface tension increases the difference between the vapor pressure over a flat surface and that over a curved surface, increasing the driving force for evaporation. Thus, all else being equal, we expect that with two particles each of a different species, the one with the larger surface tension will evaporate more readily. Since the particle density appears in the denominator of the exponential, increased density results in longer evaporation times.

## 5.6 THE SIZE DISTRIBUTION FUNCTION

An aerosol can be envisioned as a population of particles each consisting of an integral number of molecules or monomers. The smallest aerosol particle could be defined in principle as that containing two molecules or two monomers. The population can then be characterized by the concentrations of each cluster, that is, $N_k$ = concentration (per $cm^3$ of air) of particles containing $k$ molecules. Although rigorously correct, this "discrete" way of characterizing the distribution quickly becomes unwieldy because of the very large number of molecules that go to make up even the smallest of aerosol particles. For example, a sulfuric acid/water particle at equilibrium at 30% relative humidity of 0.01 μm diameter consists of approximately $10^4$ molecules.

Beyond a relatively small number of molecules, say about 100, an aerosol population can be treated as if its size variation is essentially continuous. In doing so, we replace the discrete size index $k$ by the particle diameter $D_p$ and the discrete number concentration $N_k$ by the size distribution function $n(D_p)$, defined as follows:

$$n(D_p)\, dD_p = \text{number of particles per unit volume of air having diameters in the range } D_p \text{ to } D_p + dD_p$$

The total number of particles per unit volume of air of all sizes is then just

$$N = \int_0^\infty n(D_p)\, dD_p \tag{5.102}$$

Since $n(D_p)$ is just the probability density function for particle size, a more precise term for $n(D_p)$ is a size distribution density function, although the word "density" is usually omitted in referring to $n(D_p)$. The units of $n(D_p)$ are $\mu m^{-1}\ cm^{-3}$, and those of $N$ are $cm^{-3}$. If the aerosol population is changing with time, we may write $n(D_p, t)$ and $N(t)$.

We can define a normalized size distribution function $\tilde{n}(D_p)$ by $\tilde{n}(D_p) = n(D_p)/N$, such that $\tilde{n}(D_p)\, dD_p$ = the fraction of the total number of particles per $cm^3$ of air having diameters in the range $D_p$ to $D_p + dD_p$. The units of $\tilde{n}(D_p)$ are $\mu m^{-1}$.

It is often of interest to know the distributions of both particle surface area and volume with respect to particle size. Considering all particles as spheres, we define the surface area distribution function as

$$n_s(D_p) = \pi D_p^2 n(D_p) \qquad \mu m^2\ \mu m^{-1}\ cm^{-3} \tag{5.103}$$

and the volume distribution function as

$$n_v(D_p) = \frac{\pi}{6} D_p^3 n(D_p) \qquad \mu m^3\ \mu m^{-1}\ cm^{-3} \tag{5.104}$$

The total particle surface area and volume per $cm^3$ of air are

$$S = \pi \int_0^\infty D_p^2 n(D_p)\, dD_p = \int_0^\infty n_s(D_p)\, dD_p \qquad \mu m^2\ cm^{-3} \tag{5.105}$$

$$V = \frac{\pi}{6} \int_0^\infty D_p^3 n(D_p)\, dD_p = \int_0^\infty n_v(D_p)\, dD_p \qquad \mu m^3\ cm^{-3} \tag{5.106}$$

Normalized surface area and volume distributions can then be defined by $\tilde{n}_s(D_p) = n_s(D_p)/S(\mu m^{-1})$ and $\tilde{n}_v(D_p) = n_v(D_p)/V(\mu m^{-1})$.

If the particles all have density $\rho_p$ (g $cm^{-3}$), the distribution of particle mass with respect to particle size is

$$n_m(D_p) = \frac{\rho_p}{10^6}\left(\frac{\pi}{6}\right) D_p^3\, n(D_p) = \frac{\rho_p}{10^6} n_v(D_p) \qquad \mu g\ \mu m^{-1}\ cm^{-3} \tag{5.107}$$

where the factor of $10^6$ is needed to convert $\rho_p$ from g $cm^{-3}$ to $\mu g\ \mu m^{-3}$, to maintain the units of $n_m(D_p)$ as $\mu g\ \mu m^{-1}\ cm^{-3}$.

Because particle diameters in an aerosol population typically vary over several orders of magnitude, it is often convenient to express the size distribution in terms of the logarithm of the diameter, either $\ln D_p$ or $\log D_p$.

### 5.6.1 Distributions Based on $\log D_p$

Let us define $n(\log D_p)\, d \log D_p$ = number of particles per $cm^3$ of air in the size range $\log D_p$ to $\log D_p + d \log D_p$. Note that $n(\log D_p)$ is *not* the same function as $n(D_p)$. Rather than introduce new notation for $n(\log D_p)$, to differentiate it from $n(D_p)$, we will always indicate the independent variable, either $D_p$ or $\log D_p$. Formally, we cannot take the logarithm of a dimensional quantity. Thus, when we write $\log D_p$ we must think of it as $\log (D_p/1)$, where the "reference" particle diameter is 1 $\mu m$ and is thus not explicitly indicated. The units of $n(\log D_p)$ are $cm^{-3}$ since $\log D_p$ is dimensionless. The total number concentration of particles is

$$N = \int_{-\infty}^{\infty} n(\log D_p)\, d \log D_p \qquad cm^{-3} \tag{5.108}$$

and the normalized size distribution function with respect to $\log D_p$ is $\tilde{n}(\log D_p) = n(\log D_p)/N$, which is dimensionless. Note that the limits of integration in (5.108) are now $-\infty$ to $\infty$.

Just as we defined surface area and volume distributions with respect to $D_p$, we can do so with respect to $\log D_p$,

$$n_s(\log D_p) = \pi D_p^2 n(\log D_p) \qquad \mu\text{m}^2\ \text{cm}^{-3} \tag{5.109}$$

$$n_v(\log D_p) = \frac{\pi}{6} D_p^3 n(\log D_p) \qquad \mu\text{m}^3\ \text{cm}^{-3} \tag{5.110}$$

with

$$S = \int_{-\infty}^{\infty} \pi D_p^2 n(\log D_p) d \log D_p = \int_{-\infty}^{\infty} n_s(\log D_p)\, d \log D_p \tag{5.111}$$

$$V = \int_{-\infty}^{\infty} \frac{\pi}{6} D_p^3 n(\log D_p) d \log D_p = \int_{-\infty}^{\infty} n_v(\log D_p)\, d \log D_p \tag{5.112}$$

Since $n(D_p)\, dD_p$ = the differential number of particles in the size range $D_p$ to $D_p + dD_p$, this quantity is sometimes expressed as $dN$, with similar notation for the other distributions. Thus, using this notation we have

$$dN = n(D_p)\, dD_p = n(\log D_p)\, d \log D_p \tag{5.113}$$

$$dS = n_s(D_p)\, dD_p = n_s(\log D_p)\, d \log D_p \tag{5.114}$$

$$dV = n_v(D_p)\, dD_p = n_v(\log D_p)\, d \log D_p \tag{5.115}$$

Based on this notation, the various size distribution functions are often written as

$$n(D_p) = \frac{dN}{dD_p} \qquad \tilde{n}(D_p) = \frac{dN}{N\, dD_p}$$

$$n(\log D_p) = \frac{dN}{d \log D_p} \qquad \tilde{n}(\log D_p) = \frac{dN}{Nd \log D_p}$$

$$n_s(D_p) = \frac{dS}{dD_p} \qquad \tilde{n}_s(D_p) = \frac{dS}{S\, dD_p}$$

$$n_s(\log D_p) = \frac{dS}{d \log D_p} \qquad \tilde{n}_s(\log D_p) = \frac{dS}{Sd \log D_p}$$

$$n_v(D_p) = \frac{dV}{dD_p} \qquad \tilde{n}_v(D_p) = \frac{dV}{V\, dD_p}$$

$$n_v(\log D_p) = \frac{dV}{d \log D_p} \qquad \tilde{n}_v(\log D_p) = \frac{dV}{Vd \log D_p}$$

To conform with the common notation, we will often express the distributions in this manner.

### 5.6.2 Relating Size Distributions Based on Different Independent Variables

It is important for us to be able to relate a size distribution based on one independent variable, say $D_p$, to one based on another independent variable, say $\log D_p$. The basis for such relationship is exemplified by (5.113). In a particular incremental particle size range $D_p$ to $D_p + dD_p$ the number of particles $dN$ is a certain quantity, and that quantity is the same regardless of how the size distribution function is expressed. Thus, in the particular case of $n(D_p)$ and $n(\log D_p)$ we have that

$$n(D_p)\, dD_p = n(\log D_p) d \log D_p \tag{5.116}$$

Say that we have $n(D_p)$ and wish to calculate $n(\log D_p)$ from it. Thus

$$n(\log D_p) = n(D_p) \frac{dD_p}{d \log D_p} \tag{5.117}$$

Now, since $d \log D_p = d \ln D_p/2.303 = dD_p/2.303D_p$, (5.117) becomes

$$n(\log D_p) = 2.303 D_p n(D_p) \tag{5.118}$$

which is the desired relationship between $n(D_p)$ and $n(\log D_p)$.

This procedure can be generalized to relate any two size distribution functions $n(u)$ and $n(v)$, where $u$ and $v$ are both related to diameter $D_p$. The generalization of (5.116) is [recall (C.16)]

$$n(u)\, du = n(v)\, dv \tag{5.119}$$

and if we seek to calculate $n(u)$ in terms of $n(v)$, we write

$$n(u) = n(v) \frac{dv/dD_p}{du/dD_p} \tag{5.120}$$

**Example 5.5** ***Relating Size Distributions Depending on Different Independent Variables***

We are given an aerosol size distribution function $n_m(m)$ such that $n_m(m)\, dm$ = aerosol mass per cm$^3$ of air contained in particles having masses in the range $m$ to $m + dm$. We desire to convert that distribution function to a mass distribution based on $\log D_p$, $n_m(\log D_p)$. Let us determine the relation between $n_m(\log D_p)$ and $n_m(m)$.

From (C.16) we have

$$n_m(m)\, dm = n_m(\log D_p)\, d \log D_p \tag{5.121}$$

Now $m = (\pi/6)\rho_p D_p^3$, so $dm = (\pi/6)\rho_p 3 D_p^2 dD_p$. Also, $d \log D_p = dD_p/2.303D_p$. Thus

$$n_m(m) \left(\frac{\pi}{6}\right) \rho_p 3D_p^2\, dD_p = \frac{n_m(\log D_p)\, dD_p}{2.303 D_p}$$

or

$$\begin{aligned} n_m(\log D_p) &= \left(\frac{\pi}{6}\right) \rho_p 3D_p^2 (2.303 D_p) n_m(m) \\ &= 6.9 m n_m(m) \end{aligned} \tag{5.122}$$

which is the desired relationship.

Say that we are given the *number* distribution based on particle mass, $n(m)$, and desire to obtain the mass distribution, $n_m(\log D_p)$. Since

$$n_m(m) = mn(m) \tag{5.123}$$

the desired result is just

$$n_m(\log D_p) = 6.9m^2 n(m) \tag{5.124}$$

## 5.7 THE LOG-NORMAL DISTRIBUTION

The next question that arises in our study of aerosol size distributions is, what functions are commonly used to represent aerosol size distributions? To represent particle size distributions $n(D_p)$ we need a function that is defined only for $D_p \geq 0$ and which is zero for $D_p = 0$ (clearly, no particles can exist of size zero) and approaches zero as $D_p \to \infty$ (no particles can exist with infinite size). Although many distributions with such properties exist, a popular one for representing aerosol size distributions, and one with a host of desirable properties, is the *log-normal distribution.*

If a quantity $u$ is normally distributed, the probability density function for $u$ obeys the Gaussian distribution (Table 1.13),

$$n(u) = \frac{N}{\sqrt{2\pi}\sigma_u} \exp\left[-\frac{(u - \bar{u})^2}{2\sigma_u^2}\right] \tag{5.125}$$

where $n(u)$ is defined for $-\infty < u < \infty$, $\bar{u}$ is the mean of the distribution, $\sigma_u^2$ is its variance, and

$$N = \int_{-\infty}^{\infty} n(u)\, du \tag{5.126}$$

A quantity that is log-normally distributed has its logarithm governed by a normal distribution. If the quantity of interest is particle diameter $D_p$, then saying that an aerosol population is log-normally distributed means that $u = \ln D_p$ satisfies (5.125). For now we will use the natural logarithm $\ln D_p$, but we can also express our result in terms of $\log D_p$.

Letting $u = \ln D_p$, we express (5.125) as

$$n(\ln D_p) = \frac{N}{\sqrt{2\pi} \ln \sigma_g} \exp\left[-\frac{(\ln D_p - \ln \bar{D}_{pg})^2}{2 \ln^2 \sigma_g}\right] \tag{5.127}$$

where we have let $\bar{u} = \ln \bar{D}_{pg}$ and $\sigma_u = \ln \sigma_g$. For the moment we consider $\bar{D}_{pg}$ and $\sigma_g$ to be merely the two parameters of the distribution. Shortly we will discuss the physical significance of these parameters. It is more convenient to have the size distribution function expressed in terms of $D_p$ rather than $\ln D_p$. The form of (5.118) appropriate to this transformation is $n(\ln D_p) = D_p n(D_p)$, so that (5.127) becomes

$$n(D_p) = \frac{N}{\sqrt{2\pi}\, D_p \ln \sigma_g} \exp\left[-\frac{(\ln D_p - \ln \bar{D}_{pg})^2}{2 \ln^2 \sigma_g}\right] \tag{5.128}$$

This is the conventional form of the log-normal distribution used in describing aerosol size distributions.

We now wish to examine the physical significance of the two parameters $\overline{D}_{pg}$ and $\sigma_g$. To do so, let us examine for the moment some properties of the normal distribution (5.125). The cumulative distribution function $F(u)$ is the probability that $u$ will lie in the range $-\infty$ to $u$ (C.2),

$$F(u) = \int_{-\infty}^{u} n(u')\,du' \tag{5.129}$$

so that for a normally distributed quantity,

$$F(u) = \frac{N}{\sqrt{2\pi}\,\sigma_u}\int_{-\infty}^{u} \exp\left[-\frac{(u' - \bar{u})^2}{2\sigma_u^2}\right] du' \tag{5.130}$$

To evaluate this integral we let $\eta = (u' - \bar{u})/\sqrt{2}\sigma_u$, and we obtain

$$F(u) = \frac{N}{\sqrt{\pi}}\int_{-\infty}^{(u-\bar{u})/\sqrt{2}\sigma_u} e^{-\eta^2}\,d\eta \tag{5.131}$$

The *error function* erf $z$ is defined as

$$\operatorname{erf} z = \frac{2}{\sqrt{\pi}}\int_0^z e^{-\eta^2}\,d\eta \tag{5.132}$$

where erf $\infty = 1$. If we divide the integral in (5.131) into one from $-\infty$ to 0 and the second from 0 to $(u - \bar{u})/\sqrt{2}\sigma_u$, the first integral is seen to be equal to $\sqrt{\pi}/2$ and the second to $(\sqrt{\pi}/2)$ erf $[(u - \bar{u})/\sqrt{2}\sigma_u]$. Thus, for the normal distribution

$$F(u) = \frac{N}{2} + \frac{N}{2}\operatorname{erf}\left(\frac{u - \bar{u}}{\sqrt{2}\sigma_u}\right) \tag{5.133}$$

Now, in the case of the log-normal distribution $u = \ln D_p$, so (5.133) can be expressed as

$$F(D_p) = \frac{N}{2} + \frac{N}{2}\operatorname{erf}\left[\frac{\ln (D_p/\overline{D}_{pg})}{\sqrt{2}\ln \sigma_g}\right] \tag{5.134}$$

$F(D_p)/N$ is the fraction of the total number of particles with diameters up to $D_p$. For $D_p = \overline{D}_{pg}$, since erf$(0) = 0$, $F(\overline{D}_{pg})/N = \frac{1}{2}$. Thus we see that $\overline{D}_{pg}$ is the *median diameter*, that is, that diameter for which exactly one-half the particles are smaller and one-half are larger. To understand the role of $\sigma_g$, let us consider that diameter $D_{p\sigma}$ for which $\sigma_g = D_{p\sigma}/\overline{D}_{pg}$. At that diameter

$$\frac{F(D_p)}{N} = \frac{1}{2} + \frac{1}{2}\operatorname{erf}\left(\frac{1}{\sqrt{2}}\right) = 0.841 \tag{5.135}$$

Thus $\sigma_g$ is the ratio of the diameter below which 84.1% of the particles lie to the median diameter. $D_{p\sigma}$ is one standard deviation from the median, so $\sigma_g$ is termed the *geometric standard deviation*.

In addition, 67% of all particles lie in the range $\overline{D}_{pg}/\sigma_g$ to $\overline{D}_{pg}\sigma_g$, and 95% of all particles lie in the range $\overline{D}_{pg}/2\sigma_g$ to $2\overline{D}_{pg}\sigma_g$. A monodisperse aerosol has the property that $\sigma_g = 1$.

The log-normal distribution has the useful property that when the cumulative distribution function is plotted against the logarithm of particle diameter on special graph paper with one axis scaled according to the error function of (5.134), so-called log-probability paper, a straight line results. The point at which $F(D_p) = 0.5$ occurs when $\ln D_p = \ln \overline{D}_{pg}$. The point at $F(D_p) = 0.84$ occurs for $\ln D_p = \ln \overline{D}_{pg} + \ln \sigma_g$ or $D_p = \overline{D}_{pg}\sigma_g$. The geometric mean or median is the value of $D_p$ where the straight-line plot of $F$ crosses the 50th percentile. The slope of the line is related to the geometric standard deviation $\sigma_g$, which can be calculated from the plot by dividing the 84th percentile diameter (which is one standard deviation from the mean) by the 50th percentile diameter. This property can be expressed as

$$\sigma_g = \frac{\overline{D}_{pg}}{D_{p,-\sigma}} = \frac{D_{p,+\sigma}}{\overline{D}_{pg}} = \left(\frac{D_{p,+\sigma}}{D_{p,-\sigma}}\right)^{1/2} \tag{5.136}$$

where $-\sigma$ and $+\sigma$ are minus and plus one standard deviation from the geometric mean.

We have developed the log-normal distribution for the number concentration. In addition to the number distribution, the surface area and volume distributions are of interest. Thus we wish to examine the surface area and volume distributions of an aerosol whose number distribution is log-normal and given by (5.128). Since $n_s(D_p) = \pi D_p^2 n(D_p)$ and $n_v(D_p) = (\pi/6)D_p^3 n(D_p)$, let us determine the forms of $n_s(D_p)$ and $n_v(D_p)$ when $n(D_p)$ is log-normal. From (5.128) we see that these two distributions fall within the general form of $n_\gamma(D_p) = a_\gamma D_p^\gamma n(D_p)$, where $\gamma = 2$ and 3 for the surface area and volume distributions, respectively, and $a_\gamma$ is the appropriate coefficient, either $\pi$ or $\pi/6$.

Thus we have

$$n_\gamma(D_p) = \frac{a_\gamma N D_p^\gamma}{\sqrt{2\pi}\, D_p \ln \sigma_g} \exp\left[-\frac{(\ln D_p - \ln \overline{D}_{pg})^2}{2 \ln^2 \sigma_g}\right] \tag{5.137}$$

By letting $D_p^\gamma = \exp(\gamma \ln D_p)$, expanding the exponential and completing the square in the exponent, (5.137) becomes

$$n_\gamma(D_p) = \frac{a_\gamma N}{\sqrt{2\pi}\, D_p \ln \sigma_g} \exp\left(\frac{\gamma \ln \overline{D}_{pg}}{\ln \sigma_g} + \frac{\gamma^2}{2}\right) \times \exp\left\{-\frac{[\ln D_p - (\ln \overline{D}_{pg} + \gamma \ln^2 \sigma_g)]^2}{2 \ln^2 \sigma_g}\right\} \tag{5.138}$$

Thus we see that if $n(D_p)$ is log-normal, $n_\gamma(D_p) = a_\gamma D_p^\gamma n(D_p)$ is also log-normal with the same geometric standard deviation of $\sigma_g$ as the parent distribution and with the logarithm of the median diameter given by

$$\ln \overline{D}_{pg\gamma} = \ln \overline{D}_{pg} + \gamma \ln^2 \sigma_g \tag{5.139}$$

## 5.8 GENERAL DYNAMIC EQUATION FOR AEROSOLS

The dynamic behavior of an aerosol is described by a population balance equation that can be termed the *general dynamic equation* (GDE). In the most general form of this equation the independent variables are particle size and composition, although in most applications size is the only variable characterizing the aerosol.

In the most fundamental approach to deriving the GDE, particles are represented as consisting of integer multiples of a single structural unit, typically a molecule. In these discrete equations particles differ only in the number of monomers they contain.

### 5.8.1 Discrete General Dynamic Equation

We consider the following phenomena to be occurring: (1) agglomeration of two particles, (2) evaporation or escape of a monomer from a particle, and (3) homogeneous particle generation or removal processes apart from those that occur as a result of evaporation and agglomeration. We restrict our attention to size distribution dynamics and do not consider particle composition as an independent variable. Thus the aerosol may be considered as chemically homogeneous for the purposes of deriving the governing dynamic equation.

For a spatially homogeneous aerosol the quantity of interest is the concentration of particles containing $i$ monomers, where $i \geq 1$. Assuming that an $i$-mer has a volume $v_i$, the concentration of $i$-mers, $N(v_i, t)$, will vary with time due to agglomeration, evaporation, generation, and removal processes. The rate of agglomeration of $i$-mers with $j$-mers is equal to the rate of formation of $(i + j)$-mers and is given by

$$\frac{K(v_i, v_j)N(v_i, t)N(v_j, t)}{1 + \delta_{i,j}} \qquad i, j \geq 1 \tag{5.140}$$

where $\delta_{i,j}$ is the Kronecker delta,

$$\delta_{i,j} = \begin{cases} 1 & i = j \\ 0 & \text{otherwise} \end{cases}$$

and $K(v_i, v_j)$ is the kinetic coefficient of agglomeration of two particles of volumes $v_i$ and $v_j$. The functional form of $K(v_i, v_j)$ will be discussed later. If $i$ is equal to $j$, we must divide by 2, so as not to count the agglomeration twice. The rate of evaporation of $i$-mers is $E(v_i)N(v_i, t)$, $i \geq 2$, where $E(v_i)$ is the evaporation coefficient. The rate of formation of $i$-mers by agglomeration is the sum of all agglomerations resulting in an $i$-mer and is given by

$$\frac{1}{2}\sum_{j=1}^{i-1} K(v_{i-j}, v_j)N(v_{i-j}, t)N(v_j, t) \qquad i \geq 2 \tag{5.141}$$

The factor of $\frac{1}{2}$ is introduced because $K$ is a symmetric function of its arguments, and therefore the summation counts each agglomeration twice for $i - j$ not equal to $i$. However, if $i$ is an even integer, the term $K(v_{i/2}, v_{i/2})N(v_{i/2}, t)N(v_{i/2}, t)$ is counted only once in the summation, but the factor of $\frac{1}{2}$ is still required, as given in (5.140). The rate of depletion of $i$-mers by agglomeration with all other particles is given by

$$N(v_i, t) \sum_{j=1}^{\infty} K(v_i, v_j)N(v_j, t) \qquad i \geq 1 \tag{5.142}$$

For $j$ equal to $i$, the agglomeration rate is divided by 2 as given in (5.140), but because each agglomeration removes two $i$-mers, the rate is also multiplied by 2, thereby canceling the factor of $\frac{1}{2}$. The rate of formation of $i$-mers by evaporation from $(i + 1)$-mers is $(1 + \delta_{1,i})E(v_{i+1})N(v_{i+1}, t)$, $i \geq 1$. The rate of depletion of $i$-mers due to evaporation is given by $E(v_i)N(v_i, t)$, $i \geq 2$.

The net rate of formation of monomers is thus

$$\frac{dN(v_1, t)}{dt} = -N(v_1, t) \sum_{j=1}^{\infty} K(v_1, v_j)N(v_j, t) + \sum_{j=2}^{\infty} (1 + \delta_{2,j})E(v_j)N(v_j, t) \tag{5.143}$$

and the net rate of formation of $i$-mers for $i \geq 2$ is

$$\begin{aligned} \frac{dN(v_i, t)}{dt} = {} & \frac{1}{2} \sum_{j=1}^{i-1} K(v_{i-j}, v_j)N(v_{i-j}, t)N(v_j, t) \\ & - N(v_i, t) \sum_{j=1}^{\infty} K(v_i, v_j)N(v_j, t) \\ & + E(v_{i+1})N(v_{i+1}, t) - E(v_i)N(v_i, t) \end{aligned} \tag{5.144}$$

Combined with the appropriate initial conditions [i.e., $N(v_i, 0)$, $i \geq 1$], (5.143) and (5.144) constitute the discrete GDE for a spatially homogeneous aerosol. Because agglomeration constantly produces larger particles, (5.143) and (5.144) are an infinite set of coupled ordinary differential equations.

### 5.8.2 Continuous General Dynamic Equation

Although the discrete GDE is an accurate description of aerosol dynamics, the number of equations needed to simulate actual aerosols can be immense. For $i >> 1$ the difference in size between an $i$-mer and an $(i + 1)$-mer is relatively small. Thus for particles that contain $k + 1$ or more monomers, where $k >> 1$, the discrete concentrations can be represented by $n(v, t)$, defined by

$$N(v_i, t) = \int_{v_i - v_1/2}^{v_i + v_1/2} n(v, t)\, dv \qquad i \geq k + 1 \tag{5.145}$$

If $N(v_i, t)$ is neglected for $2 \le i \le k$, (5.144) becomes, for $i \ge k$,

$$\frac{\partial n(v,t)}{\partial t} = \frac{1}{2}\int_{v_{k+1}-v_1/2}^{v-v_{k+1}+v_1/2} K(v-u,u)\,n(v-u,t)\,n(u,t)\,du$$
$$- n(v,t)\int_{v_{k+1}-v_1/2}^{\infty} K(v,u)\,n(u,t)\,du$$
$$+ n(v-v_1,t)\,K(v-v_1,v_1)\,N(v_1,t)$$
$$- n(v,t)\left[K(v,v_1)\,N(v_1,t) + E(v)\right] + E(v+v_1)\,n(v+v_1,t) \tag{5.146}$$

In the limit as $v_1/v \to 0$, the last three terms of (5.146) reduce to

$$-\frac{\partial}{\partial v}\left\{\left[K(v,v_1)\,N(v_1,t) - E(v)\right] v_1 n(v,t)\right\}$$
$$+ \frac{\partial^2}{\partial v^2}\left\{\left[K(v,v_1)\,N(v_1,t) + E(v)\right]\frac{v_1^2 n(v,t)}{2}\right\} \tag{5.147}$$

For most aerosols it has been shown that the second term of (5.147) can be neglected (Ramabhadran et al., 1976). Therefore, (5.146) becomes

$$\frac{\partial n(v,t)}{\partial t} = \frac{1}{2}\int_{v_{k+1}}^{v} K(v-u,u)\,n(v-u,t)\,n(u,t)\,du$$
$$- n(v,t)\int_{v_{k+1}}^{\infty} K(v,u)\,n(u,t)\,du - \frac{\partial[I(v,t)\,n(v,t)]}{\partial v} \tag{5.148}$$

where $I(v,t) = [K(v,v_1)\,N(v_1,t) - E(v)]\,v_1$. $I(v,t)$ is the net growth rate of a particle of volume $v$ due to condensation and evaporation of monomers and is commonly called the condensation growth rate or the *growth law*. Notice that (5.148) is defined over the domain $v \ge v_{k+1}$ and that it has been assumed that $v >> v_{k+1}$ in the upper limit of the integral. Moreover, it is generally assumed that $v_{k+1}$ may be replaced by zero in the lower limit of the integral. Thus the most common form of the continuous GDE is

$$\frac{\partial n(v,t)}{\partial t} = \frac{1}{2}\int_0^{v} K(v-u,u)\,n(v-u,t)\,n(u,t)\,du$$
$$- n(v,t)\int_0^{\infty} K(v,u)\,n(u,t)\,du$$
$$- \frac{\partial}{\partial v}[I(v,t)\,n(v,t)] + J(v^*,t) \tag{5.149}$$

The term on the left-hand side describes the evolution of the aerosol number distribution with time. The first term on the right-hand side represents the coagulation of particles of volume smaller than $v$ forming a particle of volume $v$; the second term represents the coagulation of particles of volume $v$ with other particles, forming a particle whose volume is larger than $v$; and the third term represents the change in the aerosol size distribution due to condensation, evaporation, chemical reaction, or phase transition. Such processes result in a growth or shrinkage of particles that is represented by the term $I(v, t) = dv/dt$. In some cases (e.g., for organic aerosols) the term $dv/dt$ may be negative for small particles and positive for large particles, indicating that evaporation occurs from small particles while condensation simultaneously takes place on large particles. Chemical reactions and phase transition may lead to a growth or shrinkage of particles due to changes in the molar volume of the chemical species. The fourth term represents the nucleation of particles from a gaseous chemical species. This nucleation takes place at a critical particle size at volume $v^*$. The term $J(v^*, t)$ represents the rate of formation of new particles of volume $v^*$ by nucleation. Later in this chapter we discuss how one determines the rate of and the critical particle size for nucleation.

In the absence of particle growth, $I(v, t) = 0$, and (5.149) becomes the so-called *coagulation equation*,

$$\frac{\partial n(v, t)}{\partial t} = \frac{1}{2}\int_0^v K(v-u, u)\, n(v-u, t)\, n(u, t)\, du - n(v, t)\int_0^\infty K(v, u)\, n(u, t)\, du \tag{5.150}$$

If coagulation can be neglected, the size distribution evolves due only to growth of individual particles, and $n(v, t)$ satisfies the *pure growth equation*,

$$\frac{\partial n(v, t)}{\partial t} + \frac{\partial}{\partial v}\left[I(v, t)\, n(v, t)\right] = 0 \tag{5.151}$$

If particle diameter is the size variable of interest, the continuous GDE takes the form

$$\begin{aligned}\frac{\partial n(D_p, t)}{\partial t} &= -\frac{\partial}{\partial D_p}\left[I_D(D_p, t)\, n(D_p, t)\right] \\ &+ D_p^2 \int_0^{D_p/2^{1/3}} \frac{K\left[(D_p^3 - \tilde{D}_p^3)^{1/3}, \tilde{D}_p\right] n\left[(D_p^3 - \tilde{D}_p^3)^{1/3}, t\right] n(\tilde{D}_p, t)\, d\tilde{D}_p}{(D_p^3 - \tilde{D}_p^3)^{2/3}} \\ &- n(D_p, t)\int_0^\infty K(D_p, \tilde{D}_p)\, n(\tilde{D}_p, t)\, d\tilde{D}_p \end{aligned} \tag{5.152}$$

where $I_D = dD_p/dt$, the rate of particle growth from condensation, as given by (5.98).

## 5.9 COAGULATION COEFFICIENT

Coagulation, agglomeration, or coalescence of two particles can occur as a result of several mechanisms by which two particles may be brought together. The particles may collide as a result of their mutual Brownian motion; they may be brought into contact by laminar or turbulent velocity gradients in the gas; or they may collide due to differential settling velocities. In addition, when two particles get sufficiently close to each other, electrostatic, electrodynamic, and fluid dynamical forces between particles affect the actual rate at which the particles collide.

### 5.9.1 Brownian Coagulation

Under typical atmospheric conditions and for particles smaller than a few microns in diameter, the dominant mechanism for agglomeration is Brownian motion. The form of the Brownian coagulation coefficient $K(D_p, \tilde{D}_p)$ in the continuum regime can be derived by considering the flux from the Brownian diffusion of particles of one size toward a stationary particle of the other size. In the free molecule or kinetic regime, $K$ is determined from the effusion flux (5.86) of one particle onto the area represented by the other particle. As in the case of mass transfer of vapor molecules, an appropriate interpolation formula is needed so that one has an expression for $K$ valid across the entire range of Knudsen numbers. For the derivation of these forms of $K$, we refer the reader to Seinfeld (1986, p. 391 et seq.).

A widely used form of the Brownian coagulation coefficient is that due to Fuchs (1964):

$$K(D_{pi}, D_{pj}) = 2\pi(D_i + D_j)(D_{pi} + D_{pj})\left[\frac{D_{pi} + D_{pj}}{D_{pi} + D_{pj} + 2g_{ij}} + \frac{8(D_i + D_j)}{\bar{c}_{ij}(D_{pi} + D_{pj})}\right]^{-1} \tag{5.153}$$

where $g_{ij} = (g_i^2 + g_j^2)^{1/2}$,

$$g_i = \frac{1}{3D_{pi}l_i}\left[(D_{pi} + l_i)^3 - (D_{pi}^2 + l_i^2)^{3/2}\right] - D_{pi} \tag{5.154}$$

$$l_i = \frac{8D_i}{\pi\bar{c}_i} \tag{5.155}$$

and where $D_i = kTC_c/3\pi\mu D_{pi}$ and $\bar{c}_i = (8kT/\pi m_i)^{1/2}$

In the continuum and kinetic regimes (5.153) approaches the limits

$$K_c(D_{pi}, D_{pj}) = 2\pi(D_i + D_j)(D_{pi} + D_{pj}) \tag{5.156}$$

$$K_k(D_{pi}, D_{pj}) = \frac{\pi}{4}(D_{pi} + D_{pj})^2(\bar{c}_i^2 + \bar{c}_j^2)^{1/2} \tag{5.157}$$

respectively.

Figure 5.10 shows $K(D_{pi}, D_{pj})$ as computed by (5.153). The smallest value of the Brownian coagulation coefficient occurs when both particles are of the same size. In the continuum regime for equal-sized particles, $K$ is independent of particle size and is given by $K = 8kT/3\mu$. On the other hand, when both particles are in the free molecule

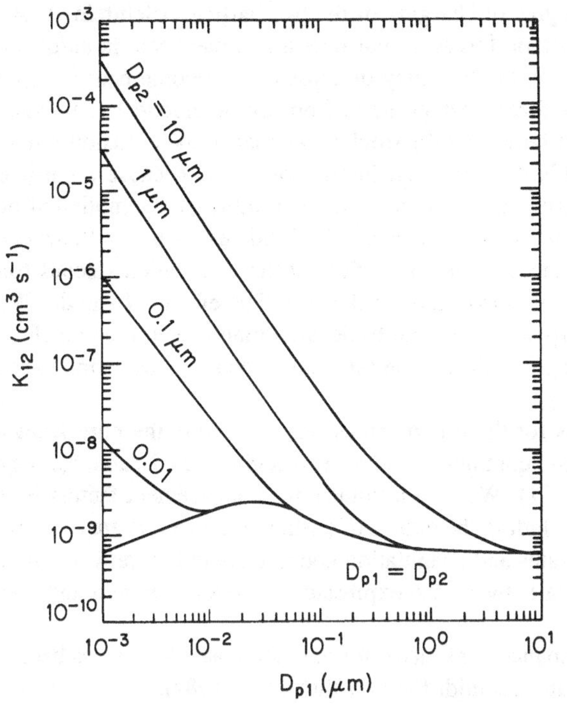


**Figure 5.10** Brownian coagulation coefficient $K(D_{pi}, D_{pj})$ for particles of density $\rho_p = 1$ g cm$^{-3}$ in air at 298 K, 1 atm.

regime and are of equal size, $K = 4(6kT/\rho_p)^{1/2}D_p^{1/2}$. A maximum coagulation coefficient for equal-sized particles is reached at about 0.02 $\mu$m diameter. The coagulation coefficient for large particles is relatively low because of their relatively small Brownian diffusivities. Very small particles, on the other hand, have relatively high velocities, but their individual target areas are small, so they tend to miss each other. The maximum in $K$ at around 0.02 $\mu$m represents a balance between particle mobility and cross-sectional area for collision. Finally, we see that the Brownian coagulation coefficient for unequal-sized particles is always larger than that for either of the two particles coagulating with another particle of its own size. This behavior is attributable to the fact that if one starts with two equal-sized particles and increases the size of one of the particles, the Brownian coagulation coefficient increases because the target area increases as $D_p^2$, whereas the Brownian diffusivity decreases only as $D_p$.

### 5.9.2 Effect of van der Waals and Viscous Forces on Brownian Coagulation

The Brownian coagulation coefficient given by (5.153) is derived by considering both aerosol particles as hard spheres in Brownian motion in the absence of interparticle forces and fluid mechanical interactions between the particles. We noted above that once particles get sufficiently close to each other, interparticle forces may act to enhance or retard the rate of collision. Electrostatic or Coulomb forces may either enhance or retard collision rates depending on the degree of charging of the two particles (Seinfeld, 1986, p. 404 et seq.). Normally, electrostatic forces do not play a dominant role in aerosol collisions, if for no other reason than the tendency of aerosols to become neutralized. On the other hand, van der Waals forces, which arise from the interaction of fluctuating dipoles in the two particles, act between all particles to enhance the collision rate over that predicted in its absence. Fluid mechanical interactions arise because a particle in motion in a fluid induces velocity gradients in the fluid that influence the motion of other particles when they approach its vicinity. Because the fluid resists being "squeezed" out from between the approaching particles, the effect of these so-called viscous forces is to retard the coagulation rate from that in their absence. The effects of van der Waals and viscous forces can be incorporated into the basic Brownian coagulation coefficient, with van der Waals forces acting to enhance the rate and viscous forces acting to retard the rate.

In order to present results for these two forces, let us express the pure Brownian collisional flux of particles in the continuum and kinetic regimes as $\Phi_c$ and $\Phi_k$, respectively (number of collisions/$\text{cm}^3$ s). We then introduce the enhancement factors $W_c$ and $W_k$, so that the collision rates including the effects of particle interactions are $F_c = W_c \Phi_c$ and $F_k = W_k \Phi_k$. The collision rates and coagulation coefficients in the transition regime will be obtained by interpolating between the expressions in the continuum and kinetic regimes.

The continuum regime enhancement factor for van der Waals forces has been presented by Friedlander (1977), and Schmidt-Ott and Burtscher (1982),

$$W_c^{-1} = (R_{pa} + R_{pb}) \int_{R_{pa}+R_{pb}}^{\infty} \exp\left[\frac{E(r)}{kT}\right] \frac{dr}{r^2} \tag{5.158}$$

where $E(r)$ is the van der Waals interaction potential. If we let $x = R_{pb}/r$, (5.158) can be expressed as

$$W_c^{-1} = \left(1 + \frac{R_{pa}}{R_{pb}}\right) \int_0^{(1+R_{pa}/R_{pb})^{-1}} \exp\left[E(x)/kT\right] dx \tag{5.159}$$

Historically, van der Waals forces between condensed media have been calculated by summing over pairwise interactions between the molecular constituents of each body. This method, called Hamaker theory, leads to the van der Waals interaction potential (Parsegian, 1975)

$$E(r) = -\frac{A}{6}\left[\frac{2R_{pa}R_{pb}}{r^2 - (R_{pa} + R_{pb})^2} + \frac{2R_{pa}R_{pb}}{r^2 - (R_{pa} - R_{pb})^2} + \ln\frac{r^2 - (R_{pa} + R_{pb})^2}{r^2 - (R_{pa} - R_{pb})^2}\right] \tag{5.160}$$

where $A$ is the so-called Hamaker constant. Values of the Hamaker constant, expressed as $A/kT$ at STP, for some substances are as follows:

| | $A/kT$ |
|---|---|
| Benzene | 63.9 |
| Ethanol | 42.6 |
| Water | 20.2 |

In deriving the pure Brownian motion collision rate, it is assumed that one particle is stationary and the second particle diffuses toward it with a diffusion coefficient that is the sum of the individual diffusion coefficients, that is, from (5.67),

$$D^{\infty} = D_a + D_b = \frac{kT}{f_a} + \frac{kT}{f_b} \tag{5.161}$$

where $f_a = 6\pi\mu R_{pa}/C_{ca}$ and $f_b = 6\pi\mu R_{pb}/C_{cb}$. In this formulation, it is assumed that each particle approaches the other oblivious of the other's existence. As we noted above, however, in reality the velocity gradients around each particle influence the motion of an approaching particle. Spielman (1970) incorporated the effect of viscous interactions between particles through a modified relative diffusion coefficient $D_{ab} = kT/f_{ab}$. The effect of viscous interactions can then be included in the enhancement factor of (5.159) through the $D^{\infty}/D_{ab}$ ratio,

$$W_c^{-1} = \left(1 + \frac{R_{pa}}{R_{pb}}\right) \int_0^{(1+R_{pa}/R_{pb})^{-1}} \frac{D^{\infty}}{D_{ab}} \exp\left[\frac{E(x)}{kT}\right] dx \tag{5.162}$$

Alam (1987) has obtained an analytical expression for $D^{\infty}/D_{ab}$ that closely approximates Spielman's solution,

$$\frac{D^{\infty}}{D_{ab}} = 1 + \frac{2.6R_{pa}R_{pb}}{(R_{pa}+R_{pb})^2}\left[\frac{R_{pa}R_{pb}}{(R_{pa}+R_{pb})(r-R_{pa}-R_{pb})}\right]^{1/2} + \frac{R_{pa}R_{pb}}{(R_{pa}+R_{pb})(r-R_{pa}-R_{pb})} \tag{5.163}$$

We note that $D^{\infty}/D_{ab}$ is unity for sufficiently large separation $r$ and increases as the particle separation decreases. The increase in the $D^{\infty}/D_{ab}$ ratio decreases the enhancement factor $W_c$. Figure 5.11 shows $W_c$ as a function of the Hamaker constant for $R_{pb}/R_{pa} = 1$ and $R_{pb}/R_{pa} = 5$. We see that in the presence of viscous forces, the enhancement factor is always less than that in their absence. With van der Waals forces $W_c$ is largest for particles of equal size, whereas viscous forces are largest when the ratio of particle radii is large. We note from Figure 5.11 that the "enhancement" factor $W_c$, including both forces, is, for some values of $A$, actually a retardation factor since $W_c < 1$.

The enhancement factor due to van der Waals forces in the free molecule or kinetic regime is (Marlow, 1980)

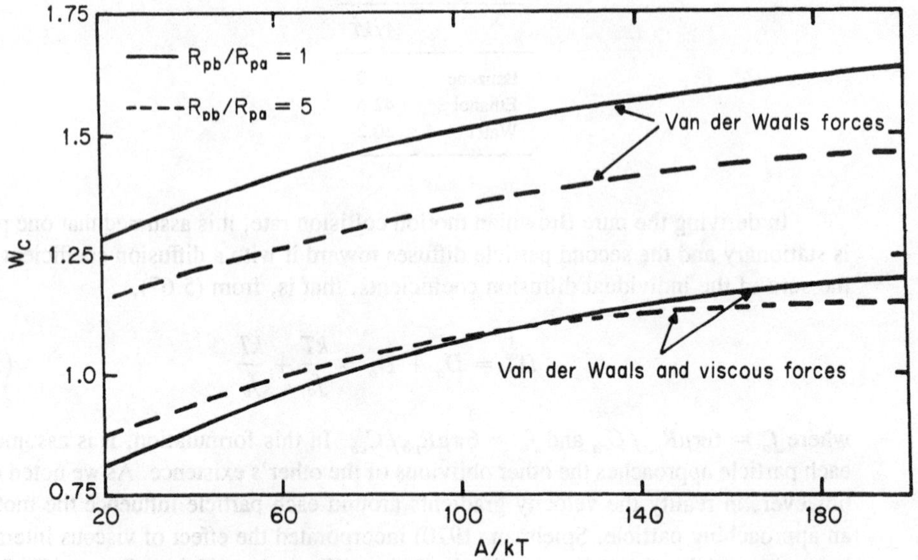


**Figure 5.11** Enhancement factor for Brownian coagulation coefficient due to van der Waals and viscous forces. (Reprinted by permission of the publisher from M. K. Alam, *Aerosol Science and Technology*, vol. 6, no. 1, pp. 41–52. Copyright 1987 by Elsevier Science Publishing Co., Inc.)

$$W_k = \frac{-1}{2(R_{pa} + R_{pb})^2 kT} \int_{R_{pa}+R_{pb}}^{\infty} \left( \frac{dE}{dr} + r\frac{d^2E}{dr^2} \right) \times \exp\left[ -\frac{1}{kT}\left( \frac{r}{2}\frac{dE}{dr} + E \right) \right] r^2\, dr \tag{5.164}$$

In the kinetic regime the particles behave as large gas molecules and the viscous interaction present when the suspending fluid can be treated as a continuum is absent.

Alam (1987) has derived an interpolation formula for the coagulation coefficient that is a function of a particle Knudsen number $\text{Kn}_p$ and that attains the proper limiting forms in the continuum and kinetic regimes. The result is

$$K(R_{pa}, R_{pb}) = 4\pi(R_{pa} + R_{pb})\,D^{\infty} \frac{W_c}{1 + (W_c/W_k)\,\alpha\, \text{Kn}_p} \tag{5.165}$$

where

$$D^{\infty} = D_a + D_b \qquad \alpha = \frac{4D^{\infty}}{(\lambda_a^2 + \lambda_b^2)^{1/2}\bar{c}}$$

$$\text{Kn}_p = \frac{(\lambda_a^2 + \lambda_b^2)^{1/2}}{R_{pa} + R_{pb}} \qquad \bar{c} = \left( \frac{8kT}{\pi m_a} + \frac{8kT}{\pi m_b} \right)^{1/2}$$

Figure 5.12 shows the coagulation coefficient $K$ at $T = 300$ K, $p = 1$ atm, $\rho_p = 1000\ \text{kg/m}^3$, and $A/kT = 20$ as a function of $\text{Kn}_p$ for particle radius ratios of 1, 2, and 5 both in the presence and in the absence of interparticle forces. Figure 5.13 shows the ratio of the coagulation coefficient to that for pure Brownian coagulation (i.e., the enhancement factor) as a function of particle Knudsen number for $A/kT = 20$. In the continuum regime, $\text{Kn} \leq 1$, for $A/kT = 20$, $W_c < 1$ due to the predominance of the retardation by viscous forces over the enhancement by van der Waals forces. As shown in Figure 5.11, for sufficiently large Hamaker constant $A$, the effect of van der Waals forces overtakes that of viscous forces to lead to an enhancement factor greater than 1. In the kinetic regime, coagulation is always enhanced due to the absence of viscous forces.

**Example 5.6** ***Coagulation of an Initially Monodisperse Aerosol***

In the absence of evaporation, (5.144) is usually expressed as

$$\frac{dN_i}{dt} = \frac{1}{2}\sum_{j=1}^{i-1} K_{i-j,j}N_{i-j}N_j - N_i\sum_{j=1}^{\infty} K_{i,j}N_j \tag{5.166}$$

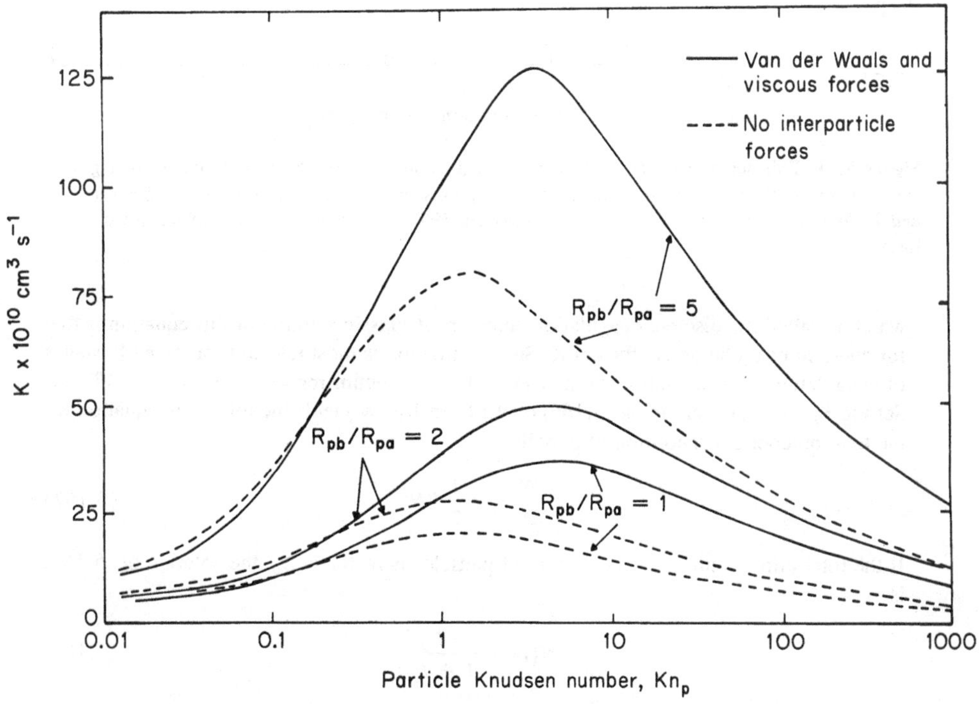


**Figure 5.12** Brownian coagulation coefficient in the presence and absence of interparticle forces. $T = 300$ K, $p = 1$ atm, $\rho_p = 1\ \text{g cm}^{-3}$, $A/kT = 20$. (Reprinted by permission of the publisher from M. K. Alam, *Aerosol Science and Technology*, vol. 6, no. 1, pp. 41–52. Copyright 1987 by Elsevier Science Publishing Co., Inc.)

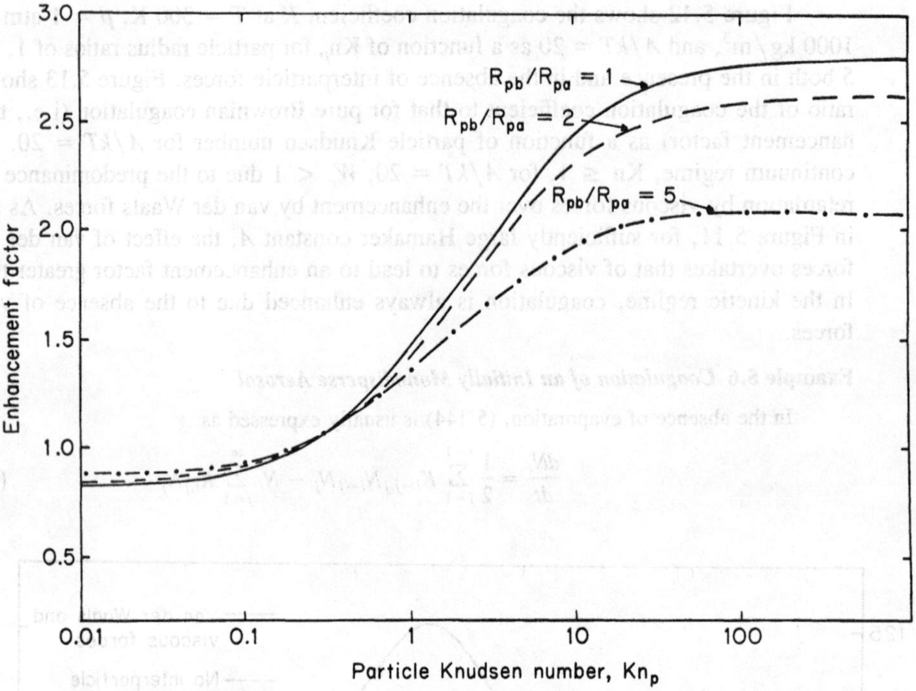


**Figure 5.13** Enhancement factor for Brownian coagulation. $T = 300$ K, $p = 1$ atm, $\rho_p = 1$ g cm$^{-3}$, $A/kT = 20$. (Reprinted by permission of the publisher from M. K. Alam, *Aerosol Science and Technology*, vol. 6, no. 1, pp. 41–52. Copyright 1987 by Elsevier Science Publishing Co., Inc.)

which is called the discrete coagulation equation. A classic solution of this equation is that for constant coagulation coefficient $K$. Such a situation is most relevant for the early stages of coagulation of a monodisperse aerosol in the continuum regime when $K_{i,j} = 8kT/3\mu$. Setting $K_{i,j} = K$ and summing (5.166) over $i$ from 1 to $\infty$ yields the following equation for the total number concentration of particles:

$$\frac{dN}{dt} = -\frac{1}{2}KN^2 \tag{5.167}$$

If the total initial number concentration of particles is $N(0) = N_0$, the solution of (5.167) is

$$N(t) = \frac{N_0}{1 + t/\tau_c} \tag{5.168}$$

where

$$\tau_c = \frac{2}{KN_0} \tag{5.169}$$

is a characteristic time for coagulation.

**Example 5.7** ***Characteristic Times for Coagulation***

The characteristic time for coagulation $\tau_c$ was defined by (5.169). It is the time needed for an intially monodisperse population of particles to coagulate to one-half the initial number concentration. Although the coagulation coefficient cannot be strictly independent of time because the average particle size of the aerosol is increasing as coagulation proceeds and the population is no longer monodisperse, $\tau_c$ represents a good approximation of the time constant for coagulation of an aerosol.

It is useful to calculate the value of $\tau_c$ for various initial particle sizes and number concentrations. For pure Brownian coagulation, the equal-size coagulation coefficients for a range of particle diameters are approximately:

| $D_p$ ($\mu$m) | $K$ (cm$^3$ s$^{-1}$) |
|---|---|
| 0.01 | $2 \times 10^{-9}$ |
| 0.1 | $1 \times 10^{-9}$ |
| 1.0 | $7 \times 10^{-10}$ |
| 10.0 | $6 \times 10^{-10}$ |

The characteristic times in seconds for coagulation of particles of these sizes with various initial number concentrations are:

| | $D_p$ ($\mu$m) | | | |
|---|---|---|---|---|
| $N_0$ (cm$^{-3}$) | 0.01 | 0.1 | 1.0 | 10.0 |
| $10^3$ | $10^6$ | $2 \times 10^6$ | $3 \times 10^6$ | $3.3 \times 10^6$ |
| $10^6$ | $10^3$ | $2 \times 10^3$ | $3 \times 10^3$ | $3.3 \times 10^3$ |
| $10^9$ | 1 | 2 | 3 | 3.3 |
| $10^{12}$ | $10^{-3}$ | $2 \times 10^{-3}$ | $3 \times 10^{-3}$ | $3.3 \times 10^{-3}$ |

**Example 5.8** ***Self-Preserving Size Distribution***

Certain aerosol coagulation and condensation situations can be solved using a similarity transformation for the size distribution function (Swift and Friedlander, 1964; Friedlander and Wang, 1966). Such solutions represent asymptotic forms approached after long times and are independent of the initial size distribution. In carrying out the similarity transformation, one assumes that the fraction of the particles in a given size range depends only on the particle volume normalized by the average particle volume,

$$\frac{n(v, t)\, dv}{N(t)} = \psi\left(\frac{v}{\bar{v}}\right) d\left(\frac{v}{\bar{v}}\right)$$

where $\bar{v} = V(t)/N(t)$ is the average particle volume at time $t$. Rearranging this equation, the dimensional size distribution function can be expressed as

$$n(v, t) = \frac{N(t)^2}{V(t)} \psi(\eta)$$

where $\eta = v/\bar{v} = vN(t)/V(t)$. Note that

$$N(t) = \int_0^\infty n(v, t)\, dv$$

$$V(t) = \int_0^\infty vn(v, t)\, dv$$

The size distribution is thus represented in terms of $N(t)$, $V(t)$, and a dimensionless function $\psi(\eta)$. The shapes of the distribution are assumed to be similar at different times, multiplied by the time-dependent factor $N(t)^2/V(t)$. Thus these solutions are called *self-preserving*.

To determine $\psi(\eta)$ the relation for $n(v, t)$ is substituted into the general dynamic equation. If the transform is consistent with the equation, an ordinary integrodifferential equation for $\psi(\eta)$ is obtained (Friedlander, 1977).

## 5.10 HOMOGENEOUS NUCLEATION

The general dynamic equation describes the evolution of a population of particles undergoing coagulation and growth or evaporation due to gas-to-particle or particle-to-gas conversion. The process by which particles form from a vapor can be viewed simply as the successive coagulation of single molecules (monomers) with molecular clusters. This process is represented explicitly in (5.143) and (5.144). When we passed to the continuous GDE (5.148) we took volume $v_{k+1}$ as the lower limit of particle volume down to which the size spectrum can be treated as if it is continuous. Presumably, coagulation among monomers and clusters of volume less than $v_{k+1}$ is still occurring that provides a flux of particles across size $v_{k+1}$ into the "continuous" portion of the size spectrum. In (5.149) and (5.150) these processes occurring at volume less than $v_{k+1}$ are simply smeared into the continuous coagulation integrals and growth term. If new particles are indeed forming from the vapor, we need to use (5.148), in which we can say something explicit about what is happening for particles of very small sizes. The formation of particles from vapor molecules may be visualized to occur as the result of the clustering of vapor molecules to form embryos of the new phase followed by the growth of some of these clusters to form macroscopic quantities of the new phase. Figure 5.14 is a schematic of the size spectrum. The passage of embryos beyond the size $k$ will be referred to as nucleation, whereas the scavenging of monomers and embryos by particles larger than $k$ is what we refer to as condensation. The major goal of nucleation theory is to predict the rate of formation of new particles from vapor molecules. We will see that a certain critical size $k$ emerges from the theory and that we will be able to predict the rate of passage of embryos across that size $k$. In this section we review briefly the theory of homogeneous nucleation. For more in-depth treatments we refer the reader to Springer (1978) and Abraham (1974).

Let us begin by considering a system composed of a spherical liquid drop of radius $r$ containing $N_l$ molecules surrounded by its vapor containing $N_v$ molecules. The temperature, pressure, and total number of molecules are assumed to be constant. The total

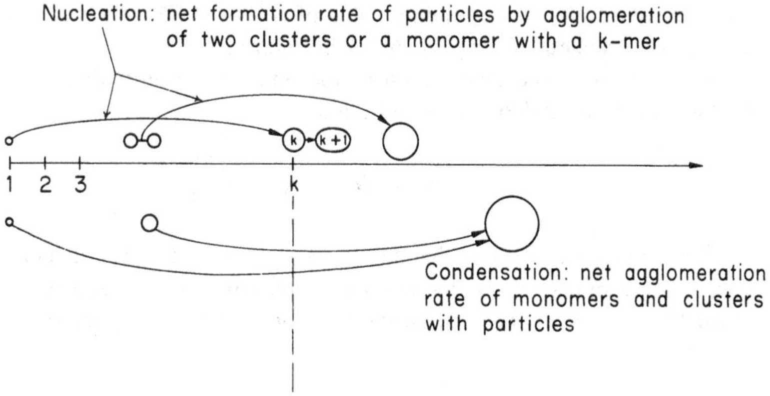


**Figure 5.14** Nucleation and condensation along the particle size spectrum.

Gibbs free energy of the system can be written as

$$G = N_v \mu_v + N_l \mu_l + 4\pi r^2 \sigma \tag{5.170}$$

where $\mu_v$ and $\mu_l$ are the chemical potentials per molecule of the vapor and liquid phases, respectively, and $\sigma$ is the surface tension of the bulk liquid. If we assume the liquid phase to be incompressible and the vapor phase to be ideal, the equilibrium state of such a system is that for which (Springer, 1978)

$$\ln \frac{p}{p_A^\circ} = \frac{2\sigma v_m}{kTr} \tag{5.171}$$

where $p$ is the pressure of the vapor in the system and $p_A^\circ$ is the equilibrium vapor pressure of the bulk liquid at temperature $T$. $v_m$ is the molecular volume. We will recognize (5.171) as equivalent to (5.97), the so-called Kelvin equation, relating the increase in vapor pressure over a curved interface to that over a flat surface in terms of the size of the droplet and its physical properties.

The Gibbs free energy of the system prior to formation of the liquid drop is $G_0 = (N_v + N_l)\mu_v$, and the difference $\Delta G = G - G_0$ is a measure of the reversible work necessary to form the drop,

$$\Delta G = g(\mu_l - \mu_v) + \xi g^{2/3} \tag{5.172}$$

where it is customary to let $g = N_l$, the number of molecules in the droplet, and $\xi = (4\pi)^{1/3}\sigma(3v_m)^{2/3}$. The first term on the right-hand side is proportional to the volume of the drop and the second to its surface area. If $\mu_l > \mu_v$, the vapor is the more stable phase, both terms in (5.172) are positive, and $\Delta G$ increases continuously as $g$ increases. If $\mu_l = \mu_v$, the vapor is saturated, but $\Delta G$ still increases continuously with $g$. A liquid drop will not form because of the additional work needed to create the droplet interface. Finally, if $\mu_l < \mu_v$, $\Delta G$ has a maximum at the point where the negative volume term begins to dominate the positive surface term. The value of $g$ at this maximum in $\Delta G$ is denoted by $g^*$ and its corresponding value of $\Delta G$ by $\Delta G^*$. We will call $g^*$ the critical size embryo or nucleus. This $g^*$ is just $k$ in Figure 5.14.

At equilibrium the distribution of the number concentrations of embryos is assumed to follow the Boltzmann distribution,

$$N_g^e = N_1 \exp\left(-\frac{\Delta G}{kT}\right) \tag{5.173}$$

The embryos can be envisioned to exist as a result of a sequence of collisions between single molecules (monomers) and embryos, embryo-embryo collisions being neglected as rare events because of the low concentration of embryos relative to single molecules,

$$A_1 + A_1 \rightleftharpoons A_2$$

$$A_2 + A_1 \rightleftharpoons A_3$$

$$\vdots$$

$$A_{g-1} + A_1 \rightleftharpoons A_g$$

where $A_1$ represents a single molecule and $A_g$ an embryo containing $g$ molecules. The nucleation rate is defined as that at which embryos pass through the critical size $g^*$.

We can write a balance on the number of embryos of size $g$ as

$$\frac{dN_g}{dt} = \alpha_{g+1}a_{g+1}N_{g+1} + \beta a_{g-1}N_{g-1} - \alpha_g a_g N_g - \beta a_g N_g \tag{5.174}$$

where $a_g$ is the surface area of a $g$-mer, $\alpha_g$ is the rate of loss of single molecules per unit surface area from a $g$-mer, and $\beta$ is the rate of impingement of vapor molecules on an embryo per unit surface area. Whereas $\alpha_g$ is generally an unknown function of embryo size, $\beta$ is usually approximated by the kinetic theory effusion flux expression, $N\bar{c}/4$, for the rate at which vapor molecules impinge on a unit area of surface per unit time.

Equation (5.174) can be written as

$$\frac{dN_g}{dt} = J_{g-1} - J_g \qquad g \geq 2 \tag{5.175}$$

where $J_g$ is the net flow of embryos from size $g$ to $g+1$,

$$J_g = \beta a_g N_g - \alpha_{g+1}a_{g+1}N_{g+1} \tag{5.176}$$

At equilibrium not only is $dN_g/dt = 0$, but $J_2 = J_3 = \cdots = J_g = 0$; that is, there is no net flow of embryos in either direction. Setting $J_g = 0$ in (5.176), and using the superscript $e$ to denote equilibrium, gives

$$\frac{N_g^e}{N_{g+1}^e} = \frac{\alpha_{g+1} a_{g+1}}{\beta a_g} \tag{5.177}$$

which allows us to calculate the evaporation rate, $\alpha_{g+1}$, in terms of $\beta$, $N_g^e$ and $N_{g+1}^e$. Equation (5.177) is a statement of the principle of microscopic reversibility.

When nucleation is occurring there is a nonzero net flow of embryos as given by (5.176), and therefore the system cannot be at equilibrium. The major assumption of classical homogeneous nucleation theory is that the embryo number concentrations $N_g$ rapidly achieve a *pseudo-steady state* in which their rate of appearance is just balanced by their rate of disappearance, that is,

$$\frac{dN_g}{dt} = 0 = J_{g-1} - J_g \tag{5.178}$$

Physically, this assumption implies that the characteristic times for the embryos to relax to a pseudo-steady-state distribution is short compared with that associated with macroscopic changes in the system. Applying (5.178) to each embryo concentration gives

$$J_{g-1} = J_g = J \tag{5.179}$$

Thus a consequence of the pseudo-steady-state approximation applied to the embryo concentrations is that the net flow through the embryo distribution is a constant, independent of $g$. This net rate is the nucleation rate that we desire to compute. It is worth reiterating at this point that the pseudo-steady-state approximation applied to the embryo concentration is *not the same* as the condition of equilibrium, since in the former all $J_g$ equal a nonzero $J$, whereas the latter all $J_g$ equal zero.

Using (5.177) and (5.179), (5.176) can be expressed as

$$J = \beta a_g N_g^e \left( \frac{N_g}{N_g^e} - \frac{N_{g+1}}{N_{g+1}^e} \right) \tag{5.180}$$

Because changes in number concentrations from one embryo to the next are small for $g \gg 1$, it is customary to replace the right-hand side of (5.180) by its different form,

$$J = -\beta a_g N_g^e \frac{d}{dg} \left( \frac{N_g}{N_g^e} \right) \tag{5.181}$$

Equation (5.181) is then integrated to give the nucleation rate,*

$$J = \frac{1}{\rho_p} \left( \frac{2\sigma M_A}{\pi} \right)^{1/2} \left( \frac{p_A^\circ}{kT} \right)^2 S^2 \exp\left( -\frac{\Delta G^*}{kT} \right) \tag{5.182}$$

where $S = p/p_A^\circ$, the saturation ratio of species $A$. $J$ can be expressed in terms of measurable quantities as

*This integration requires several approximations that can be shown to be quite valid. For elaboration, see Friedlander (1977), Springer (1978), or Seinfeld (1986).

$$J = \frac{1}{\rho_p}\left(\frac{2\sigma M_A}{\pi}\right)^{1/2}\left(\frac{p_A^\circ}{T}\right)^2 \frac{N_{av}^{3/2}}{R^2} S^2 \exp\left(-\frac{16\pi M_A^2 \sigma^3 N_{av}}{3\rho_p^2 R^3 T^3 \ln^2 S}\right) \tag{5.183}$$

where $N_{av}$ is Avogadro's number.

It is enlightening to examine the dependence of the nucleation rate $J$ on the saturation ratio $S$. Figure 5.15 shows $J$ (in number per $m^3$ per second) as a function of the saturation ratio $S$ for water and dibutylphthalate (DBP) at 293 K and 1 atm. Although the nucleation rate is a continuous function of $S$, it increases so rapidly as $S$ increases that it appears that at a certain "critical" saturation a burst of nucleation occurs. This critical saturation ratio is usually defined as that where $J = 1\ \mathrm{cm}^{-3}\ \mathrm{s}^{-1}$. Rates significantly larger than this are required to form substantial numbers of particles, although because of the extreme steepness of the $J$ versus $S$ curve, only slight increases of $S$ beyond that at $J = 1\ \mathrm{cm}^{-3}\ \mathrm{s}^{-1}$ are needed to obtain such significant rates. The critical saturation ratio for water under the conditions of Figure 5.15 occurs at about $S = 10$ ($J = 10^6\ \mathrm{m}^{-3}\ \mathrm{s}^{-1} = 1\ \mathrm{cm}^{-3}\ \mathrm{s}^{-1}$). For dibutylphthalate, the critical $S$ at these conditions is close to 800. The difference in nucleation behavior between the two compounds can be explained on the basis of their physical properties.

To show how the nucleation rate of a compound depends on its properties, we write $J$ in the form equivalent to (5.183),

$$J = \frac{p_A}{\sqrt{2\pi m_A kT}}\left(2\sqrt{\sigma}\frac{v_A}{\sqrt{kT}}\right) N_A \exp\left(-\frac{16\pi\sigma^3 v_A^2}{3k^3T^3 \ln^2 S}\right) \tag{5.184}$$

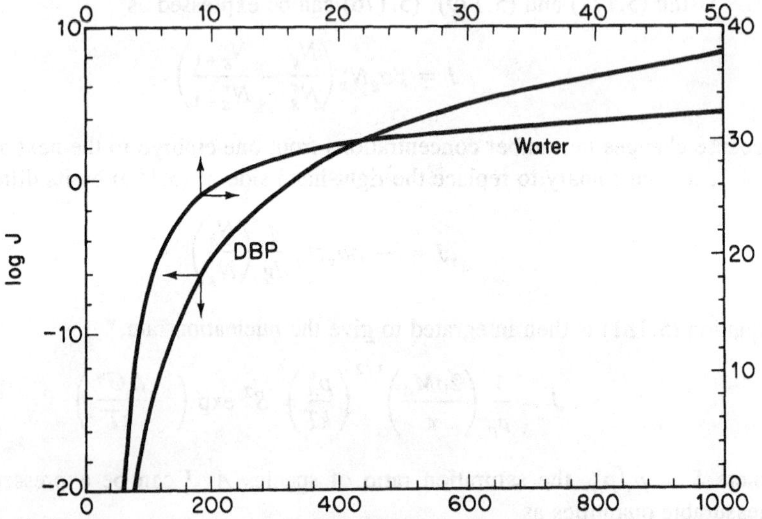


**Figure 5.15** Nucleation rates of dibutylphthalate and water at 293 K as a function of saturation ratio.

where $m_A$ and $v_A$ are the molecular mass and volume, respectively, of the monomer; and then place (5.184) in dimensionless form as follows:

$$J^* = S^2 \sqrt{\frac{\sigma^*}{6\pi}} \exp\left(-\frac{\sigma^{*3}}{2 \ln^2 S}\right) \tag{5.185}$$

where $J^* = J/\hat{J}$ and

$$\hat{J} = \left(\frac{p_A^\circ}{kT}\right)^2 \pi r_A^2 \left(\frac{8kT}{\pi m_A}\right)^{1/2}$$

$$\sigma^* = \frac{2v_A \sigma}{r_A kT}$$

Note that $\hat{J}$ is related to the monomer-monomer collision rate at saturated conditions and $\sigma^*$ is the dimemsionless surface tension.

Table 5.2 gives the relevant physical properties of water and DBP at 293 K and 1 atm, together with values of $\sigma^*$ and $\hat{J}$, the latter resulting from the extremely low saturation vapor pressure of DBP. Because of the low value of $\hat{J}$, the nucleation rate $J$ is smaller than that of water at the same value of $S$.

From (5.185) one can obtain

$$\frac{\partial \ln J}{\partial \ln S} = \frac{\partial \log J}{\partial \log S} = 2 + \left(\frac{\sigma^*}{\ln S}\right)^3 \tag{5.186}$$

We see that a material with a large $\sigma^*$, such as DBP in the present case, would require a larger value of $S$ to attain the same slope in a $\log J - \log S$ plot than a material with a smaller $\sigma^*$. This explains why DBP attains nucleation rates equivalent to those of water only at higher saturation ratios. Note that as $S$ gets very large, the asymptotic slopes of all $\log J - \log S$ plots are equal to 2.

Figure 5.16 shows $J^*$ as a function of $S$ at various values of $\sigma^*$, and Figure 5.17 presents $J^*$ as a function of $1/\sigma^*$ at various $S$ values. From Figure 5.16 we see that at low values of $S$, the dimensionless nucleation rate $J^*$ is extremely low. The larger the dimensionless surface tension $\sigma^*$, the more difficult the nucleation process. From Figure 5.17, $J^*$ has a maximum at $\sigma^* = (\ln^2 S/3)^{1/3}$ equal to $J^* = 0.1623 S^2 \ln^{1/3} S$.

TABLE 5.2 PHYSICAL PROPERTIES OF DIBUTYLPHTHALATE AND WATER AT 293 K, 1 ATM

| Properties | Dibutylphthalate (DBP) | Water |
|---|---|---|
| $m_A$ (kg) | $4.6196 \times 10^{-25}$ | $2.9890 \times 10^{-26}$ |
| $v_A$ (m$^3$) | $4.4169 \times 10^{-28}$ | $2.9890 \times 10^{-29}$ |
| $p_A^\circ$ (N m$^{-2}$) | $1.3599 \times 10^{-3}$ | 2336.468 |
| $\sigma$ (J m$^{-1}$) | $3.357 \times 10^{-2}$ | $7.275 \times 10^{-2}$ |
| $\sigma^*$ | 15.5086 | 5.5812 |
| $\hat{J}$ (m$^{-3}$ s$^{-1}$) | $1.1824 \times 10^{19}$ | $2.2786 \times 10^{31}$ |

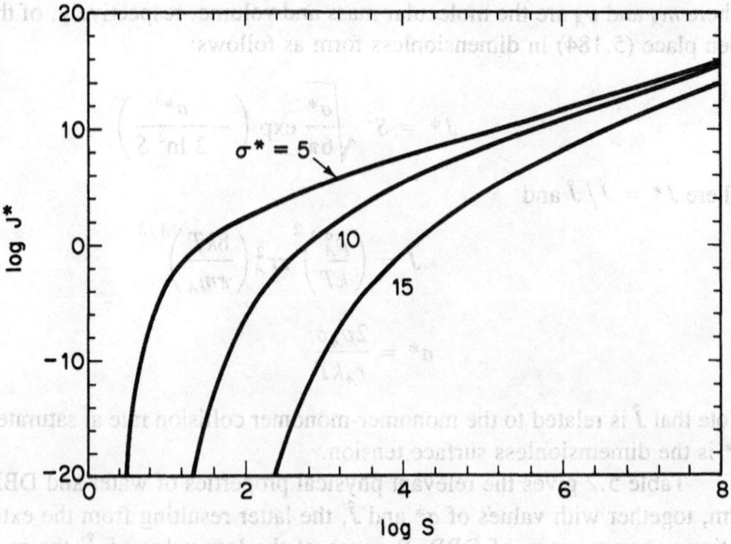


**Figure 5.16** Dimensionless nucleation rate $J^*$ as a function of saturation ratio for different values of the dimensionless surface tension $\sigma^*$.

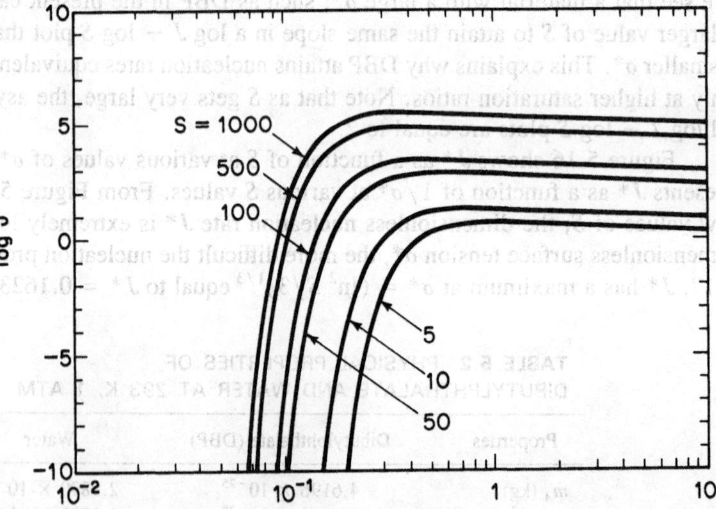


**Figure 5.17** Dimensionless nucleation rate $J^*$ as a function of $\sigma^{*-1}$ for different values of the saturation ratio S.

## 5.11 SECTIONAL REPRESENTATION OF AEROSOL PROCESSES

An aerosol size distribution is governed by a variety of physical and chemical processes comprising coagulation; gas-aerosol conversion mechanisms such as nucleation, condensation, and evaporation; gravitational settling; dry deposition on surfaces due to diffusion, impaction, and thermophoresis; phase transition; chemical reactions in the particle or at its surface; and introduction of new particles. A description of the evolution of the aerosol size distribution requires an accurate solution of the appropriate form of the general dynamic equation.

In the continuous representation of the aerosol size distribution, the aerosol population is represented by a continuous function that characterizes the number or volume concentration of aerosols as a function of aerosol size. An approximation of the continuous representation by constant number or volume concentrations of aerosols over finite aerosol size ranges leads to a so-called sectional representation of the aerosol size distribution (Gelbard et al., 1980; Gelbard and Seinfeld, 1980). The sectional approximation to the continuous size distribution leads to a simpler form of the equation since the sectional distribution is then governed by a set of ordinary differential equations.

In the sectional approach to modeling aerosol dynamics, the continuous size distribution is approximated by a series of step functions so that the aerosol characteristics (i.e., number, surface, volume, mass concentration) remain constant over each size interval. For example, the number distribution is approximated as follows:

$$N_l(t) = \int_{v_{l-1}}^{v_l} n(v, t)\, dv \qquad l = 1, \ldots, L \tag{5.187}$$

where $N_l(t)$ = aerosol number concentration in section $l$

$L$ = total number of aerosol size sections

$v_0$ and $v_L$ = aerosol volume of the lower and upper ranges of the aerosol distribution, respectively

The number distribution $n(v, t)$ is then approximated by the step functions $N_l(t)(v_l - v_{l-1})^{-1}$. The volume size distribution is approximated in a similar manner, as follows:

$$Q_l(t) = \int_{v_{l-1}}^{v_l} vn(v, t)\, dv \qquad l = 1, \ldots, L \tag{5.188}$$

The aerosol dynamic equations can be solved for any representation of the size distribution (i.e., number, volume, etc.). Generally, the sectional equations are solved for the mass or volume distribution. This solution is convenient because it allows one to check for conservation of mass or volume. In addition, it is computationally more accurate because the volume distribution covers a smaller range of concentration values than does the number distribution.

The coagulation equation for the sectional volume distribution can be deduced from the coagulation equation (5.150) to give the following ordinary differential equations (Gelbard et al., 1980):

$$\frac{dQ_l(t)}{dt} = \sum_{i=1}^{l-1} {}^a\overline{K}_{i,l-i} Q_i(t) Q_{l-i}(t) - Q_l(t) \sum_{i=1}^{L} {}^b\overline{K}_{i,l} Q_i(t) \qquad l = 1, \ldots, L \tag{5.189}$$

where the first term is not evaluated for $l = 1$.

The first term represents the formation of new aerosols in section $l$ by coagulation of aerosols from lower sections, and the second term represents the coagulation of aerosols from section $l$ with other aerosols to form particles in higher sections.

The sectional coagulation coefficients ${}^a\overline{K}_{i,l-i}$ and ${}^b\overline{K}_{i,l}$ are calculated by integrating the coagulation coefficient $K(v_1, v_2)$ over the section of interest (see Gelbard et al., 1980, for the derivation of the sectional coagulation coefficients). These coefficients generally are calculated with the constraint that $v_l \geq 2v_{l-1}$. For a fixed sectionalization of the aerosol size distribution, the sectional coagulation coefficients are computed only once, and the coagulation equation is then reduced from an integrodifferential equation to a set of ordinary differential equations.

Similarly, the pure growth equation governing the sectional volume distribution can be deduced from (5.151) to give (Gelbard and Seinfeld, 1980)

$$\frac{dQ_l(t)}{dt} = {}^a\overline{\phi}_l Q_l(t) + {}^b\overline{\phi}_{l-1} Q_{l-1}(t) - {}^b\overline{\phi}_l Q_l(t) \qquad l = 1, \ldots, L \tag{5.190}$$

where the first term is not evaluated for $l = 1$. The first term represents the growth of the aerosol volume in section $l$ due to condensation, the second term represents the growth of the aerosol volume from section $(l - 1)$ into section $l$, and the third term represents the growth of the aerosol volume from section $l$ into section $(l + 1)$. The sectional condensation coefficients are calculated by integrating the growth law over the sections. If the sectionalization of the aerosol size distribution is fixed, these sectional condensation coefficients are calculated only once, and the condensation equation is then reduced from a partial differential equation to a set of ordinary differential equations.

Warren and Seinfeld (1985) improved the accuracy of the treatment of condensation by the sectional representation by using an average value of the volume concentrations for the intersectional growth terms and by forcing conservation of the aerosol number concentration. The result is

$$\frac{dQ_l(t)}{dt} = {}^a\overline{\phi}_l Q_l(t) + {}^c\overline{\phi}_{l-1} \frac{Q_{l-1}(t) + Q_l(t)}{2} - {}^c\overline{\phi}_l \frac{Q_l(t) + Q_{l+1}(t)}{2} \qquad l = 1, \ldots, L \tag{5.191}$$

$$\sum_{l=1}^{L} \frac{1}{\overline{v}_l} \frac{dQ_l(t)}{dt} = 0 \tag{5.192}$$

where $\overline{v}_l$ is the average volume of aerosols in section $l$.

## PROBLEMS

**5.1.** What is the stop distance of a spherical particle of 10 $\mu$m diameter and density 3.0 g cm$^{-3}$ projected into still air at 298 K with an initial velocity of 100 cm s$^{-1}$?

**5.2.** Calculate and plot the terminal settling velocities of particles of density 1.0 g cm$^{-3}$ in air at 1 atm and 298 K over the particle diameter range 0.1 to 100 $\mu$m.

**5.3.** Assuming a 2% deviation to be the maximum permissible, calculate the maximum particle diameter for which the terminal settling velocity may be computed by Stokes' law in air at 1 atm and 298 K as a function of particle density over the range 0.5 to 5.0 g cm$^{-3}$.

**5.4.** A 10-$\mu$m-diameter particle of density 1 g cm$^{-3}$ is being carried by an airstream at 1 atm and 298 K in the $x$-direction with a velocity of 100 cm s$^{-1}$. The stream enters a device where the particle immediately acquires a charge of 100 electrons (the charge on a single electron, $e$, is $1.6 \times 10^{-19}$ C) and in which there exists an electric field of constant potential gradient $E_y = 10^3$ V cm$^{-1}$ perpendicular to the direction of flow.

**(a)** Determine the characteristic relaxation time of the particle.

**(b)** Determine the $x$- and $y$-coordinate locations of the particle at any downstream location in the device.

**5.5.** **(a)** Verify the relations given in Table 5.1.

**(b)** Prepare a plot of classical aerodynamic diameter versus physical diameter for physical diameter from 0.1 to 20 $\mu$m and for the four densities $\rho_p = 1, 2, 3$, and 4 g cm$^{-3}$.

**5.6.** One stage of a cascade impactor is shown in Figure 5.4. As described in Example 5.2, particles with sufficient inertia are collected on the plate; those smaller are carried around the plate to the next stage. The flow enters the stage through a cylindrical nozzle of diameter $D_j$ that terminates at a distance $s$ above the plate. The flow on the stage can be approximated as a two-dimensional, cylindrical stagnation flow near the plate,

$$u_r^* = r^* \qquad u_z^* = -2z^*$$

where the two velocity components are made dimensionless by $u_r^* = u_r/u_\infty$ and $u_z^* = u_z/u_\infty$, and $r^* = r/(D_j/2)$ and $z^* = z/(D_j/2)$.

**(a)** Show that the dimensionless equations of motion for a particle in this flow field are

$$2\text{St}\frac{d^2r^*}{dt^{*2}} + \frac{dr^*}{dt^*} = r^*$$

$$2\text{St}\frac{d^2z^*}{dt^{*2}} + \frac{dz^*}{dt^*} = -2z^* + v_t^*$$

where $\text{St} = \tau u_\infty/D_j$, $t^* = D_j t/2u_\infty$, and $v_t^* = -\tau g/u_\infty$, the dimensionless settling velocity.

**(b)** Show that appropriate initial conditions are

$$r^* = r_0^* \le 1 \qquad z^* = \frac{1}{2} \qquad \text{at } t^* = 0$$

$$\left.\frac{dr^*}{dt^*}\right|_{t^*=0} = 0 \qquad \left.\frac{dz^*}{dt^*}\right|_{t^*=0} = -1$$

(c) Show that the particle trajectory satisfies

$$r^* = r_0^* e^{-p}\left[\cosh(pq) + \frac{1}{q}\sinh(pq)\right]$$

$$z^* = \begin{cases} \frac{1}{2}(1 - v_t^*)e^{-p}\cosh(pq') + \frac{1}{2q'}\left(\frac{1+q'^2}{2} - v_t^*\right) \\ \quad \times e^{-p}\sinh(pq') + \frac{v_t^*}{2} & \text{St} \le \frac{1}{16} \\ \frac{1}{2}(1 - v_t^*)e^{-p}\cos(pq'') + \frac{1}{2q''}\left(\frac{1-q''^2}{2} - v_t^*\right) \\ \quad \times e^{-p}\sin(pq'') + \frac{v_t^*}{2} & \text{St} \ge \frac{1}{16} \end{cases}$$

where

$$p = \frac{t^*}{4\,\text{St}} \qquad q = (1 + 8\,\text{St})^{1/2}$$

$$q' = (1 - 16\,\text{St})^{1/2} \qquad q'' = (16\,\text{St} - 1)^{1/2}$$

**5.7.** At $t = 0$ a uniform concentration $N_0$ of monodisperse particles exists between two horizontal plates separated by a distance $h$. If both plates are perfect absorbers of particles and the particles settle with a settling velocity $v_t$, determine the number concentration of particles between the plates as a function of time and position. The Brownian diffusivity of the particles is $D$.

*Solution:*

$$N(z^*, t^*) = N_0 \sum_{i=1}^{\infty} \frac{2}{\pi i}\frac{1 - (-1)^i e^{2\mu^2}}{1 + 4\mu^4/\pi^2 i^2} \sin(i\pi z^*) \exp\left\{\left[-2z^* - \left(1 + \frac{\pi^2 i^2}{4\mu^4}\right)t^*\right]\mu^2\right\}$$

where

$$z^* = \frac{z}{h} \qquad t^* = \frac{t v_t}{h} \qquad \mu = \left(\frac{v_t h}{4D}\right)^{1/2}$$

Calculate the particle fluxes on the top and bottom surfaces as a function of time and obtain the limiting behavior of these fluxes for short and long times.

**5.8.** Consider a particle-sizing device in which particles are separated based on their different settling velocities. The device consists of a vertical flask that is initially filled with a fluid–particle suspension that has been vigorously mixed. Once the shaking of the flask is stopped, a sample is immediately withdrawn to determine the total suspension concentration. The particles then begin settling and a fluid–particle sample is withdrawn at predesignated times at a known distance $L$ below the surface of the fluid. In the absence of particle–particle interactions and assuming that the suspension is sufficiently dilute that the fluid can be considered as stagnant as the particles settle, show that the particle number concentration

$N_i$ of each size $i$ in the suspension as a funtion of location and time is described by

$$\frac{\partial N_i}{\partial t} + v_{t_i} \frac{\partial N_i}{\partial z} = 0$$

where $v_{t_i}$ is the terminal settling velocity of particles of size $i$. It is assumed that particles reach their terminal settling velocity immediately. We take $z$ as positive downward from the liquid surface ($z = 0$), so that $v_{t_i}$ is also a positive quantity. We neglect the effect of the fluid withdrawn for sampling on the height of the column of fluid. The appropriate initial and boundary conditions are

$$N_i(z, 0) = N_{i0} \qquad z > 0$$

$$N_i(0, t) = 0 \qquad t > 0$$

The second condition states that immediately after $t = 0$ there are no particles at the top surface of the liquid.

(a) Show that

$$N_i(z, t) = N_{i0}\left[1 - U\left(t - \frac{z}{v_{t_i}}\right)\right]$$

where $U(x)$ is the unit step function,

$$U(x) = \begin{cases} 0 & x < 0 \\ 1 & x > 0 \end{cases}$$

(b) Show that the ratio of the weight of the dried sample taken at time $t_k$ to that at $t = 0$ yields the percentage by mass of particles smaller than $D_{p_k}$, where

$$D_{p_k} = \left[\frac{18\mu L}{g t_k(\rho_p - \rho)}\right]^{1/2}$$

**5.9.** A particle sampling instrument classifies an aerosol into seven channels. A stream of particles is measured and gives the following numbers of particles in each channel of the device:

| Channel | Size range ($\mu$m) | $N_i$ |
|---|---|---|
| 1 | 0–1.5 | 80 |
| 2 | 1.5–2.3 | 140 |
| 3 | 2.3–3.2 | 180 |
| 4 | 3.2–4.5 | 220 |
| 5 | 4.5–6.0 | 190 |
| 6 | 6.0–8.0 | 60 |
| 7 | >8.0 | 0 |

(a) Plot the size distribution function as a histogram where the number of particles measured in each channel is assumed to be concentrated at the arithmetic midpoint of the size range.

(b) Assuming that all the particles have the same density, plot the mass distribution function as a histogram.

(c) Estimate the number and mass median diameters of this aerosol.

**5.10.** Show that for a log-normally distributed aerosol the volume median diameter, the diameter below which one-half of the total particle volume lies, is related to the two parameters of the distribution, $\overline{D}_{pg}$ and $\sigma_g$, by

$$\overline{D}_{pvm} = \overline{D}_{pg} \exp (3 \ln^2 \sigma_g)$$

*Note:* You may find the following integral of use:

$$\int_{L_1}^{L_2} e^{ru} \exp\left[-\frac{(u-\overline{u})^2}{2\sigma_u^2}\right] du$$

$$= \sqrt{\frac{\pi}{2}}\, \sigma_u e^{r\overline{u}} e^{r^2\sigma_u^2/2} \left\{ \text{erf}\left[\frac{L_2 - (\overline{u} + r\sigma_u^2)}{\sqrt{2}\sigma_u}\right] - \text{erf}\left[\frac{L_1 - (\overline{u} + r\sigma_u^2)}{\sqrt{2}\sigma_u}\right]\right\}$$

**5.11.** Table 5.3 gives data on the cumulative number and mass distributions of a sample of glass beads (Irani and Callis, 1963). Determine whether these data adhere to log-normal distributions. If so, determine the geometric mean diameters and standard deviations for each. What is the relationship between the parameters of the two distributions? Plot the log-normal distributions $n(D_p)$ and $n_m(D_p)$ together with their corresponding histograms.

**5.12.** The size distribution function $n(D_p)$ for the airborne particulate matter in the Pasadena, California, atmosphere in August and September 1969 is shown in Table 5.4.

(a) Plot the number distribution $n(D_p)$ versus $D_p$ on a log-log graph. If a function of the form $n(D_p) = \text{constant} \times D_p^{\alpha}$ is to be fit to the data (approximately), what is the value of $\alpha$?

(b) Determine the mass median particle diameter, the particle diameter for which one-half

TABLE 5.3 CUMULATIVE NUMBER AND MASS DISTRIBUTIONS OF A SAMPLE OF GLASS BEADS

| Diameter ($\mu$m) | Number: % smaller than | Mass: % smaller than |
|---|---|---|
| 5 | 0 | 0 |
| 10 | 1.0 | 0.1 |
| 15 | 13.8 | 1.6 |
| 20 | 42.0 | 10.5 |
| 25 | 68.0 | 28.5 |
| 30 | 85.0 | 50.0 |
| 35 | 93.0 | 67.9 |
| 40 | 97.2 | 80.8 |
| 45 | 98.8 | 89.2 |
| 50 | 99.5 | 94.0 |
| 55 | 99.85 | 97.0 |
| 60 | 99.91 | 98.1 |
| 100 | 100 | 100 |

*Source:* Irani and Callis (1963).

**TABLE 5.4** PARTICLE SIZE DISTRIBUTION FUNCTIONS AVERAGED OVER MEASUREMENTS MADE IN PASADENA, CALIFORNIA, AUGUST TO SEPTEMBER 1969

| $D_p$ ($\mu$m) | $\Delta N/\Delta D_p$ ($\mu$m$^{-1}$ cm$^{-3}$) | $\Delta V/\Delta \log D_p$ ($\mu$m$^3$ cm$^{-3}$) |
|---|---|---|
| 0.00875 | $1.57 \times 10^7$ | 0.110 |
| 0.0125 | $5.78 \times 10^6$ | 0.168 |
| 0.0175 | $2.58 \times 10^6$ | 0.289 |
| 0.0250 | $1.15 \times 10^6$ | 0.536 |
| 0.0350 | $6.01 \times 10^5$ | 1.08 |
| 0.0500 | $2.87 \times 10^5$ | 2.14 |
| 0.0700 | $1.39 \times 10^5$ | 3.99 |
| 0.0900 | $8.90 \times 10^4$ | 7.01 |
| 0.112 | $7.02 \times 10^4$ | 13.5 |
| 0.137 | $4.03 \times 10^4$ | 17.3 |
| 0.175 | $2.57 \times 10^4$ | 28.9 |
| 0.250 | $9.61 \times 10^3$ | 44.7 |
| 0.350 | $2.15 \times 10^3$ | 38.6 |
| 0.440 | $9.33 \times 10^2$ | 42.0 |
| 0.550 | $2.66 \times 10^2$ | 29.2 |
| 0.660 | $1.08 \times 10^2$ | 24.7 |
| 0.770 | $5.17 \times 10^1$ | 21.9 |
| 0.880 | $2.80 \times 10^1$ | 16.1 |
| 1.05 | $1.36 \times 10^1$ | 22.7 |
| 1.27 | 5.82 | 18.6 |
| 1.48 | 2.88 | 13.6 |
| 1.82 | 1.25 | 19.7 |
| 2.22 | $4.80 \times 10^{-1}$ | 13.4 |
| 2.75 | $2.17 \times 10^{-1}$ | 15.2 |
| 3.30 | $1.18 \times 10^{-1}$ | 13.7 |
| 4.12 | $6.27 \times 10^{-2}$ | 25.3 |
| 5.22 | $3.03 \times 10^{-2}$ | 26.9 |

*Source:* Whitby et al. (1972). Reprinted by permission of Academic Press, Inc.

the particle mass lies below and one-half lies above. Assume that all the particles have the same density.

**(c)** Figure 5.18 shows the fraction of an inhaled aerosol that will be deposited in the lungs as a function of particle size. For an aerosol density of $\rho_p = 1.5$ g cm$^{-3}$, and the 1969 Pasadena aerosol, determine how much mass would have been deposited in a person's respiratory tract daily (in $\mu$g) due to breathing the local air. What will be the distribution of this material among tracheobronchial, pulmonary, and nasal passages? Lung function data are given in Figure 5.18.

**5.13.** Show the steps to obtain (5.152) from (5.149).

**5.14.** The evolution of the number distribution based on particle volume of an aerosol undergoing

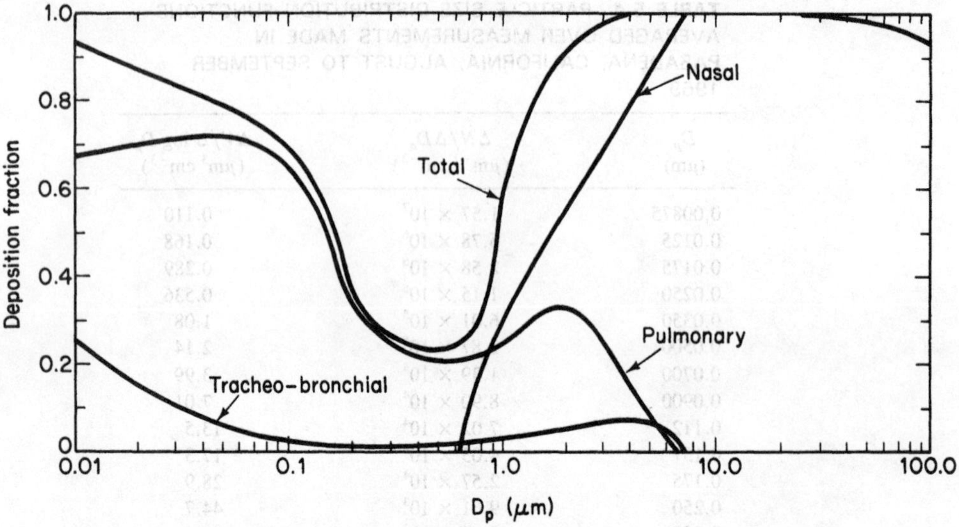


**Figure 5.18** Deposition of monodisperse aerosols as a function of aerodynamic diameter in the respiratory tract of man (assuming a respiratory rate of 15 respirations per minute and a tidal volume of 1450 cm³) (Task Group on Lung Dynamics, 1966).

coagulation is governed by

$$\frac{\partial n(v, t)}{\partial t} = \frac{1}{2}\int_0^v K(v - \tilde{v}, \tilde{v})\, n(v - \tilde{v}, t)\, n(\tilde{v}, t)\, d\tilde{v}$$

$$- n(v, t)\int_0^\infty K(v, \tilde{v})\, n(\tilde{v}, t)\, d\tilde{v}$$

For particles in the continuum regime the coagulation coefficient is

$$K(v, \tilde{v}) = \frac{2kT}{3\mu}(v^{1/3} + \tilde{v}^{1/3})(v^{-1/3} + \tilde{v}^{-1/3})$$

Let us assume that the aerosol size distribution is log-normal throughout the coagulation process (Lee, 1985). If the number distribution based on particle radius is log-normal,

$$n(R_p, t) = \frac{N}{\sqrt{2\pi}\ln\sigma_g R_p}\exp\left[-\frac{(\ln R_p - \ln \bar{R}_{pg})^2}{2\ln^2\sigma_g}\right]$$

(a) Show that the number distribution based on particle volume is

$$n(v, t) = \frac{N}{3\sqrt{2\pi}\ln\sigma_g v}\exp\left[\frac{(\ln v - \ln \bar{v}_g)^2}{18\ln^2\sigma_g}\right]$$

where $\bar{v}_g = 4\pi\bar{R}_{pg}^3/3$. In restricting the particle size distribution to be log-normal, we will allow the three parameters of the distribution, $N$, $\sigma_g$, and $\bar{v}_g$, to vary with time.

**(b)** The $k$th moment of the distribution $n(v, t)$ is

$$M_k = \int_0^\infty v^k n(v, t)\, dv$$

Show that the first three moments of the distribution satisfy

$$\frac{dM_0}{dt} = \frac{2kT}{3\mu}(M_0^2 + M_{1/3}M_{-1/3})$$

$$\frac{dM_1}{dt} = 0$$

$$\frac{dM_2}{dt} = \frac{4kT}{3\mu}(M_1^2 + M_{4/3}M_{2/3})$$

What is the physical significance of $M_0$, $M_1$, and $M_2$? Is the equation for $M_1$ consistent with your physical picture of coagulation?

**(c)** Show that the following relation exists among the moments of a log-normal distribution

$$M_k = M_1 \bar{v}_g^{k-1} \exp\left[\tfrac{9}{2}(k^2 - 1)\ln^2 \sigma_g\right]$$

*Note:* See the integral given in Problem 5.10.

**(d)** Using the relation above for $k = 0, \frac{1}{3}, -\frac{1}{3}, \frac{2}{3}, \frac{4}{3}$, and 2, and eliminating $M_1$ and $dt$ from the resulting equations, show that

$$d(\ln \bar{v}_g) = 9\left[\frac{1 - \frac{3}{2}\exp(9\ln^2\sigma_g)}{\exp(9\ln^2\sigma_g) - 2}\right] d(\ln^2\sigma_g)$$

or

$$d\left\{\ln \bar{v}_g + \tfrac{9}{2}\ln^2\sigma_g + \ln\left[\exp(9\ln^2\sigma_g) - 2\right]\right\} = 0$$

Thus

$$\bar{v}_g\left[\exp(9\ln^2\sigma_g) - 2\right]\exp\left(\tfrac{9}{2}\ln^2\sigma_g\right) = \text{constant}$$

**(e)** Show also that

$$\frac{N}{\exp(9\ln^2\sigma_g) - 2} = \text{constant}$$

We see that during the time evolution of the size distribution due to Brownian coagulation, the constant quantities are conserved. Each of the conserved quantities contains two of the three unknown parameters: $N$, $\bar{v}_g$, and $\sigma_g$. Thus, if one of the three parameters, say $N$, is measured as a function of time, from the initial condition the value of the constant quantities is fixed, and then the other two can be computed as a function of time. From the three parameters, the full size distribution is then available from the log-normal expression for $n(v, t)$.

**(f)** As $t \to \infty$, we anticipate that $\bar{v}_g \gg \bar{v}_{g0}$ and $N \ll N_0$. Show that as $t \to \infty$,

$$\exp(9\ln^2\sigma_g) = 2$$

or $\sigma_g = 1.32$. The conserved quantities break down as the size distribution attains the asymptotic geometric standard deviation of 1.32. If we make the approximation that in this asymptotic regime, $[\exp(9 \ln^2 \sigma_g) - 2]$ is constant, show that in this limit

$$\lim_{t \to \infty} N\bar{v}_g = \text{constant}$$

(g) Show that the equation for $M_0$ can be written as

$$\frac{dM_0}{dt} = -\frac{2kT}{3\mu} M_0^2[1 + \exp(\ln^2 \sigma_g)]$$

In the asymptotic regime, show that this equation can be integrated to give

$$\frac{N_\infty}{N_{\infty,0}} = \frac{1}{1 + (1 + 2^{1/9})\,\beta N_{\infty,0} t}$$

$$\frac{\bar{v}_{g\infty}}{\bar{v}_{g\infty,0}} = 1 + (1 + 2^{1/9})\,\beta N_{\infty,0} t$$

where $\beta = 2kT/3\mu$ and where the subscript $\infty,0$ represents the value of the variable at the time the distribution enters the asymptotic regime.

## REFERENCES


ABRAHAM, F. F. *Homogeneous Nucleation Theory*, Academic Press, New York (1974).

ALAM, M. K. "The Effect of van der Waals and Viscous Forces on Aerosol Coagulation," *Aerosol Sci. Technol.*, *6*, 41–52 (1987).

ALLEN, M. D., and RAABE, O. G. "Re-evaluation of Millikan's Oil Drop Data for the Motion of Small Particles in Air," *J. Aerosol Sci.*, *13*, 537–547 (1982).

BIRD, R. B., STEWART, W. E., and LIGHTFOOT, E. N. *Transport Phenomena*, Wiley, New York (1960).

CRAWFORD, M. *Air Pollution Control Theory*, McGraw-Hill, New York (1976).

DERJAGUIN, B. V., and YALAMOV, Y. I. "The Theory of Thermophoresis and Diffusiophoresis of Aerosol Particles and Their Experimental Testing," in *Topics in Current Aerosol Research*: Part 2, G. M. Hidy and J. R. Brock, Eds., Pergamon Press, New York, 1–200 (1972).

FRIEDLANDER, S. K. *Smoke, Dust and Haze—Fundamentals of Aerosol Behavior*, Wiley, New York (1977).

FRIEDLANDER, S. K., and WANG, C. S. "The Self-Preserving Particle Size Distribution for Coagulation by Brownian Motion," *J. Colloid Interface Sci.*, *22*, 126–132 (1966).

FUCHS, N. A. *Mechanics of Aerosols*, Pergamon Press, New York (1964).

FUCHS, N. A., and SUTUGIN, A. G. "High-Dispersed Aerosols," in *Topics in Current Aerosol Research*, G. M. Hidy and J. R. Brock, Eds., Pergamon Press, New York, 1–60 (1971).

GELBARD, F., and SEINFELD, J. H. "Simulation of Multicomponent Aerosol Dynamics," *J. Colloid Interface Sci.*, *78*, 485–501 (1980).

GELBARD, F., TAMBOUR, Y., and SEINFELD, J. H. "Sectional Representations for Simulating Aerosol Dynamics," *J. Colloid Interface Sci.*, *76*, 541–556 (1980).

GIESEKE, J. A. "Thermal Deposition of Aerosols" in *Air Pollution Control—Part II*, W. Strauss, Ed., Wiley-Interscience, New York, 211–254 (1972).

HINZE, J. O. *Turbulence*, McGraw-Hill, New York (1959).

IRANI, R. R., and CALLIS, C. F. *Particle Size: Measurement, Interpretation and Application*, Wiley, New York (1963).

LEE, K. W. "Conservation of Particle Size Distribution Parameters During Brownian Coagulation," *J. Colloid Interface Sci.*, *108*, 199–206 (1985).

LICHT, W. *Air Pollution Control Engineering*, Marcel Dekker, New York (1980).

MARLOW, W. H. "Derivation of Aerosol Collision Rates for Singular Attractive Contact Potentials," *J. Chem. Phys.*, *73*, 6284–6287 (1980).

MOORE, W. J. *Physical Chemistry*, 3rd Ed., Prentice-Hall, Englewood Cliffs, NJ (1962).

NEWTON, G. J., RAABE, O. G., and MOKLER, B. V. "Cascade Impactor Design and Performance," *J. Aerosol Sci.*, *8*, 339–347 (1977).

PARSEGIAN, V. A. "Long Range van der Waals Forces," in *Physical Chemistry: Enriching Topics from Colloid and Surface Science*, H. van Olphen and K. J. Mysels, Eds., Theorex, La Jolla, CA (1975).

RAMABHADRAN, T. E., PETERSON, T. W., and SEINFELD, J. H. "Dynamics of Aerosol Coagulation and Condensation," *AIChE J.*, *22*, 840–851 (1976).

REED, L. D. "Low Knudsen Number Photophoresis," *J. Aerosol Sci.*, *8*, 123–131 (1977).

SCHMIDT-OTT, A., and BURTSCHER, H. "The Effect of van der Waals Forces on Aerosol Coagulation," *J. Colloid Interface Sci.*, *89*, 353–357 (1982).

SEINFELD, J. H. *Atmospheric Chemistry and Physics of Air Pollution*, Wiley, New York (1986).

SPIELMAN, L. "Viscous Interactions in Brownian Coagulation," *J. Colloid Interface Sci.*, *33*, 562–571 (1970).

SPRINGER, G. S. "Homogeneous Nucleation," *Adv. Heat Transfer*, *14*, 281–346 (1978).

SWIFT, D. L., and FRIEDLANDER, S. K. "The Coagulation of Hydrosols by Brownian Motion and Laminar Shear Flow," *J. Colloid Sci.*, *19*, 621–647 (1964).

TALBOT, L., CHENG, R. K., SCHEFER, R. W., and WILLIS, D. R. "Thermophoresis of Particles in a Heated Boundary Layer," *J. Fluid Mech.*, *101*, 737–758 (1980).

Task Group on Lung Dynamics. "Deposition and Retention Models for Internal Dosimetry of the Human Respiratory Tract," *Health Phys.*, *12*, 173–207 (1966).

WARREN, D. R., and SEINFELD, J. H. "Simulation of Aerosol Size Distribution Evolution in Systems with Simultaneous Nucleation, Condensation, and Coagulation," *Aerosol Sci. Tech.*, *4*, 31–43 (1985).

WHITBY, K. T., HUSAR, R., and LIU, B. Y. H. "The Aerosol Size Distribution of Los Angeles Smog," *J. Colloid Interface Sci.*, *39*, 177–204 (1972).

# 6

# Particle Formation in Combustion

Combustion processes emit large quantities of particles to the atmosphere. Particles formed in combustion systems fall roughly into two categories. The first category, referred to as ash, comprises particles derived from noncombustible constituents (primarily mineral inclusions) in the fuel and from atoms other than carbon and hydrogen (so-called heteroatoms) in the organic structure of the fuel. The second category consists of carbonaceous particles that are formed by pyrolysis of the fuel molecules.

Particles produced by combustion sources are generally complex chemical mixtures that often are not easily characterized in terms of composition. The particle sizes vary widely, and the composition may be a strong function of particle size. This chapter has a twofold objective—to present some typical data on the size and chemical composition of particulate emissions from combustion processes and to discuss the fundamental mechanisms by which the particles are formed.

## 6.1 ASH

Ash is derived from noncombustible material introduced in the combustor along with the fuel and from inorganic constituents in the fuel itself. The ash produced in coal combustion, for example, arises from mineral inclusions in the coal as well as from heteroatoms, which are present in the coal molecules. Heavy fuel oils produce much less ash than coals since noncombustible material such as mineral inclusions is virtually absent in such oils and heteroatoms are the only source of ash. Fuel additives ranging from lead used to control ''knock'' in gasoline engines, to barium for soot control in diesel engines, to sulfur capture agents also contribute to ash emissions. We focus our discussion of ash on the particles produced in coal combustion.

### 6.1.1 Ash Formation from Coal

Ash particles produced in coal combustion have long been controlled by cleaning the flue gases with electrostatic precipitators (see Chapter 7). Most of the mass of particulate matter is removed by such devices, so ash received relatively little attention as an air pollutant until Davison et al. (1974) showed that the concentrations of many toxic species in the ash particles increased with decreasing particle size, as illustrated in Figure 6.1. Particle removal techniques, as we shall see in Chapter 7, are less effective for small (i.e., submicron) particles than for larger particles. Thus, even though the total

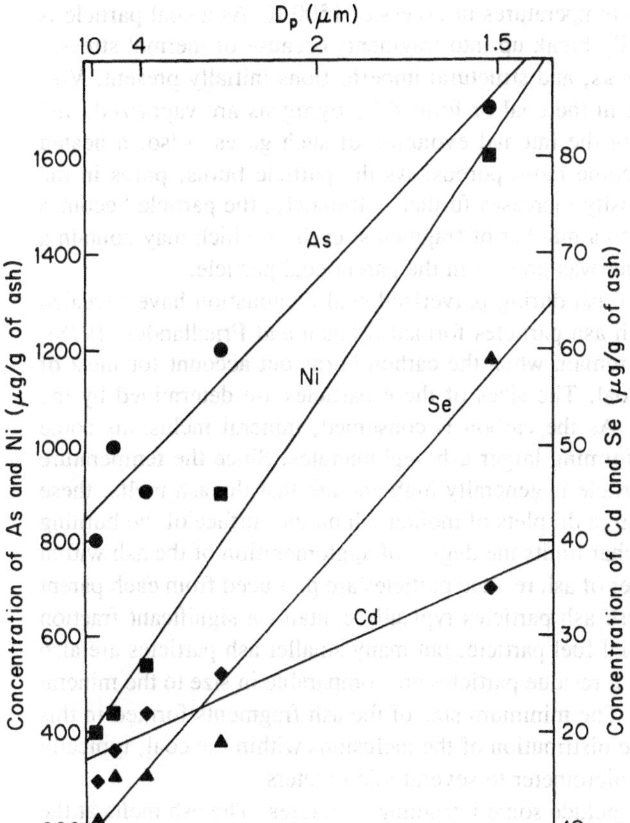


**Figure 6.1** Variation of trace element concentrations with particle size for fly ash particles produced by coal combustion (data from Davison et al., 1974). The lower axis reflects the size dependence of the condensation rate.

mass of particulate matter in the flue gases may be reduced substantially by electrostatic precipitation or another system, the particles that do escape collection are those that contain disproportionately high concentrations of toxic substances.

Coal is a complex, heterogeneous, and variable substance containing, in addition to the fossilized carbonaceous material, dispersed mineral inclusions. The sizes of these inclusions vary from one coal to another and also depend on the way the coal is prepared. The mean diameter of the mineral inclusions is typically about 1 $\mu$m. These inclusions, consisting of aluminosilicates with pyrites, calcites, and magnesites in various proportions, eventually form ash particles as the carbon burns out. The ash particles that are entrained in the combustion gases are called *fly ash*.

In the coal combustion process in most power plants, crushed coal from the mine is pulverized into a fine powder (typically, 40 to 80 $\mu$m mass median diameter) and blown into the furnace with carrier air. The coal particles are heated by radiation and conduction from hot gases up to temperatures in excess of 1500 K. As a coal particle is being heated, it may mechanically break up into fragments because of thermal stresses induced by internal fissures, cracks, and structural imperfections initially present. Volatile fractions originally present in the coal or formed by pyrolysis are vaporized, and the particle may burst open from the internal evolution of such gases. Also, a heated coal particle may swell and become more porous. As the particle burns, pores in the carbon matrix open and the porosity increases further. Ultimately, the particle becomes so porous that it disintegrates into a number of fragments, each of which may contain a fraction of the mineral matter that was present in the parent coal particle.

Studies of the evolution of ash during pulverized coal combustion have revealed two major mechanisms by which ash particles formed (Flagan and Friedlander, 1978). First, ash residue particles that remain when the carbon burns out account for most of the mineral matter in the raw coal. The sizes of these particles are determined by the combustion process as follows: As the carbon is consumed, mineral inclusions come into contact with one another, forming larger ash agglomerates. Since the temperature of a burning pulverized coal particle is generally high enough that the ash melts, these agglomerates coalesce to form large droplets of molten ash on the surface of the burning char. The fragmentation of the char limits the degree of agglomeration of the ash within a single fuel particle, so a number of ash residue particles are produced from each parent coal particle. The largest of these ash particles typically contains a significant fraction (5 to 30%) of the ash in the initial fuel particle, but many smaller ash particles are also generated. The smallest of the ash residue particles are comparable in size to the mineral inclusions in the unburned fuel. The minimum size of the ash fragments formed in this manner is determined by the size distribution of the inclusions within the coal, typically ranging from a few tenths of a micrometer to several micrometers.

The residual ash particles include some intriguing structures. The ash melts at the high temperatures encountered in pulverized coal combustion and, due to the action of surface tension, coalesces into spherical particles. Gas evolution within the molten ash leads to the formation of hollow particles, known as *cenospheres*, some of which contain large numbers of smaller particles. Figure 6.2 shows such particles produced in the

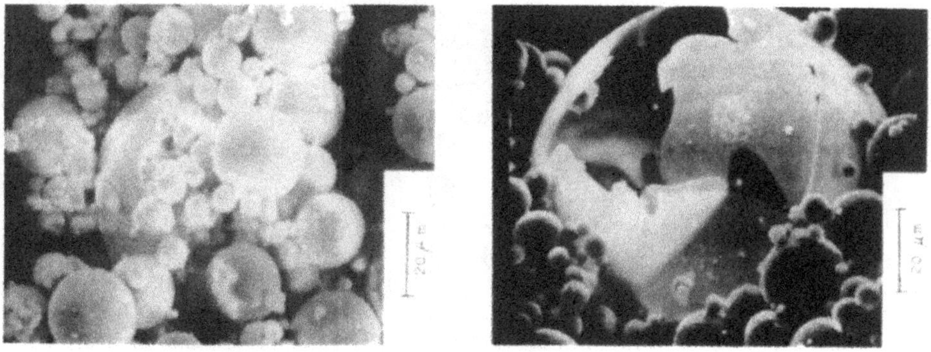

(a) (b)

**Figure 6.2** Scanning electron microscope photographs of ash particles produced in the complete combustion of a lignite coal. Photograph (a) shows a large ash cenosphere containing many smaller spheres. Photograph (b) shows a large broken cenosphere on which the thin cenosphere wall can be seen (courtesy of A. F. Sarofim).

combustion of a lignite coal. These large, picturesque particles receive disproportionate attention, since dense particles generally account for most of the mass.

The second ash particle formation mechanism involves that small fraction of the ash, on the order of 1%, that vaporizes due to the high combustion temperatures. Part of the volatilized ash homogeneously nucleates to form very small (on the order of a few angstroms in size) particles, which then grow by coagulation and condensation of additional vaporized ash. At high temperatures, the agglomerated particles may coalesce into dense spheres, but after the combustion gases cool below the melting point of the condensed material, the liquid freezes and coalescence effectively ceases. Coagulation of the resulting solid particles produces chain agglomerates such as those shown in Figure 6.3. The chain agglomerate structure is a common feature of aerosols produced by vapor nucleation in high-temperature systems. It is also observed in carbonaceous soot aerosols and in the particles produced by a variety of high-temperature combustion processes.

Particle growth by coagulation leads to a predicted narrow mode in the particle size distribution in the size range 0.01 to 0.1 $\mu$m, much smaller than the mineral inclusions present in most coals. The remaining portion of the volatilized ash heterogeneously condenses on existing particles, including those formed by homogeneous nucleation and residual ash fragments. The ash formation process is depicted in Figure 6.4.

Measured particulate size distributions in the flue gases of pulverized coal combustion systems tend to support these two mechanisms for ash particle formation. Most of the mass of ash is generally above 1 $\mu$m in size, with a broad peak in the range 3 to 50 $\mu$m diameter. This large-diameter mode can be explained by the ash residue model. Submicron ash usually comprises less than 2% of the total fly ash mass. Figure 6.5 shows the size distributions measured immediately upstream of the electrostatic precipitator on a coal-fired utility boiler. The two distinct modes are readily apparent.

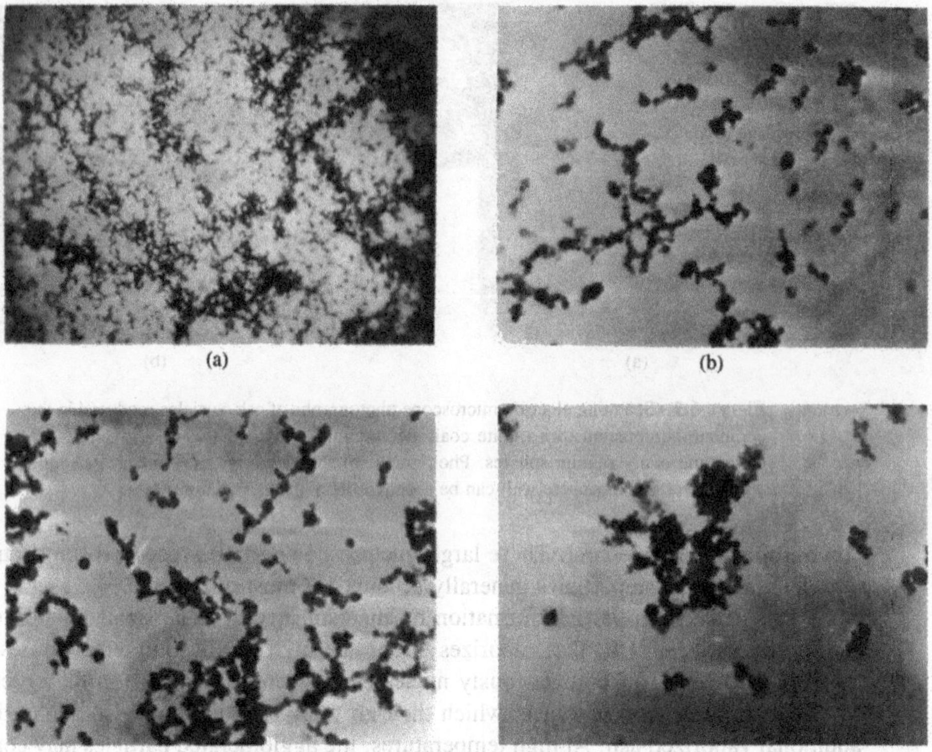

(a) (b) (c) (d)

**Figure 6.3** Transmission electron microscope photographs of the submicron ash fume produced in the combustion of entrained coal particles. The average sphere size is 20 to 30 nm (courtesy of A. F. Sarofim).

## 6.1.2 Residual Ash Size Distribution

In laboratory studies of the combustion of entrained coal particles, Sarofim et al. (1977) observed that the largest ash particles are significantly larger than the mineral inclusions in the coal from which the ash was derived, suggesting that the mineral inclusions coalesce on the surface of the burning char. Evidence for char fragmentation was also reported. Flagan and Friedlander (1978) used these observations in a first attempt to predict the size distribution of the ash residue particles, assuming that a number of equal-sized ash particles were produced from each parent coal particle and that the ash was uniformly distributed in the coal.

Neither of these assumptions is strictly true. Pulverizers used on coal-fired boilers generally employ aerodynamic classification to ensure that only particles that have been ground to a size sufficiently small to be burned completely within the combustor are sent to the burner. Because the aerodynamic diameter depends on density, mineral-rich par-

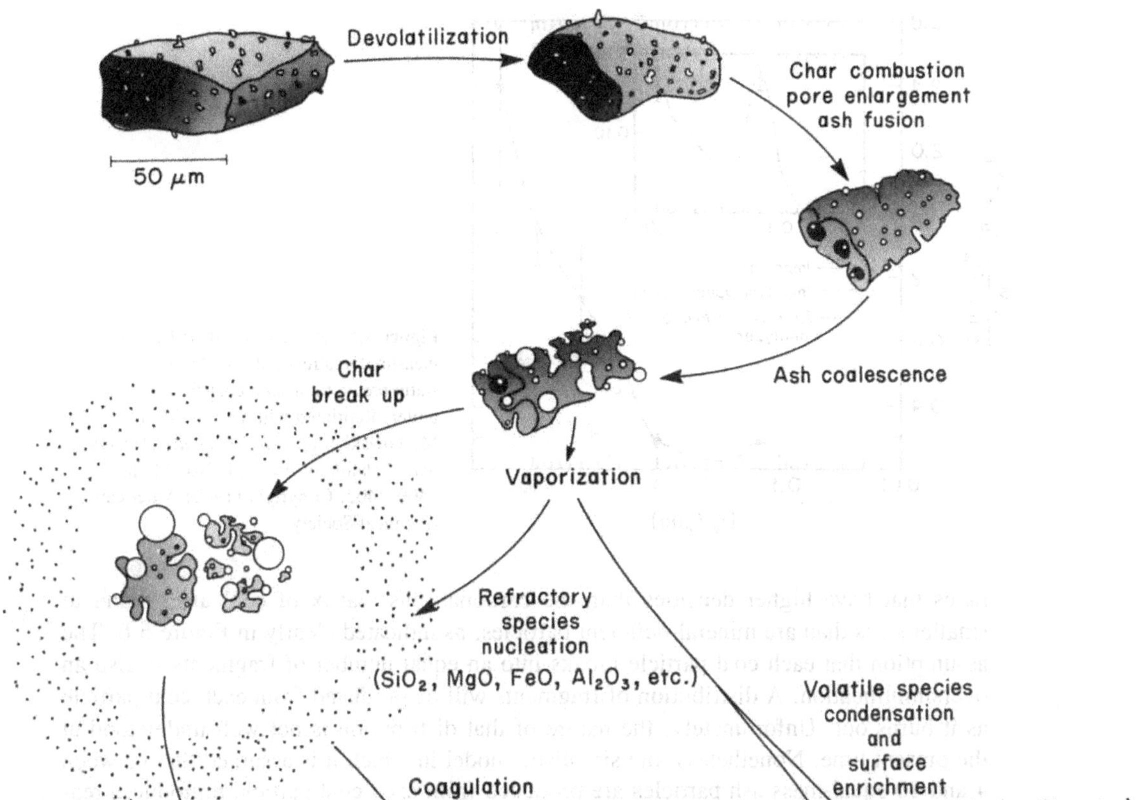


**Figure 6.4** Schematic diagram of the processes involved in ash particle formation.

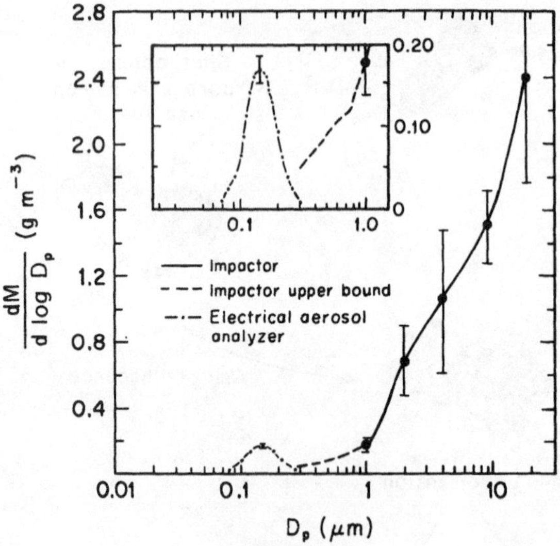


**Figure 6.5** Particle size distribution measured upstream of the electrostatic precipitator on a coal-fired utility boiler. Reprinted with permission from Markowski et al., *Environmental Science and Technology*, Vol. 14, No. 11, pp. 1400–1402. Copyright (1980) American Chemical Society.

ticles that have higher densities than the carbonaceous matrix of coal are ground to smaller sizes than are mineral-deficient particles, as indicated clearly in Figure 6.6. The assumption that each coal particle breaks into an equal number of fragments is also an oversimplification. A distribution of fragments will be produced from each coal particle as it burns out. Unfortunately, the nature of that distribution is not well understood at the present time. Nonetheless, the simplistic model in which it is assumed that between 4 and 10 equal-mass ash particles are produced from each coal particle reproduces reasonbly well the general form of observed mass distributions (Flagan and Friedlander, 1978). The distribution of smaller residual ash particles has not yet been resolved, but there are clearly many more small ash particles generated as combustion residues than this simple model would suggest. To explain the sharp peak in the submicron size distribution shown in Figure 6.5, it is necessary to examine the formation of particles from the volatilized ash. We begin with a discussion of the vaporization process.

### 6.1.3 Ash Vaporization

To understand the vaporization of ash during coal combustion, we must examine the thermodynamics and chemistry of the ash and the transport of the volatilized ash from the surface of the particle. Some components of the ash are highly volatile; examples include sodium, potassium, and arsenic. Volatile ash constituents may vaporize completely during combustion unless inhibited by diffusional resistances, either in transport through the porous structure of the char or to the surfaces of the mineral inclusions. The vaporization process can be a direct transformation from the condensed phase to the vapor phase, for example, for silica,

$$SiO_{2(l)} \rightleftharpoons SiO_{2(v)}$$

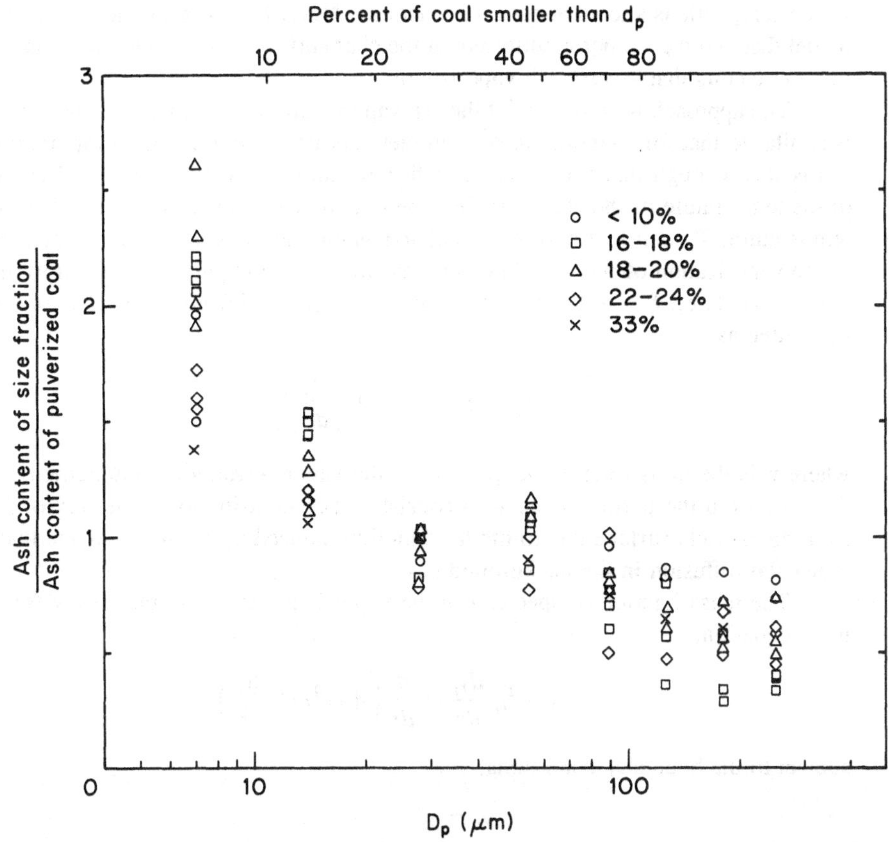


**Figure 6.6** Variation of the ash content of pulverized coal with coal particle size (data from Littlejohn, 1966).

or it may involve the production of volatile suboxides or elemental forms from the original oxides,

$$SiO_{2(l)} + CO \rightleftharpoons SiO_{(v)} + CO_2$$

The former mechanism may dominate for the more volatile ash constituents, but there is evidence that the reduction reactions play an important role in the vaporization of species with relatively low vapor pressures.

Predicting the rates of ash vaporization either by direct vaporization or by chemical reaction is difficult. The high-ionic-strength solutions produced as ash melts cannot be described by ideal solution theory, and the vapor pressures above these solutions generally are unknown. Furthermore, resistance to ash vapor diffusion in the porous char is difficult to characterize (Quann and Sarofim, 1982). For vaporization occurring as a result of chemical reactions, the rates and mechanisms of the reactions are needed. Unfortunately, little is known about such reactions. Nevertheless, a simple ash vaporization model that assumes vapor equilibrium at the char surface can provide important insights into the factors that govern ash vaporization.

The approach used to predict the ash vaporization rate from a burning char particle is similar to that for evaporation of a droplet. Let us assume that the evaporation of the ash is slow enough that it does not significantly influence either the rate of combustion or the temperature of the char particle. The rate of char combustion, $\bar{r}_c$, and the particle temperature, $T_p$, can therefore be calculated as outlined in Chapter 2 without reference to the vaporization of the ash. The net mass flux from the particle is equal to the combustion rate (i.e., $\bar{f}_c = \bar{r}_c$). The vapor flux of ash species $i$ from the particle surface may be written as

$$\bar{f}_{is} = \bar{f}_c y_{is} - \rho D \left( \frac{dy_i}{dr} \right)_{r=a} \tag{6.1}$$

where $y_i$ is the mass fraction of species $i$ in the vapor surrounding the particle. The net flux of $i$ from the particle is seen to consist of two contributions: the flux of $i$ carried from the particle surface due to the fluid motion induced by combustion and that due to molecular diffusion in the background gas.

The mass fraction of species $i$ in the vapor is governed by the steady-state continuity equation,

$$4\pi a^2 \bar{f}_c \frac{dy_i}{dr} = \frac{d}{dr}\left( 4\pi r^2 \rho D_i \frac{dy_i}{dr} \right) \tag{6.2}$$

subject to the boundary conditions,

$$y_i = y_{is} \qquad r = a$$

$$y_i = y_{i\infty} \qquad r \to \infty$$

Integrating twice, imposing the boundary conditions, and using (6.1) gives

$$\frac{a\bar{f}_c}{\rho D_i} = \ln (1 + B_i) \tag{6.3}$$

where

$$B_i = \frac{y_{is} - y_{i\infty}}{(\bar{f}_i/\bar{f}_c) - y_{is}} \tag{6.4}$$

If we assume that the diffusivity of the vapor species $i$ is equal to that for the combustion gases, $D_i = D$, and from (2.136) and (6.3) we see that $B_i = B_T = B$, the transfer number for combustion. In that case we find

$$\frac{\bar{f}_i}{\bar{f}_c} = \frac{y_{is}(1 + B) - y_{i\infty}}{B} \tag{6.5}$$

If $y_{is} >> y_{i\infty}$, (6.5) further simplifies to

$$\frac{\bar{f}_i}{\bar{f}_c} = y_{is}\left(1 + \frac{1}{B}\right) \tag{6.6}$$

The mass vaporization flux of species $i$ is then

$$\bar{f}_i = \frac{\rho D y_{is}}{a}\left(1 + \frac{1}{B}\right) \ln(1 + B) \tag{6.7}$$

The mass of an ash component that vaporizes as the particle burns is obtained by integrating (6.7) over the particle surface and burning time $\tau_B$, that is,

$$m_i = \int_0^{\tau_B} 4\pi a(t)\rho D y_{is}\left(1 + \frac{1}{B}\right) \ln(1 + B)\, dt \tag{6.8}$$

Assuming that $B$ and $y_{is}$ remain constant throughout the combustion process, (6.8) becomes

$$m_i = y_{is}\left(1 + \frac{1}{B}\right) \int_0^{\tau_B} 4\pi a(t)\rho D \ln(1 + B)\, dt \tag{6.9}$$

The integral in (6.9) is simply the total mass of carbon burned assuming complete combustion. Thus the fraction of the ash species vaporized is

$$\eta_i = \frac{m_i}{m_{i0}} = \frac{y_{is}}{\alpha_i}\left(1 + \frac{1}{B}\right)(1 - \alpha_a) \tag{6.10}$$

where $\alpha_i$ and $\alpha_a$ are the mass fractions of ash species $i$ and total ash, respectively, in the parent coal. For combustion of carbon in air the maximum value of $B$ corresponds to diffusion-limited combustion. Using (2.137) we obtain

$$B_{\max} = \frac{(0.209)(32/28.4)}{16/12} = 0.174$$

so the minimum fraction of ash vaporized is

$$\eta_{i,\min} = 6.75\, \frac{y_{is}}{\alpha_i}(1 - \alpha_a)$$

A reasonable estimate of $y_{is}$ is obtained by assuming that the vapor is at local equilibrium at the particle surface,

$$y_{is} = \frac{p_{ie} M_i}{p \bar{M}} \tag{6.11}$$

where $M_i$ and $\bar{M}$ are the molecular weight of species $i$ and the mean molecular weight of the gas, respectively. Given thermodynamic data on ash and vapor properties, the vaporization rate can thus be estimated. For a minor ash constituent ($\alpha_i << 1$), even low vapor pressure at the surface may lead to complete vaporization.

The actual rate of vaporization of an ash component will be lower than that predicted by the model above because (1) only a fraction of the external surface of the char particle is covered with ash, (2) diffusion of vapor from the inclusions distributed throughout the char volume is restricted by the small pores, and (3) diffusion of volatile species through the ash solution also limits vaporization.

Quann and Sarofim (1982) examined the vaporization of Si, Ca, and Mg during combustion of a number of coals in $O_2$–$N_2$ mixtures. In each case, they assumed that vaporization proceeded by reduction of the oxide ($MO_n$), that is,

$$MO_n + CO \rightleftharpoons MO_{n-1} + CO_2$$

Chemical equilibrium was assumed to hold at the surface of the ash,

$$K_p = \frac{p_{Me} p_{CO_2}}{p_{CO}} \tag{6.12}$$

where $p_{Me}$ is the partial pressure of the volatile form of M, and we assume a single-component ash inclusion. If there is no other source of carbon dioxide to influence the local equilibrium, at the surface, (6.12) then gives

$$p_{Me} = p_{CO_2} = \left(K_p p_{CO}\right)^{1/2} \tag{6.13}$$

The amount of each ash constituent that vaporized during combustion was estimated from measurements of the total quantity of that element in the submicron fume. Using (6.10), the species vapor pressure at the surface of the burning char can then be estimated. The temperature dependence of the vapor pressure of each species was determined using data from a number of controlled laboratory experiments.

Results obtained by Quann and Sarofim (1982) for Ca, Mg, and Si were consistent with the temperature dependence predicted using (6.13), supporting the hypothesis that these ash constituents are volatilized through chemical reduction of the mineral oxides. The net vaporization rate was significantly less than that corresponding to vapor equilibrium at the char surface, suggesting that diffusion resistances within the porous structure of the char retard vaporization, at least for chars in which the mineral matter is dispersed as discrete inclusions (Illinois number 6 and Alabama Rose). The calcium in the Montana lignite is organically bound. As the carbonaceous matter is consumed during burnout, the organically bound calcium must be mobilized. Examination of scanning electron microscope photographs of partially reacted char particles reveals that minute grains are formed on the carbon surface (Quann and Sarofim, 1982). As the burning progresses, the sizes of the ash particles on the char surface continue to increase, ultimately approaching 5 to 10 $\mu$m prior to complete combustion. Assuming that the alkaline earths form oxides very early in the combustion process and that those oxides condense as very small inclusions, one predicts that the vaporization rate should approach that predicted for external diffusion control. Figure 6.7 clearly shows that this condition is approached for calcium in the lignite in high-temperature combustion. The decrease in the partial pressure of Ca at lower temperature is thought to result from penetration of oxygen further into the pores of the burning char. The higher oxygen concentration reduces the mole fraction of metal vapors near the surface of the char.

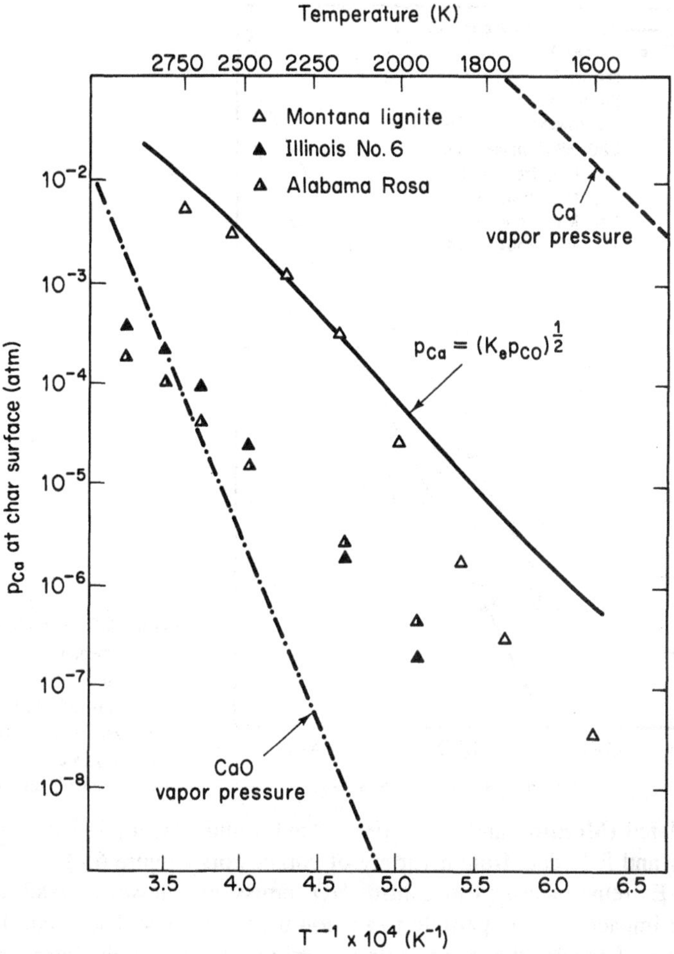


**Figure 6.7** Variation of calcium species vapor pressures at the surface of burning coal particles with temperature as inferred from total vaporization measurements. Curves represent equilibrium vapor pressures as noted (Quann and Sarofim, 1982). Reprinted by permission of The Combustion Institute.

Since the enthalpies of volatilization of typical ash constituents are high, the vaporization rates are a strong function of temperature. Fine-particle formation from volatilized ash constituents can be reduced, therefore, by combustion modifications that reduce peak flame temperatures. We recall that control of $NO_x$ formation in coal combustion involves reducing the rate at which air and fuel are brought together. This strategy reduces the peak flame temperature by allowing heat transfer to cool the gases before combustion is complete and, thereby, also reduces fine-particle formation from condensation of volatilized ash components. $NO_x$ and fine-particle formation are thus positively

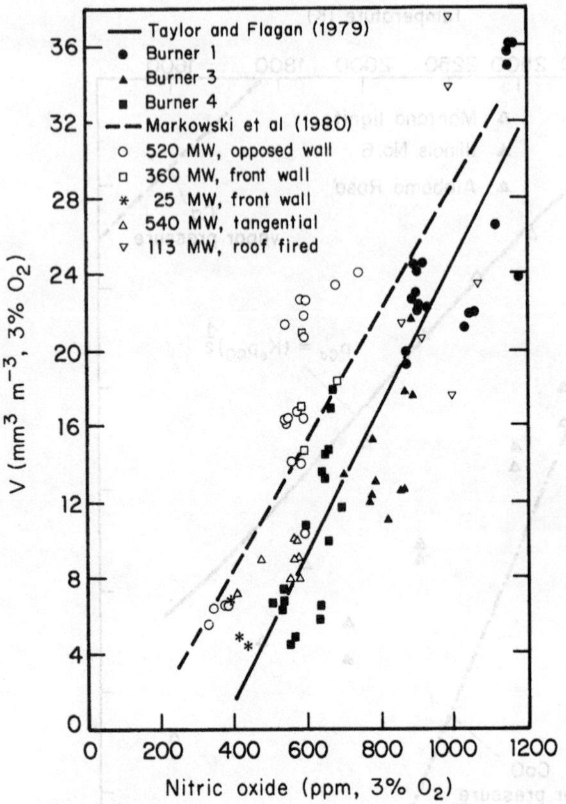


**Figure 6.8** Submicron aerosol mass loading measured upstream of control devices in coal-fired utility boilers (open points, dashed line) and in laboratory pulverized coal combustor (solid points, solid line) as a function of NO emissions (Taylor and Flagan, 1981).

correlated (McElroy and Carr, 1981; Taylor and Flagan, 1982) as illustrated by laboratory and field data from a variety of combustors (Figure 6.8).

Extreme measures to control $NO_x$ emissions, however, lead to substantially different impacts on fine-particle formation than those just discussed. The bulk of the ash consists of species that vaporize much more readily in reducing than in oxidizing conditions. Staging the combustion allows a long time in a reducing atmosphere for volatilization of the suboxides to take place, although the lower temperatures may counteract this effect (Linak and Peterson, 1984, 1987). Thus fine-particle formation may actually be favored under such conditions. In laboratory studies of staged combustion, Linak and Peterson (1987) observed an increase of as much as 50% in fine-particle emissions compared with single-stage combustion with the same fuel. As with other aspects of emission control from combustion sources, we again have encountered complex interrelationships between formation and control of different pollutants.

### 6.1.4 Dynamics of the Submicron Ash Aerosol

During the char burnout phase of coal combustion, much of the heat release occurs at the particle surface, where the carbon is oxidized to carbon monoxide. The hot reducing atmosphere at the surface provides the driving force for ash vaporization. Heat transfer from the burning particle to the surrounding gases leads to a very sharp temperature gradient in the vicinity of the burning particle. Within a few particle radii, the gas temperature approaches that of the surrounding atmosphere. The oxygen concentration approaches that of the background gas within a similar distance from the particle surface. These sharp gradients dramatically alter the ash vapor equilibrium (Senior and Flagan, 1982).

The resulting supersaturations can be quite high, leading to rapid homogeneous nucleation of some of the ash vapor species within a few radii from the surface of the burning char particle. High supersaturation means that the critical nucleus is very small, on the order of atomic dimensions in some cases. Thus, on a time scale that is comparable to that of char combustion, very large numbers of very small particles can be produced by the nucleation of volatilized ash. Coagulation dominates the growth of these small particles, although condensation can be significant at long times when the temperature is low enough that the volatile ash constituents condense.

When the number concentration is high and coagulation dominates particle growth, the dynamics of the nucleation-generated ash fume can be described by the coagulation equation (5.150). The nucleation-generated ash particles are small enough that the free molecular Brownian coagulation coefficient (5.157) can be applied during most of the growth process. The dynamics of the coagulating kinetic regime ash aerosol can be described approximately using the similarity solution to the coagulation equation that produces the self-preserving particle size distribution function (see Example 5.8).

The self-preserving size distribution model assumes that the only particles present are the small particles. The residual ash particles produced in pulverized coal combustion are sufficiently abundant, however, that the coagulation of the fine, self-preserving mode particles with the large residual ash particles is a significant sink for the former. The loss of fine particles can be estimated if we assume that the loss of fine particles to the coarse particles does not alter the form of the size distribution of the smaller particles and, further, that the accumulation of fine particles does not significantly alter the size distribution of the large particles. We assume further that the size difference between the residual ash particles and the nucleation-generated particles is large. The loss of fine particles to the larger residual ash particles is then a diffusional loss term, allowing us to separate the integrations over the coarse and the fine particles. Assuming the nucleation-generated particles to be in the free molecular size regime, the diffusional flux of the small particles to the surface of the residual ash particles is obtained using the Fuchs-Sutugin interpolation, (5.94).

Predictions of the fine particle size distribution are compared with laboratory measurements of the fine particles produced in pulverized coal combustion in Figure 6.9. For these calculations, the size distribution of the residual ash particles was estimated by assuming that five equal-mass residual ash particles were produced from each parent coal particle. The amount of ash volatilized was matched to the measured fine particle volume for the calculations. The shape of the measured size distribution closely approximates the self-preserving size distribution, and the absolute number concentrations agree quite well.

Attempts to simulate the fine-particle growth in pulverized coal-fired utility boilers

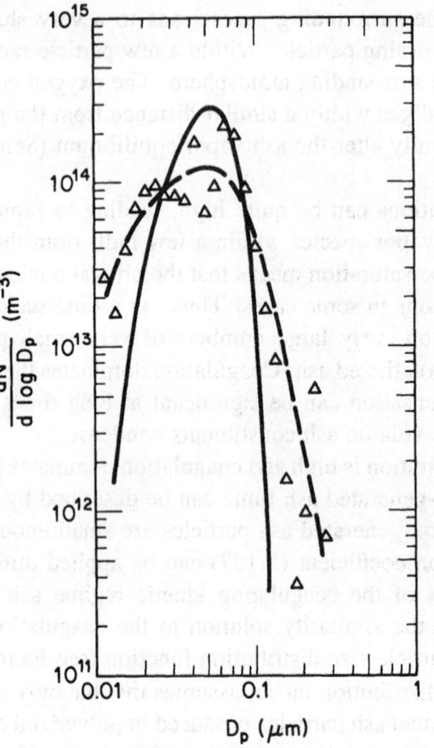


**Figure 6.9** Comparison of the submicron ash particle size distribution predicted using the self-preserving aerosol distribution model (solid curve) with size distribution measurements (broken curve) made on a laboratory pulverized coal combustor (Taylor and Flagan, 1981).

(Flagan and Friedlander, 1978; Flagan, 1979) have not been so successful. The number concentrations are generally overestimated by a significant factor. This probably indicates that the residence time during which coagulation proceeds prior to measurement is underestimated, due either to the treatment of the complex flow in the boiler as a plug flow or to coagulation in the sampling system prior to measurement of the size distribution. Nonetheless, these and other studies of fine ash particles produced in pulverized coal combustion clearly demonstrate the role of ash volatilization in fine-particle formation.

## 6.2 CHAR AND COKE

The carbonaceous char residue that remains after coal is devolatilized burns slowly by surface reactions. If the char particle is too large, mixing in the combustion is poor, or heat is transferred too quickly, char particles may not be fully consumed. Heavy fuel oils may produce similar carbonaceous particles, called *coke*. Coke particles are relatively large, 1 to 50 $\mu$m in diameter, with smaller numbers of much larger particles, and account for the majority of particulate mass emitted from boilers fired with heavy fuel oils. They are hard cenospheres, porous carbonaceous shells containing many blowholes

(Marrone et al., 1984). Fuel impurities tend to concentrate in these cenospheres (Braide et al., 1979). Coke particles are formed by liquid-phase pyrolysis of heavy fuel oil droplets. The mechanisms and rates of coke particle formation are not well understood.

In studies of the combustion of individual millimeter-sized droplets of heavy fuel oil, Marrone et al. (1984) identified two combustion times: (1) the droplet burning time, which corresponds to the time required for coke formation, and (2) the time required for the coke particle to burn. During about the first 60% of the droplet burning phase, combustion was relatively quiescent. In the final stages of droplet burning, the droplet deforms and finally appears to froth just prior to forming a small coke cenosphere. Small droplets may be ejected during the latter violent phase of coke formation. Immature coke particles were found to be tarry and soluble in organic solvents. Coke cenospheres that could not be dissolved in organic solvents were produced only at the end of the droplet lifetime. Typically, the coke particles accounted for about 3% of the mass of the residual (heavy component) oil in the original fuel oil droplet, even when the residual oil was diluted with 60% of a light distillate oil. Asphaltine conversion to coke was about 30%.

The influence of fuel droplet size on coke formation is not well documented, but there is evidence that size, at least down to 100 $\mu$m, has no effect on the quantitative behavior of burning heavy fuel oil droplets (Marrone et al., 1984). At smaller sizes, droplet lifetimes might be short enough that there would be insufficient time for the coking reactions to go to completion. Observations that the coke formation occurs within 10 to 20 ms suggest that the size of droplet required to prevent coke formation would be small indeed, less than about 20 $\mu$m in diameter. Coke particle formation appears to be almost unavoidable in the combustion of heavy fuel oils. Emission rates would be reduced substantially by reducing the time required in coke combustion. Improved atomization or dilution of the heavy fuel oil with a lighter component would decrease the initial size of the coke particles, thereby reducing the combustion time.

## 6.3 SOOT

Carbonaceous particles can also be produced in the combustion of gaseous fuels and from the volatilized components of liquid or solid fuels. The particles formed by this route, known as *soot*, differ markedly from the char and coke discussed previously. Most commonly, soot particles are agglomerates of small, roughly spherical particles such as those shown in Figure 6.10. While the size and morphology of the clusters can vary widely, the small spheres differ little from one source to another. They vary in size from 0.005 to 0.2 $\mu$m but most commonly lie in the size range 0.01 to 0.05 $\mu$m (Palmer and Cullis, 1965). The structural similarity between soot particles and the inorganic particles produced from volatilized ash (Figure 6.3) suggests a common origin. The genesis of soot, however, is much less well understood that that of the inorganic particles due to the extreme complexity of hydrocarbon chemistry in the flame, as well as to the fact that soot particles can burn if exposed to oxygen at high temperatures.

The small spheres that agglomerate together to form a soot particle consist of large numbers of lamellar crystallites that typically contain 5 to 10 sheets containing on the

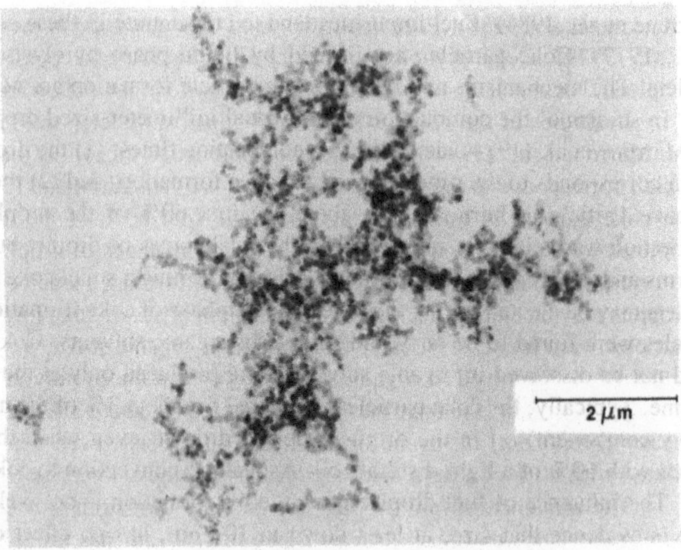


(a)

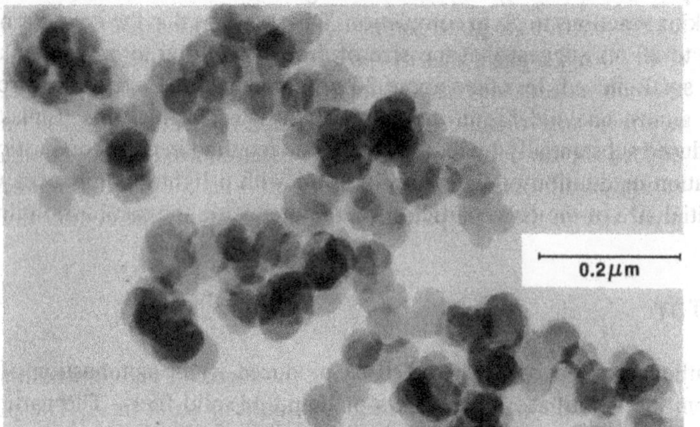


(b)

**Figure 6.10** Transmission electron microscope photographs of soot particles produced in a laminar diffusion flame with acetylene fuel illustrating the agglomerate structure (courtesy of E. Steel, National Bureau of Standards).

order of 100 carbon atoms each. The structure within each sheet is similar to that of graphite, but adjacent layers are randomly ordered in a turbostratic structure. The platelets are also randomly oriented and bound by single sheets or amorphous carbon, giving rise to the spherical particles illustrated in Figure 6.11.

Soot particles are not pure carbon. The composition of soot that has been aged in

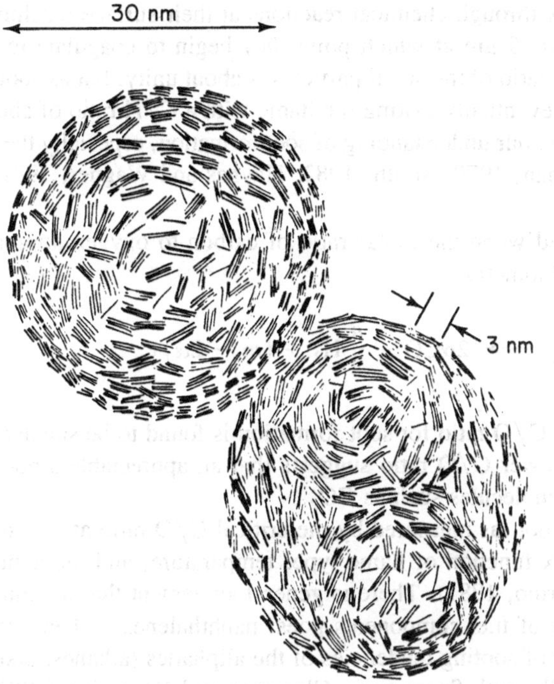


**Figure 6.11** Schematic of soot microstructure.

the high-temperature region of the flame is typically $C_8H$ (Palmer and Cullis, 1965), but soot may contain considerably more hydrogen earlier in the flame. Furthermore, soot particles adsorb hydrocarbon vapors when the combustion products cool, frequently accumulating large quantities of polycyclic aromatic hydrocarbons.

The presence of soot in a flame is actually advantageous in certain situations. Because of the high emissivity of soot particles relative to that of the gases in the flame, only a small quantity of soot is sufficient to produce an intense yellow luminosity. Soot is necessary in boiler flames to obtain good radiative heat transfer. The presence of soot is vital to give a candle flame its characteristic yellow appearance. In gas turbines, on the other hand, the intense radiative transfer from soot is to be avoided. Without soot, hydrocarbon flames appear either violet due to emissions from excited CH radicals when fuel-lean or green due to $C_2$ radical emissions when fuel-rich (Glassman, 1977). In many cases soot serves its purpose as a radiator and is consumed before it leaves the flame. If it does escape the flame, soot poses serious environmental consequences. The high emissivity of soot translates into a high absorptivity at ambient temperatures, leading to its black color in a plume.

### 6.3.1 Soot Formation

Soot forms in a flame as the result of a chain of events that begins with pyrolysis and oxidative pyrolysis of the fuel into small molecules, followed by chemical reactions that build up larger molecules that eventually get big enough to become very small particles.

The particles continue to grow through chemical reactions at their surface reaching diameters in the range 0.01 to 0.05 $\mu$m at which point they begin to coagulate to form chain agglomerates. The C/H ratio of the small particles is about unity, but as soot ages in the flame it loses hydrogen eventually exiting the flame with a C/H ratio of about 8.

Despite numerous studies, our understanding of soot formation, especially the early phases, is incomplete (Glassman, 1979; Smith, 1982; Haynes and Wagner, 1981; Lehaye and Prado, 1981).

Soot formation is favored when the molar ratio of carbon to oxygen approaches 1.0, as suggested by the stoichiometry,

$$C_mH_n + \alpha O_2 \longrightarrow 2\alpha CO + \frac{n}{2} H_2 + (m - 2\alpha) C_s$$

In premixed flames the critical C/O ratio for soot formation is found to be smaller than 1.0, actually about 0.5. The lower C/O ratio suggests that an appreciable amount of the carbon is tied up in stable molecules such as $CO_2$.

The propensity to form soot (as measured by the critical C/O ratio at which soot formation begins) is a complex function of flame type, temperature, and the nature of the fuel (Glassman and Yaccarino, 1981). There is general agreement that the rank ordering of the sooting tendency of fuel components goes: naphthalenes > benzenes > aliphatics. However, the order of sooting tendencies of the aliphatics (alkanes, alkenes, and alkynes) varies dramatically with flame type. Glassman and Yaccarino (1981) attribute much of the variability to flame temperature. In premixed flames, sooting appears to be determined by a competition between the rate of pyrolysis and growth of soot precursors and the rate of oxidative attack on these precursors (Millikan, 1962). As the temperature increases, the oxidation rate increases faster than the pyrolysis rate, and soot formation decreases.

In diffusion flames, oxygen is not present in the pyrolysis zone, so the increase in the pyrolysis rate with temperature leads to an increasing tendency to soot. Small amounts of oxygen in the pyrolysis zone appear to catalyze the pyrolysis reaction and increase sooting (Glassman and Yaccarino, 1981).

The difference between the sooting tendencies of aromatics and aliphatics is thought to result from different routes of formation. Aliphatics appear to form soot primarily through formation of acetylene and polyacetylenes at a relatively slow rate, illustrated in Figure 6.12. Aromatics might form soot by a similar process, but there is a more direct route involving ring condensation or polymerization reactions that build on the existing aromatic structure (Graham et al., 1975). The fragmentation of aromatics should occur primarily at high temperature, but such reactions may not be important even there (Frenklach et al., 1983). In flames, fuel pyrolysis generally begins at relatively low temperature as the fuel approaches the flame front, so the soot inception process may be completed well before temperatures are high enough to initiate the competitive reactions (Gomez et al., 1984).

Once soot nuclei have been formed, particle growth occurs rapidly by surface reactions (Harris and Weiner, 1983a, b). Ultimately, the soot nuclei account for only a

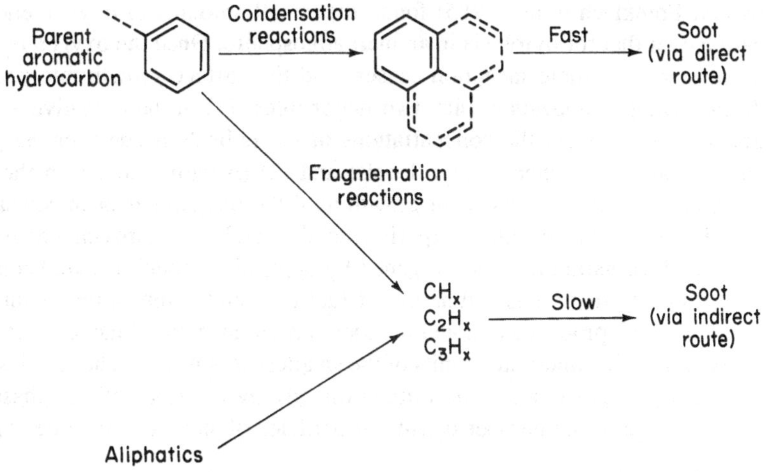


**Figure 6.12** Chemical pathways for soot formation.

small fraction of the mass of soot formed; the remainder is material that has condensed or reacted on the initial nuclei. The yield of soot increases rapidly as the C/O ratio increases beyond the sooting threshold (Haynes and Wagner, 1981). Remarkably, while the quantity of soot generated varies dramatically from one flame to another, the surface growth rate of individual particles does not appear to vary significantly with fuel composition or flame characteristics (Harris and Weiner, 1983a, b). The growth rate does vary with the age of the soot, however. Typically, the surface growth rate is 0.4 to 1 g $m^{-2}$ $s^{-1}$ early in the flame, decreases to about 0.1 to 0.2 g $m^{-2}$ $s^{-1}$ after 25 to 30 ms in the flame, and drops precipitously thereafter (Harris and Weiner, 1983a, b). The time scale of this change in growth rate corresponds to that in which the C/H ratio of the soot particles increases from an initial value of about 1 to 8 or so. Therefore, it has been suggested that the changing C/H ratio, in some as yet undefined way, is responsible for the decline in growth rate.

Since the growth rate of soot particles is not a strong function of fuel composition, the wide range of soot emissions is attributed to differences in the number of soot nuclei initially formed. The formation of the soot nuclei is poorly understood. It does not appear to involve homogeneous nucleation, as discussed in Chapter 5. Rather, it appears that a sequence of gas-phase, polymerization reactions produces hydrocarbon molecules of ever-increasing molecular weight until, at a molecular weight on the order of 1000, soot particles are formed (Wang et al., 1983). The soot precursors are large polycyclic aromatic hydrocarbons, so the chemical mechanisms of soot formation are directly related to those discussed in Chapter 3 for the formation of polycyclic aromatic hydrocarbons.

Soot particle inception takes place very early in the flame, in a region where radicals are present in superequilibrium concentrations (Millikan, 1962). In premixed flames, soot inception proceeds for only 2 to 3 ms, during which time the surface growth rate of the nuclei is approximately zero (Harris et al., 1986). In those experiments, the region of soot inception contained about 1% $O_2$. Similarly, in shock tube studies of soot formation, Frenklach et al. (1985) found that small amounts of oxygen enhance soot formation over that for pyrolysis in an inert atmosphere. Once the oxygen level in the flame declines, new particle inception ceases, and the surface growth rate rises rapidly to a plateau. Oxygen appears to play two major roles: (1) in the oxidative attack of the hydrocarbon molecules, the concentrations of the radicals needed for the polymerization reactions are maintained at high levels; and (2) oxidative attack on the incipient soot nuclei and precursors limits soot growth until the oxygen has been consumed.

In spite of the wealth of experimental data and our improving understanding of the nature and formation of soot, no generally applicable model has yet been formulated to predict soot formation as a function of fuel type and combustion conditions. Harris et al. (1986) have proposed a model for soot particle inception kinetics. Let $m_i$ be the mass of a soot particle containing $i$ units of the smallest soot nuclei. The smallest soot nucleus, of mass $m_1$, is presumed to be formed directly by reactions of gas-phase species. The rate of change of the number density of particles of size $i$ can then be expressed as

$$\frac{dN_i}{dt} = \left(\frac{dN_i}{dt}\right)_{\text{coagulation}} + \left(\frac{dN_i}{dt}\right)_{\text{inception}} + \left(\frac{dN_i}{dt}\right)_{\text{growth}} + \left(\frac{dN_i}{dt}\right)_{\text{oxidation}} \tag{6.14}$$

The rate of change of $N_i$ by coagulation is represented as outlined in Chapter 5. The inception term is nonzero only for $i = 1$, where $dN_i/dt = I_1(t)$.

In the analysis of Harris et al. (1986), an incipient soot particle was actually defined as the smallest species that absorbs light at a wavelength of 1.09 $\mu$m. Such species were argued as likely to be similar to the smallest particles actually observed by Wersborg et al. (1973) that had diameters near 1.5 nm and contained on the order of 100 carbon atoms. The inception rates are thus defined experimentally and theoretically as the flux of particles across this lower limit. The physical mechanism for growth is thought to be the thermal decomposition of acetylene on the surface of existing particles.

Although we will subsequently discuss soot oxidation in more detail, for the purposes of the soot particle dynamics model, Harris et al. (1986) assume that the ratio of surface growth and oxidation are proportional to soot surface area, so the rate of change of the mass of a particle size $i$ from growth and oxidation is

$$\frac{dm_i}{dt} = k(t)s_i \tag{6.15}$$

where $s_i$ is the surface area of a particle of size $i$, and $k(t)$ is an appropriate rate constant that, according to Harris et al. (1986), varies with time due to the aging of the carbonaceous structure of soot. Thus

$$\left(\frac{dN_i}{dt}\right)_{\text{growth}} + \left(\frac{dN_i}{dt}\right)_{\text{oxidation}} = \frac{|k(t)|}{m_1}\left[N_k s_k - N_i s_i\right] \tag{6.16}$$

where $k = i - 1$ or $i + 1$ depending on whether $k(t)$ is positive (growth rate exceeds oxidation rate) or negative (oxidation rate exceeds growth rate), respectively. When $k(t) < 0$, (6.16) still accounts for surface growth of $i = 1$ particles, but when $k(t) > 0$.

$$\frac{dN_i}{dt} = \frac{-k(t)N_1 s_1}{m_1} \tag{6.17}$$

Growth of a species smaller than size 1 is inception.

In summary, according to the model above, soot particles form by a poorly understood inception process and then grow by coagulation and the impingement of vapor species on their surface. They shrink by oxidation, which is also represented as a surface reaction process, presumably involving the collision of oxidants such as $O_2$ and OH with the soot particles.

Harris et al. (1986) estimated the soot inception rate in a premixed ethylene–argon–oxygen flame by solving the set of ordinary differential equations for the $N_i$ for diameters between 1.5 nm ($i = 1$) and 22 nm ($i = 3000$), 0.022 $\mu$m, to conform with optical data on soot particles obtained in a premixed ethylene flame.* Figure 6.13 shows the soot particle total number density, mean particle diameter, and inception rate, $J$, as functions of time over positions corresponding to $t = 1.1$ to 9.6 ms beyond the estimated reaction zone of the flame. Figure 6.13 shows the calculated rate of soot particle inception. The data, which start 1.1 ms beyond the reaction zone, missed the rise in the inception rate. Using the curve, the authors estimate that a total of about $10^{19}$ incipient particles $m^{-3}$ were created in this flame, with a peak new particle formation rate of $10^{23}$ $m^{-3}$ $s^{-1}$. This quantity corresponded to 50% of the soot volume present at 2.4 ms, 3% of the soot volume present at 9 ms, and less than 1% of the soot volume ultimately formed.

### 6.3.2 Soot Oxidation

Soot particles formed in the fuel-rich regions of diffusion flames are eventually mixed with gases sufficiently fuel-lean that oxidation becomes possible. As a result, the soot actually emitted from a combustor may represent only a small fraction of the soot formed in the combustion zone. Since the conditions leading to soot formation in diffusion flames cannot be entirely eliminated, soot oxidation downstream of the flame is very important to the control of soot emissions.

We desire to develop a useful model for soot oxidation. Park and Appleton (1973) observed that the rate of oxidation of carbon black (a commercially produced soot-like material) by oxygen could be described by a semiempirical model originally developed by Nagle and Strickland-Constable (1962) to describe the oxidation of pyrolytic graphite (2.156).

There are other potential oxidants in flames besides $O_2$. As we have seen, in fuel-rich flames, OH may be more abundant than $O_2$, and the O and $O_2$ concentrations may be comparable. Let us estimate the rate of oxidation of a soot particle by reaction with OH. The flux of molecules at a number concentration $N$ impinging on a unit area of surface in a gas is $N\bar{c}/4$, where $\bar{c}$ is the mean molecular speed of the gas molecules, $\bar{c}$

*The authors discuss those physical phenomena that influence the coagulation rate (e.g., van der Waals forces, charging, etc.) together with a variety of assumptions that must be made.

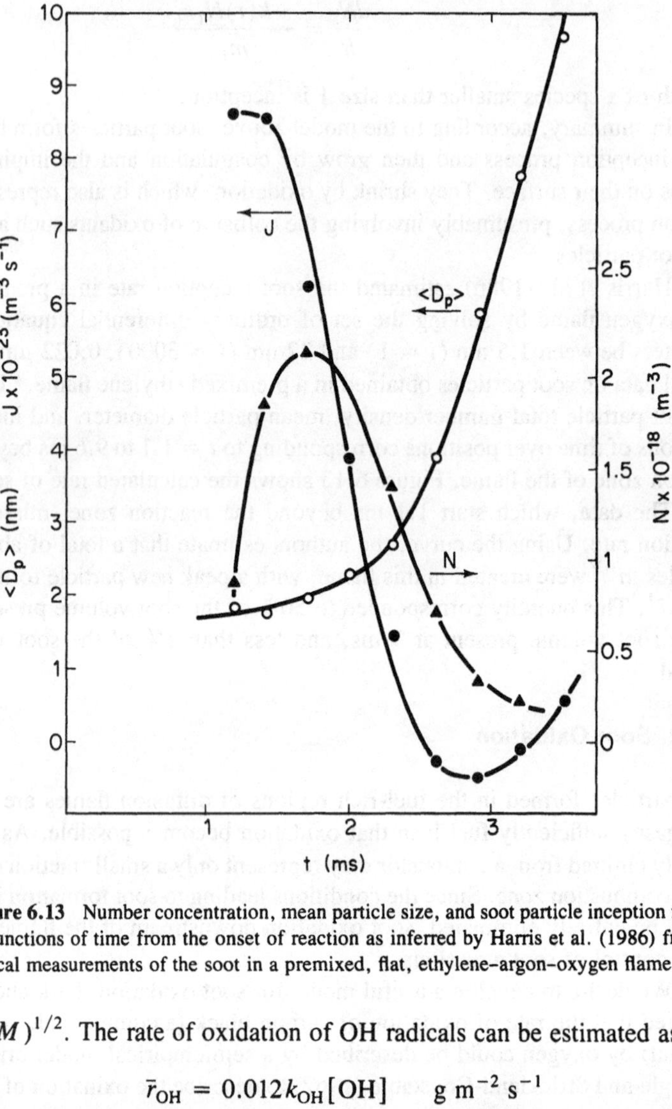


**Figure 6.13** Number concentration, mean particle size, and soot particle inception rate as functions of time from the onset of reaction as inferred by Harris et al. (1986) from optical measurements of the soot in a premixed, flat, ethylene–argon–oxygen flame.

$= (8RT/\pi M)^{1/2}$. The rate of oxidation of OH radicals can be estimated as

$$\bar{r}_{OH} = 0.012 k_{OH} [\text{OH}] \qquad \text{g m}^{-2}\,\text{s}^{-1} \tag{6.18}$$

where $k_{OH} = \gamma_{OH} (RT/2\pi M_{OH})^{1/2}$, where $\gamma_{OH}$ is the accommodation coefficient for OH on the soot particle surface, the fraction of OH radical collisions with the surface that lead to reaction. Using such a model to analyze soot oxidation by OH, Neoh et al. (1981) found that a value of $\gamma_{OH} = 0.28$ was consistent with their data. With this value of $\gamma_{OH}$, $k_{OH} = 2.47 T^{1/2}$ g m$^{-2}$ s$^{-1}$ (mol/m$^{-3}$)$^{-1}$.

Rosner and Allendorf (1968) examined the reaction of oxygen atoms with pyrolytic graphite. The accommodation coefficient was found to be weakly dependent on temperature and oxygen partial pressure, varying from about 0.15 to 0.3. Assuming that $\gamma_O =$

0.2, the rate constant for reaction of O with the carbon surface is estimated to be $k_O = 1.82\,T^{1/2}$ g m$^{-2}$ s$^{-1}$ (mol m$^{-3}$)$^{-1}$.

The total soot oxidation rate is the sum of the $O_2$ and OH oxidation rates,

$$\bar{r} = \bar{r}_{O_2} + \bar{r}_{OH} + \bar{r}_O \tag{6.19}$$

Particle burnout is described by

$$\frac{dm_p}{dt} = -\bar{r}A \tag{6.20}$$

where $A$ is the surface area. For spherical particles, the rate of change of the radius is

$$\frac{da}{dt} = \frac{-\bar{r}}{\rho_p} \tag{6.21}$$

Soot particle densities are typically $\rho_p \simeq 1800$ kg m$^{-3}$.

Figure 6.14 shows the calculated surface recession velocity as a function of equiv-

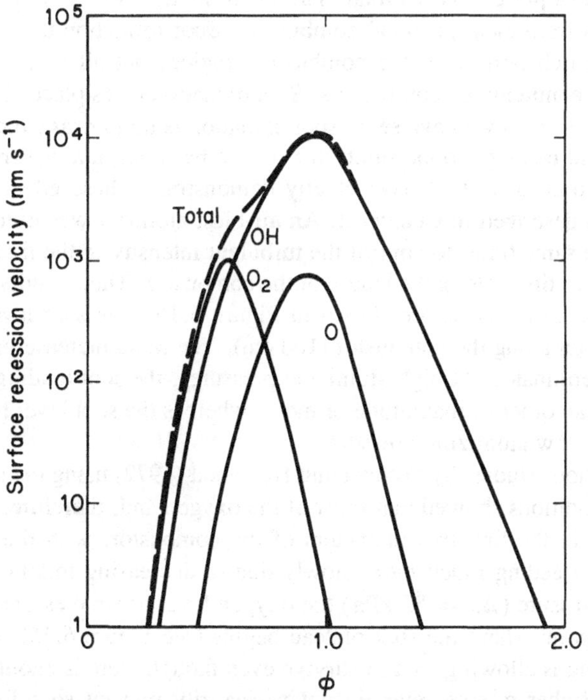


**Figure 6.14** Variation of the surface recession rate of soot particles with equivalence ratio for adiabatic combustion of a fuel with composition $CH_{1.8}$.

alence ratio for equilibrium adiabatic combustion. For $\phi < 0.7$, the $O_2$ reaction dominates, although the OH reaction is still important. As the combustion products are cooled, the equilibrium OH level will drop, reducing the contribution of that reaction. Near stoichiometric and in fuel-rich combustion products, the OH reaction clearly dominates.

The spherical soot particles are typically 10 to 50 nm in diameter. The time to burn the soot particle completely is

$$\tau_B = \frac{\rho_p}{\bar{r}} \frac{D_p}{2}$$

The time to burn a 20-nm-diameter soot particle is about 1 ms at stoichiometric conditions. Thus soot particles burn rapidly when they pass through the flame front. Very rapid dilution with cold air, however, may lead to soot emissions in spite of an abundance of oxygen.

### 6.3.3 Control of Soot Formation

The range of equivalence ratios over which reactions critical to soot formation and destruction take place is very broad. Turbulent mixing again strongly influences the emission levels from most practical combustors. Soot formation takes place only in the extreme fuel-rich portion of the combustion region, but all fuel must pass through this domain in nonpremixed combustors. Soot oxidation takes place closer to the mean composition. Since, as we have seen, soot oxidation is rapid near stoichiometric conditions, there should be ample opportunity for soot to burn out in a well-mixed combustor.

Prado et al. (1977) have clearly demonstrated these effects using the plug flow combustor described in Chapter 2. An air-blast atomizer was used to inject liquid fuels and, at the same time, to control the turbulent intensity in the first two diameters ($\sim 20$ ms residence time) along the length of the combustor. Their results for the stoichiometric combustion of kerosene are shown in Figure 6.15. The soot level rises rapidly in the first diameter along the combustor (100 cm). The rapid increase in the soot loading then abruptly terminates. At high atomizing pressures, the soot loading then drops relatively slowly by an order of magnitude or more, whereas the soot level remains approximately constant at low atomizing pressure.

Previous studies by Pompei and Heywood (1972) using the same combustor under similar conditions showed that most of the oxygen and, therefore, most of the fuel were consumed in the first two diameters of the combustor, with the final consumption of oxygen proceeding much more slowly due to decreasing mixing rates. At the low atomizing pressure ($\Delta p \simeq 82$ kPa) the oxygen level became essentially constant at about the same point where the soot plateau begins (see Figure 6.15), clearly indicating that poor mixing is allowing soot to survive even though there is enough oxygen to consume it fully. Higher mixing rates do not necessarily prevent soot formation; they simply provide the opportunity for soot formed early in the combustor to burn. While increasing

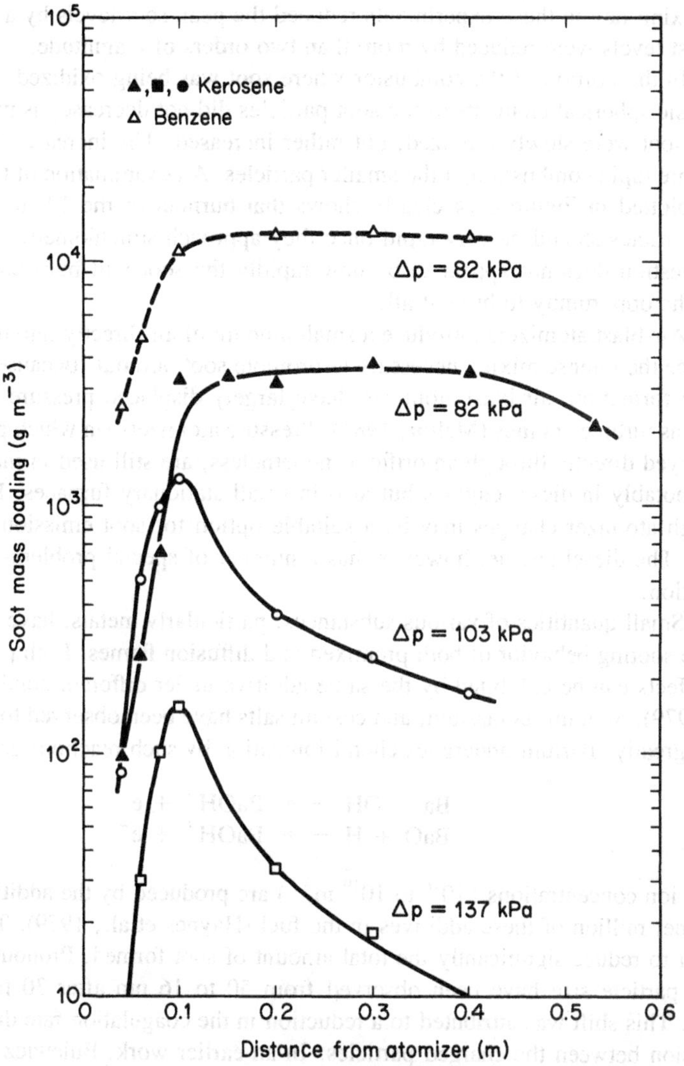


**Figure 6.15** Soot mass loadings as a function of axial position in a plug flow combustor. Data for combustion of kerosene using various pressures of an air-assist atomizer are shown as solid points. Measurements for benzene combustion are shown as open points (data from Prado et al., 1977). Reprinted by permission of The Combustion Institute.

the mixing rate in these experiments reduced the peak soot levels by a factor of 20, the exhaust levels were reduced by more than two orders of magnitude.

In the portion of the combustor where soot was being oxidized, the mean size of the basic spherical elements of the soot particles did not decrease, as might be expected if the soot were slowly oxidized, but rather increased. The increased was attributed to the more rapid combustion of the smaller particles. An examination of the soot oxidation rates plotted in Figure 6.14 clearly shows that burnout of the 19- to 25-mm-diameter soot particles should be very rapid once they approach stoichiometric conditions. Thus the question does not appear to be how rapidly the soot will burn but whether it will have the opportunity to burn at all.

Air-blast atomizers introduce a small amount of air directly into the fuel spray and produce the intense mixing necessary to promote soot burnout. Because of their reduced smoke formation, air-blast atomizers have largely displaced pressure atomizers in aircraft gas turbine engines (Mellor, 1981). Pressure atomizers (in which pressurized liquid is sprayed directly through an orifice), nonetheless, are still used in many applications, most notably in diesel engines but also in small stationary furnaces. Improved mixing through atomizer changes may be a suitable option for soot emission control in many cases. The diesel engine, however, has a number of special problems due to its cyclic operation.

Small quantities of various substances, particularly metals, have a profound effect on the sooting behavior of both premixed and diffusion flames. Both pro- and anti-sooting effects can be exhibited by the same additive under different conditions (Haynes et al., 1979). Sodium, potassium, and cesium salts have been observed to increase the soot yield greatly. Barium undergoes chemi-ionization by such reactions as

$$\mathrm{Ba + OH \longrightarrow BaOH^+ + e^-}$$
$$\mathrm{BaO + H \longrightarrow BaOH^+ + e^-}$$

Large ion concentrations ($10^{16}$ to $10^{19}$ $\mathrm{m^{-3}}$) are produced by the addition of only a few parts per million of these additives in the fuel (Haynes et al., 1979). These ions do not appear to reduce significantly the total amount of soot formed. Pronounced shifts in the mean particle size have been observed from 50 to 16 nm after 30 ms in a premixed flame. This shift was attributed to a reduction in the coagulation rate due to electrostatic repulsion between the charged particles. In an earlier work, Bulewicz et al. (1975) observed enhanced soot formation for small additions of these species, but soot inhibition at higher additive concentrations in a diffusion flame. The smaller soot particles burn faster than the larger ones produced without the additives, leading to lower soot emissions from the diffusion flame (Howard and Kausch, 1980).

Alkaline earths (Ca, Sr, Ba) form their hydroxides from the sequence (Glassman, 1977)

$$\mathrm{M + H_2O \longrightarrow MOH + H}$$

$$\mathrm{MOH + H_2O \longrightarrow M(OH)_2 + H}$$

in which M represents the alkaline earth metal. H atoms may attack water to form OH, that is,

$$H + H_2O \longrightarrow OH + H_2$$

In oxygen-deficient fuel-rich flames, this additional production of OH can dramatically increase the oxidation of soot precursors and prevent soot formation.

Transition metal additives, notably Co, Fe, Mn, and Ni, can reduce the quantity of soot emitted without significantly altering the particle size distribution. Two of the best known commercial additives, methylcylopentadienyl manganese tricarbonyl (MMT) and ferrocene (dicyclopentadrenyl iron), act by this mechanism (Howard and Kausch, 1980). These additives have no effect on the amount of soot formed in a premixed fuel-rich flame, but may increase the rate of oxidation by as much as 20%. Metal oxides may be incorporated in the soot particles and then catalyze soot oxidation by a reaction such as

$$M_xO_y + C_{(s)} \longrightarrow CO + M_xO_{y-1}$$

The emissions of the metal additives can become significant. Giovanni et al. (1972) reported an increase in particulate emissions from turbojet engines by 0.5 g (kg of fuel)$^{-1}$ above the baseline emissions of 2 to 14 g (kg of fuel)$^{-1}$ when manganese was used to reduce soot emissions. The increased emissions were attributed solely to the emissions of manganese in the aerosol.

## 6.4 MOTOR VEHICLE EXHAUST AEROSOLS

Figure 6.16 shows the volume distribution of primary automobile exhaust aerosols produced under cruise conditions with leaded fuel from a single vehicle as reported by Miller et al. (1976). At speeds up to 35 mph the mode in the volume distribution is at about 0.04 $\mu$m, whereas at 50 mph the mode is shifted to about 0.1 $\mu$m. Figure 6.17

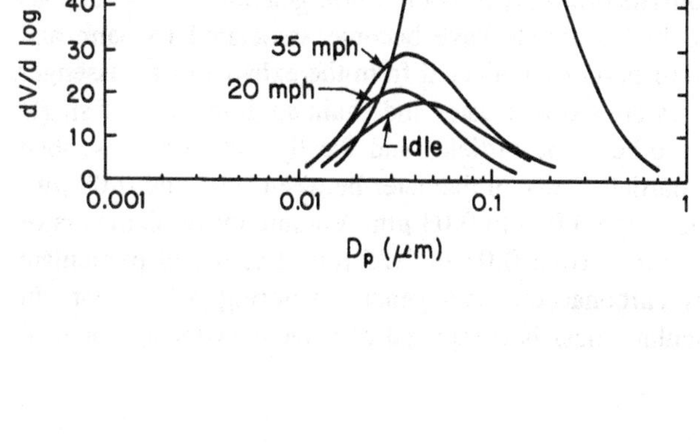


**Figure 6.16** Volume distributions of primary automobile exhaust aerosol produced under cruise conditions during combustion of leaded gasoline as reported by Miller et al. (1976). (Reprinted by permission of Air Pollution Control Association).

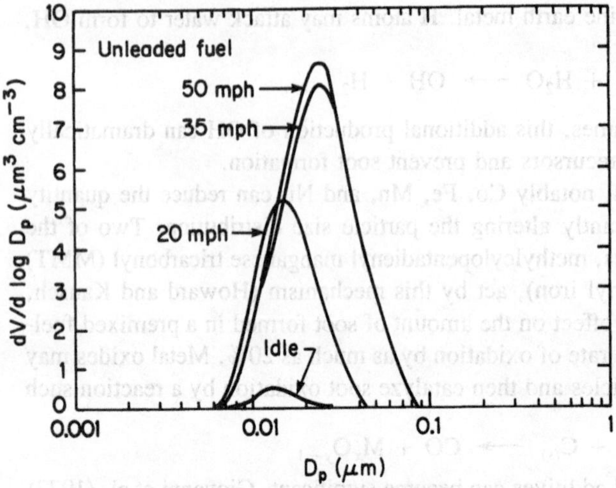


**Figure 6.17** Volume distributions of primary automobile exhaust aerosol produced under cruise conditions during combustion of unleaded gasoline as reported by Miller et al. (1976). (Reprinted by permission of Air Pollution Control Association).

shows the aerosol volume distributions from an identical vehicle equipped to run on unleaded fuel. Unlike the case with leaded fuel, the mode in the volume distribution remains between 0.01 and 0.03 $\mu$m for all cruise speeds investigated. The shift to slightly larger mean sizes between 20 and 35 mph was attributed to increase gas-to-particle conversion. We note that the conditions leading to the increase in aerosol volume at high speeds with leaded fuel appear to be absent in the case of unleaded fuel combustion. Because the data shown in Figures 6.16 and 6.17 represent a very limited sample, they should not be viewed as indicative of all motor vehicle exhaust aerosols. Nevertheless, they do exhibit the general features with respect to size distribution of automobile exhaust aerosols.

Pierson and Brachaczek (1983) reported the composition and emission rates of airborne particulate matter from on-road vehicles in the Tuscarora Mountain Tunnel of the Pennsylvania Turnpike in 1977. Particulate loading in the tunnel was found to be dominated by diesel vehicles, even though on the average they constituted only about 10% of the traffic. Diesel emission rates were of the order 0.87 g $km^{-1}$, the most abundant component of which is carbon, elemental and organic. Thirty-four elements were measured, in descending order of mass, C, Pb, H, $SO_4^{2-}$, and Br together accounting for over 90%. Size distributions of motor vehicle particulate matter exhibited a mass median diameter of 0.15 $\mu$m.

Diesel particulate matter consists primarily of combustion-generated carbonaceous soot with which some unburned hydrocarbons have become associated (Amann and Siegla, 1982). Photomicrographs of particles collected from the exhaust of a passenger car diesel indicate that the particles consist of cluster and chain agglomerates of single spherical particles, similar to the other soot particles and the fly ash fume described previously. The single spherical particles vary in diameter between 0.01 and 0.08 $\mu$m, with the most usually lying in the range 0.015 to 0.03 $\mu$m. Volume mean diameters of the particles (aggregates) tend to range from 0.05 to 0.25 $\mu$m. The diesel particulate matter is normally dominated by carbonaceous soot generated during combustion. In addition, 10 to 30% of the particulate mass is comprised of solvent-extractable hydrocarbons that adsorb or condense on the surface of the soot particles and that consist of high-boiling-point fractions and lubricating oil.

## PROBLEMS

**6.1.** The amount of ash vaporized during coal combustion is typically about 1% of the mineral matter in the coal. Reduced vapor species are oxidized as they diffuse from the surface of the burning char particle leading to rapid homogeneous nucleation and the formation of large numbers of very small particles. Assuming that the initial nuclei are 0.001 $\mu$m in diameter and the ash density is 2300 kg m$^{-3}$, compute and plot the particle number concentration as a function of residence time in the furnace, assuming the gas temperature is 1700 K for 1 second and then decreases to 400 K at a rate of 500 K s$^{-1}$. The gas viscosity may be taken as $\mu = 3.4 \times 10^{-7}\, T^{0.7}$ kg m$^{-1}$ s$^{-1}$. The aerosol may be assumed to be monodisperse.

**6.2.** A 50-$\mu$m-diameter char particle is burned in air in a furnace that is heated to 1800 K. Examine the volatilization of silica ($SiO_2$) from this particle as it burns. Silica may vaporize directly

$$SiO_{2(c)} \overset{1}{\rightleftharpoons} SiO_{2(v)} \qquad K_{p1} = 2.4 \times 10^8 e^{-69{,}000/T} \text{ atm}$$

or by means of reduction to the monoxide

$$SiO_{2(c)} + CO \overset{2}{\rightleftharpoons} SiO_{(v)} + CO_2 \qquad K_{p2} = 4.53 \times 10^8 e^{-61{,}672/T} \text{ atm}$$

The oxidation rate expression of the char is that for the Whitwick coal in Table 2.10.

(a) Calculate the equilibrium partial pressures for $SiO_2$ and SiO at the surface of the char particle.

(b) What vaporization rate corresponds to this partial pressure?

(c) Compare the vaporization rate with that predicted for a pure silica particle of the same size. Assume the silica particle temperature is the same as that of the gas.

**6.3.** For the conditions of Problem 6.2, compute the vapor concentration profile as a function of distance from the surface of the particle, assuming that no condensation takes place. The binary diffusivity may be calculated from $cD = 9.1 \times 10^7 \bar{T} + 7.4 \times 10^{-4}$ mole m$^{-1}$ s$^{-1}$, where $\bar{T}$ is the mean of gas and particle temperatures. Neglecting vapor loss, calculate and plot the supersaturation ratio and homogeneous nucleation rate. How far from the surface will nucleation occur?

Use the following properties:

| | |
|---|---|
| Surface tension: | $\sigma = 0.30$ J m$^{-2}$ |
| Density: | $\rho_c = 2200$ kg m$^{-3}$ |
| Molecular Weight: | $M = 60.9$ g mole$^{-1}$ |

**6.4.** Harris et al. (1986) observed that no growth of the soot nuclei occurred in the region of the flame where particles are formed and that the oxygen mole fraction in this region is about 0.01. Assuming that the Nagle and Strickland-Constable kinetics describe the oxidation of the soot nuclei and that the growth species is acetylene, estimate the minimum acetylene concentration that would be required to maintain the observed soot nucleus size of 1.8 nm, for temperatures ranging from 1200 K to 1500 K. The flux of acetylene reaching the particle surface may be taken as the effusion flux, and the reaction probability is unity for this estimate. The soot particle density is 1800 kg m$^{-3}$

## REFERENCES


AMANN, C. A., and SIEGLA, D.C., "Diesel Particulates—What They Are and Why," *Aerosol Sci. Technol. 1*, 73–101 (1982).

BRAIDE, K. M., ISLES, G. L., JORDAN, J. B., and WILLIAMS, A. "The Combustion of Single Droplets of Some Fuel Oils and Alternative Liquid Fuel Combinations," *J. Inst. Energy, 52*, 115–124 (1979).

BULEWICZ, E. M., EVANS, D. G., and PADLEY, P. J. "Effect of Metallic Additives on Soot Formation Processes in Flames," in *Fifteenth Symposium (International) on Combustion*, The Combustion Institute, Pittsburgh, PA, 1461–1470 (1975).

DAVISON, R. L., NATUSCH, D. F. S, WALLACE, J. R., and EVANS, C. A., JR. "Trace Elements in Fly Ash—Dependence of Concentration on Particle Size," *Environ. Sci. Tecvhnol.*, *8*, 1107–1113 (1974).

FLAGAN, R. C. "Submicron Aerosols from Pulverized Coal Combustion," *Seventeenth Symposium (International) on Combustion*, The Combustion Institute, Pittsburgh, PA, 97–104 (1979).

FLAGAN, R. C., and FRIEDLANDER, S. K., "Particle Formation in Pulverized Coal Combustion: A Review," in *Recent Developments in Aerosol Science*, D. T. Shaw, Ed., Wiley, New York, 25–59 (1978).

FRENKLACH, M., RAMACHANDRA, M. K., and MATULA, R. A. "Soot Formation in Shock Tube Oxidation of Hydrocarbons," in *Twentieth Symposium (International) on Combustion*, The Combustion Institute, Pittsburgh, PA, 871–878, (1985).

FRENKLACH, M., TAKI, S., and MATULA, R. A. "A Conceptual Model for Soot Formation in Pyrolysis of Aromatic Hydocarbons," *Combust. Flame, 49*, 275–282 (1983).

GIOVANNI, D. V., PAGNI, P. J., SAWYER, R. F., and HUGHES, L. "Manganese Additive Effects on the Emissions from a Model Gas Turbine Combustor," *Combust. Sci. Tech.*, *6*, 107–114 (1972).

GLASSMAN, I. *Combustion*, Academic Press, New York (1977).

GLASSMAN, I. "Phenomenological Models of Soot Processes in Combustion Systems," Princeton University Department of Mechanical and Aerospace Engineering Report No. 1450 (1979).

GLASSMAN, I., and YACCARINO, P. "The Temperature Effect in Sooting Diffusion Flames," in *Eighteenth Symposium (International) on Combustion*, The Combustion Institute, Pittsburgh, PA, 1175–1183, (1981).

GOMEZ, A., SIDEBOTHAM, G., and GLASSMAN, I. "Sooting Behavior in Temperature-Controlled Laminar Diffusion Flames," *Combust. Flame, 58*, 45–57 (1984).

GRAHAM, S. C., HOMER, J. B., and ROSENFELD, J. L. J. "The Formation and Coagulation of Soot Aerosols Generated in Pyrolysis of Aromatic Hydrocarbons," *Proc. R. Soc. London-A344*, 259–285 (1975).

HARRIS, S. J., and WEINER, A. M. "Surface Growth of Soot Particles in Premixed Ethylene/Air Flames," *Combust. Sci. Technol.*, *31*, 155–167 (1983a).

HARRIS, S. J., and WEINER, A. M. "Determination of the Rate Constant for Soot Surface Growth," *Combust. Sci. Technol.*, *32*, 267–275 (1983b).

HARRIS, S. J., WEINER, A. M., and ASHCRAFT, C. C. "Soot Particle Inception Kinetics in a Premixed Ethylene Flame," *Combust. Flame, 64*, 65–81 (1986).

HAYNES, B. S., and WAGNER, H. G. "Soot Formation," *Prog. Energy Combust. Sci.*, *7*, 229–273 (1981).

HAYNES, B. S., JANDER, H., and WAGNER, H. G. "The Effect of Metal Additives on the Formation of Soot in Combustion,'' in *Seventeenth Symposium (International) on Combustion,* The Combustion Institute, Pittsburgh, PA, 1365–1374, (1979).

HOWARD, J. B., and KAUSCH, W. J., JR. ''Soot Control by Fuel Additives,'' *Prog. Energy Combust. Sci. 6,* 263–276 (1980).

LEHAYE, J., and PRADO, G. ''Morphology and Internal Structure of Soot and Carbon Blacks,'' in *Particulate Carbon: Formation during Combustion,* D. C. Siegla and G. W. Smith, Eds., Plenum Press, New York, 33–56, (1981).

LINAK, W. P., and PETERSON, T. W. ''Effect of Coal Type and Residence Time on the Submicron Aerosol Distribution from Pulverized Coal Combustion,'' *Aerosol Sci. Technol., 3,* 77–96 (1984).

LINAK, W. P., and PETERSON, T. W. ''Mechanisms Governing the Composition and Size Distribution of Ash Aerosols in a Laboratory Pulverized Coal Combustor,'' in *Twenty-first Symposium (International) on Combustion,* The Combustion Institute, Pittsburgh, PA, (1987).

LITTLEJOHN, R. F. ''Mineral Matter and Ash Distribution in 'As-Fired' Samples of Pulverized Fuel,'' *J. Inst. Fuel, 39,* 59–67 (1966).

MARKOWSKI, G. R., ENSOR, D. S., HOOPER, R. G., and CARR, R. C., ''A Submicron Aerosol Mode in Flue Gas from a Pulverized Coal Utility Boiler,'' *Environ. Sci. Technol., 14,* 1400–1402 (1980).

MARRONE, N. J., KENNEDY, I. M., and DRYER, F. L. ''Coke Formation in the Combustion of Isolated Heavy Oil Droplets,'' *Combust. Sci. Tech., 36,* 149–170 (1984).

MCELROY, M. W., and CARR, R. C. in ''Proceedings of Joint Symposium on Stationary Combustion $NO_x$ Control,'' Electric Power Research Institute Report No. WS-79-220, Palo Alto, CA, 183–199 (1981).

MELLOR, A. M. ''Soot Studies in Gas Turbine Combustors and Other Turbulent Spray Flames,'' in *Particulate Carbon: Formation during Combustion,* D. C. Siegla, Ed., Plenum Press, New York, 343–349 (1981).

MILLER, D. F., LEVY, A., PUI, D. Y. H., WHITBY, K. T., and WILSON, W. E., JR. ''Combustion and Photochemical Aerosols Attributable to Automobiles,'' *J. Air Pollut. Control Assoc., 26,* 576–581 (1976).

MILLIKAN, R. C. ''Nonequilibrium Soot Formation in Premixed Flames,'' *J. Phys. Chem., 66,* 794–799 (1962).

NAGLE, J., and STRICKLAND-CONSTABLE, R. F. ''Oxidation of Carbon between 1000–2000°C,'' in *Proceedings of the Fifth Carbon Conference, 1,* 154–164 (1962).

NEOH, K. G., HOWARD, J. B., and SAROFIM, A. F. ''Soot Oxidation in Flames,'' in *Particulate Carbon: Formation during Combustion,* D. C. Siegla and G. W. Smith, Eds., Plenum Press, New York, 261–282 (1981).

PALMER, H. B., and CULLIS, C. F. ''The Formation of Carbon from Gases,'' in *Chemistry and Physics of Carbon,* P. L. Walker, Ed., Vol. 1, Marcel Dekker, New York, 265–325 (1965).

PARK, C., and APPLETON, J. P. ''Shock Tube Measurements of Soot Oxidation Rates,'' *Combust. Flame, 20,* 369–379 (1973).

PIERSON, W. R., and BRACHACZEK, W. W. ''Particulate Matter Associated with Vehicles on the Road: II,'' *Aerosol Sci. Technol., 2,* 1–40 (1983).

POMPEI, F., and HEYWOOD, J. B. ''The Role of Mixing in Burner Generated Carbon Monoxide and Nitric Oxide,'' *Combust. Flame, 19,* 407–418 (1972).

PRADO, G. P., LEE, M. L., HITES, R. A., HOULT, D. P., and HOWARD, J. B. ''Soot and Hydrocarbon Formation in a Turbulent Diffusion Flame,'' in *Sixteenth Symposium (International) on Combustion,* The Combustion Institute, Pittsburgh, PA, 649–661, (1977).

QUANN, R. J., and SAROFIM, A. F. ''Vaporization of Refractory Oxides during Pulverized Coal

Combustion,'' in *Nineteenth Symposium (International) on Combustion,* The Combustion Institute, Pittsburgh, PA, 1429–1440, (1982).

Rosner, D. E., and Allendorf, H. D. ''Comparative Studies of the Attack of Pyrolytic and Isotropic Graphite by Atomic and Molecular Oxygen at High Temperatures,'' *AIAA J., 6,* 650–654 (1968).

Sarofim, A. F., Howard, J. B., and Padia, A. S. ''The Physical Transformation of the Mineral Matter in Pulverized Coal under Simulated Combustion Conditions,'' *Combust. Sci. Tech. 16,* 187–204 (1977).

Senior, C. L., and Flagan, R. C. ''Ash Vaporization and Condensation during the Combustion of a Suspended Coal Particle,'' *Aerosol Sci. Tech., 1,* 371–383 (1982).

Smith, G. W. ''A Simple Nucleation/Depletion Model for the Spherule Size of Particulate Carbon,'' *Combust. Flame, 48,* 265–272 (1982).

Taylor, D. D., and Flagan, R. C. ''Laboratory Studies of Submicron Particles from Coal Combustion,'' *Eighteenth Symposium (International) on Combustion,* The Combustion Institute, Pittsburgh, PA, 1227–1237 (1981).

Taylor, D. D., and Flagan, R. C. ''The Influence of Combustor Operating Conditions on Fine Particles from Coal Combustion,'' *Aerosol Sci. Tech., 1,* 103–117 (1982).

Wang, T. S., Matula, R. A., and Farmer, R. C. ''Combustion Kinetics of Soot Formation from Toluene,'' in *Nineteenth Symposium (International) on Combustion,* The Combustion Institute, Pittsburgh, PA, 1149–1158 (1983).

Wersborg, B. L., Howard, J. B., and Willams, G. C. ''Physical Mechanisms in Carbon Formation in Flames,'' in *Fourteenth Symposium (International) on Combustion,* The Combustion Institute, Pittsburgh, PA, 929–940 (1973).

# 7

# Removal of Particles from Gas Streams

Particulate removal devices operate basically on the principle that a gas stream containing particles is passed through a region where the particles are acted on by external forces or caused to intercept obstacles, thereby separating them from the gas stream. When acted upon by external forces, the particles acquire a velocity component in a direction different from that of the gas stream. In order to design a separation device based on particulate separation by external forces, one must be able to compute the motion of a particle under such circumstances.

A preliminary selection of suitable particulate emission control systems is generally based on knowledge of four items: particulate concentration in the stream to be cleaned, the size distribution of the particles to be removed, the gas flow rate, and the final allowable particulate emission rate. Once the systems that are capable of providing the required efficiencies at the given flow rates have been chosen, the ultimate selection is generally made on the basis of the total cost of construction and operation. The size of a collector, and therefore its cost, is directly proportional to the volumetric flow rate of gas that must be cleaned. The operating factors that influence the cost of a device are the pressure drop through the unit, the power required, and the quantity of liquid needed (if a wet scrubbing system). In this chapter we concentrate on the design equations that are generally used for calculating efficiencies of various types of particulate emission control equipment. We shall not consider the estimation of capital or operating costs.

Devices that remove particles from gas streams rely on one or more of the following physical mechanisms:

1. *Sedimentation.* The particle-containing gas stream is introduced into a device or chamber where the particles settle under gravity to the floor of the chamber. Devices of this type are called settling chambers.

2. *Migration of charged particle in an electric field.* The particle-containing gas stream is introduced into a device in which the particles are charged and then subjected to an electric field. The resulting electrostatic force on the particles causes them to migrate to one of the surfaces of the device, where they are held and collected. Devices of this type are called electrostatic precipitators.
3. *Inertial deposition.* When a gas stream changes direction as it flows around an object in its path, suspended particles tend to keep moving in their original direction due to their inertia. Particulate collection devices based on this principle include cyclones, scrubbers, and filters.
4. *Brownian diffusion.* Particles suspended in a gas are always in Brownian motion. When the gas stream flows around obstacles, the natural random motion of the particles will bring them into contact with the obstacles, where they adhere and are collected. Because we know that Brownian motion is more pronounced the smaller the particle, we expect that devices based on diffusion as the separation mechanism will be most effective for small particles.

The key parameter that influences the choice of which device to employ in a particular case is the particle diameter $D_p$. As we will see, the physical mechanisms above vary greatly in their effectiveness depending on the size of the particle. Thus one of our major objectives in this chapter is to understand the effectiveness of particulate removal devices as a function of particle size.

There are several different classes of particulate control equipment that we consider in this chapter. The simplest particulate control device is a *settling chamber*, a large chamber in which the gas velocity is slowed, allowing the particles to settle out by gravity. A *cyclone* operates by causing the entire gas stream to flow in a spiral pattern inside a tapered tube. Because of the centrifugal force, particles migrate outward and collect on the wall of the tube. The particles slide down the wall and fall to the bottom, where they are removed. The clean gas generally reverses its flow and exits out of the top of the cyclone. An *electrostatic precipitator* utilizes the electrostatic force on charged particles in an electric field to separate particles from the gas stream. A high voltage drop is established between two electrodes, and particles passing through the resulting electric field acquire charge. The charged particles migrate to and are collected on an oppositely charged plate while the clean gas flows on through the device. Periodically, the plates are cleaned by rapping to shake off the layer of dust that has accumulated. A variety of *filters* operate on the principle that the particulate-laden gas is forced through an assemblage of collecting elements, such as a fiber or a filter mat. As the gas passes through the assemblage, particles accumulate on the collectors. Wet collection devices called *scrubbers* operate on the basis of the collision of particles with droplets of water that can easily be separated from the gas because of their large size.

Some general statements can be made about the nature of the various types of particulate gas-cleaning equipment. Mechanical collectors such as settling chambers or cyclones are typically much less expensive than the others but are generally only moderately efficient in particle removal. Since they are much better for large particles than for small ones, they often are used as precleaners for the more efficient final control devices, especially at high particulate loadings. Electrostatic precipitators can treat large volumetric flow rates of gas at relatively low pressure drops with very high removal efficiencies. However, electrostatic precipitators are expensive and are relatively inflexible to changes in process operating conditions. Fabric filters tend to have very high efficiencies but are expensive and are generally limited to dry, low-temperature conditions. Scrubbing can also achieve high efficiencies and offers the auxiliary advantage that gaseous pollutants can be removed simultaneously with particles. However, scrubbers can be expensive to operate, owing to their high pressure drop and to the fact that they produce a wet sludge that must be treated or disposed of.

We begin the chapter with a discussion of how the collection or removal efficiency of a device may be defined.

## 7.1 COLLECTION EFFICIENCY

We define the *collection efficiency* $\eta(D_p)$ of a device for particles of diameter $D_p$ as

$$\eta(D_p) = 1 - \frac{\text{number of particles of diameter } D_p \text{ per m}^3 \text{ of gas out}}{\text{number of particles of diameter } D_p \text{ per m}^3 \text{ of gas in}} \tag{7.1}$$

The *overall efficiency* of the device based on particle number is

$$\eta = 1 - \frac{\text{number of particles per m}^3 \text{ of gas out}}{\text{number of particles per m}^3 \text{ of gas in}} \tag{7.2}$$

These efficiencies can be expressed in terms of the particle size distribution functions at the inlet and outlet sides of the device,

$$\eta(D_p) = \frac{n_{\text{in}}(D_p)\,dD_p - n_{\text{out}}(D_p)\,dD_p}{n_{\text{in}}(D_p)\,dD_p} = 1 - \frac{n_{\text{out}}(D_p)}{n_{\text{in}}(D_p)} \tag{7.3}$$

and

$$\eta = \frac{\int_0^\infty \left[n_{\text{in}}(D_p) - n_{\text{out}}(D_p)\right] dD_p}{\int_0^\infty n_{\text{in}}(D_p)\,dD_p} = \frac{\int_0^\infty \eta(D_p)\,n_{\text{in}}(D_p)\,dD_p}{\int_0^\infty n_{\text{in}}(D_p)\,dD_p} \tag{7.4}$$

The definition of overall efficiency above is based on particle number. We can also define overall efficiencies based on other particle properties, such as surface area and volume (or mass). For example, the collection efficiency based on particle mass $\eta_m$ is defined as

$$\eta_m(D_p) = 1 - \frac{\text{mass of particles of diameter } D_p \text{ per m}^3 \text{ of gas out}}{\text{mass of particles of diameter } D_p \text{ per m}^3 \text{ of gas in}} \tag{7.5}$$

and the overall efficiency is

$$\eta_m = \frac{\int_0^\infty [(\pi/6)\, \rho_p D_p^3 n_{\text{in}}(D_p) - (\pi/6)\, \rho_p D_p^3 n_{\text{out}}(D_p)]\, dD_p}{\int_0^\infty (\pi/6)\, \rho_p D_p^3 n_{\text{in}}(D_p)\, dD_p}$$

$$= \frac{\int_0^\infty \eta(D_p)\, D_p^3 n_{\text{in}}(D_p)\, dD_p}{\int_0^\infty D_p^3 n_{\text{in}}(D_p)\, dD_p} \tag{7.6}$$

The overall collection efficiency by mass is usually the easiest to measure experimentally. The inlet and outlet streams may be sampled by a collection device, such as a filter, that collects virtually all of the particles.

A term that is sometimes used to express collection efficiency is the *penetration.* The penetration is based on the amount emitted rather than captured; penetration based on particle mass is just $P_m = 1 - \eta_m$. Alternatively, the penetration can be defined on the basis of particle number, $P = 1 - \eta$.

We have called the relationship between collection efficiency and particle size simply the collection efficiency. Other terms that are used for this quantity are the *grade efficiency* or the *fractional efficiency*. An important point on the collection efficiency curve is the size for which $\eta = 0.5$. The particle size at this point is called the *cut size* or the *cut diameter*.

## 7.2 SETTLING CHAMBERS

Gravitational settling is perhaps the most obvious means of separating particles from a flowing gas stream. A settling chamber is, in principle, simply a large box through which the effluent gas stream flows and in which particles in the stream settle to the floor by gravity. Gas velocities through a settling chamber must be kept low enough so that settling particles are not reentrained. The gas velocity is usually reduced by expanding the ducting into a chamber large enough so that sufficiently low velocities result. Although in principle settling chambers could be used to remove even the smallest particles, practical limitations in the length of such chambers restrict their applicability to the removal of particles larger than about 50 $\mu$m. Thus settling chambers are normally used as precleaners to remove large and possibly abrasive particles, prior to passing the gas stream through other collection devices. Settling chambers offer the advantages of (1) simple construction and low cost, (2) small pressure drops, and (3) collection of particles without need for water. The main disadvantage of settling chambers is the large space that they require.

A settling chamber is, as noted above, simply a horizontal chamber through which the particle-laden gas flows and to the floor of which the particles settle. Figure 7.1 shows a simple gravity settling chamber design. Actually, the chamber may contain a number of relatively closely spaced horizontal plates so that the distance that a particle must settle to be collected is considerably smaller than the height of the overall device.

In analyzing the performance of a settling chamber, the key feature is the nature of the gas flow through the device. We can distinguish three basic idealized flow situations: (1) laminar flow, (2) plug flow (velocity uniform across the cross section) with no vertical mixing of particles, (3) plug flow with complete vertical mixing of particles. Laminar flow is characterized by a parabolic-type velocity profile; such a flow would only be realized for Reynolds numbers below that for transition to turbulent flow. In a laminar flow, the time required for a particle at height $y$ above the floor of the chamber to settle is $y/v_t$, where $v_t$ is the particle's setting velocity, and vertical mixing of particles is absent in laminar flow. (The effect of Brownian motion is generally neglected relative to the steady downward movement due to settling.) The second flow category above, plug flow with no vertical mixing of particles, is, in a sense, an approximation to laminar flow in that vertical mixing of particles is still ignored, but a flat velocity profile is assumed and the particles all settle at their settling velocities. The third category, plug flow with thorough vertical mixing, is the model for turbulent flow. In a turbulent flow settling chamber the gas velocity is assumed to be uniform across the chamber due to the turbulent mixing. Moreover, the turbulent mixing in the core of the

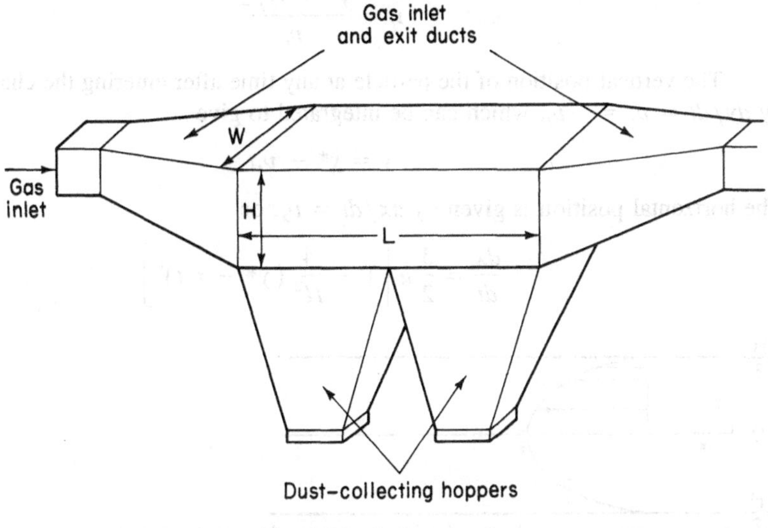


**Figure 7.1** Settling chamber.

chamber overwhelms the tendency of the particles to settle and maintains a uniform particle concentration vertically across the chamber. Removal by settling can be assumed to occur in a thin layer at the bottom of the chamber.

### 7.2.1 Laminar Flow Settling Chamber

In the laminar flow settling chamber the gas velocity profile is parabolic, as shown in Figure 7.2, and as a particle below the center streamline settles, it encounters fluid moving more slowly, and thus its residence time in the chamber increases over what it would have been on the higher streamline. Conversely, particles initially above the center streamline encounter faster moving streamlines as they fall until they pass the center streamline.

Consider the laminar flow settling chamber shown in Figure 7.2. The gas velocity profile for laminar flow between two parallel plates separated by a distance $H$ with the centerline of the chamber taken as $y = 0$ is

$$u_x = \frac{3}{2}\bar{u}\left[1 - \left(\frac{2y}{H}\right)^2\right] \tag{7.7}$$

where $\bar{u}$ is the mean velocity across the plates. We assume that particles are introduced uniformly across the entrance to the channel at concentration $N_0$.

There will be a critical height $y^*$ such that a particle of diameter $D_p$ initially at $x = 0$, $y = y^*$ will be at $y = -H/2$ at $x = L$. This particle will be the "last" particle of diameter $D_p$ collected in the device. Particles of diameter $D_p$ that entered the chamber above $y = y^*$ will not be collected; clearly, the value of $y^*$ depends on the particular $D_p$ of interest. This "last" particle collected takes time $t_f$ to fall a vertical distance $y^* + H/2$. Since $v_t$ is a constant,

$$t_f = \frac{y^* + H/2}{v_t} \tag{7.8}$$

The vertical position of the particle at any time after entering the chamber is given by $dy/dt = v_y = -v_t$, which can be integrated to give

$$y = y^* - v_t t \tag{7.9}$$

The horizontal position is given by $dx/dt = v_x$, or

$$\frac{dx}{dt} = \frac{3}{2}\bar{u}\left[1 - \frac{4}{H^2}(y^* - v_t t)^2\right] \tag{7.10}$$

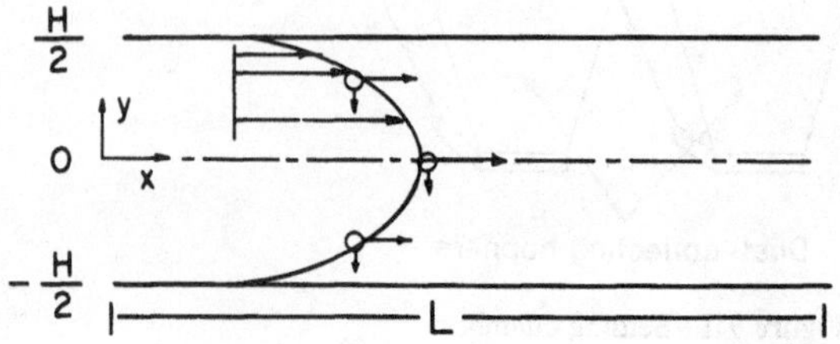


**Figure 7.2** Laminar flow settling chamber.

where the local horizontal velocity of the particle is that of the gas (7.7). Integrating (7.10) from the entrance to the exit of the chamber, we obtain

$$\frac{LH^2}{6\bar{u}} = t_f\left(\frac{H^2}{4} - y^{*2}\right) + v_t y^* t_f^2 - \frac{v_t^2}{3} t_f^3 \tag{7.11}$$

Using (7.8) and (7.11) gives

$$\beta = \left[1 - 4\left(\frac{y^*}{H}\right)^2\right]\left(\frac{\alpha}{2} + \alpha\frac{y^*}{H}\right) + 4\alpha\frac{y^*}{H}\left(\frac{1}{2} + \frac{y^*}{H}\right)^2 - \frac{4}{3}\alpha\left(\frac{1}{2} + \frac{y^*}{H}\right)^3 \tag{7.12}$$

where $\beta = 2v_t/3\bar{u}$ and $\alpha = H/L$.

To determine the expression for the collection efficiency, we need to compute the fraction of particles of a size $D_p$ that is collected over a length $L$. The flow of particles into the chamber, in number of particles per unit time, for a chamber of width $W$, is

$$\int_{-H/2}^{H/2} N_0 u_x(y) W \, dy = N_0 \bar{u} WH$$

The number of particles collected per unit time is that portion of the inlet flow of particles between $y = -H/2$ and $y = y^*$,

$$\int_{-H/2}^{y^*} N_0 u_x(y) W \, dy = N_0 W \int_{-H/2}^{y^*} u_x(y) \, dy$$

Therefore, the collection efficiency is just the ratio of the flow of particles collected to the total inlet flow,

$$\begin{aligned} \eta(D_p) &= \frac{N_0 W \int_{-H/2}^{y^*} u_x(y) \, dy}{N_0 \bar{u} WH} \\ &= \frac{1}{H\bar{u}} \int_{-H/2}^{y^*} u_x(y) \, dy \end{aligned} \tag{7.13}$$

Using (7.7), (7.13) becomes

$$\eta(D_p) = \frac{1}{2} + \frac{3}{2}\frac{y^*}{H} - 2\left(\frac{y^*}{H}\right)^3 \tag{7.14}$$

We now have two equations, (7.12) and (7.14), for the two unknowns $(y^*/H)$ and $\eta$. We can simplify these further by letting $z = \frac{1}{2} + (y^*/H)$. In doing so, (7.12) becomes

$$\frac{\beta}{\alpha} = 2z^2 - \tfrac{4}{3}z^3 \tag{7.15}$$

Similarly, (7.14) can be expressed as

$$\eta(D_p) = 3z^2 - 2z^3 \tag{7.16}$$

Combining (7.15) and (7.16), we see immediately that

$$\eta(D_p) = \frac{3\beta}{2\alpha} = \frac{v_t L}{\bar{u} H} \tag{7.17}$$

This is the equation governing the collection efficiency of a laminar flow settling chamber that consists of two parallel plates of length $L$ separated by a distance $H$, with a mean gas velocity of $\bar{u}$.

To evaluate the efficiency of the laminar flow settling chamber, we need only to determine the settling velocity $v_t$. If the particle is sufficiently small to be in the Stokes law regime, then $v_t = \rho_p g D_p^2/18\mu$, as derived in (5.30). Because of the large particle sizes of interest, we need not include the slip correction factor. For particles that are too large for Stokes' law to apply, the terminal settling velocity can be determined using the drag coefficient, as outlined in Section 5.3.4.

**Example 7.1** ***Efficiency of a Laminar Flow Settling Chamber in the Stokes Law Regime***

Consider a settling chamber for which $H = 0.1$ m, $L = 10$ m, $\bar{u} = 0.1$ m s$^{-1}$, and $\rho_p = 1$ g cm$^{-3}$. At 298 K, $\nu_{air} = 0.15$ cm$^2$ s$^{-1}$ and $\mu = 1.8 \times 10^{-4}$ g cm$^{-1}$ s$^{-1}$. Under these conditions the Reynolds number for the channel flow is 667, so laminar flow conditions exist. From (7.17) and (5.30) we find that $\eta = 0.03024 D_p^2$, with $D_p$ in $\mu$m. Thus, for these particular conditions, the collection efficiency depends on particle diameter as follows:

| $D_p$ ($\mu$m) | $\eta(D_p)$ |
|---|---|
| 1.0 | 0.03 |
| 2.0 | 0.12 |
| 3.0 | 0.27 |
| 4.0 | 0.48 |
| 5.0 | 0.76 |
| 5.75 | 1.0 |

Thus all particles with diameter exceeding 5.75 $\mu$m are totally collected in this chamber.

### 7.2.2 Plug Flow Settling Chamber

The second type of flow situation we consider is that of plug flow with no vertical mixing of particles. We assume that the particles are distributed uniformly across the entrance to the chamber. Whether a particle is collected is determined solely by the height $y$ at its entrance above the collecting surface. A critical height $y^*$ can be defined such that all particles entering with $y \le y^*$ are collected and those for which $y > y^*$ escape collection. The collection efficiency is then just

$$\eta(D_p) = \frac{v_t L}{\bar{u} H}$$

which is precisely the expression (7.17) obtained for the laminar flow settling chamber. Thus, in the parabolic velocity profile case, even though the particle falls across streamlines with different velocities, the overall effect is as if the particle were simply falling across streamlines all having a velocity equal to the mean velocity of the flow.

### 7.2.3 Turbulent Flow Settling Chamber

The flow in a rectangular channel can be assumed to be turbulent if the Reynolds number $\mathrm{Re}_c > 4000$ (McCabe and Smith, 1976, p. 52). For a duct the Reynolds number can be defined as $\mathrm{Re}_c = 4r_H\bar{u}\rho/\mu$, where $r_H$ is the hydraulic radius, defined as the ratio of the cross-sectional area to the perimeter. Thus, for a duct of height $H$ and width $W$, $r_H = HW/[2(H + W)]$. The average velocity $\bar{u}$ is just the volumetric flow rate $Q$ divided by the cross-sectional area $HW$. If the duct contains $N$ horizontal plates, each space receives a volumetric flow of $Q/N$ and has a height $H/N$ (neglecting the effect of plate thickness). The Reynolds number for the flow in each space is then

$$\mathrm{Re}_c = \frac{2Q}{\nu(H + NW)}$$

The turbulent flow settling chamber is shown schematically in Figure 7.3. In the laminar flow settling chamber just considered, particles settle at all heights above the floor of the chamber, the key to the analysis being to calculate the overall residence time of the particles as they fall across streamlines. The mechanism of collection in a turbulent flow settling chamber is, although ultimately based on the settling of particles under gravity, rather different from that in the laminar flow chamber. The difference is due to the turbulent flow in the chamber. In the bulk flow in the chamber, turbulent mixing is vigorous enough so that particles are overwhelmed by the flow and do not settle. We shall assume that the turbulent mixing maintains a uniform particle concentration over the height of the chamber. Very near the floor of the chamber a thin layer can be assumed to exist across which particles settle the short distance to the floor. Thus, once a particle, vigorously mixed in the core of the flow, enters this layer, it settles to the floor.

Consider a particle close to the wall. In time $dt$ the particle travels forward a distance $dx = \bar{u}\, dt$, where $\bar{u}$ is the mean velocity of the flow in the chamber. (Thus, we assume that the mean velocity $\bar{u}$ extends into the layer in spite of the absence of turbulent mixing in the layer.) During the time interval $dt$ the particle settles a distance $dy = v_t\, dt$. Therefore, the distances $dx$ and $dy$ are related by $dy = v_t\, dx/\bar{u}$.

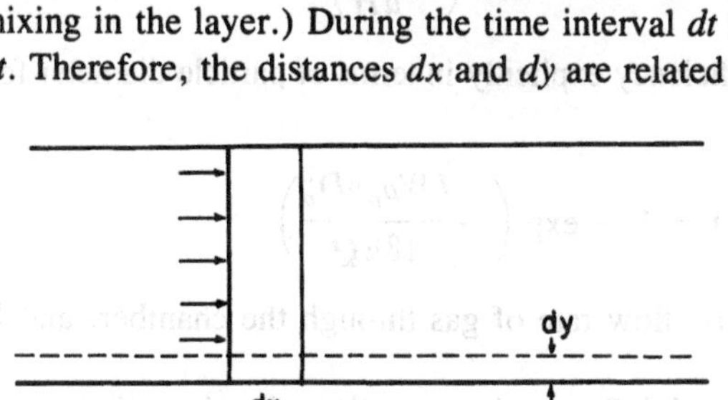

**Figure 7.3** Turbulent flow settling chamber.

In order to develop an overall design equation for the turbulent flow settling chamber, let us form a particle balance over the vertical section $dx$ in Figure 7.3. At the entrance to the section $dx$ there is a uniform distribution of particles across the entire chamber. The fraction of particles in the thin layer of thickness $dy$ is just $dy/H$. Since $dy$ was defined in terms of $dx$ such that $dx$ is just the distance a particle moves in the horizontal direction while it falls the distance $dy$, all particles in $dy$ are collected over the distance $dx$. Thus the fraction of particles collected in $dx$ is $dy/H = v_t\, dx/\bar{u}H$.

If the cross-sectional area of the device is $A_c$, a particle number balance over the section $dx$ is

$$\bar{u}A_c(N|_x - N|_{x+dx}) = \left(\frac{v_t\, dx}{\bar{u}H}\right)\bar{u}A_c N|_x \tag{7.18}$$

The left-hand side of (7.18) is the difference in flows in particles $s^{-1}$ into and out of the volume $A_c\, dx$, and the right-hand side is the number of particles $s^{-1}$ removed in that volume. Dividing by $dx$ and taking the limit as $dx \to 0$ yields

$$\frac{dN}{dx} = -\frac{v_t}{\bar{u}H}N \tag{7.19}$$

If the particle number concentration at the entrance to the chamber is $N_0$, then

$$N(x) = N_0 \exp\left(\frac{-v_t x}{\bar{u}H}\right) \tag{7.20}$$

Note that this equation holds for particles of each diameter since the particles are assumed not to interact with each other. Particle size dependence enters through the settling velocity $v_t$. Thus, if desired, we can indicate the particle size dependence of $N$ explicitly by $N(x; D_p)$, where $N$ is strictly the number of particles in the diameter range $(D_p, D_p + dD_p)$.

The collection efficiency of a settling chamber of length $L$ is

$$\eta(D_p) = 1 - \frac{N(L)}{N_0}$$

$$= 1 - \exp\left(-\frac{v_t L}{\bar{u}H}\right) \tag{7.21}$$

We can express the collection efficiency explicitly in terms of particle diameter for Stokes law settling as

$$\eta(D_p) = 1 - \exp\left(-\frac{LW\rho_p g D_p^2}{18\mu Q}\right) \tag{7.22}$$

where $Q = \bar{u}HW$, the volumetric flow rate of gas through the chamber, and $W$ is the width of the chamber.

We note a rather fundamental difference between the collection efficiencies for the settling chamber for laminar (and plug) and turbulent flows. The laminar flow collection

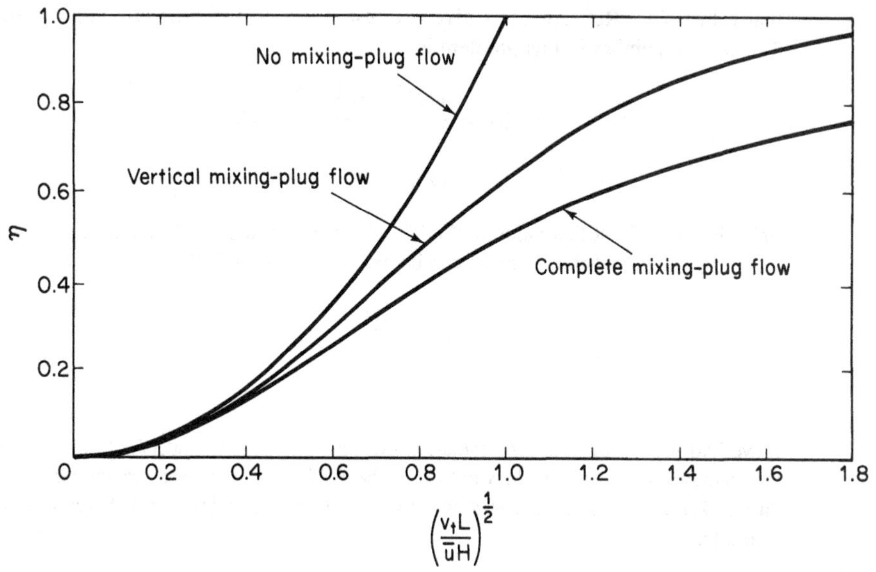


**Figure 7.4** General collection efficiency curves for settling chambers (Licht, 1980).

efficiency (7.17) predicts that $\eta(D_p) = 1.0$ for all particles large enough that $v_t \geq \bar{u}H/L$. If $v_t \sim D_p^2$, the Stokes law case, then $\eta(D_p)$ versus $D_p$ is a parabolic curve. On the other hand, in the case of turbulent flow (7.22), $\eta(D_p)$ approaches 1.0 asymptotically as $D_p \rightarrow \infty$. These features are illustrated schematically in Figure 7.4. The abscissa of Figure 7.4 is the group $(v_t L/\bar{u}H)^{1/2}$, which for Stokes law settling is directly proportional to $D_p$. Collection efficiency curves for actual chambers tend to have the S-shaped behavior of the turbulent flow curve in Figure 7.4 since any real unit will exhibit some degree of mixing in the flow.

**Example 7.2** ***Design of a Turbulent Flow Settling Chamber***

Determine the length of a settling chamber required to achieve 90% efficiency for 50-$\mu$m particles of density 2.0 g $cm^{-3}$ from an airstream of 1 $m^3$ $s^{-1}$ at 298 K, 1 atm. The chamber is to be 1 m wide and 1 m high.

We first evaluate the Reynolds number for the chamber to determine if the flow will be laminar or turbulent.

$$\mathrm{Re}_c = \frac{H\bar{u}\rho}{\mu} = \frac{Q}{W\nu}$$

Using $\nu = 0.15$ $cm^2$ $s^{-1}$, $Q = 10^6$ $cm^3$ $s^{-1}$, $W = 100$ cm, we find $\mathrm{Re}_c = 6.67 \times 10^4$. Thus the flow will be turbulent.

Now we need to determine the settling velocity of a 50-$\mu$m particle under the conditions of operation. We do not know ahead of time whether Stokes law will be valid for particles of this size, so to be safe we will determine the settling velocity using the drag coefficient. From (5.54), $\mathrm{Ga} = C_D \mathrm{Re}^2$, where Re is the particle Reynolds number, we can determine the value of Ga and then from Figure 5.6 we can determine the value of Re at that value of $C_D$ Re$^2$. An alternative is to use (5.55) and (5.56). We will use Figure 5.6. The Galileo number in this problem is

$$\text{Ga} = \frac{4gD_p^3\rho(\rho_p - \rho)}{3\mu^2} = C_D \text{Re}^2$$

$$= 12.1$$

From Figure 5.6, at this value of $C_D$ Re$^2$, Re = 0.7 and $v_t$ = 21 cm s$^{-1}$.

The length of the chamber can be determined from (7.21),

$$L = -\frac{\bar{u}H \ln(1 - \eta)}{v_t}$$

$$= 11 \text{ m}$$

If we had used Stokes law to calculate the settling velocity and (7.22) for the efficiency, the chamber length predicted would have been 15.2 m. Thus we see the effect of the fact that Stokes law is no longer strictly valid for 50-$\mu$m particles under the conditions of this example.

## 7.3 CYCLONE SEPARATORS

Cyclone separators are gas cleaning devices that utilize the centrifugal force created by a spinning gas stream to separate particles from a gas. A standard tangential inlet vertical reverse flow cyclone separator is shown in Figure 7.5. The gas flow is forced to follow the curved geometry of the cyclone while the inertia of particles in the flow causes them to move toward the outer wall, where they collide and are collected. A particle of mass $m_p$ moving in a circular path of radius $r$ with a tangential velocity $v_\theta$ is acted on by a centrifugal force of $m_p v_\theta^2/r$. At a typical value of $v_\theta$ = 10 m s$^{-1}$, $r$ = 0.5 m, this force is 20.4 times that of gravity on the same particle. Thus we see the substantially enhanced force on the particle over that of settling alone that can be achieved in a cyclone geometry. In a cyclone the particles in the spinning gas stream move progressively closer to the outer wall as they flow through the device. As shown in Figure 7.5, the gas stream may execute several complete turns as it flows from one end of the device to the other. One way to pose the question of the design of a cyclone separator is: For a given gas flow rate and inner and outer radii, how long must the body of the cyclone be to ensure that a desired collection efficiency for particles of a given size be attained? Since the length of the body of a cyclone is related through the gas flow rate to the number of turns executed by the gas stream, the design problem is often posed in terms of computing the number of turns needed to achieve a specified collection efficiency.

There are a variety of designs of cyclone separators, differing in the manner in which the rotating motion is imparted to the gas stream. Conventional cyclones can be placed in the following categories:

1. Reverse-flow cyclones (tangential inlet and axial inlet)

2. Straight-through-flow cyclones
3. Impeller collectors

Figure 7.5 shows a conventional reverse-flow cyclone with a tangential inlet. The dirty gas enters at the top of the cyclone and is given a spinning motion because of its tangential entry. Particles are forced to the wall by centrifugal force and then fall down the wall due to gravity. At the bottom of the cyclone the gas flow reverses to form an inner core that leaves at the top of the unit. In a reverse-flow axial-inlet cyclone, the inlet gas is introduced down the axis of the cyclone, with centrifugal motion being imparted by permanent vanes at the top.

In straight-through-flow cyclones the inner vortex of air leaves at the bottom (rather than reversing direction), with initial centrifugal motion being imparted by vanes at the top. This type is used frequently as a precleaner to remove fly ash and large particles. The chief advantages of this unit are low pressure drop and high volumetric flow rates.

In the impeller collector, gases enter normal to a many-bladed impeller and are swept out by the impeller around its circumference while the particles are thrown into an annular slot around the periphery of the device. The principal advantage of this unit is its compactness; its chief disadvantage is a tendency toward plugging from solid buildup in the unit.

Cyclones can be constructed of any material, metal or ceramic, for example, that is capable of withstanding high temperatures, abrasive particles, or corrosive atmo-

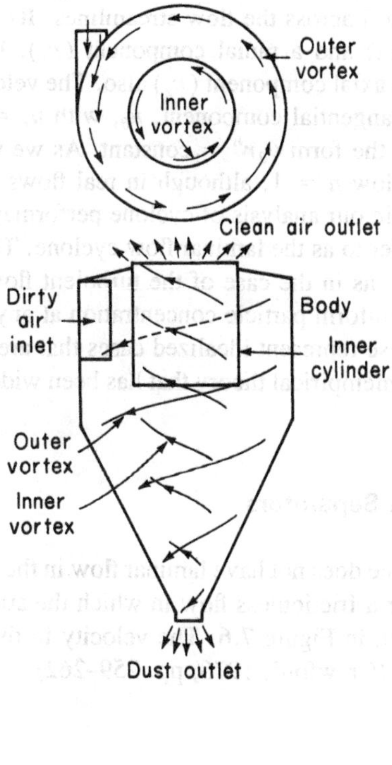


**Figure 7.5** Tangential inlet vertical reverse flow cyclone.

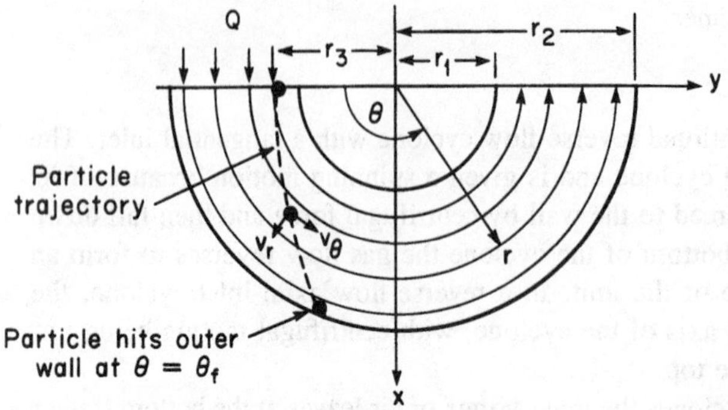


**Figure 7.6** Trajectory of a particle in one-half complete turn of an ideal flow cyclone.

spheres. It is necessary that the interior surface be smooth so that the collected particles may slide easily down the wall to the hopper. There are no moving parts to a cyclone, so operation is generally simple and relatively free of maintenance. Their low capital cost and maintenance-free operation make them ideal for use as precleaners for more efficient final control devices, such as electrostatic precipitators. Although cyclones have traditionally been regarded as relatively low efficiency collectors, some cyclones currently available from manufacturers can achieve efficiencies greater than 98% for particles larger than 5 $\mu$m. Generally, cyclones routinely achieve efficiencies of 90% for particles larger than 15 to 20 $\mu$m.

Consider a particle entering tangentially onto a horizontal plane of a spinning gas stream at $r_3$, as shown in Figure 7.6. Because of a centrifugal force of $m_p v_\theta^2/r$, the particle will follow a path outward across the flow streamlines. Its velocity vector will have a tangential component ($v_\theta$) and a radial component ($v_r$). Because the flow is actually into the page, there is an axial component ($v_z$) also. The velocity of the spinning gas is assumed to have only a tangential component, $u_\theta$, with $u_r = 0$. Tangential gas flows of this type usually are of the form $u_\theta r^n$ = constant. As we will see shortly, for an ideal fluid in such a vortex flow $n = 1$, although in real flows the value of $n$ may range downward to 0.5. We begin our analysis of cyclone performance with the case of the ideal flow, which we will refer to as the laminar flow cyclone. Then we consider the turbulent flow cyclone in which, as in the case of the turbulent flow settling chamber, mixing in the flow maintains a uniform particle concentration at any tangential position in the cyclone. Since both of these represent idealized cases that are not attained in real cyclones, we turn finally to a semiempirical theory that has been widely used in practical cyclone design.

### 7.3.1 Laminar Flow Cyclone Separators

The so-called laminar flow cyclone does not have laminar flow in the sense of the laminar flow settling chamber, but rather a frictionless flow in which the streamlines follow the contours of the cyclone as shown in Figure 7.6. The velocity in the case of ideal flow is given in $(r, \theta)$ coordinates as (Crawford, 1976, pp. 259–262)

$$u_r = 0 \qquad u_\theta = u = \frac{Q}{Wr \ln (r_2/r_1)} \tag{7.23}$$

where the entering flow is through a rectangular slot of area $W(r_2 - r_1)$.

To determine the collection efficiency consider a particle entering the cyclone at $r = r_3$ that strikes the wall at $\theta_f$. The particle's velocity components at any point on its trajectory are $v_r$ and $v_\theta$. The radial velocity component is the terminal velocity of the particle when acted on by the centrifugal force $F_c = m_p v_\theta^2/r$, which, in the case in which the drag force can be given by Stokes law, is

$$v_r = \frac{F_c}{3\pi\mu D_p} \tag{7.24}$$

Since the $\theta$-component of the particle's velocity is that of the fluid, $v_\theta = u_\theta$, and

$$F_c = \frac{m_p v_\theta^2}{r} = \frac{\pi}{6}\rho_p D_p^3 \frac{v_\theta^2}{r}$$

$$= \frac{\pi}{6}\rho_p D_p^3 \frac{Q^2}{W^2 r^3 (\ln r_2/r_1)^2} \tag{7.25}$$

Thus, combining (7.24) and (7.25), we obtain

$$v_r = \frac{\rho_p Q^2 D_p^2}{18\mu r^3 W^2 (\ln r_2/r_1)^2} \tag{7.26}$$

We now want to obtain an equation for the trajectory of a particle in the cyclone. The distance traveled in the $\theta$-direction in a time interval $dt$ is $v_\theta\, dt = r\, d\theta$. Also, the distance the particle moves in the $r$-direction in time $dt$ is $dr = v_r\, dt$. Then over a time interval $dt$, $r\, d\theta/v_\theta = dr/v_r$. From this relation we have

$$\frac{r\, d\theta}{dr} = \frac{v_\theta}{v_r} \tag{7.27}$$

and substituting the expressions for $v_\theta$ and $v_r$ gives us

$$\frac{d\theta}{dr} = \frac{18\mu W \ln (r_2/r_1) r}{\rho_p Q D_p^2} \tag{7.28}$$

a differential equation describing the particle's trajectory. If the particle enters the device at $r = r_3$ and hits the outer wall at $\theta = \theta_f$, then integrating (7.28) gives

$$\theta_f = \frac{9\mu W \ln (r_2/r_1)}{\rho_p Q D_p^2}(r_2^2 - r_3^2) \tag{7.29}$$

Conversely, we can solve (7.29) for $r_3$ to find the entrance position $r_3$ of a particle that hits the outer wall at $\theta = \theta_f$,

$$r_3 = \left[r_2^2 - \frac{\rho_p Q D_p^2 \theta_f}{9\mu W \ln (r_2/r_1)}\right]^{1/2} \tag{7.30}$$

We can now determine an expression for the collection efficiency of a cyclone. Assume that the cyclone has an angle $\theta_f$. All particles that enter the cyclone at $r \geq r_3$ hit the wall over $0 \leq \theta \leq \theta_f$. If the entering particle concentration and gas velocity are uniform across the cross section, the collection efficiency is just that fraction of the particles in the entering flow that hits the outer wall before $\theta = \theta_f$,

$$\eta = \frac{r_2 - r_3}{r_2 - r_1} \tag{7.31}$$

which is

$$\eta(D_p) = \frac{1 - \left[1 - \dfrac{\rho_p Q D_p^2 \theta_f}{9 \mu W r_2^2 \ln(r_2/r_1)}\right]^{1/2}}{1 - r_1/r_2} \tag{7.32}$$

The value of $\theta_f$ at which $\eta = 1$ is the value of $\theta_f$ when $r_3 = r_1$,

$$\theta_f = \frac{9 \mu W \ln(r_2/r_1)}{\rho_p Q D_p^2}(r_2^2 - r_1^2) \tag{7.33}$$

We had earlier noted a comparison of the centrifugal force acting on a particle to that due to gravity. Using $F_c$ from (7.25) and $F_g = (\pi/6)\,\rho_p D_p^3 g$, we obtain the ratio of the centrifugal to gravity force as

$$\frac{F_c}{F_g} = \frac{Q^2}{g r^3 W^2 \ln(r_2/r_1)^2} \tag{7.34}$$

which, as can be shown, for typical cyclones, $F_c/F_g \gg 1$.

### 7.3.2 Turbulent Flow Cyclone Separators

The model of the turbulent flow cyclone separator is shown in Figure 7.7. Because of turbulent mixing the particle concentration is assumed to be uniform across the cyclone, and, as in the case of the turbulent flow settling chamber, removal occurs across a thin layer at the outer wall. For lack of a better approximation, we continue to use the inviscid gas velocity components given by (7.23) to represent the fluid velocity field in the turbulent flow cyclone. Thus the key difference between the laminar and turbulent cyclones relates to the assumption made concerning particle behavior in the cyclone. The distance a particle travels in the $\theta$-direction in the laminar sublayer over a time interval $dt$ is $v_\theta\, dt = r_2\, d\theta$, where we can evaluate $v_\theta$ at $r = r_2$. For the particle to be captured across the layer of thickness $dr$, $dr = v_r\, dt = v_r r_2\, d\theta / v_\theta$, where $v_r$ is also evaluated at $r = r_2$.

To derive an expression for the change in particle number concentration with $\theta$, we perform a particle balance over the sector of angle $d\theta$. The fractional number of particles removed over $d\theta$ is just the fraction of particles that are in the boundary layer,

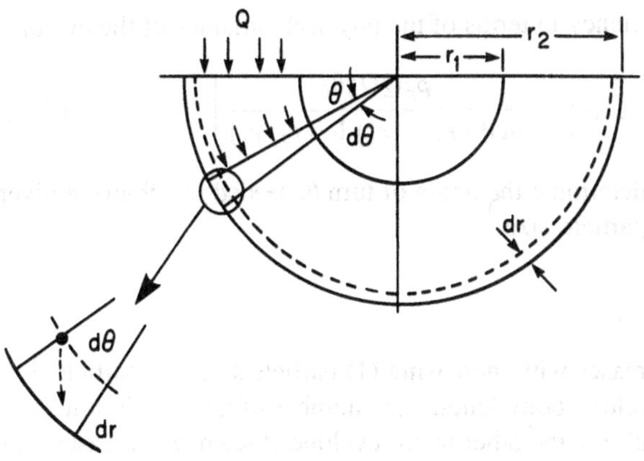


**Figure 7.7** One-half complete turn of a turbulent flow cyclone.

$2r_2dr/(r_2^2 - r_1^2)$. Thus

$$N\Big|_\theta - N\Big|_{\theta+d\theta} = \frac{2r_2\,dr}{(r_2^2 - r_1^2)}N\Big|_\theta \tag{7.35}$$

where we need not include the product of mean velocity and cross-sectional area since it appears on both sides of the equation. Using $dr = v_{r_2}r_2\,d\theta/v_\theta$, dividing by $d\theta$, and taking the limit as $d\theta$ approaches zero gives us

$$\frac{dN}{d\theta} = -\frac{v_{r_2}}{v_{\theta_2}}\frac{2r_2^2}{r_2^2 - r_1^2}N \tag{7.36}$$

where $v_{\theta_2}$ is $v_\theta$ at $r = r_2$. This equation is to be integrated subject to $N = N_0$ at $\theta = 0$. The result is

$$N(\theta) = N_0 \exp\left(-\frac{v_{r_2}}{v_{\theta_2}}\frac{2r_2^2}{r_2^2 - r_1^2}\theta\right) \tag{7.37}$$

The collection efficiency of a cyclone that has an angle $\theta_f$ is

$$\eta(D_p) = 1 - \frac{N(\theta_f)}{N_0}$$

$$= 1 - \exp\left[-\frac{2v_{r_2}r^2\theta_f}{v_{\theta_2}(r_2^2 - r_1^2)}\right] \tag{7.38}$$

Using the explicit expressions for the two velocity components,

$$v_{\theta_2} = \frac{Q}{Wr_2 \ln(r_2/r_1)} \tag{7.39}$$

$$v_{r_2} = \frac{\rho_p Q^2 D_p^2}{18\mu r_2^3 W^2 (\ln r_2/r_1)^2} \tag{7.40}$$

we can express the collection efficiency in terms of the physical variables of the cyclone,

$$\eta(D_p) = 1 - \exp\left[-\frac{\rho_p Q D_p^2 \theta_f}{9\mu W(r_2^2 - r_1^2)\ln(r_2/r_1)}\right] \tag{7.41}$$

This equation can be inverted to determine the angle of turn $\theta_f$ needed to achieve a given collection efficiency for a given particle size.

### 7.3.3 Cyclone Dimensions

Cyclone collection efficiency increases with increasing (1) particle size, (2) particle density, (3) inlet gas velocity, (4) cyclone body length, (5) number of gas revolutions, and (6) smoothness of the cyclone wall. On the other hand, cyclone efficiency decreases with increasing (1) cyclone diameter, (2) gas outlet duct diameter, and (3) gas inlet area. For any specific cyclone whose ratio of dimensions is fixed, the collection efficiency increases as the cyclone diameter is decreased. The design of a cyclone separator represents a compromise among collection efficiency, pressure drop, and size. Higher efficiencies require higher pressure drops (i.e., inlet gas velocities) and larger sizes (i.e., body length).

The dimensions required to specify a tangential-entry, reverse-flow cyclone are shown in Figure 7.8. In classic work that still serves as the basis for cyclone design, Shepherd and Lapple determined "optimum" dimensions for cyclones. All dimensions were related to the body diameter $D_c$. A common set of specifications is given on the right-hand side of Figure 7.8. Other standard cyclone dimensions are given by Licht (1984) and Cooper and Alley (1986). The number of revolutions that the gas makes in the outer vortex can be approximated by

$$n_r = \frac{1}{H_c}\left(L_c + \frac{Z_c}{2}\right)$$

where the dimensions are shown in Figure 7.8.

Besides collection efficiency the other major consideration in cyclone specification is pressure drop. While higher efficiencies are obtained by forcing the gas through the cyclone at higher velocities, to do so results in an increased pressure drop. Since increased pressure drop requires increased energy input into the gas, there is ultimately an economic trade-off between collection efficiency and operating cost. A simple pressure-drop equation for cyclones is given by Cooper and Alley (1986). Cyclone pressure drops range from 250 to 4000 Pa.

### 7.3.4 Practical Equation for Cyclone Efficiency

We have analyzed the collection efficiency of a cyclone assuming that the particles behave as if they are in either a laminar or a turbulent flow. Actually, the flow pattern in a cyclone is a complex one, and the two models that we have presented represent extremes in cyclone performance. Although a Reynolds number for a cyclone can be de-

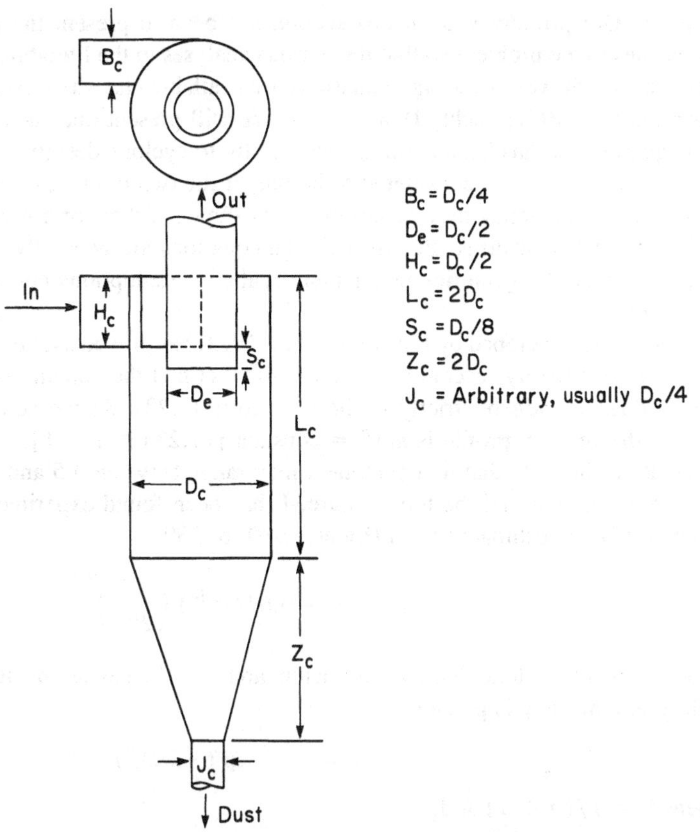


**Figure 7.8** Geometric specifications for the design of a cyclone separator. The dimensions given on the right-hand side of the figure are those of the classic design of Shepherd and Lapple. This particular set of specifications appears in *Perry's Handbook* (Perry and Chilton, 1973; Figure 20-96, p. 20-82); reprinted by permission of McGraw-Hill Publishing Company.

fined as $Re_{cy} = (\rho u/\mu)(4A_c/\pi)^{1/2}$, where $A_c$ is the cross-sectional area so that $(4A_c/\pi)^{1/2}$ is an equivalent diameter, and for the velocity it is sufficient to use $u = Q/W(r_2 - r_1)$, a characteristic velocity in the cyclone, a precise criterion for transition from laminar to turbulent flow in a cyclone does not exist. The laminar flow theory predicts a well-defined critical value for the smallest particle size that may be collected completely, whereas the turbulent flow result gives an asymptotic approach to complete collection as particle size increases. Experimentally determined collection efficiency curves generally approach 100% efficiency asymptotically and thus appear to conform more closely to turbulent than to laminar flow conditions. Since operating cyclones do not conform to either of these limiting cases, one must resort to semiempirical design equations to predict cyclone performance.

There has been a great deal of effort devoted to predicting the performance of cyclones. Our primary goal in this section has been to present the general theoretical approaches to the problem so that the various analyses in the literature will be accessible to the reader. Surveys of design equations are available elsewhere (see, e.g., Bhatia and Cheremisinoff, 1977; Licht, 1980, 1984). We will present one such semiempirical design equation that has been applied successfully to cyclone design.

If the flow can be considered to be one of the two limiting cases analyzed above, the collection efficiency may be computed as shown earlier for a given geometry, flow rate, and number of turns. Practical design equations are generally derived by considering the particle trajectories under more realistic assumptions concerning the flow in the cyclone.

A theory developed by Leith and Licht (1972) has proved useful in practical cyclone design. In that theory, account is taken of the fact that the velocity profile in a cyclone usually does not adhere strictly to the ideal form (7.23). As we noted, a more general form of the velocity profile is $u_\theta r^n$ = constant [(7.23) is $n = 1$], where experimental observations indicate that in a cyclone $n$ may range between 0.5 and 0.9, depending on the size of the unit and the temperature. It has been found experimentally that the exponent $n$ may be estimated from (Licht, 1980, p. 239)

$$n = 1 - (1 - 0.67D_c^{0.14})\left(\frac{T}{283}\right)^{0.3}$$

where $D_c$ is the cyclone diameter in meters and $T$ is the gas temperature in kelvin. The collection efficiency is given by

$$\eta(D_p) = 1 - \exp(-MD_p^N) \tag{7.42}$$

where $N = 1/(n + 1)$ and

$$M = 2\left[\frac{KQ}{D_c^3}\frac{\rho_p(n+1)}{18\mu}\right]^{N/2}$$

where $D_p$ is in cm, $\rho_p$ is in g cm$^{-3}$, $Q$ is the gas volumetric flow rate in m$^3$ s$^{-1}$, $\mu$ is in g cm$^{-1}$ s$^{-1}$ and $K$ is a geometric configuration parameter that depends only on the relative dimensions of the unit. For the relative dimensions suggested in Figure 7.8, $K$ = 402.9; for other dimensions the values of $K$ are given by Licht (1980, 1984). The calculation of $K$ is explained by Leith and Licht (1972) and Licht (1980).

**Example 7.3 *Cyclone Collection Efficiency***

Three design equations for cyclone collection efficiency were presented in this section. We wish to compare the collection efficiencies predicted by each approach. To do so, consider a cyclone having $W$ = 4 m and $Q$ = 20 m$^3$ s$^{-1}$, inner and outer radii of 0.5 m and 1 m, respectively, and an angle of turn of $12\pi$. Assume that the particle size range of interest is from 1 to 30 $\mu$m and that the particles have a density of 2 g cm$^{-3}$. The relative dimensions of the cyclone are those suggested in Figure 7.8. Assume $T$ = 293 K.

Figure 7.9 shows the collection efficiencies for this cyclone predicted by the laminar flow theory (7.32), the turbulent flow theory (7.41), and the theory of Leith and Licht (7.42). We see that the laminar flow theory, which is based on computing particle trajec-

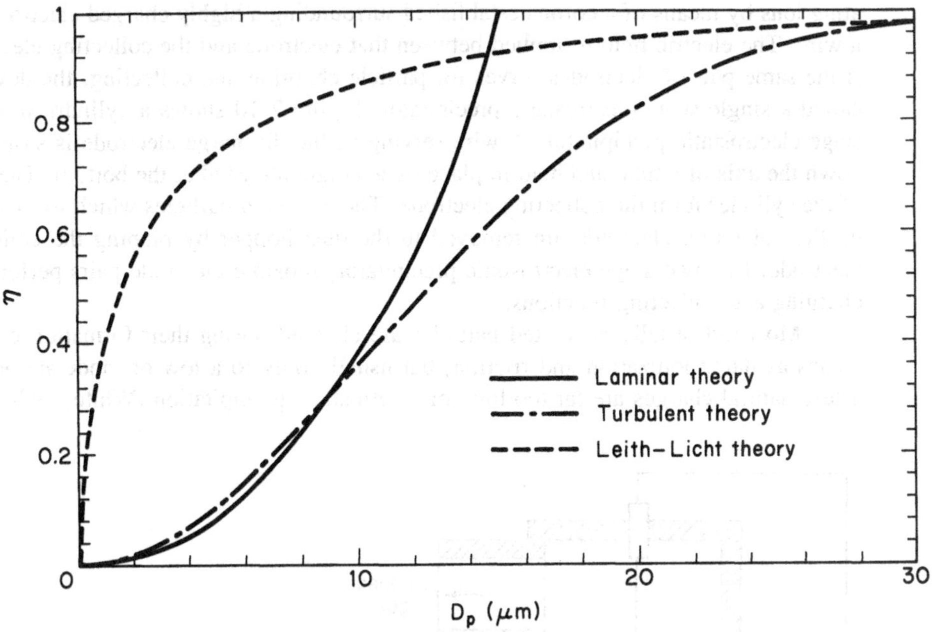


**Figure 7.9** Collection efficiency curves for the conditions of Example 7.3 based on assuming laminar flow, turbulent flow and using the Leith and Licht equation.

tories across the entire device, predicts that particles larger than about 15 μm are totally collected. The other two theories predict an asymptotic approach to complete collection with increasing particle diameter, a type of behavior that conforms to that observed in operating cyclones.

## 7.4 ELECTROSTATIC PRECIPITATION

Electrostatic precipitators are one of the most widely used particulate control devices, ranging in size from those installed to clean the flue gases from the largest power plants to those used as small household air cleaners. The basic principle of operation of the electrostatic precipitator is that particles are charged, then an electric field is imposed on the region through which the particle-laden gas is flowing, exerting an attractive force on the particles and causing them to migrate to the oppositely charged electrode at right angles to the direction of gas flow. Electrostatic precipitation differs from mechanical methods of particle separation in that the external force is applied directly to the individual particles rather than indirectly through forces applied to the entire gas stream (e.g., in a cyclone separator). Particles collect on the electrode. If the particles collected are liquid, then the liquid flows down the electrode by gravity and is removed at the bottom of the device. If the particles are solid, the collected layer on the electrode is removed periodically by rapping the electrode. Particle charging is achieved by generating ions by means of a corona established surrounding a highly charged electrode like a wire. The electric field is applied between that electrode and the collecting electrode. If the same pair of electrodes serves for particle charging and collecting, the device is called a single-stage electrostatic precipitator. Figure 7.10 shows a cylindrical single-stage electrostatic precipitator. A wire serving as the discharge electrode is suspended down the axis of a tube and held in place by a weight attached at the bottom. The sides of the cylinder form the collecting electrode. The collected particles which form a layer on the collecting electrode are removed to the dust hopper by rapping the collecting electrode. In a two-stage electrostatic precipitator, separate electrode pairs perform the charging and collecting functions.

Most industrially generated particles are charged during their formation by such means as flame ionization and friction, but usually only to a low or moderate degree. These natural charges are far too low for electrostatic precipitation (White, 1984). The

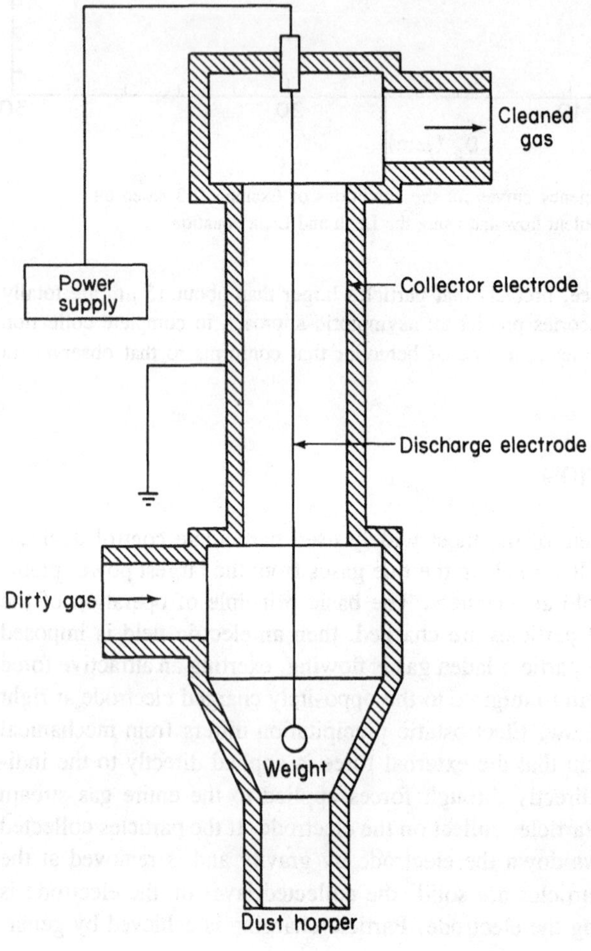


**Figure 7.10** Cylindrical single-stage electrostatic precipitator.

high-voltage dc corona is the most effective means for particle charging and is universally used for electrostatic precipitation. The corona is formed between an active high-voltage electrode such as a fine wire and a passive ground electrode such as a plate or pipe. The corona surrounding the discharge electrode can lead to the formation of either positive or negative ions that migrate to the collecting electrode. The ions, in migrating from the discharging to the collecting electrode, collide with the particulate matter and charge the particles. Because the gas molecule ions are many orders of magnitude smaller than even the smallest particles and because of their great number, virtually all particles that flow through the device become charged. The charged particles are then transported to the collecting electrode, to which they are held by electrostatic attraction. The particles build a thickening layer on the collecting electrode. The charge slowly bleeds from the particles to the electrode. As the layer grows, the charges on the most recently collected particles must be conducted through the layer of previously collected particles. The resistance of the dust layer is called the dust resistivity.

As the particle layer grows in thickness, the particles closest to the plates lose most of their charge to the electrode. As a result, the electrical attraction between the electrode and these particles is weakened. However, the newly arrived particles on the outside layer have a full charge. Because of the insulating layer of particles, these new particles do not lose their charge immediately and thus serve to hold the entire layer against the electrode. Finally, the layer is removed by rapping, so that the layer breaks up and falls into a collecting hopper.

Of direct interest is the determination of the collection efficiency of a given precipitator as a function of precipitator geometry, gas flow rate, particle size, and gas properties. Flow in commercial electrostatic precipitators is turbulent. The prediction of the migration of particles therefore requires consideration of the motion of particles in turbulent flow subject to both electrostatic and inertial forces. Because one cannot describe exactly the motion of particles in turbulent flow, even in the absence of electric forces, there does not exist a rigorous general theory for the design of turbulent-flow electrostatic precipitators. In order to obtain design equations for collection efficiency, we resort, as we have been doing, to an idealized representation of the turbulent mixing process in the device and removal in a thin layer at the collector wall.

Electrostatic precipitators are commonly employed for gas cleaning when the volumetric throughput of gas is high. Such units are used routinely for fly ash removal from power plant flue gases. Electrostatic precipitators are also widely employed for the collection of particles and acid mists in the chemical and metallurgical process industries.

### 7.4.1 Overall Design Equation for the Electrostatic Precipitator

Figure 7.11 depicts the wall region of an electrostatic precipitator, a chamber of perimeter $P$ and cross-sectional area $A_c$ through which a gas containing charged particles is flowing. Turbulent flow will be assumed so that, as before, the particle number concentration is uniform at any point across the device. Again, we will suppose the existence of a thin layer adjacent to the walls of the device across which the particle migration

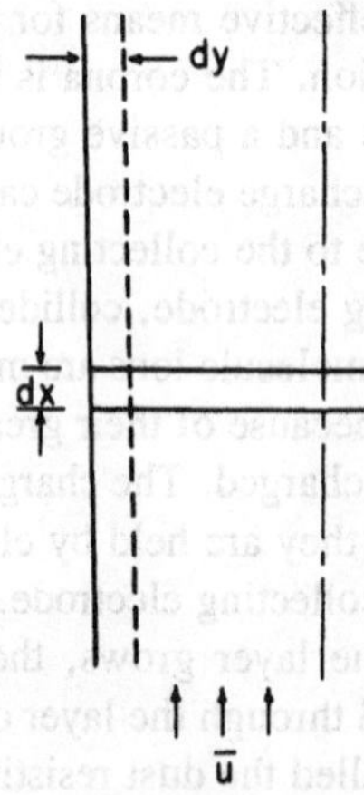


**Figure 7.11** Wall region of an electrostatic precipitator.

and collection occur. To reiterate, because the turbulent mixing in the core of the flow overwhelms the tendency of particles to migrate, the only migration occurs across a layer close to the wall. Thus, from the point of view of the overall design equation, the turbulent flow electrostatic precipitator is quite analogous to the turbulent flow settling chamber and cyclone; only the physical mechanism leading to particle migration differs.

Assume for the moment that the charge on a particle and the electric field between the electrodes are known. As we noted in Chapter 5, the electrostatic force on a particle with charge $q$ in an electric field of strength $E$ is $F_{el} = qE$. The electrical migration velocity of a particle of diameter $D_p$ in such a field is that given by (5.43)

$$v_e = \frac{qEC_c}{3\pi\mu D_p} \tag{7.43}$$

where we now retain the slip correction factor $C_c$ because we will be dealing with submicron-sized particles. The charge $q$ is equal to the product of the number of charges $z_p$ and the charge on an electron $e$.

As we have done before, we will define the wall layer thickness $dy$ such that all particles in $dy$ are captured over the distance $dx$, that is, $dy = v_e\, dt = v_e\, dx/\bar{u}$. The fraction of particles captured in distance $dx$ is just the ratio of the cross-sectional area of the wall layer to the overall cross-sectional area of the device, $P\, dy/A_c$. A balance on particle number over the section $dx$ gives

$$\bar{u}A_c\left(N\big|_x - N\big|_{x+dx}\right) = \left[\left(\frac{P\, dy}{A_c}\right)N\big|_x\right]\bar{u}A_c \tag{7.44}$$

Using $dy = v_e\, dx/\bar{u}$ and then taking the limit as $dx \rightarrow 0$ give

$$\frac{dN}{dx} = -\frac{Pv_e}{A_c\bar{u}}N \tag{7.45}$$

where the electrical migration velocity $v_e$ is evaluated at conditions at the collector surface. Equation (7.45) is to be integrated subject to $N(0) = N_0$.

The migration velocity $v_e$ depends on the number of charges on the particle, which, as we will see, is a function of particle size as well as the electric field and ion density conditions in the precipitator, and on the local electric field strength. Both $q$ and $E$ are in general a function of distance $x$ down the precipitator. If it can be assumed that $v_e$ is independent of the number concentration $N$, integration over a unit of length $L$ yields*

$$N(L) = N_0 \exp\left(-\frac{P}{A_c\bar{u}}\int_0^L v_e\,dx\right)$$

$$= N_0 \exp\left(-\frac{A/L}{Q}\int_0^L v_e\,dx\right) \tag{7.46}$$

where $A_c\bar{u} = Q$, the volumetric flow rate of gas through the unit, and $PL = A$, the collector surface area. Furthermore, if the electrical migration velocity can be assumed to be constant, then (7.46) gives $N(L) = N_0 \exp(-Av_e/Q)$, and the collection efficiency is given by

$$\eta = 1 - \exp\left(-\frac{Av_e}{Q}\right) \tag{7.47}$$

Equation (7.47) is seen to be analogous to that derived for a turbulent flow settling chamber, with only the physical mechanism leading to particle migration differing. This equation was first used in an empirical form in 1919 by Evald Anderson and derived theoretically by W. Deutsch in 1922 (White, 1984). It has generally been referred to as the Deutsch equation and sometimes as the Deutsch-Anderson equation.

Although the Deutsch equation can be used to estimate the collection efficiency of an electrostatic precipitator, the assumption of constant $v_e$ is overly restrictive. In the remainder of this section, we take into account the variation in migration velocity with position in the precipitator. Our development will focus on the cylinder and wire configuration, although it can be carried through in a similar fashion for other geometries, such as parallel plates. We should point out, however, that even though it is possible to derive theoretically the electric fields and migration velocities in devices with well-defined geometry, the idealized conditions corresponding to the theory seldom exist in actual practice. Factors such as particle reentrainment and gas channeling around the collecting zones cannot be accounted for theoretically. Because of these uncertainties, industrial precipitator design is often based on empirical migration velocities for use in the Deutsch equation (White, 1984). Nevertheless, it is important to understand the underlying fundamental relationships among the variables in an electrostatic precipitator, and we will develop these relationships subsequently.

### 7.4.2 Generation of the Corona

The mechanism for particle charging in an electrostatic precipitator is the generation of a supply of ions that attach themselves to the particles. The corona is the mechanism for

*We will see subsequently that the migration velocity is, in fact, a function of the local number concentration.

forming ions. The corona can be either positive or negative. A gas usually has a few free electrons and an equal number of positive ions, a situation that is exploited in generating a corona. When a gas is placed between two electrodes, small amount of current results as the free electrons migrate to the positive electrode and the positive ions migrate to the negative electrode.

In the positive corona the discharge electrode, the wire in the cylindrical electrostatic precipitator, is at a positive potential. The few free electrons normally present in the gas migrate toward the wire. As the electrons approach the wire, their energy increases due to an increased attractive force. These free electrons collide with gas molecules, the collision leading in some cases to the ejection of an electron from the molecule, producing two free electrons and a positive ion. The two free electrons continue toward the positive electrode, gaining energy, until they collide with two more gas molecules, producing four free electrons and two positive ions. This process is referred to as an electron avalanche. The positive ions formed migrate to the negative electrode. It is these positive ions that must migrate across the entire device to the negative electrode that collide with and attach to the particles in the gas. The region immediately surrounding the wire in which the electron avalanche is established is the corona. Thus, with a positive corona the particles become positively charged. The term "corona" arises from the fact that the electron avalanche is often accompanied by the production of light.

In the negative corona the discharge electrode is maintained at a negative potential. The electron avalanche begins at the outer surface of the wire and proceeds radially outward. Close to the wire the electrons are sufficiently energetic to form positive ions upon collision with gas molecules, thus initiating the electron avalanche. The positive ions formed migrate the short distance to the wire. As the electrons migrate outward into a region of lower electric field strength, they are slowed down by collisions with gas molecules. These electrons eventually have lower energy than those that are accelerated toward the positive electrode in the positive corona. These relatively low energy electrons, rather than ejecting an electron from the gas molecule upon collision, are absorbed by the gas molecules to produce negative ions. The formation of negative ions, which begins to occur at the outer edge of the corona, essentially absorbs all the free electrons produced in the electron avalanche at the wire surface. These negative ions then migrate to the positive electrode, in the course of which attaching to gas molecules and forming negative ions. For a negative corona to be effective it is necessary that the gas molecules can absorb free electrons to form negative ions. Sulfur dioxide is one of the best electron-absorbing gases of those present in flue gases. Oxygen, $CO_2$ and $H_2O$ are also effective electron absorbers. The negative corona is generally more stable than the positive corona, so it is preferred in most industrial applications. A by-product of the negative corona is the production of $O_3$, which is an undesirable feature of the household use of an electrostatic precipitator with a negative corona. Moreover, since the positive corona does not need an electron-absorbing gas, it is more suitable for domestic application.

A few comments are in order about the collecting, or passive, electrode. As the electrostatic precipitator is operated, a layer of the collected material builds up on the collecting electrode. Particle deposits on the precipitator collection surface must possess at least a small degree of electrical conductivity in order to conduct the ion currents from the corona to ground. The minimum conductivity required is about $10^{-10}$ ($\Omega$ cm)$^{-1}$, which is the inverse of resistivity. A conductivity of $10^{-10}$ ($\Omega$ cm)$^{-1}$, or a resistivity of $10^{10}$ $\Omega$ cm, is small compared to that of ordinary metals but is much greater than that of good insulators such as silica and most plastics. The resistivity of a material is determined by establishing a current flow through a slab of known thickness of the material. As long as the resistivity of the collected dust layer is less than about $10^{10}$ $\Omega$ cm, the layer will surrender its charge to the electrode. A typical dust has a resistivity of about $10^{8}$ $\Omega$ cm at room temperature, due to a layer of water on the surface of the particles. As the temperature is increased beyond 373 K, the water is evaporated and the resistivity increases to a value characteristic of the collected solids. Fly ash resistivities can vary from $10^{8}$ to $10^{13}$ $\Omega$ cm. When the resistivity of the layer exceeds about $10^{10}$ $\Omega$ cm, the potential across the layer increases so that the voltage that can be maintained across the electrostatic precipitator decreases and the collection efficiency decreases. The electrical resistivity of collected particulate matter depends on its chemical composition, the constituents of the gas, and the temperature (Bickelhaupt, 1979; Han and Ziegler, 1984). The resistivity of fly ash is dependent on the content of $SO_3$, $Na_2O$, and to a lesser extent, hydrophilic compounds ($Fe_2O_3$, $K_2$, $Li_2O$) in the ash and on the water content in the flue gas. When sulfur-containing coal is burned, from 1 to 5% of the $SO_2$ is oxidized to $SO_3$ in the combustion process. The $SO_3$ condenses on the fly ash as $H_2SO_4$ and lowers its resistivity. Materials with very low resistivities, such as carbon black with a resistivity in the range $10^{-3}$ $\Omega$ cm, are difficult to collect because these materials assume the charge of the collecting electrode upon contact and are repelled toward the discharge electrode.

### 7.4.3 Particle Charging

Particle charging in an electrostatic precipitator occurs in the gas space between the electrodes where the gas ions generated by the corona bombard and become attached to the particles. The gas ions may reach concentrations as high as $10^{15}$ ions m$^{-3}$. The level of charge attained by a particle depends on the gas ion concentration, the electric field strength, the conductive properties of the particle, and the particle size. A 1-$\mu$m particle typically acquires the order of 300 electron charges, whereas a 10-$\mu$m particle can attain 30,000 electron charges. Predicting the level of charge acquired by a particle is necessary in order to predict the particle's migration velocity, on the basis of which the collection efficiency can be calculated for a given set of operating conditions.

There are actually two mechanisms by which particles become charged in an electrostatic precipitator. In the first mechanism particle charging occurs when ions that are migrating toward the collecting electrode encounter particles to which they become attached. In migrating between the electrodes the ions follow the electric flux lines, which are curves everywhere tangent to the electric field vector. When the particle first enters the device and is uncharged, the electric flux lines deflect toward the particle, resulting in the capture of even a larger number of ions than would be captured if the ions followed their normal path between the electrodes. As the particle becomes charged, ions begin to be repelled by the particle, reducing the rate of charging. Eventually, the particle will acquire a saturation charge and charging will cease. This mechanism is called ion bombardment or *field charging*. The second mode of particle charging is *diffusion charging*, in which the particle acquires a charge by virtue of the random thermal motion of ions and their collision with and adherence to the particles.

The theories of both field and diffusion charging, in their full generality, are quite complex and have received a great deal of attention. Strictly speaking, field and diffusion charging occur simultaneously once a particle enters an electrostatic precipitator, and hence to predict the overall charge acquired by a particle, one should consider the two mechanisms together. However, because, as we shall see, diffusion charging is predominant for particles smaller than about 1 $\mu$m in diameter and field charging is predominant for particles larger than about 1 $\mu$m, the two mechanisms often are treated in electrostatic precipitator design as if they occur independently. In doing so, one estimates the total charge on a particle as the sum of the charges resulting from each of the two separate mechanisms.

### 7.4.4 Field Charging

When a dielectric particle of radius $R_p$ containing charge $q$ is placed in a preexisting, uniform electric field $E_\infty$ with an initially unipolar ion density $N_{i\infty}$, the electric potential at point $(r, \theta, \phi)$ in the region outside the sphere is (Stratton, 1941)

$$V = \frac{q}{4\pi\epsilon_0 r} + \left(\frac{r}{R_p} - \frac{\kappa - 1}{\kappa + 2}\frac{R_p^2}{r^2}\right)E_\infty R_p \cos\theta$$

where $\kappa$ is the dielectric constant of the sphere and $\epsilon_0$ is the permittivity of free space ($8.85 \times 10^{-12}$ F m$^{-1}$). The range of values of the dielectric constant $\kappa$ is $\kappa = 1$ for a perfect insulator and $\kappa = \infty$ for a perfect conductor. The dielectric constants of insulating particles of mineral origin commonly are of order 2 to 10. The value of $\kappa$ for air is approximately 1.

The first term in $V$ is the Coulombic potential, and the second term combines the $r$-component external potential uniformly built by $E_\infty$ and the $r$-component image potential resulting from the sphere dielectric polarization in $E_\infty$. The electric field around the sphere is just the negative gradient of the potential $V$. For the $r$-component of the electric field,

$$E_r = -\frac{\partial V}{\partial r} = -E_\infty \cos\theta\left(1 + 2\frac{\kappa - 1}{\kappa + 2}\frac{R_p^3}{r^3}\right) + \frac{q}{4\pi\epsilon_0 r^2}$$

At the surface of the sphere

$$E_r\big|_{r=R_p} = -E_\infty \cos\theta \frac{3\kappa}{\kappa + 2} + \frac{q}{4\pi\epsilon_0 R_p^2}$$

Field charging occurs as the ions that are migrating in the field $E_\infty$ become close to a particle (the sphere) and follow the distorted electric field lines around the particle and impinge on it. The electric field at the surface of the particle has a zero-potential circle at $\theta = \theta_0$ such that for $\theta \leq \theta_0$ ions will impinge on the surface and for $\theta \geq \theta_0$ ions will drift past the particle. We find $\theta_0$ by setting $E_r|_{r=R_p} = 0$,

$$\theta_0 = \cos^{-1}\left(\frac{\kappa + 2}{3\kappa}\frac{q}{4\pi\epsilon_0 R_p^2}\frac{1}{E_\infty}\right)$$

To determine the impingement rate of ions (in number per second), we need to integrate the ion flux,

$$\mp B_i E_r\big|_{r=R_p} N_{i\infty}$$

where $B_i$ is the ion mobility, as given in (5.44) (minus sign for positive charging, plus sign for negative charging) from $\theta = 0$ to $\theta = \theta_0$ and from $\phi = 0$ to $\phi = 2\pi$,

$$J_{fc} = \int_0^{2\pi}\int_0^{\theta_0} \left(\mp B_i E_r\big|_{r=R_p} N_{i\infty}\right) R_p^2 \sin\theta \, d\theta \, d\phi$$

which gives

$$J_{fc} = \pm\frac{q_s B_i N_{i\infty}}{4\epsilon_0}\left(1 - \frac{q}{q_s}\right)^2 \tag{7.48}$$

where $q_s$ is the saturation charge,

$$q_s = 4\pi\epsilon_0\left(\frac{3\kappa}{\kappa + 2}\right)E_\infty R_p^2 \tag{7.49}$$

Since the rate of charging of the particle for singly charged ions equals the ion impingement rate $J_{fc}$ multiplied by the unit charge, $\pm e$, we obtain for the rate of charging of the particle,

$$\frac{dq}{dt} = \begin{cases} \dfrac{q_s e B_i N_{i\infty}}{4\epsilon_0}\left(1 - \dfrac{q}{q_s}\right)^2 & |q| < |q_s| \\ 0 & q = q_s \end{cases} \tag{7.50}$$

which can be integrated subject to $q = 0$ at $t = 0$ to give the time-dependent field charge,

$$q = \frac{q_s e B_i N_{i\infty} t}{e B_i N_{i\infty} t + 4\epsilon_0} \tag{7.51}$$

Under usual operating conditions in an electrostatic precipitator, the saturation charge is attained soon after the particles enter the device (White, 1984). For our purposes, then, it suffices to assume that the field-charging contribution to total particle charge is given by (7.49)

$$q_{fc} = \left(\frac{3\kappa}{\kappa + 2}\right)\pi\epsilon_0 E D_p^2 \tag{7.49}$$

and that this charge is attained by particles immediately upon entrance into the precipitator.

We can examine the validity of this approximation from (7.51). In order for $q$ from (7.51) to be approximated by $q_s$, it is necessary that $t >> 4\epsilon_0/eB_iN_{i\infty}$. Now, $\epsilon_0$ is the order of $10^{-11}$ C m$^{-1}$ V$^{-1}$, $e$ is the order of $10^{-19}$ C, $B_i$ is the order of $10^{-4}$ m$^2$ V$^{-1}$ s$^{-1}$, and $N_{i\infty}$ is the order of $10^{13}$ m$^{-3}$. Thus we find that under usual conditions $q \simeq q_s$ for $t > 0.1$ s, and therefore approximating the field-charging contribution by (7.49) is valid for electrostatic precipitators since the residence time of the particles in the precipitator will generally exceed 1 s or so.

**Example 7.4** ***Field Charging***

The saturation charge on a particle attained by field charging in an electric field of strength $E$ is given by (7.49). Charging electric fields in an electrostatic precipitator are typically in the range 300 to 600 kV m$^{-1}$, but may exceed 1000 kV m$^{-1}$ in special cases. Let us calculate the magnitude of this charge for the following conditions: $D_p = 1\ \mu$m, $E = 500$ kV m$^{-1}$, $\kappa >> 1$ (conducting particle). Then from (7.49)

$$q_{fc} = 4.17 \times 10^{-17}\ \text{C}$$

The number of electronic charges to which this charge corresponds is

$$z_p = \frac{q_{fc}}{e} = \frac{4.17 \times 10^{-17}}{1.60 \times 10^{-19}}$$

$$= 260 \text{ electronic charges}$$

### 7.4.5 Diffusion Charging

Diffusion charging occurs as the ions in their random thermal motion collide with a particle and surrender their charge to it. In that sense the mechanism of diffusion charging is identical to that of the diffusion of uncharged vapor molecules to the surface of a particle (Section 5.5). However, because both the particle and the ions are charged, the random thermal motion of the ions in the vicinity of a particle is influenced by an electrostatic force. This force gives rise to a tendency of the ions to migrate away from the particle as the particle charge increases. The overall flux of ions to a particle thus must include both the random diffusive motion and the electrical migration. As in the case of diffusion of gas molecules to a particle, the particular flux expression depends on the ratio of the ion mean free path, $\lambda_i$, to the particle radius, that is, the ion Knudsen number. We neglect the effect of the background electric field in the precipitator in analyzing the flux of ions to a particle.

In the free molecule regime a kinetic theory argument can be used to deduce the rate of diffusion charging. If the particle has a charge $q$, the radial distribution of ions around the particle should be given by a Boltzmann expression (White, 1963)

$$N_{i,s} = N_i\Big|_{r=R_p} = N_{i\infty} \exp\left(\mp \frac{qe}{2\pi\epsilon_0 kTD_p}\right)$$

(minus sign for positive charging; plus sign for negative charging). The rate at which ions strike the surface of the particle per unit surface area is given by the effusion flux,

$\frac{1}{4} N_{i,s}\bar{c}_i$. Thus the rate of accumulation of charge is

$$\frac{dq}{dt} = \pm \frac{\pi D_p^2}{4} N_{i\infty} \bar{c}_i e \exp\left(\mp \frac{qe}{2\pi\epsilon_0 kTD_p}\right)$$

which can be integrated subject to $q = q_0$ at $t = 0$ to give

$$q = \pm \frac{2\pi\epsilon_0 kTD_p}{e} \ln\left[\exp\left(\pm \frac{q_0 e}{2\pi\epsilon_0 kTD_p}\right) + \frac{N_{i\infty} e^2 D_p \bar{c}_i t}{8\epsilon_0 kT}\right] \tag{7.52}$$

which in the case of $q_0 = 0$ becomes

$$q = \pm \frac{2\pi\epsilon_0 kTD_p}{e} \ln\left(1 + \frac{N_{i\infty} e^2 D_p \bar{c}_i t}{8\epsilon_0 kT}\right) \tag{7.53}$$

where the plus sign is for positive charging and the minus sign for negative charging.

In the continuum regime the flux of ions toward the particle at any distance $r$ is given by

$$J_i = 4\pi r^2 \left(D_i \frac{dN_i}{dr} \mp B_i E N_i\right) \tag{7.54}$$

The first term of the right-hand side is the diffusive contribution to the flux and the second is that due to the field-induced migration in the vicinity of the particle. The steady-state ion concentration profile cannot be prescribed to be Boltzmann equilibrium distributed since it is now influenced by the presence of the particle. The local electric field around the particle is, since we are neglecting the overall field in the precipitator, the Coulombic field

$$E = \frac{q}{4\pi\epsilon_0 r^2} \tag{7.55}$$

At steady state $J_i$ is a constant independent of $r$. We substitute (7.55) into (7.54), and solve the differential equation subject to $N_i = N_{i\infty}$ as $r \to \infty$ to get

$$N_i = \mp \frac{J_i \epsilon_0}{B_i q} + \left(\frac{\pm J_i \epsilon_0}{B_i q} + N_{i\infty}\right) \exp\left(\mp \frac{qe}{2\pi\epsilon_0 kTD_p}\right)$$

(upper sign for positive charging; lower sign for negative charging). Note that if $J_i = 0$, we recover the Boltzmann distribution.

To determine $J_i$ we assume that $N_i = 0$ at $r = R_p$,

$$J_i = \pm \frac{B_i q N_{i\infty}}{\epsilon_0} \left[\exp\left(\pm \frac{qe}{2\pi\epsilon_0 kTD_p}\right) - 1\right]^{-1}$$

The rate of accumulation of charge is just

$$\frac{dq}{dt} = \frac{B_i q e N_{i\infty}}{\epsilon_0} \left[\exp\left(\pm \frac{qe}{2\pi\epsilon_0 kTD_p}\right) - 1\right]^{-1}$$

This equation can be integrated subject to $q = q_0$ at $t = 0$ to give the following implicit expression for $q$ as a function of time:

$$\sum_{m=1}^{\infty} \frac{1}{mm!} \left( \frac{e}{2\pi\epsilon_0 kTD_p} \right)^m [(\pm q)^m - (\pm q_0)^m] = \frac{B_i e N_{i\infty} t}{\epsilon_0} \tag{7.56}$$

We have now developed diffusion charging results in the free molecule, (7.53), and continuum regimes, (7.56). Lassen (1961) obtained an expression that spans the two regimes,

$$\frac{dq}{dt} = \frac{B_i q e N_{i\infty}}{\epsilon_0} \left[ \left( 1 \pm \frac{2qe \text{ Kn}}{a\pi\epsilon_0 kTD_p} \right) \exp\left( \pm \frac{qe}{2\pi\epsilon_0 kTD_p} \right) - 1 \right]^{-1} \tag{7.57}$$

where the ion Knudsen number $\text{Kn} = 2\lambda_i/D_p$. The ion mean free path is related to its molecular diffusivity by

$$D_i = \frac{1}{a} \lambda_i \bar{c}_i$$

Lassen used the value of $a = 3$, which we recall is used in conjunction with the Fuchs-Sutugin interpolation formula [i.e., (5.95)].

In 1918 Enskog obtained the following expression for the binary diffusivity of species $i$ in a background gas $j$,

$$D_i = \frac{3\pi}{32} \left( \frac{30z^2 + 16z + 13}{30z^2 + 16z + 12} \right) (1 + z)\lambda_i \bar{c}_i$$

where $z = m_i/m_j$, the ratio of the mass ($m_i$) to that of the background gas ($m_j$). This equation results from the second-order Chapman-Enskog solution to the Boltzmann equation for a hard sphere model. The term in the first parentheses is the correction factor to the first-order solution [recall (5.11)].

Equation (7.57) can be integrated subject to $q = q_0$ at $t = 0$ to give

$$\sum_{m=1}^{\infty} \left( \frac{4 \text{ Kn}}{a} + \frac{1}{m} \right) \frac{1}{m!} \left( \frac{e}{2\pi\epsilon_0 kTD_p} \right)^m [(\pm q)^m - (\pm q_0)^m] = \frac{B_i e N_{i\infty} t}{\epsilon_0} \tag{7.58}$$

For Kn >> 1 and $D_i = kTB_i/e$, (7.58) reduces to (7.52).

Many treatments of electrostatic precipitation confine their analysis of the diffusion charging contribution to particle charge to the free molecule result (7.53). One relevant question concerns the difference in charge predicted by that equation as compared with the more complete result (7.58). We will examine that difference in Example 7.5.

The classical diffusion charging equations derived above are based on the absence of an external electric field and the neglect of the electrostatic image force between the ions and the dielectric particles. Diffusion charging can be enhanced by an external electric field, so-called "field-enhanced diffusion." Results on combined field and diffusion charging have been obtained by Liu and Yeh (1968), Brock and Wu (1973), Smith and McDonald (1976), Liu and Kapadia (1978), and Withers and Melcher (1981). The effect of the electrostatic image force has been considered by Marlow and Brock (1975), Liu and Pui (1977), and Davison and Gentry (1985). In the transition regime Fuchs obtained the flux for diffusion charging by flux matching. Fuchs' formula includes the electrostatic force on the trajectory of an ion in the vicinity of a particle and has shown good agreement with recent experimental results (Adachi et al., 1985).

**Example 7.5** ***Particle Charging by Field and Diffusion Charging Mechanisms***

Let us compute the charge acquired by particles as a function of $D_p$ by field and diffusion charging mechanisms separately. We consider the following conditions:

$$T = 293 \text{ K} \qquad E = 2 \times 10^5 \text{ V m}^{-1}$$

$$N_{i\infty} = 10^{13} \text{ m}^{-3}$$

$$\kappa = 1$$

We need to select an ion mass. The ion masses are difficult to determine accurately, as the ions tend to form clusters that may change with time. Adachi et al. (1985) have considered the available data and have recommended:

$$m_i^+ = 109 \text{ to } 130 \text{ amu}$$

$$m_i^- = 50 \text{ to } 100 \text{ amu}$$

$$B_i^+ = 1.4 \times 10^{-4} \text{ m}^2 \text{ V}^{-1} \text{ s}^{-1} \qquad B_i^- = 1.9 \times 10^{-4} \text{ m}^2 \text{ V}^{-1} \text{ s}^{-1}$$

Using these values, we obtain:

$$\bar{c}_i^+ = 2.18 \times 10^2 \text{ to } 2.38 \times 10^2 \text{ m s}^{-1} \qquad \bar{c}_i^- = 2.48 \times 10^2 \text{ to } 3.52 \times 10^2 \text{ m s}^{-1}$$

$$\lambda_i^+ = 1.44 \times 10^{-8} \text{ to } 1.46 \times 10^{-8} \text{ m} \qquad \lambda_i^- = 1.79 \times 10^{-8} \text{ to } 1.94 \times 10^{-8} \text{ m}$$

$$D_i^+ = 3.54 \times 10^{-6} \text{ m}^2 \text{ s}^{-1} \qquad D_i^- = 4.80 \times 10^{-6} \text{ m}^2 \text{ s}^{-1}$$

Figure 7.12 shows the number of elementary charges as a function of particle diameter for both field and diffusion charging. The field charging line is (7.49), which, since this is a log-log plot, is a straight line of slope 2. The saturation charge from field charging depends, in addition to size, only on the dielectric constant of the particle and the field strength. The diffusion charging contribution varies with time. That contribution as given by (7.58) assuming no initial charge is shown at $t = 1$ and 10 s. Also, we show by the dashed line the classic free molecule result (7.53). We note that the free molecule result is quite close to the more complete equation (7.58) for particle diameters less than about 1 $\mu$m, the regime where diffusion charging dominates. For this reason and because the field charging contribution reaches saturation very quickly, it will suffice henceforth to use (7.52) as an approximation to the diffusion charging contribution in electrostatic precipitation with $q_0$ equal to the field charge.

**Example 7.6** ***Migration Velocity***

The charged particle migration velocity in an electric field was given by (7.43). Let us compute the migration velocities of the particles of Figure 7.12 at a charging time of $t$ = 1 s in air at 298 K. At this temperature, the mean free path of the air molecules is $\lambda_{\text{air}}$ = 0.065 $\mu$m. The migration velocity is shown in Figure 7.13, and the individual contributions

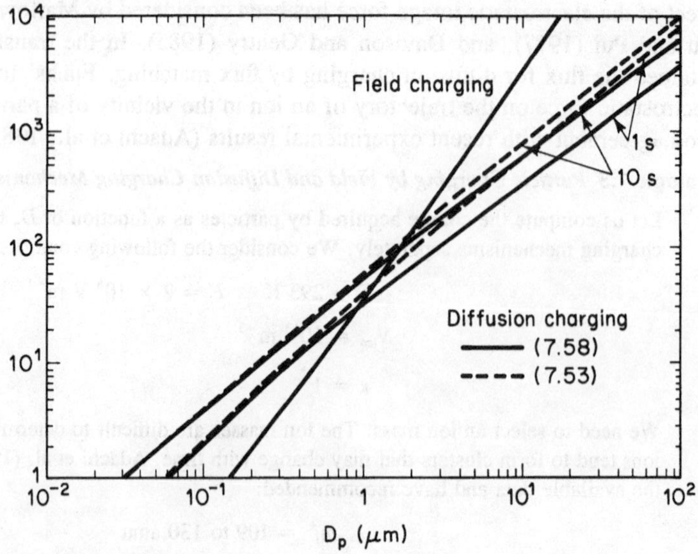


**Figure 7.12** Particle charging by field and diffusion charging mechanisms. The following conditions are assumed: $T$ = 293 K, $m_i^+$ = 130 amu, $B_i^+ = 1.4 \times 10^{-4}$ m$^2$ V$^{-1}$s$^{-1}$, $\bar{c}_i^+ = 218$ m s$^{-1}$, $\lambda_i^+ = 1.46 \times 10^{-8}$ m, $D_i^+ = 3.54 \times 10^{-6}$ m$^2$s$^{-1}$, $a$ = 0.905, $E = 2 \times 10^5$ V m$^{-1}$, $N_{i\infty} = 10^{13}$ m$^{-3}$, $\kappa = 1$.

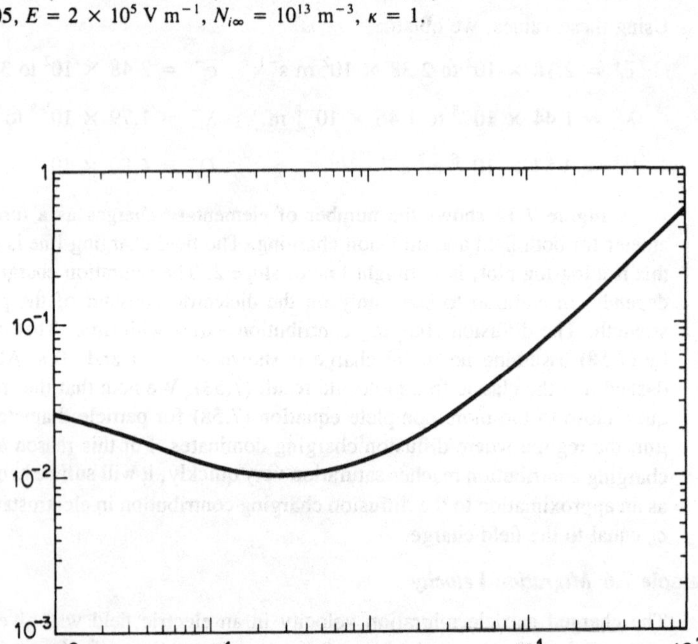


**Figure 7.13** Particle migration velocity. Conditions are the same as Figure 7.12.

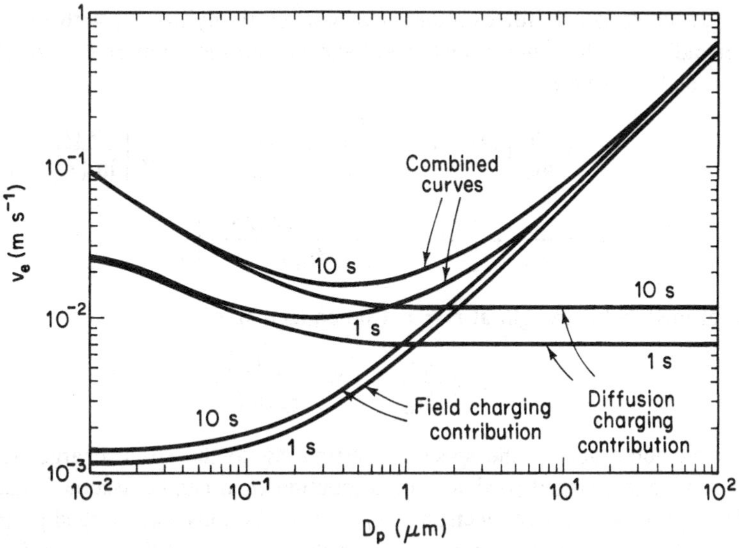


**Figure 7.14** Contributions to particle migration velocity from field and diffusion charging. Conditions are the same as in Figure 7.12.

are shown in Figure 7.14. We note that the migration velocity is relatively large for very small particles and very large particles. Even though the particle charge is lower for small particles, the mobility of such small particles is large enough to more than compensate for the relatively lower level of charge. Large particles are able to acquire such a substantial charge that even the increased Stokes drag cannot overcome the charge effect. This can be seen simply from the fact that the particle charge by field charging increases as $D_p^2$, whereas the Stokes drag increases only as $D_p$, leading to an overall increase in the migration velocity with $D_p$ as particle diameter increases.

### 7.4.6 The Electric Field

Our final step in developing the information needed to design an electrostatic precipitator is to calculate the electric field in the device. The chamber consists of two electrodes, the discharge and the collecting electrodes. Between the electrodes the gas contains free electrons, ions, and charged particles.

The electric field intensity, $\boldsymbol{E}$, is defined in terms of the potential $V$ by

$$\boldsymbol{E} = -\nabla V \tag{7.59}$$

and in a medium with space charge density $q_v$ is governed by Poisson's equation

$$\nabla \cdot \boldsymbol{E} = \frac{q_v}{\epsilon_0} \tag{7.60}$$

[Actually, the denominator on the right-hand side of (7.60) is the product of the dielectric constant of the medium and $\epsilon_0$. Since $\kappa$ is essentially unity for gases under all realistic precipitation conditions, we simply use $\epsilon_0$ in (7.60).]

If we consider, for example, two concentric cylinders, with inner radius $r_0$ and outer radius $r_c$, the inner radius at voltage $V_0$ and the outer at $V = 0$, the solutions of (7.59) and (7.60) are

$$V = \frac{q_v}{4\epsilon_0}(r_c^2 - r^2) + \left[V_0 - \frac{q_v}{4\epsilon_0}(r_c^2 - r_0^2)\right]\frac{\ln(r_c/r)}{\ln(r_c/r_0)} \tag{7.61}$$

$$E = \frac{q_v r}{2\epsilon_0} + \left[\frac{V_0 - q_v(r_c^2 - r_0^2)/4\epsilon_0}{\ln(r_c/r_0)}\right]\frac{1}{r} \tag{7.62}$$

The electric field strength at $r = r_c$ (the collector) is

$$E_c = \frac{q_v r_c}{2\epsilon_0} + \frac{V_0 - q_v(r_c^2 - r_0^2)/4\epsilon_0}{r_c \ln(r_c/r_0)} \tag{7.63}$$

As noted above, the species contributing to the space charge density are ions, electrons, and charged particles. In computing $q_v$ it can be assumed that the gas molecules capture all the free electrons so that only the ions and charged particles contribute to $q_v$. Actually, an ionic current flows in the direction of the electric field consisting of ions charged with the same polarity as the charging electrode and moving to the collecting electrode. The ions migrate to the collecting electrode with a velocity large enough to be unaffected by the turbulent flow in the chamber. The space charge density due to the ionic current depends on the local value of $E$. Accounting for the dependence of $q_v$ on $E$ leads to a nonlinear equation for $E$ that cannot be solved exactly. Under usual conditions of electrostatic precipitation operation, the effect of the ionic current on $q_v$ can be neglected (Crawford, 1976).

The edge of the corona, at $r = r_0$ in the cylindrical case, is defined by the electric field strength $E_0$. An empirical expression for the electric field strength at the edge of the corona is given by White (1963) as

$$E_0 = \pm 3 \times 10^6 f\left[\frac{T_0 p}{T p_0} + 0.03\left(\frac{T_0 p}{T p_0 r_0}\right)^{1/2}\right] \quad \text{V m}^{-1} \tag{7.64}$$

where $T_0 = 293$ K, $p_0 = 1$ atm, $r_0$ is the radius at the edge of the corona (m), and $f$ is a roughness factor that accounts for rough spaces on the wire surface. The effect of roughness is to reduce the field strength needed to form the corona. For a clean smooth wire, $f = 1$; for practical applications $f = 0.6$ is a reasonable value to use in the absence of other information. For a positive corona, the positive sign is used in (7.64); the negative sign for a negative corona.

We are now in a position to summarize the basic design equations for the electrostatic precipitator. To do so, we follow the treatment of Crawford (1976). The electric field strength at the edge of the corona is fixed and given by (7.64). The electric field strength (7.62) can be written in terms of $E_0$ as

$$E = \frac{q_v}{2\epsilon_0}\left(r - \frac{r_0^2}{r}\right) + \frac{r_0 E_0}{r} \tag{7.65}$$

The voltage at the edge of the corona is

$$V_0 = \frac{q_v}{4\epsilon_0}\left(r_c^2 - r_0^2 - 2r_0^2 \ln\frac{r_c}{r_0}\right) + r_0 E_0 \ln\frac{r_c}{r_0} \tag{7.66}$$

Since it is usually the case that $r_c >> r_0$, (7.66) may be approximated as

$$V_0 = \frac{q_v r_c^2}{4\epsilon_0} + r_0 E_0 \ln\frac{r_c}{r_0} \tag{7.67}$$

Even though we have been evaluating $E$ as a function of radial position, keep in mind that if $q_v$ varies down the length of the precipitator, $E$ is also a function of axial position $x$.

At this point we need to specify $q_v$. As we noted above, we will not include the ion current as a contribution to the space charge density, only the charge on the particles. Field charging occurs rapidly after the particles enter the precipitator, so that if the particle concentration is substantially less than the ion concentration, it is reasonable to assume that every particle acquires the saturation charge given by (7.49) corresponding to the field strength at the entrance. For this purpose it is sufficient to use the mean electric field strength across the entrance,

$$\hat{E}(0) = \frac{\int_{r_0}^{r_c} 2\pi r E(0, r)\, dr}{\pi(r_c^2 - r_0^2)} \tag{7.68}$$

Thus the saturation charge from field changing is

$$q_{fc} = \frac{3\kappa}{\kappa + 2}\pi\epsilon_0 D_p^2 \hat{E}(0) \tag{7.69}$$

where in evaluating $\hat{E}(0)$ we invoke the approximation that $r_c >> r_0$,

$$\hat{E}(0) = \frac{q_v(0) r_c}{3\epsilon_0} + \frac{2E_0 r_0 (r_c - r_0)}{r_c^2} \tag{7.70}$$

Let us recapitulate. Particles at number concentration $N_0$ enter the precipitator and are assumed to be immediately charged to $q_{fc}$, corresponding to the mean electric field strength $\hat{E}(0)$ across the entrance to the precipitator. The space charge density at the entrance to the precipitator is then just the product of the charge on each particle and the number concentration of the particles,

$$q_v(0) = q_{fc} N_0 \tag{7.71}$$

where $q_{fc}$ is given by (7.69). If we combine (7.69)–(7.71), we can eliminate $\hat{E}(0)$ and obtain a relation for $q_{fc}$ in terms of known quantities,

$$q_{fc} = \frac{2E_0 r_0 (r_c - r_0)\epsilon_0}{r_c^2[(\kappa + 2)/3\kappa\pi D_p^2 - r_c N_0/3]} \tag{7.72}$$

This is the relation for the charge/particle, due to field charging, at the entrance to the precipitator. We noted above that we are assuming that each particle immediately acquires the saturation field charge upon entrance into the precipitator, and that for this assumption to hold, the number concentration of particles must be substantially less than the ion concentration. The mean electric field at the entrance $\hat{E}(0)$ is computed taking into account the space charge density due to the particles that are charged due to $\hat{E}(0)$. The contribution to $\hat{E}(0)$ from the space charge density [i.e., the first term on the right-hand side of (7.70)] cannot exceed $\hat{E}(0)$ itself. This restriction is reflected in the fact that for (7.72) to be valid, it is necessary that the denominator of (7.72) be positive. Given $D_p$, $\kappa$, and $r_c$, this condition places an upper limit on the value of $N_0$ for which the approach is valid. In fact, for the theory to be applicable, we require that

$$\frac{\kappa + 2}{3\kappa\pi D_p^2} \gg \frac{r_c N_0}{3}$$

For example, if $\kappa = 1$, $r_c = 0.3$ m, and $D_p = 1\ \mu$m, we require that $N_0 \ll 3 \times 10^{12}$ m$^{-3}$. For $D_p = 0.1\ \mu$m, $N_0 \ll 3 \times 10^{14}$ m$^{-3}$.

As the particles flow through the precipitator, each particle retains its charge $q_{fc}$ and may gain additional charge due to diffusion charging. The space charge density at any point is the product of the charge $q$ on each particle and the number concentration of particles,

$$q_v(x) = qN(x) \tag{7.73}$$

If we neglect the charging contribution from diffusion charging, an assumption valid for particles larger than about 0.5 $\mu$m, the charge on each particle is just $q_{fc}$, as given by (7.72). The space charge density $q_v(x)$ decreases down the precipitator as particles are deposited on the collecting electrode,

$$q_v(x) = \frac{2E_0 r_0 (r_c - r_0)\,\epsilon_0 N(x)}{r_c^2[(\kappa + 2)/3\kappa\pi D_p^2 - r_c N_0/3]} \tag{7.74}$$

The electric field strength at any radial and axial position is found by combining (7.65) and (7.74),

$$E(x, r) = \frac{E_0 r_0 (r_c - r_0)(r - r_0^2/r) N(x)}{r_c^2[(\kappa + 2)/3\kappa\pi D_p^2 - r_c N_0/3]} + \frac{r_0 E_0}{r} \tag{7.75}$$

The electric field strength, through its dependence on $q_v$, and therefore on $N$, also varies down the length of the precipitator.

The electric field strength at the collector is obtained from (7.75), using the approximation $r_c \gg r_0$, as

$$E_c(x) = \frac{E_0 r_0 (r_c - r_0)\, N(x)}{r_c[(\kappa + 2)/3\kappa\pi D_p^2 - r_c N_0/3]} + \frac{r_0 E_0}{r_c} \tag{7.76}$$

We are now ready to return to the basic design equation for the electrostatic precipitator. We recall that we performed a balance on the number of particles of a given size across a differential section of the precipitator that led to (7.44). For the concentric cylinder geometry we have been considering, the wall layer of thickness $dr$ is defined by $dx/\bar{u} = dr/v_e$. The fraction of the cross-sectional area occupied by the wall layer is $2\pi r_c\, dr/(\pi(r_c^2 - r_0^2))$. If we assume that $r_c >> r_0$, this fraction is just $2\, dr/r_c$. Then the form of (7.44) appropriate to this geometry is

$$\bar{u}A_c(N\big|_x - N\big|_{x+dx}) = \bar{u}A_c \frac{2dr}{r_c} N\big|_x$$

$$= \bar{u}A_c \left(\frac{2v_e}{r_c \bar{u}} dx\right) N\big|_x \tag{7.77}$$

Taking the limit of (7.77) as $dx \to 0$ and using (7.43) gives

$$\frac{dN}{dx} = -\frac{2qE_cC_c}{3\pi\mu D_p \bar{u} r_c} N \tag{7.78}$$

Using the volumetric flow rate $Q = \bar{u}\pi(r_c^2 - r_0^2) \cong \bar{u}\pi r_c^2$, (7.78) can be written as

$$\frac{dN}{dx} = -\frac{2qE_c r_c C_c}{3\mu D_p Q} N \tag{7.79}$$

Now, substituting (7.76) for $E_c(x)$ in (7.79) gives

$$\frac{dN}{dx} = -a(bN + 1)N \tag{7.80}$$

where

$$a = \frac{2qr_0E_0C_c}{3\mu D_p Q}$$

$$b = \frac{3(r_c - r_0)}{(\kappa + 2)/\kappa\pi D_p^2 - r_c N_0}$$

Integrating (7.80) from $N = N_0$ at $x = 0$ to $N$ at $x = L$ gives

$$L = \frac{1}{a} \ln \frac{N_0/(bN_0 + 1)}{N/(bN + 1)} \tag{7.81}$$

Since the collection efficiency is $\eta(D_p) = 1 - N/N_0$, (7.81) can be written in terms of the collection efficiency as

$$L = \frac{1}{a} \ln \frac{bN_0 + (1 - \eta)^{-1}}{bN_0 + 1} \tag{7.82}$$

or as the collection efficiency achieved by a given length,

$$\eta = 1 - \left[(bN_0 + 1)e^{aL} - bN_0\right]^{-1} \tag{7.83}$$

In our derivation of the design equations for the electrostatic precipitator we assumed that the particle charge in the device is that due solely to field charging at the entrance to the precipitator. We know, however, that for particles smaller than about 0.5 $\mu$m diameter, diffusion charging dominates field charging. Let us see how the precipitator design equations are modified to include diffusion charging.

The space charge density at any point is given by (7.73), now written as

$$q_v(x) = q(x)N(x) \tag{7.84}$$

Since we assume that field charging occurs immediately at the entrance and that diffusion charging begins only for $x > 0$, $q_v(0) = q_{fc}N_0$. The value of $q_{fc}$ is that given by (7.72). From (7.53) the net charge is

$$q(x) = \pm \frac{2\pi\epsilon_0 kTD_p}{e} \ln\left[\exp\left(\pm \frac{q_0 e}{2\pi\epsilon_0 kTD_p}\right) + \frac{N_{i\infty}e^2 D_p}{2\epsilon_0 kT}\left(\frac{kT}{2\pi m_i}\right)^{1/2}\frac{x}{\bar{u}}\right] \tag{7.85}$$

where time $t$ is replaced by $x/\bar{u}$.

The electric field strength at any position in the unit is given by (7.65)

$$E(x, r) = \frac{q_v(x)}{2\epsilon_0}\left(r - \frac{r_0^2}{r}\right) + \frac{r_0 E_0}{r} \tag{7.86}$$

and that at the collector surface, assuming that $r_c >> r_0$, is

$$E_c(x) = \frac{r_c q(x)N(x)}{2\epsilon_0} + \frac{r_0 E_0}{r_c} \tag{7.87}$$

Finally, (7.79) becomes

$$\frac{dN}{dx} = -\frac{2r_c C_c}{3\mu D_p Q} q(x)\left\{\frac{r_c}{2\epsilon_0} q(x)N(x) + \frac{r_0 E_0}{r_c}\right\}N(x) \tag{7.88}$$

which must be solved to obtain the collection efficiency.

In this section we have focused on developing the basic equations for predicting electrostatic precipitator collection efficiency. In the design of an actual electrostatic precipitator one must specify the configuration (e.g., parallel plates or wire in tube), the plate area and spacing, the corona power and current, and the pressure drop. These will depend on the gas velocity, the particle loading, the required removal efficiency, and the resistivity of the particulate matter. White (1977) presents an analysis of all of these factors in the design of an electrostatic precipitator.

**Example 7.7** ***Electrostatic Precipitator Design***

An airstream flowing at 1.5 m s$^{-1}$ at 573 K, 1 atm containing a particle mass concentration of $3 \times 10^{-3}$ g m$^{-3}$ with a particle density of 1.75 g cm$^{-3}$ is to be treated by a cylindrical electrostatic precipitator. All particles can be assumed to have $\kappa = 3$. The electrostatic precipitator is to consist of a cylinder of dimensions $r_0 = 0.005$ m and $r_c = 0.1$ m. A value of $f = 0.7$ can be assumed. Assume a negative corona. We want to determine the efficiency of the preciptator as a function of particle diameter and precipitator length.

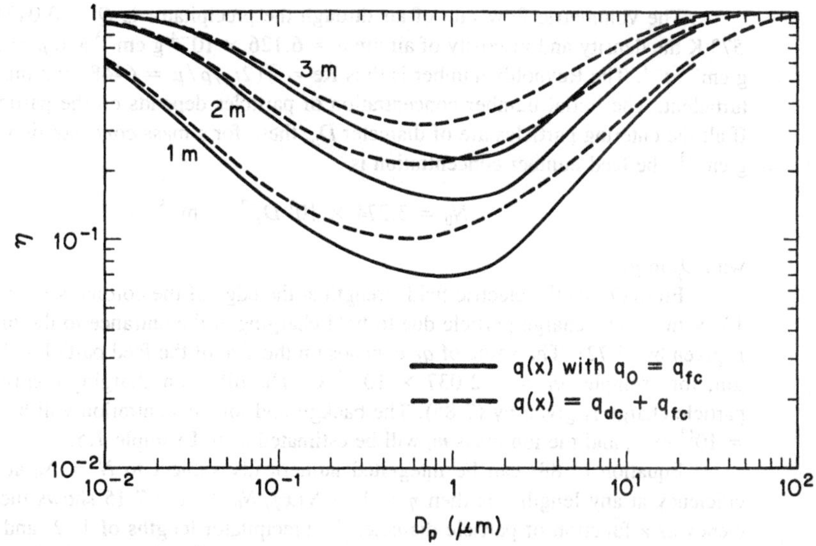


**Figure 7.15** Overall efficiency of the electrostatic precipitator in Example 7.7 as a function of particle diameter and length.

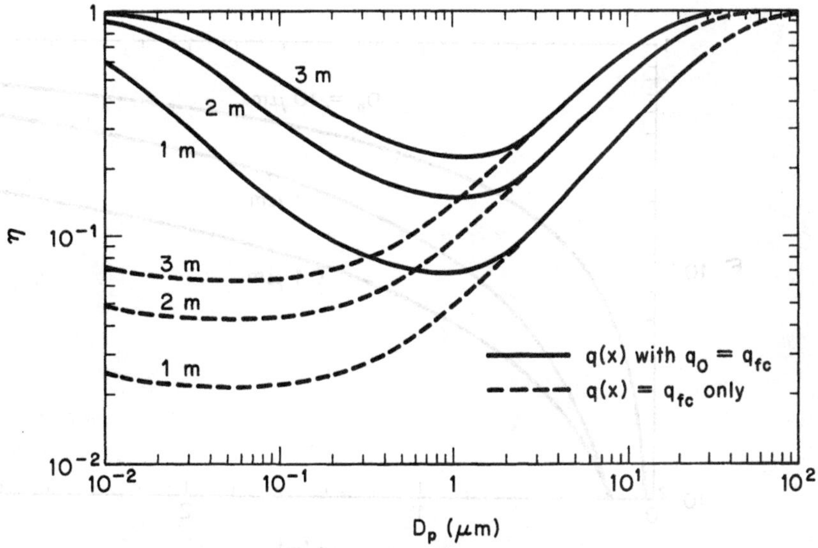


**Figure 7.16** Overall efficiency of the electrostatic precipitator in Example 7.7 as a function of particle diameter and length. Comparison of efficiencies calculated with field and diffusion charging and field charging only.

The volumetric flow rate of air through the precipitator is $Q = 0.0471\ \mathrm{m^3\ s^{-1}}$. At 573 K the density and viscosity of air are $\rho = 6.126 \times 10^{-4}\ \mathrm{g\ cm^{-3}}$ and $\mu = 2.85 \times 10^{-4}\ \mathrm{g\ cm^{-1}\ s^{-1}}$. The Reynolds number is thus $\mathrm{Re} = \bar{u}(2r_c)\rho/\mu = 6448$, and the flow will be turbulent. The initial number concentration of particles depends on the particle diameter. If all the entering particles are of diameter $D_p$, then, for a mass concentration of $3 \times 10^{-3}\ \mathrm{g\ cm^{-3}}$, the feed number concentration is

$$N_0 = 3.274 \times 10^9\, D_p^{-3} \quad \mathrm{m^{-3}}$$

with $D_p$ in $\mu$m.

From (7.64) the electric field strength at the edge of the corona is $E_0 = -1.7109 \times 10^6\ \mathrm{V\ m^{-1}}$. The charge/particle due to field charging at the entrance to the precipitator $q_{fc}$ is given by (7.72). The value of $q_{fc}$ depends on the size of the feed particles. For $D_p = 0.5$ $\mu$m, for example, $q_{fc} = -2.037 \times 10^{-18}$ C. The diffusion charging contribution to the particle charge is given by (7.85). The background ion concentration will be taken as $N_{i\infty} = 10^{13}\ \mathrm{m^{-3}}$, and the ion mass $m_i$ will be estimated as in Example 7.5.

Equation (7.88) can be integrated numerically subject to $N = N_0$ at $x = 0$. The efficiency at any length $x$ is then $\eta = 1 - N(x)/N_0$. Figure 7.15 shows the overall efficiency as a function of particle diameter for precipitator lengths of 1, 2, and 3 m. In this figure we also show the efficiency calculated assuming that the particle charge is the sum of independent field and diffusion charging contributions. This assumption is seen to lead to substantial errors especially in the region of minimum efficiency. Figure 7.16 gives the same result compared to that considering field charging only. We see that for particles of diameter smaller than 1 $\mu$m, diffusion charging cannot be neglected. Figure 7.17 shows the overall efficiency as a function of precipitator length at various particle diameters.

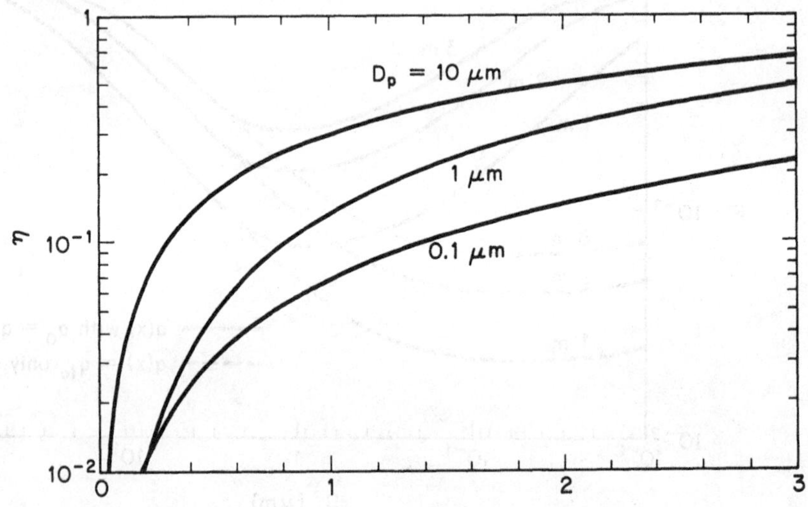


**Figure 7.17** Overall efficiency of the electrostatic precipitator in Example 7.7 as a function of precipitator length.

## 7.5 FILTRATION OF PARTICLES FROM GAS STREAMS

A major class of particulate air pollution control devices relies on the filtration of particles from gas streams. A variety of filter media is employed, including fibrous beds, packed beds, and fabrics. Fibrous beds used to collect airborne particles are typically quite sparsely packed, usually only about 10% of the bed volume being fibers. Packed-bed filters consist of solid packing in, say, a tube and tend to have higher packing densities than do fibrous filters. Both fibrous and packed beds are widely used in ventilation systems. Fabric filters are frequently used to remove solid particles from industrial gases, whereby the dusty gas flows through fabric bags and the particles accumulate on the cloth.

The physical mechanisms by which the filtration is accomplished vary depending on the mode of filtration. Conventional sparsely packed fibrous beds can be viewed as assemblages of cylinders. In such a filter the characteristic spacing between fibers is much larger than the size of the particles being collected. Thus the mechanism of collection is not simply sieving, in which the particles are trapped in the void spaces between fibers; rather, the removal of particles occurs by the transport of particles from the gas to the surface of a single collecting element. Because the filtration mechanisms in a fibrous bed can be analyzed in terms of a single collector, it is possible to describe them in considerable theoretical detail. Packed-bed filters are sometimes viewed as assemblages of interacting, but essentially separate, spherical collectors, although the close proximity of individual packing elements casts doubt as to the validity of this approach. Because of the relatively closer packing in packed-bed filters, and the resulting difficulty of describing the particle collection process in clean theoretical terms, predicting collection in such systems is more empirically based than for fibrous filters. Fabric filter efficiencies must be predicted strictly empirically since the accumulated particle layer actually does the collecting. We will devote most of our attention in this section to filtration by fibrous filters wherein theoretical predictions may be made.

We begin with an analysis of the overall collection efficiency of a fibrous filter bed. Then we consider the mechanisms of collection by a single cylinder placed in a particulate-laden gas flow. Finally, we discuss briefly industrial fabric filters and packed-bed filters.

### 7.5.1 Collection Efficiency of a Fibrous Filter Bed

A fibrous filter bed is viewed as a loosely packed assemblage of single cylinders. Even though the fibers are oriented in all directions in the bed, from a theoretical point of view the bed is treated as if every fiber is normal to the gas flow through the bed. Since, as we have noted, the solid fraction of the filter, $\alpha$, is generally the order of only 10%, we assume, in addition, that each fiber acts more or less independently as a collector. (As we will see later, there is assumed to be an effect of the other fibers on the flow field around an individual fiber.) Thus, to compute the particle removal by a filter bed, we basically need to determine the number of fibers per unit volume of the bed and then multiply that quantity by the efficiency of a single fiber.

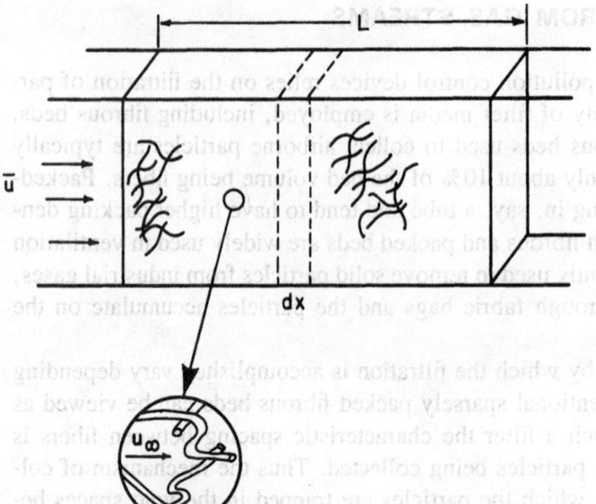


**Figure 7.18** Filter bed composed of an assemblage of single fibers.

Figure 7.18 shows a schematic of a filter bed. Let $D_f$ be the uniform diameter of each fiber comprising the bed. We will perform a balance on the number concentration of particles of diameter $D_p$ across the bed, and, as usual, to do so we consider the balance over a slice of thickness $dx$. Let the cross-sectional area of the bed be $A_c$, and let $L_f$ be the total length of fiber per unit volume of the bed. Then the solid fraction of the filter can be expressed in terms of $D_f$ and $L_f$ as

$$\alpha = \frac{\pi D_f^2 L_f}{4} \tag{7.89}$$

The gas velocity inside the filter is greater than that approaching the filter, $\bar{u}$, due to the volume of flow excluded by the fibers. The volumetric flow rate of air through the filter is $Q = A_c\bar{u}$, so the velocity inside the bed, $u_\infty$, is related to that upstream of the bed, $\bar{u}$, by

$$u_\infty = \frac{Q}{A_c(1-\alpha)} = \frac{\bar{u}}{1-\alpha} \tag{7.90}$$

The particle flows into and out of the element $dx$ are $QN|_x$ and $QN|_{x+dx}$, respectively. The number of particles removed per unit time in the element $dx$ is the product of the flow of particles into the element and the fractional removal of particles by fibers. Let the collection efficiency of a single fiber $\eta$ be defined as the ratio of the number of particles collected to the total number of particles in the projected upstream area $(D_f L_f)$ of the fiber. Thus the particle balance over $dx$ is

$$A_c\bar{u}\left(N\big|_x - N\big|_{x+dx}\right) = (D_f L_f \eta)\, u_\infty N\big|_x (A_c\, dx)$$

Taking the limit as $dx \to 0$ and using (7.89) and (7.90), we obtain

$$\frac{dN}{dx} = -\frac{4\alpha\eta}{\pi(1-\alpha)D_f}N \tag{7.91}$$

which, when integrated over a bed of length $L$, subject to $N(0) = N_0$, gives

$$\frac{N(L)}{N_0} = \exp\left[-\frac{4\alpha\eta L}{\pi(1-\alpha)D_f}\right] \tag{7.92}$$

The overall efficiency of the bed is

$$\eta_b = 1 - \frac{N(L)}{N_0} = 1 - \exp\left[-\frac{4\alpha\eta L}{\pi(1-\alpha)D_f}\right] \tag{7.93}$$

The quantity $\pi(1-\alpha)D_f/4\alpha\eta$ can be viewed as a characteristic depth of penetration of suspended particles in the bed. Since experiments on collection by an isolated fiber are difficult, the isolated fiber collection efficiency $\eta$ is sometimes determined from (7.92) by measuring $N(L)$ and $N_0$ over a bed of length $L$ and known $\alpha$ and $D_f$.

### 7.5.2 Mechanics of Collection by a Single Fiber

As we have just seen, the basis of predicting the collection efficiency of a filter bed is the collection efficiency of a single filter element in the bed. That filter element is taken as an isolated cylinder normal to the gas flow. Three distinct mechanisms can be identified whereby particles in the gas reach the surface of the cylinder:

1. Particles in a gas undergo *Brownian diffusion* that will bring some particles in contact with the cylinder due to their random motion as they are carried past the cylinder by the flow. A concentration gradient is established after the collection of a few particles and acts as a driving force to increase the rate of deposition over that which would occur in the absence of Brownian motion. Because the Brownian diffusivity of particles increases as particle size decreases, we expect that this removal mechanism will be most important for very small particles. When analyzing collection by Brownian diffusion, we treat the particles as diffusing massless points.
2. *Interception* takes place when a particle, following the streamlines of flow around a cylinder, is of a size sufficiently large that its surface and that of the cylinder come into contact. Thus, if the streamline on which the particle center lies is within a distance $D_p/2$ of the cylinder, interception occurs.
3. *Inertial impaction* occurs when a particle is unable to follow the rapidly curving streamlines around an obstacle and, because of its inertia, continues to move toward the obstacle along a path of less curvature than the flow streamlines. Thus, collision occurs because of the particle's momentum. Note that the mechanism of inertial impaction is based on the premise that the particle has mass but no size, whereas interception is based on the premise that the particle has size but no mass.

Collection may also result from electrostatic attraction when either particles or fiber or both possess a static charge. These electrostatic forces may be either direct, when both particle and fiber are charged, or induced, when only one of them is charged. Such charges are usually not present unless deliberately introduced during the manufacture of the fiber. We will not discuss the mechanisms of electrostatic attraction here. Such a discussion is presented by Strauss (1966).

The size ranges in which the various mechanisms of collection are important are:

Inertial impaction: $>1$ μm
Interception: $>1$ μm
Diffusion: $<0.5$ μm
Electrostatic attraction: 0.01 to 5 μm

It is common to analyze the mechanisms of collection separately and then combine the individual efficiencies to give the overall collection efficiency for the cylinder or other obstacle. To see how to combine efficiencies, let us consider two independent mechanisms of collection: one with efficiency $\eta_1$, the other with efficiency $\eta_2$. The probability that a particle will escape collection by mechanism 1 is $(1 - \eta_1)$. If it escapes collection by mechanism 1, the probability that it will escape collection altogether is that probability times the probability of escaping collection by mechanism 2, $(1 - \eta_1)(1 - \eta_2)$. Thus the probability that it will be collected is $1 - (1 - \eta_1)(1 - \eta_2)$, or $\eta_1 + \eta_2 - \eta_1\eta_2$. With $n$ independent mechanisms, the probability of collection is $1 - (1 - \eta_1)(1 - \eta_2) \cdots (1 - \eta_n)$. For two independent mechanisms of collection, we see that the overall collection efficiency is

$$\eta = \eta_1 + \eta_2 - \eta_1\eta_2 \tag{7.94}$$

Because collection efficiencies of two independent mechanisms, such as those listed above, are frequently such that one mechanism is dominant in a particular range of particle size, the overall efficiency is often calculated simply as $\eta = \eta_1 + \eta_2$. Later when we present collection efficiencies for impaction/interception (mechanism 1) and Brownian diffusion (mechanism 2), we will use this approximation.

Most developments of particle collection assume, for lack of better information, that particles transported to the surface of a fiber are retained by the fiber. Experiments have shown, however, that for a variety of substances and filter media, the fraction of particles striking the collector surface that adhere is generally less than unity and may in some cases be as low as 0.5. All of the results we will present can be modified by including an accommodation coefficient if one is known, although we will not discuss this factor further here.

### 7.5.3 Flow Field around a Cylinder

We begin our analysis of the collection of particles by a cylinder with a brief discussion of the velocity field around a cylinder placed normal to the flow. The Reynolds number, based on the cylinder diameter, $\text{Re} = D_f u_\infty \rho / \mu$, for the flows of interest to us is usually of order unity or smaller. It is customary to determine the flow field around the cylinder based on the assumption of creeping flow (i.e., $\text{Re} \ll 1$). There exists no solution of the creeping flow equations of motion that satisfies simultaneously the condition of zero velocity at the cylinder surface and that of $u_\infty$ far from the cylinder. The solution that diverges least rapidly when $r \to \infty$ is (Rosenhead, 1963)

$$u_r = u_\infty A_f \cos\theta \left(1 - 2\ln\frac{2r}{D_f} - \frac{D_f^2}{4r^2}\right) \tag{7.95}$$

$$u_\theta = u_\infty A_f \sin\theta \left(1 + 2\ln\frac{2r}{D_f} - \frac{D_f^2}{4r^2}\right) \tag{7.96}$$

where $A_f = [2(2.0 - \ln \text{Re})]^{-1}$. The velocity field defined by (7.95) and (7.96) is accurate at distances for which the following condition holds: $A_f \text{Re}\,(2r/D_f)\ln(2r/D_f) \ll 1$. For $\text{Re} = 0.1$, this condition is satisfied as long as $2r/D_f \leq 10$.

The velocity field (7.95) and (7.96) pertains to low Reynolds number flow around an isolated cylinder. Our ultimate interest is in cylinders that are elements of a filter bed. Experimental pressure drop data for fibrous beds show that the drag force per filter element increases as the packing density is increased (Happel and Brenner, 1965). Thus it is advantageous to develop a velocity field that depends on the fiber solid fraction $\alpha$. A number of investigators have derived velocity fields around a cylinder assuming that the cylinder is contained in a fluid "cell" with a radius determined by requiring the volume of fluid to be in the same ratio to the cylinder volume as the fluid-to-fiber volume ratio in the fibrous medium (Happel, 1959; Kuwabara, 1959; Spielman and Goren, 1968). These cell models endeavor to account for the interference effect of neighboring cylinders on the flow field near a representative cylinder in an approximate way, and the resulting velocity fields are expected to apply best near the cylinder surface. Fortunately, since most mechanisms of particle capture are dominated by phenomena near the collector surface, these models are useful for providing flow fields within which to compute particle removal. The Kuwabara solution is

$$u_r = \frac{u_\infty}{2\,\text{Ku}}\left[1 - 2\ln\frac{2r}{D_f} - \alpha - \frac{D_f^2}{4r^2}\left(1 - \frac{\alpha}{2}\right) + \frac{2\alpha r^2}{D_f^2}\right]\cos\theta \tag{7.97}$$

$$u_\theta = \frac{u_\infty}{2\,\text{Ku}}\left[1 + 2\ln\frac{2r}{D_f} + \alpha - \frac{D_f^2}{4r^2}\left(1 - \frac{\alpha}{2}\right) + \frac{6\alpha r^2}{D_f^2}\right]\sin\theta \tag{7.98}$$

where $\text{Ku} = \alpha - 3/4 - \alpha^2/4 - \frac{1}{2}\ln\alpha$. The stream function for this velocity field is

$$\Psi = \frac{u_\infty r}{2\,\text{Ku}}\left[1 - 2\ln\frac{2r}{D_f} - \alpha - \frac{D_f^2}{4r^2}\left(1 - \frac{\alpha}{2}\right) + \frac{2\alpha r^2}{D_f^2}\right]\sin\theta \tag{7.99}$$

The effect of crowding by neighboring fibers is to compress the streamlines and increase the fluid speed close to the central cylinder since Ku decreases as $\alpha$ increases. Streamlines of the Kuwabara flow field, expressed as $\overline{\Psi} = \Psi / u_\infty D_f$, are shown in Figure

7.19 for $\alpha = 0.001, 0.01$, and $0.1$. Streamlines corresponding to constant values of $\overline{\Psi} = 0.01$ and $0.3$ are plotted. The packing density $\alpha$ plays a role similar to that of the Reynolds number: that is, as $\alpha$ increases, the streamlines are compressed toward the cylinder and toward the line of symmetry at $y = 0$.

At present there is considerable disagreement as to which of the available flow models for fibrous beds is best suited to predict particle capture. For our purposes it will be sufficient to employ the Kuwabara solution above. For more in-depth discussions of the flow fields, we refer the reader to Spielman (1977) and Adamczyk and van de Ven (1981).

### 7.5.4 Deposition of Particles on a Cylindrical Collector by Brownian Diffusion

When analyzing the transport of particles by Brownian diffusion, the particles are treated as if they are gas molecules (see Section 5.4), and under steady-state conditions the number concentration of particles obeys the convective diffusion equation,

$$\mathbf{u} \cdot \nabla N = D \nabla^2 N \tag{7.100}$$

where $D$ is the Brownian diffusivity. By defining $\mathbf{u}^* = \mathbf{u}/u_\infty$, $N^* = N/N_\infty$, and the Peclet number $\text{Pe} = D_f u_\infty / D$, (7.100) can be placed in dimensionless form,

$$\mathbf{u}^* \cdot \nabla N^* = \frac{1}{\text{Pe}} \nabla^2 N^* \tag{7.101}$$

The Peclet number is the product of the Reynolds number, $\text{Re} = D_f u_\infty \rho / \mu$, and the Schmidt number, $\text{Sc} = \mu / \rho D$, and represents the ratio of convective to diffusive transport. The boundary conditions on (7.101) are $N^* = 1$ far upstream of the cylinder and $N^* = 0$ at its surface.

Since the dimensionless velocity field $\mathbf{u}^*$ is itself a function of Re from the dimensionless Navier-Stokes equations, dimensional analysis implies immediately that $N^* = N^*(\mathbf{r}^*, \text{Re}, \text{Pe})$, where $\mathbf{r}^*$ denotes the dimensionless position. Our interest is in the dependence of the flux of particles to the surface of the cylinder. The local flux of particles to the cylinder surface is

$$-D\left(\frac{\partial N}{\partial r}\right)_{r = D_f/2} = -\frac{D N_\infty}{D_f}\left(\frac{\partial N^*}{\partial r^*}\right)_{r^* = 1/2} \tag{7.102}$$

The deposition of particles over the entire surface of the cylinder can be represented in terms of an average mass transfer coefficient $k_{av}$, such that the product of the mass transfer coefficient, the surface area of the cylinder ($\pi D_f L_f$) and the "driving force" ($N_\infty - 0$) is equal to the local flux from the solution of the convective diffusion equation, (7.101), integrated over the surface of the cylinder.

From the Brownian diffusion coefficients shown in Figure 5.7 we calculate that for particles in air larger than about 0.01 $\mu$m in diameter, $\text{Sc} \gg 1$. (The Schmidt number for particles in air is the same order of magnitude as for molecules of a liquid.) Thus, even though Re is generally of order 1 or smaller, Sc is sufficiently large that in

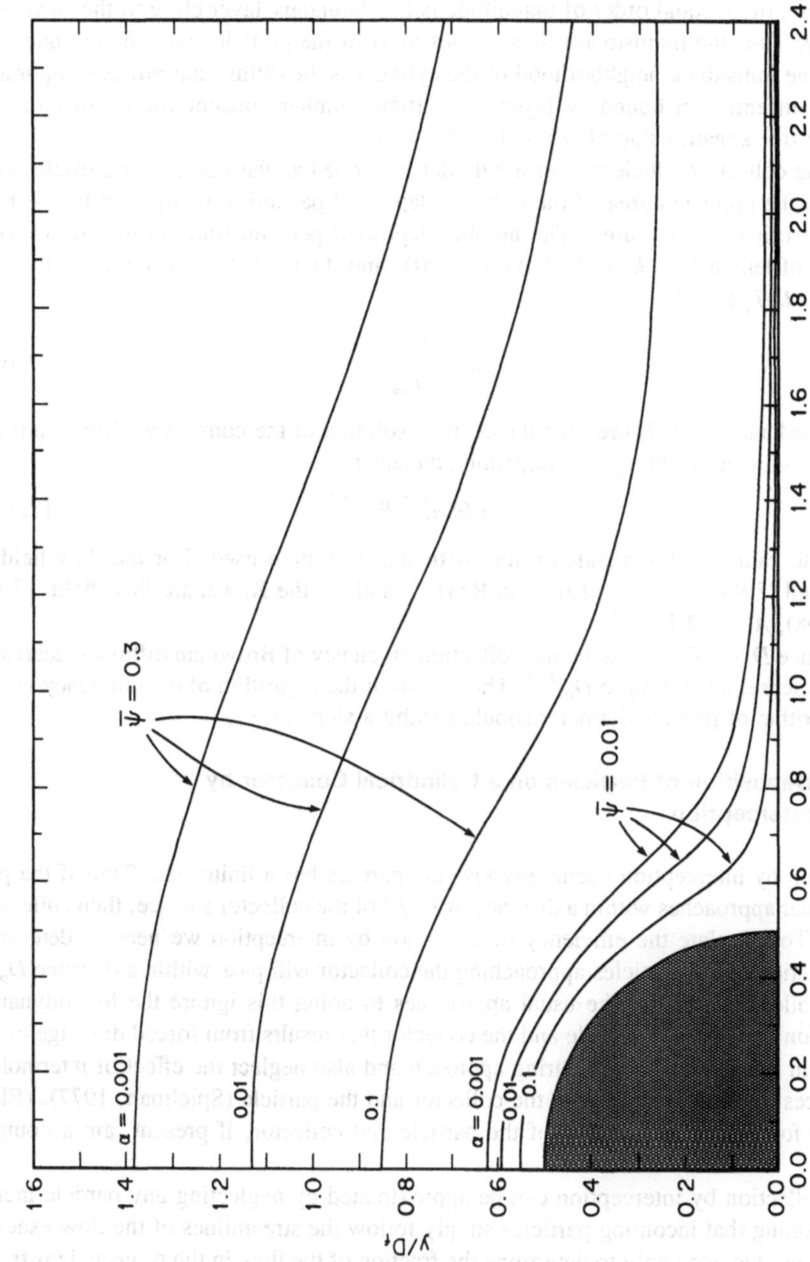

y/D_f
x/D_f
α = 0.001
0.01
0.1
α = 0.001
0.01
0.1
Ψ̄ = 0.3
Ψ̄ = 0.01
0.0
0.2
0.4
0.6
0.8
1.0
1.2
1.4
1.6
1.8
2.0
2.2
2.4

the cases of interest to us, Pe >> 1. Physically, the large Peclet number implies that convective transport greatly exceeds diffusive transport, and the only region in which the two are of an equal order of magnitude is in a boundary layer close to the surface of the body. Thus the mainstream flow carries most of the particles past the cylinder and only in the immediate neighborhood of the cylinder is the diffusional process important. In this concentration boundary layer the particle number concentration drops sharply from the free stream value of $N^* = 1$ to $N^* = 0$.

The collection efficiency for a cylinder is defined as the number of particles contained in the projected area of the cylinder deposited per unit time divided by the total flow of particles in that area. The number deposited per unit time on the surface of a cylinder of length $L_f$ is $k_{av}(\pi D_f L_f)(N_\infty - 0)$, and the total flow in the projected area is $u_\infty N_\infty (D_f L_f)$. Thus

$$\eta = \frac{k_{av}\pi}{u_\infty} \tag{7.103}$$

Friedlander (1977) presents the detailed solution of the convective diffusion problem to a cylinder, yielding the collection efficiency

$$\eta = 3.68 A_f^{1/3} \text{ Pe}^{-2/3} \tag{7.104}$$

where the value of $A_f$ depends on the particular flow field used. For the flow field of (7.95) and (7.96), $A_f = [2(2.0 - \ln \text{Re})]^{-1}$, and for the Kuwabara flow field, (7.97) and (7.98), $A_f = (2 \text{ Ku})^{-1}$.

Since $D = kTC_c/3\pi\mu D_p$, the collection efficiency of Brownian diffusion decreases as $D_p$ increases according to $D_p^{-2/3}$. Thus a plot of the logarithm of the efficiency versus the logarithm of particle diameter should exhibit a slope of $-\frac{2}{3}$.

### 7.5.5 Deposition of Particles on a Cylindrical Collector by Interception

Collection by interception occurs because the particle has a finite size. Thus if the particle center approaches within a distance of $D_p/2$ of the collector surface, then collection occurs. To calculate the efficiency of collection by interception we need to determine what fraction of the particles approaching the collector will pass within a distance $D_p/2$ of the collector surface. The usual approaches to doing this ignore the hydrodynamic interaction between the particle and the collector that results from forced drainage of the fluid from the narrowing gap during approach and also neglect the effect of intermolecular forces of attraction between the collector and the particle (Spielman, 1977). (Electrostatic forces due to charging of the particle and collector, if present, are accounted for.)

Collection by interception can be approximated by neglecting any particle inertia and assuming that incoming particles simply follow the streamlines of the flow exactly. In so doing, we need only to determine the fraction of the flow in the projected upstream area of the collector that passes within a distance $D_p/2$ of the collector surface. For the velocity fields we have been considering, the collection efficiency is just

$$\eta = 2A_f \left(\frac{D_p}{D_f}\right)^2 \tag{7.105}$$

where, due to the neglect of particle inertia, this expression is most applicable if $D_p/D_f \ll 1$.

Actually, Brownian diffusion and interception can be treated simultaneously using (7.100) with the modified boundary condition that $N = 0$ at the collision envelope $r = D_f/2 + D_p/2$ rather than at $r = D_f/2$. Analysis of (7.100) under this condition gives (Friedlander, 1977)

$$\eta \frac{D_p}{D_f} \text{Pe} = f\left[A_f \left(\frac{D_p}{D_f}\right)^3 \text{Pe}\right] \tag{7.106}$$

which includes both (7.104) and (7.105) as special cases, since (7.104) can be expressed as $\eta (D_p/D_f)\,\text{Pe} = 3.68[A_f (D_p/D_f)^3\,\text{Pe}]^{1/3}$, and (7.105) can be rewritten as $\eta (D_p/D_f)\,\text{Pe} = 2[A_f (D_p/D_f)^3\,\text{Pe}]$. We reiterate that both (7.105) and (7.106) neglect any particle inertia (except that inherent in the concept of a particle's Brownian motion).

### 7.5.6 Deposition of Particles on a Cylindrical Collector by Inertial Impaction and Interception

The final mechanism of particle collection that we consider is inertial impaction. As we described earlier, inertial impaction results because sufficiently massive particles are unable to follow curvilinear fluid motion and tend to continue along a straight path as the fluid curves around the collector. Therefore, when one accounts for particle inertia, the collection efficiency will exceed that calculated for interception alone without particle inertia (7.105), because some particles assumed to follow the flow streamlines around the collector cannot do so because of this mass. The basic approach to analyzing inertial impaction is to compute the trajectories of particles that approach the collecting cylinder and to determine those upstream locations from which particles are collected. Figure 7.20 shows the geometry of the collection of a particle by inertial impaction and interception. The trajectory of a particle initially at a distance $y_1$ from the centerline is shown. If all particles between the centerline and $y_1$ are captured and all particles farther from the centerline than $y_1$ escape collection, the flow streamline through $y_1$ is the *limiting streamline* and the particle trajectory through $y_1$ is the *limiting* or *critical trajectory*. Once $y_1$ has been determined, the collection efficiency is just $\eta = 2y_1/D_f$. If the critical trajectory is taken as that passing within a distance $D_p/2$ of the cylinder surface rather than touching the surface, interception is automatically included within the analysis. Our object, then, is, by calculating particle trajectories, to identify the distance $y_1$ and thereby the collection efficiency for combined impaction and interception.

In the absence of external forces on the particle, its equation of motion is given

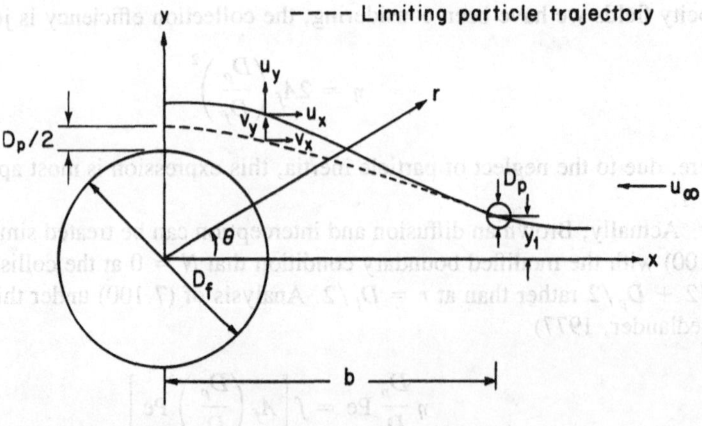


**Figure 7.20** Collection of a particle by a cylinder placed transverse to the flow carrying the particles by the mechanisms of inertial impaction and interception.

by (5.19)

$$\tau \frac{d\mathbf{v}}{dt} = \mathbf{u} - \mathbf{v} \tag{7.107}$$

where $\tau = \rho_p C_c D_p^2/18\mu$. The particle trajectory is determined by integrating (7.107). It is advantageous to address the problem in Cartesian coordinates, so the form of (7.107) that we integrate to determine the particle trajectory, $(x(t), y(t))$, is

$$\tau \frac{d^2x}{dt^2} + \frac{dx}{dt} = u_x \tag{7.108}$$

$$\tau \frac{d^2y}{dt^2} + \frac{dy}{dt} = u_y \tag{7.109}$$

With reference to Figure 7.20, the initial conditions on (7.108) and (7.109) are

$$x(0) = b \qquad \left.\frac{dx}{dt}\right|_{t=0} = u_x(b, y_1) \tag{7.110}$$

$$y(0) = y_1 \qquad \left.\frac{dy}{dt}\right|_{t=0} = u_y(b, y_1) \tag{7.111}$$

At this point we need to specify the velocity field in Cartesian coordinates. Upon transforming the Kuwabara velocity field (7.97) and (7.98) into Cartesian coordinates, it is clear that a numerical solution of (7.108) and (7.109) is necessary. We can obtain an approximate solution by replacing the exact expressions for $u_x$ and $u_y$ by appropriate average velocities between points 1 and 2 in Figure 7.20 (Crawford, 1976). The average velocity in the $x$-direction between points 1 and 2 is

$$\bar{u}_x = \tfrac{1}{2}\left(u_x\big|_1 + u_x\big|_2\right) \tag{7.112}$$

Now, $b$ is chosen so that $u_x|_1 = -u_\infty$. Equating mass flows at planes 1 and 2, $-u_\infty y_1 = u_x|_2 y_2$. Thus

$$\bar{u}_x = -\frac{u_\infty}{2}\left(1 + \frac{y_1}{y_2}\right) \tag{7.113}$$

The average velocity in the $y$-direction is obtained by noting that the fluid must travel the vertical distance from 1 to 2 in the time during which it travels horizontally,

$$\begin{aligned}\bar{u}_y &= \frac{y_2 + D_f/2 - y_1}{-b/\bar{u}_x} \\ &= \frac{u_\infty}{2b}\left[\left(y_2 + \frac{D_f}{2} - y_1\right)\left(1 + \frac{y_1}{y_2}\right)\right]\end{aligned} \tag{7.114}$$

Now we need to obtain a relation between $y_1$ and $y_2$. We see that $y_1$ and $y_2$ lie on the same fluid streamline. Streamlines in a flow are defined by the stream function $\Psi$, such that lines of constant $\Psi$ are the streamlines. The stream function for the Kuwabara flow field was given by (7.99). To obtain a relation between $y_1$ and $y_2$, we note that the stream function at point 2 where $\theta = \pi/2$ and $r = D_f/2 + y_2$,

$$\begin{aligned}\Psi|_2 = -\frac{u_\infty[(D_f/2) + y_2]}{2\mathrm{Ku}}&\left[2\ln\frac{D_f + 2y_2}{D_f} - 1 + \alpha\right. \\ &\left. + \frac{D_f^2}{4[(D_f/2) + y_2]^2}\left(1 - \frac{\alpha}{2}\right) - \frac{2\alpha}{D_f^2}\left(\frac{D_f}{2} + y_2\right)^2\right]\end{aligned} \tag{7.115}$$

The value of the stream function at point 1 is $\Psi|_1 = -u_\infty y_1$.* Equating the values of $\Psi$ at the two points gives the desired relation between $y_1$ and $y_2$,

$$\begin{aligned}\frac{2y_1}{D_f} = \frac{1}{2\mathrm{Ku}}\left(1 + \frac{2y_2}{D_f}\right)&\left[2\ln\left(1 + \frac{2y_2}{D_f}\right) - 1\right. \\ &\left. + \alpha + \frac{1 - \alpha/2}{(1 + 2y_2/D_f)^2} - \frac{\alpha}{2}\left(1 + \frac{2y_2}{D_f}\right)^2\right]\end{aligned} \tag{7.116}$$

This equation gives us the relationship between any two points at planes 1 and 2, as expressed by the distance $y_1$ from the $y = 0$ line at plane 1 and by the distance $y_2$ from the cylinder surface at plane 2, that lie on the *same streamline* of the Kuwabara flow field. Now we need to find that *particular* streamline on which a particle starting at position $y_1$ at plane 1 is just captured, that is such that its trajectory comes within a distance $D_p/2$ of the cylinder surface at plane 2. To find the starting location $y_1$ such that a particle is just captured, we turn to the equations of motion of the particle.

As noted, in order to make the problem more tractable we will replace $u_x$ and $u_y$

*To see this, note that $u_x = -\partial\Psi/\partial y$. At point 1, $u_x = -u_\infty$. Integrating from the $y = 0$ streamline along which $\Psi = 0$ to $y = y_1$ gives $\Psi = u_\infty y_1$.

in (7.108) and (7.109) by $\bar{u}_x$ and $\bar{u}_y$,

$$\tau \frac{d^2x}{dt^2} + \frac{dx}{dt} = \bar{u}_x \tag{7.117}$$

$$\tau \frac{d^2y}{dt^2} + \frac{dy}{dt} = \bar{u}_y \tag{7.118}$$

to be solved subject to

$$x(0) = b \qquad \left.\frac{dx}{dt}\right|_{t=0} = \bar{u}_x \tag{7.119}$$

$$y(0) = y_1 \qquad \left.\frac{dy}{dt}\right|_{t=0} = 0 \tag{7.120}$$

The solution of (7.117) and (7.118) subject to (7.119) and (7.120) is

$$x(t) = b + \bar{u}_x t \tag{7.121}$$

$$y(t) = y_1 - \bar{u}_y \tau (1 - e^{-t/\tau}) + \bar{u}_y t \tag{7.122}$$

For capture to occur, $y = (D_f + D_p)/2$ when $x = 0$. This occurs at $t = -b/\bar{u}_x$. Thus (7.122) becomes

$$y_1 = \frac{D_f + D_p}{2} + \bar{u}_y \tau (1 - e^{b/\bar{u}_x \tau}) + \frac{\bar{u}_y b}{\bar{u}_x} \tag{7.123}$$

This equation provides a relationship among the starting position $y_1$, the particle properties, $D_p$ and $\tau$, and the average flow field, $\bar{u}_x$ and $\bar{u}_y$, for a particle that is just captured at $x = 0$. Equation (7.123) therefore defines the *limiting trajectory*. All particles that start out at plane 1 with $y \leq y_1$ are collected, and vice versa.

We now have (7.116), which relates $y_1$ and $y_2$ along any streamline of the flow, and (7.123), which specifies that particular $y_1$ for which capture is just attained. Note that these two equations are coupled since $\bar{u}_x$ and $\bar{u}_y$ depend on both $y_1$ and $y_2$ through (7.113) and (7.114). Thus we now have two equations (7.116) and (7.123), in the two unknowns $y_1$ and $y_2$. (We are not really interested in the value of $y_2$.)

At this point we need to address the upstream distance $b$. Recall that we related the fiber volume fraction $\alpha$ to the length of fiber per unit volume of filter by (7.89). Let the cylinder of radius $b$ be the void region around the cylinder. This idea is consistent with the choice of $b$ as that point where the approach velocity can be taken as $u_\infty$. Thus $\alpha = D_f^2/4b^2$ or

$$b = \frac{D_f}{2\sqrt{\alpha}} \tag{7.124}$$

It is useful to place our results in dimensionless form so that the problem need not be resolved for every different combination of variables. The natural length scale to use is the diameter of the cylinder $D_f$, and the characteristic velocity is the approach velocity

$u_\infty$. Thus the two components of the average velocity can be expressed as

$$\frac{\bar{u}_x}{u_\infty} = -\frac{1}{2}\left(1 + \frac{2y_1/D_f}{2y_2/D_f}\right) \tag{7.125}$$

$$\frac{\bar{u}_y}{u_\infty} = \frac{\sqrt{\alpha}}{2}\left[\left(1 + \frac{2y_1/D_f}{2y_2/D_f}\right)\left(1 + \frac{2y_2}{D_f} - \frac{2y_1}{D_f}\right)\right] \tag{7.126}$$

Using (7.125) and (7.126) in (7.123), we obtain

$$\frac{2y_1}{D_f} = \left(1 + \frac{D_p}{D_f}\right) + \mathrm{St}\sqrt{\alpha}\left[\left(1 + \frac{2y_1/D_f}{2y_2/D_f}\right)\left(1 + \frac{2y_2}{D_f} - \frac{2y_1}{D_f}\right)\right]$$
$$\times\left\{1 - \exp\left[-\frac{1}{\mathrm{St}\sqrt{\alpha}}\left(1 + \frac{2y_1/D_f}{2y_2/D_f}\right)^{-1}\right]\right\} - \left(1 + \frac{2y_2}{D_f} - \frac{2y_1}{D_f}\right) \tag{7.127}$$

where the dimensionless distances, $2y_1/D_f$ and $2y_2/D_f$, are seen to depend on the interception parameter, $D_p/D_f$, the packing fraction $\alpha$, and the Stokes number, $\mathrm{St} = u_\infty \tau/D_f = \rho_p C_c D_p^2 u_\infty/18\mu D_f$, the ratio of the stop distance to the diameter of the fiber. The collection efficiency is just $\eta = 2y_1/D_f$. To determine the efficiency, (7.116) and (7.127) must be solved simultaneously for $2y_1/D_f$ and $2y_2/D_f$. It is easiest to eliminate $2y_1/D_f$ from (7.127) using (7.116) to obtain a single equation for $2y_2/D_f$ that can be solved numerically.

We see that the collection efficiency depends on three dimensionless parameters, $D_p/D_f$, $\alpha$, and St. Actually, we could have anticipated the dependence of the collection efficiency on these three dimensionless parameters by initially making the particle equation of motion (7.107) dimensionless at the start of the analysis by letting $\mathbf{u}^* = \mathbf{u}/u_\infty$, $\mathbf{v}^* = \mathbf{v}/u_\infty$, and $t^* = tu_\infty/D_f$. The result is

$$\mathrm{St}\frac{d\mathbf{v}^*}{dt^*} = \mathbf{u}^* - \mathbf{v}^* \tag{7.128}$$

Since $\mathbf{u}^*$ is a function of $\alpha$, and the solution of (7.128) is evaluated at $y^* = (D_p/D_f + 1)/2$, the dimensionless particle trajectory is seen to depend only on St, $\alpha$, and $D_p/D_f$.

The collection efficiency for combined impaction and interception is a function of the Stokes number, $\mathrm{St} = \rho_p C_c D_p^2 u_\infty/18\mu D_f$, the packing density $\alpha$, and the ratio of particle to fiber diameter, $D_p/D_f$. Figure 7.21 shows $\eta$ as a function of St for $\alpha = 0.001$, 0.01, and 0.1, and $D_p/D_f = 0.1$ calculated based on our approximate analysis. The results clearly show the effect of changing $\alpha$. For the larger values of $\alpha$ the streamlines lie closer to the cylinder than at smaller $\alpha$ (recall Figure 7.19). Thus, at fixed Stokes number, increasing $\alpha$ leads to increasing collection efficiency since the streamlines are crowded closer and closer to the cylinder, allowing fewer particles to escape past the cylinder. Conversely, to attain the same collection efficiency at a smaller value of $\alpha$ requires greater particle inertia (i.e., Stokes number).

The parameter $b$ is related to $\alpha$ by (7.124) and is treated as the distance along the $x$-axis at which the flow is undisturbed by the presence of the cylinder. For $\alpha = 0.1$,

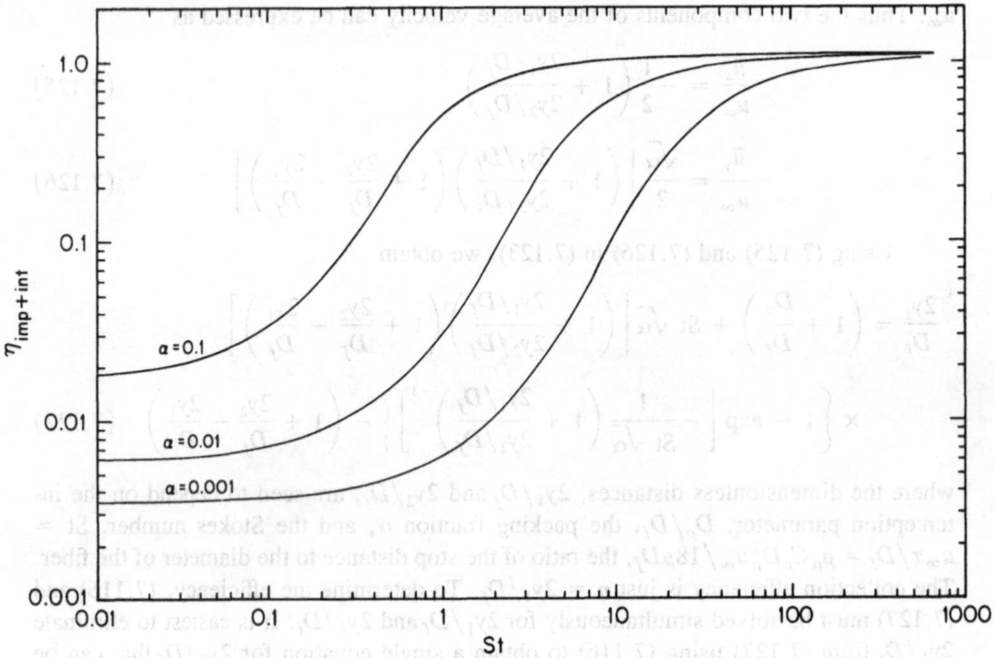


**Figure 7.21** Collection efficiency for combined impaction and interception for a cylinder placed transverse to the flow as a function of Stokes number for $D_p/D_f = 0.1$.

the value of $b/D_f$ is calculated to be 1.58. The solution for the stream function actually begins to break down for $x/D_f$ values exceeding this value; fortunately, we only need the flow field in the vicinity of the cylinder. For $\alpha = 0.01$, $b/D_f = 5$, and for $\alpha = 0.001$, $b/D_f = 15.8$.

Continuing with Figure 7.21, we see that at a fixed value of $\alpha$ the collection efficiency increases with increasing Stokes number, eventually reaching a value of unity. Physically, a convenient way to think of increasing St is to imagine the particle density $\rho_p$ increasing at fixed size $D_p$ and approach velocity $u_\infty$. Thus, as St increases the particle becomes heavier and heavier and is less able to follow any changes in the flow field. A point is eventually reached as St increases where all the particles contained in the upstream projected area of the cylinder are collected; in fact, we see that $\eta$ values slightly larger than 1.0 are obtained, reflecting the interception contribution from particles even initially outside the upstream projected area of the cylinder but within the collection envelope at $r = (D_f + D_p)/2$.

**Example 7.8** ***Collection Efficiency by Inertial Impaction and Interception***

In this section we presented an approximate solution to the determination of the collection efficiency of a cylinder by combined inertial impaction and interception. The approximation arose in using the average velocity components $\bar{u}_x$ and $\bar{u}_y$ as given by (7.113) and (7.114), and by assuming that the critical particle trajectory for capture is that which passes within a distance $D_p/2$ of the cylinder surface at $\theta = \pi/2$. In this example we want to integrate

(7.108) and (7.109) subject to (7.110) and (7.111) using the exact Kuwabara velocity field and compare the calculated collection efficiencies to those obtained using the approximate analysis.

The Kuwabara velocity field in Cartesian coordinates is found from (7.97) and (7.98) by using

$$u_x = u_r \cos\theta - u_\theta \sin\theta$$

$$u_y = u_r \sin\theta + u_\theta \cos\theta$$

and $r^2 = x^2 + y^2$. Doing so, we find

$$u_x = -\frac{u_\infty}{2}\,\mathrm{Ku}\left[\ln\frac{4(x^2+y^2)}{D_f^2} + \alpha - \alpha\frac{x^2+3y^2}{D_f^2/2} + \frac{y^2-x^2}{x^2+y^2} + \frac{D_f^2}{4}\left(1-\frac{\alpha}{2}\right)\frac{x^2-y^2}{(x^2+y^2)^2}\right]$$

$$u_y = \frac{u_\infty}{2}\,\mathrm{Ku}\left[\frac{2xy}{x^2+y^2} - \frac{D_f^2\,xy}{2(x^2+y^2)^2}\left(1-\frac{\alpha}{2}\right) - \frac{4\alpha xy}{D_f^2}\right]$$

To determine the trajectory of a particle we solve (7.108) and (7.109) using this $u_x$ and $u_y$ subject to (7.110) and (7.111). At this point it is advantageous to put everything in dimensionless form. Let $z_1 = x/D_f$, $z_2 = y/D_f$, $t^* = t/\tau$, and $\mathrm{St} = u_\infty\tau/D_f$, and we obtain

$$\frac{d^2z_1}{dt^{*2}} + \frac{dz_1}{dt^*} = -\frac{\mathrm{St}}{2\,\mathrm{Ku}}\left[\ln\left(4(z_1^2+z_2^2)\right) + \alpha - 2\alpha(z_1^2+3z_2^2) + \frac{z_2^2-z_1^2}{z_1^2+z_2^2} + \frac{1}{4}\left(1-\frac{\alpha}{2}\right)\frac{z_1^2-z_2^2}{(z_1^2+z_2^2)^2}\right]$$

$$\frac{d^2z_2}{dt^{*2}} + \frac{dz_2}{d^*} = \frac{\mathrm{St}}{2\,\mathrm{Ku}}\left[\frac{2z_1z_2}{z_1^2+z_2^2} - \left(1-\frac{\alpha}{2}\right)\frac{z_1z_2}{2(z_1^2+z_2^2)^2} - 4\alpha z_1 z_2\right]$$

to be solved subject to

$$z_1(0) = \frac{1}{2\sqrt{\alpha}} \qquad \left.\frac{dz_1}{dt^*}\right|_{t^*=0} = \frac{u_x(b, y_1)\,\tau}{D_f}$$

$$z_2(0) = \frac{y_1}{D_f} \qquad \left.\frac{dz_2}{dt^*}\right|_{t^*=0} = 0$$

These two second-order ordinary differential equations must be solved numerically. A convenient way to do so is to convert the two second-order ordinary differential equations into four first-order ordinary differential equations by letting $w_1 = z_1$, $w_2 = z_2$, $w_3 = dz_1/dt^*$, and $w_4 = dz_2/dt^*$. The four new dependent variables, $w_1$, $w_2$, $w_3$, and $w_4$ now satisfy four first-order ordinary differential equations, the first two of which are

$$\frac{dw_1}{dt^*} = w_3$$

$$\frac{dw_2}{dt^*} = w_4$$

and the second two of which are obtained from the original equations by noting that $dw_3/dt^* = d^2z_1/dt^{*2}$ and $dw_4/dt^* = d^2z_2/dt^{*2}$. We have solved these four coupled differential equations using a fourth-order Runga–Kutta method.

The solution gives the trajectory of a particle starting at $t^* = 0$ at $z_1 = (2\alpha^{1/2})^{-1}$ and $z_2 = y_1/D_f$. To determine the collection efficiency it is necessary to solve the equations repeatedly for ever-increasing values of $y_1$ to find that value of $y_1$ for which the particle just escapes capture. This entire procedure is then repeated for a series of different Stokes numbers.

Figure 7.22 shows the trajectory of a particle and the streamline on which it starts for St = 1, $\alpha = 0.1$, $D_p/D_f = 0.1$, and $y_1/D_f = 0.2$. The approximate and exact collection efficiencies are compared in Figure 7.23 for $\alpha = 0.1$ and $D_p/D_f = 0.1$. The maximum difference between the two efficiencies is about 75%, occurring in the vicinity of St = 0.1.

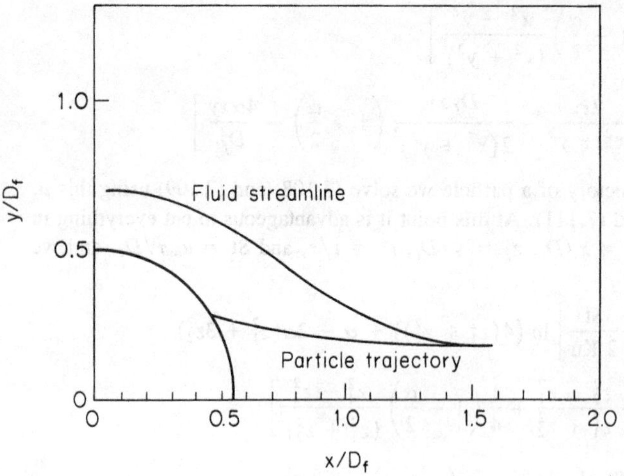


**Figure 7.22** Particle trajectory approaching a cylinder in the Kuwabara flow field for St = 1, $\alpha = 0.1$, $D_p/D_f = 0.1$, and $y_1/D_f = 0.2$.

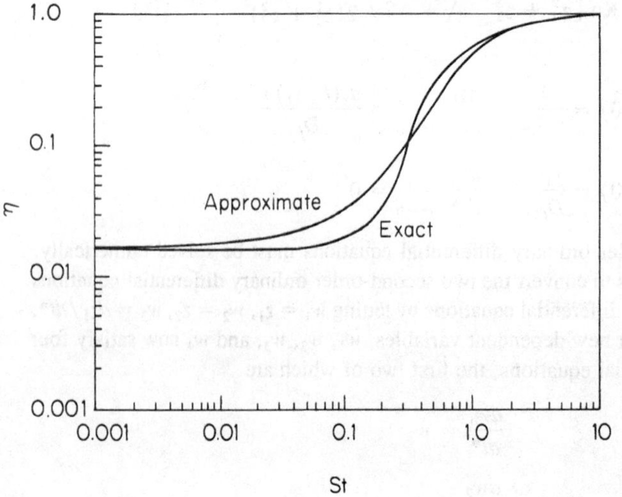


**Figure 7.23** Approximate and exact collection efficiencies for inertial impaction and interception by a cylinder. The Kuwabara flow field is assumed with a filter solid fraction $\alpha = 0.1$. The approximate efficiency is that already given in Figure 7.21; the exact is that determined from numerical solution of the particle trajectories.

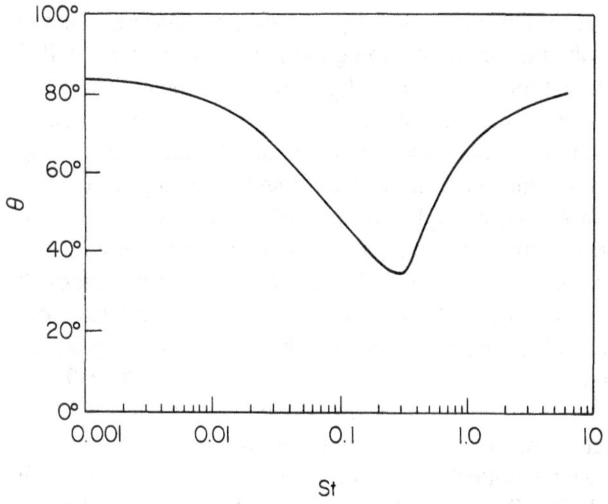


**Figure 7.24** Critical impaction angle as a function of Stokes number for exact solution of particle trajectories using Kuwabara flow field and filter solid fraction $\alpha = 0.1$.

In view of the approximate nature of the Kuwabara velocity field, this discrepancy is probably not large enough to invalidate the approximate result.

In deriving the collection efficiency by the approximate approach, we assumed that the critical angle for capture is $\theta = \pi/2$. We can evaluate this assumption by plotting the angle of impaction obtained from the exact solution as a function of Stokes number (Figure 7.24). The results of the exact solution show that the critical angle for impaction can get as low as 30° instead of the 90° assumed in the approximate solution. We see that $\theta \rightarrow 90°$ as $\text{St} \rightarrow 0$ and also as $\text{St} \rightarrow \infty$. In the case of $\text{St} \rightarrow 0$, the particle merely follows the flow streamlines, and the critical angle for collection will be 90°. On the other hand, in the limit of $\text{St} \rightarrow \infty$, the details of the flow streamlines around the cylinder are irrelevant since the particle proceeds on a straight line to the cylinder. Thus the critical angle for capture will also be 90°, reflecting particles starting out at $y_1 = (D_p + D_f)/2$.

The analysis we have presented strictly applies when the size of the particle is much smaller than the diameter of the collecting cylinder. The Kuwabara velocity field has been modified in the case where the particle and cylinder are the same order in size by Yeh and Liu (1974).

### 7.5.7 Collection Efficiency of a Cylindrical Collector

We can now summarize and evaluate the collection efficiencies by Brownian diffusion, interception, and inertial impaction. The collection efficiency for deposition by Brownian diffusion is given by (7.104). As we noted, by this mechanism the efficiency decreases as $D_p$ increases according to the two-thirds power of $D_p$. Interception efficiency was calculated simply by determining those flow streamlines that fall within a "collision envelope" at a distance of $D_p/2$ from the front half of the cylinder. In doing so, particle diffusion and inertia are neglected; only the velocity field is needed. Such an analysis gives the interception collection efficiency of (7.105). We see that this efficiency increases as particle size increases according to $D_p^2$. Finally, a combined interception and inertial impaction efficiency was obtained by determining the particle trajectory, called the limiting trajectory, that comes within a distance of $D_p/2$ of the cylinder. Since the latter analysis includes both interception and inertial impaction, the collection efficiency will be larger than that predicted on the basis of interception alone. Similarly, by setting $D_p/D_f = 0$, a pure inertial impaction efficiency can be determined. Although there does not exist a closed-form expression for the inertial impaction efficiency, we expect that impaction efficiency should increase with increasing particle size and fluid velocity.

Figure 7.25 shows $\eta$ for combined impaction and interception as a function of St for $\alpha = 0.001$, 0.01, and 0.1, and $D_p/D_f = 0.001$, 0.01, and 0.1 as calculated from the approximate analysis in the preceding section. In studying the effect of varying $D_p/D_f$ we see clearly the influence of interception on the collection efficiency. As St $\rightarrow$ 0, the collection efficiency reflects pure interception only. (Again, think of St $\rightarrow$ 0 as reflecting $\rho_p \rightarrow 0$.) Assuming that the effect of impaction is largely negligible at St = 0.001, the intercepts of the $\eta$ curves show how the collection efficiency by interception varies with both $\alpha$ and $D_p/D_f$. At very small $D_p/D_f$ (e.g., 0.001) we expect very low collection efficiency due to interception. The efficiency from pure interception was given by (7.105), so that $\eta$ increases as $(D_p/D_f)^2$. The intercept values shown for $D_p/D_f = 0.001$ and 0.1 can be confirmed to adhere to (7.105). At large enough Stokes number the efficiency curves for the different values of $D_p/D_f$ converge as impaction replaces interception as the principal mechanism of collection.

It is of interest to compare the three mechanisms of collection: Brownian diffusion, impaction, and interception. The collection efficiency for Brownian diffusion (7.104) depends on the Peclet number. To represent the collection efficiency on a single plot as a function of the Stokes number, we need only specify the approach velocity $u_\infty$. Figure 7.26 shows $\eta$ as a function of St for $\alpha = 0.1$, $D_p/D_f = 0.1$, and $u_\infty = 1.0 \text{ cm s}^{-1}$. The curve for diffusion efficiency shows the expected decrease in efficiency for increasing Stokes number (heavier, less mobile particles).

Figure 7.27 is the more commonly used representation of collection efficiency versus particle diameter. The values of $\alpha$ and $u_\infty$ are the same as in Figure 7.26, but $D_p$ is now allowed to vary, with $D_f$ fixed at 1.0 $\mu$m. Thus, now the ratio $D_p/D_f$ varies along the curve. As $D_p/D_f$ approaches unity, the assumptions made in the impaction/interception theory begin to break down. For example, hydrodynamic interactions between the particle and the cylinder become important. Therefore, the portion of the curve that exceeds $\eta = 1.0$ is not correct. (For example, for $\alpha = 0.1$, $D_f = 1.0$ $\mu$m, $b = 1.58$ $\mu$m, and the particle actually starts at a distance from the cylinder less than its diameter.) The overall collection efficiency versus particle diameter curve shown in Figure 7.27 exhibits a minimum in the efficiency between 0.1 and 1.0 $\mu$m in diameter. In this range the particle is large enough so that its Brownian diffusivity is too small to lead to a substantial collection efficiency by that mechanism, and at the same time, it is too small for its inertia to be large enough so that impaction can be a strong contribution. In fact, by now we are not surprised to see this type of aerosol collection behavior, wherein a minimum in the collection occurs in the range 0.1 to 1.0 $\mu$m.

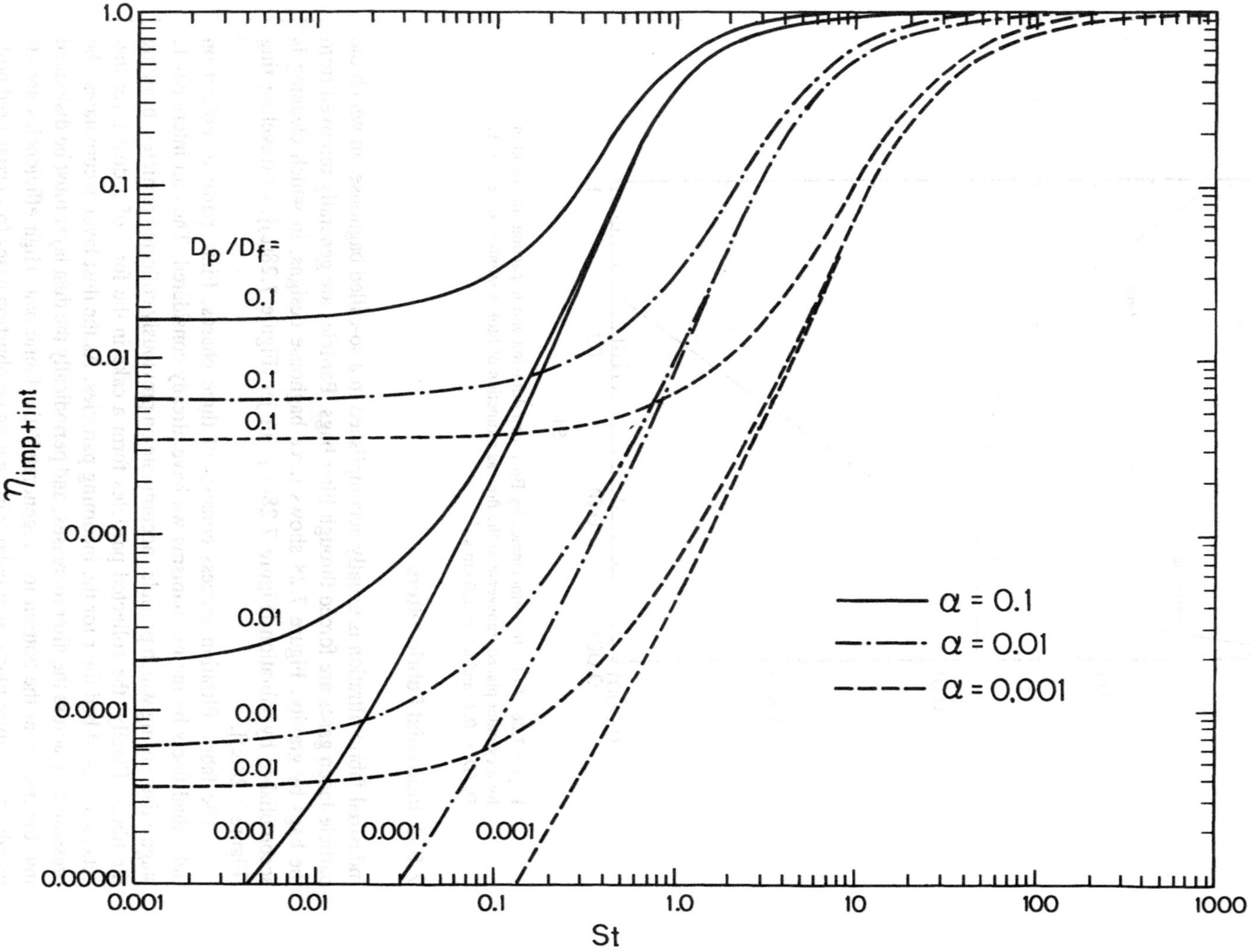


**Figure 7.25** Collection efficiency for combined impaction and interception for a cylinder placed transverse to the flow as a function of Stokes number for $D_p/D_f = 0.001$, 0.01, and 0.1 and $\alpha =$ 0.001, 0.01, and 0.1.

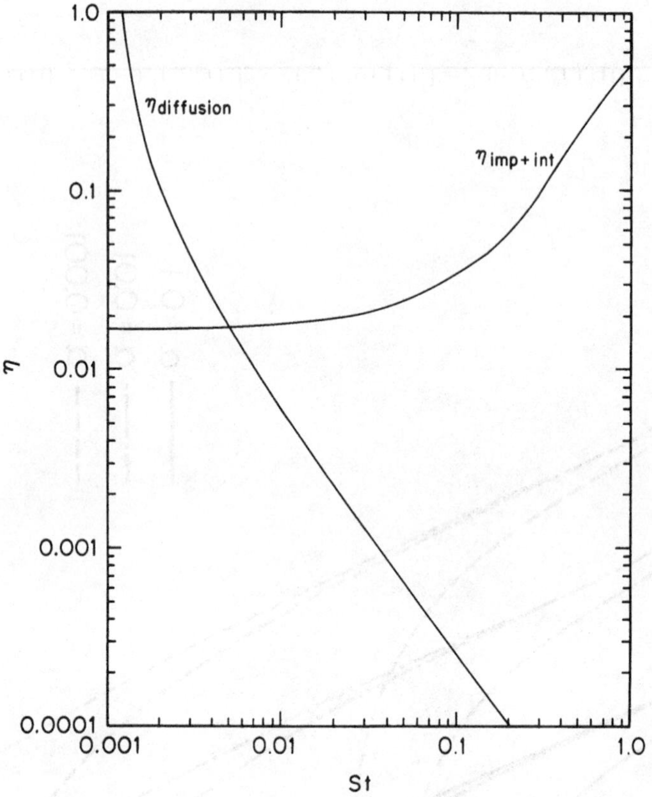


**Figure 7.26** Collection efficiencies by Brownian diffusion and impaction/interception for a cylinder placed transverse to the flow as a function of Stokes number for $\alpha = 0.1$, $D_p/D_f = 0.1$, and $u_\infty = 1.0 \text{ cm s}^{-1}$.

### 7.5.8 Industrial Fabric Filters

Industrial fabric filtration is usually accomplished in a so-called baghouse, in which the particle-laden gases are forced through filter bags. Particles are generally removed from the bags by gravity. Figure 7.28 shows three baghouse designs, in which cleaning is accomplished by vibration [Figure 7.28(a)], air jet [Figure 7.28(b)], or traveling ring [Figure 7.28(c)].

The fabric filtration process consists of three phases. First, particles collect on individual fibers by the mechanisms we have already considered. Then an intermediate stage exists during which particles accumulate on previously collected particles, bridging the fibers. Finally, the collected particles form a cake in the form of a dust layer that acts as a packed bed filter for the incoming particles. As the dust layer accumulates, the pressure drop across the filter increases, and periodically the dust layer must be dislodged into the hopper at the bottom to "regenerate" the fabric bag. High efficiencies are attainable with fabric filters, particularly in treating combustion gases from coal-fired boil-

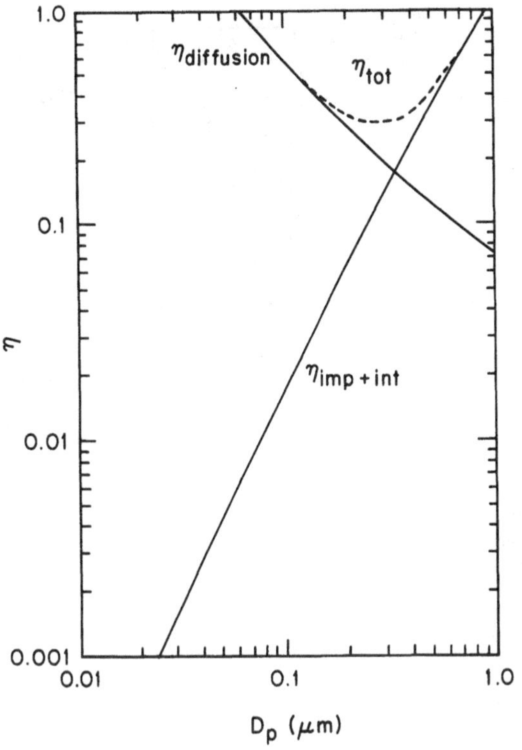


**Figure 7.27** Individual collection efficiencies due to Brownian diffusion and impaction/interception, together with total collection efficiency as a function of particle diameter. The other parameters are $\alpha = 0.1$, $u_\infty = 1.0 \text{ cm s}^{-1}$, and $D_f = 1.0\ \mu\text{m}$.

ers. To the extent that effective operation of an electrostatic precipitator depends on the presence of $SO_2$ in the gas as an ionizable species, fabric filters can operate with no loss of efficiency with low-sulfur fuel.

Fabric filters consist of semipermeable woven or felted materials that constitute a support for the particles to be removed. A brand-new woven filter cloth has fibers roughly 100 to 150 μm in diameter with open spaces between the fibers of 50 to 75 μm. Initially, the collection efficiency of such a cloth is low because most of the particles will pass directly through the fabric. However, deposited particles quickly accumulate, and it is the deposited particle layer that enables the high-efficiency removal once a uniform surface layer has been established. Although fiber mat filters are similar in some respects to fabric filters, they do not depend on the layer of accumulated particles for high efficiency. Fiber mat filters generally are not cleaned but are discarded. They are ordinarily used when particle concentrations are low, so that resonable service life can be achieved before discarding.

Fabric filters offer the following advantages: (1) they can achieve very high collection efficiencies even for very small particles; (2) they can be used for a wide variety of particles; (3) they can operate over a wide range of volumetric flow rates; and (4) they require only moderate pressure drops. The limitations of fabric filters are: (1) operation must be carried out at temperatures lower than that at which the fabric is destroyed, or its life is shortened to an uneconomical degree; (2) gas or particle constituents

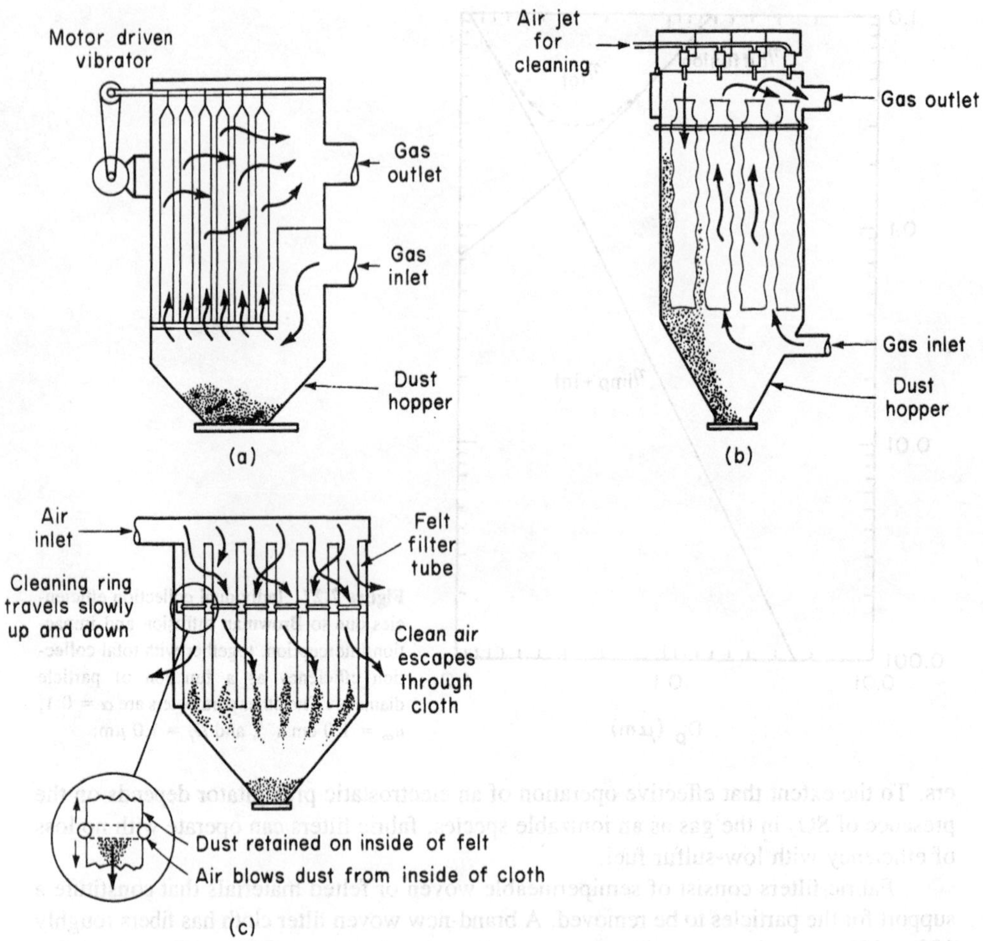


**Figure 7.28** Three designs for a baghouse (a) motor-driven vibrator, (b) air jet, (c) cleaning ring for removing particles from fabric filters.

that attack the fabric or prevent proper cleaning, such as sticky particles difficult to dislodge, are to be avoided; and (3) baghouses require large floor areas. The advantages of fabric filter baghouses clearly outweigh their limitations, as they currently represent close to 50% of the industrial gas-cleaning market.

In a fabric filter the particle layer performs the removal task. As the layer of collected particles grows in thickness, the pressure drop across the particle layer and the underlying fabric increases. The two major considerations in the design of a fabric filter assembly are the collection efficiency and the pressure drop as a function of time of operation (since the last cleaning). Dennis and Klemm (1979) (see also Turner and McKenna, 1984) developed a series of equations for predicting outlet concentration through a fabric filter. The collection efficiency depends on the local gas velocity and the particle loading on the fabric. Empirical correlations for the pressure drop through a combined fabric-dust layer are available in Turner and McKenna (1984) and Cooper and Alley (1986).

### 7.5.9 Filtration of Particles by Granular Beds

An alternative to filtration in fibrous beds is the use of granular beds. The granular bed can be a fixed (packed), fluidized, or moving assemblage of inert particles. In the analysis of a granular bed filter, the bed is usually assumed to consist of an array of spherical elements through which the particle-laden gas flows. As before, the essential component of determining overall collection efficiency is the efficiency for particle capture by a single filter element, in this case a sphere. And, as before, collection occurs by the mechanisms of inertial impaction, interception, and diffusion. Gravity may also be important. A comprehensive experimental study of packed-bed filtration was reported by Gebhart et al. (1973), and their data were subsequently correlated by Balasubramanian and Meisen (1975). Given the single-sphere collection efficiency $\eta$, the overall collection efficiency of a granular bed of length $L$ can be derived as follows.

Let $D_s$ be the uniform diameter of each sphere comprising the bed. The collection efficiency of a single sphere is defined as the ratio of the number of particles collected per unit time to that in the projected upstream area, $\pi D_s^2/4$, of the sphere. As in the case of the fibrous bed, the interstitial gas velocity in the bed, $u_\infty$, is greater than that approaching the filter, $\bar{u}$, due to the volume of the flow excluded by the spheres. The volumetric flow rate of air through the filter is $Q = \bar{u}A_c$, so, as before, $Q = u_\infty A_c(1 - \alpha)$, so $u_\infty = \bar{u}/(1 - \alpha)$. If the number of spheres per unit volume of the bed is $N_s$, the solid fraction $\alpha$ of the bed is $\alpha = (\pi/6)D_s^3 N_s$.

We now perform the customary balance on the number concentration of particles over a differential element of bed depth $dx$. The flows into and out of the element $dx$ are $QN|_x$ and $QN|_{x+dx}$, respectively. The number of particles removed per unit time in the element $dx$ is the product of the flow rate of particles into the element and the fraction that is removed,

$$\left(\frac{\pi}{4}D_s^2\eta N_s\right)(u_\infty N|_x)(A_c\,dx)$$

Thus the balance over $dx$ is

$$A_c\bar{u}\left(N|_x - N|_{x+dx}\right) = \left(\frac{\pi}{4}D_s^2\eta N_s\right)(u_\infty N|_x)(A_c\,dx)$$

Eliminating $N_s$ in terms of $\alpha$, using the relation between $\bar{u}$ and $u_\infty$, and taking the limit of $dx \rightarrow 0$, we obtain

$$\frac{dN}{dx} = -\frac{3}{2}\left(\frac{\alpha}{1-\alpha}\right)\frac{\eta}{D_s}N \tag{7.129}$$

to be solved subject to $N(0) = N_0$. The overall bed efficiency $\eta_b = 1 - N(L)/N_0$, so

$$\eta_b = 1 - \exp\left[-\frac{3}{2}\left(\frac{\alpha}{1-\alpha}\right)\frac{\eta L}{D_s}\right] \tag{7.130}$$

A number of authors have considered the efficiency of collection of particles by spheres (Michael and Norey, 1969; Paretsky et al., 1971; Nielsen and Hill, 1976a, b; Rajagopalan and Tien, 1976; Tardos et al., 1976, 1978; Tardos and Pfeffer, 1980). Tardos and Pfeffer (1980) have derived an expression for the collection efficiency of a single sphere by interception and gravitational effects when $D_p/D_s << 1$,

$$\eta = \left(1 + \frac{D_p}{D_s}\right)^2 \eta_G + \frac{\eta_R}{1 + \text{Gr St}} \tag{7.131}$$

where the efficiency for gravitational collection,

$$\eta_G = \frac{\text{Gr St}}{1 + \text{Gr St}} \tag{7.132}$$

with $\text{Gr} = D_s g/2u_\infty^2$ and $\text{St} = \rho_p u_\infty C_c D_p^2/9\mu D_s$, and where the efficiency for interception is

$$\eta_R = \frac{3}{2}\left(\frac{1.31}{1 - \alpha}\right)^3 \left(\frac{D_p}{D_s}\right)^2 \tag{7.133}$$

Note that the collection efficiency due to gravitational effects is independent of the flow field and is therefore independent of the bed solid fraction $\alpha$. The efficiency expression (7.131) has been shown by Tardos and Pfeffer (1980) to be applicable for values of the Stokes number smaller than about St = 0.05. For larger values of St, a combined inertial, interception, and gravitational efficiency must be computed using the limiting trajectory.

## 7.6 WET COLLECTORS

Wet collectors, or scrubbers, employ water washing to remove particles directly from a gas stream. Scrubbers may be grouped broadly into two main classes: (1) those in which an array of liquid drops (sprays) form the collecting medium, and (2) those in which wetted surfaces of various types constitute the collecting medium. The first class includes spray towers and venturi scrubbers, while the second includes plate and packed towers. In this book we concentrate on the first class of devices.

Scrubbing is a very effective means of removing small particles from a gas. Removal of particles results from collisions between particles and water drops. In the humid environment of a scrubber, small, dry particles also grow in size by condensation of water and thereby become easier to remove. Reentrainment of particles is avoided since the particles become trapped in droplets or in a liquid layer. A scrubber also provides the possibility of simultaneously removing soluble gaseous pollutants. The particle-laden scrubbing liquid must be disposed of—a problem not encountered in dry methods of gas cleaning.

A spray scrubber is a device in which a liquid stream is broken into drops, approximately in the range 0.1 to 1.0 mm in diameter, and introduced into the particle-laden gas stream. The array of moving drops becomes a set of targets for collection of the particles in the gas stream. Collection efficiency is computed by considering the efficiency of a single spherical collector and then summing over the number of drops per unit volume of gas flow. The relative motion between the drops and particles is an important factor in the collection efficiency because capture occurs by impaction and direct interception. (Diffusion is also important for smaller particles.)

There are two general types of spray scrubbers. The first class comprises those having a preformed spray where drops are formed by atomizer nozzles and sprayed into the gas stream. These include:

1. Countercurrent gravity tower, where drops settle vertically against the rising gas stream
2. Cross-current tower, where drops settle through a horizontal gas stream
3. Cocurrent tower, where spray is horizontal into a horizontal gas stream

The second class comprises those in which the liquid is atomized by the gas stream itself. Liquid is introduced more or less in bulk into a high-velocity gas flow that shatters the liquid into drops. Devices in this class are called venturi scrubbers since the high-velocity gas flow is achieved in a venturi (a contraction).

Figure 7.29 illustrates four types of wet collection equipment. The simplest type of wet collector is a spray tower into which water is introduced by means of spray nozzles [Figure 7.29(a)]. Gas flow in a spray chamber is countercurrent to the liquid, the configuration leading to maximum efficiency. Collection efficiency can be improved over the simple spray chamber with the use of a cyclonic spray tower, as shown in Figure 7.29(b). The liquid spray is directed outward from nozzles in a central pipe. An unsprayed section above the nozzles is provided so that the liquid drops with the collected particles will have time to reach the walls of the chamber before exit of the gas. An impingement plate scrubber, as shown in Figure 7.29(c), consists of a tower containing layers of baffled plates with holes (5000 to 50,000 $m^{-2}$) through which the gas must rise and over which the water must fall. Highest collection efficiencies of wet collectors are obtained in a venturi scrubber, shown in Figure 7.29(d), in which water is introduced at right angles to a high-velocity gas flow in a venturi tube, resulting in the formation of very small water droplets by the flow and high relative velocities of water and particles. The high gas velocity is responsible for the breakup of the liquid. Aside from the small droplet size and high impingement velocities, collection is enhanced through particle growth by condensation. Table 7.1 summarizes particle scrubbing devices.

The collection efficiency of wet collectors can be related to the total energy loss in the equipment; the higher the scrubber power, per unit volume of gas treated, the better is the collection efficiency. Almost all the energy is introduced in the gas, and thus the energy loss can be measured by the pressure drop of gas through the unit.

The major advantage of wet collectors is the wide variety of types, allowing the selection of a unit suitable to the particular removal problem. As disadvantages, high-pressure drops (and therefore energy requirements) must be maintained, and the handling and disposal of large volumes of scrubbing liquid must be undertaken.

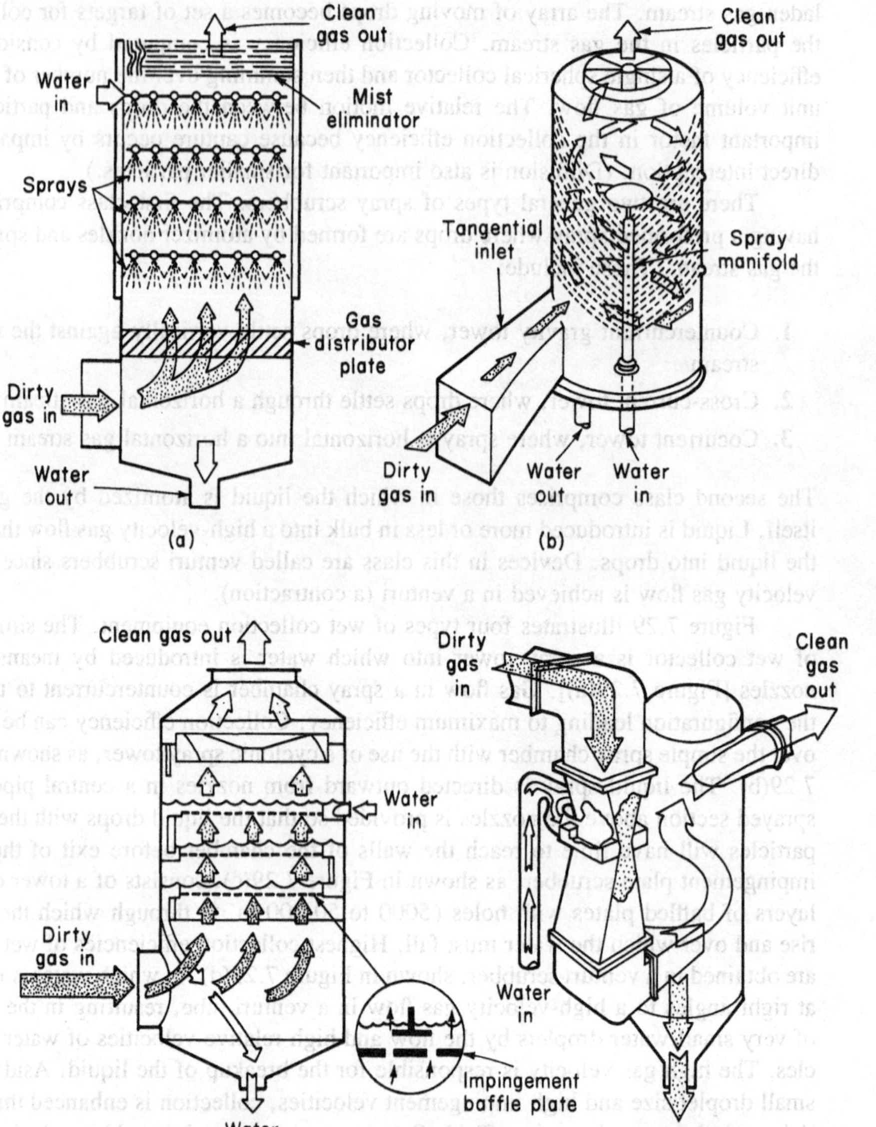


**Figure 7.29** Wet collectors: (a) spray tower; (b) cyclone spray tower; (c) impingement scrubber; (d) venturi scrubber.

**TABLE 7.1 PARTICLE SCRUBBERS**

| Type | Description |
|---|---|
| Plate scrubber | A vertical tower containing one or more horizontal plates (trays). Gas enters the bottom of the tower and must pass through perforations in each plate as it flows countercurrent to the descending water stream. Plate scrubbers are usually named for the type of plates they contain (e.g., sieve plate tower). Collection efficiency increases as the diameter of the perforations decreases. A cut diameter, that collected with 50% efficiency, of about 1 $\mu$m aerodynamic diameter can be achieved with 3.2-mm-diameter holes in a sieve plate. |
| Packed-bed scrubber | Operates similarly to packed-bed gas absorber (see Chapter 8). Collection efficiency increases as packing size decreases. A cut diameter of 1.5 $\mu$m aerodynamic diameter can be attained in columns packed with 2.5-cm elements. |
| Spray scrubber | Particles are collected by liquid drops that have been atomized by spray nozzles. Horizontal and vertical gas flows are used, as well as spray introduced cocurrent, countercurrent, or cross-flow to the gas. Collection efficiency depends on droplet size, gas velocity, liquid/gas ratio, and droplet trajectories. For droplets falling at their terminal velocity, the optimum droplet diameter for fine-particle collection lies in the range 100 to 500 $\mu$m. Gravitational settling scrubbers can achieve cut diameters of about 2.0 $\mu$m. The liquid/gas ratio is in the range 0.001 to 0.01 $m^3$ $m^{-3}$ of gas treated. |
| Venturi scrubber | A moving gas stream is used to atomize liquids into droplets. High gas velocities (60 to 120 m $s^{-1}$) lead to high relative velocities between gas and particles and promote collection. |
| Cyclone scrubber | Drops can be introduced into the gas stream of a cyclone to collect particles. The spray can be directed outward from a central manifold or inward from the collector wall. |
| Baffle scrubber | Changes in gas flow velocity and direction induced by solid surfaces. |
| Impingement-entrainment scrubber | The gas is forced to impinge on a liquid surface to reach a gas exit. Some of the liquid atomizes into drops that are entrained by the gas. The gas exit is designed so as to minimize the loss of entrained droplets. |
| Fluidized-bed scrubber | A zone of fluidized packing is provided where gas and liquid can mix intimately. Gas passes upward through the packing, while liquid is sprayed up from the bottom and/or flows down over the top of the fluidized layer of packing. |

*Source:* Calvert (1984).

### 7.6.1 Spray Chamber

We begin our analysis of spray scrubbing with the conceptually simplest of the devices, a gravity spray chamber. Water droplets are introduced at the top of an empty chamber through atomizing nozzles and fall freely at their terminal settling velocities countercurrently through the rising gas stream. The particle-containing liquid collects in a pool at the bottom and must be pumped out for treatment to remove the solids, and the cleaned liquid is usually recycled to the tower. A schematic of a spray chamber is given in Figure

7.30. We assume that all the falling drops have the same diameter $D_s$. The volumetric flow rate of water fed to the top of the chamber is $W\,(\mathrm{m^3\,s^{-1}})$. If every drop has diameter $D_s$, the number of drops per second fed to the top of the chamber and passing any point in the chamber is $W/[(\pi/6)D_s^3]$. The drop concentration at any point in the chamber is $W/[(\pi/6)D_s^3 A_c v]$, where $A_c$ is the cross-sectional area of the chamber and $v$ is the fall velocity of the drops. We assume that $D_s$ remains constant in the chamber.

If a drop of diameter $D_s$ is falling in still air, its terminal velocity $v_t$ is such that the drag force is just balanced by the gravitational force on the drop. Now in the spray chamber the drop is falling at a fall velocity $v$ relative to a fixed coordinate system in the presence of a rising gas velocity $v_g$. Thus $v$ is not the same as $v_t$, due to the influence of the rising gas. The new fall velocity $v$ is still determined by the equality of the drag and gravity forces. The drag force can be expressed from (5.8) as

$$F_{\text{drag}} = \frac{\pi}{8} C_D \rho_g D_s^2 (v + v_g)^2 \tag{7.134}$$

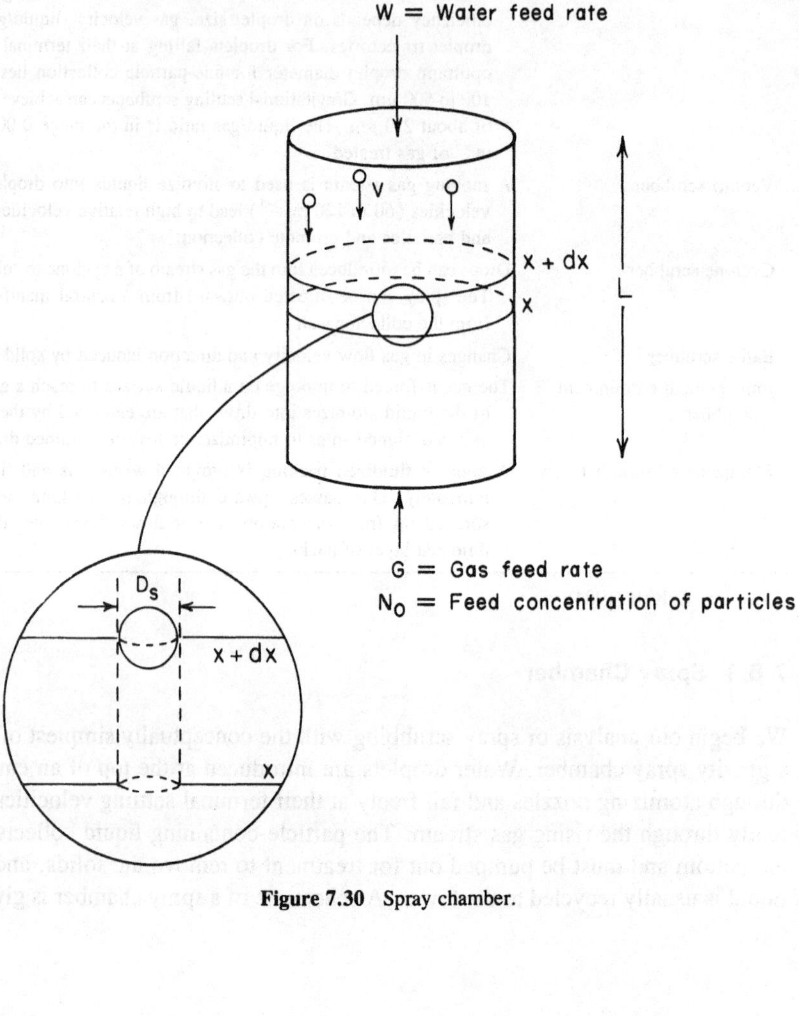


**Figure 7.30** Spray chamber.

where the drag force depends on the *relative* velocity between the drop and the gas, $v + v_g$. By equating this drag force to the gravity force,

$$F_{gravity} = \frac{\pi}{6} D_s^3 (\rho_p - \rho_g) g \tag{7.135}$$

we find that the sum of the new fall velocity and the rising gas velocity equals the terminal velocity in still air,

$$v + v_g = v_t \tag{7.136}$$

Given the drop size, we can compute (or find correlations for) $v_t$, and given the volumetric flow rate of gas through the unit, we can calculate $v_g$. Thus we can compute the water drop fall velocity from (7.136).

We now wish to derive an equation governing the overall collection efficiency of a spray tower. Let $\eta$ be the collection efficiency of particles on an individual droplet, defined as the ratio of the cross-sectional area of the hypothetical tube of gas from which the particles are all removed to the frontal area of the droplet. Consider a differential section of chamber height as shown in Figure 7.30. The number of particles removed per second from the gas stream over $dx$ is just

$$v_g A_c (N|_x - N|_{x+dx})$$

This quantity is equated to the product of the fraction of the volumetric flow of gas through $dx$ from which all particles are removed and the total incoming number of particles per second. The total incoming number of particles per second is $N|_x v_g A_c$. Thus we need to obtain an expression for the fraction of the volumetric flow of gas through $dx$ from which all particles are removed.

The distance $dx$ is fixed as the distance a drop falls in time $dt$ relative to the chamber,

$$dx = v\, dt$$

During the time $dt$ the volume of air that flows through the hypothetical tube having the frontal area of the droplet is

$$\left(\frac{\pi}{4} D_s^2\right) v_t\, dt$$

where $v_t$ is the relative velocity between the droplet and the gas. This quantity can be expressed in terms of $dx$ as

$$\left(\frac{\pi}{4} D_s^2\right) \frac{v_t}{v}\, dx$$

Now the volume from which all particles are removed by the single drop is

$$\eta \left(\frac{\pi}{4} D_s^2\right) \frac{v_t}{v}\, dx$$

Thus the total volume of gas swept clean per second by all the droplets in $dx$ is

$$\eta\left(\frac{\pi}{4}D_s^2\right)\frac{v_t}{v}\,dx\,\frac{W}{(\pi/6)D_s^3}$$

and the fraction of the volumetric flow of gas through $dx$ from which all particles are removed is

$$\frac{[\eta(\pi/4)D_s^2(v_t/v)\,dx]\left\{W/[(\pi/6)D_s^3]\right\}}{v_g A_c}$$

Then the number of particles removed per second from $dx$ is

$$\frac{[\eta(\pi/4)D_s^2(v_t/v)\,dx]\left\{W/[(\pi/6)D_s^3]\right\}}{v_g A_c}N\Big|_x v_g A_c$$

Thus

$$N\Big|_x - N\Big|_{x+dx} = \frac{[\eta(\pi/4)D_s^2(v_t/v)\,dx]\left\{W/[(\pi/6)D_s^3]\right\}}{v_g A_c}N\Big|_x$$

Taking the limit as $dx \to 0$ gives

$$\frac{dN}{dx} = -\left[\frac{3}{2}\eta\left(\frac{v_t}{v}\right)\frac{W}{v_g A_c}\frac{1}{D_s}\right]N \tag{7.137}$$

Integrating (7.137) subject to $N(0) = N_0$ gives

$$N(L) = N_0 \exp\left[-\frac{3}{2}\eta\left(\frac{v_t}{v}\right)\frac{W}{v_g A_c}\frac{L}{D_s}\right] \tag{7.138}$$

The overall spray chamber efficiency is

$$\begin{aligned}\eta_t &= 1 - \frac{N(L)}{N_0}\\ &= 1 - \exp\left[-\frac{3}{2}\eta\left(\frac{v_t}{v}\right)\frac{W}{v_g A_c}\frac{L}{D_s}\right]\end{aligned} \tag{7.139}$$

The quantity $(W/v_g A_c)$ is the ratio of the volumetric flow rate of water to the volumetric flow rate of air. This ratio is of basic importance in the operation of a scrubber. Liquid-to-gas volumetric ratios usually fall in the range 2 to 20 gal of liquid per thousand cubic feet of gas, or $0.27 \times 10^{-3}$ to $2.7 \times 10^{-3}$ m$^3$ of liquid per m$^3$ of gas. This ratio determines the number of drops per unit volume of gas, $6W/\pi D_s^3 G$. For example, the number of droplets per cm$^3$ at the two limits of the flow rate ratio for different droplet diameters is

| | $W/G$ (m$^3$ liquid/m$^3$ gas) | |
|---|---|---|
| $D_s$ (mm) | 0.00027 | 0.0027 |
| 0.1 | 516 | 5159 |
| 1 | 0.516 | 5.16 |

If too small a $D_s$ is attempted at a high $W/G$ ratio, the drop concentration would be so large that collision and coalescence would probably occur, driving the droplet population to larger sizes and lower concentration.

In summary, the overall efficiency of a spray tower increases as the collection efficiency of a single drop increases, as the length of the chamber increases, and as the ratio of the volumetric flow rate of water to that of air increases. It increases as the diameter of the drops decreases.

### 7.6.2 Deposition of Particles on a Spherical Collector

The collection efficiency of a sphere is equal to the ratio of the total number of collisions per second occurring between particles and the spherical collector to the total number of particles per second flowing into the tube having the cross-sectional area of the sphere. We can follow exactly the same approach as we did in determining the collection efficiency of a cylinder by Brownian motion, impaction, and interception; only here we need the flow field around a sphere. However, the current problem is somewhat more complicated than just being the spherical analog of the cylindrical collector. The collecting spheres are falling water drops, which may develop internal circulations that influence the flow field of the gas in their vicinity. Also, drops of sufficiently large size may no longer be spherical as they fall, although we will not include this aspect in our analysis. An alternative to the approaches in Section 7.5 is to rely on dimensional analysis to suggest the dimensionless variables on which the collection efficiency should depend. To formulate a correlation for $\eta$ based on dimensional analysis, we must identify the dimensionless groups that arise in the dimensionless equations of motion of a particle. We are interested specifically in the case of falling water droplets. Allowing for the possibility of internal circulations in the drop that may affect the flow field around it, we find that $\eta$ depends on eight variables: $D_p$, $D_s$, $v$, $v_g$, $\mu_w$, $\mu_{air}$, $D$, $\rho_{air}$. These eight variables have three dimensions. By the Buckingham pi theorem, there are eight minus three, or five, independent dimensionless groups. The actual groups can be obtained by nondimensionalizing the equations of motion for the fluid and the particles. The five dimensionless groups are:

$$\text{Re} = \frac{D_s v_t \rho_{air}}{\mu_{air}} \qquad \text{Reynolds number of sphere}$$

$$\text{Sc} = \frac{\mu_{air}}{\rho_{air} D} \qquad \text{Schmidt number of particles}$$

$$\text{St} = \frac{C_c \rho_p D_p^2 v_t}{18 \mu D_s} \qquad \text{Stokes number of particles}$$

$$\kappa = \frac{D_p}{D_s} \qquad \text{Ratio of diameters of particle and drop}$$

$$\omega = \frac{\mu_w}{\mu_{\text{air}}} \qquad \text{Viscosity ratio of water to air}$$

Note that the Reynolds and Stokes numbers of the falling drop are based on the relative velocity to the air, which is just its terminal settling velocity, $v_t$.

Slinn (1983) has presented the following general equation for the collection efficiency of a sphere

$$\eta = \frac{8}{\text{Re Sc}}\left[1 + \frac{0.4}{\sqrt{2}} \text{Re}^{1/2}\,\text{Sc}^{1/3} + \frac{0.16}{\sqrt{2}} \text{Re}^{1/2}\,\text{Sc}^{1/2}\right]$$
$$+ 4\kappa\left[\omega^{-1} + (1 + \sqrt{2}\,\text{Re}^{1/2})\,\kappa\right] + \left[\frac{2\,\text{St} - S_*}{2\,\text{St} - S_* + \frac{2}{3}}\right]^{3/2} \tag{7.140}$$

$$S_* = \frac{1.2 + \frac{1}{12} \ln (1 + \text{Re}/2)}{1 + \ln (1 + \text{Re}/2)}$$

The first term in (7.140) is the contribution from Brownian diffusion, the second is that due to interception, and the third accounts for impaction. In (7.140) it is assumed that both the collector drop and the collected particles have unit density. For particles of density different from 1.0 g $\text{cm}^{-3}$, the last term in (7.140) should be multiplied by $(\rho_w/\rho_p)^{1/2}$.

Figure 7.31 shows the single-sphere collection efficiency $\eta$ as a function of $D_p$ for $D_s = 0.5$, 1.0, 2.0, and 4.0 mm as predicted by (7.140). At a fixed value of $D_s$, at the lower end of the size spectrum $\eta$ decreases with increasing $D_p$ due to the decreased importance of Brownian diffusion. At the large particle end of the spectrum $\eta$ increases as $D_p$ increases due to the predominant role of inertial impaction and interception. A minimum in the collection efficiency is seen to exist between 0.5 and 1.0 $\mu$m diameter. At a fixed value of $D_p$, $\eta$ decreases as $D_s$ increases due to the decreased importance of interception. The empirical nature of (7.140) is evident in the rather abrupt increase in the efficiency at about $D_p = 4\ \mu$m due to the impaction contribution. This abrupt change is the result of attempting to fit two different physical phenomena into a single empirical equation. Figure 7.32 shows $\eta$ as a function of $D_s$ for $D_p$ ranging from 1 to 7 $\mu$m. The increase in $\eta$ for $D_p$ larger than 4 $\mu$m is due to the impaction contributions as predicted by (7.140).

Equation (7.140) is quite general in that it accounts for all three collection mechanisms. In many scrubber applications inertial impaction is the predominant removal mechanism, especially for particles larger than 1 $\mu$m in diameter. In that case Calvert (1984) has suggested an alternative to (7.140) for the collection efficiency due to im-

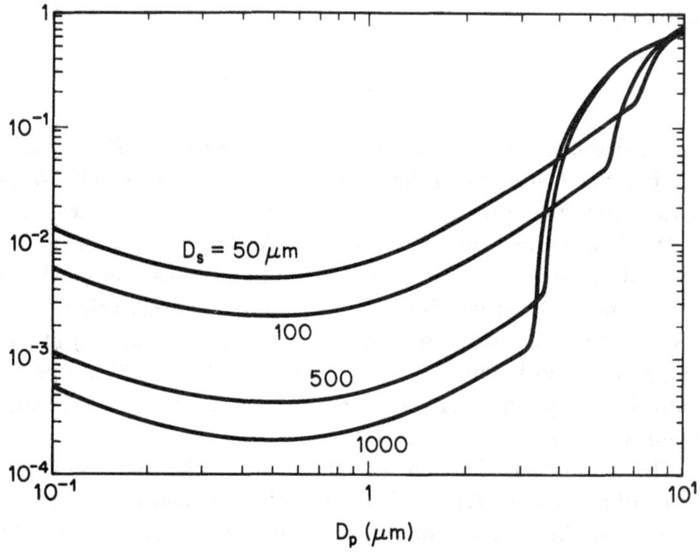


**Figure 7.31** Collection efficiency (7.140) for a single sphere as a function of collected particle diameter $D_p$ at collector water droplet diameter $D_s$ = 50, 100, 500, and 1000 μm. Conditions are for water droplets falling in still air at 298 K, 1 atm, collecting particles of $\rho_p$ = 1 g cm$^{-3}$.

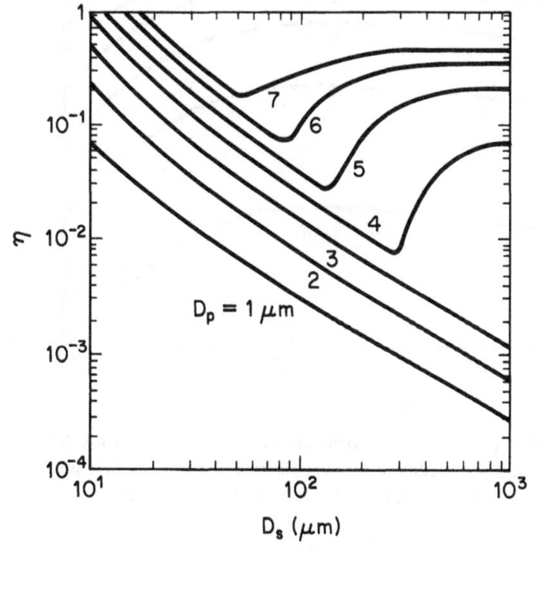


**Figure 7.32** Collection efficiency (7.140) for a single sphere as a function of collector water droplet diameter $D_s$ at collected particle diameters $D_p$ = 1, 2, 3, 4, 5, 6, 7 μm. Conditions are for water droplets falling in still air at 298 K, 1 atm, $\rho_p$ = 1 g cm$^{-3}$.

paction only,

$$\eta = \left(\frac{\text{St}}{\text{St} + 0.35}\right)^2 \tag{7.141}$$

Figure 7.33 shows $\eta$ from (7.141) as a function of $D_s$ for $D_p = 1$ to 7 $\mu$m for $\rho_p = 1$ g cm$^{-3}$. For small droplets in the Stokes law regime, $v_t \sim D_s^2$, so as $D_s$ increases, St increases proportional to $D_s$. Thus, as $D_s$ increases, $\eta$ increases. At intermediate sizes $v_t \sim D_s$, St is constant, and $\eta$ is constant. For large sizes $v_t$ increases less rapidly than $D_s$, so St decreases as $D_s$ increases, leading to a decrease in $\eta$. Thus there is a value of droplet diameter $D_s$ for which $\eta$ is a maximum. The peak value of $\eta$ occurs at about $D_s$ = 600 $\mu$m regardless of particle diameter $D_p$. The value of $\eta$ at its peak is larger for larger particles and is rather flat, extending for 200 or 300 $\mu$m on either side of $D_s \simeq$ 600 $\mu$m. By comparing (7.141) to (7.140), we see that (7.141) should be valid for $D_p \geq 6$ $\mu$m when 100 $\mu$m $\leq D_s \leq$ 1000 $\mu$m.

The total effect of $D_s$ on the overall spray chamber efficiency is a result of the variation of $v_t$ and $\eta$ in (7.139). Since $\eta$ is relatively constant with $D_s$ in the range around $D_s$ = 600 $\mu$m, the net effect on the factor $\eta v_t/(v_t - v_g) D_s$ is to make $\eta_t$ a maximum at the low end of the $D_s$ range, around 300 to 400 $\mu$m.

**Example 7.9 *Overall Efficiency of a Spray Chamber***

We desire to calculate the overall efficiency of a spray scrubber as a function of water droplet diameter, particle diameter, and ratio of water to gas volumetric flow rates. Assume

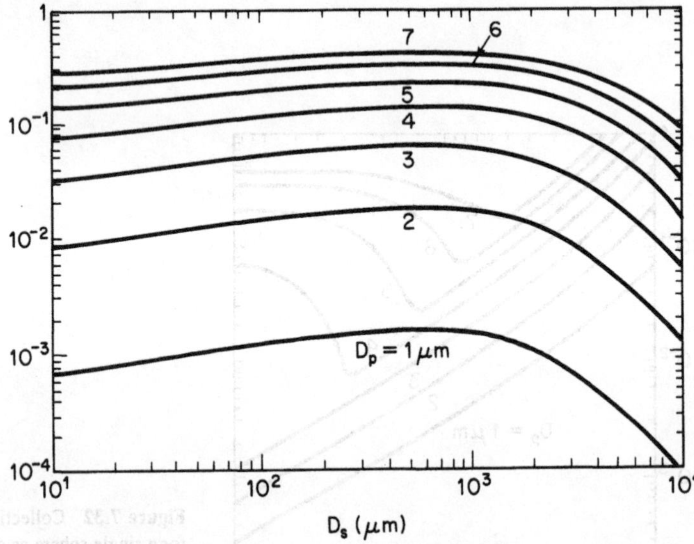


**Figure 7.33** Collection efficiency (7.141) for a single sphere as a function of collector water droplet diameter $D_s$ at collected particle diameters $D_p$ = 1, 2, 3, 4, 5, 6, 7 $\mu$m. Conditions are for water droplets falling in still air at 298 K, 1 atm, $\rho_p$ = 1 g cm$^{-3}$.

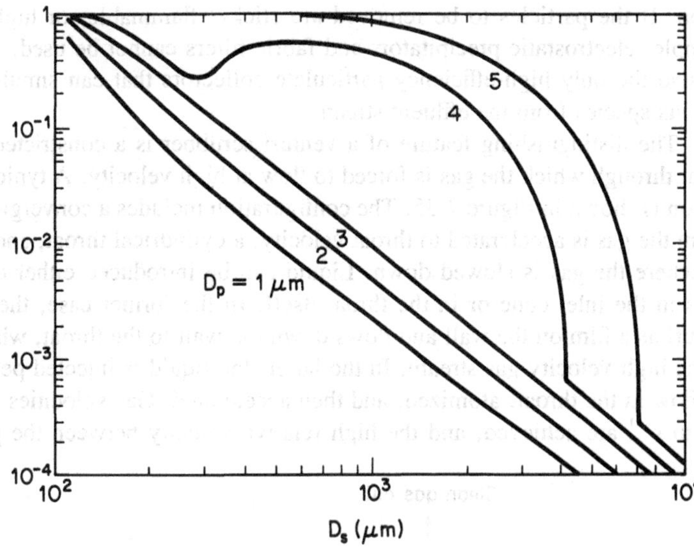


**Figure 7.34** Overall collection efficiency of a spray chamber as a function of droplet diameter for the conditions of Example 7.10.

that the collected particles have a density of 1 g $cm^{-3}$. The chamber has a diameter of 1 m, is 5 m high, and operates at 298 K and 1 atm. Plot the overall collection efficiency as a function of droplet diameter over the range $D_s$ = 50 $\mu$m (0.05 mm) to $10^4$ $\mu$m (10 mm) for particle sizes from $D_p$ = 1 to 5 $\mu$m and for water and gas flow rates of $W = 0.001$ $m^3$ $s^{-1}$ and $G = 1$ $m^3$ $s^{-1}$.

The following empirical equation can be used for the terminal velocity of water droplets for $D_s \geq 50$ $\mu$m,

$$v_t = 958\left\{1 - \exp\left[-\left(\frac{D_s}{0.171}\right)^{1.147}\right]\right\}$$

where $v_t$ is in cm $s^{-1}$ and $D_s$ is in cm. Figure 7.34 shows the overall efficiency $\eta_t$ for the spray chamber as a function of $D_s$ and $D_p$ ranging from 1 to 5 $\mu$m. The individual sphere collection efficiency is that predicted by (7.140). Although droplet diameters exceeding 1 mm (1000 $\mu$m) are unlikely, we have calculated $\eta_t$ for $D_s$ values up to 10 mm (1 cm) to show that a maximum in efficiency is achieved for a particular range of $D_s$ values when impaction is the controlling collection mechanism. The explanation for that maximum is that at a fixed $W/G$, larger droplets imply fewer droplets and thus a decreased target area for particle collection. As the droplets get very small, on the other hand, the Stokes number decreases and the individual sphere impaction contribution decreases.

### 7.6.3 Venturi Scrubbers

Venturi scrubbers are employed when high collection efficiencies are required and when most of the particles are smaller than 2 $\mu$m in diameter. There are a number of instances, in fact, where a venturi scrubber is the only practical device for a gas-cleaning application. If the particles to be removed are sticky, flammable, or highly corrosive, for example, electrostatic precipitators and fabric filters cannot be used. Venturi scrubbers are also the only high-efficiency particulate collectors that can simultaneously remove gaseous species from the effluent stream.

The distinguishing feature of a venturi scrubber is a constricted cross section or throat through which the gas is forced to flow at high velocity. A typical venturi configuration is shown in Figure 7.35. The configuration includes a converging conical section where the gas is accelerated to throat velocity, a cylindrical throat, and a conical expander where the gas is slowed down. Liquid can be introduced either through tangential holes in the inlet cone or in the throat itself. In the former case, the liquid enters the venturi as a film on the wall and flows down the wall to the throat, where it is atomized by the high-velocity gas stream. In the latter, the liquid is injected perpendicular to the gas flow in the throat, atomized, and then accelerated. Gas velocities in the range 60 to 120 m $s^{-1}$ are achieved, and the high relative velocity between the particle-laden gas

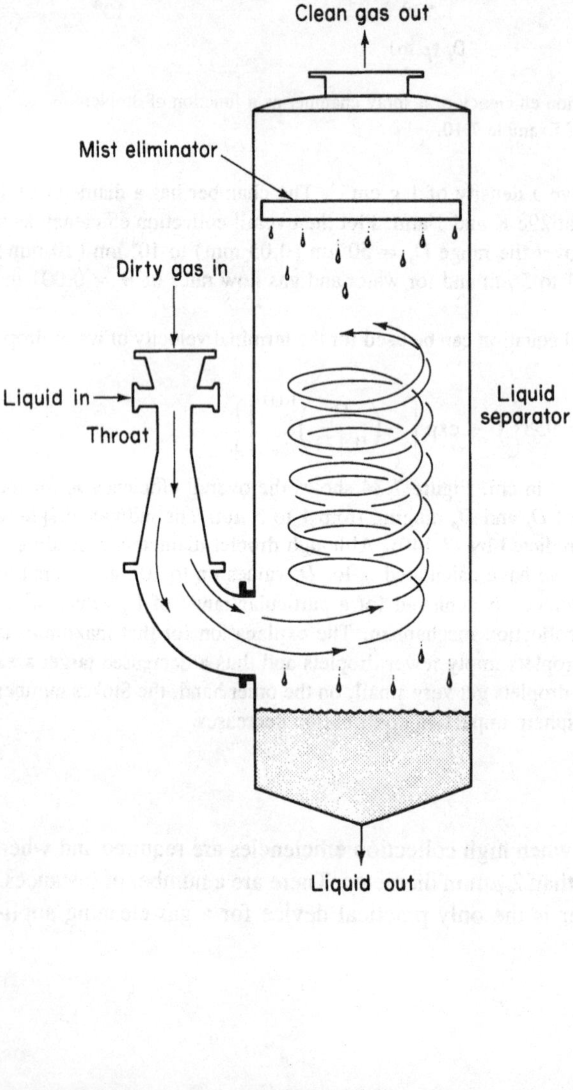


**Figure 7.35** Venturi scrubber.

flow and the droplets promotes collection. The collection process is essentially complete by the end of the throat. Because they operate at much higher velocities than electrostatic precipitators or baghouses, venturi scrubbers are physically smaller and can be economically made of corrosion-resistant materials. Venturis have the simplest configuration of the scrubbers and are the smallest in size.

A typical range of liquid to gas flow rate ratios for a venturi scrubber is 0.001 to 0.003 $m^3$ liquid per $m^3$ gas. At the higher liquid/gas ratios, the gas velocity at a given pressure drop is reduced, and at lower ratios, the velocity is increased. For gas flow rates exceeding about 1000 $m^3 \ min^{-1}$ venturi scrubbers are generally constructed in a rectangular configuration in order to maintain an equal distribution of liquid over the throat area.

In essence, venturi scrubbers are cocurrent flow devices for which the incremental collection along the axis of the venturi can be described by

$$\frac{dN}{dx} = -\left[\frac{3}{2}\eta\left(\frac{W}{G}\right)\frac{v_g - v_d}{v_d}\frac{1}{D_s}\right]N \qquad (7.142)$$

where $v_g$ and $v_d$ are the gas (particle) and droplet velocities, respectively. One may integrate (7.142) together with an inertial impaction expression for $\eta$ to obtain the overall efficiency of a venturi scrubber as (Calvert, 1984)

$$\eta_t = 1 - \exp\left[\frac{1}{55}\left(\frac{W}{G}\right)\frac{v_g \rho_l D_s}{\mu_g} F(K_p f)\right] \qquad (7.143)$$

where

$$F(K_p f) = \frac{1}{K_p}\left[-0.7 - K_p f + 1.4 \ln\left(\frac{K_p f + 0.7}{0.7}\right) + \frac{0.49}{0.7 + K_p f}\right]$$

and $K_p = 2$ St. $f$ is an empirical parameter that accounts for collection by means other than impaction, such as particle growth due to condensation. It has been found that the performance of a variety of large-scale venturi and other gas-atomized spray scrubbers can be correlated with (7.143) using $f = 0.5$. For hydrophobic particles in smaller units $f = 0.25$.

## 7.7 SUMMARY OF PARTICULATE EMISSION CONTROL TECHNIQUES

Table 7.2 presents a summary of particulate emission control techniques, including minimum particle sizes, ranges of efficiency, and advantages and disadvantages of each type of unit. In selecting a method to meet a particular gas cleaning need, the most important consideration is the total cost (operating and equipment) of the method. The advantages and disadvantages listed in Table 7.2 give an indication of the considerations that enter into a determination of the cost of a particular device. Figure 7.36 shows typical collection efficiency curves for the devices considered in this chapter.

**TABLE 7.2 SUMMARY OF PARTICULATE EMISSION CONTROL TECHNIQUES**

| Device | Minimum particle size ($\mu m$) | Efficiency (%) (mass basis) | Advantages | Disadvantages |
|---|---|---|---|---|
| Gravitational settler | >50 | <50 | Low-pressure loss<br>Simplicity of design and maintenance | Much space required<br>Low collection efficiency |
| Cyclone | 5–25 | 50–90 | Simplicity of design and maintenance<br>Little floor space required<br>Dry continuous disposal of collected dusts<br>Low-to-moderate pressure loss<br>Handles large particles<br>Handles high dust loadings<br>Temperature independent | Much head room required<br>Low collection efficiency of small particles<br>Sensitive to variable dust loadings and flow rates |
| Wet collectors | | | Simultaneous gas absorption and particle removal<br>Ability to cool and clean high-temperature, moisture-laden gases<br>Corrosive gases and mists can be recovered and neutralized<br>Reduced dust explosion risk<br>Efficiency can be varied | Corrosion, erosion problems<br>Added cost of wastewater treatment and reclamation<br>Low efficiency on submicron particles<br>Contamination of effluent stream by liquid entrainment<br>Freezing problems in cold weather<br>Reduction in buoyancy and plume rise |
| Spray towers | >10 | <80 | | |
| Cyclonic | >2.5 | <80 | | |
| Impingement | >2.5 | <80 | | |
| Venturi | >0.5 | <99 | | |

| | | | | |
|---|---|---|---|---|
| | | | | Water vapor contributes to visible plume under some atmospheric conditions |
| Electrostatic precipitator | <1 | 95–99 | 99+% efficiency obtainable<br>Very small particles can be collected<br>Particles may be collected wet or dry<br>Pressure drops and power requirements are small compared with other high-efficiency collectors<br>Maintenance is nominal unless corrosive or adhesive materials are handled<br>Few moving parts<br>Can be operated at high temperatures (573 to 723 K) | Relatively high initial cost<br>Precipitators are sensitive to variable dust loadings or flow rates<br>Resistivity causes some material to be economically uncollectable<br>Precautions are required to safeguard personnel from high voltage<br>Collection efficiencies can deteriorate gradually and imperceptibly |
| Fabric filtration | <1 | >99 | Dry collection possible<br>Decrease of performance is noticeable<br>Collection of small particles possible<br>High efficiencies possible | Sensitivity to filtering velocity<br>High-temperature gases must be cooled<br>Affected by relative humidity (condensation)<br>Susceptibility of fabric to chemical attack |

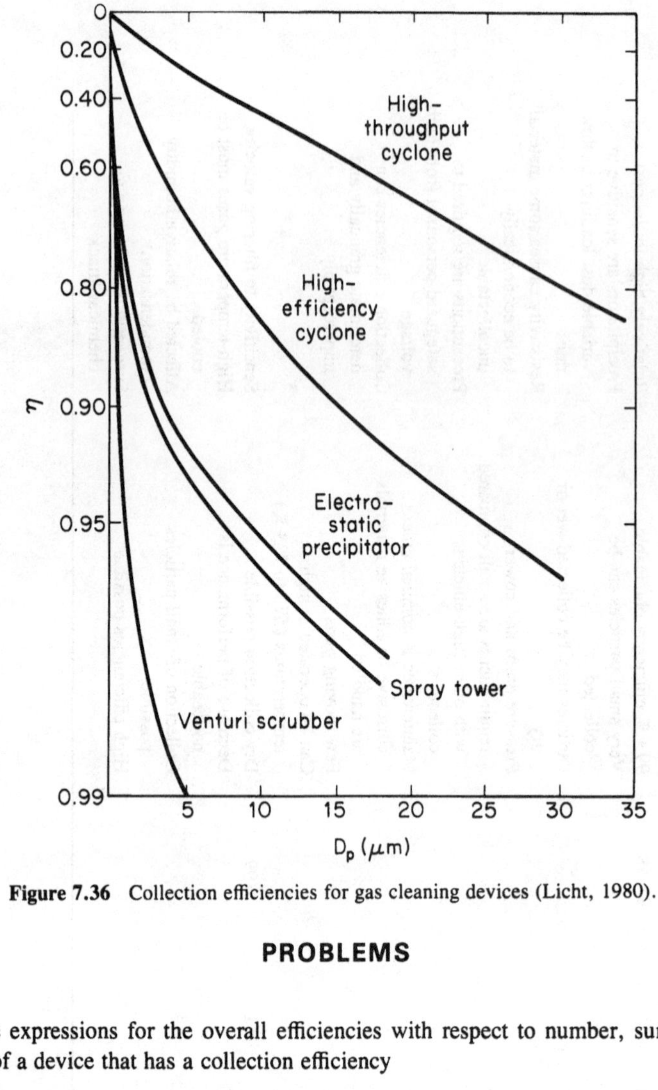


**Figure 7.36** Collection efficiencies for gas cleaning devices (Licht, 1980).

## PROBLEMS

**7.1.** Derive expressions for the overall efficiencies with respect to number, surface area, and mass of a device that has a collection efficiency

$$\eta(D_p) = \begin{cases} 1 & D_p > D_{p0} \\ 0 & D_p \leq D_{p0} \end{cases}$$

with a log-normal size distribution entering the device. *Note:* You may find the following integral useful:

$$\int_{L_1}^{L_2} e^{ru} \exp\left[-\frac{(u-\bar{u})^2}{2\sigma_u^2}\right] du = (\pi/2)^{1/2}\, \sigma_u e^{r\bar{u}} e^{r^2\sigma_u^2/2} \left\{ \operatorname{erf}\left[\frac{L_2 - (\bar{u} + r\sigma_u^2)}{\sqrt{2}\sigma_u}\right] - \operatorname{erf}\left[\frac{L_1 - (\bar{u} + r\sigma_u^2)}{\sqrt{2}\sigma_u}\right]\right\}$$

**7.2.** A particulate collector has a collection efficiency

$$\eta(D_p) = \begin{cases} 1 - D_p^{-1} & D_p > 1 \ \mu\text{m} \\ 0 & D_p \leq 1 \ \mu\text{m} \end{cases}$$

An aerosol with a log-normal size distribution with $\overline{D}_{pg} = 2 \ \mu$m and $\sigma_g = 1.35$ is passed through the device. Compute the overall efficiency of the device with respect to particle number, surface area, and mass. *Note:* The formula given in Problem 7.1 is needed.

**7.3.** Several particulate collection devices are often operated in series, with each succeeding device used to collect smaller and smaller particles. Consider $n$ particulate removal devices connected in series, such that the outlet stream from unit 1 is the inlet stream to unit 2, and so on. If the efficiencies of the $n$ devices are $\eta_1(D_p), \eta_2(D_p), \ldots, \eta_n(D_p)$, show that the total efficiency of the $n$ units is

$$E_n = \int_0^\infty \Big\{\eta_1(D_p) + \eta_2(D_p)[1 - \eta_1(D_p)] + \cdots + \eta_n(D_p)[1 - \eta_{n-1}(D_p)] \times [1 - \eta_{n-2}(D_p)] \cdots [1 - \eta_1(D_p)]\Big\} n(D_p)\, dD_p$$

**7.4.** Design a plate-type settling chamber for a stream of 100 m$^3$ s$^{-1}$ of air at 298 K and 1 atm to collect particles of density 2.0 g cm$^{-3}$. The chamber must not exceed 5 m in width or 6 m in height and must collect particles of 50 $\mu$m with 99.5% efficiency. Determine the length of the chamber required if 100 trays are used. For simplicity, assume that Stokes' law may be used to calculate the settling velocity of 50-$\mu$m particles.

**7.5.** Air at 298 K and 1 atm laden with acid fog is led from a process to a square horizontal settling chamber 8 m long and 50 cm high. The fog can be considered to consist of spherical droplets of diameter 0.8 mm and density 1 g cm$^{-3}$. It is desired to remove 90% of the fog from the stream. Find the volumetric flow rate, in cubic meters per hour, which will allow 90% removal.

**7.6.** Settling chambers are commonly used in a sinter plant to remove large particles of quartz and iron oxide from effluent gas streams. A settling chamber 3 m high and wide and 6 m long is available. The volumetric flow rate of air through the chamber is 5000 m$^3$ h$^{-1}$. The densities of quartz and iron oxide particles are 2.6 and 4.5, respectively. Compute and plot efficiency curves for this unit at the given gas flow rate at 298 K for both types of particles as a function of particle diameter over the diameter range 1 to 60 $\mu$m.

**7.7.** Consider a settling chamber that is so well mixed internally that at any instant all uncollected particles are uniformly mixed throughout the entire volume of the chamber. Show that the collection efficiency for such a device is given by

$$\eta(D_p) = \frac{v_t WL/Q}{1 + v_t WL/Q}$$

**7.8.** Consider a cyclone flow having an inlet velocity of 10 m s$^{-1}$, an angle of turn of $15\pi$, and inner and outer radii of 5.0 and 10.0 cm, respectively. Assume that the particles have density $\rho_p = 1.5$ g cm$^{-3}$. Plot the collection efficiency as a function of particle size over the range 0.5 $\mu$m $\leq D_p \leq$ 10 $\mu$m for $T = 323$ K and $W = 0.4$ m.

**(a)** assuming laminar flow conditions.

**(b)** assuming turbulent flow conditions.

**(c)** using (7.42).

Discuss your results.

**7.9.** Design a cyclone to remove 99% of particles of 20 $\mu$m diameter and density 1.5 g cm$^{-3}$ from a stream of 20 m$^3$ s$^{-1}$ of air at 298 K and 1 atm. Determine suitable values for the major dimensions of the unit. Plot the collection efficiency as a function of particle size over the range 1 $\mu$m $\le D_p \le$ 25 $\mu$m.

**7.10.** Determine the collection efficiency versus $D_p$ curve over the range 0.1 $\mu$m $\le D_p \le$ 10 $\mu$m for the following single-stage electrostatic precipitator:

$$Q = 0.1 \text{ m}^3 \text{ s}^{-1} \qquad T = 300 \text{ K}$$
$$\rho_p = 1.5 \text{ g cm}^{-3} \qquad \kappa = 4$$
$$r_c = 0.25 \text{ m} \qquad \mu = 1.8 \times 10^{-4} \text{ g cm}^{-1} \text{ s}^{-1}$$
$$r_0 = 0.002 \text{ m}$$
$$f = 0.6$$
$$L = 30 \text{ m}$$

**7.11.** A cylindrical single-stage electrostatic precipitator for gas sampling is to provide 95% efficiency for a flow rate of 0.01 m$^3$ s$^{-1}$ for particles of 3 $\mu$m diameter, $\rho_p$ = 1.5 g cm$^{-3}$, $\kappa$ = 4, and inlet concentrations of 5 $\times$ 10$^4$ $\mu$g m$^{-3}$. Use a velocity of 100 cm s$^{-1}$ in a single tube with $r_0 = 0.1 r_c$ and $f$ = 0.65. Determine the required length of the tube.

**7.12.** Consider a filter of packing density 0.04 consisting of fibers of diameter 7 $\mu$m in an airstream with an approach velocity of 0.55 m s$^{-1}$ at 298 K and 1 atm, containing particles of $\rho_p$ = 1.5 g cm$^{-3}$. Compute the collection efficiency of the fiber as a function of $D_p$ from 0.05 to 1.0 $\mu$m from diffusion, interception, and impaction.

**7.13.** A filter bed of packing density 0.1 and fiber diameter 4 $\mu$m for use in removing radioactive particles from a gas stream must provide an overall collection efficiency of at least 99.99% for particles of any size. Given a flow rate of air at 298 K, 1 atm of 10 m$^3$ s$^{-1}$, $\rho_p$ = 1.2 g cm$^{-3}$, and filter width and height of 1.5 and 1.0 m, respectively, determine the necessary depth of the filter.

**7.14.** Compute the collection efficiency of a cigarette filter which is a fiber layer of thickness 1 cm and void fraction 0.5. Assume that the smoke particles are a monodisperse aerosol of diameter 0.2 $\mu$m and density 1 g cm$^{-3}$ and that the fiber filaments have a diameter of 50 $\mu$m. Smoke is inhaled at a velocity of 3 cm s$^{-1}$ and at 298 K and 1 atm.

**7.15.** Determine the overall efficiency of a spray chamber as a function of particle size for particles of density 1.5 g cm$^{-3}$ over the range 0.1 $\mu$m $\le D_p \le$ 10 $\mu$m. The cylindrical chamber has a diameter of 1 m, is 5 m high, and operates with air and water flow rates of 1 m$^3$ s$^{-1}$ and 0.01 m$^3$ s$^{-1}$, respectively. The air is at 298 K, 1 atm, and the water drop diameter is 1 mm, for which the terminal velocity in still air is 4 m s$^{-1}$.

**7.16.** Derive an expression for the overall efficiency of a settling chamber in which a horizontal flow of gas is contacted by a vertical settling spray.

**7.17.** With increasingly stringent fuel economy standards, diesel engines appear to be attractive alternative power plants for passenger vehicles. However, despite its superior fuel economy, a diesel engine presents a challenging emission control problem. As the engine is operated in a manner to maintain low $NO_x$ emissions, particulate emissions exceed the exhaust emission standard. Thus an additional scheme for controlling particulate emissions is necessary. One such scheme that has received extensive study is the filtration of exhaust particulate matter by a filter bed placed in the exhaust system. In this problem we consider the design of a fibrous filter bed for this purpose (Oh et al., 1981). (Disposal of the collected particulate matter is a key problem, although we do not consider it here.) We will focus on the initial performance of the filter, that is, on the period during which the deposition of particles is not influenced by those already collected.

The following conditions can be assumed for the exhaust filtration problem:

$$\alpha = 0.05 \qquad T = 473 \text{ K}$$

$$D_f = 10\ \mu\text{m} \qquad u_\infty = 8 \text{ cm s}^{-1}$$

$$\rho_p = 1 \text{ g cm}^{-3}$$

**(a)** Calculate the Stokes number for particles ranging in size from 0.02 to 1.0 $\mu$m in diameter. Show that the Stokes number is sufficiently small that inertial impaction can be neglected as a significant mechanism of collection.

**(b)** Plot the single fiber collection efficiency as a function of particle diameter over the range 0.02 to 1.0 $\mu$m showing the individual contributions of diffusion and interception. Assume the Kuwabara flow field to be applicable.

**(c)** For a given set of operating conditions there are four filter design parameters that can be varied: filter face area, filter thickness, fiber size, and packing density. These design parameters can be divided into two groups; the first two parameters are related to the size and shape of the filter, and the last two refer to the properties of the filter medium. Consider the operating conditions corresponding to an Oldsmobile 5.7-liter diesel engine automobile being driven at 40 mph (64 km h$^{-1}$):

$$\text{Exhaust flow rate} = 2832 \text{ l min}^{-1} \text{ (at 293 K, 1 atm)}$$

$$\text{Exhaust temperature} = 473 \text{ K}$$

$$\text{Mass median particle diameter} = 0.2\ \mu\text{m}$$

$$\text{Particle density} = 1 \text{ g cm}^{-3}$$

With filter thickness as the ordinate (in cm) and filter face area as the abscissa (in cm$^2$), assuming that $\alpha = 0.05$ and $D_f = 10\ \mu$m, plot the curve of overall bed efficiency = 90% based on particle mass. (If we require at least 90% efficiency, all points above this line are candidates for a filter design.)

**(d)** Now assuming a filter face area of 2500 cm$^2$ and a thickness of 3.4 cm, with the operating conditions above, plot the overall bed efficiency as a function of packing density $\alpha$ over the range $0 < \alpha < 0.1$ for $D_f$ = 5, 10, 20, and 40 $\mu$m. Discuss your results. (Note that an important design consideration that we have not included here is the pressure drop across the filter bed. Just as it is desired to maintain the efficiency higher than a certain level, it is sought to keep the pressure drop across the bed below a certain level.)

**7.18.** In most real fibrous filtration problems there is not only a distribution of particles by size to be filtered but also a distribution of sizes of the filter elements. In this problem we wish to extend the treatment given in the text on the collection efficiency of a filter bed to include distributions in size of both the particles and the cylindrical filter elements.

**(a)** Assuming that the particle volume distribution is log-normally distributed by particle diameter

$$n_v(D_p) = \frac{V_t}{\sqrt{2\pi}\log\sigma_g D_p}\exp\left[-\frac{(\log D_p - \log \bar{D}_{pg})^2}{2\log^2\sigma_g}\right]$$

where $V_t$ is the total particle volume concentration; and assuming that the fiber size distribution is normally distributed, such that $n_f(D_p)dD_p$ is the fraction of fibers having diameters in the range $[D_p, D_p + dD_p]$,

$$n_f(D_p) = \frac{1}{\sqrt{2\pi}\sigma_f} \exp\left[\frac{(D_f - \overline{D}_f)^2}{2\sigma_f^2}\right]$$

derive an equation for the overall filter efficiency in terms of particle mass. (Since fibers cannot have negative diameters, in using the normal distribution for $n_f$ one needs to assume that the distribution is rather sharply peaked about $\overline{D}_f$.)

**(b)** Let us apply the result of part (a) to the filtration of diesel exhaust particles (Oh et al., 1981). The diesel exhaust particles are characterized by $\rho_p = 1$ g cm$^{-3}$, $V_t = 50{,}700$ $\mu$m$^3$ cm$^{-3}$, $\overline{D}_{pg} = 0.17$ $\mu$m, and $\sigma_g = 1.74$. The filter medium consists of medium-fine grade commercial steel wool for which $\overline{D}_f = 22.5$ $\mu$m and $\sigma_f = 7.5$ $\mu$m. Assuming a cylindrical bed with a face diameter of 20 cm, a depth of 5.1 cm, and a packing density of 0.03, calculate the mass removal efficiency of the bed when operated at 473 K and an exhaust flow rate of 300 l min$^{-1}$ (at 293 K, 1 atm).

## REFERENCES


ADACHI, M., KOUSAKA, Y., and OKUYAMA, K. "Unipolar and Bipolar Diffusion Charging of Ultrafine Aerosol Particles," *J. Aerosol Sci.*, *16*, 109–123 (1985).

ADAMCZYK, Z., and VAN DE VEN, T. G. M. "Deposition of Brownian Particles onto Cylindrical Collectors," *J. Colloid Interface Sci.*, *84*, 497–518 (1981).

BALASUBRAMANIAN, M., and MEISEN, A. "A Note on the Diffusional Deposition of Aerosol Particles in Packed Beds," *J. Aerosol Sci.*, *6*, 461–463 (1975).

BHATIA, M. V., and CHEREMISINOFF, P. N. "Cyclones," in *Air Pollution Control and Design Handbook: Part 1*, P. N. Cheremisinoff and R. A. Young, Eds., Marcel Dekker, New York, 281–316 (1977).

BICKELHAUPT, R. E. "A Technique for Predicting Fly Ash Resistivity," U.S. Environmental Protection Agency Report No. EPA-600/7-79-204, Research Triangle Park, NC (1979).

BROCK, J. R., and WU, M. "Field Charging of Aerosol Particles," *J. Colloid Interface Sci.*, *45*, 106–114 (1973).

CALVERT, S. "Particle Control by Scrubbing" in *Handbook of Air Pollution Technology*, S. Calvert and H. M. Englund, Eds., Wiley, New York, 215–248 (1984).

COOPER, C. D., and ALLEY, F. C. *Air Pollution Control: A Design Approach*, PWS Engineering, Boston (1986).

CRAWFORD, M. *Air Pollution Control Theory*, McGraw-Hill, New York (1976).

DAVISON, S. W., and GENTRY, J. W. "Differences in Diffusion Charging of Dielectric and Conducting Ultrafine Aerosols," *Aerosol Sci. Technol.*, *4*, 157–163 (1985).

DENNIS, R., and KLEMM, H. A. "Fabric Filter Model Format Change," Vol. 1, "Detailed Technical Report," U.S. Environmental Protection Agency Report No. EPA-600/7-79-043a, Research Triangle Park, NC (1979).

FRIEDLANDER, S. K. *Smoke, Dust and Haze: Fundamentals of Aerosol Behavior*, Wiley, New York (1977).

GEBHART, J., ROTH, C., and STAHLHOFEN, W. "Filtration Properties of Glass Bead Media for Aerosol Particles in the 0.1–2 $\mu$m Size Range," *J. Aerosol Sci.*, *4*, 355–371 (1973).

HAN, H., and ZIEGLER, E. N. "The Performance of High Efficiency Electrostatic Precipitators," *Environ. Prog.*, *3*, 201–206 (1984).

HAPPEL, J. "Viscous Flow Relative to Arrays of Cylinders," *AIChE J.*, *5*, 174–177 (1959).

HAPPEL, J., and BRENNER, H. *Low Reynolds Number Hydrodynamics*, Prentice-Hall, Englewood Cliffs, NJ (1965).

KUWABARA, S. "The Forces Experienced by Randomly Distributed Parallel Circular Cylinders or Spheres in a Viscous Flow at Small Reynolds Numbers," *J. Phys. Soc. Jpn.*, *14*, 527–532 (1959).

LASSEN, L. "Die Anlagerung von Zerfallsprodukten der naturlichen Emanationen an elektrisch geladene Aerosole (Schwebstoffe), *Z. Phys.*, *163*, 363–376 (1961).

LEITH, D., and LICHT, W. "The Collection Efficiency of Cyclone Type Particle Collectors—A New Theoretical Approach," in *AIChE Symposium Series*, No. 126, Vol. 68, 196–206 (1972).

LICHT, W. *Air Pollution Control Engineering*, Marcel Dekker, New York (1980).

LICHT, W. "Control of Particles by Mechanical Collectors," in *Handbook of Air Pollution Technology*, S. Calvert and H. M. Englund, Eds., Wiley, New York, 319–329 (1984).

LIU, B. Y. H., and KAPADIA, A. "Combined Field and Diffusion Charging of Aerosol Particles in the Continuum Regime," *J. Aerosol Sci.*, *9*, 227–242 (1978).

LIU, B. Y. H., and PUI, D. Y. H. "On Unipolar Diffusion Charging of Aerosols in the Continuum Regime," *J. Colloid Interface Sci.*, *58*, 142–149 (1977).

LIU, B. Y. H., and YEH, H. C. "On the Theory of Charging of Aerosol Particles in an Electric Field," *J. Appl. Phys.*, *39*, 1396–1402 (1968).

LIU, B. Y. H., WHITBY, K. T., and YU, H. H. S. "Diffusion Charging of Aerosol Particles at Low Pressures," *J. Appl. Phys.*, *38*, 1592–1597 (1967).

MARLOW, W. H., and BROCK, J. R. "Unipolar Charging of Small Aerosol Particles," *J. Colloid Interface Sci.*, *50*, 32–38 (1975).

MCCABE, W. L., and SMITH, J. C. *Unit Operations of Chemical Engineering*, 3rd ed., McGraw-Hill, New York (1976).

MICHAEL, D. H., and NOREY, P. W. "Particle Collision Efficiencies for a Sphere," *J. Fluid Mech.*, *37*, 565–575 (1969).

NIELSEN, K. A., and HILL, J. C. "Collection of Inertialess Particles on Spheres with Electrical Forces," *Ind. Eng. Chem. Fund.*, *15*, 149–156 (1976a).

NIELSEN, K. A., and HILL, J. C. "Capture of Particles on Spheres by Inertial and Electrical Forces," *Ind. Eng. Chem. Fundam.*, *15*, 157–163 (1976b).

OH, S. H., MACDONALD, J. S., VANEMAN, G. L., and HEGEDUS, L. L. "Mathematical Modeling of Fibrous Filters for Diesel Particulates—Theory and Experiment," SAE Paper No. 810113, Society of Automotive Engineers, Warrendale, PA (1981).

PARETSKY, L., THEODORE, L., PFEFFER, R., and SQUIRES, A. M. "Panel Bed Filters for Simultaneous Removal of Fly Ash and Sulfur Dioxide: II. Filtration of Dilute Aerosols by Sand Beds," *J. Air Pollut. Control Assoc.*, *21*, 204–209 (1971).

PERRY, R. H., and CHILTON, C. H., eds., *Chemical Engineers' Handbook*, 5th ed., McGraw-Hill, New York (1973).

RAJAGOPALAN, R., and TIEN, C. "Trajectory Analysis of Deep Bed Filtration with the Sphere in Cell Porous Media Model," *AIChE J.*, *22*, 523–533 (1976).

ROSENHEAD, L. *Laminar Boundary Layers*, Oxford University Press, Oxford (1963).

SLINN, W. G. N. "Precipitation Scavenging," in *Atmospheric Sciences and Power Production—1979*, Div. of Biomedical and Environmental Research, U.S. Dept. of Energy, Chap. 11 (1983).

SMITH, W. B., and MCDONALD, J. R. "Development of a Theory for the Charging of Particles by Unipolar Ions," *J. Aerosol Sci.*, *7*, 151–166 (1976).

SPIELMAN, L. A. "Particle Capture from Low-Speed Laminar Flows," *Annu. Rev. Fluid Mech.*, *9*, 297–319 (1977).

SPIELMAN, L. A., and GOREN, S. L. "Model for Predicting Pressure Drop and Filtration Efficiency in Fibrous Media," *Environ. Sci. Technol.*, 2, 279–287 (1968).

STRATTON, J. A. *Electromagnetic Theory*, McGraw Hill, New York, 205 (1941).

STRAUSS, W. *Industrial Gas Cleaning*, Pergamon Press, New York (1966).

TARDOS, G., and PFEFFER, R. "Interceptional and Gravitational Deposition of Inertialess Particles on a Single Sphere and in a Granular Bed," *AIChE J.*, *26*, 698–701 (1980).

TARDOS, G. I., GUTFINGER, C., and ABUAF, N. "High Peclet Number Mass Transfer to a Sphere in a Fixed or Fluidized Bed," *AIChE J.*, *22*, 1147–1150 (1976).

TARDOS, G. I., ABUAF, N., and GUTFINGER, C. "Dust Deposition in Granular Bed Filters—Theories and Experiments," *J. Air Pollut. Control. Assoc.*, *18*, 354–363 (1978).

TURNER, J. H., and MCKENNA, J. D. "Control of Particles by Filters," in *Handbook of Air Pollution Technology*, S. Calvert and H. M. Englund, Eds., Wiley, New York, 249–281 (1984).

WHITE, H. J. *Industrial Electrostatic Precipitation*, Addison-Wesley, Reading, MA (1963).

WHITE, H. J. "Electrostatic Precipitation of Fly Ash: Parts I, II, III, and IV," *J. Air Pollut. Control Assoc.*, *27*, Nos. 1, 2, 3, 4 (1977).

WHITE, H. J. "Control of Particulates by Electrostatic Precipitation," in *Handbook of Air Pollution Technology*, S. Calvert and H. M. Englund, Eds., Wiley, New York, 283–317 (1984).

WITHERS, R. S., and MELCHER, J. R. "Space-Charge Effects in Aerosol Charging and Migration," *J. Aerosol Sci.*, *12*, 307–331 (1981).

YEH, H. C., and LIU, B. Y. H. "Aerosol Filtration by Fibrous Filter: I. Theoretical," *J. Aerosol Sci.*, *5*, 191–204 (1974).

# 8

# Removal of Gaseous Pollutants from Effluent Streams

There are a variety of approaches to removing gaseous pollutants from effluent streams: absorption, adsorption, condensation, chemical reaction, incineration, and selective diffusion through a membrane.

Absorption is an operation involving mass transfer of a soluble vapor component to a solvent liquid in a device that promotes intimate contact between the gas and the liquid. The driving force for absorption is the difference between the partial pressure of the soluble gas in the gas mixture and its vapor pressure just above the surface of the liquid. It is necessary to employ a liquid solvent within which the gas to be removed is soluble. Water is, by itself, quite efficient for removing soluble acidic gases such as HCl and HF and the soluble basic gas $NH_3$. Gases of more limited solubility, such as $SO_2$, $Cl_2$, and $H_2S$, can be absorbed readily in an alkaline solution such as dilute NaOH. Thus, when water is used as the solvent, it may contain added species, such as acids, alkalines, oxidants, or reducing agents to react with the gas being absorbed and enhance its solubility. Nonaqueous, organic liquids of low volatility can be used for absorption of gases with low water solubility, such as hydrocarbons. Examples of such solvents are dimethylaniline and amines. Organic solvents are often limited to treating particle-free gases to avoid sludge formation. To provide a large liquid surface area for mass transfer, a means of breaking the liquid stream into small droplets or thin films is provided in the gas absorber. The most commonly used devices are columns containing packing or regularly spaced plates, open spray chambers and towers, and combinations of sprayed and packed chambers. Countercurrent contact of liquid and gas is employed to maximize the driving forces.

Adsorption is employed to remove low concentration gases from exhaust streams by causing the gaseous solutes to intimately contact a porous solid to which the solute will adhere. Gas adsorption is used industrially for odor control and for the removal of volatile solvents such as benzene, ethanol, trichloroethylene, and so on, from effluent streams.

Condensation can be used to remove species with relatively low vapor pressure and is carried out in a device that appropriately cools the gas stream and provides a means to remove the layer of condensed liquid. We will not consider condensation in this book.

The fourth and fifth methods of removing gaseous species from an effluent stream listed at the outset were chemical reaction and incineration. Although many of the separation processes involve chemical reaction, such as absorption of an acidic gas in an alkaline solvent, the chemical reaction category refers to those where the key element of the separation is the reaction itself. For example, even though $SO_2$ scrubbing by an aqueous solution containing lime involves a chemical reaction, the absorption is the key removal step. Incineration involves the combustion of the species and is an important process for the treatment of toxic species, where virtually complete removal is necessary. The waste gas is fed to a combustor where the unwanted species are burned at sufficiently high temperature to convert them to harmless products such as $CO_2$ and $H_2O$. Selection of the combustion temperature is determined by the combustion chemistry of the particular substances to be removed.

The final separation method listed was selective diffusion through a membrane. Membrane processes have found application in removing gases such as $CO_2$, $H_2S$, and $H_2$ from natural gas streams (Cooley and Dethloff, 1985). Due to its specialized nature and applications, we do not consider membrane diffusion here.

## 8.1 INTERFACIAL MASS TRANSFER

In the process of gas absorption the gaseous effluent stream containing the pollutant to be removed is brought into contact with a liquid in which it will dissolve. The mechanism by which the species is removed from the gas consists of three steps that occur in series: (1) diffusion of the pollutant molecules through the gas to the surface of the absorbing liquid, (2) dissolution into the liquid at the interface, and (3) diffusion of the dissolved species from the interface into the bulk of the liquid. To predict the extent to which a compound can be removed by gas absorption, we must be able to compute the rates of these three mass transfer processes.

The key process of mass transfer of a dissolving species through a gas to a liquid surface is that of diffusion of component $A$ through a nondissolving background gas $B$. Fick's law in this case, referred to as diffusion of $A$ through stagnant $B$, $N_{B_z} = 0$, becomes

$$N_{A_z} = -cD_{AB}\frac{dx}{dz} + N_{A_z}x \tag{8.1}$$

where $x$ is the mole fraction of $A$. Integrating (8.1), in accordance with the situation in Figure 8.1(a), with $N_{A_z}$ = constant, we obtain the flux of $A$ as

$$N_{A_z} = \frac{-cD_{AB}}{l} \ln \frac{1 - x_0}{1 - x_1} \tag{8.2}$$

We can rewrite (8.2) as

$$N_{A_z} = \frac{-cD_{AB}}{l(x_B)_{lm}} (x_1 - x_0) \tag{8.3}$$

where $(x_B)_{lm}$, the log mean mole fraction of component $B$, is defined as

$$(x_B)_{lm} = \frac{(1 - x)_1 - (1 - x)_0}{\ln [(1 - x)_1/(1 - x)_0]} \tag{8.4}$$

Partial pressures may be used instead of mole fractions, in which case (8.3) becomes

$$N_{A_z} = \frac{-pD_{AB}}{RTl(p_B)_{lm}} (p_{A_1} - p_{A_0}) \tag{8.5}$$

where $p$ is the total pressure.

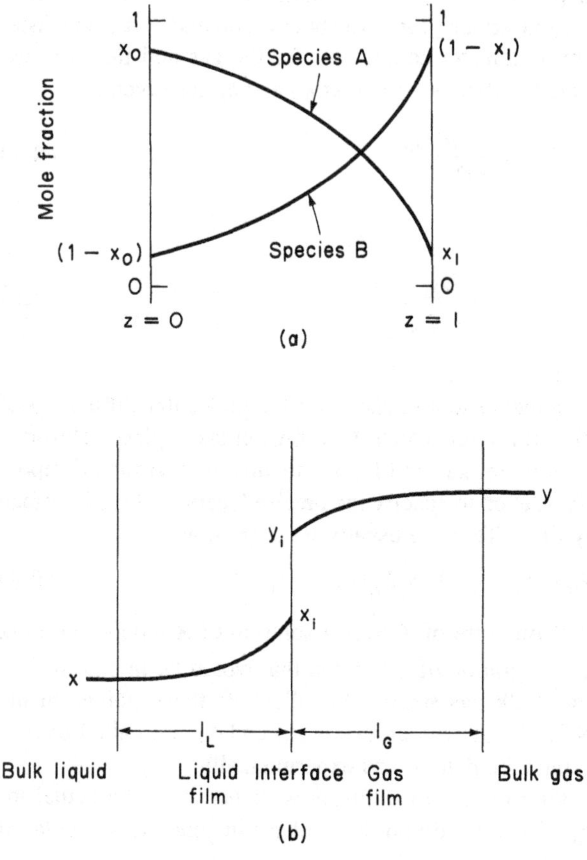


**Figure 8.1** Binary mass transfer: (a) diffusion of $A$ through stagnant $B$; (b) the two-film model of interfacial mass transfer.

In gas absorption, the gas is in turbulent flow and the transport of species occurs across a gas–liquid interface. Most situations of mass transfer in turbulent flow near an interface are too complicated to allow an exact evaluation of profiles and fluxes. Thus certain idealized models are postulated for the mass transfer in such a situation, models that enable the solution for the flux of a species in terms of readily measurable empirical coefficients.

Turbulent motion maintains a fairly uniform composition in the bulk gas. Close to the surface of the liquid, a laminar boundary layer exists in the gas across which species in the bulk gas must diffuse to reach the liquid surface. Similarly, on the liquid side, the bulk liquid is at a uniform composition with a thin layer near the surface of the liquid through which species diffuse from the interface into the bulk liquid. At steady state it can be assumed that the flux of species *A* from the bulk gas to the interface equals the flux of *A* from the interface to the bulk liquid. The simplest model one can envision for this situation is two stagnant layers on either side of the interface, as shown in Figure 8.1(b). Based on the form of (8.5) we assume that the flux of *A* is given by

$$N_{A_z} = k_G(p_A - p_{A_i}) = k_L(c_{A_i} - c_A) \tag{8.6}$$

where $p_A$ and $p_{A_i}$ are the partial pressures of *A* in the bulk gas and at the interface, respectively; $c_A$ and $c_{A_i}$ are the concentrations of *A* in the bulk liquid and at the interface, respectively; and $k_G$ and $k_L$ are mass transfer coefficients for the gas and liquid films, respectively. For the case considered earlier, we see that $k_G$ and $k_L$ are given by

$$k_G = \frac{pD_{AB}}{RTl_G(p_B)_{lm}} \tag{8.7}$$

and

$$k_L = \frac{D_{AB}}{l_L(x_B)_{lm}} \tag{8.8}$$

For dilute mixtures of *A* in *B*, $(x_B)_{lm} \simeq 1.0$.

Thus the mass transfer coefficients should depend on the molecular diffusivity of *A* in *B* and on the thickness of the film over which the diffusion takes place. Unfortunately, in mass transfer between turbulent gas and liquid streams, it is virtually impossible to specify $l_G$ and $l_L$ and, in fact, even to specify the precise location of the interface or the values of $p_{A_i}$ and $c_{A_i}$ at any time. Thus we usually write (8.6) as

$$N_{A_z} = K_G(p_A - p_A^*) = K_L(c_A^* - c_A) \tag{8.9}$$

where $p_A^*$ is the equilibrium partial pressure of *A* over a solution of *A* having the bulk concentration $c_A$, and $c_A^*$ is the concentration of a solution that would be in equilibrium with the partial pressure $p_A$ of the bulk gas stream. We illustrate these points on the equilibrium diagram in Figure 8.2. The new coefficients $K_G$ and $K_L$ are called overall mass transfer coefficients. These must be determined experimentally.

Originally, in (8.6), the driving forces for diffusion were based on the actual interfacial compositions, $p_{A_i}$ and $c_{A_i}$. Since we do not know these in general, we replaced

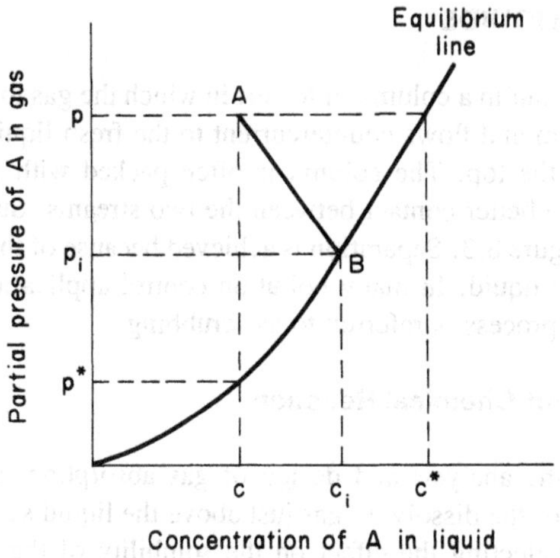


**Figure 8.2** Driving forces in the two-film model of interfacial mass transfer.

(8.6) with (8.9), in which the new overall mass transfer coefficients $K_G$ and $K_L$ were defined. The new driving forces for diffusion, $p_A - p_A^*$ and $c_A^* - c_A$ are shown in Figure 8.2. The point $B$ on the equilibrium curve represents the interfacial composition, which we assume to be $(p_{A_i}, c_{A_i})$. The line $AB$ has a slope $-k_L/k_G$ and is given by

$$\frac{p_A - p_{A_i}}{c_{A_i} - c_A} = \frac{k_L}{k_G} \tag{8.10}$$

When the equilibrium line is given by Henry's law,

$$p_A^* = H_A c_A \tag{8.11}$$

we can explicitly relate $k_G$ and $k_L$ and $K_G$ and $K_L$ through (8.6) and (8.9) by

$$\frac{1}{K_G} = \frac{1}{k_G} + \frac{H_A}{k_L} \tag{8.12}$$

and

$$\frac{1}{K_L} = \frac{1}{k_L} + \frac{1}{H_A k_G} \tag{8.13}$$

If $H_A \ll 1$, species $A$ is very soluble, $K_G \simeq k_G$, and the overall process is controlled by diffusion through the gas film. On the other hand, if $H_A \gg 1$, species $A$ is sparingly soluble and overall rate of mass transfer is liquid-film controlled. Note that we have drawn the equilibrium line in Figure 8.2 as curved, since $H_A$ is not usually constant [and thus (8.12) and (8.13) are not generally valid].

It is also possible to express (8.6) in terms of mole fractions, in which case

$$N_{A_z} = k_y(y - y_i) = k_x(x_i - x) \tag{8.14}$$

where $y$ and $x$ refer to the gas and liquid phases, respectively.

## 8.2 ABSORPTION OF GASES BY LIQUIDS

Gas absorption is usually carried out in a column or tower, in which the gas to be cleaned (the rich gas) enters at the bottom and flows countercurrent to the fresh liquid (the lean liquid), which is introduced at the top. The column is often packed with inert solids (e.g., ceramic beads) to promote better contact between the two streams. Such a tower is shown diagrammatically in Figure 8.3. Separation is achieved because of the solubility of the species in question in the liquid. In many pollution control applications the absorbing liquid is water, and the process is referred to as scrubbing.

### 8.2.1 Gas Absorption without Chemical Reaction

The conventional approach to the analysis and design of gas absorption towers is to assume that the partial pressure of the dissolving gas just above the liquid surface is that calculated from Henry's law neglecting the effect on the solubility of the gas of any further chemical reactions in the liquid. Such an approach can be termed "gas absorption without chemical reaction." The basic gas absorption design problem is the following. Given:

1. A rich gas stream entering at a rate $G$ (mol $h^{-1}$ $m^{-2}$ of empty tower) containing a known mole fraction of component, $A$, $y_0$
2. A desired exit gas mole fraction $y_1$

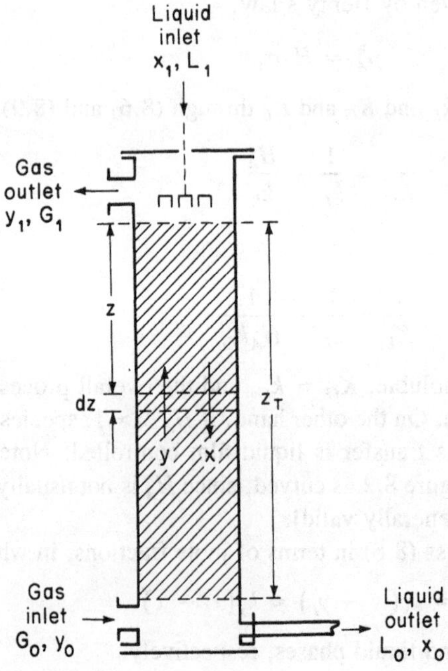


**Figure 8.3** Countercurrent gas absorption tower.

3. A specified mole fraction of $A$ in the inlet liquid, $x_1$
4. The equilibrium curve of $y$ versus $x$ for the system

we wish to compute the height of the tower required to carry out the separation.*

We assume that the gas and liquid phases are immiscible. For example, $SO_2$ is removed from air by absorption in a liquid amine of low vapor pressure. The low vapor pressure of the amine ensures that virtually no amine evaporates into the gas phase, and operation at atmospheric pressure ensures that no air dissolves in the amine. Thus, even though $SO_2$ is transferred between phases, the assumption of immiscibility refers to the fact that the two carrier streams, in this case air and amine, do not dissolve in each other to an appreciable extent.

We note that although the molal gas flow rate $G$ is usually specified, that for the liquid phase is not. A little reflection will show that there is no maximum to the value of the liquid flow rate $L$, but, indeed, there is a minimum value of $L$ below which the required $A$ cannot be separated from the gas. Actually, the total gas and liquid flows are not constant through the tower. The gas flow $G$ consists of $A$ + inert gas, and $L$ consists of $A$ + inert liquid. By our assumption of immiscibility, the flow rates of inert gas and inert liquid always remain constant down the tower. We denote these flow rates by $G'$ and $L'$, respectively. Thus, when we want to select the liquid flow rate, we really want to determine first the minimum $L'$ and then add a comfortable operating margin to that minimum.

In Figure 8.4, point $(x_1, y_1)$ denotes the top of the tower and point $(x_0, y_0)$ the bottom. The driving force for mass transfer is proportional to the line $AB$, as shown in Figure 8.2. Point $A$ must always lie above the equilibrium line; however, as $A$ approaches $B$, the driving force for mass transfer approaches zero. When $A$ actually coincides with $B$ at any point in the tower, mass transfer ceases, because, of course, the

*One would normally also determine the column diameter based on the liquid flow rate and desired pressure drop characteristics. We do not consider this aspect of the design here; rather, we simply assume that the column diameter is constant and known.

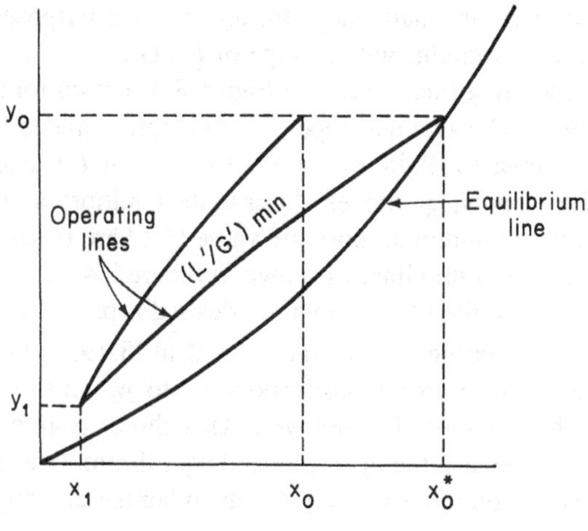


**Figure 8.4** Equilibrium and operating lines for a gas absorption tower.

two phases are in equilibrium at that point. Clearly, if a point is reached in an actual column where $A$ and $B$ coincide, no more mass transfer can take place past that point regardless of the height of the column.

To determine the minimum value of $L'$, we must perform a material balance on species $A$ for the absorption tower. A balance on species $A$ over the whole tower gives

$$L_1x_1 + G_0y_0 = L_0x_0 + G_1y_1 \tag{8.15}$$

where $(G_0, L_0)$ and $(G_1, L_1)$ represent the flows at the bottom and top of the column, respectively. At any point in the tower, where the flow rates are $G$ and $L$, a balance around the top of the column gives

$$L_1x_1 + Gy = Lx + G_1y_1 \tag{8.16}$$

Rearranging (8.16), we have

$$y = \frac{L}{G}x + \frac{1}{G}(G_1y_1 - L_1x_1) \tag{8.17}$$

On a plot of $y$ versus $x$, (8.17) represents a line, not necessarily straight (unless $L$ and $G$ are constant through the whole column), that relates the compositions of passing streams at any point. Such a line is called an *operating line*. The two ends of the column are represented by points $(x_0, y_0)$ and $(x_1, y_1)$.

In order to draw the operating line, we need to know $L$ and $G$ at each point in the column. In the case of gas absorption, in which only component $A$ is transferred between phases, we know that

$$G = \frac{G'}{1-y} \qquad L = \frac{L'}{1-x} \tag{8.18}$$

where $G'$ and $L'$ are constant. Thus (8.16) becomes

$$L'\left(\frac{x_1}{1-x_1} - \frac{x}{1-x}\right) = G'\left(\frac{y_1}{1-y_1} - \frac{y}{1-y}\right) \tag{8.19}$$

If the mole fraction of $A$ in each phase is small, then, for all practical purposes, $G \simeq G'$, $L \simeq L'$, and the operating line is straight, with a slope of $L'/G'$.

We now consider the two operating lines shown in Figure 8.4, drawn for the case in which $L/G$ varies over the tower. The average slope of the operating line is $L/G$, so that as $L$ is decreased, the slope decreases. Point $(x_1, y_1)$ is fixed, so as $L$ is decreased, the upper end of the operating line, that is, $x_0$, moves closer to the equilibrium line. The maximum possible value of $x_0$ and the minimum possible value of $L'$ are reached when the operating line just touches the equilibrium line, as shown in Figure 8.4. At this point, an infinitely long column would be required to achieve the desired separation. We can and the minimum value of $L'/G'$ by setting $y = y_0$ and $x = x_0^*$ in (8.19), where $x_0^*$ is the abscissa of the point on the equilibrium line corresponding to $y_0$. Customarily, a value of $L'/G'$ about 1.5 times the minimum is employed. This choice is an economic one. If $L'/G'$ is large, the distance between the operating and equilibrium lines is large, the driving force is large, and a short column is needed. On the other hand, a high liquid flow rate may be costly. Thus the optimum $L'/G'$ results from a balance between capital equipment costs and operating costs.

Assuming that $L'$ has been specified, we wish to determine the required column height. Let us consider a differential height of the column $dz$. If the interfacial area per unit volume is $a$, a balance on component $A$ in the gas phase over the height $dz$, using (8.14), yields

$$Gy\Big|_{z+dz} - Gy\Big|_{z} = k_y a(y - y_i)\,dz \tag{8.20}$$

which, upon division by $dz$, and letting $dz \to 0$, gives

$$\frac{d(Gy)}{dz} = k_y a(y - y_i) \tag{8.21}$$

where $y_i$ is a point on the equilibrium curve. Using (8.18), we see that

$$d(Gy) = G'd\left(\frac{y}{1-y}\right) = G'\frac{dy}{(1-y)^2} = G\left(\frac{dy}{1-y}\right) \tag{8.22}$$

Integrating (8.21) with the aid of (8.22) produces

$$\int_0^{z_T} dz = \int_{y_1}^{y_0} \frac{G}{k_y a}\frac{dy}{(1-y)(y-y_i)} \tag{8.23}$$

To determine the total height $z_T$, we must evaluate the integral in (8.23). The method of integration depends on the shape of the equilibrium line, the variation in $G$, and the relative importance of the two mass transfer coefficients $k_x a$ and $k_y a$.

From (8.14) we note that

$$\frac{y - y_i}{x_i - x} = \frac{k_x a}{k_y a} \tag{8.24}$$

Thus, at any point, (8.24) describes a straight line with slope $-k_x a/k_y a$, passing through $(x, y)$ and $(x_i, y_i)$. From a knowledge of $k_x a/k_y a$ we can determine $x_i$ and $y_i$ corresponding to any $(x, y)$ on the operating line. Then (8.23) can be integrated. It is common to express (8.23) as

$$z_T = \overline{\left(\frac{G}{k_y a}\right)} \int_{y_1}^{y_0} \frac{dy}{(1-y)(y-y_i)} \tag{8.25}$$

where $\overline{(G/k_y a)}$ is the average value of this group over the column. (Since $G$ decreases from bottom to top, and $k_y a$ also decreases from bottom to top, these changes somewhat compensate each other.) The functional dependence of $k_x a$ and $k_y a$ on the molal flow rates must be determined experimentally.

In deriving an expression for $z_T$ we could have considered a liquid-side balance, in which case the equation corresponding to (8.25) is

$$z_T = \overline{\left(\frac{L}{k_x a}\right)} \int_{x_1}^{x_0} \frac{dx}{(1-x)(x_i-x)} \tag{8.26}$$

Either (8.25) or (8.26) is suitable for carrying out calculations.

The design method embodied in (8.25) and (8.26) is applicable to an equilibrium line of arbitrary shape. A strongly curved equilibrium line is often due to a significant temperature variation over the height of the tower. Appreciable temperature differences result from the heat of solution of a highly concentrated solute in the rich gas. If the rich gas contains a rather dilute concentration of solute, the temperature gradient in the column is small, and the equilibrium line is approximately straight. When the equilibrium line is straight, overall mass transfer coefficients, which are easier to determine experimentally than $k_x a$ and $k_y a$ can be used. The overall coefficients $K_x a$ and $K_y a$ are defined on the basis of the fictitious driving forces $(x^* - x)$ and $(y - y^*)$.

The design equations analogous to (8.25) and (8.26) are, in this case,

$$z_T = \left(\frac{\overline{G}}{k_y a}\right) \int_{y_1}^{y_0} \frac{dy}{(1 - y)(y - y^*)} \tag{8.27}$$

and

$$z_T = \left(\frac{\overline{L}}{K_x a}\right) \int_{x_1}^{x_0} \frac{dx}{(1 - x)(x^* - x)} \tag{8.28}$$

which can be evaluated given the $y^*$ versus $x^*$ equilibrium line.

It has been customary in gas absorption design to express the equations for $z_T$, that is, (8.25)–(8.28), as the product of a number of *transfer units* and the depth of packing required by a single of these units (the height of a transfer unit). Then $z_T$ is written

$$z_T = NH \tag{8.29}$$

where $N$ is the number of transfer units and $H$ is the height of a transfer unit (HTU). For example, using (8.25) and (8.27), we define

$$N_y = \int_{y_1}^{y_0} \frac{dy}{(1 - y)(y - y_i)} \tag{8.30}$$

and

$$N_{0y} = \int_{y_1}^{y_0} \frac{dy}{(1 - y)(y - y^*)} \tag{8.31}$$

where $N_y$ and $N_{0y}$ are based on the individual and overall driving forces, respectively. Of course, $N_y$ and $N_{0y}$ are different, and, in order to produce the same $z_T$ in (8.29), they are compensated for by the corresponding $H$s. Thus

$$H_y = \frac{\overline{G}}{k_y a} \tag{8.32}$$

and

$$H_{0y} = \frac{\overline{G}}{K_y a} \tag{8.33}$$

Similar relations hold for the liquid-side equations. When the equilibrium line is straight, and $G$ and $L$ are constant throughout the tower, $H_{0y}$ (and $H_{0x}$) are constant.

Clearly, the concept of an HTU merely represents a different manner of viewing $z_T$. Its advantage is that the HTU is usually fairly constant for a particular type of tower (usually, with a value in the range 0.1 to 1.5 m), and data are often reported in terms of the HTU. Correlations for HTUs for packed absorption towers are presented by McCabe and Smith (1976).

**Example 8.1** ***Absorption of $SO_2$ from Air by Water***

A packed tower is to be designed for absorption of $SO_2$ from air by contact with fresh water. The entering gas has a mole fraction of $SO_2$ of 0.10, and the exit gas must contain a mole fraction of $SO_2$ no greater than 0.005. The water flow rate used is to be 1.5 times the minimum, and the inlet airflow rate (on an $SO_2$-free basis) is 500 kg m$^{-2}$ h$^{-1}$. The column is to be operated at 1 atm and 303 K. We wish to determine the required depth of the packed section for such a tower.

The following correlations are available for absorption of $SO_2$ at 303 K in towers packed with 1-in. rings (McCabe and Smith, 1976):

$$k_x a = 0.6634\tilde{L}^{0.82} \qquad k_y a = 0.09944\tilde{L}^{0.25}\tilde{G}^{0.7}$$

where $\tilde{L}$ and $\tilde{G}$ are the mass flow rates of liquid and gas, respectively, in kg m$^{-2}$ h$^{-1}$, and $k_x a$ and $k_y a$ are in kg-mol m$^{-3}$ h$^{-1}$ mole fraction$^{-1}$.

Equilibrium data for $SO_2$ in air and water at this temperature are available:

| $p_{SO_2}$ (mmHg) | $c$($gSO_2$/100 g $H_2O$) |
|---|---|
| 0.6 | 0.02 |
| 1.7 | 0.05 |
| 4.7 | 0.10 |
| 8.1 | 0.15 |
| 11.8 | 0.20 |
| 19.7 | 0.30 |
| 36.0 | 0.50 |
| 52.0 | 0.70 |
| 79.0 | 1.00 |

From these data we can calculate the equilibrium curve:

$$y = \frac{p_{SO_2}}{760} \qquad x = \frac{c/64}{c/64 + 100/18}$$

The equilibrium curve is shown in Figure 8.5.

The first step in the solution is calculation of the minimum water flow rate. Using (8.19) with $y_0 = 0.10$, $x_1 = 0$, $y_1 = 0.005$, and $x_0^* = 0.0027$, we obtain $L'_{\min} = 667$ kg-mol m$^{-2}$ h$^{-1}$. Thus the actual water rate to be used is $667 \times 1.5 = 1000$ kg-mol m$^{-2}$ h$^{-1}$.

The equation for the operating line is

$$\frac{x}{1-x} = 0.0172\frac{y}{1-y} - 0.000086$$

This line is shown in Figure 8.5.

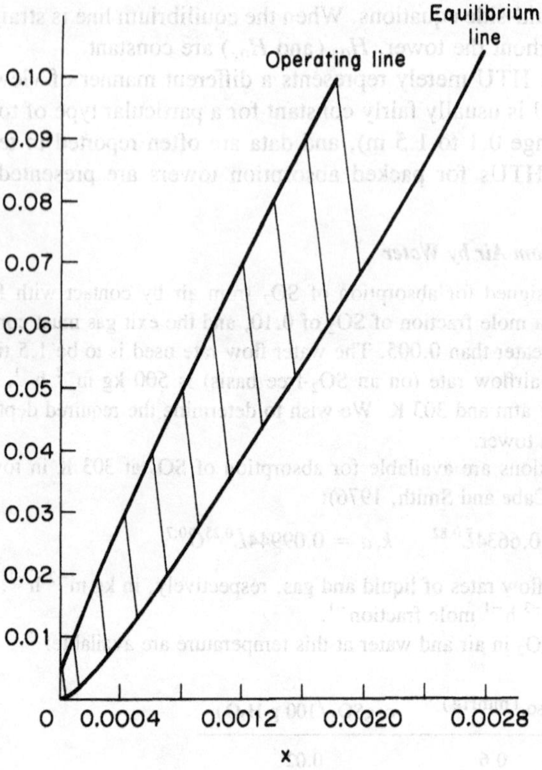


**Figure 8.5** Equilibrium and operating lines for $SO_2$ absorption in water.

The $SO_2$ enters at a rate of 122 kg $m^{-2}$ $h^{-1}$ and leaves at a rate of 5.5 kg $m^{-2}$ $h^{-1}$. The total exit gas rate is 505.5 kg $m^{-2}$ $h^{-1}$. The freshwater feed at the top is 18,000 kg $m^{-2}$ $h^{-1}$, and the rich liquor leaving at the bottom is 18,116.5 kg $m^{-2}$ $h^{-1}$.

The liquid-side mass transfer coefficient will not change appreciably from the top to the bottom since $L$ is nearly constant. We can calculate $k_x a$ from the average mass velocity of 18,058 kg $m^{-2}$ $h^{-1}$:

$$k_x a = 2052.6$$

Because of the change of the total gas velocity from the top to the bottom, $k_y a$ will change somewhat over the tower. The values at the top and bottom are

$$(k_y a)_0 = 104.17$$

$$(k_y a)_1 = 89.48$$

We shall use the average value of 96.82.

Therefore, from any point $(x, y)$ on the operating line, we can determine $x_i$, $y_i$ by drawing a straight line with slope $-2052.6/96.82 = -21.2$. The integral in (8.25) can be evaluated graphically. Table 8.1 shows the calculation of the quantity $1/(1 - y)(y - y_i)$ and the graphical integration (8.25). The value of the integral in (8.25) is found to be 5.72.

**TABLE 8.1** EVALUATION OF INTEGRAND IN (8.25)

| $y$ | $1-y$ | $y_i$ | $y-y_i$ | $(1-y)(y-y_i)$ | $\frac{1}{(1-y)(y-y_i)}$ | $\Delta I$ | $\Delta I \Delta Y$ |
|---|---|---|---|---|---|---|---|
| 0.005 | 0.995 | 0.0005 | 0.0045 | 0.0048 | 223 | | |
| 0.01 | 0.99 | 0.002 | 0.0080 | 0.00792 | 126.5 | 164 | 0.82 |
| 0.02 | 0.98 | 0.0075 | 0.0125 | 0.01225 | 81.7 | 102 | 1.02 |
| 0.03 | 0.97 | 0.014 | 0.0160 | 0.01552 | 64.5 | 72 | 0.72 |
| 0.04 | 0.96 | 0.0215 | 0.0185 | 0.01775 | 56.4 | 60 | 0.60 |
| 0.05 | 0.95 | 0.0285 | 0.0215 | 0.0204 | 49 | 52.5 | 0.525 |
| 0.06 | 0.94 | 0.036 | 0.0240 | 0.0226 | 44.2 | 46.5 | 0.465 |
| 0.07 | 0.93 | 0.044 | 0.0260 | 0.0242 | 41.4 | 42.8 | 0.428 |
| 0.08 | 0.92 | 0.0520 | 0.0280 | 0.0258 | 38.8 | 40 | 0.400 |
| 0.09 | 0.91 | 0.0605 | 0.0295 | 0.0268 | 37.3 | 38 | 0.380 |
| 0.10 | 0.90 | 0.0685 | 0.0315 | 0.0283 | 35.3 | 36 | 0.360 |
| | | | | | | | 5.718 |

Finally, we evaluate the quantity $k_y a/G$ at the two ends of the tower,

$$\left(\frac{k_y a}{G}\right)_0 = 5.448$$

$$\left(\frac{k_y a}{G}\right)_1 = 5.202$$

and use the average value of 5.325 to calculate $z_T$ as 1.08 m.

### 8.2.2 Gas Absorption with Chemical Reaction

The equilibrium vapor pressure of $SO_2$ over the liquid depends on the concentration of dissolved $SO_2$. In the previous analysis the concentration of dissolved $SO_2$ is just equal to that which has been absorbed into the liquid at that point in the column. However, the equilibrium vapor pressure of the dissolved solute can be decreased almost to zero by adding a reagent to the absorbing liquid that reacts with the dissolved solute, effectively "pulling" more of the solute gas into solution. Some examples of the use of chemically enhanced absorption are the removal of acid gases (such as $SO_2$) by alkaline solutions (see Section 8.4), the removal of odorous gases in oxidizing solutions, and the absorption of $CO_2$ and $H_2S$ in amine solutions.

We wish to consider the same situation as in the preceding section, except that to enhance the solubility of $SO_2$ in the water an alkaline reagent will be assumed to have been added to the fresh liquid feed at the top of the column. It will be necessary to account for the chemical state of the dissolved $SO_2$ in order to compute its equilibrium vapor pressure and, therefore, the gas-phase driving force for absorption. Let

$G_0$ = total molar flow rate of entering gas, kg-mol $m^{-2}$ $h^{-1}$ of empty tower

$y_0$ = mole fraction of $SO_2$ in the entering gas

$y_1$ = desired mole fraction of $SO_2$ in the exiting gas

$pH_0$ = initial pH of water feed

$W$ = volumetric flow rate of liquid fed to the top of the column per unit cross-sectional area of the tower, $m^3$ of liquid $m^{-2}$ of column $h^{-1}$

In the design we need to derive equations for the compositions of the gas and liquid phases as a function of position in the unit. We let the position in the chamber be denoted by $z$, where $z$ is the distance measured from the top of the unit, and $z_T$ is the total height of the chamber (to be determined). We derive these equations by considering balances on $SO_2$ both over a slice of differential depth and over the unit as a whole.

The total molar flow rate of gas introduced at the bottom of the unit is $G_0$ (kg-mol $m^{-2}$ $h^{-1}$), which consists of mole fractions $y_0$ of $SO_2$ and $(1 - y_0)$ of air. Thus, the molar flow rate of air in the unit is $(1 - y_0)\,G_0$. If the mole fraction of $SO_2$ at any depth in the chamber is $y$, the total molar flow rate of gas per unit cross-sectional area at that point is $G = (1 - y_0)\,G_0/(1 - y)$.

We now perform a balance on gas-phase $SO_2$ over a section of depth $dz$. At steady state:

$$\text{flow in with gas at } z + dz = \text{flow out with gas at } z + \text{amount transferred to water} \tag{8.34}$$

The first two terms of the balance are:

$$\text{flow in with gas at } z + dz = (1 - y_0)\,G_0 \left.\frac{y}{1 - y}\right|_{z+dz}$$

$$\text{flow out with gas at } z = (1 - y_0)\,G_0 \left.\frac{y}{1 - y}\right|_{z}$$

where we see that these two terms are each just $yG$, the molar flow rate of $SO_2$ in the gas.

Thus, the balance on gas-phase $SO_2$ becomes

$$(1 - y_0)\,G_0 \left.\frac{y}{1 - y}\right|_{z+dz} - (1 - y_0)\,G_0 \left.\frac{y}{1 - y}\right|_{z} = k_y a(y - y_i)\,dz \tag{8.35}$$

Dividing by $dz$ and taking the limit as $dz \to 0$ gives us

$$\frac{d}{dz}\left(\frac{y}{1 - y}\right) = \frac{k_y a(y - y_i)}{G_0(1 - y_0)} \tag{8.36}$$

which is to be solved subject to

$$y(z_T) = y_0 \tag{8.37}$$

All the quantities in (8.36) are known except $y_i$, the $SO_2$ mole fraction just above the liquid surface. To determine $y_i$ we must consider the behavior of the liquid phase as a function of position in the chamber. Note that $y_i = p^s_{SO_2}/p$, where $p^s_{SO_2}$ is the partial pressure of $SO_2$ just above the liquid surface.

The absorption of $SO_2$ by water leads to the equilibria given in Table 8.2. The concentrations of the dissolved sulfur species in the liquid, in units of kg-mol m$^{-3}$, given $p^s_{SO_2}$, are found from the equilibrium constant expressions in Table 8.2 to be

$$[SO_2 \cdot H_2O] = K_{hs} p^s_{SO_2} \tag{8.38}$$

$$[HSO_3^-] = \frac{K_{hs} K_{s1} p^s_{SO_2}}{[H^+]} \tag{8.39}$$

$$[SO_3^{2-}] = \frac{K_{hs} K_{s1} K_{s2} p^s_{SO_2}}{[H^+]^2} \tag{8.40}$$

As we noted, to enhance the solubility of $SO_2$ in the water it is customary to raise the pH of the feed drops over that for pure water through the addition of an alkaline substance. Thus let us presume that an amount of nonvolatile salt MOH that dissociates in solution into $M^+$ and $OH^-$ has been added to the feed water such that the initial pH is $pH_0$. Electroneutrality must always be maintained locally in the liquid, so the concentration of the ion $M^+$ is found from the specified initial pH,

$$[H^+]_0 + [M^+]_0 = [OH^-]_0 \tag{8.41}$$

and since MOH is assumed to be nonvolatile,

$$[M^+] = [M^+]_0 = \frac{K_w}{[H^+]_0} - [H^+]_0 \tag{8.42}$$

**TABLE 8.2** EQUILIBRIUM CONSTANTS FOR AQUEOUS ABSORPTION OF $SO_2$

| Reaction | Equilibrium constant expression | Equilibrium constant[a,b] |
|---|---|---|
| $SO_{2(g)} + H_2O \rightleftharpoons SO_2 \cdot H_2O$ | $K_{hs} = \dfrac{[SO_2 \cdot H_2O]}{p_{SO_2}}$ M atm$^{-1}$ | $\log K_{hs} = \dfrac{1376.1}{T} - 4.521$ |
| $SO_2 \cdot H_2O \rightleftharpoons H^+ + HSO_3^-$ | $K_{s1} = \dfrac{[H^+][HSO_3^-]}{[SO_2 \cdot H_2O]}$ M | $\log K_{s1} = \dfrac{853}{T} - 4.74$ |
| $HSO_3^- \rightleftharpoons H^+ + SO_3^{2-}$ | $K_{s2} = \dfrac{[H^+][SO_3^{2-}]}{[HSO_3^-]}$ M | $\log K_{s2} = \dfrac{621.9}{T} - 9.278$ |
| $H_2O \rightleftharpoons H^+ + OH^-$ | $K_w = [H^+][OH^-]$ M$^2$ | $\log K_w = \dfrac{4470.99}{T} + 6.0875 - 0.01706T$ |

[a]Values of $K_{hs}$, $K_{s1}$, and $K_{s2}$ from Maahs (1982).
[b]Value of $K_w$ from Harned and Owen (1958, pp. 643-646).

Electroneutrality at any time is expressed as

$$[H^+] + [M^+] = [OH^-] + [HSO_3^-] + 2[SO_3^{2-}] \tag{8.43}$$

which can be written in terms of $[H^+]$ as

$$[H^+] + [M^+] = \frac{K_w}{[H^+]} + \frac{K_{hs}K_{s1}p^s_{SO_2}}{[H^+]} + \frac{2K_{hs}K_{s1}K_{s2}p^s_{SO_2}}{[H^+]^2} \tag{8.44}$$

or

$$[H^+]^3 + [M^+][H^+]^2 - (K_w + K_{hs}K_{s1}p^s_{SO_2})[H^+] - 2K_{hs}K_{s1}K_{s2}p^s_{SO_2} = 0 \tag{8.45}$$

The local hydrogen ion concentration in the liquid is related to the $SO_2$ partial pressure just above the drop surface by (8.42) and (8.45). Since there are two unknowns, $[H^+]$ and $p^s_{SO_2}$, we need to obtain another equation relating these two quantities. At the top of the tower $[H^+] = [H^+]_0$; however, as soon as the falling liquid encounters $SO_2$, absorption takes place and the hydrogen ion concentration begins to change. To calculate how $[H^+]$ changes with $z$, we perform an overall material balance on $SO_2$ between the bottom of the tower ($z = z_T$) and any level $z$. That overall balance takes the form:

flow in with gas at $z_T$ + flow in with water at $z$

= flow out with water at $z_T$ + flow out with gas at $z$

Two of the terms in this balance are:

$$\text{flow in with gas at } z_T = G_0 y_0$$

$$\text{flow out with gas at } z = \frac{(1 - y_0) G_0 y}{1 - y}$$

The quantity of $SO_2$ flowing out with the water at $z = z_T$ is just the difference between that fed to the unit in the gas and that allowed to leave with the cleaned gas:

$$\text{flow out with water at } z_T = G_0 y_0 - \frac{(1 - y_0) G_0 y_1}{1 - y_1}$$

The final term in the material balance is the quantity of $SO_2$ flowing with the liquid across the plane at $z$. This flow is given by

$$W([SO_2 \cdot H_2O] + [HSO_3^-] + [SO_3^{2-}])$$

We can use the equilibrium relations to express this quantity as

$$WK_{hs}p^s_{SO_2}\left(1 + \frac{K_{s1}}{[H^+]} + \frac{K_{s1}K_{s2}}{[H^+]^2}\right)$$

Collecting terms in the entire balance, we have

$$G_0 y_0 + WK_{hs}p^s_{SO_2}\left(1 + \frac{K_{s1}}{[H^+]} + \frac{K_{s1}K_{s2}}{[H^+]^2}\right) = G_0 y_0 - \frac{(1 - y_0) G_0 y_1}{1 - y_1} + \frac{(1 - y_0) G_0 y}{1 - y} \tag{8.46}$$

or

$$WK_{hs}p_{SO_2}^s\left(1 + \frac{K_{s1}}{[H^+]} + \frac{K_{s1}K_{s2}}{[H^+]^2}\right)$$

$$= (1 - y_0)\, G_0\left(\frac{y}{1 - y} - \frac{y_1}{1 - y_1}\right) \tag{8.47}$$

Now, to place this equation in a more compact form, let

$$\eta = [H^+] \qquad A = \frac{K_w}{[H^+]_0} - [H^+]_0 \qquad B = K_{hs}K_{s1} \qquad C = K_{hs}K_{s1}K_{s2}$$

$$D = \frac{(1 - y_0)\, G_0}{W} \qquad E = \frac{y_1}{1 - y_1}$$

and (8.47) can be written as

$$p_{SO_2}^s = \frac{D\eta^2}{K_{hs}\eta^2 + B\eta + C}\left(\frac{y}{1 - y} - E\right) \tag{8.48}$$

Also, (8.45) becomes

$$p_{SO_2}^s = \frac{\eta^3 + A\eta^2 - K_w\eta}{B\eta + 2C} \tag{8.49}$$

Equating (8.48) and (8.49) yields a single nonlinear algebraic equation relating $[H^+]$ (i.e., $\eta$) and $y$,

$$y = \frac{f(\eta)}{1 + f(\eta)} \tag{8.50}$$

where

$$f(\eta) = \frac{(\eta^3 + A\eta^2 - K_w\eta)(K_{hs}\eta^2 + B\eta + C)}{D\eta^2(B\eta + 2C)} + E \tag{8.51}$$

Now return to (8.36). Using (8.49) and (8.50), (8.36) becomes

$$m_2(\eta)\frac{d\eta}{dz} = m_1(\eta) \tag{8.52}$$

where

$$m_1(\eta) = \frac{k_y a}{G_0(1 - y_0)}\left[\frac{f(\eta)}{1 + f(\eta)} - \frac{\eta^3 + A\eta^2 - K_w\eta}{p(B\eta + 2C)}\right]$$

$$m_2(\eta) = \frac{df(\eta)}{d\eta}$$

Integrating (8.52) over the tower gives

$$\int_{\eta_0}^{\eta_T} \frac{m_2(\eta)}{m_1(\eta)}\, d\eta = \int_0^{z_T} dz = z_T \tag{8.53}$$

The upper limit of the integral in (8.53), $\eta_T$, is the solution of (8.50) at $y(z_T) = y_0$, that is,

$$y_0 = \frac{f(\eta_T)}{1 + f(\eta_T)} \tag{8.54}$$

To determine the tower height $z_T$ for a given set of operating conditions, we must first solve (8.54) for $\eta_T$, then evaluate the integral in (8.53) to find $z_T$.

**Example 8.2** ***Absorption of*** $\mathbf{SO_2}$ ***with Chemical Reaction***

Let us calculate the tower height $z_T$ as a function of the water feed rate for gas molar flow rates ranging from 0.006 to 0.015 kg-mol $m^{-2}$ $s^{-1}$. Let $y_0 = 0.2$, $y_1 = 0.01$, and $pH_0 = 11$. Consider water volumetric flow rates from 20 to 30 $m^3$ $m^{-2}$ $h^{-1}$. We use the correlation for $k_y a$ from Example 8.1.

Figure 8.6 shows $z_T$ as a function of $W$ for $G_0$ values ranging from 0.006 to 0.015 kg-mol $m^{-2}$ $s^{-1}$. We see that at each gas flow rate, there is a liquid flow rate below which the specified separation cannot be achieved. As $G_0$ increases, the value of this liquid flow rate also increases. Figure 8.7 shows $z_T$ as a function of $pH_0$ and $G_0$. For all gas flow rates, there is a $pH_0$ value below which the required length does not change (e.g., $pH_0$ 11 at $G_0 = 0.009$ kg-mol $m^{-2}$ $s^{-1}$, since the capacity of the water is saturated). Very high values of $pH_0$ promote more absorption and decrease the required scrubber length.

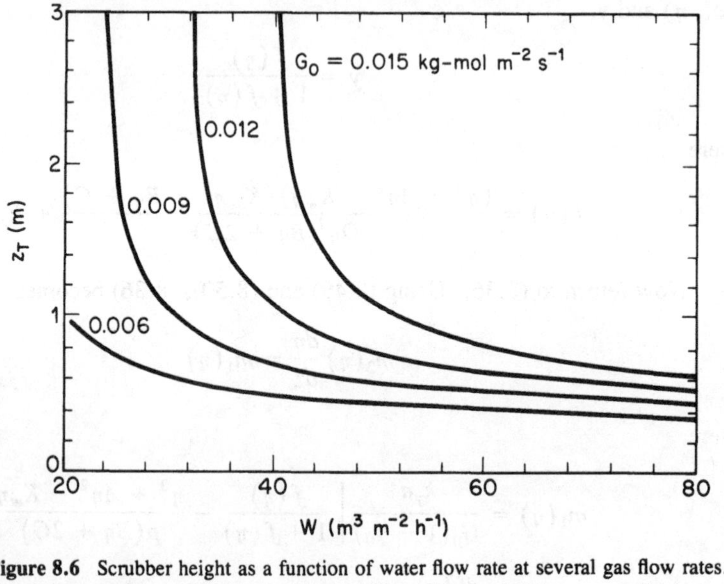


**Figure 8.6** Scrubber height as a function of water flow rate at several gas flow rates. Conditions are: $T = 303$ K, $p = 1$ atm, $pH_0 = 11$, $y_0 = 0.2$, $y_1 = 0.01$, $k_y a = 0.09944 \tilde{L}^{0.25} \tilde{G}^{0.7}$.

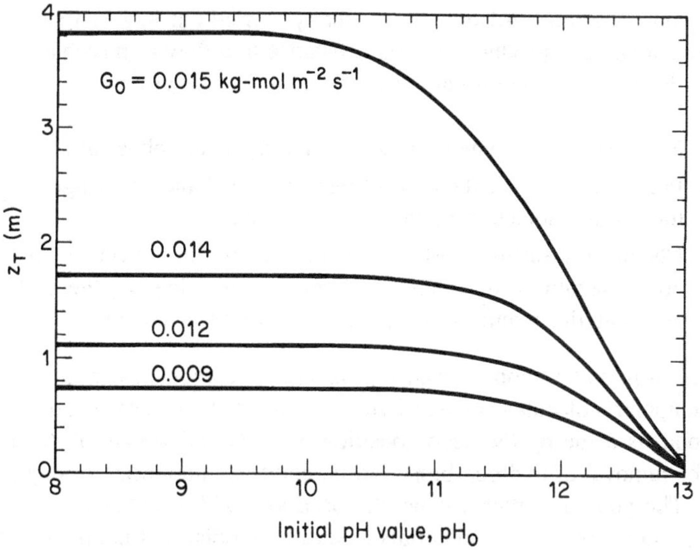


**Figure 8.7** Scrubber height as a function of $pH_0$ at several gas flow rates. $W = 40$ m$^3$ m$^{-2}$ h$^{-1}$.

## 8.3 ADSORPTION OF GASES ON SOLIDS

Adsorption involves the use of a solid substrate to remove the contaminant. The intermolecular attractive forces in the bulk of a solid are, at its surface, available for holding other materials, such as gases and liquid. In adsorption, the thermal motion of a gas molecule is converted to heat as the molecule becomes bound to the surface. Adsorption is, therefore, an exothermic process. Conversely, regeneration of the adsorbent (or sorbent) by desorption of the adsorbed gas (the adsorbate or sorbate) is endothermic and energy must be supplied. Some adsorption processes occur so strongly that the adsorbed material can only be desorbed by removal of some of the solid substrate. In such a case, chemical bonds form between the adsorbent and the adsorbed species, and the adsorption process is referred to as chemisorption. For example, oxygen chemisorbed on activated carbon can only be removed as CO or $CO_2$. Adsorption is particularly well suited to treating large volumes of gases with very dilute pollutant levels and to removing contaminants down to trace levels. The solids best suited for use as adsorbents are those with large surface/volume ratios, that is, very porous. Polar adsorbents such as activated aluminas, silica gels, and molecular sieves have high selectivity for polar gases. Such adsorbents also effectively remove water, a polar molecule, and thus in the presence of moisture can become ineffective due to the relatively large amount of water that is adsorbed. Activated carbon is a second class of adsorbent that is commonly used. Since activated carbon is composed largely of neutral atoms of a single species, there are no significant potential gradients to attract and orient polar molecules in preference to nonpolar molecules. Activated carbon tends to adsorb all gases roughly in proportion to their concentrations. Adsorption is usually carried out in a bed packed with a granular adsorbent sized to produce as little resistance to a flow as possible.

Potential adsorbents can be classified into three groups:

1. Nonpolar solids, where the adsorption is mainly physical
2. Polar solids, where the adsorption is chemical and no change in the chemical structure of the molecules or the surface occurs
3. Chemical adsorbing surfaces, which adsorb the molecules and then release them after reaction, which may be either catalytic, leaving the surface unchanged, or noncatalytic, requiring replacement of the surface atoms

The most important nonpolar adsorbing solid is carbon, which is very effective in binding nonpolar molecules, such as hydrocarbons. Activated carbon (charcoal, if the source is wood) is made by the decomposition of coals and woods. Activated carbon is used for the removal of hydrocarbons, odors, and trace impurities from gas streams.

The polar adsorbents generally used are oxides, either of silicon or other metals (e.g., aluminum). These materials adsorb both polar and nonpolar molecules, but they exhibit preference for polar molecules. Thus silicon and aluminum oxides are used to adsorb polar molecules such as water, ammonia, hydrogen sulfide, and sulfur dioxide.

The equilibrium characteristics of a solid–gas system are described by a curve of the concentration of adsorbed gas on the solid as a function of the equilibrium partial pressure of the gas at constant temperature. Such a curve is called an *adsorption isotherm*. In the case in which only one component of a binary gas mixture is adsorbed, the adsorption of that species is relatively uninfluenced by the presence of the other gas, and the adsorption isotherm for the pure vapor is applicable as long as the equilibrium pressure is taken as the partial pressure of the adsorbing gas.

Separation of one component from a gaseous mixture by adsorption on a solid may be carried out in a batchwise or continuous manner of operation. Continuous operation can, in turn, be employed in a series of distinct stages or in continuous contact, such as in gas absorption. When one component is being adsorbed, the design of the operation is, from the point of view of the calculational procedure, analogous to gas absorption, in that only one component is transferred between two essentially immiscible phases. A rather thorough treatment of gas adsorption operations is given by Treybal (1968). We consider here only the process of adsorption of a species as the gas is passed through a stationary (fixed) bed of adsorbent.

The key difference between gas absorption with two continuous countercurrent streams and gas adsorption in a fixed bed is that the former is a steady-state process, whereas the latter, due to the accumulation of adsorbed gas on the solid, is an unsteady-state process.

We will consider a mixture of two gases, one strongly adsorbed, which is to be passed through a bed initially free of adsorbent. When the mixture first enters the fresh bed, the solid at the entrance to the bed at first adsorbs the gas almost completely. Thus, initially the gas leaving the bed is almost completely free of the solute gas. As the layers of solid near the entrance to the bed become saturated with adsorbed gas, the zone of solid in which the major portion of the adsorption takes place moves slowly through the bed, at a rate generally much slower than the actual gas velocity through the bed. Finally, the so-called adsorption zone reaches the end of the bed. At this point, the exit concentration of solute gas rises sharply and approaches its inlet concentration, since, for all practical purposes, the bed is saturated and at equilibrium with the inlet gas. The curve of effluent concentration as a function of time thus has an S-shaped appearance that may be steep or relatively flat, depending on the rate of adsorption, the nature of the adsorption equilibrium, the fluid velocity, the inlet concentration, and the length of the bed. The time at which the breakthrough curve first begins to rise appreciably is called the breakpoint. The passage of an adsorption wave through a stationary bed during an adsorption cycle is depicted in Figure 8.8.

When a bed reaches saturation, the adsorbed material must be removed from the solid. Desorption of an adsorbed solute by passing a solvent through the bed is called *elution*. The process of gas chromatography is based on the elution of a bed that contains small quantities of several adsorbed gases. As a suitable eluent is passed through such a bed, the adsorbed solutes are desorbed at different rates and pass out of the bed at different times, enabling their identification by comparison with eluent curves previously established for known species.

The design of a fixed-bed adsorption column would normally require that one predict the breakthrough curve, and thus the length of the adsorption cycle between elutions of the bed, given a bed of certain length and equilibrium data. Alternatively, one could seek the bed depth required for operation over a specified period of time to achieve a desired degree of separation. Because of the different types of equilibrium relationships that can be encountered and the unsteady nature of the process, prediction of the solute breakthrough curve is, in general, quite difficult. We present here a design method applicable only when the solute concentration in the feed is small, the adsorption isotherm

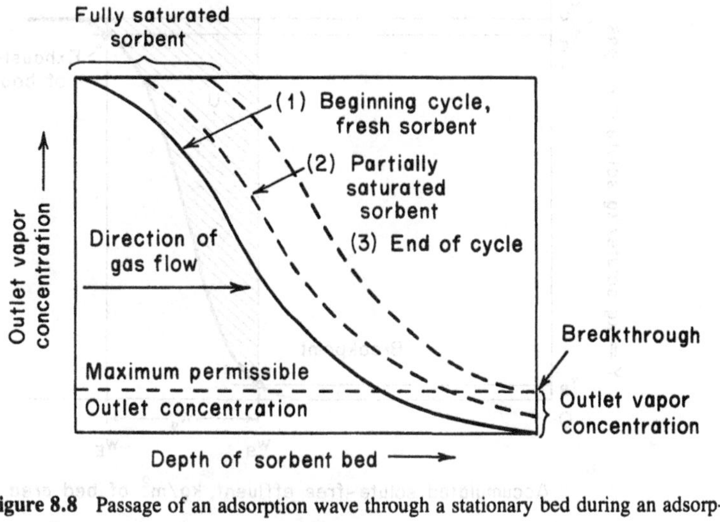


**Figure 8.8** Passage of an adsorption wave through a stationary bed during an adsorption cycle.

is concave to the gas-phase concentration axis, the adsorption zone is constant in height as it travels through the column, and the length of the column is large compared with the height of the adsorption zone (Treybal, 1968).

Let us consider the idealized breakthrough curve shown in Figure 8.9 resulting from flow of an inert gas through a bed with a rate $\tilde{G}'$ kg m$^{-2}$ h$^{-1}$ containing an inlet solute concentration of $Y_0$ kg solute/kg inert gas. The total amount of solute-free gas that has passed through the bed up to any time is $w$ kg m$^{-2}$ of bed cross section. Values $Y_B$ and $Y_E$, shown in Figure 8.9, mark the breakpoint and equilibrium concentrations, respectively; $w_B$ and $w_E$ denote the values of $w$ at $Y_B$ and $Y_E$, respectively. The adsorption zone, taken to be of constant height $z_a$, is that part of the bed in which the concentration profile from $Y_B$ to $Y_E$ exists at any time.

If $\theta_a$ and $\theta_E$ are the times required for the adsorption zone to move its own length and down the entire bed, respectively, then

$$\theta_a = \frac{w_a}{\tilde{G}'} \tag{8.55}$$

and

$$\theta_E = \frac{w_E}{\tilde{G}'} \tag{8.56}$$

If $\theta_F$ is the time required for the adsorption zone to form, and if $z$ is the length of the bed,

$$z_a = z\frac{\theta_a}{\theta_E - \theta_F} \tag{8.57}$$

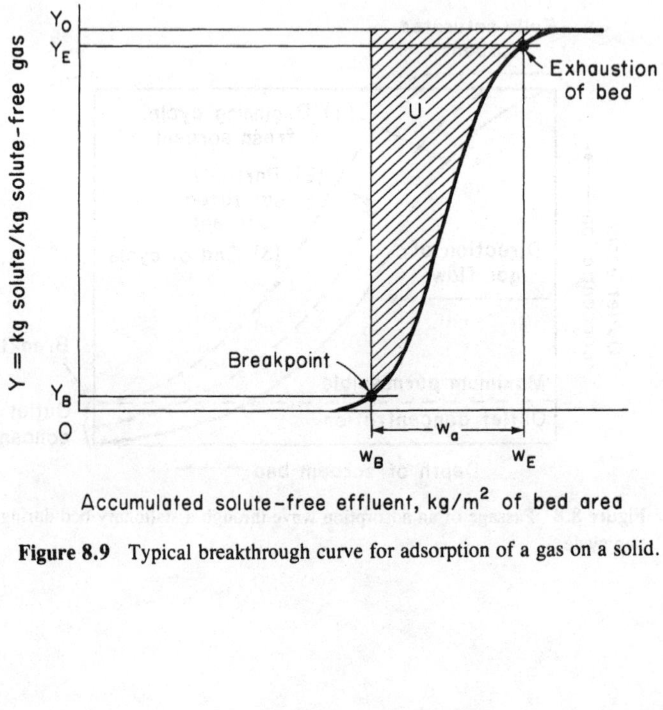


**Figure 8.9** Typical breakthrough curve for adsorption of a gas on a solid.

The solute removed from the gas in the adsorption zone is $U$ kg m$^{-2}$ of bed cross section; $U$ is shown in Figure 8.9 by the shaded area, which is

$$U = \int_{w_B}^{w_E} (Y_0 - Y)\, dw \tag{8.58}$$

If all the adsorbent in the zone were saturated, the solid would contain $Y_0 w_a$ kg solute m$^{-2}$. Thus the fractional capacity of the adsorbent in the zone to continue adsorbing solute is $f = U/Y_0 w_a$. The shape of the breakthrough curve is thus characterized by $f$. If $f = 0$, the time of formation $\theta_F$ of the zone should be the same as the time required for the zone to travel its own thickness, $\theta_a$, since the breakthrough curve will be a vertical line. If $f = 1$, the time to establish the zone should be zero. To satisfy these two limiting cases, one sets $\theta_F = (1 - f)\theta_a$. Thus (8.57) becomes

$$z_a = z \frac{\theta_a}{\theta_E - (1 - f)\theta_a} = z \frac{w_a}{w_E - (1 - f) w_a} \tag{8.59}$$

If the column contains $z A_c \rho_s$ kg of adsorbent, where $A_c$ is the cross-sectional area of the bed and $\rho_s$ is the solid density in the bed, at complete saturation the bed would contain $z A_c \rho_s X_T$ kg of solute, where $X_T$ is the solute concentration on the solid in equilibrium with the feed. At the breakpoint, $z - z_a$ of the bed is saturated, and $z_a$ of the bed is saturated to the extent of $1 - f$. The degree of overall bed saturation at the breakpoint is thus

$$\alpha = \frac{(z - z_a)\rho_s X_T A_c + z_a \rho_s (1 - f) X_T A_c}{z \rho_s X_T A_c} = \frac{z - f z_a}{z} \tag{8.60}$$

The determination of the breakthrough curve can be carried out in the following way. Let us consider the adsorption column in Figure 8.10, where the adsorption zone $z_a$ is in the column, and the solute composition in the gas is $Y_0$ and 0 at the entrance and

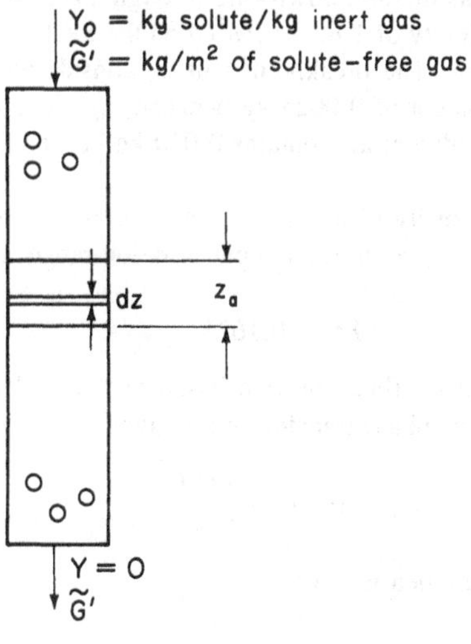


**Figure 8.10** Fixed-bed adsorber with adsorption zone of depth $z_a$.

exit, respectively. Corresponding to these gas-phase compositions, we assume that those on the solid are $X_T$ (saturation at the entrance to the column) and 0 (no adsorbed solute at the exit). If the column is considered to be infinitely long, the situation depicted in Figure 8.10 is applicable. This point will not really concern us since our only real interest is in the adsorption zone $z_a$. The operating line, which relates $Y$ and $X$ at any point in the column, is then a straight line connecting the origin with the point $(Y_0, X_T)$ on the equilibrium curve.

Over a differential depth $dz$ in $z_a$ the rate of adsorption is

$$\tilde{G}'\,dY = K_Y a(Y - Y^*)\,dz \tag{8.61}$$

where $K_Y a$ is the overall mass transfer coefficient for transfer from gas to solid phase. Thus, over the adsorption zone,

$$z_a = \frac{\tilde{G}'}{K_Y a}\int_{Y_B}^{Y_E} \frac{dY}{Y - Y^*} \tag{8.62}$$

and for any value of $z$ less than $z_a$, but within the zone,

$$\frac{z}{z_a} = \frac{w - w_B}{w_a} = \frac{\int_{Y_B}^{Y} dY/(Y - Y^*)}{\int_{Y_B}^{Y_E} dY/(Y - Y^*)} \tag{8.63}$$

The breakthrough curve can be plotted directly from (8.63).

Table 8.3 lists the various types of adsorption equipment, together with brief comments on their operation and use.

**Example 8.3** ***Adsorption of Benzene from Air***

Benzene vapor present to the extent of 0.025 kg benzene/kg air (benzene-free basis) is to be removed by passing the gas mixture downward through a bed of silica gel at 298 K and 2 atm pressure at a linear velocity of 1 m s$^{-1}$ (based on the total cross-sectional area). It is desired to operate for 90 min. The breakpcint will be considered as that time when the effluent air has a benzene content of 0.0025 kg benzene/kg air, and the bed will be considered exhausted when the effluent air contains 0.020 kg benzene/kg air. Determine the depth of bed required.

Silica gel has a bulk density of 625 kg m$^{-3}$ and an average particle diameter $D_p$ of 0.60 cm. For this temperature, pressure, and concentration range, the adsorption isotherm is

$$Y^* = 0.167X^{1.5}$$

where $Y^*$ = kg benzene/kg air (benzene-free basis) and $X$ = kg benzene/kg gel. We assume that the height of a gas-phase transfer unit is given by

$$H_{0Y} = 0.00237\left(\frac{D_p\tilde{G}'}{\mu_{air}}\right)^{0.51}$$

The cross-sectional area of the bed is 1 m$^2$.

**TABLE 8.3 TYPES OF ADSORPTION EQUIPMENT**

| Type | Operation and use |
|---|---|
| Disposable and rechargeable canisters | Small flow; effluent with low sorbate concentration. |
| Fixed regenerable beds | When volume of flow or sorbate concentration is high enough to make recovery attractive, or when cost of fresh sorbent is expensive. |
| Shallow beds | Large gas volumes of low pollutant concentration. |
| Deep beds | When pollutant concentrations exceed 100 ppm or flow exceeds 4.7 $m^3$ $s^{-1}$; typically, 0.3–1 m thick. |
| Traveling bed | Freshly regenerated adsorbent is added continuously to the top of the bed at a rate to maintain a constant solid depth. Saturated sorbent is continuously removed from the bottom of the bed and regenerated before return to the top. Gas to be treated enters the bottom and passes countercurrent to the slowly moving sorbent. Used for high concentrations of sorbates requiring high sorbent to gas ratio. |
| Fluid bed | Particles continuously removed and regenerated to maintain bed particles relatively unsaturated. Use for adsorption or organics from a moist stream where continuous carbon regeneration is needed. |
| Chromatographic baghouse | Granular absorbent introduced continuously into the gas stream which conveys the particles through a line of sufficient length to provide appreciable contact. Sorbent removed in a baghouse. |

First, we can compute $H_{OY}$. The density of air at 298 K and 2 atm is 2.38 kg $m^{-3}$, and so $\tilde{G}' = 2.38$ kg $m^{-2}$ $s^{-1}$. The viscosity of air at 298 K is $1.8 \times 10^{-5}$ kg $m^{-1}$ $s^{-1}$. Thus $H_{OY} = 0.071$ m.

The adsorption isotherm is shown in Figure 8.11. The operating line has been drawn to intersect the equilibrium curve at $Y_0 = 0.025$. From the problem specifications, $Y_B = 0.0025$ and $Y_E = 0.020$. From Figure 8.11 we see that $X_T = 0.284$.

The integral in (8.62) can be evaluated numerically (see Table 8.4) as 5.925. Thus

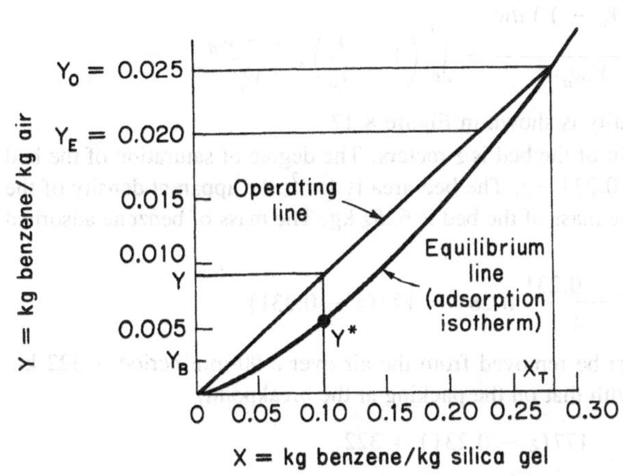


**Figure 8.11** Equilibrium and operating lines for adsorption of benzene on silica gel.

TABLE 8.4 NUMERICAL EVALUATION OF INTEGRAL IN (8.62)

| $Y$ | $Y^*$ | $Y - Y^*$ | $\dfrac{1}{Y - Y^*}$ | $\displaystyle\int_{Y_B}^{Y_E} \frac{dY}{Y - Y^*}$ | $\dfrac{w - w_B}{w_a}$ | $\dfrac{Y}{Y_0}$ |
|---|---|---|---|---|---|---|
| 0.0025 | 0.0009 | 0.0016 | 625 | | 0 | 0.01 |
| 0.0050 | 0.0022 | 0.0028 | 358 | 1.1375 | 0.192 | 0.2 |
| 0.0075 | 0.0042 | 0.0033 | 304 | 1.9000 | 0.321 | 0.3 |
| 0.0100 | 0.0063 | 0.0037 | 270 | 2.6125 | 0.441 | 0.4 |
| 0.0125 | 0.0089 | 0.0036 | 278 | 3.3000 | 0.556 | 0.5 |
| 0.0150 | 0.0116 | 0.0034 | 294 | 4.0125 | 0.676 | 0.6 |
| 0.0175 | 0.0148 | 0.0027 | 370 | 4.8375 | 0.815 | 0.7 |
| 0.0200 | 0.0180 | 0.0020 | 500 | 5.9250 | 1.00 | 0.8 |

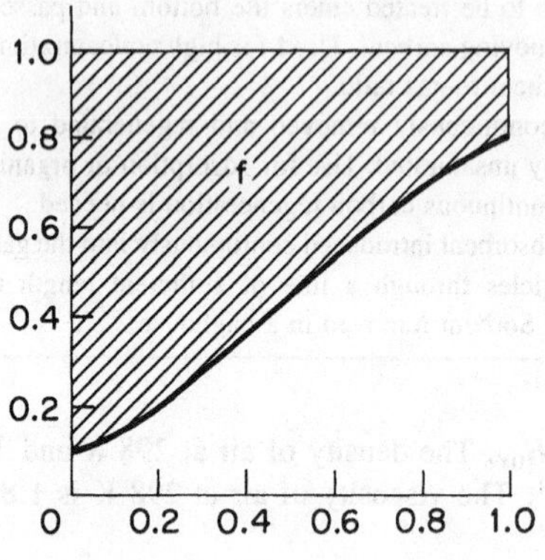


**Figure 8.12** Breakthrough curve for adsorption of benzene on silica gel.

the height of the adsorption zone $z_a$ is $0.071 \times 5.925 = 0.42$ m. The extent of saturation,

$$f = \frac{\int_{w_B}^{w_E} (Y_0 - Y)\, dw}{Y_0 w_a} = \int_0^1 \left(1 - \frac{Y}{Y_0}\right) d\,\frac{w - w_B}{w_a}$$

is found to be 0.55. This quantity is shown in Figure 8.12.

Let us suppose the height of the bed is $z$ meters. The degree of saturation of the bed at the breakpoint is $\alpha = (z - 0.231)/z$. The bed area is 1 m$^2$; the apparent density of the packing is 625 kg m$^{-3}$; thus the mass of the bed is $625z$ kg. The mass of benzene adsorbed on the gel is then

$$625z\,\frac{z - 0.231}{z}\,0.284 = 177(z - 0.231)$$

The mass of benzene that must be removed from the air over a 90-min period is 322 kg. Equating this mass removed with that on the packing at the breakpoint,

$$177(z - 0.231) = 322$$

we obtain the required bed depth of 2.04 m.

## 8.4 REMOVAL OF $SO_2$ FROM EFFLUENT STREAMS

As noted in Chapter 1, $SO_2$ is emitted from coal-fired power plants (about two-thirds of U.S. emissions), from industrial fuel combustion, sulfuric acid manufacturing, and smelting of nonferrous metals. The two basic approaches to $SO_2$ emission control are (1) to remove the sulfur from the fuel before it is burned, or (2) to remove $SO_2$ from the exhaust gases. There has been a significant amount of effort expended in the United States and worldwide on the development of processes in both categories. We will concentrate here on methods for the removal of $SO_2$ from exhaust gases.

The technical and economic feasibility of an $SO_2$ removal process depends on the type and quantity of effluent gases that must be cleaned. With regard to $SO_2$ removal, there are essentially two types of effluent gas treatment problems. The first is the problem of removing $SO_2$ from power plant flue gases. Power plant flue gases generally contain low concentrations of $SO_2$ ( $<0.5\%$ by volume), but emitted at tremendous volumetric flow rates. For example, a coal-fired power plant burning 2% sulfur coal (by weight) will produce 40,000 kg of $SO_2$ for every $10^6$ kg of coal burned. The second class of $SO_2$ effluent gas treatment problems comprises those resulting from the need to remove $SO_2$ from streams containing relatively high concentrations of $SO_2$ at low flow rates. Streams of this type are typical of those emitted from smelter operations. A smelter emission gas typically contains $SO_2$ at a concentration of about 10% by volume (100,000 ppm).

In this section we concentrate largely on the problem of $SO_2$ removal from power plant flue gases, so-called flue gas desulfurization (FGD), since it represents a more prevalent and, in many respects, the more difficult problem than that of $SO_2$ removal from smelting and other industrial operations. Elliot et al. (1982) have reviewed a number of processes for the cleaning of smelter gases, and we refer the reader to this source for those applications.

There are two ways of classifying flue gas desulfurization systems. The first is based on what is done with the $SO_2$-absorbing or $SO_2$-reacting medium, and by this means processes are categorized as *throwaway* or *regenerative*. In a throwaway process, the sulfur removed, together with the absorbing or reacting medium, is discarded. A process is regenerative if the sulfur is recovered in a usable form and the medium is reused. The second way of classifying FGD processes is by the phase in which the main removal reactions occur. By this means processes are categorized as *wet* or *dry*. Both wet and dry processes can be throwaway or regenerative, so there are, in effect, four categories of FGD processes.

In the majority of the throwaway processes an alkaline agent reacts with the $SO_2$, leading to a product that is discarded. Commonly used agents in this type of process are limestone ($CaCO_3$) and lime ($CaO$). In another type of throwaway process the agent is injected directly into the furnace, and the sulfated product is subsequently scrubbed out of the flue gas with water. Part of the $SO_2$ is captured chemically within the furnace, the rest in the scrubbing step.

In the regenerative alkaline processes, an alkaline agent strips $SO_2$ from the flue gas stream, combining chemically with the $SO_2$. In a separate regeneration step, the agent is reconstituted and sulfur is recovered, usually as liquid $SO_2$ or sulfuric acid.

Some of the agents used include MgO, $Na_2SO_3$, and metal carbonates. Regenerative solid adsorption comprises several activated char processes, in which $SO_2$ is adsorbed on char and desorbed to lead to the production of sulfuric acid.

### 8.4.1 Throwaway Processes: Lime and Limestone Scrubbing

The most prevalent throwaway processes involve lime and limestone. Approximately 75% of all installed flue gas desulfurization systems use a lime or limestone slurry as the scrubbing liquor (Joseph and Beachler, 1981; Beachler and Joseph, 1984). In this process $SO_2$ reacts with the lime or limestone slurry to form a $CaSO_3/CaSO_4$ sludge that must be disposed of in a pond or landfill. Most wet scrubbing flue gas desulfurization systems are capable of reducing $SO_2$ emissions by 90%. In dry scrubbing an alkaline slurry is injected in a spray dryer with dry particle collection. Spray dryers are units where hot flue gases are contacted with a fine, wet, alkaline spray, which absorbs the $SO_2$. The high temperature of the flue gas (393 to 573 K) evaporates the water from the alkaline spray, leaving a dry product that can be collected in a baghouse or electrostatic precipitator. Dry scrubbing can remove 75 to 90% of $SO_2$ emissions.

In conventional limestone or lime scrubbing, a limestone/water or lime/water slurry is contacted with the flue gas in a spray tower. The essence of the process lies in the absorption equilibrium of $SO_2$ in water as given in Table 8.2:

$$SO_{2(g)} + H_2O \rightleftharpoons SO_2 \cdot H_2O$$

$$SO_2 \cdot H_2O \rightleftharpoons H^+ + HSO_3^-$$

$$HSO_3^- \rightleftharpoons H^+ + SO_3^{2-}$$

Limestone consists of a mixture of $CaCO_3$ and inert siliceous compounds. Although limestone is very plentiful, it has been estimated that only about 2% of the deposits are of "chemical grade," that is, containing 95% or more $CaCO_3$. Calcium carbonate is relatively insoluble in water (0.00153 g per 100 g of $H_2O$ at 273 K), and its solubility increases only slightly with increasing temperature. Although its low solubility is one of its main drawbacks for use in wet scrubbing, it can be finely pulverized to produce a limestone/water slurry.

Lime (CaO, calcium oxide) can be obtained by heating (calcining) $CaCO_3$ at about 1100 K:

$$CaCO_3 \longrightarrow CaO + CO_2$$

When added to water, lime produces calcium hydroxide (slaked lime):

$$CaO + H_2O \longrightarrow Ca(OH)_2$$

which dissociates according to

$$Ca(OH)_2 \rightleftharpoons Ca^{2+} + 2OH^-$$

While still relatively insoluble (0.185 g per 100 g of $H_2O$ at 273 K), $Ca(OH)_2$ is considerably more soluble than $CaCO_3$. The solubility of $Ca(OH)_2$ decreases as the temperature increases (0.078 g per 100 g of $H_2O$ at 373 K).

A chemical mechanism that is consistent with the overall stoichiometry of limestone scrubbing is

$$SO_{2(g)} + H_2O \rightleftharpoons SO_2 \cdot H_2O$$

$$SO_2 \cdot H_2O \rightleftharpoons H^+ + HSO_3^-$$

$$H^+ + CaCO_3 \rightleftharpoons Ca^{2+} + HCO_3^-$$

$$Ca^{2+} + HSO_3^- + 2H_2O \rightleftharpoons CaSO_3 \cdot 2H_2O + H^+$$

$$H^+ + HCO_3^- \rightleftharpoons CO_2 \cdot H_2O$$

$$CO_2 \cdot H_2O \rightleftharpoons CO_{2(g)} + H_2O$$

The overall reaction corresponding to this mechanism is

$$CaCO_3 + SO_2 + 2H_2O \longrightarrow \underset{\text{calcium sulfite dihydrate}}{CaSO_3 \cdot 2H_2O} + CO_2$$

Two routes have been proposed for the mechanism of lime scrubbing. The first simply involves the conversion of the CaO to $CaCO_3$ by reacting with $CO_2$ in the flue gas,

$$CaO + CO_2 \longrightarrow CaCO_3$$

in which case the mechanism for limestone above would also apply for lime. The second route involves the chemistry of lime itself:

$$SO_{2(g)} + H_2O \rightleftharpoons SO_2 \cdot H_2O$$

$$SO_2 \cdot H_2O \rightleftharpoons H^+ + HSO_3^-$$

$$CaO + H_2O \rightleftharpoons Ca(OH)_2$$

$$Ca(OH)_2 \rightleftharpoons Ca^{2+} + 2OH^-$$

$$Ca^{2+} + HSO_3^- + 2H_2O \longrightarrow CaSO_3 \cdot 2H_2O + H^+$$

$$H^+ + OH^- \rightleftharpoons H_2O$$

The overall reaction corresponding to this mechanism is

$$CaO + SO_2 + 2H_2O \longrightarrow CaSO_3 \cdot 2H_2O$$

Lime is more reactive toward $SO_2$ than is limestone, although both are highly favorable reactants. In spite of the fact that lime scrubbing can achieve higher $SO_2$ removal efficiencies than limestone scrubbing, lime is more expensive and hence is not as widely in use as limestone. The critical step in both mechanisms is the formation of the calcium ion, which reacts with the bisulfite ion to remove $SO_2$ from the solution. We note that in the limestone system the formation of $Ca^{2+}$ depends on the $H^+$ concentration, whereas in the lime system the $Ca^{2+}$ formation step is independent of pH. Thus, in order to drive the $H^+—CaCO_3$ reaction to the right, the limestone system must operate at a fairly high $H^+$ concentration (low pH). The optimal operating pH for limestone scrubbing is between 5.8 and 6.2, whereas that for lime scrubbing is about 8.0.

There is an additional problem related to the setting of the pH in lime and limestone scrubbing—the formation of calcium sulfate. Although the reaction was not used in either of the chemical mechanisms above, we know that the bisulfite ion is in equilibrium with the sulfite ion:

$$HSO_3^- \rightleftharpoons SO_3^{2-} + H^+$$

(Although it was not necessary to consider this reaction to explain the mechanism of $SO_2$ removal, for lime scrubbing this reaction should be added due to the high pH employed.) Excess oxygen in the flue gas can lead to some dissolved oxygen in the slurry. The sulfite ion can be oxidized by dissolved $O_2$ to the sulfate ion:

$$SO_3^{2-} + \tfrac{1}{2}O_2 \longrightarrow SO_4^{2-}$$

When this reaction occurs, the net result is conversion of $CaSO_3$ to $CaSO_4$:

$$CaSO_3 \cdot 2H_2O + \tfrac{1}{2}O_2 \longrightarrow CaSO_4 \cdot 2H_2O$$

Calcium sulfate (gypsum) forms a hard, stubborn scale on the surface of the scrubber, and its formation must be avoided.

The solubility of $CaSO_3$ increases markedly as the pH decreases (100 ppm at pH 5.8 and 1000 ppm at pH 4.4). Thus the rate of oxidation of $SO_3^{2-}$ to $SO_4^{2-}$ increases as pH decreases. The $CaSO_4$ formed by the oxidation has a solubility that decreases slightly as pH decreases. Because of both of these factors, but primarily the increased solubility of $CaSO_3$ at low pH, $CaSO_4$ precipitation occurs at low pH. Thus the pH must be kept sufficiently high to prevent $CaSO_4$ scale formation. Limestone scrubbing systems operating at a pH around 6.0 can successfully avoid $CaSO_4$ scale formation.

We noted that the optimal pH from the point of view of $Ca^{2+}$ formation in lime scrubbing is about 8.0. At high pHs, however, the low solubility of $CaSO_3$ leads to a phenomenon known as soft pluggage, the formation of large leafy masses of $CaSO_3$ inside the scrubber. The soft pluggage can be dissolved by lowering the pH to promote $CaSO_3$ solubility. As long as a pH of 8.0 is not exceeded, lime scrubbing can avoid soft pluggage.

Because neither lime nor limestone is particularly soluble, the liquid/gas ratio must be relatively high. For limestone scrubbing this ratio must exceed 65 gal per 1000 $ft^3$ of gas (0.0088 $m^3$ water per $m^3$ of gas). For lime systems a liquid/gas ratio of 35 gal per 1000 $ft^3$ is adequate (0.0047 $m^3$ water per $m^3$ of gas) due to the higher solubility of lime.

The scrubbing solution is sent from the tower to a retention tank where the precipitation of $CaSO_3$, $CaSO_4$, and unreacted $CaCO_3$ occurs. The residence time needed in the retention tank is about 5 min for a lime system and 10 min for limestone. The crystallized products from the retention tank constitute the waste sludge. Typical waste sludge compositions are given in Table 8.5.

**TABLE 8.5** TYPICAL COMPOSITIONS OF LIME AND LIMESTONE SCRUBBING WASTE SLUDGES

| Compound | Percent dry weight |
|---|---|
| Limestone systems | |
| $CaCO_3$ | 33 |
| $CaSO_3 \cdot 2H_2O$ | 58 |
| $CaSO_4 \cdot 2H_2O$ | 9 |
| | 100 |
| Lime systems | |
| $CaCO_3$ | 5 |
| $CaSO_3 \cdot 2H_2O$ | 73 |
| $CaSO_4 \cdot 2H_2O$ | 11 |
| $Ca(OH)_2$ | 11 |
| | 100 |

*Source:* Fellman and Cheremisinoff (1977).

The main problems with lime and limestone scrubbing are scaling and plugging inside the scrubber unit. The *dual alkali* system eliminates these problems. A solution of sodium sulfite ($Na_2SO_3$)/sodium hydroxide (NaOH) is sprayed in the tower. Sulfur dioxide is absorbed and neutralized in the solution, and since both $Na_2SO_3$ and $Na_2SO_4$ are soluble in water, no precipitation occurs in the scrubber. The $Na_2SO_3/Na_2SO_4$ solution from the scrubber cannot simply be discarded because of water pollution problems and because NaOH is relatively expensive. Thus, in a separate tank, lime or limestone and some additional NaOH are added to the scrubbing effluent. The lime or limestone precipitates the sulfite and sulfate and regenerates the NaOH.

There are several dry throwaway processes. As mentioned earlier, a wet lime slurry can be injected into the tower, and $SO_2$ is absorbed by the droplets forming $CaSO_3$ and $CaSO_4$. If the liquid/gas ratio is low enough, the water will evaporate before the droplets reach the bottom of the tower. The dry particles are subsequently collected, usually in a baghouse.

Direct injection of pulverized lime or limestone into the boiler has been demonstrated as an effective means of $SO_2$ removal. The $SO_2$ is adsorbed on the dry particles, and the dry $SO_2$-laden particles are collected in a baghouse. The smaller the particle size, the more efficient the removal process.

**Example 8.4** ***pH Control in Lime Scrubbing (Shinskey, 1977)***

We noted above that the optimal pH for lime scrubbing of $SO_2$ is about 8.0. Let us see how this value can be determined. The idea is to determine the pH at which all of the $SO_2$ is converted to product.

The ions present in the lime scrubbing system must obey charge neutrality,

$$[H^+] + 2[Ca^{2+}] = [OH^-] + [HSO_3^-] + 2[SO_3^{2-}]$$

where we have included the sulfite ion, $SO_3^{2-}$, for completeness. The equilibria involving sulfur compounds and water were given in Table 8.2. The additional one needed is the solubility product of $CaSO_3$,

$$[Ca^{2+}][SO_3^{2-}] = K_s$$

Substituting the equilibrium expressions into the electroneutrality relation yields

$$\left(1 + \frac{2K_{s2}}{[H^+]}\right)[HSO_3^-]^2 + \left(\frac{K_w}{[H^+]} - [H^+]\right)[HSO_3^-] - \frac{2K_s[H^+]}{K_{s2}} = 0$$

This quadratic equation can be solved for $[HSO_3^-]$ given a value of $[H^+]$, or pH, since $[H^+] = 10^{-pH}$.

Shinskey (1977) has carried out this solution and obtained

| pH | $[HSO_3^-]$ (M) |
|---|---|
| 3 | 0.159 |
| 4 | 0.050 |
| 5 | 0.0156 |
| 6 | $4.4 \times 10^{-3}$ |
| 7 | $7.7 \times 10^{-4}$ |
| 8 | $8.8 \times 10^{-5}$ |
| 9 | $9.0 \times 10^{-6}$ |
| 10 | $9.0 \times 10^{-7}$ |

We now have the bisulfite ion concentration as a function of pH, and we need to connect the pH to the material balance. The total sulfur concentration in solution at any time is

$$[S] = [SO_2 \cdot H_2O] + [HSO_3^-] + [SO_3^{2-}] + [CaSO_3]$$

Similarly, the total calcium concentration is

$$[Ca] = [Ca^{2+}] + [CaSO_3]$$

We can combine these two and eliminate $[CaSO_3]$,

$$[Ca] - [S] = [Ca^{2+}] - [SO_2 \cdot H_2O] - [HSO_3^-] - [SO_3^{2-}]$$

The $[Ca^{2+}]$ and $[SO_3^{2-}]$ terms may be removed by using the electroneutrality relation, to give

$$[Ca] - [S] = \tfrac{1}{2}([OH^-] - [H^+] - [HSO_3^-]) - [SO_2 \cdot H_2O]$$

The right-hand side may be brought completely in terms of only $[HSO_3^-]$ and $[H^+]$ using the equilibrium constant expressions.

The difference $[Ca] - [S]$ represents the difference between the reagent added and the absorbed $SO_2$. The difference has been computed by Shinskey (1977) as a function of pH:

| pH | $[Ca] - [S]$ (M) |
|---|---|
| 3 | $-0.09$ |
| 4 | $-0.025$ |
| 5 | $-7.8 \times 10^{-3}$ |
| 6 | $-2.2 \times 10^{-3}$ |
| 7 | $-3.9 \times 10^{-4}$ |
| 8 | $-4.3 \times 10^{-5}$ |
| 9 | $5.0 \times 10^{-7}$ |
| 10 | $5.0 \times 10^{-5}$ |

At a pH between 8.0 and 9.0 all the sulfur is predicted to be consumed, and this represents, therefore, the optimal pH for full utilization of the calcium.

### 8.4.2 Regenerative Processes

In regenerative processes the sulfur is recovered in a usable form. One of the oldest regenerative FGD processes is the *Wellman-Lord process*. In this process the flue gas is contacted with aqueous sodium sulfite, and the dissolved $SO_2$ reacts to form sodium bisulfite:

$$Na_2SO_3 + SO_2 + H_2O \longrightarrow 2NaHSO_3$$

If excess oxygen is present in the flue gas, some of the $Na_2SO_3$ is oxidized to sodium sulfate:

$$Na_2SO_3 + \tfrac{1}{2}O_2 \longrightarrow Na_2SO_4$$

Part of the liquid stream leaving the bottom of the absorber is sent to a crystallizer where $Na_2SO_4$, which is less soluble than $Na_2SO_3$, crystallizes. The $Na_2SO_4$ solids are removed and discarded. The remaining liquid is recycled to the process. The remainder of the liquid stream from the absorber is sent to a unit where it is heated:

$$2NaHSO_3 \longrightarrow Na_2SO_3 + SO_2 + H_2O$$

The $SO_2$ gas produced is quite concentrated (approximately 85% $SO_2$ and 15% $H_2O$) and in that form can be reduced to elemental sulfur or oxidized to sulfuric acid. Finally, because some of the feed sodium is discarded with the $Na_2SO_4$, soda ash ($Na_2CO_3$) is added to the absorption tower to produce more sodium sulfite:

$$Na_2CO_3 + SO_2 \longrightarrow Na_2SO_3 + CO_2$$

The magnesium oxide (MgO) process involves scrubbing the flue gas with a slurry of MgO and recycled $MgSO_3$ and $MgSO_4$. Absorption takes place by the reactions

$$MgO + SO_2 + 6H_2O \longrightarrow MgSO_3 \cdot 6H_2O$$

$$MgO + SO_2 + 3H_2O \longrightarrow MgSO_3 \cdot 3H_2O$$

The absorbate enters a centrifuge system where the hydrated crystals of $MgSO_3$ and $MgSO_4$ are separated from the mother liquor. The liquor is returned to the absorber and the centrifuged wet cake is sent to a dryer. Regeneration takes place upon heating:

$$MgSO_3 \longrightarrow MgO + SO_2$$

$$MgSO_4 + \underset{\text{(coke)}}{\frac{1}{2}C} \longrightarrow MgO + SO_2 + \frac{1}{2}CO_2$$

The flue gas from the heating step contains about 15 to 16% $SO_2$, which can then be used for sulfuric acid production.

The catalytic process converts $SO_2$ to $H_2SO_4$ by passing the flue gases over a vanadium pentoxide ($V_2O_5$) catalyst, which oxidizes $SO_2$ to $SO_3$, followed by contacting the $SO_3$ with water to form $H_2SO_4$. In the process, gas enters the catalyst bed, after particulate removal, at temperatures of 698 to 728 K. After the catalyst bed, the $SO_3$ is contacted with water, and $H_2SO_4$ is condensed. The advantages of the process are that the system is basically simple and catalyst recycle is not necessary. Disadvantages are that expensive, corrosion-resistant materials are needed, the catalyst is easily deactivated by certain particles, and the sulfuric acid produced is usually too dilute to be salable.

## 8.5 REMOVAL OF $NO_x$ FROM EFFLUENT STREAMS

Stationary source $NO_x$ control is based on both modifications in combustion conditions (Chapter 3) and removal of $NO_x$ from exhaust gases. Combustion and design modification techniques appear to be the most economical means of achieving substantial $NO_x$ emission reductions. It is uncertain, however, whether the $NO_x$ emission reductions attainable by use of combustion modification techniques alone can provide the overall level of $NO_x$ control necessary to meet ambient air quality standards.

Typical uncontrolled and controlled $NO_x$ concentrations in utility boiler flue gases are given in Table 8.6. Flue gas treatment (FGT) methods for $NO_x$ removal are generally used together with combustion modifications. In Table 8.6 the flue gas treatment method indicated is selective catalytic reduction, a method that we will discuss shortly. To achieve an excess of 90% $NO_x$ reduction, the combination of combustion modifications (to reduce 35 to 50% of the $NO_x$ emissions) and FGT, such as by selective catalytic reduction (to remove 80 to 85% of the remaining $NO_x$) is generally more economical than FGT alone.

For a number of reasons, $NO_x$ removal from flue gases is more difficult than $SO_2$ removal, and, as a result, technology for $NO_x$ cleaning of flue gases is not as advanced as that for $SO_2$. The key problem is that NO, the principal $NO_x$ species in flue gas, is relatively insoluble and unreactive. In addition, flue gases containing NO often also contain $H_2O$, $CO_2$, and $SO_2$ in greater concentrations than NO. These species are more reactive than NO and interfere with its removal.

**TABLE 8.6** UNCONTROLLED AND CONTROLLED $NO_x$ CONCENTRATIONS (ppm) IN UTILITY BOILER FLUE GAS

| Fuel | Without control | With combustion modifications | With combustion modifications and selective catalytic reduction |
|---|---|---|---|
| Gas | 200 | 50 | 10 |
| Oil | 300 | 100 | 20 |
| Coal | 600 | 250 | 50 |

*Source:* Ando (1983).

### 8.5.1 Shell Flue Gas Treating System

There is at this time at least one commercially demonstrated dry simultaneous $NO_x/SO_x$ removal process, the Shell Flue Gas Treating System. This process was originally designed for $SO_2$ control but was found also to be adaptable for $NO_x$ control (Mobley, 1979). Flue gas is introduced at 673 K into two or more parallel passage reactors containing copper oxide (CuO) supported on alumina ($Al_2O_3$), where the $SO_2$ reacts with the copper oxide to form copper sulfate ($CuSO_4$):

$$CuO + \tfrac{1}{2}O_2 + SO_2 \longrightarrow CuSO_4$$

The $CuSO_4$ and, to a lesser extent, the CuO act as catalysts in the reduction of NO with added ammonia,

$$4NO + 4NH_3 + O_2 \longrightarrow 4N_2 + 6H_2O$$

When the reactor is saturated with copper sulfate, the flue gas is switched to a fresh reactor, and the spent reactor is regenerated. In the regeneration cycle, hydrogen is used to reduce the copper sulfate to copper,

$$CuSO_4 + 2H_2 \longrightarrow Cu + SO_2 + 2H_2O$$

producing an $SO_2$ stream of sufficient concentration for conversion to sulfur or sulfuric acid. The copper is then oxidized back to copper oxide,

$$Cu + \tfrac{1}{2}O_2 \longrightarrow CuO$$

The process can be operated in the $NO_x$-only mode by eliminating the regeneration cycle or in the $SO_x$-only mode by eliminating the ammonia injection.

### 8.5.2 Wet Simultaneous $NO_x/SO_x$ Processes

Although wet $NO_x$ removal processes do not as yet compete economically with dry $NO_x$ processes, wet simultaneous $NO_x/SO_x$ processes may be competitive with the sequential installation of dry $NO_x$ control followed by $SO_2$ control by flue gas desulfurization (FGD). The first wet simultaneous $NO_x/SO_x$ systems, called oxidation/absorption/reduction processes, evolved from FGD systems (Mobley, 1979). Since the NO is relatively insoluble in aqueous solutions, a gas-phase oxidant, such as ozone ($O_3$) or chlorine dioxide ($ClO_2$), is injected before the scrubber to convert NO to the more soluble $NO_2$. The absorbent then forms, with $SO_2$, a sulfite ion that reduces a portion of the absorbed $NO_x$ to $N_2$. The remaining $NO_x$ is removed from the wastewater as nitrate salts, while the remaining sulfite ions are oxidized to sulfate by air and removed as gypsum. Oxidation/absorption/reduction processes have the potential to remove 90% of both $SO_x$ and $NO_x$ from combustion flue gas. However, the use of a gas-phase oxidant is expensive. Chlorine dioxide, although cheaper than ozone, adds to the wastewater problems created by the nitrate salts.

Absorption/reduction processes circumvent the need for use of a gas-phase oxidant

TABLE 8.7 COMPARISON OF DRY $NO_x$ AND WET SIMULTANEOUS $NO_x/SO_2$ SYSTEMS

| | Advantages | Disadvantages |
|---|---|---|
| Dry $NO_x$ | Low capital investment | Sensitive to inlet particulate levels |
| | Simple process | Requires ammonia |
| | High $NO_x$ removal efficiency (>90%) | Possible emission of $NH_3$ and $(NH_4)_2SO_4$ |
| | Extensive tests in large units | Relative high temperatures (573 to 673 K) |
| | No waste stream generated | |
| Wet $NO_x/SO_2$ | Simultaneous $NO_x/SO_2$ removal | Expensive to process due to complexity and NO insolubility |
| | Insensitive to particulate levels | |
| | High $SO_2$ removal (>95%) | Formation of nitrates ($NO_3^-$) and other potential water pollutants |
| | | Extensive equipment requirements |
| | | Formation of low-demand byproducts |
| | | Flue gas reheat required |
| | | Only moderate $NO_x$ removal |
| | | High $SO_2$ to $NO_x$ ratios in feed required |

through the addition of a chelating compound, such as ferrous-EDTA (ethylenediamine tetracetic acid), which has an affinity for the relatively insoluble NO. The NO is absorbed into a complex with the ferrous ion, and the $SO_2$ is absorbed as the sulfite ion. Then the NO complex is reduced to $N_2$ by reaction with the sulfite ion. A series of regeneration steps recovers the ferrous chelating compound and oxidizes the sulfite to sulfate, which is removed as gypsum. Although absorption/reduction processes also have the potential to remove 90% of both $SO_x$ and $NO_x$ from combustion flue gas, a large absorber is required, and the process is sensitive to the flue gas composition of $SO_2$, $NO_x$, and $O_2$. The molar ratio of $SO_2$ to $NO_x$ must remain above approximately 2.5, and the oxygen concentration must remain low. Table 8.7 presents a comparison of the advantages and disadvantages of dry $NO_x$ and wet simultaneous $NO_x/SO_x$ systems.

### 8.5.3 Selective Noncatalytic Reduction

There are two promising routes for $NO_x$ control involving the use of ammonia, one noncatalytic homogeneous reduction and the other selective catalytic (heterogeneous) reduction. Noncatalytic ammonia injection removes $NO_x$ from effluent gases by reducing NO to $N_2$ and $H_2O$ in the presence of oxygen. This process has been discussed in detail in Chapter 3. The overall reactions are:

$$4NO + 4NH_3 + O_2 \xrightarrow{1} 4N_2 + 6H_2O$$

$$4NH_3 + 5O_2 \xrightarrow{2} 4NO + 6H_2O$$

The first reaction dominates at temperatures ranging from 1070 to 1270 K; above 1370 K the second reaction becomes significant, leading to the undesirable formation of NO.

The ammonia injection process is thus highly temperature sensitive, with maximum NO reduction occurring in the range 1200 to 1300 K. The ammonia injection $NO_x$ control system is commercially available and has been demonstrated on a number of boilers and furnaces.

### 8.5.4 Selective Catalytic Reduction

*Selective catalytic reduction* (SCR) refers to the process wherein $NO_x$ is reduced by $NH_3$ over a heterogeneous catalyst in the presence of $O_2$. The process is termed selective because the $NH_3$ preferentially reacts with $NO_x$ rather than with $O_2$. The oxygen, however, enhances the reaction and is a necessary component of the process. Because of the need for oxygen to be present, SCR is most applicable to flue gases from flue-lean firing combustion systems. The predominant reaction is 1 above. Note that, in theory, a stoichiometric amount of $NH_3$ sufficient to reduce all the NO according to reaction 1 is in a 1 : 1 ratio to the NO. In practice, molar $NH_3$ : NO ratios in noncatalytic $NH_3$ injection range from 1.5 for NO levels below 200 ppm to approaching 1.0 as the NO level increases. In selective catalytic reduction a ratio of 1.0 has typically reduced $NO_x$ emissions by 80 to 90%.

The SCR processes are relatively simple, requiring only a reactor, a catalyst, and an ammonia storage and injection system. The optimum temperature for the noncatalyzed reaction is about 1300 K. The catalyst effectively reduces the reaction temperature to the range 570 to 720 K. To avoid the need to reheat the flue gas, the reactor is usually located just after the boiler, either before or after the particulate control device.

Many different types of catalyst compositions and configurations have been developed for SCR. Initially, catalysts were developed for flue gases without particles, such as those from natural gas firing. For these applications a catalyst of platinum (Pt) on an alumina ($Al_2O_3$) support was used. Alumina is poisoned by $SO_x$, particularly $SO_3$, so titanium dioxide ($TiO_2$), which is resistant to $SO_x$ poisoning, was found to be an acceptable catalyst support. Vanadium compounds are resistant to $SO_x$ attack and also promote the reduction of $NO_x$ with ammonia. A common catalyst support is thus $TiO_2$ and $V_2O_5$.

A problem with SCR processes is the formation of solid ammonium sulfate, $(NH_4)_2SO_4$, and liquid ammonium bisulfate, $NH_4HSO_4$, both of which are highly corrosive and interfere with heat transfer. The problem is most severe with high sulfur oil firing. With low sulfur oils, the $SO_3$ is not present in sufficient quantity. Tests with coal indicate that $(NH_4)_2SO_4$ and $NH_4HSO_4$ may deposit on the fly ash or be removed from the heat exchanger surface by the erosive action of the fly ash. The formation of these two substances is minimized by reducing the $SO_3$ and $NH_3$ in the effluent and by increasing the exhaust temperature of the flue gas.

Selective catalytic reduction has achieved widespread use in Japan. By the beginning of 1985 about 160 SCR plants were in operation in Japan. About 60% of these SCR plants are being used with oil-fired utility boilers, 21% with coal-fired boilers, and 19% with gas-fired boilers (Ando, 1983, 1985).

### 8.5.5 $NO_x$ and $SO_x$ Removal by Electron Beam

In the electron beam process, an electron beam is caused to penetrate into the effluent gas stream where collisions between the electrons and gas molecules produce ions that, in turn, interact with the gas to create free atoms and radicals that will react with pollutants in the gas stream (Bush, 1980). Primary reactions induced by the electron beam are the decomposition of water and oxygen by the electrons:

$$H_2O \xrightarrow{e} H + OH$$

$$O_2 \xrightarrow{e} 2O$$

followed by reactions of the free atoms and radicals with pollutant species:

$$OH + NO \longrightarrow HNO_2$$

$$O + NO \longrightarrow NO_2$$

$$OH + NO_2 \longrightarrow HNO_3$$

$$SO_2 + O \longrightarrow SO_3$$

The essence of the process is thus that the hydroxyl radicals and oxygen atoms formed by the irradiation oxidize $NO_x$ and $SO_2$ to form the corresponding acids, which are then removed by appropriate neutralization of the acids with added basic substances, such as $Ca(OH)_2$:

$$2HNO_3 + Ca(OH)_2 \longrightarrow Ca(NO_3)_2 + 2H_2O$$

$$SO_3 + H_2O + Ca(OH)_2 \longrightarrow CaSO_4 \cdot 2H_2O$$

An electron beam is generated by accelerating electrons through a potential field (Gleason and Helfritch, 1985). The depth of penetration of the electron beam into a gas stream is proportional to the electron energy and inversely proportional to the gas density. The yield of a particular radical, such as OH, is proportional to the absorbed beam energy.

Electron beam treatment can be combined with conventional spray dryer alkali absorption and particulate collection (fabric filter or electrostatic precipitator). Flue gas exiting a boiler is first reacted in the spray dryer with lime and recycled fly ash. A typical operation would capture 50 to 60% of the $SO_2$ across the spray dryer, with the moisture content of the gas stream increasing from about 9 to approximately 13% by volume. The electron beam reactor then converts the NO and $NO_2$ to nitric acid ($HNO_3$). The acid formed in the electron beam reactor is neutralized by the dispersed alkali particles in the gas stream and further neutralized in the downstream filter. The dry scrubbing step can be focused on $SO_2$ removal with only incidental $NO_x$ reduction, due possibly to $NO-NO_2-SO_2$ reactions, since the electron beam step removes the $NO_x$.

If the electron beam reactor is operated at reduced levels of irradiation, $HNO_3$ production can be minimized, and it is theoretically possible to produce a 50/50 mixture of $NO_2$ and NO from the effluent NO. Gleason and Helfritch (1985) have proposed that reduced irradiation can be combined with subsequent contacting of the gas with an aqueous $NH_4OH$ solution:

$$NO_{(g)} + NO_{2(g)} + 2NH_4OH_{(aq)} \rightleftharpoons 2NH_4NO_2 + H_2O$$

In this process, the reduction in power consumption associated with a lower-energy beam is offset by the need to introduce ammonia.

Although the electron beam reactor is a relatively new concept, simultaneous $NO_x$ and $SO_x$ removal at efficiencies exceeding 90% have been demonstrated for high-sulfur fuels.

## PROBLEMS

**8.1.** A flue gas containing 3% $SO_2$ by volume is to be scrubbed by a fresh absorbent to remove 90% of the $SO_2$. At equilibrium, the dissolved $SO_2$ mole fraction in the absorbent is 0.0027 when the mole fraction in the gas phase is 0.03. What is the minimum $L/G$ for the absorber? Assume that in this region the equilibrium line is straight.

**8.2.** Ninety-five percent of the $SO_2$ in a process effluent stream of $SO_2$ and air is to be removed by gas absorption with water. The entering gas contains a mole fraction of $SO_2$ of 0.08; the entering water contains no $SO_2$. The water flow rate is to be twice the minimum. The entering gas flow rate is 100 mol min$^{-1}$.

**(a)** Assume for the purposes of the calculation that the equilibrium line for $SO_2$ is straight with a slope of 35. Determine the depth of the packing needed. Use the relation for $H_{0y}$ given in Problem 8.3.

**(b)** In the case in which both the operating and equilibrium lines are straight, that is, when the concentration of solute is lean ($y \ll 1$, $x \ll 1$), the integral in (8.27) can be approximated by

$$\int_{y_1}^{y_0} \frac{dy}{y - y^*}$$

which can be integrated analytically. Show that in this case

$$z_T = \frac{G}{K_y a} \frac{y_0 - y_1}{(y - y^*)_{lm}}$$

where

$$(y - y^*)_{lm} = \frac{(y_0 - y_0^*) - (y_1 - y_1^*)}{\ln[(y_0 - y_0^*)/(y_1 - y_1^*)]}$$

Repeat case (a), assuming that the operating and equilibrium lines are both straight.

**8.3.** An absorber is to be used to remove acetone from an airstream by contact with water. The entering air contains an acetone mole fraction of 0.11, and the entering water is acetone-free. The inlet gas flow rate is 10 m$^3$ min$^{-1}$. The mole fraction of acetone in the air leaving the column is to be 0.02. The equilibrium curve for acetone–water at 1 atm and 299.6 K, the conditions of operation of the tower, is given by (McCabe and Smith, 1976)

$$y = 0.33x\, e^{1.95(1-x)^2}$$

**(a)** What is the water flow rate if it is to be 1.75 times the minimum?

**(b)** What is the required height of the tower if the gas-phase HTU is given by

$$H_{0y} = 3.3\tilde{G}^{0.33}\tilde{L}^{-0.33} \text{ meters}$$

where $\tilde{G}$ and $\tilde{L}$ are the mass velocities, in kg m$^{-2}$ h$^{-1}$?

**8.4.** Benzene vapor present at a concentration of 0.030 kg benzene/kg air is to be removed by passing the gas mixture downward through a bed of silica gel at 323 K and 2 atm pressure at a linear velocity of 0.5 m s$^{-1}$ (based on the total cross-sectional area of 1 m$^2$). The bed has a packing depth of 3 m. The breakpoint will be considered that time when the effluent air has a benzene content of 0.0030 kg benzene/kg air, and the bed will be considered exhausted when the effluent air contains 0.024 kg benzene/kg air. Determine the time required to reach the breakpoint. Pertinent data are given in Example 8.3.

Under these conditions the adsorption isotherm is

$$Y^* = 0.1167X^{1.3}$$

**8.5.** A coal containing 3% sulfur by weight is burned at a rate of 50 kg s$^{-1}$ in a 500-MW power plant. Ninety percent of the $SO_2$ in the flue gas is to be removed by limestone scrubbing. Assume the limestone to be pure $CaCO_3$. Calculate the limestone feed rate needed to achieve the 90% removal assuming perfect stoichiometric reaction.

The stoichiometric ratio can be defined as the weight of reagent actually needed divided by the theoretical stoichiometric weight to remove the same quantity of $SO_2$. Stoichiometric ratios for 90% $SO_2$ removal for lime systems range from 1.05 to 1.15, while those for limestone range from 1.25 to 1.6. Using actual ratios of 1.10 and 1.40, calculate the ratio of the weight of limestone to that of lime for this flue gas.

**8.6.** A power plant flue gas contains 1000 ppm of NO and is emitted at a rate of 1000 m$^3$ s$^{-1}$ at 573 K and 1 atm. A selective catalytic reduction system is to be used to achieve 75% removal of the NO. Calculate the quantity of ammonia needed in kg h$^{-1}$.

**8.7.** Section 8.2.2 considered the design of a packed $SO_2$ absorber when the aqueous-phase $SO_2$ equilibria are explicitly accounted for. For the conditions of Example 8.2, compute and plot the tower height as a function of the percentage of $SO_2$ removed over the range of 90% to 97.5% removal. Assume $G_0 = 30$ kg-mol m$^{-2}$ h$^{-1}$, $W = 50$ m$^3$ m$^{-2}$ h$^{-1}$, $pH_0 = 10$, and $y_0 = 0.2$.

Note that to carry out this calculation it will be necessary to solve the nonlinear algebraic equation (8.54) to determine $\eta_T$ and the numerically evaluate the integral in (8.53) to find $z_T$. The nonlinear algebraic equation can be solved by Newton's method, for example with the IBM Scientific Subroutine Package (SSP) RTNI, and the integral can be evaluated numerically by the IBM SSP Gaussian quadrature subroutine DQG32.

**8.8.** In the dual alkali processes $SO_2$ is absorbed from the flue gas by an $Na_2SO_3$ solution. The spent solution is then sent to a regenerating system where lime is added to precipitate $CaSO_3$ and regenerate $Na_2SO_3$. In this problem we want to determine the optimum pH at which to carry out the scrubbing step. The overall reaction in the system is

$$Na_2SO_3 + H_2O + SO_2 + CO_2 \rightleftharpoons Na^+ + H^+ + OH^- + HSO_3^- + SO_3^{2-} + HCO_3^- + CO_3^{2-}$$

Equilibrium constants for the carbonate systems at 333 K are:

$$\frac{[CO_2 \cdot H_2O]}{p_{CO_2}} = K_{hc} = 0.0163 \text{ M atm}^{-1}$$

$$\frac{[HCO_3^-][H^+]}{[CO_2 \cdot H_2O]} = K_{c1} = 10^{-6.35} \text{ M}$$

$$\frac{[CO_3^{2-}][H^+]}{[HCO_3^-]} = K_{c2} = 10^{-10.25} \text{ M}$$

The sodium mass in the system is just

$$[Na] = [Na^+]$$

whereas that for sulfur is

$$[S] = [SO_2 \cdot H_2O] + [HSO_3^-] + [SO_3^{2-}]$$

Derive an equation for [Na] as a function of pH, $p_{CO_2}$, and [S]. Calculate and plot [Na] − [S], in M, as a function of pH over the range pH = 2 to 7 for [S] = 0.01 M at $p_{CO_2}$ = 0.16 atm. For these conditions, what is the pH of complete sodium utilization?

## REFERENCES


ANDO, J. "$NO_x$ Abatement for Stationary Sources in Japan," U.S. Environmental Protection Agency Project Summary No. EPA-600/57-83-027 (1983).

ANDO, J. "Recent Developments in $SO_2$ and $NO_x$ Abatement Technology for Stationary Sources in Japan," U.S. Environmental Protection Agency Project Summary No. EPA-600/57-85-040 (1985).

BEACHLER, D. S., and JOSEPH, G. T. "Emission Regulations and Air Pollution Control Equipment for Industrial and Utility Boilers," *Environ. Prog.*, *3*, 44–50 (1984).

BUSH, J. R. "Removal of $NO_x$ and $SO_x$ from Flue Gases Using Electron Beam Irradiation," Final Report to U.S. Department of Energy, Contract No. EP-78-C-02-4902 (1980).

COOLEY, T. E., and DETHLOFF, W. L. "Field Tests Show Membrane Processing Attractive," *Chem. Eng. Prog.*, 45–50 (October 1985).

ELLIOT, R. A., MATYAS, A. G., GOODFELLOW, H. D., and NENNINGER, E. H. "$SO_2$ Emission Control in Smelters," *Environ. Prog.*, *1*, 261–267 (1982).

FELLMAN, R. T., and CHEREMISINOFF, P. N. "A Survey of Lime/Limestone Scrubbing for $SO_2$ Removal," in *Air Pollution Control and Design Handbook: Part 2*, P. N. Cheremisinoff and R. A. Young, Eds., Marcel Dekker, New York, 813–834 (1977).

GLEASON, R. J., and HELFRITCH, D. J. "High-Efficiency $NO_x$ and $SO_x$ Removal by Electron Beam," *Chem. Eng. Prog.*, 33–38 (October 1985).

HARNED, H. S., and OWEN, B. B. *The Physical Chemistry of Electrolyte Solutions*, Van Nostrand Reinhold, New York (1958).

JOSEPH, G. T., and BEACHLER, D. S. "Control of Gaseous Emissions," U.S. Environmental Protection Agency Report No. EPA-450/2-81-005 (1981).

MAAHS, H. G. "Sulfur Dioxide/Water Equilibrium between 0° and 50°C. An Examination of

Data at Low Concentrations,'' in *Heterogeneous Atmospheric Chemistry*, D. R. Schryer, Ed., American Geophysical Union, Washington, DC, 187–195 (1982).

McCabe, W. L., and Smith J. C. *Unit Operations of Chemical Engineering*, 3rd ed., McGraw-Hill, New York (1976).

Mobley, J. D. ''Flue Gas Treatment Technology for $NO_x$ Control,'' in *Proceedings of the Third Stationary Source Combustion Symposium*, Vol. 2, *Advanced Processes and Special Topics*. U.S. Environmental Protection Agency Report No. EPA-600/7-79-0506, 245–281 (1979).

Shinskey, F. G. ''pH Controls for $SO_2$ Scrubbers,'' in *Air Pollution Control and Design Handbook: Part 2*, P. N. Cheremisinoff and R. A. Young, Eds., Marcel Dekker, New York, 899–913 (1977).

Treybal, R. E. *Mass Transfer Operations*, 2nd ed., McGraw-Hill, New York (1968).

# 9

# Optimal Air Pollution Control Strategies

In general, the goal of air pollution abatement is the meeting of a set of air quality standards (see Table 1.9). Air pollution abatement programs can be divided into two categories:

1. Long-term control
2. Short-term control (episode control)

Long-term control strategies involve a legislated set of measures to be adopted over a multiyear period. Short-term (or episode) control involves shutdown and slowdown procedures that are adopted over periods of several hours to several days under impending adverse meteorological conditions. An example of a short-term strategy is the emergency procedures for fuel substitution by coal-burning power plants in Chicago when $SO_2$ concentrations reach certain levels (Croke and Booras, 1969).

Figure 9.1 illustrates the elements of a comprehensive regional air pollution control strategy, consisting of both long- and short-term measures. Under each of the two types of measures are listed some of the requirements for setting up the control strategy. The air quality objectives of long- and short-term strategies may be quite different. For long-term control, a typical objective might be to reduce to a specified value the expected number of days per year that the maximum hourly average concentration of a certain pollutant exceeds a given value. On the other hand, a goal of short-term control is ordinarily to keep the maximum concentration of a certain pollutant below a given value on that particular day.

The alternatives for abatement policies depend on whether long- or short-term

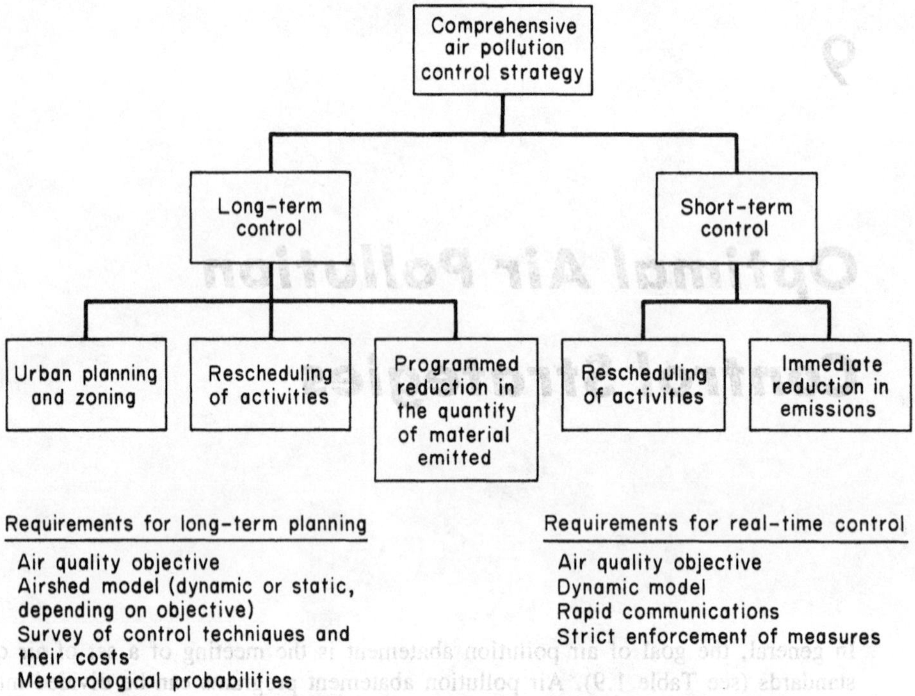


**Figure 9.1** Elements of a comprehensive air pollution control strategy for a region.

control measures are being considered. Some examples of long-term air pollution control policies are:

- Enforcing standards that restrict the pollutant content of combustion exhaust
- Requiring used motor vehicles to be outfitted with exhaust control devices
- Requiring new motor vehicles to meet certain emissions standards
- Prohibiting or encouraging the use of certain fuels in power plants
- Establishing zoning regulations for the emission of pollutants
- Encouraging the use of vehicles powered by electricity or natural gas for fleets

Short-term controls are of an emergency nature and are more stringent than long-term controls that are continuously in effect. Examples of short-term control strategies are:

- Prohibiting automobiles with fewer than three passengers from using certain lanes of freeways
- Prohibiting the use of certain fuels in some parts of the city
- Prohibiting certain activities, such as incineration of refuse

The objectives of a short-term control system are to continuously monitor concentrations at a number of stations (and perhaps also at the stacks of a number of important emission sources) and, with these measurements and weather predictions as a basis, to prescribe actions that must be undertaken by sources to avert dangerously high concentrations. Figure 9.2 shows in schematic, block-diagram form a possible real-time control system for an airshed. Let us examine each of the loops. The innermost loop refers to an automatic stack-monitoring system of major combustion and industrial sources. If the stack emissions should exceed the emission standards, the plant would automatically curtail its processes to bring stack emissions below the standard. The emission standards would normally be those legislated measures currently in force. The next loop represents a network of automatic monitoring stations that feed their data continuously to a central computer that compares current readings with air quality "danger" values. These values are not necessarily the same as the air quality standards discussed earlier. For example, if the air quality standard for $SO_2$ is 0.14 ppm for a 24-h average, the alert level might be 0.5 ppm for a 1-h average. In such a system one would not rely entirely on measurements to initiate action, since once pollutants reach dangerous levels it is difficult to restore the airshed quickly to safe levels. Thus we would want to predict the weather to 3 to 48 h in advance, say, and use the information from this prediction combined with the feedback system in deciding what action, if any, to take.

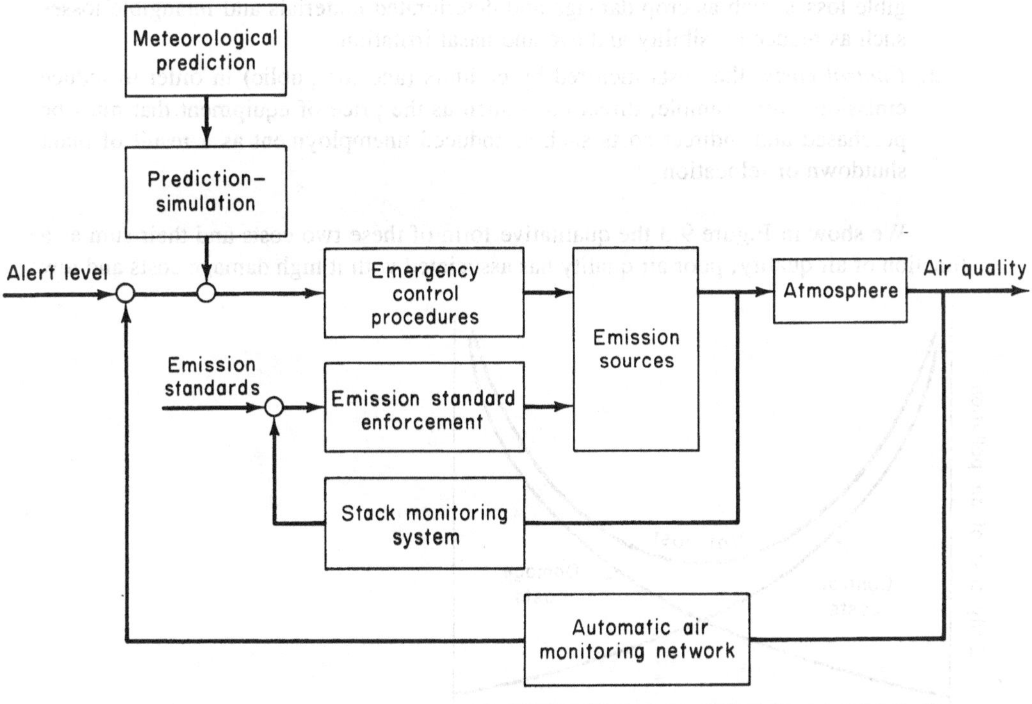


**Figure 9.2** Elements of a real-time air pollution control system involving automatic regulation of emission sources based on atmospheric monitoring.

We refer the interested reader to Rossin and Roberts (1972), Kyan and Seinfeld (1973), and Akashi and Kumamoto (1979), for studies of short-term air pollution control.

## 9.1 LONG-TERM AIR POLLUTION CONTROL

Let us focus our attention primarily on long-term control of air pollution for a region. It is clear that potentially there are a number of control policies that could be applied by an air pollution control agency to meet desired air quality goals. The question then is: How do we choose the "best" policy from among all the possibilities? It is reasonable first to establish criteria by which the alternative strategies are to be judged.

Within the field of economics, there is a hierarchy of techniques called cost/benefit analysis, within which all the consequences of a decision are reduced to a common indicator, invariably dollars. This analysis employs a single measure of merit, namely the total cost, by which all proposed programs can be compared. A logical inclination is to use total cost as the criterion by which to evaluate alternative air pollution abatement policies. The total cost of air pollution control can be divided into a sum of two costs:

1. *Damage costs:* the costs to the public of living in polluted air, for example, tangible losses such as crop damage and deteriorated materials and intangible losses such as reduced visibility and eye and nasal irritation
2. *Control costs:* the costs incurred by emitters (and the public) in order to reduce emissions, for example, direct costs such as the price of equipment that must be purchased and indirect costs such as induced unemployment as a result of plant shutdown or relocation

We show in Figure 9.3 the qualitative form of these two costs and their sum as a function of air quality; poor air quality has associated with it high damage costs and low

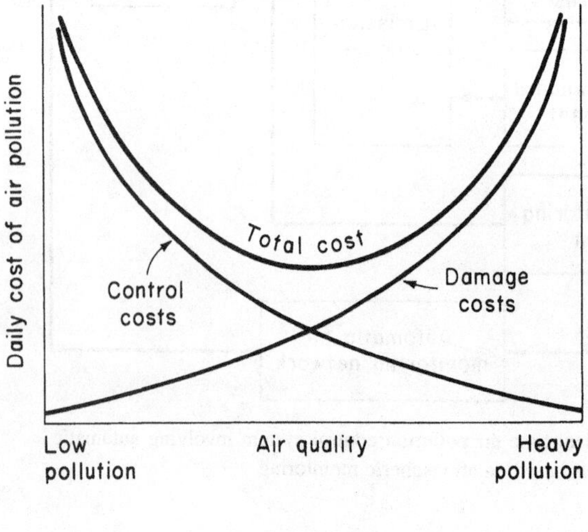


**Figure 9.3** Total cost of air pollution as a sum of control and damage costs.

control costs, whereas good air quality is just the reverse. Cost/benefit principles indicate that the optimal air quality level is at the minimum of the total cost curve. The key problem is: How do we compute these curves as a function of air quality? Consider first the question of quantifying damage costs.

Damage costs to material and crops, cleaning costs due to soiling, and so on, although not easy to determine, can be estimated as a function of pollutant levels (Ridker, 1967). However, there is the problem of translating into monetary value the effects on health resulting from air pollution. One way of looking at the problem is to ask: How much are people willing to spend to lower the incidence of disease, prevent disability, and prolong life? Attempts at answering this question have focused on the amount that is spent on medical care and the value of earnings missed as a result of sickness or death. Lave and Seskin (1970) stated that "while we believe that the value of earnings foregone as a result of morbidity and mortality provides a gross underestimate of the amount society is willing to pay to lessen pain and premature death caused by disease, we have no other way of deriving numerical estimates of the dollar value of air pollution abatement." Their estimates are summarized in Table 9.1. These estimates are so difficult to make that we must conclude that it is generally not possible to derive a quantitative damage-cost curve such as that shown in Figure 9.3.

There are actually other reasons why a simple cost/benefit analysis of air pollution control is not feasible. Cost is not the only criterion for judging the consequences of a control measure. Aside from cost, social desirability and political acceptability are also important considerations. For example, a policy relating to zoning for high and low emitting activities would have important social impacts on groups living in the involved areas, and it would be virtually impossible to quantify the associated costs.

It therefore appears that the most feasible approach to determining air pollution abatement strategies is to treat the air quality standards as constraints not to be violated and to seek the combination of strategies that achieves the required air quality at minimum cost of control. In short, we attempt to determine the minimum cost of achieving a given air quality level through emission controls (i.e., to determine the control cost curve in Figure 9.3).

In the case of the control cost curve, it is implicitly assumed that *least-cost* control

**TABLE 9.1** ESTIMATED HEALTH COSTS OF AIR POLLUTION IN 1970

| Disease | Total annual estimated cost (millions of dollars) | Estimated percentage decrease in disease for a 50% reduction in air pollution | Estimated savings incurred for a 50% reduction in air pollution (millions of dollars) |
|---|---|---|---|
| Respiratory disease | 4887 | 25 | 1222 |
| Lung cancer | 135 | 25 | 33 |
| Cardiovascular disease | 4680 | 10 | 468 |
| Cancer | 2600 | 15 | 390 |
| | | | 2100 |

*Source:* Lave and Seskin (1970).

strategies are selected in reaching any given abatement level. There will usually be a wide assortment of potential control strategies that can be adopted to reduce ambient pollution a given amount. For instance, a given level of $NO_x$ control in an urban area could be achieved by reducing emissions from various types of sources (e.g., power plants, industrial boilers, automobiles, etc.). The range of possible strategies is further increased by alternative control options for each source (e.g., flue gas recirculation, low-excess-air firing, or two-stage combustion for power plant boilers). Out of all potential strategies, the control cost curve should represent those strategies that attain each total emission level at minimum control cost.

## 9.2 A SIMPLE EXAMPLE OF DETERMINING A LEAST-COST AIR POLLUTION CONTROL STRATEGY

Let us now consider the formulation of the control method–emission-level problem for air pollution control, that is, to determine that combination of control measures employed that will give mass emissions not greater than prescribed values and do so at least cost. Let $E_1, \ldots, E_N$ represent measures of the mass emissions* of $N$ pollutant species (e.g., these could be the total daily emissions in the entire airshed in a particular year or the mass emissions as a function of time and location during a day); then we can express the control cost $C$ (say in dollars per day) as $C = C(E_1, \ldots, E_N)$. To illustrate the means of minimizing $C$, we take a simple example (Kohn, 1969).

Let us consider a hypothetical airshed with one industry, cement manufacturing. The annual production is $2.5 \times 10^6$ barrels of cement, but this production is currently accompanied by 2 kg of particulate matter per barrel lost into the atmosphere. Thus the uncontrolled particulate emissions are $5 \times 10^6$ kg yr$^{-1}$. It has been determined that particulate matter emissions should not exceed $8 \times 10^5$ kg yr$^{-1}$. There are two available control measures, both electrostatic precipitators: type 1 will reduce emissions to 0.5 kg bbl$^{-1}$ and costs 0.14 dollars bbl$^{-1}$; type 2 will reduce emissions to 0.2 kg bbl$^{-1}$ but costs 0.18 dollar bbl$^{-1}$. Let

$$X_1 = \text{bbl yr}^{-1} \text{ of cement produced with type 1 units installed}$$

$$X_2 = \text{bbl yr}^{-1} \text{ of cement produced with type 2 units installed}$$

The total cost of control in dollars is thus

$$C = 0.14X_1 + 0.18X_2 \tag{9.1}$$

We would like to minimize $C$ by choosing $X_1$ and $X_2$. But $X_1$ and $X_2$ cannot assume any values; their total must not exceed the total cement production,

$$X_1 + X_2 \leq 2.5 \times 10^6 \tag{9.2}$$

and a reduction of at least $4.2 \times 10^6$ kg of particulate matter must be achieved,

$$1.5X_1 + 1.8X_2 \geq 4.2 \times 10^6 \tag{9.3}$$

*Note that $E_i$ is 0 if $i$ is purely a secondary pollutant.

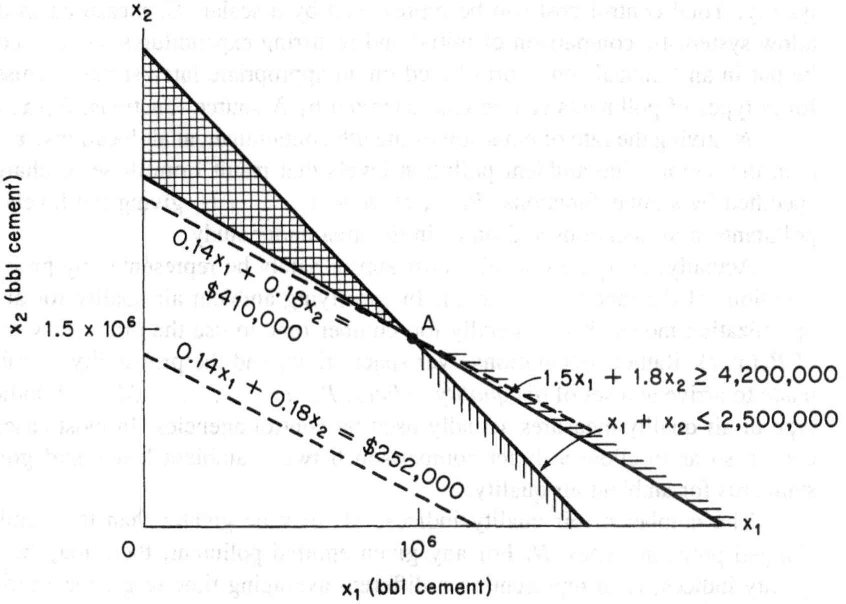


**Figure 9.4** Least-cost strategy for cement industry example (Kohn, 1969).

and both $X_1$ and $X_2$ must be nonnegative,

$$X_1, X_2 \geq 0 \tag{9.4}$$

The complete problem is to minimize $C$ subject to (9.2)–(9.4). In Figure 9.4 we have plotted lines of constant $C$ in the $X_1$-$X_2$ plane. The lines corresponding to (9.2) and (9.3) are also shown. Only $X_1$, $X_2$ values in the crosshatched region are acceptable. Of these, the minimum cost set is $X_1 = 10^6$ and $X_2 = 1.5 \times 10^6$ with $C = 410{,}000$ dollars. If we desire to see how $C$ changes with the allowed particulate emissions, we solve this problem repeatedly for many values of the emission reduction (we illustrated the solution for a reduction of $8 \times 10^5$ kg of particulate matter per year) and plot the minimum control cost $C$ as a function of the amount of reduction (see Problem 9.1).

The problem that we have described falls within the general framework of *linear programming* problems. Linear programming refers to minimization of a linear function subject to linear equality or inequality constraints. Its application requires that control costs and reductions remain constant, independent of the level of control.

## 9.3 GENERAL STATEMENT OF THE LEAST-COST AIR POLLUTION CONTROL PROBLEM

The first step in formulating the least-cost control problem mathematically is to put the basic parameters of the system into symbolic notation. There are three basic sets of variables in the environmental control system: control cost, emission levels, and air quality. Total control cost can be represented by a scalar, $C$, measured in dollars. To allow systematic comparison of initial and recurring expenditures, control costs should be put in an "annualized" form based on an appropriate interest rate. Emission levels for $N$ types of pollutants can be characterized by $N$ source functions, $E_n(\boldsymbol{x}, t)$, $n = 1, \ldots, N$, giving the rate of emission of the $n$th contaminant at all locations, $\boldsymbol{x}$, and times, $t$, in the region. The ambient pollution levels that result from these discharges can be specified by similar functions, $P_h(\boldsymbol{x}, t)$, $h = 1, \ldots, H$, giving the levels of $H$ final pollutants at all locations and times in the area under study.

Actually, air quality would most appropriately be represented by probability distributions of the functions $P_h(\boldsymbol{x}, t)$. In specifying ambient air quality for an economic optimization model, it is generally too cumbersome to use the probability distributions of $P_h(\boldsymbol{x}, t)$. Rather, integrations over space, time, and the probability distributions are made to arrive at a set of *air quality indices*, $P_m$, $m = 1, \ldots, M$. Such indices are the type of air quality measures actually used by control agencies. In most cases, they are chosen so as to allow a direct comparison between ambient levels and governmental standards for ambient air quality.

The number of air quality indices, $M$, may be greater than the number of discharged pollutant types, $N$. For any given emitted pollutant, there may be several air quality indices, each representing a different averaging time (e.g., the yearly average, maximum 24-h, or maximum 1-h ambient levels). Multiple indices will also be used to represent multiple receptor locations, seasons, or times of day. Further, a single emitted pollutant may give to rise to more than one type of ambient species. For instance, sulfur dioxide emissions contribute to both sulfur dioxide and sulfate air pollution.

Among the three sets of variables, two functional relationships are required to define the least-cost control problem. First, there is the control cost–emission function that gives the minimum cost of achieving any level and pattern of emissions. It is found by taking each emission level, $E_n(\boldsymbol{x}, t)$, $n = 1, \ldots, N$, technically determining the subset of controls that exactly achieves that level, and choosing the specific control plan with minimum cost, $C$. This function, the minimum cost of reaching various emission levels, will be denoted by $G$,

$$C = G[E_1(\boldsymbol{x}, t), \ldots, E_N(\boldsymbol{x}, t)] \tag{9.5}$$

Second, there is the discharge–air quality relationship. This is a physicochemical relationship that gives expected air quality levels, $P_m$, as functions of discharge levels, $E_n(\boldsymbol{x}, t)$. For each air quality index, $P_m$, this function will be denoted by $F_m$,

$$P_m = F_m[E_1(\boldsymbol{x}, t), \ldots, E_N(\boldsymbol{x}, t)], \quad m = 1, \ldots, M \tag{9.6}$$

With the definitions above, we can make a general mathematical statement of the minimal-cost air pollution control problem. To find the minimal cost of at least reaching air quality objectives $P_m^\circ$, choose those

$$E_n(\boldsymbol{x}, t) \qquad n = 1, \ldots, N$$

that minimize

$$C = G[E_1(\boldsymbol{x}, t), \ldots, E_N(\boldsymbol{x}, t)] \tag{9.7}$$

subject to

$$F_m[E_1(x, t), \ldots, E_N(x, t)] \leq P_m^\circ \qquad m = 1, \ldots, M$$

Thus one chooses the emission levels and patterns that have the minimum control cost subject to the constraint that they at least reach the air quality goals.

## 9.4 A LEAST-COST CONTROL PROBLEM FOR TOTAL EMISSIONS

The problem (9.7), though simply stated, is extremely complex to solve, because, as stated, one must consider all possible spatial and temporal patterns of emissions as well as total emission levels. It is therefore useful to remove the spatial and temporal dependence of the emissions and air quality. Let us consider, therefore, minimizing the cost of reaching given levels of total regional emissions. We assume that:

- The spatial and temporal distributions of emissions can be neglected. Accordingly, the discharge functions, $E_n(x, t)$, $n = 1, \ldots, N$, can be more simply specified by, $E_n$, $n = 1, \ldots, N$, that are measures of total regionwide emissions.
- The air quality constraints can be linearly translated into constraints on the total magnitude of emissions in the region of interest.
- The problem is static (i.e., the optimization is performed for a fixed time period in the future).
- There are a finite number of emission source types. For each source type, the available control activities have constant unit cost and constant unit emission reductions.

With these assumptions, the problem of minimizing the cost of reaching given goals for total emissions can be formulated in the linear programming framework of Section 9.2. Table 9.2 summarizes the parameters for this linear programming problem. The mathematical statement of the problem is as follows: Find $X_{ij}$, $i = 1, \ldots, I$ and $j = 1, \ldots, J_i$ that minimize

$$C = \sum_{i=1}^{I} \sum_{j=1}^{J_i} c_{ij} X_{ij} \tag{9.8}$$

subject to

$$\sum_{i=1}^{I} \sum_{j=1}^{J_i} e_{in}(1 - b_{ijn}) X_{ij} \leq E_n \qquad \text{for } n = 1, \ldots, N \tag{9.9}$$

$$\sum_{j=1}^{J_i} A_{ij} X_{ij} \leq S_i \qquad \text{for } i = 1, \ldots, I \tag{9.10}$$

and

$$X_{ij} \geq 0 \qquad \text{for } i = 1, \ldots, I; \ \ j = 1, \ldots, J_i \tag{9.11}$$

**TABLE 9.2 PARAMETERS FOR THE LEAST-COST PROBLEM FOR TOTAL EMISSIONS**

| Parameter | | Definition |
|---|---|---|
| $X_{ij}$ | $i = 1, \ldots, I$<br>$j = 1, \ldots, J_i$ | The number of units of the $j$th control activity applied to source type $i$ (e.g., the number of a certain control device added to 1980 model year vehicles or the amount of natural gas substituted for fuel oil in power plant boilers). The total number of source types is $I$; the number of control alternatives for the $i$th source type is $J_i$. |
| $c_{ij}$ | $i = 1, \ldots, I$<br>$j = 1, \ldots, J_i$ | The total annualized cost of one unit of control type $j$ applied to source type $i$. |
| $C$ | | The total annualized cost for the control strategy as specified by all the $X_{ij}$. |
| $E_n$ | $i = 1, \ldots, N$ | The uncontrolled (all $X_{ij} = 0$) emission rate of the $n$th pollutant as specified by all $X_{ij}$ (e.g., the resultant total $NO_x$ emission level in kg day$^{-1}$). There are $N$ pollutants. |
| $e_{in}$ | $i = 1, \ldots, I$<br>$n = 1, \ldots, N$ | The uncontrolled (all $X_{ij} = 0$) emission rate of the $n$th pollutant from the $i$th source (e.g., the $NO_x$ emissions from power plant boilers under no controls). |
| $b_{ijn}$ | $i = 1, \ldots, I$<br>$j = 1, \ldots, J_i$<br>$n = 1, \ldots, N$ | The fractional emission reduction of the $n$th pollutant from the $i$th source attained by applying one unit of control, type $j$ (e.g., the fractional $NO_x$ emission reduction from power plant boilers attained by substituting one unit of natural gas for fuel oil). |
| $S_i$ | $i = 1, \ldots, I$ | The number of units of source type $i$ (e.g., the number of 1980 model year vehicles or the number of power plant boilers). |
| $A_{ij}$ | $i = 1, \ldots, I$<br>$j = 1, \ldots, J_i$ | The number of units of source type $i$ controlled by one unit of control type $j$ (e.g., the number of power plants controlled by substituting one unit of natural gas for fuel oil). |

In this linear programming problem, (9.8) is the objective function, and (9.9)–(9.11) are the constraints. Equation (9.9) represents the constraint of at least attaining the specified emission levels, $E_n$. Equations (9.10) and (9.11) represent obvious physical restrictions, namely not being able to control more sources than those that exist and not using negative controls.

Solution techniques are well developed for linear programming problems, and computer programs are available that accept numerous independent variables and constraints. Thus the solution to the problem is straightforward once the appropriate parameters have been chosen. The results are the minimum cost, $C$, and the corresponding set of control methods, $X_{ij}$, associated with a least-cost strategy for attaining any emission levels, $E_n$.

More generality is introduced if we do not translate the air quality constraints linearly into emission constraints. Rather, we may allow for nonlinear relationships between air quality and total emissions and can include atmospheric interaction between emitted pollutants to produce a secondary species. The general least-cost control problem can then be restated as: Choose

$$E_n \qquad n = 1, \ldots, N$$

to minimize

$$C = G(E_n) \tag{9.12}$$

subject to

$$F_m(E_n) \leq P_m \qquad m = 1, \ldots, M$$

Here $G(E_n)$ represents the minimum cost of attaining various total emission levels. This function can be found by linear programming. The functions, $F_m(E_n)$, represent the air quality–emission relationships. These can be found by a variety of means, such as empirical/statistical or physicochemical models (Seinfeld, 1986). If linear functions are adopted for the $f_m(E_n)$, this case degenerates into that above. In general, however, the air quality–emission relationships can be nonlinear and can involve interactions between two or more types of emissions.

A hypothetical example of the solution to (9.12) for two emitted contaminants ($E_1$, $E_2$) and two final pollutants ($P_1$, $P_2$) is illustrated in Figure 9.5. The axes of the graph measure total emission levels of the two contaminants, $E_1$ and $E_2$. The curves labeled $C_1$, $C_2$, and so on, are iso-cost curves determined by repeated application of a linear programming submodel. Along any curve labeled $C_k$, the minimum cost of reaching any point on that curve is $C_k$. As emission levels fall (downward and to the left in the graph), control costs rise. Thus $C_1 < C_2 < \ldots < C_5$. The air quality constraints are represented by the two curves, $P_1$ and $P_2$, derived from a nonlinear air quality–emission level relationship. The constraint of at least reaching air quality level $P_1$ for the first pollutant requires that emissions be reduced below the curve. The constraint that air quality be at least as good as $P_2$ for the second pollutant requires that emissions be reduced to the left of the $P_2$ curve. The emission levels that satisfy both air quality

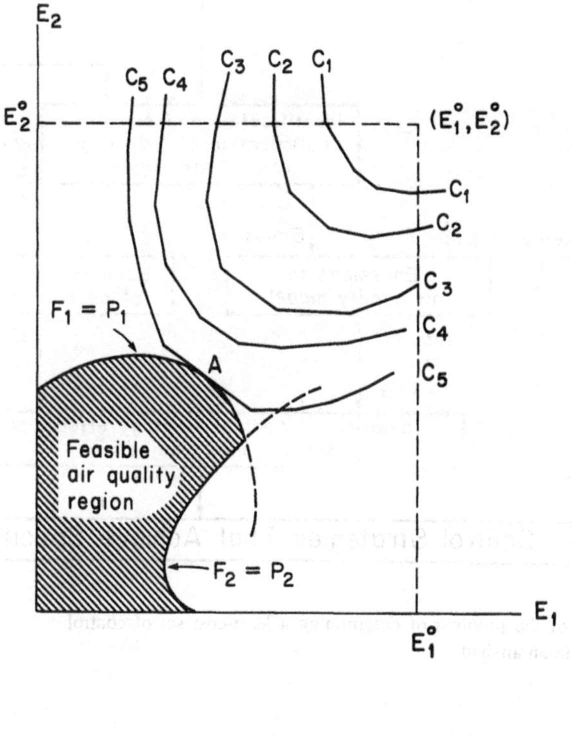


**Figure 9.5** Iso-cost lines in the plane of emission levels of two pollutants, $E_1$ and $E_2$, showing a feasible region of air quality defined by the curves $F_1(E_1, E_2) \leq P_1$ and $F_2(E_1, E_2) \leq P_2$.

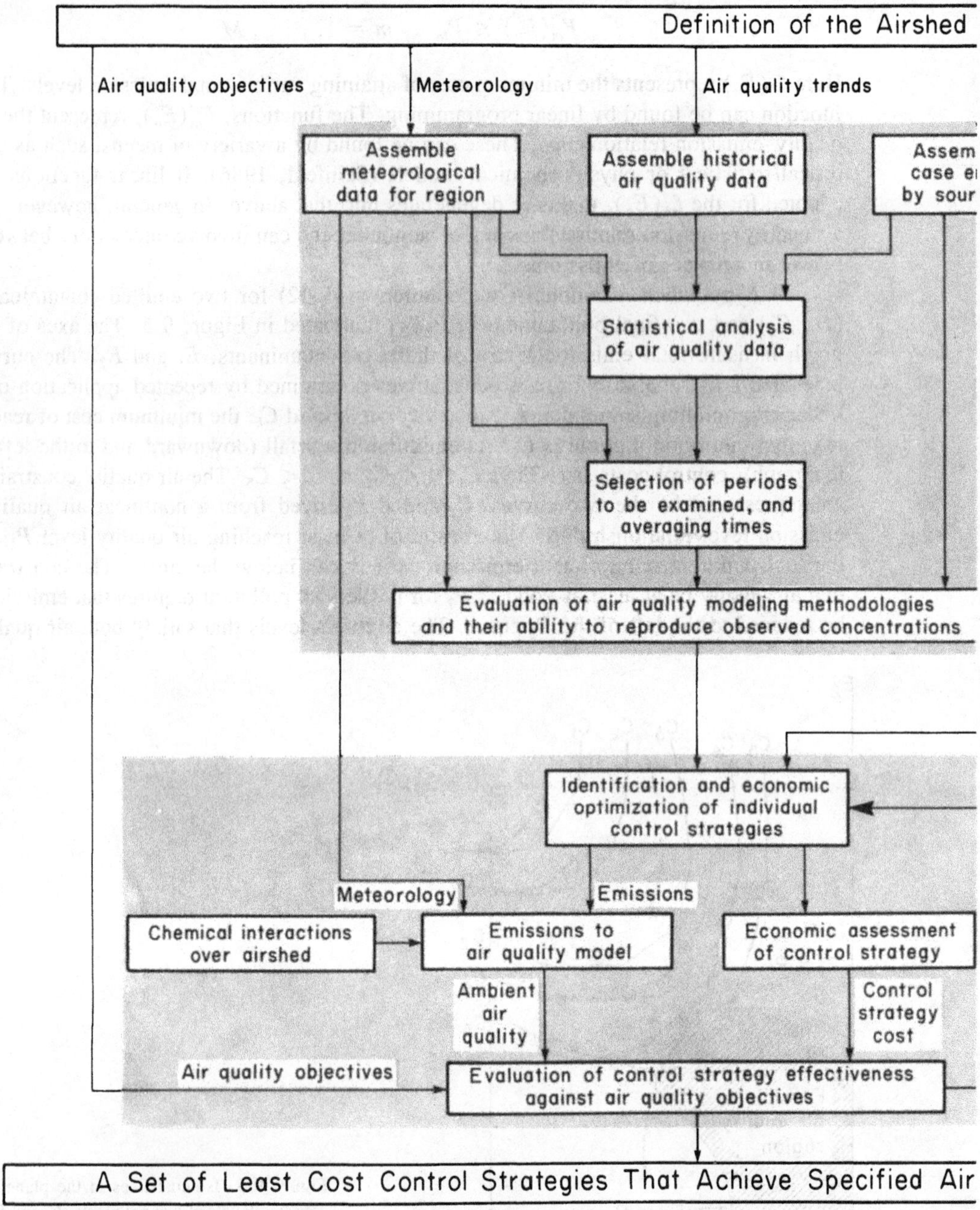


**Figure 9.6** Comprehensive structure of the problem of determining a least-cost set of control actions to achieve specified air quality in an airshed.

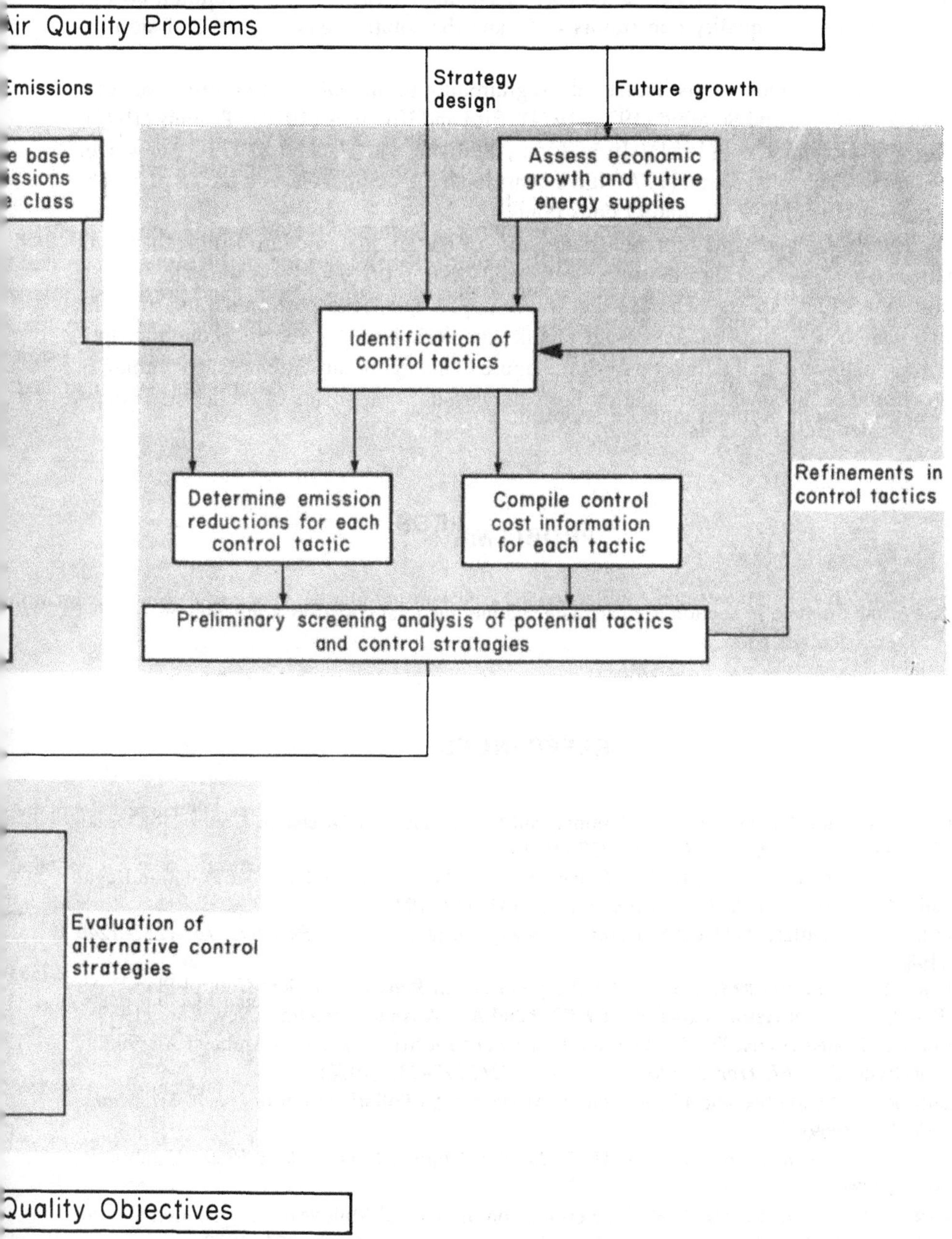

\ir Quality Problems
:missions
Strategy design
Future growth
e base
issions
class
Assess economic growth and future energy supplies
Identification of control tactics
Refinements in control tactics
Determine emission reductions for each control tactic
Compile control cost information for each tactic
Preliminary screening analysis of potential tactics and control stratagies
Evaluation of alternative control strategies
Quality Objectives

constraints lie in the crosshatched admissible air quality region. The minimum cost of meeting the two air quality constraints is $C_5$ and the solution is to reduce emissions to point A.

For applications of mathematical programming to air pollution control, we refer the reader to Kyan and Seinfeld (1972, 1974), Bird and Kortanek (1974), Trijonis (1974), Kohn (1978), and Cass (1981). In addition, Sullivan and Hackett (1973), Schweizer (1974), and Dejax and Gazis (1976) have considered the optimal electric power dispatching problem to achieve air quality constraints.

Figure 9.6 gives a comprehensive picture of the air quality control problem for an airshed. The large block in the upper left-hand portion of Figure 9.6 indicates the air quality modeling aspects, whereas that in the upper right-hand portion summarizes the identification of control tactics. Both inputs then feed into the overall economic optimization in the box in the lower part of the figure. This figure thus attempts to summarize the material in this chapter indicating how the various components of the airshed abatement problem must be attacked.

## PROBLEMS

**9.1.** For the example in Section 9.2 calculate and plot the total cost $C$ as a function of the level of emission reduction.

## REFERENCES

AKASHI, H., and KUMAMOTO, H. "Optimal Sulfur Dioxide Gas Discharge Control for Osaka City, Japan," *Automatica, 15*, 331–337 (1979).

BIRD, C. G., and KORTANEK, K. O. "Game Theoretic Approaches to Some Air Pollution Regulation Problems," *Socio-Econ. Plan. Sci., 8*, 141–147 (1974).

CASS, G. R. "Sulfate Air Quality Control Strategy Design," *Atmos. Environ., 15*, 1227–1249 (1981).

CROKE, E. J., and BOORAS, S. G. "The Design of an Air Pollution Incident Control Plan," Air Pollution Control Assoc. Paper No. 69-99, 62nd APCA Annual Meeting, New York (1969).

DEJAX, P. J., and GAZIS, D. C. "Optimal Dispatch of Electric Power with Ambient Air Quality Constraints," *IEEE Trans. Autom. Control, AC21*, 227–233 (1976).

KOHN, R. E. "A Mathematical Programming Model for Air Pollution Control," *Sch. Sci. Math.*, 487–494 (1969).

KOHN, R. E. *A Linear Programming Model for Air Pollution Control*, MIT Press, Cambridge, MA (1978).

KYAN, C. P., and SEINFELD, J. H. "Determination of Optimal Multiyear Air Pollution Control Policies," *J. Dyn. Syst. Meas. and Control Trans. ASME, G94*, 266–274 (1972).

KYAN, C. P., and SEINFELD, J. H. "Real-Time Control of Air Pollution," *AIChE J., 19*, 579–589 (1973).

KYAN, C. P., and SEINFELD, J. H. "On Meeting the Provisions of the Clean Air Act," *AIChE J.*, *20*, 118–127 (1974).

LAVE, L. B., and SESKIN, E. P. "Air Pollution and Human Health," *Science*, *169*, 723–733 (1970).

RIDKER, R. *Economic Costs of Air Pollution: Studies in Measurement*, Praeger, New York (1967).

ROSSIN, A. D., and ROBERTS, J. J. "Episode Control Criteria and Strategy for Carbon Monoxide," *J. Air Pollut. Control Assoc.*, *22*, 254–259 (1972).

SCHWEIZER, P. F., "Determining Optimal Fuel Mix for Environmental Dispatch," *IEEE Trans. Autom. Control*, *AC19*, 534–537 (1974).

SEINFELD, J. H. *Atmospheric Chemistry and Physics of Air Pollution*, Wiley-Interscience, New York (1986).

SULLIVAN, R. L., and HACKETT, D. F. "Air Quality Control Using a Minimum Pollution-Dispatching Algorithm," *Environ. Sci. Technol.*, *7*, 1019–1022 (1973).

TRIJONIS, J. C. "Economic Air Pollution Control Model for Los Angeles County in 1975—General Least Cost—Air Quality Model," *Environ. Sci. Technol.*, *8*, 811–826 (1974).

# Appendix A

# ELECTROPHORETIC MIGRATION FOR AEROSOL MEASUREMENT

An electrostatic force of

$$F_e = qE \tag{A.1}$$

acts on a particle that carries charge $q$ when exposed to an electric field of strength $E$. At steady-state, that force is opposed by the aerodynamic drag so that the particle migrates at velocity

$$v_E = qBE = ZE \tag{A.2}$$

where $B = C_c/(3\pi\mu d_p)$ is the mechanical mobility of the particle under the prevailing atmospheric conditions. $Z = qB$ is known as the *electrical mobility* of the particle, i.e., the migration velocity per unit of electric field strength.

The charge on the particle is $q = n_e e$, where $n_e$ is the number of charges that the particle carries, and $e$ is the elementary unit of charge. Particles smaller than about 1 $\mu$m in diameter can be charged so that most have $q = \pm e$, with only a small fraction carrying more than one elementary charge. The electrical mobility of such singly-charged submicron particles is a simple function of particle size, at least for spherical particles. The so-called *differential mobility analyzer* (DMA) takes advantage of the resulting unique relationship between electrical mobility and particle size to enable high resolution measurements of particle size distributions in this important size range.

The DMA consists of a flow channel that is confined by two electrodes. These are usually coaxial cylinders (Knutson and Whitby, 1974), as illustrated in Fig. A.1, although parallel plates, or disks with flow in the radial direction have also been used. We shall consider the coaxial cylinder design, since that is most common. A small flow of aerosol, which includes a fraction of charged particles, is introduced near one electrode while a larger, particle-free sheath flow fills most of the gap between the two electrodes. An applied voltage between the two electrodes creates an electric field of strength

$$E(r) = \frac{V}{r \ln \frac{R_2}{R_1}} = E_2 \frac{R_2}{r} \tag{A.3}$$

that causes charged particles to migrate across the gap between the two electrodes as the gas flow carries them along the length of the electrodes. Here, we introduce the shorthand, $E_2 = E(R_2)$, which is the electric field strength at the outer electrode surface. Particles that migrate across the channel quickly deposit on the counter electrode; those that migrate too slowly exit from the downstream end of the classifier in the so-called excess air flow. Those that migrate across the channel in the time it takes to be carried to a downstream sample extraction port leave the flow channel in the classified aerosol flow. Typically, the volumetric flow rate of the classified aerosol is the same as that of the incoming aerosol sample flow. The classified sample contains particles with electrical mobilities centered in a narrow range about a target mobility. We seek to derive the transfer function, which is defined as the mobility dependence of the fraction of particles that are transmitted from the aerosol flow to the classified sample flow, i.e., $\Omega(Z, Z^*)$ is the fraction of particles of mobility $Z$ when the DMA operating parameters are set to target particles of mobility $Z^*$. First, we find the dependence of $Z^*$ on the flow rates and applied voltage for an idealized DMA. Then we will examine the range of mobilities that is transmitted under ideal conditions, and derive an expression for the DMA transfer function. Finally, we will explore the conditions under which diffusion broadens the range of mobilities that is transmitted and, thereby, degrades the performance of the DMA.

### Target Mobility

We consider an ideal cylindrical DMA in which the flow velocity of the gas is assumed to be uniform, i.e., we assume plug flow. The DMA is assumed to be radially symmetric. The equations of particle motion are

$$\frac{dx}{dt} = u(r) \tag{A.4}$$

$$\frac{dr}{dt} = \frac{Z_p E_2 R_2}{r} \tag{A.5}$$

Since we are interested in the position of the particle, not the time, we apply the chain rule to note

$$\frac{dr}{dt} = \frac{dr}{dx}\frac{dx}{dt} \tag{A.6}$$

from which,

$$\frac{dr}{dx} = \frac{Z_p E_2 R_2}{r u(r)}. \tag{A.7}$$

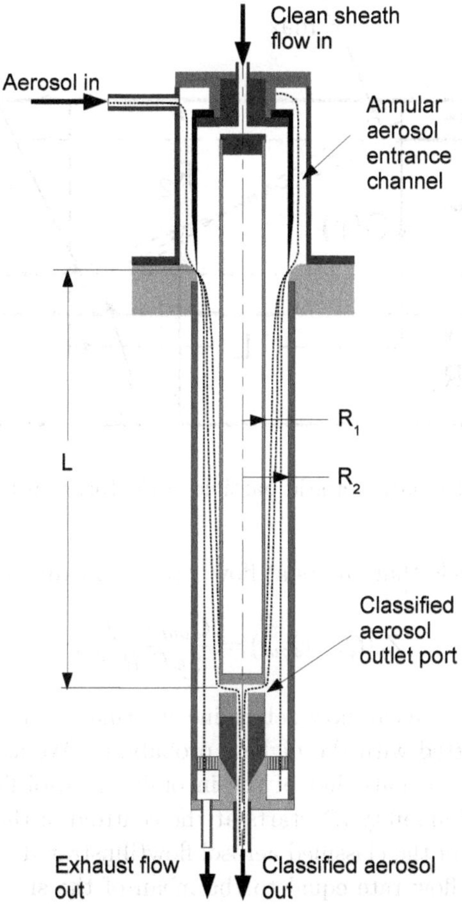


Figure A.1: Schematic diagram of the Knudsen and Whitby (1974) cylindrical differential mobility analyzer.

For the plug flow, $u(r) = \bar{U} =$ constant, so the radial positions at the inlet and outlet of the classification column are related through

$$\int_{R_{in}}^{R_{out}} 2\pi r u(r) dr = 2\pi Z_p E_2 R_2 L, \tag{A.8}$$

where the factor $2\pi$ has been introduced to relate the integral to the volumetric flow rate. It is convenient to define the cumulative flow rate as an integral beginning at the inner electrode, i.e.,

$$Q(r) = -\int_{R_2}^{r} 2\pi r u(r) dr. \tag{A.9}$$

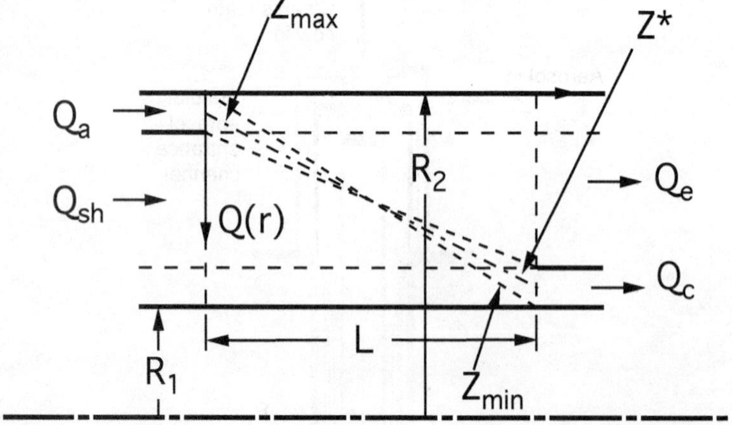


Figure A.2: Characteristic particle trajectories in the DMA.

The mobility of the particle that enters at flow, $Q_{in}$, and exits at flow $Q_{out}$ is then

$$Z_p(Q_{in}, Q_{out}) = \frac{Q_{out} - Q_{in}}{2\pi E_2 R_2 L}. \tag{A.10}$$

With this information, we can now determine the characteristic mobility of the particles that will be transmitted with the highest probability. We assume that the outgoing classified aerosol flow rate equals that of the incoming aerosol flow, i.e., $Q_c = Q_a$. Assuming that a particle of mobility $Z^*$ starts at the centroid of the incoming aerosol flow, and exits at the centroid of the classified aerosol flow(illustrated in Fig. A.2), the particle must cross a volumetric flow rate equal to the mean of the sheath and excess air flows (which, in the case of balanced flows, is equal to the sheath flow rate), then,

$$Z^* = \frac{Q_{sh}}{2\pi E_2 R_2 L} \tag{A.11}$$

is the target mobility for the given flow rates, DMA dimensions, and applied voltage.

### Minimum and maximum transmitted mobilities

The minimum particle mobility that will allow a particle to be transmitted is that which corresponds to the maximum value of $Q_{in} = Q_a$ and the minimum value of $Q_{out} = Q_a + Q_{sh} - Q_e$, which, for balanced aerosol and classified sample flows is

$$Q_{out} = Q_{sh}. \tag{A.12}$$

This trajectory is illustrated in Fig. A.2. Using Eq. (A.10), this yields

$$Z_{min} = \frac{Q_{sh} - Q_a}{2\pi E_2 R_2 L} = Z^* \left(1 - \frac{Q_{sh} - Q_a}{Q_{sh}}\right) = Z^* (1 - \beta), \tag{A.13}$$

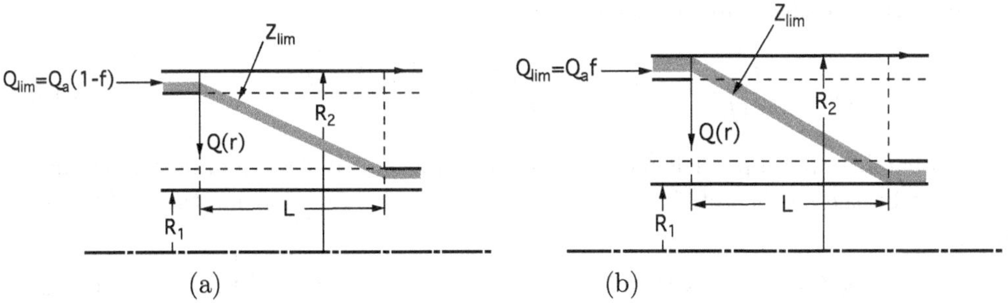


Figure A.3: Geometry used to calculate the fraction of particles transmitted for (a) $Z > Z_{min}$; and (b) $Z < Z_{max}$.

where the flow rate ratio, $\beta \equiv \frac{Q_a}{Q_{sh}}$, becomes one of the key operating parameters for the DMA. Similarly, the maximum transmitted particle mobility is that of a particle that enters at $Q_{in} = 0$ and exits at $Q_{out} = Q_a + Q_{sh}$, or

$$Z_{max} = \frac{Q_{sh} + Q_a}{2\pi E_2 R_2 L} = Z^* (1 + \beta) \tag{A.14}$$

### Fractional transmission

Particles can be transmitted if the mobility lies in the range $Z_{min} \leq Z \leq Z_{max}$. Those with mobility $Z = Z^*$ will be transmitted with 100% efficiency. Only a fraction of the other particles will be transmitted. We begin by considering the fraction of particles with mobilities $Z > Z_{min}$ that is transmitted, $f_{min}$. As illustrated in Fig. 1, only a fraction of the particles in the incoming flow is transmitted to the classified aerosol flow.The limiting particle trajectory is the one that exits at $Q_{out} = Q_a + Q_{sh} - Q_c = Q_{sh}$. Particles on that limiting trajectory entered the DMA at

$$Q_{lim} = Q_{sh} - 2\pi E_2 R_2 Z. \tag{A.15}$$

The fraction of the incoming aerosol that is included in the classified sample flow is then

$$f_{min} = \frac{Q_a - Q_{lim}}{Q_a} = \frac{2\pi E_2 R_2 Z - Q_{sh}(1-\beta)}{Q_a}. \tag{A.16}$$

Applying the definition of $Z^*$ (Eq. A.11), $f_{min}$ becomes

$$f_{min} = \frac{\frac{Z}{Z^*} - 1 + \beta}{\beta}. \tag{A.17}$$

For $Z^* < Z < Z^*(1+\beta)$, the limiting particle trajectory is one that enters at $Q_{lim} = Q_{in} < Q_a$ and exits at $Q_{out} = Q_{sh} + Q_a$. The fraction of particles in the incoming flow that exits in the classified sample flow is then

$$f_{max} = \frac{Q_{lim}}{Q_a} = \frac{Q_{sh} + Q_a - 2\pi E_2 R_2 L}{Q_a} = \frac{1 + \beta - \frac{Z}{Z^*}}{\beta}. \tag{A.18}$$

The probability that a particle will be transmitted cannot be less than 0, greater than 1, and must be the smaller of $f_{min}$ or $f_{max}$. The transfer function for the DMA thus becomes

$$\Omega(Z, Z^*) = \max\left[0, \min\left(\frac{\frac{Z}{Z^*} - 1 + \beta}{\beta}, \frac{1 + \beta - \frac{Z}{Z^*}}{\beta}, 1\right)\right]. \tag{A.19}$$

## Effect of Diffusion on DMA performance

The analysis presented above assumes that particles exactly follow the trajectories predicted by the kinematic equations of motion. Small particles also undergo Brownian diffusion, which will allow particles outside the range $Z^*(1 - \beta) \leq Z \leq Z^*(1 + \beta)$ to be transmitted, and will cause some particles within this range to deposit on the DMA electrodes or to exit in the exhaust flow. We will not undertake a detailed analysis of diffusive transport (which may be found in Flagan, 1999). Instead, we will apply scaling principles to understand when diffusion must be taken into account.

The relative importance of particle transport by electrophoretic migration to that due to Brownian diffusion is characterized by a migration Péclet number,

$$\mathrm{Pe}_{mig} = \frac{\text{electrophoretic migration}}{\text{diffusive transport}} = \frac{b v_E}{\mathcal{D}}, \tag{A.20}$$

where $b = R_2 - R_1$ is the spacing between the two electrodes, $v_E$ is the electrical migration velocity, and $\mathcal{D}$ is the particle diffusivity. From the Sutherland-Stokes-Einstein relation,

$$\mathcal{D} = B k_B T. \tag{A.21}$$

Combining Eqs. (A.2), (A.3), (A.20), and (A.21), we find

$$\mathrm{Pe}_{mig} = \frac{qV}{k_B T} f_{geom}, \tag{A.22}$$

where

$$f_{geom} = \frac{1 - \frac{R_1}{R_2}}{\ln \frac{R_2}{R_1}} \tag{A.23}$$

is a geometry factor which is of order unity and accounts for nonuniformities in the electric field. DMAs are generally applied to singly charged particles, for which $q = \pm e$. The migration Péclet number thus becomes

$$\mathrm{Pe}_{mig} = \frac{eV}{k_B T} f_{geom}. \tag{A.24}$$

The group $eV/k_B T$ is simply the ratio of the electrostatic potential energy of the charged

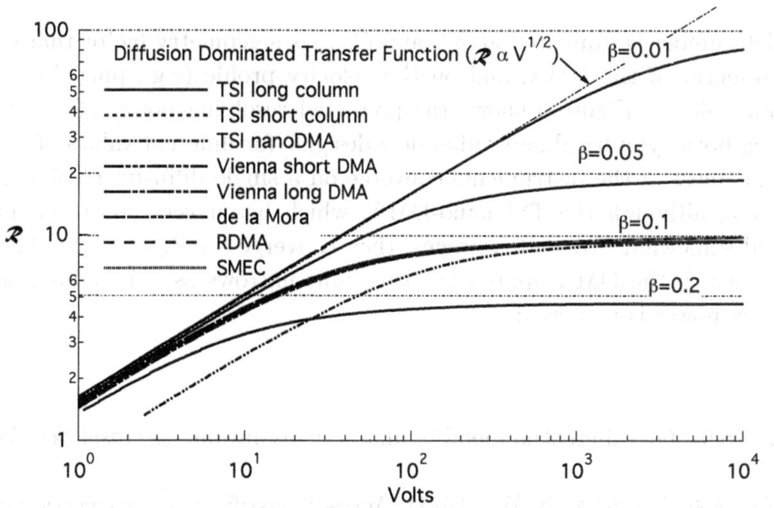


Figure A.4: Resolving power of the DMA. Theoretical predictions are shown for several different cylindrical DMAs, and two in which classification is performed in radial flow between parallel disk electrodes (the RDMA and the SMEC). Horizontal lines show the asymptotic limit for non-diffusive particles; the $V^{\frac{1}{2}}$ line shows the slope of the asymptotic limit for diffusion dominated separation. (After Flagan, 1999)

particle in the imposed electric field to its thermal energy. At normal ambient conditions at which most DMA measurements are made, $k_BT/e = 26$ mV.

Most DMAs have electrodes that are separated by a distance of about 10 mm. The maximum voltage is limited by electrostatic breakdown (arcing), typically ~10 kV. Generally, in the high voltage limit, the transfer function approaches the nondiffusive limit. This can be characterized by the resolving power of the DMA, which is defined as the ratio of the mobility of the particle that is transmitted with the highest efficiency, to the full width of the transfer function cooresponding to transmission with half of that maximum efficiency, i.e.,

$$\mathcal{R} = \frac{Z^*}{\Delta Z_{fwhm}}. \tag{A.25}$$

In the high voltage limit Eq. (A.19) yields

$$\mathcal{R}_{nd} = \beta^{-1}, \tag{A.26}$$

i.e., the nondiffusive transfer function is determined by the ratio of the sheath flow rate to that of the aerosol. At low voltages, where diffusion dominates, the diffusive resolving power has been empirically found to be

$$\mathcal{R}_d \approx 0.425 \left(\frac{\mathrm{Pe}^*_{mig}}{G_{mig}}\right)^{\frac{1}{2}}, \tag{A.27}$$

where the Péclet number is evaluated at $Z^*$, and $G_{mig}$ is a geometry factor that depends both on the geometry of the DMA, and on the velocity profile (e.g., plug flow or fully developed laminar flow). Figure 3 shows the predicted resolving power of a number of DMAs, including both cylindrical and radial flow designs, for different values of $\beta$. In the low voltage limit, most of the instruments converge on a single diffusive resolving power that scales as $V^{\frac{1}{2}}$, although the TSI nano-DMA, which has a very short classification region, is shifted somewhat. At high voltages, they converge on $\mathcal{R}_{nd} = \beta^{-1}$. The range of voltages over which the DMA approaches this limit narrows as $\beta$ is reduces and the targeted resolving power is increased.

## References

FLAGAN, R.C. (1999) "On differential mobility analyzer resolution." *Aerosol Sci. Technol.* **30,** 556-570.

KNUTSON, E.O., AND WHITBY, K.W. (1974) "Aerosol classification by electric mobility: Apparatus, theory, and applications." *J. Aerosol Sci.* **6,** 443-451.

# Appendix B

# WALL DEPOSITION OF AEROSOL PARTICLES IN STEADY LAMINAR FLOW IN A TUBE

A classic problem is heat or mass transfer between a fluid and the wall in steady laminar flow in a tube; this is sometimes referred to as the Graetz problem, as this problem was first analyzed by Graetz (1883). Here we consider the mathematically equivalent problem of the deposition of aerosol particles to the wall of a tube in steady laminar flow. The system is shown in Figure B.1. Air containing particles at number concentration $N_o$ flows into a tube of radius $R$. Steady state, fully-developed laminar flow is assumed. Particles undergo Brownian diffusion with Brownian diffusivity as a function of particle size given by (5.67). When a particle encounters the wall, it is removed permanently from the flow. The problem is to determine the concentration of particles emerging from the tube at a distance $L$ from the entrance.

The velocity distribution of axi-symmetric laminar flow in a tube (so-called Poiseuille flow) is given by

$$v_r = 0 \qquad v_\theta = 0 \qquad v_z(r) = 2U\left(1 - \frac{r^2}{R^2}\right) \tag{B.1}$$

where $U$ is the mean velocity of the fluid.

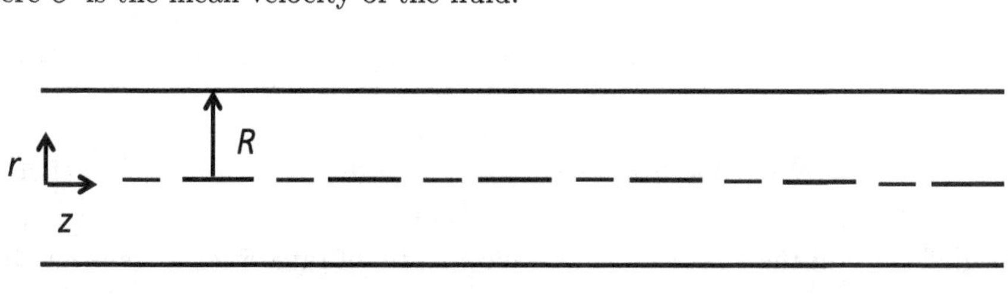


Figure B.1: Geometry for Laminar Tube Flow

The convection/diffusion equation governing particle number concentration $N(r, z)$, accounting for the fluid velocity field (B.1) is

$$2U\left(1-\frac{r^2}{R^2}\right)\frac{\partial N}{\partial z}=D\left[\frac{1}{r}\frac{\partial}{\partial r}\left(r\frac{\partial N}{\partial r}\right)+\frac{\partial^2 N}{\partial z^2}\right] \tag{B.2}$$

The obvious boundary conditions on (B.1) are

$$\text{Inlet:}\quad N(r,0)=N_0 \tag{B.3}$$

$$\text{Wall:}\quad N(R,z)=0 \tag{B.4}$$

$$\text{Centerline:}\quad N(0,Z)\text{ is finite or }\left.\frac{\partial N}{\partial r}\right|_{r=0}=0 \tag{B.5}$$

The presence of the axial diffusion term, $\partial^2 N/\partial z^2$, in (B.2) means that another boundary condition in the $z$-coordinate is needed. Actually, the inlet condition $N(r, 0) = N_0$ is mathematically incompatible with the inclusion of diffusion in the axial direction, because axial diffusion will lead to some of the information about the step change in wall removal of particles to propagate backward.

To assess the possible importance of the axial diffusion term, it is useful to place the variables in dimensionless form. Let us define

$$\tilde{N}=\frac{N}{N_0}\qquad Y=\frac{r}{R}\qquad Z=\frac{z}{R\mathrm{Pe}} \tag{B.6}$$

where $\mathrm{Pe} = 2UR/D$ is the Peclet number. With these definitions, (B.1) becomes

$$(1-Y^2)\frac{\partial \tilde{N}}{\partial Z}=\frac{1}{Y}\frac{\partial}{\partial Y}\left(Y\frac{\partial \tilde{N}}{\partial Y}\right)+\frac{1}{\mathrm{Pe}^2}\frac{\partial^2 \tilde{N}}{\partial Z^2} \tag{B.7}$$

The boundary conditions (B.3) – (B.5) become

$$\tilde{N}(Y,0)=1 \tag{B.8}$$

$$\tilde{N}(1,Z)=0 \tag{B.9}$$

$$\tilde{N}(0,Z)\text{ is finite or }\left.\frac{\partial \tilde{N}}{\partial Y}\right|_{Y=0}=0 \tag{B.10}$$

The physical significance of the Peclet number is Pe $\sim$ ratio of rate of mass transport by convection to rate of mass transport by diffusion.

$$\mathrm{Pe}\sim\frac{\text{Rate of mass transport by convection}}{\text{Rate of mass transport by diffusion}}$$

The numerator represents the order of magnitude of the convective flux of particles in the main flow direction; the denominator represents the order of magnitude of the diffusive flux in the radial direction. Physically, there are two mechanisms for transporting particles in the axial direction, convection and diffusion, whereas in the radial direction diffusion is the only transport mechanism.

Let us estimate the order of magnitude of Pe for a typical laminar flow particle deposition problem. We assume $R = 1\,\text{cm}, U = 5\,\text{cm s}^{-1}$, and particles of diameter 0.1 μm, for which $D = 7 \times 10^{-6}\text{cm}^2\,\text{s}^{-1}$. We find that Pe >> 100. Consequently,

$$\frac{1}{\text{Pe}^2} << 1$$

Since each of the terms in (B.7) besides the final term on the R.H.S. is of order unity, the last term on the R.H.S. can be neglected. The physical interpretation of this result is that convection (flow of the fluid) is much more effective in transporting particles in the axial direction than Brownian diffusion.

The final form of the convection/diffusion equation is

$$\left(1 - Y^2\right) \frac{\partial \tilde{N}}{\partial Z} = \frac{1}{Y} \frac{\partial}{\partial Y} \left( Y \frac{\partial \tilde{N}}{\partial Y} \right) \tag{B.11}$$

for which the boundary conditions (B.8) – (B.10) are complete.

Equation (B.11) can be solved by separation of variables, which yields an infinite series solution of the form,

$$\tilde{N}(Y, Z) = \sum_{n=1}^{\infty} A_n\, e^{-\lambda_n^2 Z} \quad \phi_n(Y) \tag{B.12}$$

where $\lambda_n$ and $\phi_n(Y)$ are the eigenvalues and eigenfunctions of the corresponding Sturm-Liouville system. A number of solutions of this problem exist in the literature, e.g. Gormley and Kennedy (1949), Brown (1960), and Davies (1973).

The first five eigenvalues are:

| $n$ | $\lambda_n$ | $\lambda_n^2$ |
|---|---|---|
| 1 | 2.7044 | 7.3136 |
| 2 | 6.6790 | 44.609 |
| 3 | 10.673 | 113.92 |
| 4 | 14.671 | 215.24 |
| 5 | 18.670 | 348.57 |

The eigenfunctions are given by

$$\phi_n(Y) = \exp\left( - \frac{\lambda_n Y^2}{2} \right) W_n(\lambda_n Y^2) \tag{B.13}$$

where $W_n$ is the confluent hypergeometric function (Abramowitz and Stegun, 1965). The coefficients $A_n$ in (B.12) are determined from

$$A_n = \frac{\int_0^1 \phi_n(Y)Y(1-Y^2)dY}{\int_0^1 \phi_n^2(Y)Y(1-Y^2)dY} \tag{B.14}$$

One is interested in the mean concentration of particles at a distance down the tube, that is, the fraction of entering particles that survive diffusional deposition to the wall,

$$\langle N(z)\rangle = \frac{\int_0^R N(r,z)2\pi r v(r)dr}{\pi R^2 U} \tag{B.15}$$

from which

$$\left\langle \tilde{N}(Z)\right\rangle = 4\int_0^1 \tilde{N}(Y,Z)Y(1-Y^2)dY \tag{B.16}$$

Approximate analytical solutions to this problem are given by Gormley and Kennedy (1949) and Alonso et al. (2010). These have been shown to closely match the exact solution and are useful for practical computations.

As an illustrative application, we consider the following parameter values:

$$\begin{aligned} R &= 0.002\,\text{m} \\ L &= 1\,\text{m} \\ D &= 6.914\times10^{-10}\ \text{m}^2\,\text{s}^{-1} \qquad (D_p = 1\times10^{-7}\text{m}) \\ C_c &= 2.85 \\ U &= 0.1\,\text{m}\,\text{s}^{-1} \end{aligned}$$

Considering $n = 11$ terms in the eigenfunction expansion, we obtain $\left\langle \tilde{N}(L)\right\rangle = 0.965$. Thus, under these conditions, 3.5% of the entering particles deposit on the tube wall owing to diffusion. Note that for this set of parameters $\text{Pe} = 2UR/D = 5.79\times10^5$. Neglect of the axial diffusion term in the convection/diffusion equation is clearly justified. The flow Reynolds number for this example,

$$\text{Re} = \frac{2RU\rho}{\mu} \tag{B.17}$$

can be determined. Using an air density $\rho = 1.8\,\text{kg}\,\text{m}^{-3}$ and viscosity $\mu = 1.8\times10^{-5}\,\text{kg}\,\text{m}^{-1}\,\text{s}^{-1}$, $\text{Re} = 26.2$, clearly well within the laminar range ($< 2100$) for flow in a tube at 298 K.

## B.1 WALL DEPOSITION OF AEROSOL PARTICLES IN STEADY TURBULENT FLOW IN A TUBE

Under turbulent flow conditions, the velocity profile at any axial location can be taken as uniform across the tube. Thus, the mean velocity $U$ exists across the entire flow. The approach that must be taken to derive the equation governing the particle number concentration at any axial distance down the tube is a macroscopic balance over a thin

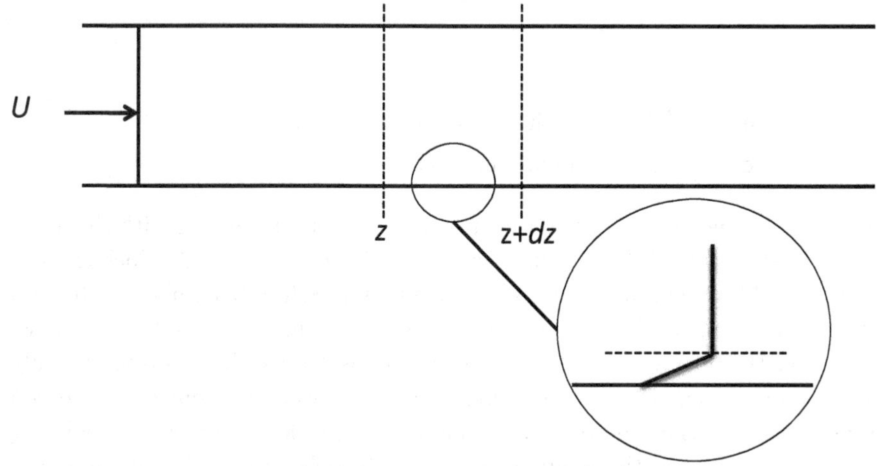


Figure B.2: Particle Deposition in Turbulent Flow in a Tube

slice of thickness $dz$ (Figure B.2). Owing to the intense mixing in the core of the flow, the profile of particle concentration across any tube cross-section is flat, except for a thin layer adjacent to the tube wall, across which the particle concentration drops from its value in the core of the flow to zero at the wall (see inset in Figure B.2).

The flux $F$ of particles to the wall, expressed in units of number of particles per unit area per unit time, is written in terms of an empirical turbulent mass transfer coefficient, $k$, times the driving force for transport, $N(z) - 0$, i.e.

$$F = k\,N\,(z) \tag{B.18}$$

Performing a steady state balance on particle number concentration over the slice between $z$ and $z + dz$ gives

$$0 = N(z)U\pi R^2\big|_z - N(z)U\pi R^2\big|_{z+dz} - kN(2\pi R dz)\big|_z \tag{B.19}$$

Dividing (B.19) by $dz$ and letting $dz \to 0$ yields

$$\frac{dN}{dz} = \frac{2k}{UR}N \tag{B.20}$$

with entrance condition $N(0) = N_0$. The solution of (B.21) is

$$N(z) = N_0 \exp\left(-\frac{2k}{UR}z\right) \tag{B.21}$$

A dimensionless form of the mass transfer coefficient for turbulent transfer between a fluid and the wall of a tube is (Bird et al., 1960)

$$\mathrm{Nu} = 0.023\,\mathrm{Re}^{0.8}\,\mathrm{Sc}^{1/3} \tag{B.22}$$

where

$$\text{Nu} = \frac{2Rk}{D} \quad \text{(Nusselt number for mass transfer}$$
$$\text{Sc} = \frac{\mu}{\rho D} \quad \text{(Schmidt number)}$$

Let us consider the same example as in the laminar flow case above, but with the mean velocity increased to $10\,\text{m}\,\text{s}^{-1}$. With this increase in air velocity, Re = 2622, which exceeds the value of 2100 at which the flow becomes turbulent. The Schmidt number for 0.1 μm particles is Sc=22076. From (B.22) one obtains $k = 6.05 \times 10^{-5}\text{m}\,\text{s}^{-1}$. At a distance down the tube of $L = 1\,\text{m}$, (B.21) gives $N(L)/N_0 = 0.994$. Thus, under these conditions only 0.6% of the particles entering the tube are deposited on the wall. Of course, one reason for this small percentage is that the residence time in the 1 m long tube is now only 0.1 s, as compared with the 10 s in the laminar flow case. To achieve comparable residence times, the tube length in the turbulent flow case would have to be increased to 100 m.

## REFERENCES

6
ABRAMOWITZ, M., AND STEGUN, I. (1965) Handbook of Mathematical Functions, Dover, New York.

ALONSO, M., ALGNACIL, F. J., AND HUANG, C. H.(2010) "Analytical Approximation to the Fully Developed Concentration Profile of Diffusive Aerosol Particles in Laminar Flow in a Circular Tube." *J. Aerosol Sci.* **41**, 413-417.

BROWN, G. M. (1960) "Heat or Mass Transfer in a Fluid in Laminar Flow in a Circular or Flat Conduit." *AIChE J.* **6**, 179-183.

DAVIS, C. N. (1973) "Diffusion and Sedimentation of Aerosol Particles from Poiseuille Flow in Pipes." *J. Aerosol Sci.* **4**, 317-328.

GORMLY, P. G. AND KENNEDY, M. (1949) "Diffusion from a Stream Flowing through a Cylindrical Tube." *Proc. Royal Irish Acad.* **52A**, 163-169.

GRAETZ, L. (1883) Über die Wärmeleitungsfähigkeit von Flüssigkeiten, *Annalen der Physik und Chemie*, **18**, 79.

# Appendix C

# EVAPORATION OF DIESEL EXHAUST PARTICLES DUE TO DILUTION

Particles emitted from a diesel-fueled vehicle comprise a mixture of organic compounds of different vapor pressures. In the concentrated exhaust, it can be assumed that the particles are in equilibrium with the vapor phase; that is, the amount of each compound in the particle phase is exactly that which is in equilibrium with its concentration in the gas phase. As the exhaust is diluted with clean air, the gas-phase concentrations of the organic compounds are reduced, and, as a result, particle-phase compounds will evaporate in an attempt to re-establish equilibrium. Here we model the rate of evaporation of such compounds from the exhaust particles when the gas phase in which the particles are suspended suddenly undergoes dilution. For computational purposes, we assume that the gas-phase concentrations are suddenly reduced by a factor of 10.

We assume that the fresh diesel exhaust consists of monodisperse particles of diameter 15 nm. We also assume that the entire evaporation process occurs at a constant temperature of 293 K. Thus, any effect of evaporation on the particle temperature is neglected. We assume, for simplicity, that fresh diesel exhaust particles consist of the following compounds in equal mole fractions (pure component vapor pressure and molecular weight given for each):

$$\begin{aligned} 2-\text{methyl naphthalene} \quad & 5.75\times 10^{-5}\,\text{atm} \quad M_w = 142 \\ \text{pentadecane} \quad & 3.1\times 10^{-6}\,\text{atm} \quad M_w = 212 \\ \text{tetradecane} \quad & 9.33\times 10^{-6}\,\text{atm} \quad M_w = 198 \end{aligned}$$

Assume that the condensed mixture of these three compounds can be taken as ideal, that is, the vapor pressure of each compound over a particle (if the particle had infiniate diameter) is the product of this pure compound vapor pressure and its mole fraction (Raoult's law). The surface tension of the particles is equal to 0.028 $N\text{m}^{-1}$. Overall particle density can be taken as $p_p = 1\,\text{g cm}^{-3}$.

Since all particles are initially of the same size and composition, we need consider only the dynamics of a single particle. The molar diffusion flux of species $i$ from the particle is given by

$$J_i = \frac{4\pi R_p D_i}{RT} \left( p_{\infty i} - p_{si} \right) \beta_{FS} \left( Kn \right)$$

where $D_i$ is the molecular diffusion coefficient of species $i$ in air, $p_{\infty i}$ is the partial pressure of $i$ far from the particle, and $p_{si}$ is the vapor pressure of $i$ just above the particle surface. The non-continum correction fator $\beta_{FS}(Kn)$ is given by (5.94).

The total number of moles of species $i$ in the particle, $N_i$, is

$$N_i = \frac{\rho_p x_i}{\bar{M}} \frac{4}{3} \pi R_p^3 \tag{C.1}$$

where $\bar{M}$ is the average molar mass and $x_i$ is the mole fraction of species $i$, $x_i = N_i/N_{tot}$. The particle diameter at any time is

$$D_p = \left[ \frac{6}{\pi \rho_p} \sum_{i=1}^{3} M_i N_i \right]^{1/3} \tag{C.2}$$

The vapor pressure of species $i$ over the particle, using Raoult's law and (5.97) is

$$p_{si} = p_i^0 x_i \exp \left[ \frac{4\sigma \bar{M}}{\rho_\rho R T D_p} \right] \tag{C.3}$$

Combining (??) and (C.3),

$$\frac{dN_i}{dt} = \frac{2\pi D_p D_i}{RT} \left[ p_{\infty i} - p_i^0 x_i \exp \left[ \frac{4\sigma \bar{M}}{\rho_p R T D_p} \right] \right] \beta_{FS} \left( Kn \right) \tag{C.4}$$

At $t<0$, the particle is at equilibrium with the gas. Since the initial particle diameter is specified as 15 nm, and each $x_i = 0.33$ initially, the equilibrium is given by

$$p_{\infty i}^{t<0} = p_i^0 x_i \exp \left[ \frac{4\sigma \bar{M}}{\rho_p R T D_p} \right] \qquad t < 0 \tag{C.5}$$

The R.H.S. of (C.5) is completely determined, so that (C.5) gives $p_{\infty i}$ for $t < 0$. At $t = 0$, the gas is diluted by a factor of 10, so that

$$p_{\infty i} = \frac{p_{\infty i}^{t<0}}{10} \qquad t \geq 0 \tag{C.6}$$

Equation (C.4) can be solved numerically for $i = 1, 2, 3$, with $p_{\infty i}$ given by (C.5) and (C.6). For the numerical evaluation, we use $D_i = 0.1\,\text{cm}^2\,\text{s}^{-1}$ for all 3 species. For these parameter values, the evaporation is complete in 400 $\mu s$ (Figures C.1 - C.3 ). The most volatile species, 2-methyl naphthalene, evaporates most readily, followed by tetradecane and then pentadecane (Figures C.1 and C.2). At the point at which the

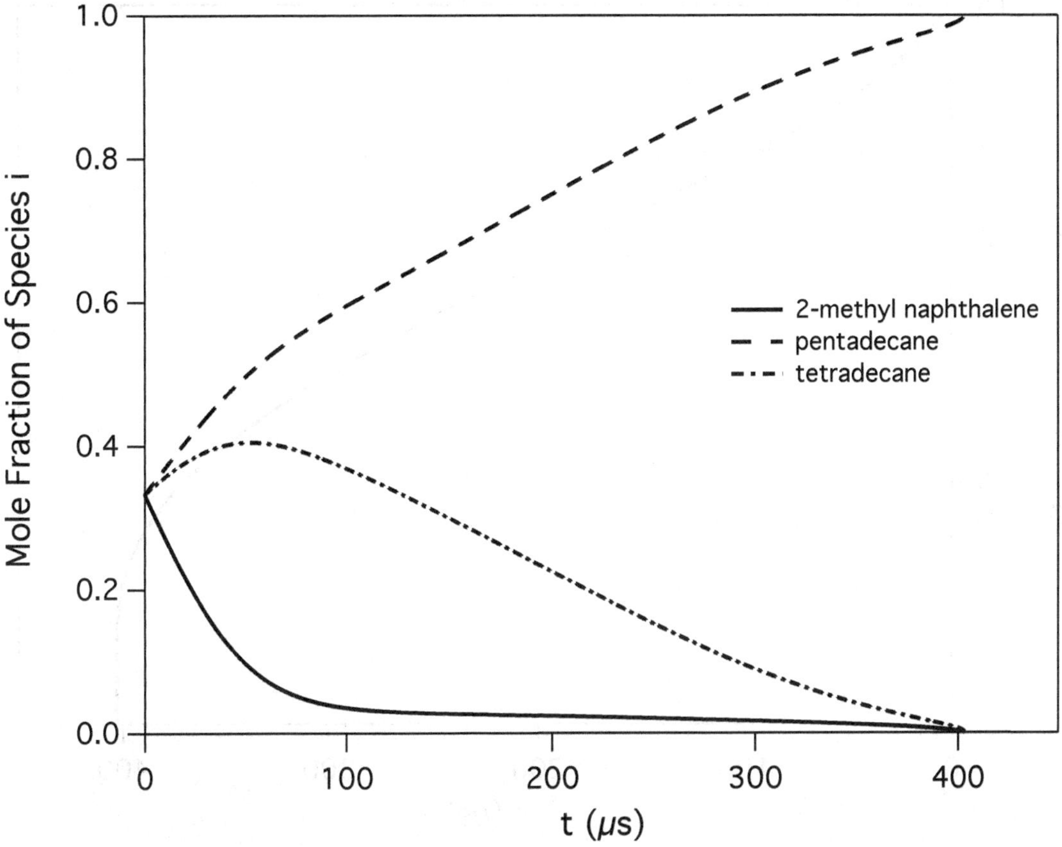


Figure C.1:

particle fully evaporates, it consists essentially entirely of pentadecane. Once the particle diameter reaches about 5 nm, the Kelvin effect is so dominant that the particle evaporates essentially instantaneously (Figure C.3).

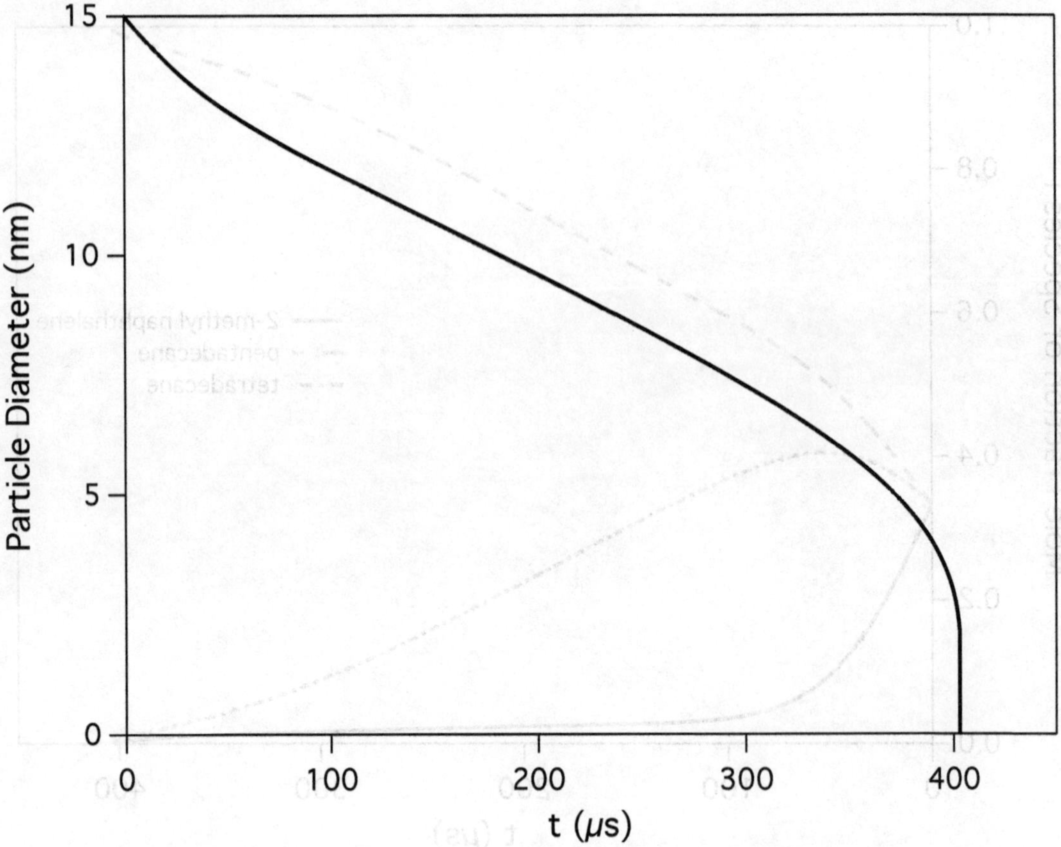


Figure C.2:

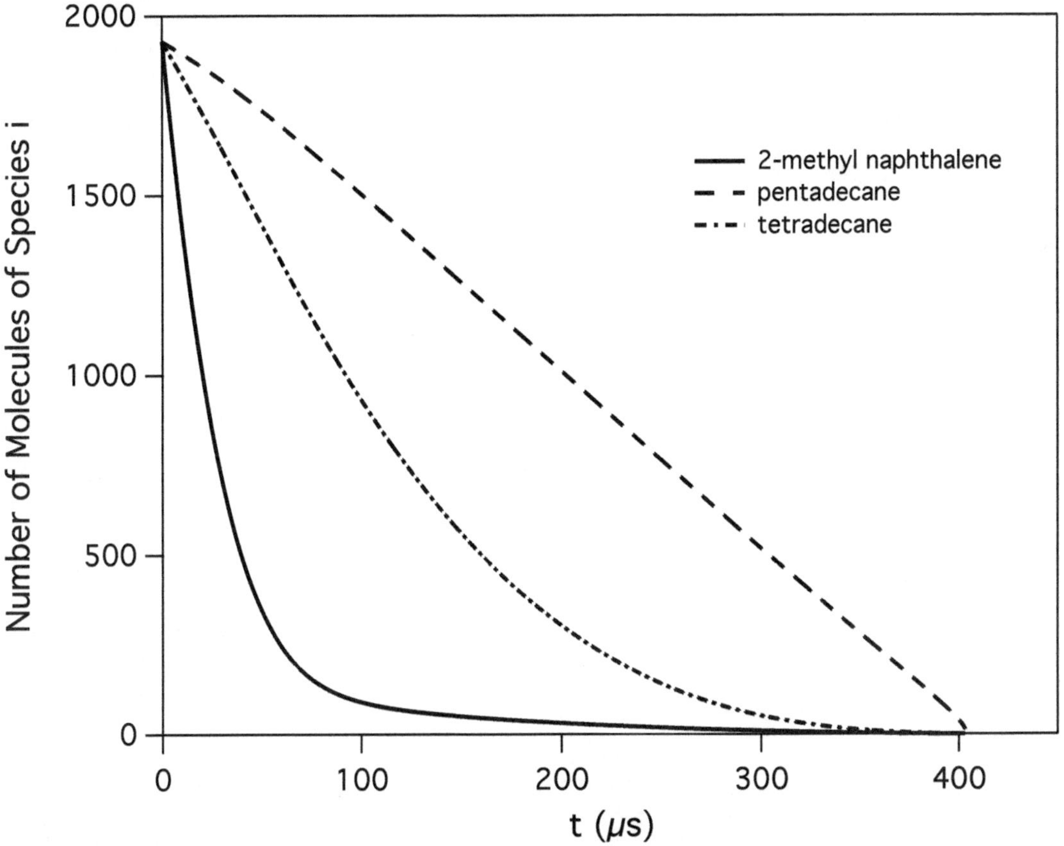
2000
1500
1000
500
0
Number of Molecules of Species i
2-methyl naphthalene
pentadecane
tetradecane
0
100
200
300
400
t (μs)

Figure C.3:

# Index